MW00836871

ELECTROMAGNETIC WAVE PROPAGATION, RADIATION, AND SCATTERING

ELECTROMAGNETIC WAVE PROPAGATION, RADIATION, AND SCATTERING
From Fundamentals to Applications

Second Edition

AKIRA ISHIMARU
University of Washington, Seattle, WA, USA

 IEEE Antennas and Propagation Society, *Sponsor*

The IEEE Press Series on Electromagnetic Wave Theory
Andreas C. Cangellaris, *Series Editor*

IEEE PRESS

WILEY

For general information on our other products and services or for technical support, please contact our Customer Care Department within the United States at (800) 762-2974, outside the United States at (317) 572-3993 or fax (317) 572-4002.

Wiley also publishes its books in a variety of electronic formats. Some content that appears in print may not be available in electronic formats. For more information about Wiley products, visit our web site at www.wiley.com.

Library of Congress Cataloging-in-Publication Data is available.

ISBN: 978-1-118-09881-3

Printed in the United States of America.

10 9 8 7 6 5 4 3 2 1

To

YUKO

and

Emily, Ben, Joyce, Gene, Aaron, Eric, Andrew, Karen, Jim,
Grace, Patty, Chad, Katie, Will, Jane, Chuck, John

CONTENTS

ABOUT THE AUTHOR

Akira Ishimaru was born in Fukuoka, Japan. He became a naturalized US citizen in 1963. He received the B.S. degree from the University of Tokyo, Tokyo, Japan, in 1951 and the Ph.D. degree in electrical engineering from the University of Washington, Seattle, in 1958. He is a registered engineer in the state of Washington.

From 1951 to 1952, he was with the Electrotechnical Laboratory, Tanashi, Tokyo, and in 1956, he was with Bell Laboratories, Holmdel, NJ. In 1958, he joined the faculty of the Department of Electrical Engineering, University of Washington, where he was a Professor of electrical engineering, and an Adjunct Professor of applied mathematics. He is currently Professor Emeritus there. He has also been a Visiting Associate Professor at the University of California, Berkeley. His current research includes waves in random media, remote sensing, object detection, and imaging in clutter environment, inverse problems, millimeter wave, optical propagation and scattering in the atmosphere and the terrain, rough surface scattering, and optical diffusion and ultrasound scattering in tissues. He is the author of *Wave Propagation and Scattering in Random Media* (New York: Academic Press, 1978; IEEE-Oxford University Press classic reissue, 1997) and *Electromagnetic Wave Propagation, Radiation, and Scattering* (Englewood Cliffs, NJ: Prentice Hall, 1991). He was Editor (1979–1983) of *Radio Science* and Founding Editor of *Waves in Random Media* (Institute of Physics, UK), and *Waves in Random and Complex Media* (Taylor & Francis, UK).

Dr. Ishimaru has served as a member-at-large of the US National Committee (USNC) and was chairman (1985–1987) of Commission B of the USNC/International Union of Radio Science. He is a fellow of the IEEE, the Optical Society of America, the Acoustical Society of America, and the Institute of Physics, UK. He was the recipient of the 1968 IEEE Region VI Achievement Award and the IEEE Centennial Medal in 1984. He was appointed as Boeing Martin Professor in the College of

Engineering in 1993. In 1995, he was awarded the Distinguished Achievement Award from the IEEE Antennas and Propagation Society. In 1998, he was awarded the Distinguished Achievement Award from the IEEE Geoscience and Remote Sensing Society. He is the recipient of the 1998 IEEE Heinrich Hertz Medal and the 1999 URSI Dellinger Gold Medal. In 2000, he received the IEEE Third Millennium Medal. He was elected to the National Academy of Engineering in 1996.

PREFACE

The original edition of this book, *Electromagnetic Wave Propagation, Radiation, and Scattering*, was published by Prentice Hall in 1991. When IEEE Press/AP-S Liaison Committee expressed interest in updating the original edition by including additional chapters on new topics that reflect changing interests in the field since the original publication, I was delighted and excited by the opportunity to include some new materials and ideas about the state of electromagnetics.

Electromagnetics is fundamental to all aspects of electrical engineering and in fact most of engineering and science. Electromagnetics theory describes and expresses all electromagnetic fields in space and time, as well as in vector formulations, using both real and complex mathematical and analytical methods.

Analytical and computational electromagnetics are two aspects of the fundamental work of pursuing unique and novel ideas in the field, with the same purpose and objectives. This book is focused on an analytical approach, and emphasizes new ideas and applications. The important topic of computational electromagnetics is not included in this book; there are many excellent texts on the subject for the interested reader, such as *Computational Methods for Electromagnetics* by Peterson, Ray, and Mittra (Wiley-IEEE Press, 1998).

The newly revised material in this edition is focused on the following topics:

1. Statistical wave theories; these theories have been extensively applied to topics such as geophysical remote sensing, bioelectromagnetics, bio-optics, and bioultrasound imaging;
2. Integration of several distinct yet related disciplines such as statistical wave theories, communications, signal processing, and time-reversal imaging;

3. New phenomena of multiple scattering, such as coherent scattering and memory effects;

4. Multiphysics applications that combine theories for different physical phenomena such as seismic coda waves (which are acoustic waves in elastic solid earth), stochastic wave theory, heat diffusion, and temperature rise in biological and other media; and

5. Metamaterials and solitons in optical fibers, nonlinear phenomena, and porous media.

Chapters 1 to 17 of this revised book are from the original edition, and emphasize the fundamentals of electromagnetic wave propagation. The new chapters 18 to 26 are largely focused on new topics in electromagnetics. The new chapters describe the following topics:

Chapter 18 presents an overview of the stochastic Sommerfeld problem, stochastic Green's function, channel capacity in random media, and related topics.

Chapter 19 covers geophysical applications, including polarimetric radars, space–time vector radiative transfer, the Wigner distribution, and the Stokes vector. Bioelectromagnetics, heat diffusion in tissues, optical diffusion and ultrasonic scattering in tissues, and OCT (optical coherence tomography) and low coherence interferometry are included in Chapter 20.

Chapter 21 presents fundamentals of metamaterials, including wave packets in a dispersive negative index medium, backward lateral and surface waves, negative Goos–Hanchen shift, perfect lens, evanescent waves, electromagnetic Brewster's angle, acoustic Brewster's angle, and transformation electromagnetics.

Chapter 22 describes "time-reversal" imaging, including the relation to SVD (singular value decomposition), MUSIC (multiple signal classification) imaging, and DORT (decomposition of time-reversal operator).

Scattering in random media is not normally discussed in electromagnetic texts. However, there is extensive literature on the subject, including my IEEE Press book *Wave Propagation and Scattering in Random Media* (1997). This topic is summarized in Chapter 23, with an emphasis on angular spectrum, pulse propagation, coherence time and bandwidth, and spectral broadening.

Multiple scattering theories in random media are fundamental not only in electrical engineering but in astrophysics and condensed matter physics. Chapter 24 gives a sketch of this important and difficult problem, and formulations of the Dyson and Bethe–Salpeter equations and Feynman diagram. Included are recent studies of backscattering enhancement, Anderson localization, super resolution, and memory effects.

The interesting history of the discovery of solitons and the KdV equation are discussed in Chapter 25, with special emphasis on optical fibers.

The mixing formula was originally discussed in Chapter 8 and used to find the permittivity of the mixture. The same material can alternatively be viewed as a "porous medium," where the medium contains "pores." The porous medium approach can be used to find the mixture's permittivity and the flow rate of fluid through the pores (fluid permeability). This is a situation encountered in oil shale, and a discussion of that application is included in Chapter 26.

Chapter 26 also discusses the topic of seismic waves. A seismic wave is an acoustic wave in the earth, which is considered as an elastic solid. The initial part of the wave pulse is important and is used to determine the seismic wave (earthquake) magnitude. However, the subsequent wave train due to multiple scattering can reveal useful additional information about the total seismic energy and the medium's characteristics. This wave train, called the "seismic coda," is discussed in Chapter 26. This chapter also includes a discussion of P-waves (pressure), S-waves (shear), and Rayleigh surface waves, all of which are important for seismic study.

The revised book is intended as a textbook for graduate courses in electrical engineering. It should also be of interest to graduate students in bioengineering, geophysics, ocean engineering, and geophysical remote sensing. The book is also a useful reference for engineers and scientists working in fields such as geophysical remote sensing, biomedical engineering in optics and ultrasound, and new materials and integration with signal processing. Please note that the solutions manual for the problems in Chapters 1 through 17 is available to anyone who requests it. To obtain the manual, please email your request to ieeeproposals@wiley.com.

I am grateful to IEEE Press, in particular Bob Mailloux, Gary Brown, and Ross Stone, and to Mary Hatcher at Wiley, for their help and encouragement. John Ishimaru has been of enormous help in preparing the manuscript. I thank my colleagues, in particular Yasuo Kuga, Leung Tsang, and Radha Poovendran, and graduate students, including T.-H. Liao and many others. Some of the work described in this book was supported by the National Science Foundation and the Office of Naval Research.

AKIRA ISHIMARU

PREFACE TO THE FIRST EDITION

Numerous new advances have been made in electromagnetic theory in recent years. This is due, in part, to new applications of the theory to many practical problems. For example, in microwave and millimeter wave applications, there is an increasing need to investigate the electromagnetic problems of new guiding structures, phased arrays, microwave imaging, polarimetric radars, microwave hazards, frequency-sensitive surfaces, composite materials, and microwave remote sensing. In the field of optics, applications involving fiber optics, integrated optics, atmospheric optics, light diffusion in tissues, and optical oceanography are among many problems whose solutions require the use of electromagnetic theory as an essential element. The mathematical techniques used in electromagnetic theory are also equally applicable to ocean acoustics, acoustic scattering in the atmosphere, and ultrasound imaging of tissues. In this book, we attempt to present a cohesive account of these advances with sufficient background material.

This book has three overall objectives. First, we present the fundamental concepts and formulations in electromagnetic theory. Second, advanced mathematical techniques and formulations such as the saddlepoint technique, integral equations, GTD, and the T-matrix method are presented. Third, new topics, such as inverse scattering, diffraction tomography, vector radiative transfer, and polarimetry, are discussed.

The book includes several topics that are often not adequately covered in many electromagnetics texts. Examples are the Maxwell–Garnett and Polder–van Santen mixing formulas for the effective dielectric constant of mixtures and waves in chiral media. Also included are London's equations and the two-fluid model of the complex conductivity of superconductors at high frequencies. Other examples are Radon transform, diffraction tomography, holographic and physical optics inverse problems, Abel's inversion formula, and polarimetry.

The book also includes more conventional but important topics such as the spectral method, strip lines, patches, apertures, Stokes vectors, modified saddlepoint techniques, Watson transforms, residue series, T-matrix, GTD, and UTD. Also covered are Zenneck waves, plasmons, Goos–Hanchen effects, construction of Green's functions, Green's dyadic, radiometry, and noise temperatures. An important current topic that is not included due to space limitations is that of wave propagation and scattering in random media. However, this has been covered adequately in the author's previous book (Academic Press, 1978). Other topics that have been omitted are nonlinear electromagnetics, solitons, and chaos.

This book is based on a set of lecture notes prepared for a three-quarter graduate course on electromagnetic waves and applications given in the Department of Electrical Engineering at the University of Washington. The course has been offered yearly for students primarily from within the department of electrical engineering, but also including those from bioengineering, atmospheric sciences, geophysics, and oceanography. The course is also offered via cable television to students in remote sites located at nearby aerospace and bioengineering companies. The interest of the students vary considerably, from microwave antennas, radomes, and radar cross sections, to ultrasound imaging, optical scattering, ocean acoustics applications, microwave hazards, and bioelectromagnetics. This book is therefore intended to concentrate on fundamentals as well as recent analytical techniques, so that students may be prepared to handle a variety of old and new applications.

It is assumed that readers are familiar with material normally covered in undergraduate courses on electromagnetic theory, differential equations, Fourier and Laplace transforms, vector analysis, and linear algebra. However, brief reviews of these topics are given whenever needed. The time convention used in this book is $\exp(j\omega t)$. For some topics such as Stokes parameters, which are used extensively by the optics and acoustics community, the convention of $\exp(-i\omega t)$ is used and its use is clearly noted.

This book is intended as a textbook for a first-year graduate course, and the material can be covered in two semesters or three-quarters with appropriate selections depending on the students' interests. The book is also a useful reference for engineers working on electromagnetic, optical, or acoustics problems with aerospace, bioengineering, remote sensing, and ocean engineering applications.

I am grateful to Noel Henry for her expert typing and her assistance in organizing the manuscript. I also thank my graduate students and colleagues who have made many valuable comments. Some of the work in this book was supported by the National Science Foundation, the US Army Research Office, the Office of Naval Research, and the US Army Engineer Waterways Experiment Station. I also wish to thank Elizabeth Kaster, Editor, Prentice Hall, for her encouragement and editorial assistance.

AKIRA ISHIMARU

ACKNOWLEDGMENTS

This book was based on a set of lecture notes originally prepared for first- and second-year graduate courses on electromagnetics and applications in the Department of Electrical Engineering at the University of Washington. The first 17 chapters were the original text published in 1991. Chapters 18 to 26 are new chapters in this revised edition, reflecting new topics and interests that had not been covered in the 1991 edition. The revised edition is aimed at a broader audience including students in bioengineering, geophysics, and atmospheric and earth sciences. I wish to express my appreciation to my graduate students and colleagues for their encouragement and interest. I wish to thank Y. Kuga, L. Tsang, and I. C. Peden at the University of Washington. I am grateful to my good friends Bob Mailloux, Gary Brown, Ross Stone, and all members of the IEEE Press/AP-S Liaison Committee. John Ishimaru has been of enormous and steady help. A large part of the research described in this book was supported by the National Science Foundation, the Army Research Office, the Office of Naval Research, and the Air Force Office of Scientific Research.

I am grateful to IEEE Press and AP-S for sponsoring this project, and to Mary Hatcher and Divya Narayanan at Wiley for their help and encouragement and their efficient work on the production of this book.

PART I

FUNDAMENTALS

CHAPTER 1

INTRODUCTION

Many advances in electromagnetic theory were made in recent years in response to new applications of the theory to microwaves, millimeter waves, optics, and acoustics; as a result, there is a need to present a cohesive account of these advances with sufficient background material. In this book we present the fundamentals and the basic formulations of electromagnetic theory as well as advanced analytical theory and mathematical techniques and current new topics and applications.

In Chapter 2, we review the fundamentals, starting with Maxwell's equations and covering such fundamental concepts and relationships as energy relations, potentials, Hertz vectors, and uniqueness and reciprocity theorems. The chapter concludes with linear acoustic-wave formulation. Plane-wave incidence on dielectric layers and wave propagation along layered media are often encountered in practice. Examples are microwaves in dielectric coatings, integrated optics, waves in the atmosphere, and acoustic waves in the ocean. Chapter 3 deals with these problems, starting with reviews of plane waves incident on layered media, Fresnel formulas, Brewster's angle, and total reflection. The concepts of complex waves, trapped surface waves, and leaky waves are presented with examples of surface-wave propagation along dielectric slabs, and this is followed by discussion on the relation between Zenneck waves and plasmons. The chapter concludes with Wentzel–Kramers–Brillouim (WKB) solutions and the Bremmer series for inhomogeneous media and turning points, and WKB solutions for the propagation constant of guided waves in inhomogeneous media such as graded-index fibers.

Electromagnetic Wave Propagation, Radiation, and Scattering: From Fundamentals to Applications, Second Edition. Akira Ishimaru.
© 2017 by The Institute of Electrical and Electronic Engineers, Inc. Published 2017 by John Wiley & Sons, Inc.

Chapter 4 deals with microwave waveguides, dielectric waveguides, and cavities. Formulations for transverse magetic (TM), transverse electric (TE), and transverse electromagnetic (TEM) waves, eigenfunctions, eigenvalues, and the k–β diagram are given, followed by pulse propagation in dispersive media. Dielectric waveguides, step-index fibers, and graded-index fibers are discussed next with due attention to dispersion. It concludes with radial and azimuthal waveguides, rectangular and cylindrical cavities, and spherical waveguides and cavities. This chapter introduces Green's identities, Green's theorem, special functions, Bessel and Legendre functions, eigenfunctions and eigenvalues, and orthogonality.

One of the most important and useful tools in electromagnetic theory is Green's functions. They are used extensively in the formulation of integral equations and radiation from various sources. Methods of constructing Green's functions are discussed in Chapter 5. First, the excitation of waves by electric and magnetic dipoles is reviewed. Three methods of expressing Green's functions are discussed. The first is the representation of Green's functions in a series of eigenfunctions. The second is to express them using the solutions of homogeneous equations. Here, we discuss the important properties of Wronskians. The third is the Fourier transform representation of Green's functions. In actual problems, these three methods are often combined to obtain the most convenient representations. Examples are shown for Green's functions in rectangular waveguides and cylindrical and spherical structures.

Chapter 6 deals with the radiation field from apertures. We start with Green's theorem applied to the field produced by the sources and the fields on a surface. Here, we discuss the extinction theorem and Huygens' formula. Next, we consider the Kirchhoff approximation and Fresnel and Fraunhofer diffraction formulas. Spectral representations of the field are used to obtain Gaussian beam waves and the radiation from finite apertures. The interesting phenomenon of the Goos–Hanchen shift of a beam wave at an interface and higher-order beam waves are also discussed. The chapter concludes with the electromagnetic vector Green's theorem, Stratton–Chu formula, Franz formula, equivalence theorem, and electromagnetic Kirchhoff approximations.

The periodic structures discussed in Chapter 7 are used in many applications, such as optical gratings, phased arrays, and frequency-selective surfaces. We start with the Floquet-mode representation of waves in periodic structures. Guided waves along periodic structures and plane-wave incidence on periodic structures are discussed using integral equations and Green's function. An interesting question regarding the Rayleigh hypothesis for scattering from sinusoidal surfaces is discussed. Also included are the coupled-mode theory and co-directional and contra-directional couplers.

Chapter 8 deals with material characteristics. We start with the dispersive characteristics of dielectric material, the Sellmeier equation, plasma, and conductors. It also includes the Maxwell–Garnett and Polder–van Santen mixing formulas for the effective dielectric constant of mixtures. Wave propagation characteristics in magnetoplasma, which represents the ionosphere and ionized gas, and in ferrite, used in microwave networks, are discussed as well as Faraday rotation, group velocity, warm plasma, and reciprocity relations. This is followed by wave propagation in chiral material. The chapter concludes with London's equations and the two-fluid model of superconductors at high frequencies.

Chapter 9 presents selected topics on antennas, apertures, and arrays. Included in this chapter are radiation from current distributions, dipoles, slots, and loops. Also discussed are arrays with nonuniform spacings, microstrip antennas, mutual couplings, and the integral equation for current distributions on wire antennas. Chapter 10 starts with a general description of the scattering and absorption characteristics of waves by dielectric and conducting objects. Definitions of cross sections and scattering amplitudes are given, and Rayleigh scattering and Rayleigh–Debye approximations are discussed. Also included are the Stokes vector, the Mueller matrix, and the Poincaré sphere for a description of the complete and partial polarization states. Techniques discussed for obtaining the cross sections of conducting objects include the physical optics approximation and the moment method. Formal solutions for cylindrical structures, spheres, and wedges are presented in Chapter 11, including a discussion of branch points, the saddle-point technique, the Watson transform, the residue series, and Mie theory. Also discussed is diffraction by wedges, which will be used in Chapter 13.

Electromagnetic scattering by complex objects is the topic of Chapter 12. We present scalar and vector formulations of integral equations. Babinet's principle for scalar and electromagnetic fields, electric field integral equation (EFIE), and magnetic field integral equation (MFIE) are discussed. The T-matrix method, also called the extended boundary condition method, is discussed and applied to the problem of sinusoidal surfaces. In addition to the surface integral equation, also included are the volume integral equation for two- and three-dimensional dielectric bodies and Green's dyadic. Discussions of small apertures and slits are also included.

Geometric theory of diffraction (GTD) is one of the powerful techniques for dealing with high-frequency diffraction problems. GTD and UTD (uniform geometric theory of diffraction) are discussed in Chapter 13. Applications of GTD to diffraction by slits, knife edges, and wedges are presented, including slope diffraction, curved wedges, and vertex and surface diffractions.

Chapter 14 deals with excitation and scattering by sources, patches, and apertures embedded in planar structures. Excitation of a dielectric slab is discussed, followed by the WKB solution for the excitation of waves in inhomogeneous layers. An example of the latter is acoustic-wave excitation by a point source in the ocean. Next, we give general spectral formulations for waves in patches, strip lines, and apertures embedded in dielectric layers. Convenient equivalent network representations are presented that are applicable to strip lines and periodic patches and apertures.

The Sommerfeld dipole problem is that of finding the field when a dipole is located above the conducting earth. This classical problem, which dates back to 1907, when Zenneck investigated what is now known as the Zenneck wave, is discussed in Chapter 15, including a detailed study of the Sommerfeld pole, the modified saddle-point technique, lateral waves, layered media, and mode representations.

The inverse scattering problem in Chapter 16 is one of the important topics in recent years. It deals with the problem of obtaining the properties of an object by using the observed scattering data. First, we present the Radon transform, used in computed tomography or X-ray tomography. The inverse Radon transform is obtained by using the projection slice theorem and the back projection of the filtered projection. Also included is an alternative inverse Radon transform in terms of the Hilbert transform. For ultrasound and electromagnetic imaging problems, it is necessary to include the diffraction effect. This leads to diffraction tomography, which makes use of back propagation rather than back projection. Also discussed are physical optics inverse scattering and the holographic inverse problem. Abel's integral equations are frequently used in inverse problems. Here, we illustrate this technique by using it to find the electron density profile in the ionosphere. Polarimetric radars are becoming increasingly more important because of the advances in polarimetric measurement techniques. We present the fundamentals of polarimetry and optimization as well as polarization signatures.

Chapter 17 presents fundamentals of radiometry and noise. The definitions of antenna temperature, radiative transfer theory, and the scattering cross section of surfaces are given, followed by consideration of system noise temperature and minimum detectable temperature. Also included is a discussion on the determination of sky brightness distribution by interferometry used in radio astronomy. Here, we discuss the Fourier transform and convolution relationships among the aperture distribution, the radiation pattern, and sky brightness distributions.

Chapters 18 through 26 were added for the second edition of this book. The contents of those chapters are summarized in the Preface to the Second Edition.

The appendices give many formulas and detailed derivations of equations that are too lengthy to be included in the text. They should be helpful in understanding the material in the text.

Useful reference books on electromagnetics at the intermediate undergraduate level include Ramo et al. (1965), Jordan and Balmain (1968), Wait (1986), Shen and Kong (1987), and Cheng (1983). Among books at the advanced level are Stratton (1941), Harrington (1961), Collin (1966), Felsen and Marcuvitz (1973), Schelkunoff (1965), Balanis (1989), Kong (1981, 1986), Jones (1964, 1979), and Van Bladel (1964).

CHAPTER 2

FUNDAMENTAL FIELD EQUATIONS

The fundamental field equations for electromagnetic fields are Maxwell's equations. In this chapter, we review differential and integral forms of these equations and discuss boundary conditions, energy relations, Poynting's theorem, the uniqueness theorem, and the reciprocity theorem. Vector and scalar potentials and Hertz vectors are discussed, as they give alternative and often simpler formulations of the problem. Although electromagnetic waves are vector fields, they can sometimes be represented or approximated by scalar fields. We present formulations for scalar acoustic waves as examples of scalar fields.

It will be assumed that the reader is familiar with the electromagnetic field theory normally covered in undergraduate courses, and therefore, this book starts with a review of the fundamental field equations. A detailed historical development of electromagnetic theory is given in Elliott (1966) and Born and Wolf (1970).

2.1 MAXWELL'S EQUATIONS

The fundamental differential equations governing the behavior of electromagnetic fields were given by Maxwell in 1865:

$$\nabla \times \bar{H} = \frac{\partial \bar{D}}{\partial t} + \bar{J}, \tag{2.1}$$

$$\nabla \times \bar{E} = -\frac{\partial \bar{B}}{\partial t}, \tag{2.2}$$

Electromagnetic Wave Propagation, Radiation, and Scattering: From Fundamentals to Applications,
Second Edition. Akira Ishimaru.
© 2017 by The Institute of Electrical and Electronic Engineers, Inc. Published 2017 by John Wiley & Sons, Inc.

$$\nabla \cdot \bar{D} = \rho, \tag{2.3}$$
$$\nabla \cdot \bar{B} = 0. \tag{2.4}$$

Here $\bar{E}$ is the electric field vector in volts/meter, $\bar{H}$ the magnetic field vector in amperes/meter, $\bar{D}$ the electric displacement vector in coulombs/meter^2, $\bar{B}$ the magnetic flux density vector in webers/meter^2, $\bar{J}$ the current density vector in amperes/meter^2, and ρ the volume charge density in coulombs/meter^3.

The physical meanings of Maxwell's equations are often easier to understand if expressed in alternative integral form. The first two equations, (2.1) and (2.2), can be converted into integral form by employing Stokes' theorem:

$$\int_a \nabla \times \bar{A} \cdot d\bar{a} = \oint_l \bar{A} \cdot d\bar{l}, \tag{2.5}$$

where $\bar{A}$ is a vector, $d\bar{a}$ is a differential surface element vector with magnitude da pointed in the normal direction $\hat{n}$ ($d\bar{a} = \hat{n}\, da$), and $d\bar{l}$ is a differential line element vector with magnitude dl pointed in the direction $\hat{l}$ (see Fig. 2.1). The directions of $\hat{n}$ and $\hat{l}$ are chosen so that $\hat{l}$ is in the direction of the rotation of the right-handed screw advancing in the direction $\hat{n}$.

Using (2.5), (2.1) and (2.2) are expressed as follows:

$$\int_l \bar{H} \cdot d\bar{l} = \oint_a \left(\frac{\partial \bar{D}}{\partial t} + \bar{J} \right) \cdot d\bar{a}, \tag{2.6}$$
$$\int_l \bar{E} \cdot d\bar{l} = \oint_a \frac{\partial \bar{B}}{\partial t} \cdot d\bar{a}. \tag{2.7}$$

Equation (2.6) represents Ampère's law that the line integral of a magnetic field around a closed path is equal to the total current, including the displacement current, $\partial \bar{D}/\partial t$, through the loop. Equation (2.7) represents Faraday's law of induction that the line integral of an electric field around a closed path is equal to the negative of time rate of change of the total magnetic flux through the loop.

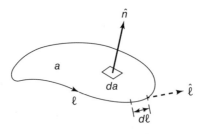

FIGURE 2.1   Stoke's theorem.

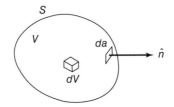

FIGURE 2.2   Divergencen theorem.

Equations (2.3) and (2.4) can be expressed in integral form by using the divergence theorem

$$\int_V \nabla \cdot \bar{A} dV = \int_S \bar{A} \cdot d\bar{a}, \tag{2.8}$$

where S is the closed surface surrounding a volume V and $\hat{n}$ is pointed outward (Fig. 2.2). With (2.8), (2.3) and (2.-4) can be written as

$$\int_S \bar{D} \cdot d\bar{a} = \int_V \rho \, dV, \tag{2.9}$$

$$\int_S \bar{B} \cdot d\bar{a} = 0. \tag{2.10}$$

Equation (2.9) is Gauss's law that the total electric flux flowing out of any closed surface is equal to the total charge enclosed by the surface. Equation (2.10) states that there is no magnetic charge and that there is no net magnetic flux flowing in or out of a closed surface. In addition to Maxwell's equations, the following force law holds concerning the force on a charge q moving with velocity $\bar{v}$ through an electric field $\bar{E}$ and a magnetic field $\bar{B}$:

$$\bar{F} = q(\bar{E} + \bar{v} \times \bar{B}). \tag{2.11}$$

The conservation of charge is embodied in the following continuity equation:

$$\nabla \cdot \bar{J} + \frac{\partial \rho}{\partial t} = 0. \tag{2.12}$$

The integral form of (2.12) is obtained by employing the divergence theorem:

$$\int_S \bar{J} \cdot d\bar{a} + \frac{\partial}{\partial t} \int_V \rho dV = 0. \tag{2.13}$$

This states that the outward flow of the current through a closed surface S must be accompanied by the decrease per unit time of the total charge inside the volume V.

The continuity equation (2.12) can be derived from Maxwell's equations by taking the divergence of (2.1) and using (2.3) and the identity

$$\nabla \cdot \nabla \times \bar{A} = 0. \tag{2.14}$$

Some vector formulas and theorems and the gradient, the divergence, the curl, and the Laplacian in Cartesian, cylindrical, and spherical coordinate systems are shown in Appendix 2.A.

Equations (2.1)–(2.4) exhibit a mathematical similarity. $\bar{E}$ and $\bar{H}$ appear on the left sides of (2.1) and (2.2) under the same operator, and $\bar{D}$ and $\bar{B}$ appear similarly on the right sides. This may appear to suggest that $\bar{E}$ and $\bar{H}$ belong to one class and $\bar{D}$ and $\bar{B}$ to another. However, (2.11) shows that the force depends on $\bar{E}$ and $\bar{B}$, not $\bar{E}$ and $\bar{H}$. In fact, physically, $\bar{E}$ and $\bar{B}$ are the fundamental fields and $\bar{D}$ and $\bar{H}$ are the derived fields related to $\bar{E}$ and $\bar{B}$ through constitutive relations (Section 2.3).

2.2 TIME-HARMONIC CASE

Since the general behavior of a wave as a function of time can always be expressed as a superposition of waves at different frequencies through a Fourier transform, it is sufficient to investigate the characteristics of a wave at a single frequency. The wave at a single frequency is often called a *time-harmonic* or *monochromatic wave* and is most conveniently described by the real part of the phasor field. For example, the vector field $\bar{E}(\bar{r}, t)$, a real function of position $\bar{r}$ and time t, is given by

$$\bar{E}(\bar{r},\ t) = \mathrm{Re}[\bar{E}_{\mathrm{ph}}(\bar{r})e^{j\omega t}], \tag{2.15}$$

where $\bar{E}_{\mathrm{ph}}(\bar{r})$ is a phasor field and, in general, complex. Here we use the convention that $\bar{E}_{\mathrm{ph}}$ is the peak value rather than the rms value, and thus $\bar{E}_{\mathrm{ph}} = \sqrt{2}\bar{E}(\mathrm{rms})$. The x component of $\bar{E}(\bar{r}, t)$ is then given by

$$\begin{aligned} E_x(\bar{r}, t) &= \mathrm{Re}[E_{\mathrm{ph}x}(\bar{r})e^{j\omega t}] \\ &= A_x(\bar{r})\ \cos[\omega t + \phi_x(\bar{r})], \end{aligned} \tag{2.16}$$

where A_x and ϕ_x are the amplitude and phase, respectively, of the x component of the phasor $E_{\mathrm{ph}x} = A_x \exp[j\phi_x]$. In the following, we omit the subscript ph for the phasor whenever no confusion is expected to arise.

Let us rewrite Maxwell's equations for the time-harmonic case:

$$\nabla \times \bar{H}(\bar{r}) = j\omega \bar{D}(\bar{r}) + \bar{J}(\bar{r}), \tag{2.17}$$

$$\nabla \times \bar{E}(\bar{r}) = j\omega \bar{B}(\bar{r}), \tag{2.18}$$

$$\nabla \cdot \bar{D}(\bar{r}) = \rho(r), \tag{2.19}$$

$$\nabla \cdot \bar{B}(\bar{r}) = 0, \tag{2.20}$$

with the continuity equation

$$\nabla \cdot \bar{J}(\bar{r}) + j\omega\rho(\bar{r}) = 0. \tag{2.21}$$

Note that $\exp(j\omega t)$ is dropped, as it is common to all terms and that all the field quantities in (2.17) to (2.21) are phasors. Equation (2.19) can be obtained by taking the divergence of (2.17) and using (2.21). Equation (2.20) is obtained by taking the divergence of (2.18). Thus, for the time-harmonic case, (2.17), (2.18), and (2.21) constitute the complete set of differential equations. For the static field, $\omega = 0$, however, we need (2.18) and (2.19) for electrostatics and (2.17) and (2.20) for magnetostatics.

Once the time-harmonic fields in terms of the phasor [e.g., $\bar{E}(\bar{r})$] are obtained, the general transient field $\bar{E}(\bar{r}, t)$ as a function of time can be obtained by the inverse Fourier transform

$$\bar{E}(\bar{r}, t) = \frac{1}{2\pi} \int \bar{E}(\bar{r}) e^{j\omega t} d\omega. \tag{2.22}$$

Let us next consider the current $\bar{J}$ and charge ρ. It is often convenient to separate the current into the source current and the induced current. For example, in a radio broadcast the current on the transmitting antenna at a radio station is the *source current*, but the currents induced in receiver antennas and nearby metallic walls are considered the *induced current*. Similarly, we can separate the charges into the source charge and the induced charge. The current density $\bar{J}$ and the charge density ρ in Maxwell's equations refer to all currents and charges, source and induced. It is more convenient, however, to express separately the source currents $\bar{J}_s$ and the source charges ρ_s in the following manner:

$$\nabla \times \bar{H}(\bar{r}) = j\omega\bar{D}(\bar{r}) + \bar{J}(\bar{r}) + \bar{J}_s(\bar{r}), \tag{2.23}$$

$$\nabla \times \bar{E}(\bar{r}) = -j\omega\bar{B}(\bar{r}), \tag{2.24}$$

$$\nabla \cdot \bar{D}(\bar{r}) = \rho + \rho_s, \tag{2.25}$$

$$\nabla \cdot \bar{B}(\bar{r}) = 0. \tag{2.26}$$

In this form, $\bar{J}$ and ρ are the induced current density and the induced charge density, respectively, and as will be shown in Section 2.3, $\bar{J}$ and ρ will be incorporated into the medium characteristics.

2.3 CONSTITUTIVE RELATIONS

Let us consider Maxwell's equations (2.23)–(2.26) for time-harmonic electromagnetic fields. The relationships among $\bar{D}, \bar{E}, \bar{B}, \bar{H}$, and $\bar{J}$ depend on the characteristics of the medium, and they are expressed by the *constitutive* parameters. In a linear passive medium, $\bar{D}$ and $\bar{B}$ are linearly related to $\bar{E}$ and $\bar{H}$, respectively. Furthermore, if the constitutive relationships between $\bar{D}$ and $\bar{E}$, and $\bar{B}$ and $\bar{H}$, do not depend on the direction of $\bar{E}$ and $\bar{H}$, the medium is called *isotropic*.

Let us first consider a lossless medium where $\bar{J} = 0$. For a linear, passive, isotropic medium, we have

$$\bar{D} = \varepsilon \bar{E}, \tag{2.27}$$

$$\bar{B} = \mu \bar{H}, \tag{2.28}$$

where ε is the permittivity or dielectric constant (farads/meter), μ is the permeability (henries/meter), and both ε and μ are real and scalar. For free space, the dielectric constant and the permeability are

$$\varepsilon_0 = 8.854 \times 10^{-12} \simeq \frac{10^{-9}}{36\pi} \text{ F/m},$$

$$\mu_0 = 4\pi \times 10^{-7} \text{ H/m}. \tag{2.29}$$

Note that the light velocity in free space is given by

$$c = \frac{1}{(\mu_0 \varepsilon_0)^{1/2}} \simeq 3 \times 10^8 \text{ m/s}. \tag{2.30}$$

It is often convenient to use the dimensionless quantities, the relative dielectric constant ε_r, and the relative permeability μ_r.

$$\varepsilon_r = \frac{\varepsilon}{\varepsilon_0} \quad \text{and} \quad \mu_r = \frac{\mu}{\mu_0}. \tag{2.31}$$

If ε and μ are constant from point to point, the medium is called *homogeneous*.

In an anisotropic medium, where the relationship between $\bar{D}$ and $\bar{E}$ (or $\bar{B}$ and $\bar{H}$) depends on the direction of $\bar{E}$ (or $\bar{H}$) and, therefore, $\bar{D}$ and $\bar{E}$ are in general not parallel, the constitutive parameters should be expressed by the tensor dielectric constant $\bar{\bar{\varepsilon}}$ (*or permeability* $\bar{\bar{\mu}}$):

$$\bar{D} = \bar{\bar{\varepsilon}} \bar{E}, \tag{2.32}$$

$$\bar{B} = \bar{\bar{\mu}} \bar{H}. \tag{2.33}$$

For example, (2.32) in the Cartesian coordinate system can be expressed in the following matrix form:

$$\begin{bmatrix} D_x \\ D_y \\ D_z \end{bmatrix} = \begin{bmatrix} \varepsilon_{11} & \varepsilon_{12} & \varepsilon_{13} \\ \varepsilon_{21} & \varepsilon_{22} & \varepsilon_{23} \\ \varepsilon_{31} & \varepsilon_{32} & \varepsilon_{33} \end{bmatrix} \begin{bmatrix} E_x \\ E_y \\ E_z \end{bmatrix}. \tag{2.34}$$

In (2.32) and (2.33), we called $\bar{\bar{\varepsilon}}$ and $\bar{\bar{\mu}}$ the tensor without defining them. In Chapter 8, we present a detailed discussion of anisotropic media as well as chiral media in which $\bar{D}$ and $\bar{B}$ are coupled to both $\bar{E}$ and $\bar{H}$.

Let us consider a lossy medium. In a linear lossy medium, the current density $\bar{J}$ is proportional to the electric field $\bar{E}$, and this relationship is called *Ohm's law*:

$$\bar{J} = \sigma\bar{E}, \tag{2.35}$$

where σ is the conductivity (siemens/meter) of the medium. For low frequencies up to the microwave region, the conductivity is often essentially real and independent of frequency. We can then rewrite one of Maxwell's equations (2.23) in the following manner:

$$\nabla \times \bar{H} = j\omega\varepsilon\bar{E} + \sigma\bar{E} + \bar{J}_s$$
$$= j\omega\varepsilon_c\bar{E} + \bar{J}_s, \tag{2.36}$$

where $\varepsilon_c = \varepsilon - j(\sigma/\omega)$ is called the complex dielectric constant. Note that in (2.36), the conductivity term $\sigma\bar{E}$ is absorbed into the dielectric constant as the imaginary part. The relative complex dielectric constant is then given by

$$\varepsilon_r = \frac{\varepsilon_c}{\varepsilon_0} = \frac{\varepsilon}{\varepsilon_0} - j\frac{\sigma}{\omega\varepsilon_0} = \varepsilon' - j\varepsilon''. \tag{2.37}$$

The ratio $\varepsilon''/\varepsilon'$ is called the loss tangent,

$$\tan\delta = \frac{\varepsilon''}{\varepsilon'}, \tag{2.38}$$

and the complex index of refraction n is given by

$$n = (\varepsilon_r)^{1/2} = n' - jn''. \tag{2.39}$$

For most materials, the permeability μ is equal to that of free space ($\mu = \mu_0$). However, in magnetic material, μ is different from μ_0 and may be lossy. This will be discussed in more detail later when we discuss ferrite materials. We note here that in general the relative permeability $\mu_r = \mu/\mu_0$ can be complex ($\mu_r = \mu' - j\mu''$). The complex refractive index n is then given by $n = (\varepsilon_r \mu_r)^{1/2}$. Some examples of the relative dielectric constant ε' and the conductivity σ are shown in Table 2.1. Note that using the relative dielectric constant ε_r in (2.37), the displacement vector $\bar{D}$ is now given by

$$\bar{D} = \varepsilon_c\bar{E}.$$

In the frequency ranges above the microwave region, ε and σ in (2.37) are no longer independent of frequency and it is often more convenient simply to use the relative complex dielectric constant ε_c and not separate it into ε and σ. Therefore, whenever

TABLE 2.1 Relative Dielectric Constants and Conductivities for Low Frequencies

	ε'	σ (S/m)
Wet earth	10	10^{-3}
Dry earth	5	10^{-5}
Fresh water	81	10^{-3}
Seawater	81	4
Copper	1	5.8×10^{7}
Silver	1	6.17×10^{7}
Brass	1	1.57×10^{7}

no confusion is expected to arise, we use ε to indicate the complex dielectric constant. Maxwell's equations can then be written as

$$\nabla \times \bar{E} = -j\omega\mu\bar{H}, \tag{2.40}$$

$$\nabla \times \bar{H} = -j\omega\varepsilon\bar{E} + \bar{J}_s, \tag{2.41}$$

$$\nabla \cdot \bar{D} = \rho_s, \tag{2.42}$$

$$\nabla \cdot \bar{B} = 0, \tag{2.43}$$

where $\bar{J}_s$ and $\bar{\rho}_s$ are the source current and charge density, respectively, and ε and μ are, in general, complex and dependent on frequency. The medium is called *dispersive* if ε or μ is dependent on frequency.

In (2.40) to (2.43), the constitutive equations should be written with the complex ε and μ:

$$\bar{D} = \varepsilon\bar{E}, \quad \bar{B} = \mu\bar{H}. \tag{2.44}$$

Equations (2.40)–(2.44) constitute a complete description of the fundamental field equations for a lossy medium. Note that the medium is now expressed in terms of the complex dielectric constant and the complex permeability. Alternatively, we can express the medium in terms of the real dielectric constant and the conductivity.

Instead of the dielectric constant and the permeability, we can use two vectors, the electric polarization $\bar{P}$ and the magnetic polarization $\bar{M}$ defined by

$$\bar{D} = \varepsilon_0\bar{E} + \bar{P},$$
$$\bar{B} = \mu_0(\bar{H} + \bar{M}). \tag{2.45}$$

The vectors $\bar{P}$ and $\bar{M}$ represent the electric and magnetic dipole distributions in the medium and vanish in free space. In a linear medium, they are related to $\bar{E}$ and $\bar{H}$ by the electric and magnetic susceptibilities, χ_e and χ_m:

$$\bar{P} = \chi_e\varepsilon_0\bar{E},$$
$$\bar{M} = \chi_m\bar{H}. \tag{2.46}$$

The dielectric constant ε in (2.44) is defined for a time-harmonic electromagnetic phasor field at a certain frequency ω. If the dielectric constant is independent of frequency, the relationship (2.44) in time is simply given by

$$\bar{D}(\bar{r}, t) = \varepsilon \bar{E}(\bar{r}, t). \tag{2.47}$$

However, if the dielectric constant is a function of frequency, the temporal relationship is a Fourier transform of the product of $\varepsilon(\omega)$ and $\bar{E}(\omega)$ and, therefore, is given by the following convolution integral:

$$\bar{D}(\bar{r}, t) = \int_{-\infty}^{t} h(t - t')\bar{E}(\bar{r}, t')dt', \tag{2.48}$$

where

$$h(t) = \frac{1}{2\pi} \int \varepsilon(\omega)e^{j\omega t}d\omega, \tag{2.49}$$

$$\bar{E}(\bar{r}, t) = \frac{1}{2\pi} \int \bar{E}(\bar{r}, \omega)e^{j\omega t}d\omega. \tag{2.50}$$

The medium whose dielectric constant is a function of frequency $\varepsilon(\omega)$ is called dispersive. Although strictly speaking, all media are dispersive, a medium can often be treated as nondispersive within a frequency range used for a particular problem.

If the medium is linear but time varying, the relationship between $\bar{D}$ and $\bar{E}$ cannot be expressed as the convolution integral (2.48). The general relationship should then be given by

$$\bar{D} = (\bar{r}, t) = \int_{-\infty}^{t} h(t, t - t')\bar{E}(\bar{r}, t')dt'. \tag{2.51}$$

Substituting (2.50) into (2.51), we get

$$\bar{D}(\bar{r}, t) = \frac{1}{2\pi} \int \varepsilon(t, \omega)\bar{E}(\bar{r}, \omega)e^{j\omega t}\, d\omega, \tag{2.52}$$

where $\varepsilon(t,\omega)$ is the time-varying dielectric constant given by

$$\varepsilon(t, \omega) = \int_{0}^{\infty} h(t, t'')e^{-j\omega t''}\, dt''. \tag{2.53}$$

We will, however, not discuss time-varying media in this book.

2.4 BOUNDARY CONDITIONS

At an interface between two media, the field quantities must satisfy certain conditions. Consider an interface between two media with complex dielectric constants

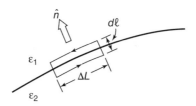

FIGURE 2.3   Boundary conditions for tangential components.

ε_1 and ε_2. In the absence of the source current $\bar{J}_s$, we write (2.41) in integral form,

$$\int_l \bar{H} \cdot d\bar{l} = \int_a j\omega\varepsilon\bar{E} \cdot d\bar{a}, \qquad (2.54)$$

and apply it to a line integral shown in Fig. 2.3. As $dl \to 0$, the right side of (2.54) vanishes and the left side becomes

$$(H_{t1} - H_{t2})\Delta L = 0, \qquad (2.55)$$

where H_{t1} and H_{t2} are the tangential components of the magnetic field in media 1 and 2, respectively. Similarly, from (2.40) we get

$$(E_{t1} - E_{t2})\Delta L = 0, \qquad (2.56)$$

where E_{t1} and E_{t2} are the tangential components of the electric field in media 1 and 2. We therefore state that the tangential components of the electric field and magnetic fields must be continuous, respectively, across the boundary. Mathematically, we write

$$\hat{n} \times \bar{E}_1 = \hat{n} \times \bar{E}_2, \qquad (2.57a)$$
$$\bar{n} \times \bar{H}_1 = \hat{n} \times \bar{H}_2, \qquad (2.57b)$$

where $\hat{n}$ is the unit vector normal to the interface and (E_1, H_1) and (E_2, H_2) are the fields in the medium with ε_1 and ε_2, respectively. If a surface current J_{sf} (amperes/meter) exists on the boundary, we have

$$\hat{n} \times \bar{H}_1 - \hat{n} \times \bar{H}_2 = \bar{J}_{sf}. \qquad (2.58)$$

We can also obtain the conditions on the normal components of the electric and magnetic fields on the boundary. In the absence of the source ($\bar{J}_s = 0$ and $\rho_s = 0$), we apply the divergence theorem to (2.42) and (2.43) for a pillbox volume shown in Fig. 2.4, and in the limit $dl \to 0$ we get

$$(D_{n1} - D_{n2})\,\Delta a = 0,$$
$$(B_{n1} - B_{n2})\,\Delta a = 0. \qquad (2.59)$$

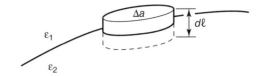

FIGURE 2.4   Boundary conditions for normal components.

his states that the normal components of $\bar{D}$ and $\bar{B}$ must be continuous across the boundary. Mathematically, we write

$$\bar{D}_1 \cdot \hat{n} = \bar{D}_2 \cdot \hat{n},$$
$$\bar{B}_1 \cdot \hat{n} = \bar{B}_2 \cdot \hat{n}. \tag{2.60}$$

If a surface charge exists on the boundary, we have

$$(D_{n1} - D_{n2})\Delta a = \rho_{sf}\Delta a, \tag{2.61}$$

or

$$\bar{D}_1 \cdot \hat{n} - \bar{D}_2 \cdot \hat{n} = \rho_{sf}. \tag{2.62}$$

Here ρ_{sf} is the surface charge density (coulombs/meter^2).

Let us examine how we use the boundary conditions (2.57), (2.58), (2.60), and (2.62) to solve electromagnetic problems. According to the uniqueness theorem, which is discussed in Section 2.10, only one unique solution exists among all the possible solutions of Maxwell's equations that satisfy both Maxwell's equations and the boundary conditions. Thus it is important to determine what constitutes the necessary and sufficient boundary conditions in order to yield the unique solution. (See Morse and Feshbach (1953), Chapter 6, for more details.)

Here, we first note that (2.60) is not independent of (2.57a) and (2.57b) for a time-harmonic case since (2.42) and (2.43) can be derived by taking the divergence of (2.40) and (2.41), respectively. Therefore, the necessary and sufficient boundary conditions are the continuity of the tangential electric and magnetic fields as given in (2.57a) and (2.57b).

If the second medium is a perfect conductor, the fields inside the conductor vanish, and the necessary and sufficient boundary condition is that the tangential component of the electric field is zero on the boundary:

$$\hat{n} \times \bar{E}_1 = 0. \tag{2.63}$$

The surface current density is then given by $\bar{J}_{sf} = \hat{n} \times \bar{H}_1$.

If the second medium is a good conductor, so that $\varepsilon' \ll \varepsilon''$, the wave can only penetrate a distance of the skin depth given by $(2/\omega\mu\sigma)^{1/2}$ (Chapter 3). If the radius of curvature of the surface is much greater than the skin depth, the following condition, called the *Leontovich impedance boundary condition*, holds approximately (Brekhovskikh, 1960):

$$\bar{E}_t = Z_s(\hat{n} \times \bar{H}), \tag{2.64}$$

where $\bar{E}_t$ is the electric field tangential to the surface and $Z_s = (\mu/\varepsilon)^{1/2} \approx (j\mu\omega/\sigma)^{1/2}$. This means that the ratio of the tangential electric field to the tangential magnetic field is constant at the surface. The Leontovich boundary condition eliminates the need to consider the field in the second medium and thus leads to considerable mathematical simplification.

In addition to the foregoing conditions at the boundary, if the region under consideration extends to infinity, the wave must be outgoing at infinity. This requirement at infinity is called the *Sommerfeld radiation condition* (Sommerfeld, 1949). For a scalar field ψ, the radiation condition is given by

$$\lim_{r \to \infty} r \left(\frac{\partial \psi}{\partial r} + jk\psi \right) = 0, \tag{2.65}$$

where $k = \omega/c$. For electromagnetic fields, they are given by (Collin and Zucker, 1969)

$$\lim_{r \to \infty} r \left[\left(\frac{\mu}{\varepsilon} \right)^{1/2} \hat{r} \times \bar{H} + \bar{E} \right] = 0,$$

$$\lim_{r \to \infty} r \left[\hat{r} \times \bar{E} - \left(\frac{\mu}{\varepsilon} \right)^{1/2} \bar{H} \right] = 0. \tag{2.66}$$

This means that the field is outgoing and the Poynting vector is pointed outward and decreases as r^{-2}. The radial components E_r and H_r must decrease faster than r^{-1}.

If the region includes a sharp edge, the field can become infinite, but the energy stored around the edge must be finite. Thus the field must satisfy the edge condition, which is discussed in Appendix 7.C. In general, therefore, the complete mathematical description of the electromagnetic problem includes Maxwell's equations, boundary conditions, the radiation condition, and the edge condition.

2.5 ENERGY RELATIONS AND POYNTING'S THEOREM

Let us consider the general time-varying electromagnetic fields satisfying Maxwell's equations (2.1)–(2.4). To obtain the energy relations, we use the vector identity

$$\nabla \cdot (\bar{A} \times \bar{B}) = \bar{B} \cdot \nabla \times \bar{A} - \bar{A} \cdot \nabla \times \bar{B} \tag{2.67}$$

and let $\bar{A} = \bar{E}$ and $\bar{B} = \bar{H}$. We then substitute $\nabla \times \bar{E}$ from (2.2) and $\nabla \times \bar{H}$ from (2.1) into the right side of (2.67) and obtain

$$\nabla \cdot (\bar{E} \times \bar{H}) + \bar{H} \cdot \frac{\partial \bar{B}}{\partial t} + \bar{E} \cdot \frac{\partial \bar{D}}{\partial t} + \bar{E} \cdot \bar{J} = 0, \tag{2.68}$$

where all field quantities are real functions of position $\bar{r}$ and time t. The vector $\bar{S} = \bar{E} \times \bar{H}$ is called the *Poynting vector* and represents the flow of the power flux per unit area. It is the power flux density and its unit is watts/meter^2.

Equation (2.68) is a mathematical representation of Poynting's theorem. Let us examine its physical meaning. In a nondispersive, lossless medium, $\bar{D} = \varepsilon \bar{E}$ and + and ε and μ are real constants. We can then identify the electromagnetic energy density W:

$$W = W_e + W_m = \frac{1}{2}\varepsilon\bar{E} \cdot \bar{E} + \frac{1}{2}\mu\bar{H} \cdot \bar{H}, \tag{2.69}$$

where W is the sum of the electric W_e and magnetic W_m energy densities. Poynting's theorem (2.68) can then be stated as

$$\nabla \cdot \bar{S} + \frac{\partial}{\partial t}W + \bar{E} \cdot \bar{J} = 0. \tag{2.70}$$

The physical meaning of this can be seen more clearly in the following integral form obtained by applying the divergence theorem to (2.70):

$$-\int_S \bar{S} \cdot d\bar{a} = \frac{\partial}{\partial t}\int_V WdV + \int_V \bar{E} \cdot \bar{J}dV. \tag{2.71}$$

Here the left side of (2.71) is the total power flow into the volume V. The first term on the right represents the time rate of increase of the total electromagnetic energy inside the volume and the second term represents the total power dissipation in the volume. Thus (2.71) states that the total energy flow into a volume per unit time is equal to the sum of the increase in the total electromagnetic energy and the energy dissipation per unit time in the volume.

In many practical problems, we deal with time-harmonic electromagnetic fields, and it is necessary to consider Poynting's theorem for the phasor field quantities. In a time-harmonic case, we use the complex Poynting vector

$$\bar{S} = \frac{1}{2}\bar{E} \times \bar{H}^*, \tag{2.72}$$

where $\bar{E}$ and $\bar{H}$ are phasors. The magnitudes $|\bar{E}|$ and $|\bar{H}|$ are peak values, and therefore the rms values are $(1/\sqrt{2})|\bar{E}|$ and $(1/\sqrt{2})|\bar{H}|$. $\bar{S}$ gives the direction and the rms value of the complex power flux density. The real part of $\bar{S}$ represents the real power flux density, and the imaginary part represents the reactive power flux density. Using the identity (2.67), we write

$$\nabla \cdot \bar{S} = \frac{1}{2}\bar{H}^* \cdot \nabla \times \bar{E} - \frac{1}{2}\bar{E} \cdot \nabla \times \bar{H}^*. \tag{2.73}$$

Substituting Maxwell's equations (2.40) and (2.41) into this, we get Poynting's theorem in the following complex form:

$$\nabla \cdot \bar{S} + 2j\omega(W_m - W_e) + L + \frac{1}{2}\bar{E} \cdot \bar{J}_s^* = 0, \tag{2.74}$$

where

$$W_e = \frac{\varepsilon_0 \varepsilon'}{4}|\bar{E}|^2,$$

$$W_m = \frac{\mu_0 \mu'}{4}|\bar{H}|^2,$$

$$L = \frac{\omega \varepsilon_0 \varepsilon''}{2}|\bar{E}|^2 + \frac{\omega \mu_0 \mu''}{2} + \frac{\omega \mu_0 \mu''}{2}|\bar{H}|^2,$$

$$\varepsilon = \varepsilon_0(\varepsilon' - j\varepsilon''), \quad \varepsilon'' = \frac{\sigma}{\omega \varepsilon_0}, \quad \mu = \mu_0(\mu' - j\mu'').$$

Here ε and μ are complex dielectric constant and permeability, respectively (μ is generalized to be complex). W_e and W_m are the time-averaged electric and magnetic stored energy densities and are equal to the average of $\frac{1}{2}\varepsilon_0\varepsilon'|\bar{E}(t)|^2$ and $\frac{1}{2}\mu_0\mu'|\bar{H}(t)|^2$ over the period $T = 2\pi/\omega$.

$$\frac{1}{T}\int_0^T \frac{\varepsilon_0\varepsilon'}{2}|\bar{E}(t)|^2 dt = \frac{1}{T}\frac{\varepsilon_0\varepsilon'}{2}\int_0^T |\text{Re}[\bar{E}(\text{phasor})e^{j\omega t}]|^2 dt$$

$$= \frac{\varepsilon_0\varepsilon'}{4}|\bar{E}(\text{phasor})|^2. \tag{2.75}$$

The third term L in (2.74) is real and positive and represents the power dissipation per unit volume in a lossy medium. The last term in (2.74) is the power absorbed by the source current $\bar{J}_s$. The power emitted from the source $\bar{J}_s$ is therefore given by $-\frac{1}{2}\bar{E} \cdot \bar{J}_s^*$. This can be seen by taking the volume integral of (2.74) over the source volume only. Then W_e, W_m, and L are zero and the total power emitted is equal to the volume integral of $-\frac{1}{2}\bar{E} \cdot \bar{J}_s^*$.

The Poynting vector $\bar{S}$ is in general complex, and its real part represents the real power flow and its imaginary part represents the reactive power. Taking the real and imaginary part of (2.74), we write

$$-\nabla \cdot \bar{S}_r - \frac{1}{2}\text{Re}(\bar{E} \cdot \bar{J}_s^*) = L, \tag{2.76a}$$

$$-\nabla \cdot \bar{S}_i - \frac{1}{2}\text{Im}(\bar{E} \cdot \bar{J}_s^*) = 2\omega(W_m - W_e), \tag{2.76b}$$

where $\bar{S}_r$ and $\bar{S}_i$ are the real and imaginary parts of $\bar{S}$, respectively.

TABLE 2.2   Dielectric Constants and Conductivities of (a) Muscle, Skin, and Tissues with High Water Content and (b) Fat, Bone, and Tissues with Low Water Content

Frequency (MHz)	Dielectric Constant		Conductivity (S/m)	
	(a)	(b)	(a)	(b)
27.12	113	20	0.612	10.9–43.2
40.68	97.3	14.6	0.693	12.6–52.8
433	53	5.6	1.43	37.9–118
915	31	5.6	1.60	55.6–147
2450	47	5.5	2.21	96.4–213
5000	43.3	5.05	4.73	186–338

Source: Johnson and Guy (1972).

Equation (2.76a) states that the real power flowing into a unit volume $(-\nabla \cdot \bar{S}_r)$ plus the power supplied by the source $-\frac{1}{2}\text{Re}(\bar{E} \cdot \bar{J}_s^*)$ per unit volume is equal to the power loss per unit volume L. Similarly, (2.76b) represents the reactive power per unit volume due to the power flow, the source, and the stored energy densities.

The specific absorption rate (SAR) is used to represent the power loss per unit mass of biological media when the incident power flux density is 1 mW/cm^2 (Table 2.2). If the density of the medium is given by ρ (kg/m^3), the SAR is given by

$$SAR = \frac{L}{\rho} \quad W/kg. \tag{2.77}$$

For biological media, $\mu'' = 0$, and therefore,

$$SAR = \frac{\omega \varepsilon_0 \varepsilon'' |E|^2}{2\rho} = \frac{\sigma |E|^2}{2\rho}. \tag{2.78}$$

The density ρ is usually taken to be approximately equal to that of water ($\rho = 10^3$ kg/m^3).

The definition of the time-averaged electric and magnetic stored energy densities as given in (2.74) and (2.75) is valid if ε and μ are independent of frequency. For a dispersive medium, the time-averaged electric and magnetic stored energy densities are given by

$$W_e = \frac{1}{4}\text{Re}\left[\frac{\partial}{\partial \omega}(\omega\varepsilon)|\bar{E}|^2\right],$$
$$W_m = \frac{1}{4}\text{Re}\left[\frac{\partial}{\partial \omega}(\omega\mu)|\bar{H}|^2\right]. \tag{2.79}$$

Note that if ε and μ are constant, (2.79) reduces to those given in (2.74). The derivation of (2.79) requires consideration of $\omega\varepsilon$ in the neighborhood of the operating frequency and is given in Landau and Lifshitz (1960) and Yeh and Liu (1972).

2.6 VECTOR AND SCALAR POTENTIALS

Maxwell's equations are vector differential equations and each equation represents three scalar equations for each of three orthogonal components. It would be more convenient, therefore, if the vector problem was reduced to a scalar problem with a fewer number of equations. This has been done in electrostatics and magnetostatics by using electrostatic and vector potentials to describe electric and magnetic fields, respectively. The concept of these potentials can be extended to electromagnetic fields in the following manner.

We assume that the medium is isotropic, homogeneous, and nondispersive, and therefore, μ and ε are scalar and constant. First, we note from (2.4) that the divergence of $\bar{B}$ is zero, and recalling that the divergence of the curl of any vector is zero, $\bar{B}$ can be expressed by the curl of an arbitrary vector $\bar{A}$, called the *vector potential*.

$$\nabla \cdot \bar{B} = 0, \tag{2.80}$$

$$\bar{B} = \nabla \times \bar{A}. \tag{2.81}$$

Then the second Maxwell's equation (2.2) becomes

$$\nabla \times \left(\bar{E} + \frac{\partial \bar{A}}{\partial t} \right) = 0. \tag{2.82}$$

Since the curl of the gradient of any scalar function is zero, the bracketed factor is represented by the gradient of an arbitrary scalar function ϕ, which is called the *scalar potential*.

$$\bar{E} + \frac{\partial \bar{A}}{\partial t} = -\nabla \phi. \tag{2.83}$$

Substituting (2.81) and (2.83) into the first of Maxwell's equation (2.1), we get

$$-\nabla \times \nabla \times \bar{A} - \mu\varepsilon\frac{\partial^2 \bar{A}}{\partial t^2} - \mu\varepsilon\nabla\frac{\partial \phi}{\partial t} = -\mu\bar{J}. \tag{2.84}$$

Now, substituting (2.83) into (2.3), we get

$$\nabla^2 \phi + \frac{\partial}{\partial t}\nabla \cdot \bar{A} = -\frac{\rho}{\varepsilon}. \tag{2.85}$$

Alternatively, we can get (2.85) by taking the divergence of (2.84) and using the continuity equation (2.12). Equations (2.84) and (2.85) are the two equations, the vector and scalar potentials must satisfy.

The vector potential $\bar{A}$ above is defined only through $\nabla \times \bar{A}$ in (2.81). In general, a vector field $\bar{A}$ consists of a curl-free component $\bar{A}_1$ and a divergence-free component $\bar{A}_2$.

$$\bar{A} = \bar{A}_1 + \bar{A}_2, \quad \nabla \times \bar{A}_1 = 0, \quad \text{and} \quad \nabla \cdot \bar{A}_2 = 0. \tag{2.86}$$

Since $\nabla \times \bar{A} = \nabla \times \bar{A}_2$ and $\nabla \cdot \bar{A} = \nabla \cdot \bar{A}_1$, we can still choose any $\nabla \cdot \bar{A}$ without affecting $\bar{E}$ and $\bar{H}$. If we choose $\nabla \cdot \bar{A}$ so that it satisfies the *Lorentz condition*

$$\nabla \cdot \bar{A} + \mu\varepsilon \frac{\partial \phi}{\partial t} = 0, \tag{2.87}$$

(2.84) and (2.85) become

$$\nabla^2 \bar{A} - \mu\varepsilon \frac{\partial^2 \bar{A}}{\partial t^2} = -\mu \bar{J}, \tag{2.88}$$

$$\nabla^2 \phi - \mu\varepsilon \frac{\partial^2 \phi}{\partial t^2} = -\frac{\rho}{\varepsilon}, \tag{2.89}$$

where $\nabla^2 \bar{A} = -\nabla \times \nabla \times \bar{A} + \nabla(\nabla \cdot \bar{A})$, and $\bar{J}$ and ρ are related through the continuity equation:

$$\nabla \cdot \bar{J} + \frac{\partial \rho}{\partial t} = 0. \tag{2.90}$$

Once (2.88) and (2.89) are solved for $\bar{A}$ and ϕ, the fields are given by (2.81) and (2.83).

$$\bar{E} = -\frac{\partial \bar{A}}{\partial t} - \nabla\phi, \tag{2.91}$$

$$\bar{B} = \nabla \times \bar{A}. \tag{2.92}$$

Another useful choice of $\nabla \cdot \bar{A}$ is the *Coulomb gauge*, for which $\nabla \cdot \bar{A} = 0$. This is particularly useful for a source-free region ($\bar{J} = 0, \rho = 0$). In this case, (2.84) and (2.85) become

$$\nabla^2 \bar{A} - \mu\varepsilon \frac{\partial^2 \bar{A}}{\partial t} = 0 \quad \text{and} \quad \nabla^2 \phi = 0. \tag{2.93}$$

The fields are given by

$$\bar{E} = -\frac{\partial \bar{A}}{\partial t} \quad \text{and} \quad \bar{B} = \nabla \times \bar{A}. \tag{2.94}$$

For a dispersive, isotropic, and homogeneous medium, the time-harmonic Maxwell's equations (2.40)–(2.43) must be used. We then obtain

$$\nabla^2 \bar{A} + \omega^2 \mu \varepsilon \bar{A} = -\mu \bar{J}_s,$$
$$\nabla^2 \phi + \omega^2 \mu \varepsilon \phi = -\frac{\rho_s}{\varepsilon_c}, \tag{2.95}$$

and the Lorentz condition

$$\nabla \cdot \bar{A} + j\omega\mu\varepsilon\phi = 0. \tag{2.96}$$

Equations (2.95) and (2.96) constitute the basic formulations of electromagnetic problems in terms of the vector and scalar potentials, and the fields are obtained by (2.91) and (2.92) from A and ϕ with $\partial/\partial t$ replaced by $j\omega$. If the medium is inhomogeneous, however, the equations above do not hold, and it is more convenient to go back and start with the original Maxwell's equations.

2.7 ELECTRIC HERTZ VECTOR

It is possible to combine the vector and scalar potentials and the Lorentz condition and form a single vector called the Hertz vector, from which all the field components can be derived. This useful formulation has been applied to many engineering problems.

Let us define the electric Hertz vector $\bar{\pi}$ such that

$$\bar{A} = \mu\varepsilon\frac{\partial\bar{\pi}}{\partial t} \quad \text{and} \quad \phi = -\nabla \cdot \bar{\pi}. \tag{2.97}$$

Then the Lorentz condition is satisfied automatically. Furthermore, we combine $\bar{J}$ and ρ consistent with the continuity equation by using $\bar{P}$:

$$\bar{J} = \frac{\partial\bar{P}}{\partial t} \quad \text{and} \quad \rho = -\nabla \cdot \bar{P}. \tag{2.98}$$

Then we get a single vector equation

$$\nabla^2\bar{\pi} - \mu\varepsilon\frac{\partial^2\bar{\pi}}{\partial t^2} = -\frac{\bar{P}}{\varepsilon}, \tag{2.99}$$

from which all the electromagnetic fields can be derived.

$$\bar{E} = \nabla(\nabla \cdot \bar{\pi}) - \mu\varepsilon\frac{\partial^2}{\partial t^2}\bar{\pi} = \nabla \times \nabla \times \bar{\pi} - \frac{\bar{P}}{\varepsilon} \quad \text{and} \quad \bar{H} = \varepsilon\nabla \times \frac{\partial\bar{\pi}}{\partial t}. \tag{2.100}$$

Equations (2.99) and (2.100) constitute the basic formulation in terms of the electric Hertz vector $\bar{\pi}$. The vector $\bar{P}$ is called the electric polarization vector and is equal to the dipole moment per unit volume of the exciting source.

For a time-harmonic case (2.40) to (2.43) applicable to a dispersive medium, we have

$$\nabla^2\bar{\pi} + \omega^2\mu\varepsilon\bar{\pi} = -\frac{\bar{J}_s}{j\omega\varepsilon}, \tag{2.101}$$

$$\bar{E} = \nabla(\nabla \cdot \bar{\pi}) + \omega^2\mu\varepsilon\bar{\pi} = -\nabla \times \nabla \times \bar{\pi} - \frac{\bar{J}_s}{j\omega\varepsilon}, \tag{2.102}$$

$$\bar{H} = j\omega\varepsilon\nabla \times \bar{\pi}, \tag{2.103}$$

where $\bar{J}_s$ is the source current density and ε is the complex dielectric constant. Equation (2.101) gives scalar wave equation for each Cartesian component π_x, π_y, and π_z. Care should be exercised in the use of (2.101) for the component of $\bar{\pi}$ in other coordinate systems. See Section 3.1.

2.8 DUALITY PRINCIPLE AND SYMMETRY OF MAXWELL'S EQUATIONS

At present, no magnetic charge has been found to exist in nature and Maxwell's equations contain only electric charges and currents. In practice, however, it is often convenient to use the concept of fictitious magnetic currents and charges. For example, we show later that a small current loop is equivalent to a magnetic current. If we include the fictitious magnetic current density $\bar{J}_m$ and charge density ρ_m, Maxwell's equations take the following symmetric form:

$$\nabla \times \bar{H} = \varepsilon\frac{\partial\bar{E}}{\partial t} + \bar{J}, \tag{2.104}$$

$$\nabla \times \bar{E} = -\mu\frac{\partial\bar{H}}{\partial t} - \bar{J}_m, \tag{2.105}$$

$$\nabla \cdot \bar{B} = \rho_m, \tag{2.106}$$

$$\nabla \cdot \bar{D} = \rho. \tag{2.107}$$

Because of this symmetry, we can interchange $\bar{E}$ and $\bar{H}, \bar{J}$ and J_m, ρ and ρ_m, and ε and μ from the unprimed fields to the new primed fields in the following manner:

$$\begin{matrix} \bar{E} \to \bar{H} & \bar{J} \to \bar{J}'_m & \rho \to \rho'_m & \mu \to \varepsilon' \\ \bar{H} \to -\bar{E} & \bar{J}_m \to -\bar{J}' & \rho_m \to -\rho' & \varepsilon \to \mu' \end{matrix} \tag{2.108}$$

Then the primed fields satisfy the same Maxwell's equations. Using this duality principle, when a solution is known for the unprimed fields, the solution for the primed fields can easily be obtained.

The duality relations above are not the only ones to transform the unprimed fields to the primed fields. We can also use the following transformation without affecting Maxwell's equations:

$$\bar{E} \to \sqrt{\frac{\mu}{\varepsilon}}\bar{H} \quad \bar{J} \to \sqrt{\frac{\varepsilon}{\mu}}\bar{J}'_m \quad \rho \to \sqrt{\frac{\varepsilon}{\mu}}\rho'_m$$

$$\bar{H} \to -\sqrt{\frac{\varepsilon}{\mu}}\bar{E} \quad \bar{J}'_m \to -\sqrt{\frac{\mu}{\varepsilon}}\bar{J}' \quad \rho_m \to -\sqrt{\frac{\varepsilon}{m}}\rho'.$$

(2.109)

The transformation does not require the interchange of ε and μ, and therefore it is useful in dealing with the field relationship for the same medium.

2.9   MAGNETIC HERTZ VECTOR

The symmetric Maxwell's equations (2.104)–(2.107) contain the electric current $\bar{J}$ and the magnetic current $\bar{J}_m$. Therefore, the total field consists of the field due to $\bar{J}$ and the field due to $\bar{J}_m$. The field due to $\bar{J}$ was already obtained in Section 2.7 in terms of the electric Hertz vector $\bar{\pi}$. Similarly, the field due to $\bar{J}_m$ can be obtained by using the transformation (2.108) and by replacing the electric Hertz vector $\bar{\pi}$ with the magnetic Hertz vector $\bar{\pi}_m$. Thus we have the following vector equation corresponding to (2.99):

$$\nabla^2 \bar{\pi}_m - \mu\varepsilon\frac{\partial^2}{\partial t^2}\bar{\pi}_m = -\bar{M} \quad \text{and} \quad \bar{M} = \frac{\bar{J}_{ms}}{j\omega\mu},$$

(2.110)

where $\bar{M}$ is the magnetic polarization vector. The fields corresponding to (2.100) are given by

$$\bar{E} = -\mu\nabla \times \frac{\partial\bar{\pi}_m}{\partial t} \quad \text{and} \quad \bar{H} = \nabla(\nabla \cdot \bar{\pi}_m) - \mu\varepsilon\frac{\partial^2}{\partial t^2}\bar{\pi}_m = \nabla \times \nabla s\,\bar{\pi}_m - \bar{M}.$$

(2.111)

The magnetic polarization vector $\bar{M}$ is the magnetic dipole moment per unit volume.
   For a time-harmonic case, we have

$$\nabla^2 \bar{\pi}_m + \omega^2\mu\varepsilon\,\bar{\pi}_m = -\frac{\bar{J}_{ms}}{j\omega\mu},$$

(2.112)

$$\bar{E} = -j\omega\mu\nabla \times \bar{\pi}_m,$$

(2.113)

$$\bar{H} = \nabla(\nabla \cdot \bar{\pi}_m) + \omega^2\mu\varepsilon\,\bar{\pi}_m = \nabla \times \nabla \times \bar{\pi}_m - \frac{\bar{J}_{ms}}{j\omega\mu},$$

(2.114)

where $\bar{J}_{ms}$ is the magnetic source current density.

2.10 UNIQUENESS THEOREM

For a passive network with N terminals, if N voltages $v_1, v_2, \ldots, v_N$ are applied at these N terminals, all the voltages and currents inside the network are uniquely determined. Similarly, if N currents $I_1, I_2, \ldots, I_N$ are applied at the terminals, this also uniquely determines all the voltages and currents. Or, we can specify the voltages at some of the N terminals and specify the currents at the rest of the terminals. This also gives a unique distribution of voltages and currents. It is also obvious that we cannot specify *both* voltages $v_1, \ldots, v_N$ and currents $I_1, \ldots, I_N$ at the N terminals. These conditions appear to be obvious for network problems. However, in electromagnetic problems, these conditions are not obvious since we need to consider a volume V surrounded by a surface S and ask: What field quantities should be specified on the surface S in order to uniquely determine all fields inside?

The quantities could be tangential or normal, electric or magnetic fields, or displacement or flux vectors. Among all these quantities, which can give the unique field inside the surface S? We will show that one of the following three conditions is necessary and sufficient to determine uniquely all the fields inside.

1. The tangential electric field $\hat{n} \times \bar{E}$ is specified on S.
2. The tangential magnetic field $\hat{n} \times \bar{H}$ is specified on S.
3. The tangential electric field $\hat{n} \times \bar{E}$ is specified on a part of S and the tangential magnetic field $\hat{n} \times \bar{H}$ is specified on the rest of S.   (2.115)

Note that these conditions correspond to the three conditions mentioned above for network problems.

We shall prove (2.115) for a time-harmonic case. Let us consider two different fields, $(\bar{E}_1, \bar{H}_1)$ and $(\bar{E}_2, \bar{H}_2)$, both of which satisfy Maxwell's equations. We will show that if both satisfy one of the conditions in (2.115) on S, these two fields are identical within V, and thus the field in V is unique. To show this, consider the fields $\bar{E}_d = \bar{E}_1 - \bar{E}_2$ and $\bar{H}_d = \bar{H}_1 - \bar{H}_2$. Since both $(\bar{E}_1, \bar{H}_1)$ and $(\bar{E}_2, \bar{H}_2)$ satisfy Maxwell's equations, $(\bar{E}_d, \bar{H}_d)$ also satisfies Maxwell's equations and consequently, satisfies Poynting's theorem (2.74). Noting that in a passive medium, the source current $\bar{J}_s$ is zero, we write (2.74) in the following integral form for the volume V:

$$\int_S \frac{1}{2}\bar{E}_d \times \bar{H}_d \times \bar{H}_d^* \cdot d\bar{a} = -2j\omega \int_V \left(\frac{\varepsilon_0 \varepsilon'}{4}|\bar{E}_d|^2 - \frac{\mu_0 \mu'}{4}|\bar{E}_d|^2 \right) dV$$
$$- \int_V \left(\frac{\omega\varepsilon_0\varepsilon''}{2}|\bar{E}_d|^2 + \frac{\omega\mu_0\mu''}{2}|\bar{E}_d|^2 \right) dV. \quad (2.116)$$

If both $(\bar{E}_1, \bar{H}_1)$ and $(\bar{E}_2, \bar{H}_2)$ satisfy one of the conditions (2.115), the left side of (2.116) is zero, since $\bar{E}_d \times \bar{H}_d^* \cdot \hat{n} = \hat{n} \times \bar{E}_d \cdot \bar{H}_d^* = 0$, where $d\bar{a} = \hat{n}da$. Therefore, the right side of equation (2.116) must be zero. The first integral is purely imaginary. The second integral on the right side of (2.116) is always positive and real whenever $\varepsilon'' \neq 0$ and $\mu'' \neq 0$ unless $\bar{E}_d = 0$ and $\bar{H}_d = 0$. For any physical medium, ε'' and μ''

are always nonzero and positive, and therefore $\bar{E}_d$ and $\bar{H}_d$ must be zero, proving that $\bar{E}_1 = \bar{E}_2$ and $\bar{H}_1 = \bar{H}_2$ and the field inside the surface S is unique as long as one of (2.115) is satisfied.

Note that if a lossless cavity is inside S, the field inside the cavity is independent of the field on S and cannot be determined uniquely by the field on S. However, a completely lossless cavity does not exist and thus the uniqueness theorem can be applied to any physical medium. For a general time-varying case, a similar proof can be given to show that (2.115) is also a necessary and sufficient condition (Stratton, 1941, p. 486).

2.11 RECIPROCITY THEOREM

It is well understood that the reciprocity theorem holds for any linear passive network. For example, if a voltage V_a applied at the input terminal produces a short circuit current I_a at the output terminal, a voltage V_b applied at the output terminal produces the short circuit current I_b at the input satisfying the following reciprocity relationship:

$$\frac{I_a}{V_a} = \frac{I_b}{V_b}. \tag{2.117}$$

In electromagnetics, the equivalent relationship is called the "Lorentz reciprocity theorem."

Let us consider time-harmonic electromagnetic fields. We consider the field $(\bar{E}_a, \bar{H}_a)$ produced by $\bar{J}_a$ and $\bar{J}_{ma}$. We also consider the field $(\bar{E}_b, \bar{H}_b)$ produced by $\bar{J}_b$ and $\bar{J}_{mb}$ in the same medium (Fig. 2.5). We first note the following:

$$\nabla \cdot (\bar{E}_a \times \bar{H}_b) = \bar{H}_b \cdot \nabla \times \bar{E}_a - \bar{E}_a \cdot \nabla \times \bar{H}_b. \tag{2.118}$$

We then use Maxwell's equations for $(\bar{E}_a, \bar{H}_a)$:

$$\begin{aligned} \nabla \times \bar{E}_a &= -j\omega\mu\,\bar{H}_a - \bar{J}_{ma}, \\ \nabla \times \bar{H}_a &= j\omega\varepsilon\,\bar{E}_a + \bar{J}_a. \end{aligned} \tag{2.119}$$

We also use the similar equations for $(\bar{E}_b, \bar{H}_b)$. We then get

$$\nabla \cdot (\bar{E}_a \times \bar{H}_b) = -j\omega\mu\,\bar{H}_a \cdot \bar{H}_b - \bar{H}_b \cdot \bar{J}_{ma} - j\omega\varepsilon\,\bar{E}_a \cdot \bar{H}_b - \bar{E}_a \cdot \bar{J}_b. \tag{2.120}$$

Interchanging a and b, we get a similar equation for $\nabla \cdot (\bar{E}_b \times \bar{H}_a)$. Subtracting one from another, we get the *Lorentz reciprocity theorem*.

$$\nabla \cdot (\bar{E}_a \times \bar{H}_b - \bar{E}_b \times \bar{H}_a) = -(\bar{E}_a \cdot \bar{J}_b - \bar{H}_a \cdot \bar{J}_{mb}) + \bar{E}_b \cdot \bar{J}_a - \bar{H}_b \cdot \bar{J}_{ma}. \tag{2.121}$$

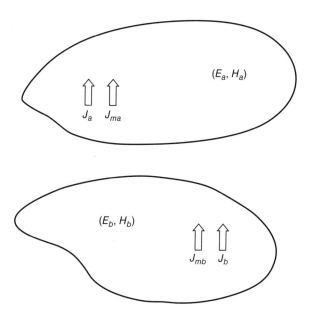

FIGURE 2.5 Lorentz reciprocity theorem.

In order to examine the meaning of the Lorentz reciprocity theorem, let us consider two special cases. If the volume V contains no exciting sources, $\bar{J} = \bar{J}_m = 0$, we have the following Lorentz reciprocity theorem applicable to any closed surface enclosing a source-free region:

$$\int_S (\bar{E}_a \times \bar{H}_b - \bar{E}_b \times \bar{H}_a) \cdot d\bar{a} = 0. \qquad (2.122)$$

Next, we consider a case where the volume V extends to infinity. In this case, at a large distance r from the source, the field component behaves as a spherical wave and is proportional to $(1/r)\exp(-jkr)$ and since $d\bar{a}$ is proportional to r^2, the left side of (2.121) is proportional to $\exp(-j2kr)$. Since $k = k_0 n$ has a negative imaginary part for any physical medium, the left side of (2.121) tends to zero as $r \to \infty$. Therefore, the volume integral of the right side of (2.121) taken over the entire space extending to infinity is zero.

We rewrite this reciprocity relationship using the following form:

$$\langle a, b \rangle = \langle b, a \rangle, \qquad (2.123)$$

where $\langle a, b \rangle = \int_\infty (\bar{E}_a \cdot \bar{J}_b - \bar{H}_a \cdot \bar{J}_{mb}) dV$. This shows that the electric field $\bar{E}_a$ at a point $\bar{r}_1$ due to the current $\bar{J}_a$ at $\bar{r}_2$ is the same as the electric field $\bar{E}_b$ at $\bar{r}_2$ due to the same current $\bar{J}_b = \bar{J}_a$ at $\bar{r}_1$. It should be easy to see the correspondence between (2.117) and (2.123). The quantity $\langle a, b \rangle$ given in (2.123) is called the *reaction* by Rumsey (1954) and is useful in solving boundary value problems.

2.12 ACOUSTIC WAVES

Although electromagnetic waves are vector fields, there are several reasons why we wish to examine scalar fields. First, electromagnetic problems can often be approximated by scalar problems because they reveal many important features of the problem without excessive mathematical complexities. Also, in cases such as two-dimensional problems, vector electromagnetic problems can be reduced to scalar problems. In addition, the study of scalar acoustic problems is of practical interest in many applications, including ultrasound in medicine and ocean acoustics.

For acoustic waves, the pressure variation $p(\bar{r}, t)$ replaces the electric field $\bar{E}(\bar{r}, t)$ in Sections 2.1 and 2.2. In this section, we present a brief description of basic formulations for acoustic waves propagating in material media. We may classify material media into fluids and solids. *Fluids* include both gases and liquids and are in general *viscous*. However, often the effects of viscosity are negligible, and we consider *nonviscous* fluids, which are also called *perfect* fluids. Fluids are in general *compressible*, but when the density of the medium can be assumed constant, they are called *incompressible* fluids. Propagation of an acoustic wave takes place in elastic solids, viscous compressible fluids, and perfect compressible fluids. In this section, we outline the formulation for acoustic-wave propagation in perfect compressible fluids such as gases and liquids. Examples include fog particles in air and bubbles in water (Morse and Ingard, 1968).

There are two fundamental equations: the equation of motion and the equation of conservation of mass. The equation of motion for a small elementary volume of the medium is

$$\rho \frac{d\bar{V}}{dt} + \mathrm{grad}\, p = 0, \tag{2.124}$$

where ρ is density (kg/m^3), $\bar{V}$ is the particle velocity (m/s), and p is pressure (N/m^2). The particle velocity $\bar{V}$ is the velocity of the elementary volume of the fluid and should be distinguished from the velocity c of an acoustic wave in the medium. The derivative d/dt is the time rate of change in a coordinate system attached to a particular portion of a fluid and moving with the fluid. This is called the Lagrangian description. In contrast, $\partial/\partial t$ is the time derivative at a fixed point in space as the fluid flows past that point and is called the Eulerian description. They are related through

$$\frac{d}{dt} = \frac{\partial}{\partial t} + (\bar{V} \cdot \mathrm{grad}). \tag{2.125}$$

The conservation of mass is given by

$$\nabla \cdot (\rho \bar{V}) + \frac{\partial \rho}{\partial t} = 0. \tag{2.126}$$

Let us decompose p, ρ, and $\bar{V}$ into their average values p_0, ρ_0, and $\bar{V}_0$ and the small vibrating acoustic-wave components p_1, ρ_1, and $\bar{V}_1$.

First, we assume that the fluid is stationary, $\bar{V}_0 = 0$, and the magnitudes of acoustic pressure p_1 and density ρ_1 are small compared with the average values p_0 and ρ_0. Then keeping the linear terms, (2.124) and (2.126) become

$$\rho_0 \frac{\partial \bar{V}_1}{\partial t} + \text{grad} p_1 = 0,$$

$$\rho_0 \nabla \cdot \bar{V}_1 + \frac{\partial \rho_1}{\partial t} = 0. \tag{2.127}$$

In general, pressure p is a function of density ρ and therefore for small p_1 and ρ_1, we can expand p in Taylor's series and keep the first two terms.

$$p = p_0 + p_1 = p_0 + \left(\frac{\partial p}{\partial \rho}\right)_{p0} \rho_1. \tag{2.128}$$

Since p_1 is linearly related to ρ_1, we write

$$p_1 = c^2 \rho_1, \quad \text{where} \quad c^2 = \left(\frac{\partial p}{\partial \rho}\right)_{p0}. \tag{2.129}$$

The value of the constant c depends on the material under consideration, and it is the velocity of an acoustic wave in that medium. Note the difference between the particle velocity $\bar{V}_1$ and the acoustic velocity c. Using (2.129), we express ρ_1 in (2.127) by p_1.

$$\nabla \cdot \bar{V}_1 + \frac{1}{c^2 \rho_0} \frac{\partial p_1}{\partial t} = 0. \tag{2.130}$$

Taking the divergence of $\bar{V}_1$ in (2.127) and substituting into (2.130), we get

$$\nabla \cdot \left(\frac{1}{\rho_0} \nabla p_1\right) - \kappa \frac{\partial^2 p_1}{\partial t^2} = 0, \tag{2.131}$$

where $\kappa = 1/c^2 \rho_0$ is called the *compressibility*.

Equation (2.131) is the basic acoustic wave equation for a stationary medium. Note that $(1/\rho_0)$ is inside the divergence operation. For a uniform medium, this can be taken outside the divergence,

$$\nabla^2 p_1 - \frac{1}{c^2} \frac{\partial^2 p_1}{\partial t^2} = 0. \tag{2.132}$$

For a time-harmonic case $\exp(j\omega t)$, we have

$$(\nabla^2 + k^2) p_1 = 0,$$

$$\bar{V}_1 = -\frac{1}{j\omega\rho_0} \nabla p_1. \tag{2.133}$$

The boundary conditions at an interface between two media are:

1. Continuity of pressure p_1
2. Continuity of the normal component of the particle velocity $\bar{V}_1$ [or $(1/\rho_0)$ $(\partial p_1/\partial n)$, where $\partial/\partial n$ is the normal derivative]

The power flux density vector $\bar{S}$ is given by

$$\bar{S} = \tfrac{1}{2} p_1 \bar{V}_1^* . \tag{2.134}$$

Alternatively, we can use the velocity potential ψ. For a uniform medium, the curl of $\bar{V}_1$ is zero from (2.127), and thus it is possible to express $\bar{V}_1$ in terms of a scalar function ψ.

$$\bar{V}_1 = -\text{grad } \psi. \tag{2.135}$$

The function ψ is called the *velocity potential*, and it satisfies the wave equation

$$\nabla^2 \psi - \frac{1}{c^2} \frac{\partial^2 \psi}{\partial t^2} = 0, \tag{2.136}$$

and p_1 and ρ_1 are expressed by

$$p_1 = \rho_0 \frac{\partial \psi}{\partial t}, \quad \rho_1 = \frac{\rho_0}{c^2} \frac{\partial \psi}{\partial t}. \tag{2.137}$$

For a time-harmonic case [$\exp(j\omega t)$], we have

$$(\nabla^2 + k^2)\psi = 0, \quad k = \frac{\omega}{c},$$

$$\bar{V}_1 = -\text{grad } \psi, \quad p_1 = j\omega \rho_0 \psi. \tag{2.138}$$

The boundary conditions at an interface are:

1. Continuity of $\rho_0 \psi$
2. Continuity of $\partial \psi/\partial n$

For a plane acoustic wave propagating in the x direction, we have from (2.133)

$$p_1 = A_0 \exp(-jkx)$$

$$\bar{V}_1 = \hat{x} \frac{p_1}{\rho_0 c}. \tag{2.139}$$

The ratio $\rho_0 c$ of p_1 to V_1 is called the characteristic impedance. Its MKS unit is rayl (after Rayleigh) and 1 rayl = 1 kg/m²s. The MKS unit for acoustic pressure is Pa (pascal) and 1 Pa = 1 N/m² = 10 μbar.

In seawater at 13°C and with a salinity of 35 (in parts by weight per thousand), the standard sound velocity is taken to be 1500 m/s, the standard characteristic impedance is $\rho_0 c = 1.54 \times 10^6$ rayl, and the density $\rho_0 = 1026$ kg/m³. Air at a temperature of

20°C and at standard atmospheric pressure (1 atm $= 1.013 \times 10^5$ N/m^2 $= 1013.25$ mbar) has the density $\rho_0 = 1.21$ kg/m^3, the sound velocity $c = 343$ m/s, and the characteristic impedance $\rho_0 c = 415$ rayl. Oil has a density of 900 kg/m^3, sound velocity of 1300 m/s, and characteristic impedance of 1.117×10^6 rayl.

PROBLEMS

2.1 Consider a circular parallel-plate capacitor with dielectric material shown in Fig. P2.1. A dc current I_0 is switched on at $t = 0$. Find the fields $\bar{E}$, $\bar{H}$, and the Poynting vector $\bar{S} = \bar{E} \times \bar{H}$ as functions of position $\bar{r}$ and time inside the capacitor. Assume that the fringe field is negligible and that the fields are confined within the capacitor. Also find the electromagnetic energy density W defined in (2.69). Show that

$$\nabla \cdot \bar{S} + \frac{\partial}{\partial t} W = 0.$$

2.2 A microwave propagating in tissue is expressed by

$$E_x = E_0 e^{-jkz},$$

$$H_y = \frac{nE_0}{Z_0} e^{-jkz},$$

$$k = \frac{\omega}{c} n, \quad Z_0 = \left(\frac{\mu_0}{\varepsilon_0} \right)^{1/2}.$$

The power flux density at $z = 0$ is 1 mW/cm^2. Calculate E_0, W_e, W_m, L, and SAR as functions of z. Show that they satisfy (2.76a) and (2.76b). Use Table 2.2a and 915 MHz.

2.3 The magnetic Hertz vector $\bar{\pi}_m$ for a small loop antenna is given by

$$\bar{\pi}_m = \frac{Ae^{-jkr}}{4\pi r} \hat{z}, \quad A = \text{constant}.$$

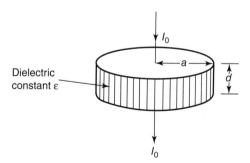

FIGURE P2.1   Circular capacitor.

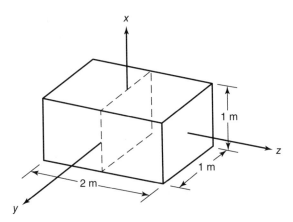

FIGURE P2.6   Current sheet is located at z = 0. The valume is 1 m × 1 m × 2 m.

Find $\bar{E}$ and $\bar{H}$ and express all components (E_θ, E_ϕ, E_r, H_θ, H_ϕ, and H_r) in the spherical coordinate system.

2.4   The electric Hertz vector $\bar{\pi}$ for a small wire antenna in free space is given by

$$\bar{\pi} = \frac{Ae^{-jk_0 r}}{4\pi r}\hat{z}, \quad A = \text{constant}.$$

Find $\bar{E}$ and $\bar{H}$ and express all components (E_θ, E_ϕ, E_r, H_θ, H_ϕ, and H_r) in the spherical coordinate systems.

2.5   Show that the following $\bar{\pi}$ satisfies the wave equation in free space:

$$\bar{\pi} = A\sin k_1 x \sin k_2 y \exp[-j\sqrt{k_0^2 - k_1^2 - k_2^2}\,z]\hat{z}.$$

Find all components of $\bar{E}$ and $\bar{H}$ derived from this $\bar{\pi}$.

2.6   A sheet of uniform current I_0 (A/m) flowing in the x direction is located in free space at $z = 0$. This is represented by the source current density $\bar{J}_s = \hat{x}I_0\delta(z)$. The Hertz vector $\bar{\Pi} = \Pi_x\hat{x}$ is given by

$$\Pi_x = -\frac{I_0\exp(-jk_0|z|)}{2k_0\omega\varepsilon_0}, \quad k_0 = \omega(\mu_0\varepsilon_0)^{1/2}.$$

Find $\bar{E}$ and $\bar{H}$ at z.

   Consider the volume shown in Fig. P2.6 and integrate (2.76) over this volume. Show that the real power flowing out of the volume is equal to the power supplied by the source. Also calculate the reactive power shown in (2.76b).

CHAPTER 3

WAVES IN INHOMOGENEOUS AND LAYERED MEDIA

In this chapter, we start with a review of plane-wave propagation in homogeneous media and reflection and refraction of plane waves by layered media. We then consider the propagation characteristics of guided waves along layered media. This gives us an opportunity to discuss various types of guided waves and to introduce the idea of complex waves. The chapter concludes with the problem of wave propagation in a medium whose refractive index varies as a function of z only. Here we discuss the WKB approximation and the turning point.

3.1 WAVE EQUATION FOR A TIME-HARMONIC CASE

Maxwell's equations are the coupled first-order vector differential equations for the two field quantities $\bar{E}$ and $\bar{H}$. We can combine these two equations, eliminate one of the field quantities, and obtain the uncoupled second-order vector differential equation for one field quantity in the following manner.

Let us write Maxwell's equations for a time-harmonic wave

$$\nabla \times \bar{H} = j\omega\varepsilon\bar{E}, \tag{3.1}$$

$$\nabla \times \bar{E} = -j\omega\mu\bar{H}, \tag{3.2}$$

where $\varepsilon = \varepsilon_0\,\varepsilon_r$ is the complex dielectric constant (permittivity) and $\mu = \mu_0 \cdot \mu_r$ is the permeability. $\mu_r = 1$ for most materials except for magnetic material. We consider a homogeneous medium where ε and μ are constant.

Electromagnetic Wave Propagation, Radiation, and Scattering: From Fundamentals to Applications,
Second Edition. Akira Ishimaru.
© 2017 by The Institute of Electrical and Electronic Engineers, Inc. Published 2017 by John Wiley & Sons, Inc.

To combine (3.1) and (3.2), we take the curl of (3.2) and substitute into (3.1). We get

$$\nabla \times (\nabla \times \bar{E}) - \omega^2 \mu \varepsilon \bar{E} = 0. \tag{3.3}$$

This is a vector wave equation and can be decomposed into three components in three orthogonal directions.

For the special case of a Cartesian coordinate system, we use the following identity:

$$-\nabla \times \nabla \times \bar{E} + \nabla(\nabla \cdot \bar{E}) = \nabla^2 \bar{E}. \tag{3.4}$$

Together with $\nabla \cdot \bar{D} = 0$, we get the following scalar wave equation for each component E_x, E_y, and E_z:

$$(\nabla^2 + k^2)E_x = 0, \tag{3.5}$$

where ∇^2 is the Laplacian, $k = k_0 n$ is the wave number in the medium, $k_0 = \omega(\mu_0 \varepsilon_0)^{1/2} = \omega/c = (2\pi)/\lambda_0$ is the wave number in free space, c is the speed of light in free space, λ_0 is the wavelength in free space, and $n = (\mu_r \varepsilon_r)^{1/2}$ is the refractive index of the medium.

The identity (3.4) is true for the Cartesian system, but care should be exercised in the use of (3.4) for other coordinate systems. For example, in the cylindrical coordinate system (ρ, ϕ, z), the z component of the left side of (3.4) is equal to $\nabla^2 E_z$, the right side. However, the ρ (or ϕ) component of the left side is *not* equal to $\nabla^2 E_\rho$ (or $\nabla^2 E_\phi$).

Once we solve (3.5) and obtain $\bar{E}$, we can easily obtain $\bar{H}$ by (3.2):

$$\bar{H} = -\frac{1}{j\omega\mu}\nabla \times \bar{E}. \tag{3.6}$$

3.2 TIME-HARMONIC PLANE-WAVE PROPAGATION IN HOMOGENEOUS MEDIA

Let us consider a time-harmonic plane wave propagating in the z direction. First, we note that all the fields are functions of z only and since $\nabla \cdot \bar{E} = \partial E_z/\partial z = 0$, E_z is independent of z. Therefore, E_z varying in z is zero, and we conclude that $E_z = 0$ for a plane wave propagating in the z direction. Similarly, we have $H_z = 0$ for a plane wave from $\nabla \cdot \bar{H} = 0$. Since all the components of $\bar{E}$ and $\bar{H}$ are transverse to the direction of the wave propagation, the plane electromagnetic field is called the *transverse electromagnetic* (TEM) wave.

Let us assume that the field is linearly polarized in the x direction ($E_x \neq 0$, $E_y = 0$). Solving (3.5) and using (3.6), we get

$$E_x = E_0 e^{-jkz} \tag{3.7}$$

$$H_y = \frac{E_0}{\eta} e^{-jkz}, \tag{3.8}$$

where $\eta = (\mu/\varepsilon)^{1/2} = \eta_0(\mu_r/\varepsilon_r)^{1/2}$ is the characteristic impedance of the medium, and $\eta_0 = (\mu_0/\varepsilon_0)^{1/2} \approx 120\pi$ ohms is the characteristic impedance of free space.

In a lossy medium, using $n = n' - jn''$, we get

$$E_x = E_0 \exp(-jkz) = E_0 \exp(-j\beta z - \alpha z),$$

$$H_y = \frac{E_z}{\eta}, \tag{3.9}$$

where $\beta = k_0 n' = 2\pi/\lambda = \omega/V$ is called the *phase constant*, λ the wavelength in the medium, V the phase velocity, and α the *attenuation constant*. Also note that at the distance $\delta = \alpha^{-1}$, the wave attenuates to the value of $\exp(-1)$ of the magnitude at $z = 0$. This distance δ is called the *skin depth* of the medium, because the wave cannot penetrate in the medium much farther than this depth. For example, if the medium is highly conducting,

$$\varepsilon_r = \varepsilon' - j\varepsilon'' \approx -j\frac{\sigma}{\omega \varepsilon 0}. \tag{3.10}$$

Therefore, we get

$$\delta \approx \left(\frac{2}{\omega \mu_0 \sigma}\right)^{1/2}, \tag{3.11}$$

In general, however, the skin depth δ is given by $\alpha^{-1} = (k_0 n'')^{-1}$.

A plane wave propagating in an arbitrary direction $\hat{\imath}$ is given by

$$\bar{E} = \bar{E}_0 e^{-j\bar{K}\cdot\bar{r}}, \tag{3.12}$$

We also note that for a plane wave,

$$\nabla e^{-j\bar{K}\cdot\bar{r}} = -j\bar{K}e^{-j\bar{K}\cdot\bar{r}}. \tag{3.13}$$

Therefore, $\nabla \cdot \bar{D} = 0$ means that $\bar{K} \cdot \bar{D} = 0$. Similarly $\bar{K} \cdot \bar{B} = 0$, showing that $\bar{D}$ and $\bar{B}$ are perpendicular to the direction of propagation $\hat{\imath}$.

3.3 POLARIZATION

In Section 3.2, we considered a linearly polarized plane wave polarized in the x direction ($E_x \neq 0$, $E_y = 0$, and $E_z = 0$) and propagating in the z direction. In general, however, the electric field vector $\bar{E}$ can be polarized in any direction ($E_x \neq 0$, $E_y \neq 0$, and $E_z = 0$). If E_x and E_y are in phase, the electric field vector $\bar{E}$ at a given location is always pointed in a fixed direction, and the wave is said to be *linearly polarized*. For a linearly polarized wave, the direction of the electric vector is called the *direction of polarization*, and the plane containing the electric vector and the direction of wave propagation is called the *plane of polarization* (Fig. 3.1a).

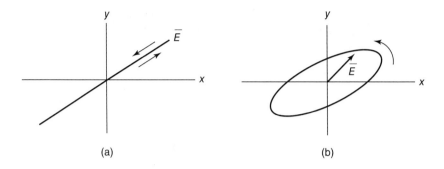

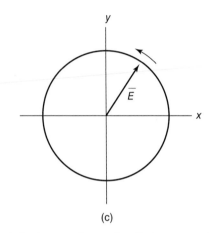

FIGURE 3.1   Polarizations: (a) linear; (b) elliptic; (c) circular (right handed).

In general, when E_x and E_y are not in phase, we write

$$
\begin{aligned}
E_x &= E_1 e^{j\delta_1}, \\
E_y &= E_2 e^{j\delta_2},
\end{aligned}
\tag{3.14}
$$

where $\delta = \delta_1 - \delta_2$ is the phase difference. As a function of time, E_x and E_y are given by

$$
\begin{aligned}
E_x &= E_1 \cos(\omega t + \delta_1), \\
E_y &= E_2 \cos(\omega t + \delta_2).
\end{aligned}
\tag{3.15}
$$

If we eliminate ωt from (3.15), we get an equation to describe the locus of the tip of the electric vector $\bar{E}$. To do so, we write

$$
\frac{E_y}{E_2} = \cos A, \quad A = \omega t + \delta_2,
\tag{3.16}
$$

$$
\frac{E_x}{E_1} = \cos(A + \delta) = \cos A \cos \delta - \sin A \sin \delta.
\tag{3.17}
$$

Eliminating A from (3.16) and (3.17), we get

$$\left(\frac{E_x}{E_1}\right)^2 + \left(\frac{E_y}{E_2}\right)^2 - 2\left(\frac{E_x}{E_1}\right)\frac{E_y}{E_2}\cos\delta = \sin^2\delta. \qquad (3.18)$$

This is an equation of an ellipse, and thus the wave is said to be *elliptically polarized* (Fig. 3.1b).

If the phase difference is exactly 90° or –90° and the magnitudes E_1 and E_2 are the same ($E_1 = E_2 = E_0$), then (3.18) becomes an equation of a circle, and the wave is said to be *circularly polarized* (Fig. 3.1c). If $\delta = +90°$, the rotation of the electric field vector and the direction of the wave propagation form a right-handed screw, and this is called the *right-handed circular polarization* (RHC), and E_x and E_y are related by

$$E_x = +jE_y \qquad (3.19)$$

If $\delta = -90°$, this is called the *left-handed circular polarization* (LHC) and E_x and E_y satisfy

$$E_x = -jE_y. \qquad (3.20)$$

This definition of RHC and LHC is normally used in electrical engineering and IEEE publications. However, in physics, RHC refers to the wave whose electric vector is rotating in the clockwise direction as observed by the receiver, while the engineering definition of RHC refers to the wave whose electric vector is rotating in the clockwise direction as viewed by the transmitter. Therefore, RHC in engineering definition is LHC in physics.

The *polarized waves* described above are deterministic, which means that the field quantities are definite functions of time and position. In contrast, if the field quantities are completely random and E_x and E_y are uncorrelated, the wave is called *unpolarized*. The light from the sun can be considered unpolarized. In many situations the waves may be partially polarized. We include a detailed discussion of this topic, including Stokes' vectors and the Poincaré sphere, in Chapter 10.

3.4 PLANE-WAVE INCIDENCE ON A PLANE BOUNDARY: PERPENDICULAR POLARIZATION (s POLARIZATION)

Let us consider the reflection and refraction of a plane wave incident on a single boundary separating two media with dielectric constants ε_1 and ε_2 and permeabilities μ_1 and μ_2, respectively (Fig. 3.2). Both media can be lossy, and therefore, ε_1, ε_2, μ_1, and μ_2 can be complex. We define the *plane of incidence* as the plane including the direction of the wave propagation and the normal to the boundary, and we choose the x–z plane to be the plane of incidence. The *angle of incidence* θ_i is defined as the angle between the direction of propagation and the normal to the boundary (Fig. 3.2). The refractive indexes n_1 and n_2 are given by $n_1 = [(\varepsilon_1\,\mu_1)/(\varepsilon_0\,\mu_0)]^{1/2}$ and $n_2 = [(\varepsilon_2\,\mu_2)/$

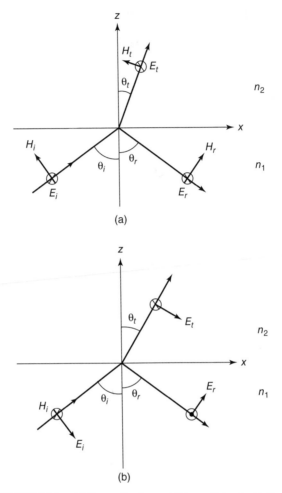

FIGURE 3.2 (a) Perpendicular and (b) parallel polarizations.

$(\varepsilon_0\ \mu_0)]^{1/2}$, and the characteristic impedance of the medium is $\eta_1 = (\mu_1/\varepsilon_1)^{1/2}$ and $\eta_2 = (\mu_2/\varepsilon_2)^{1/2}$, respectively (see Section 3.2).

In Fig. 3.2 we show two cases: (a) perpendicular polarization (electric field perpendicular to the plane of incidence) and (b) parallel polarization (electric field parallel to the plane of incidence). Case (a) is also called the *s polarization* (*senkrecht*, "perpendicular" in German) or the *TE wave* (electric field is transverse to the direction of wave propagation). Case (b) is then called the *p polarization* (parallel) or the *TM wave* (magnetic field is transverse to the direction of wave propagation). These two waves are independent. In fact, for a two-dimensional problem where the dielectric constant and permeability are functions of x and z only $(\partial/\partial y = 0)$, the electromagnetic field can be separated into two independent waves: TE and TM.

Let us consider the perpendicular polarization. We use (3.12) and note that $\hat{\imath} = \hat{x} \sin \theta_i + \hat{z} \cos \theta_i$. We then write the incident field E_{yi} with magnitude E_0.

$$E_{yi} = E_0 \exp(-jq_i z - j\beta_i x), \tag{3.21}$$

where $q_i = k_1 \cos \theta_i$, $\beta_i = k_1 \sin \theta_i$, $k_1 = \omega(\mu_1 \varepsilon_1)^{1/2} = k_0 n_1$ is the wave number in medium I, and $k_0 = \omega(\mu_0 \varepsilon_0)^{1/2}$ is a free-space wave number. The reflected field E_{yr} also satisfies the wave equation, and therefore we write

$$E_{yr} = R_s E_0 \exp(+jq_r z - j\beta_r x), \tag{3.22}$$

where R_s is the reflection coefficient for perpendicular polarization (s-polarization), $q_r = k_1 \cos \theta_r$ and $\beta_r = k_1 \sin \theta_r$, and θ_r is the angle of reflection. Similarly, for the transmitted field we have

$$E_{yt} = T_s E_0 \exp(-jq_t z - j\beta_t x), \tag{3.23}$$

where T_s is the transmission coefficient, $q_t = k_2 \cos \theta_t$, $\beta_t = k_2 \sin \theta_t$, $k_2 = k_0 n_2$, and θ_t is the angle of transmission.

Now we apply the boundary conditions at $z = 0$, the continuity of the tangential electric field E_y and the magnetic field H_x. First we apply the continuity of the tangential electric field:

$$E_{yi} + E_{yr} = E_{yt} \quad at z = 0 \tag{3.24}$$

This yields the following:

$$\exp(-j\beta_i x) + R_s \exp(-j\beta_r x) = T_s \exp(-j\beta_t x) \tag{3.25}$$

For this to hold at all x, all exponents must be the same, thus we get the *phase-matching condition*:

$$\beta_i = \beta_r = \beta_t \tag{3.26}$$

From this, we obtain the *law of reflection*,

$$\theta_i = \theta_r, \tag{3.27}$$

and Snell's law,

$$n_1 \sin \theta_i = n_2 \sin \theta_t. \tag{3.28}$$

With (3.26), (3.25) becomes

$$1 + R_s = T_s. \tag{3.29}$$

To apply the boundary condition for the tangential magnetic field H_x, we use

$$H_x = \frac{1}{j\omega\mu} \frac{\partial}{\partial z} E_y, \tag{3.30}$$

and write

$$H_{xi} = -\frac{E_0}{Z_1} \exp(-jq_i z - j\beta_i x),$$

$$H_{xr} = -\frac{R_s E_0}{Z_1} \exp(+jq_i z - j\beta_i x), \tag{3.31}$$

$$H_{xt} = -\frac{T_s E_0}{Z_2} \exp(-jq_t z - j\beta_i x),$$

where $Z_1 = \omega \mu_1/q_i$ and $Z_2 = \omega \mu_2/q_t$ represent the ratio of the tangential components E_y and H_x and are called the *wave impedance*. We also use Snell's law and write

$$q_t = (k_2^2 - \beta_i^2)^{1/2} = k_2 \cos \theta_t$$

and

$$\cos \theta_t = \left[1 - \left(\frac{n_1}{n_2} \right)^2 \sin^2 \theta_i \right]^{1/2}. \tag{3.32}$$

Since the transmitted wave is propagating in the $+z$ direction, the sign for the square root must be chosen such that Re $q_t > 0$ and Im $q_t < 0$.

We now apply the continuity of the tangential magnetic field

$$H_{xi} + H_{xr} = H_{xt} \quad at\ z = 0. \tag{3.33}$$

We then get

$$\frac{1 - R_s}{Z_1} = \frac{T_s}{Z_2}. \tag{3.34}$$

From (3.29) and (3.34), we get the following *Fresnel formula*:

$$R_s = \frac{Z_2 - Z_1}{Z_2 + Z_1}, \quad T_s = \frac{2Z_2}{Z_2 + Z_1}, \tag{3.35}$$

where Z_1 and Z_2 are as given in (3.31). Note that since $\exp(-jq_t z)$ in (3.23) should represent the wave propagating in the $+z$ direction, if $\cos \theta_t$ becomes complex, the choice of the plus or minus sign in the square root must be made such that the imaginary part of $\cos \theta_t$ is negative. If $\mu_1 = \mu_2 = \mu_0$, (3.35) is reduced to a more familiar form:

$$R_s = \frac{n_1 \cos \theta_i - n_2 \cos \theta_t}{n_1 \cos \theta_i + n_2 \cos \theta_t} \quad \text{and} \quad \frac{2n_1 \cos \theta_i}{n_1 \cos \theta_i + n_2 \cos \theta_t}. \tag{3.36}$$

3.5 ELECTRIC FIELD PARALLEL TO A PLANE OF INCIDENCE: PARALLEL POLARIZATION (p POLARIZATION)

We now consider the parallel polarization. We let E_0 be the magnitude of the incident electric field. The x component of the incident electric field is given by

$$E_{xi} = E_0 \cos \theta_i \exp(-jq_i z - j\beta_i x) \tag{3.37}$$

The reflected and the transmitted waves are

$$E_{xr} = R_p E_0 \cos \theta_i \exp(+jq_i z - j\beta_i x), \tag{3.38}$$
$$E_{xt} = T_p E_0 \cos \theta_t \exp(-jq_t z - j\beta_i x), \tag{3.39}$$

where R_p and T_p are the reflection and transmission coefficients and the law of reflection $\theta_i = \theta_r$ has already been used. As explained in Section 3.4, Snell's law holds here:

$$n_1 \sin \theta_i = n_2 \sin \theta_t \tag{3.40}$$

The magnetic field H_y is related to E_x by

$$E_x = -\frac{1}{j\omega\varepsilon} \frac{\partial}{\partial z} H_y. \tag{3.41}$$

Now we apply the boundary conditions:

$$E_{xi} + E_{xr} = E_{xt} \quad \text{and} \quad H_{yi} + H_{yr} = H_{yt}, \tag{3.42}$$

and obtain the following *Fresnel formula*:

$$R_p = \frac{Z_2 - Z_1}{Z_2 + Z_1}, \quad T_p = \frac{2Z_2}{Z_2 + Z_1} \frac{\cos \theta_i}{\cos \theta_t}. \tag{3.43}$$

Here the wave impedance Z_1 and Z_2 are given by

$$Z_1 = \frac{q_i}{\omega\varepsilon_1} \quad \text{and} \quad Z_2 = \frac{q_t}{\omega\varepsilon_2} \tag{3.44}$$

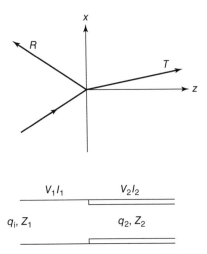

FIGURE 3.3 Equivalent transmission line.

If $\mu_1 = \mu_2 = \mu_0$, (3.43) is reduced to a more familiar form:

$$R_p = \frac{(1/n_2)\cos\theta_t - (1/n_1)\cos\theta_i}{(1/n_2)\cos\theta_t - (1/n_1)\cos\theta_i} \quad \text{and} \quad \frac{(2/n_2)\cos\theta_i}{(1/n_2)\cos\theta_t - (1/n_1)\cos\theta_i}, \quad (3.45)$$

where

$$\cos\theta_t = \left[1 - \left(\frac{n_1}{n_2}\right)^2 \sin^2\theta_i\right]^{1/2}.$$

The factor $\cos\theta_i/\cos\theta_t$ of T_p in (3.43) comes from the definition of T_p as the ratio of the total electric field. The total transmitted and incident electric fields are $E_{xt}/\cos\theta_t$ and $E_{xi}/\cos\theta_i$, respectively.

If we take the tangential components E_{xi}, E_{xr}, and E_{xt}, (3.43) should be identical to (3.35). Equations (3.43) and (3.35) can be recognized as the same as the reflection and transmission coefficients for a junction of two transmission lines. In fact, if we take the tangential electric field as the voltage and the tangential magnetic field as the current, we have a completely equivalent transmission-line problem (Fig. 3.3).

3.6 FRESNEL FORMULA, BREWSTER'S ANGLE, AND TOTAL REFLECTION

Let us examine the Fresnel formula (3.35) and (3.43), when n_1 and n_2 are real and positive. If medium 2 is denser than medium 1 ($n_1 < n_2$), the variations of these coefficients as functions of the incident angle θ_i are as shown in Fig. 3.4a.

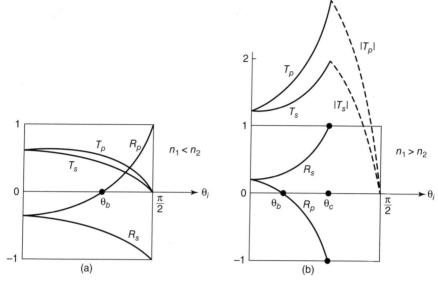

FIGURE 3.4   Reflection and transmission coefficients.

The reflection coefficient R_p for parallel polarization becomes zero when

$$\cos\theta_i = \frac{n_1}{n_2}\cos\theta_t = \frac{n_1}{n_2}\left[1-\left(\frac{n_1}{n_2}\right)^2\sin^2\theta_i\right]^{1/2}.$$

Solving this for θ_i, we get

$$\theta_i = \theta_b = \sin^{-1}\left(\frac{n_2^2}{n_1^2+n_2^2}\right)^{1/2} = \tan^{-1}\frac{n_2}{n_1}. \qquad (3.46)$$

This incident angle θ_b is called *Brewster's angle* (Fig. 3.5), and at this angle all the incident power passes into the second medium. Note that this occurs only when the incident field is polarized in the plane of incidence. If the wave consisting of both parallel and perpendicular polarizations is incident on the boundary at Brewster's angle, the component with parallel polarization is all transmitted into the second medium and the reflected wave consists only of the perpendicular polarization. Note also that the Brewster angle occurs for both $n_1 > n_2$ and $n_1 < n_2$. It is also easy to prove that at Brewster's angle, $\theta_b + \theta_t = \pi/2$. This means that the electric field in the second medium is polarized such that all the polarization vectors $\bar{P}$ are pointed in the direction of the reflected wave (Fig. 3.4). It will be shown that the far-field radiation from each electric dipole in the direction of the dipole axis is zero, and since the reflected wave is a sum of the radiation from all electric dipoles corresponding to the polarization vectors $\bar{P}$, the reflection should be zero at Brewster's angle (Sommerfeld, 1954).

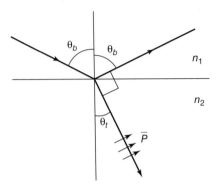

FIGURE 3.5   Brewster's angle.

Let us next examine the case when medium 2 is rarer than medium 1, $n_1 > n_2$. The reflection coefficients are shown in Fig. 3.4b. Note that for the parallel polarization, there is a Brewster angle. Also, note that since $n_1 > n_2$, $\theta_t > \theta_i$, and there is a critical incident angle $\theta_i = \theta_c$ at which θ_t becomes $\pi/2$, and the incident wave $\theta_i > \theta_c$ is totally reflected. The critical angle θ_c is given by

$$\theta_c = \sin^{-1} \frac{n_2}{n_1}. \tag{3.47}$$

When the wave is totally reflected ($\theta_i > \theta_c$), there is no real power transmitted through the boundary. However, this does not mean that there is no field in medium 2. To examine this phenomenon, consider the transmitted wave for the perpendicular polarization.

$$E_{yt} = T_s E_0 \exp(-jk_2 z \cos \theta_t - jk_1 x \sin \theta_i), \tag{3.48}$$

where

$$\cos \theta_t = (1 - \sin^2 \theta_t)^{1/2} = \left[1 - \left(\frac{n_1}{n_2} \right)^2 \sin^2 \theta_i \right]^{1/2}.$$

When $\theta_i > \theta_c = \sin^{-1}(n_2/n_1)$, $\cos \theta_t$ becomes purely imaginary. In fact, as θ_i varies from 0 to θ_c, θ_t varies from 0 to $\pi/2$, and when θ_i becomes greater than θ_c, θ_t becomes complex ($\pi/2 + j\delta$), $\delta > 0$ (Fig. 3.6).

$$\cos \theta_t = \cos \left(\frac{\pi}{2} + j\delta \right) = -j \sin h\delta = -j \left[\left(\frac{n_1}{n_2} \right)^2 \sin^2 \theta_i - 1 \right]^{1/2}.$$

Note that $\cos \theta_t$ must have a negative imaginary part, as noted in Section 3.4. The transmitted field is then given by

$$E_{yt} = T_0 E_0 \exp(-k_2 z \sinh \delta - jk_1 x \sin \theta_i). \tag{3.49}$$

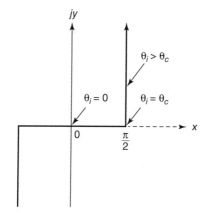

FIGURE 3.6   Complex θ_t plane ($\theta_t = x + jy$).

This shows that the field decays exponentially in the positive z direction. This field, (3.49), carries no real power in the $+z$ direction. However, the reactive power given by the imaginary part of the Poynting vector in the $+z$ direction is not zero and represents the stored energy. This wave is called the *evanescent wave* and occurs in many situations involving total reflections. This is in contrast with the exponentially attenuating wave in a lossy medium, which does carry real power.

Let us next consider the conservation of power across a boundary. We showed at the end of Section 3.5 that the reflection and transmission of a wave at a boundary is equivalent to a transmission-line junction, if the tangential electric and magnetic fields are defined as the voltage and current, respectively. Therefore, the Poynting vector normal to the surface is equal to $\frac{1}{2} VI^*$. Now we let the incident, reflected, and transmitted Poynting vectors be $\bar{S}_i, \bar{S}_r$, and $\bar{S}_t$, respectively. Then we have

$$
\begin{aligned}
P_i &= \bar{S}_i \cdot \hat{z} = \tfrac{1}{2}\bar{E}_i \times \bar{H}_i^* \cdot \hat{z} = \tfrac{1}{2} V_i I_i^*, \\
P_r &= \bar{S}_r \cdot (-\hat{z}) = \tfrac{1}{2}\bar{E}_r \times \bar{H}_i^* \cdot (-\hat{z}) = \tfrac{1}{2} V_i I_i^*, \\
P_t &= \bar{S}_t \cdot \hat{z} = \tfrac{1}{2} E_t \times \bar{H}_i^* \cdot \hat{z} = \tfrac{1}{2} V_t I_t^*.
\end{aligned}
$$

Using $1 + R = T$ and $(1 - R)/Z_1 = T/Z_2$, we can show the conservation of real power:

$$
\operatorname{Re} P_i = \operatorname{Re} P_r + \operatorname{Re} P_t, \tag{3.50}
$$

if Z_1 or R is real.

3.7   WAVES IN LAYERED MEDIA

At the end of Section 3.5, we noted that if we take the tangential electric field to be the voltage and the tangential magnetic field to be the current, we get an equivalent

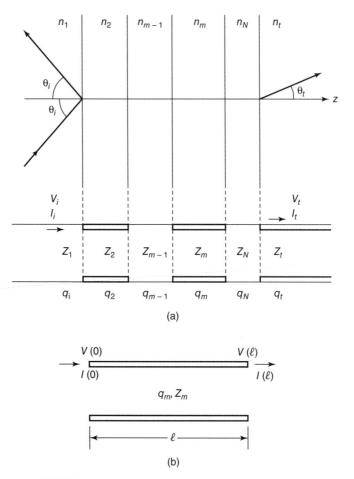

FIGURE 3.7   (a) Layered medium; (b) transmission line.

transmission line. We can easily generalize this result to wave propagation in a layered medium.

Let us first consider the perpendicular polarization. If a wave is incident on a layer as shown in Fig. 3.7, we can obtain the equivalent transmission line as shown. Here we have, from Section 3.4,

$$E_y = V(z)e^{-j\beta x}, \quad -H_x = I(z)e^{-j\beta x},$$

$$q_m = k_m \cos\theta_m = k_m \left[1 - \left(\frac{n_1}{n_m}\right)^2 \sin^2\theta_i\right]^{1/2},$$

$$Z_m = \frac{\omega\mu_m}{q_m}, \quad k_m = k_0 n_m,$$

$$q_i = k_1 \cos\theta_i, \quad q_t = k_t \cos\theta_t. \tag{3.51}$$

The transmission-line problem can be solved by noting that the voltage and current at $z = 0$ and $z = l$ are related by (Fig. 3.7b)

$$\begin{bmatrix} V(0) \\ I(0) \end{bmatrix} = \begin{bmatrix} A_m & B_m \\ C_m & D_m \end{bmatrix} \begin{bmatrix} V(l) \\ I(l) \end{bmatrix}.$$

$$A_m = D_m = \cos q_m l, \quad B_m = j Z_m \sin q_m l, \tag{3.52}$$

$$C_m = \frac{j \sin q_m l}{Z_m}, \quad A_m D_m - B_m C_m = 1.$$

The total layer can be expressed by the total *ABCD* matrix.

$$\begin{bmatrix} V_i \\ I_i \end{bmatrix} = \begin{bmatrix} A & B \\ C & D \end{bmatrix} \begin{bmatrix} V_t \\ I_t \end{bmatrix}, \tag{3.53}$$

where

$$\begin{bmatrix} A & B \\ C & D \end{bmatrix} = \begin{bmatrix} A_2 & B_2 \\ C_2 & D_2 \end{bmatrix} \begin{bmatrix} A_3 & B_3 \\ C_3 & D_3 \end{bmatrix} \cdots \begin{bmatrix} A_N & B_N \\ C_N & D_N \end{bmatrix}.$$

At the input we have

$$V_i = E_0(1 + R_s) \quad \text{and} \quad I_i = \frac{E_0}{Z_1}(1 - R_s). \tag{3.54}$$

At the other end we have

$$V_t = T_s E_0 \quad \text{and} \quad I_t = \frac{T_s E_0}{Z_t}. \tag{3.55}$$

From these we get the solution

$$R_s = \frac{A + B/Z_t - Z_1(C + D/Z_t)}{A + B/Z_t - Z_1(C + D/Z_t)},$$

$$T_s = \frac{2}{A + B/Z_t - Z_1(C + D/Z_t)}. \tag{3.56}$$

For the parallel polarization, we have the identical transmission line, but V, I, and Z are different.

$$V = E_x, \quad I = H_y,$$

$$Z_m = \frac{q_m}{\omega \varepsilon_m}, \quad k_m = k_0 n_m, \tag{3.57}$$

$$q_m = k_m \cos \theta_m.$$

The total electric field E is, however, not E_x, but $E_x / \cos \theta_m$.

3.8 ACOUSTIC REFLECTION AND TRANSMISSION FROM A BOUNDARY

The fundamental equations for acoustic waves in fluid are given in Section 2.12. In terms of the acoustic pressure p_1 and the particle velocity $\bar{V}_1$, we have

$$(\nabla^2 + k^2)p_1 = 0,$$
$$\bar{V}_1 = \frac{1}{j\omega\rho_0}\nabla p_1. \tag{3.58}$$

The boundary conditions are the continuity of p_1 and $(1/\rho_0)(\partial/\partial n)p_1$.

For the wave incident on a boundary at $z = 0$ from the medium 1 with the acoustic velocity c_1 and the density ρ_1 on the medium 2 with c_2 and ρ_2, we write the incident p_i, the reflected p_r, and the transmitted waves p_t.

$$p_i = A_0 \exp(-jq_i z - j\beta_i x),$$
$$p_r = RA_0 \exp(+jq_i z - j\beta_i x), \tag{3.59}$$
$$p_t = TA_0 \exp(-jq_t z - j\beta_i x),$$

where A_0 is a constant, R and T are the reflection and transmission coefficients, respectively, and

$$k_1 = \frac{\omega}{c_1}, \quad k_2 = \frac{\omega}{c_2},$$
$$\beta_i = k_1 \sin\theta_i = k_2 \sin\theta_t \quad \text{(Snells law)},$$
$$q_i = k_1 \cos\theta_i,$$
$$q_t = k_2 \cos\theta_t = (k_2^2 - \beta_i^2)^{1/2},$$
$$\cos\theta_t = \left[1 - \left(\frac{c_2}{c_2}\right)^2 \sin^2\theta_i\right]^{1/2}.$$

Now we satisfy the boundary conditions at $z = 0$:

$$1 + R = T,$$
$$\frac{1}{Z_1}(1 - R) = \frac{1}{Z_2}T, \tag{3.60}$$

where $Z_1 = \rho_1 c_1/\cos\theta_i$ and $Z_2 = \rho_2 c_2/\cos\theta_t$, and $\rho_1 c_1$ and $\rho_2 c_2$ are the characteristic impedances of medium 1 and medium 2, respectively. From (3.60), we get

$$R = \frac{Z_2 - Z_1}{Z_2 + Z_1}, \quad T = \frac{2Z_2}{Z_2 + Z_1}. \tag{3.61}$$

This is identical to electromagnetic problems. There are some differences, however. For the electromagnetic case, μ_1 and μ_2 for nonmagnetic materials are equal to μ_0, and therefore the medium is characterized by the refractive index $n = \sqrt{\varepsilon}$ only. For the acoustic case, the medium is characterized by two parameters, the sound velocity and the density (or compressibility; see Section 2.12). This is equivalent to having different μ and ε for the electromagnetic case. Therefore, even though there is no Brewster's angle for electromagnetic perpendicular polarization, Brewster's angle θ_b is possible for the acoustic case and is given by setting $R = 0$.

$$\sin \theta_b = \left[\frac{(\rho_2/\rho_1)^2 - (c_1/c_2)^2}{(\rho_2/\rho_1)^2 - 1} \right]^{1/2}. \tag{3.62}$$

It is clear that Brewster's angle exists only when the right side of (3.62) is real and its magnitude is less than unity. Thus for Brewster's angle to exist, c_1, c_2, ρ_1, and ρ_2 must satisfy the condition that either $\rho_2/\rho_1 < c_1/c_2 < 1$ or $\rho_2/\rho_1 > c_1/c_2 > 1$.

3.9 COMPLEX WAVES

Up to this point we have discussed the reflection and transmission of a plane wave incident on plane boundaries. We also noted that the reflection coefficient becomes zero when the angle of incidence is equal to Brewster's angle. Can the reflection coefficient become infinite under certain conditions? The infinite reflection coefficient is equivalent to having a finite field with a vanishing incident field and thus represents a wave that is guided along the dielectric surface. This is often called the *surface wave* or the *trapped surface wave*. Important applications of this wave type include guided waves on optical fibers and thin films, microwaves on Goubau lines, and artificial dielectric, and thin dielectric coatings on metal surfaces.

In studying the surface-wave mode and other guided waves, it is important first to recognize various wave types and their mathematical representations. The general wave types are usually called *complex waves*, and the clear understanding of their characteristics is important in the study of surface waves, leaky waves, Zenneck waves, and many other wave types.

To clarify the relationships among various wave types, it is instructive to examine all possible waves propagating along a plane boundary and their physical significance. Let us examine a wave $u(x, z)$ propagating in the z direction in free space. We consider the two-dimensional problem $\partial/\partial y = 0$ and assume that the dielectric or other guiding structure is located below $x = 0$ and that free space extends to infinity in the $+x$ direction. This field $u(x, z)$ satisfies a scalar wave equation.

$$\left(\frac{\partial^2}{\partial x^2} + \frac{\partial^2}{\partial z^2} + k^2 \right) u(x, z) = 0. \tag{3.63}$$

We write a solution in the following form:

$$u(x, z) = e^{-jpx - j\beta z} \tag{3.64}$$

Substituting this into (3.63), we get the following condition:

$$p^2 + \beta^2 = k^2 \tag{3.65}$$

In general, p and β can be complex and thus we let

$$p = p_r - j\alpha_t \quad \text{and} \quad \beta = \beta_r - j\alpha. \tag{3.66}$$

Substituting (3.66) into (3.65) and equating the real and imaginary parts of both sides, we get

$$p_r^2 - \alpha_t^2 + \beta_r^2 - \alpha^2 = k^2, \tag{3.67}$$

$$p_r\alpha_r + \beta_r\alpha = 0. \tag{3.68}$$

Using (3.68), we can first show that the constant-amplitude and constant-phase planes are perpendicular to each other. To prove this, we write

$$u(x, z) = e^{-j(p_r x + \beta_r z) - (\alpha_t x + \alpha z)} = e^{-j\bar{k}_r \cdot \bar{r} - \bar{\alpha} \cdot \bar{r}} \tag{3.69}$$

The constant-phase plane is given by

$$\bar{k}_r \cdot \bar{r} = \text{constant}, \quad \bar{k}_r = p_r\hat{x} + \beta_r\hat{z}, \quad \hat{r} = x\hat{x} + z\hat{z}, \tag{3.70}$$

and the constant-amplitude plane by

$$\bar{\alpha} \cdot \bar{r} = \text{constant}, \quad \bar{\alpha} = \alpha_t\hat{x} + \alpha\hat{z}. \tag{3.71}$$

Thus the constant-phase plane is perpendicular to $\bar{k}_r$ and the constant-amplitude plane is perpendicular to $\bar{\alpha}$. But according to (3.68), $\bar{k}_r$ is perpendicular to $\bar{\alpha}$

$$\bar{k}_r \cdot \bar{\alpha} = p_r\alpha_t + \beta_r\alpha = 0, \tag{3.72}$$

and therefore the constant-amplitude plane is perpendicular to the constant-phase plane (Fig. 3.8).

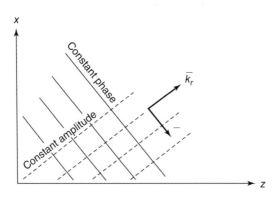

FIGURE 3.8 Complex waves.

TABLE 3.1   Proper and Improper Waves

	Case	β_r	α	p_r	α_t	
	A	+	0	+	0	Fast wave (waveguide modes)
Proper wave	B	+	–	+	+	Backward leaky wave
	C	+	0	0	+	Trapped surface wave
	D	+	+	–	+	Zenneck wave
	E	+	0	–	0	Plane-wave incidence
Improper wave	F	+	–	–	–	
	G	+	0	0	–	Untrapped surface wave
	H	+	+	+	–	Forward leaky wave

Next, we consider the behavior of the wave as $x \to +\infty$. We take the wave whose phase progression along the surface is in the $+z$ direction ($\beta_r > 0$) and then consider the magnitude of the wave in the transverse (x) direction, which is given by

$$|e^{-jpx}| = e^{-\alpha_i x}. \tag{3.73}$$

From this we note that if $\alpha_t > 0$, the wave attenuates exponentially in the $+x$ direction and is called the *proper wave*. If $\alpha_t < 0$, the wave has the amplitude increasing exponentially in the $+x$ direction and is called the *improper wave*. It is obvious that depending on the signs of β_r, α, p_r, and α_t, a variety of wave types can result.

Let us start with the following combinations of p and β shown in Table 3.1. In terms of $\bar{k}_r = p_r \hat{x} + \beta_r \hat{z}$ and $\bar{\alpha} = \alpha_t \hat{x} + \alpha \hat{z}$, they are pictured in Fig. 3.9. In terms of the complex $p = p_r + jp_i$ and $\beta = \beta_r + j\beta_i$ planes, they are shown in Fig. 3.10. Note that for a given β, there can be two values of $p = \pm(k^2 - \beta^2)^{1/2}$. For example, two waves at D and H have the same β but different p. D is the Zenneck wave and H is the leaky wave. Those waves in the upper half-plane of the complex p plane are the improper waves, whereas those in the lower half are the proper waves (see Section 3.12 for Zenneck

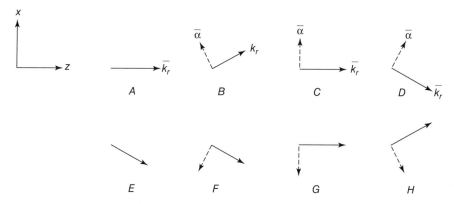

FIGURE 3.9   Proper and improper waves.

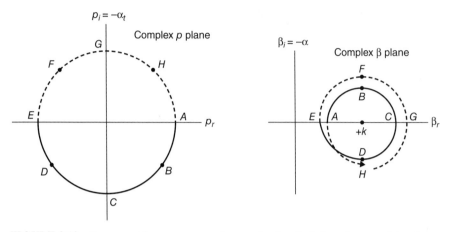

FIGURE 3.10   Proper and improper waves in complex $\beta = \beta_r + j\beta_i$ and $p = p_r + jp_i$ planes, where $\beta_i = -\alpha$ and $p_i = -\alpha_t$.

waves). A more complete discussion must be postponed to Chapter 15, where these waves are interpreted as a portion of the complete spectrum representation of a wave. In this chapter, we concentrate on trapped surface waves and leaky waves.

3.10 TRAPPED SURFACE WAVE (SLOW WAVE) AND LEAKY WAVE

If a wave propagates along a surface with the phase velocity lower than the velocity of light, the wave can be trapped near the surface and can propagate without attenuation. This characteristic is useful for guiding waves over a long distance. Examples are optical communication through optical fiber, microwave transmission through Goubau wire, and surface waves on a dielectric coating. Other examples are waves along Yagi antennas and slow waves along helical structures in traveling-wave tubes.

Let us assume that a wave propagates along a surface without attenuation. If the phase velocity of this wave is slow, $\beta = \beta_r = \omega/v_p > k$. Thus, from (3.65), we get

$$p^2 = k^2 - \beta_r^2 < 0,$$

and therefore, p must be purely imaginary. Writing $p = -j\alpha_t$, the wave is expressed by

$$u(x, z) = e^{-\alpha_t x - j\beta_r z}, \tag{3.74}$$

where $\alpha_t = \sqrt{\beta_r^2 - k^2}$. Obviously, in this case, $\bar{k}_r$ is directed in the z direction and $\bar{\alpha}$ in the x direction (Fig. 3.11).

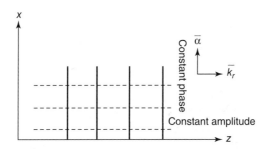

FIGURE 3.11 Trapped surface waves.

We note that the slower the wave, the greater the value of β, and therefore, the greater the value of α_t. Because of the attenuation due to α_t in the $+x$ direction, the wave is mostly concentrated near the surface and the total power is finite:

$$\int_0^\infty |wu(x,y)|^2\, dx = \text{finite}. \tag{3.75}$$

Thus the wave is said to be *trapped* near the surface. The *trapped surface wave* propagates the finite amount of power along the surface without attenuation, and it decays exponentially in the transverse $+x$ direction. In this chapter, we examine several slow-wave structures that support the trapped surface wave.

Let us consider a fast wave propagating along the surface, $\beta_r = \omega/v_p < k$. In this case, in general, the wave must attenuate in the $+z$ direction, $\alpha > 0$, and according to (3.68), $p_r \alpha_t$ must be negative. If we consider the outgoing wave in the $+x$ direction, p_r is positive, and therefore, α_t must be negative. Now $\alpha_t < 0$ represents the wave whose amplitude increases exponentially in the $+x$ direction, and thus this is the improper wave. Note that the amplitude decays in the z direction ($A_1 > A_2 > A_3 > A_4$), but in the x direction at a given z, the amplitude increases exponentially (Fig. 3.12). This wave is called a *leaky wave*, as the energy is constantly leaked out from the surface.

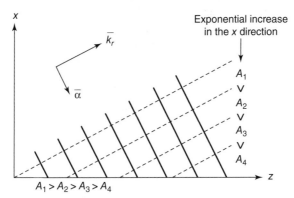

FIGURE 3.12 Leaky waves.

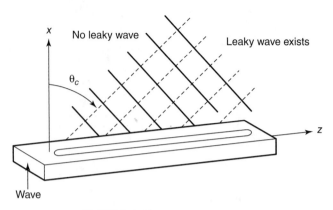

FIGURE 3.13 Leaky waveguides.

Since the leaky wave is an improper wave, it cannot exist by itself, but it can exist within a portion of a space. A typical example is the radiation from a narrow slit cut along a waveguide. The energy leaks out through this slit, and thus the wave attenuates in the z direction. But the wave farther from the surface has greater amplitude because it is originated at the point on the waveguide where the amplitude is greater. The angle θ_c is approximately given by the propagation constant β_z for the waveguide (Fig. 3.13).

$$\beta_z = k \sin \theta_c.$$

Note that the leaky wave exists only in a portion of the space, and thus in the transverse (x) direction the amplitude increases up to a point and then decreases. More rigorous analysis is necessary to analyze this problem completely and will be done later in terms of a Fourier transform technique (see Section 14.1).

Another important example is the coupling of an optical beam into a thin film as shown in Fig. 3.14. A slow wave propagates along a thin film. A prism is placed close to the thin film, and the wave in the prism becomes a leaky wave extracting the power from the surface wave. The leaky wave is then taken out from the prism at Brewster's angle. This process can be reversed and a beam can enter into a prism in the opposite direction and the power can be converted into a surface wave.

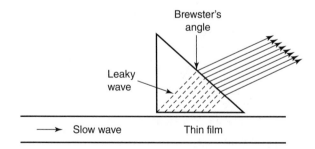

FIGURE 3.14 Coupling of a beam and surface wave.

3.11 SURFACE WAVES ALONG A DIELECTRIC SLAB

A dielectric slab can support a trapped surface wave. This occurs in a variety of situations, such as optical beam propagation along a thin film and optical fibers. Also, radar cross sections of spacecraft and rockets are severely affected by the existence of surface waves on the surface coated with thin dielectric materials. Furthermore, the study of this dielectric slab requires a basic mathematical technique that is common to other wave propagation problems for planar stratified media, such as the wave propagation along the earth–ionosphere waveguide. In this section, we consider the characteristics of trapped surface-wave modes on a dielectric slab.

Let us consider a wave propagating in the z direction along a dielectric slab of thickness d placed on a perfectly conducting plane (Fig. 3.15). Since this is a two-dimensional problem with $\partial/\partial y = 0$, there are two independent modes. One consists of H_y, E_x, and E_z and since the magnetic field H_y is transverse to the direction of propagation (the z direction), we may call them TM (transverse magnetic) modes. The other consists of E_y, H_x, and H_z and may be called TE (transverse electric) modes.

The TM modes can be derived from H_y, which satisfies the wave equation

$$\left(\frac{\partial^2}{\partial x^2} + \frac{\partial^2}{\partial z^2} + k^2 \right) H_y = 0, \tag{3.76}$$

and E_x and E_z are given in terms of H_y:

$$E_x = j\frac{1}{\omega\varepsilon}\frac{\partial}{\partial z}H_y, \quad E_z = -j\frac{1}{\omega\varepsilon}\frac{\partial}{\partial x}H_y. \tag{3.77}$$

Similarly, the TE mode is given by E_y:

$$\left(\frac{\partial^2}{\partial x^2} + \frac{\partial^2}{\partial z^2} + k^2 \right) E_y = 0, \tag{3.78}$$

$$H_x = -j\frac{1}{\omega\mu}\frac{\partial}{\partial z}E_y, \quad H_z = j\frac{1}{\omega\mu}\frac{\partial}{\partial x}E_y, \tag{3.79}$$

where $k = k_0$ and $\varepsilon = \varepsilon_0$ in free space and $k = k_1$ and $\varepsilon = \varepsilon_1$ in the dielectric slab.

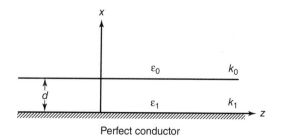

FIGURE 3.15 Dielectric slab on a conducting surface.

Let us first consider the TM mode. This mode can easily be excited by a horn or a waveguide. We solve (3.76) in free space and in the dielectric slab.

$$H_{y0} = Ae^{-jp_0x - j\beta z} \quad \text{for} \quad x > d \text{ (free space)}, \tag{3.80}$$

$$H_{y1} = Be^{-jpx} - j\beta z + Ce^{+jpx - j\beta z} \quad \text{for} \quad d > x > 0 \quad \text{(in dielectric)}, \tag{3.81}$$

where

$$p_0^2 + \beta^2 = k_0^2 \quad \text{and} \quad p^2 + \beta^2 = k_1^2.$$

We note that the propagation constant β along the surface must be the same for both free space and dielectric to ensure the same phase progression along the surface. This is necessary to satisfy the boundary condition. We also note that (3.80) has only one term because this region extends to infinity and only the outgoing wave exists. Equation (3.81), on the other hand, consists of two terms representing the waves propagating in the $+x$ and the $-x$ direction.

Now the boundary conditions at $x = 0$ and $x = d$ are applied. At $x = 0$, we have

$$E_{z1} = -j\frac{1}{\omega\varepsilon_1}\frac{\partial}{\partial x}H_{y1} = 0, \tag{3.82}$$

and at $x = d$, E_z and H_y must be continuous.

$$E_{z0} = E_{z1} \quad \text{and} \quad H_{y0} = H_{y1}, \tag{3.83}$$

where

$$E_{z0} = -j\frac{1}{\omega\varepsilon_0}\frac{\partial}{\partial x}H_{y0} \quad \text{and} \quad E_{z1} = -j\frac{1}{\omega\varepsilon_1}\frac{\partial}{\partial x}H_{y1}.$$

Equations (3.82) and (3.83) may be written as

$$-B + C = 0, \quad Be^{-jpd} + Ce^{jpd} = Ae^{jp_0d},$$

$$\frac{p}{\varepsilon_1}(Be^{-jpd} - Ce^{jpd}) = \frac{p_0}{\varepsilon_0}Ae^{-jp_0d}.$$

From these, we eliminate A, B, and C and obtain the transcendental equation to determine the propagation constant β.

$$\frac{p}{\varepsilon_1}\tan pd = j\frac{p_0}{\varepsilon_0}. \tag{3.84}$$

Once this equation is solved for p, the propagation constant β is given by $\beta = (k_1^2 - p^2)^{1/2}$. Therefore, p is a number that is characteristic of the geometry of the problem and is called the *eigenvalue*. Equation (3.84) is the *eigenvalue equation*.

The ratio of the amplitudes can also be determined.

$$\frac{B}{A} = \frac{C}{A} = \frac{e^{-jp_0 d}}{2 \cos pd}.$$ (3.85)

Equations (3.84) and (3.85) should give the propagation constant and the field configurations for the TM mode.

3.11.1 Slow-Wave Solution for the TM Mode

Let us consider the most important case of a trapped surface-wave propagation. As noted in Section 3.10, we look for a solution of the form

$$H_{y0} = Ae^{-\alpha_t x - j\beta z},$$ (3.86)

with

$$\beta > k_0 \quad \text{and} \quad p_0 = -j\alpha_t.$$

In this case, $\beta^2 - \alpha_t^2 = k_0^2$ and $\beta^2 - p^2 = k_1^2$ and thus, eliminating β from these two, we get

$$\alpha_t^2 = (k_1^2 - = k_0^2) - p^2.$$ (3.87)

Thus the eigenvalue equation (3.84) becomes, for $\alpha_t > 0$,

$$\frac{\varepsilon_0}{\varepsilon_1} X \tan X = \sqrt{V^2 - X^2},$$ (3.88)

where $X = pd$ and $V^2 = (k_1^2 - k_0^2)d^2$. If the dielectric constant ε of the material is greater than ε_0, V^2 is real and positive. If we plot the left side and the right side of (3.88) as functions of X, we get a curve Y_l for the left side and a circle Y_r for the right side. The solution is obtained at the intersection of Y_r and Y_l (Fig. 3.16). Note that when $0 < V < \pi$, there is only one solution. Similarly, when $(N-1)\pi < V < N\pi$, there are N different modes. Therefore, there exist N trapped surface-wave modes if the thickness of the slab d is given by

$$\frac{N-1}{2(\varepsilon_r - 1)^{1/2}} < \frac{d}{\lambda_0} < \frac{N}{2(\varepsilon_r - 1)^{1/2}}, \quad \varepsilon_r = \frac{\varepsilon_1}{\varepsilon_0},$$ (3.89)

and in particular, there exists only one mode if

$$0 < \frac{d}{\lambda_0} < \frac{1}{2(\varepsilon_r - 1)^{1/2}}.$$ (3.90)

The parameter $V = k_0 d(n_1^2 - 1)^{1/2}$ determines the number of modes and is called the *normalized frequency*. The total number of propagating modes N is given by the *mode volume formula*:

$$(N - 1)\pi < V < N\pi \tag{3.91}$$

Once we find $X = pd$ from (3.88), the propagation constant β is obtained by $\beta^2 = k_1^2 - p^2$. As is seen for $V = V_2$ (Fig. 3.16), there are also some solutions for $\alpha_t < 0$. This is called *untrapped surface wave*. When this structure is excited by a source, this wave has some influence over the total field. In practice, however, this untrapped surface wave contributes very little to the total field and is insignificant.

The propagation constant β can be found from the value of X in (3.88) using $\beta^2 = k_1^2 - p^2$. A general shape of k_0–β diagram may be seen by examining a few important points (Fig. 3.17). For example, as $k_0 \to 0$, the curve approaches $k_0 = \beta$, and as $k_0 \to \infty$, the curve approaches $k_1 = \beta$. Also, note that at each cutoff frequency (k_c), the curve is tangent to $k_0 = \beta$. It is clear then that β is always between k_0 and k_1, and thus the phase velocity is always faster than that of a plane wave in dielectric $1/\sqrt{\mu_0 \varepsilon_1}$, but slower than that in free space $1/\sqrt{\mu_0 \varepsilon_0}$. At the cutoff frequency k_c, $V = X = N\pi$ in (3.88) and α_t becomes zero. This means that at the cutoff, the magnitude of the wave outside the slab becomes uniform, extending to infinity instead of decaying exponentially.

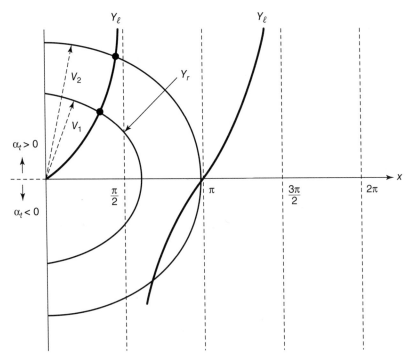

FIGURE 3.16   The plot of the left side Y_l and the right side Y_r of (3.88). The values of X at which Y_l and Y_r intersect are the solution. $V^2 = k_0^2(\varepsilon_r - 1) d^2$.

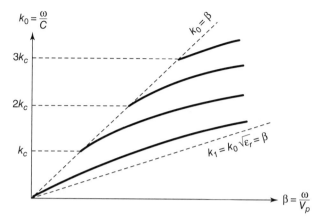

FIGURE 3.17 $k_0 - \beta$ diagram for the wave along a dielectric slab. The cutoff wavenumber $k_c = (\pi/d)(1/\sqrt{\epsilon_1 - 1})$

The field in dielectric is given using (3.85)

$$H_{y1} = Ae^{-\alpha_t d}\frac{\cos px}{\cos pd}e^{-j\beta z},\qquad(3.92)$$

and other field components are easily obtained from (3.77).

$$E_{x1} = \left(\frac{\beta}{k}\eta_0\right)Ae^{-\alpha_t x - j\beta z}\qquad\text{in dielectric slab,}$$

$$E_{x0} = \left(\frac{\beta}{k_1}\eta_1\right)Ae^{-\alpha_t d}\frac{\cos px}{\cos pd}e^{-j\beta z}\quad\text{in dielectric slab,}$$

where $\eta_0 = (\mu_0/\varepsilon_0)^{1/2}$ and $\eta_1 = (\mu_0/\varepsilon_1)^{1/2}$ are the characteristic impedances of free space and the dielectric medium. Note that the transverse electric and magnetic fields are perpendicular to each other and have the same configuration. The ratio of these two is the wave impedance $Z_e = (\beta/k)\eta$. Also, the field has sinusoidal distribution (cos px) inside dielectric, but it has exponential behavior ($e^{-\alpha_t x}$) in free space (Fig. 3.18).

Next, let us examine the impedance looking down on the surface of the dielectric slab. Noting that the Poynting vector $E_z\hat{z} \times H_y^*\hat{y}$ is pointed into the dielectric, the impedance is given by

$$\frac{E_z}{H_y} = j\frac{\alpha_t}{k_0}\eta_0.\qquad(3.93)$$

This is purely inductive. It is clear that the surface supporting the trapped surface wave must be purely reactive. The TM mode requires a purely inductive surface, whereas the TE mode requires a purely capacitive surface.

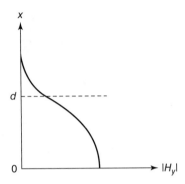

FIGURE 3.18   Field distribution for a trapped surface wave.

3.11.2 Slow-Wave Solution for the TE Mode

In a similar manner, we obtain the solution for the TE mode. We get

$$E_{y0} = Ae^{-\alpha_t x - j\beta z},$$

$$H_{x0} = -\left(\frac{\beta}{k_0\eta_0}\right)Ae^{-\alpha_t x - j\beta z},$$

$$H_{z0} = -j\left(\frac{\alpha_t}{k_0\eta_0}\right)Ae^{-\alpha_t x - j\beta z},$$

(3.94)

in free space $x > d$, and

$$E_{y1} = Ae^{-\alpha_t d}\frac{\sin px}{\sin pd}e^{-j\beta z},$$

$$H_{x1} = -\left(\frac{\beta}{k_1\eta_1}\right)Ae^{-\alpha_t d}\frac{\sin px}{\sin pd}e^{-j\beta z},$$

$$H_{z1} = j\left(\frac{p}{k_1\eta_1}\right)Ae^{-\alpha_t d}\frac{\cos px}{\sin pd}e^{-j\beta z},$$

(3.95)

in the dielectric $0 < x < d$, where $\eta_0 = (\mu_0/\varepsilon_0)^{1/2}$ and $\eta_1 = (\mu_0/\varepsilon_1)^{1/2}$. The eigenvalue equation is then, using $X = pd$,

$$-X\cot X = \sqrt{V^2 - X^2} \quad \text{where} \quad V^2 = k_0^2(\varepsilon_r - 1)d^2.$$

(3.96)

The left and right sides of (3.96) are pictured in Fig. 3.19. Note that there is no solution for $V < \pi/2$ and there are N solutions for $(N - \frac{1}{2})\pi < V < (N + \frac{1}{2})\pi$. This is the mode volume equation for TE modes. Thus the minimum thickness d required for a trapped wave is

$$\frac{d}{\lambda_0} > \frac{1}{4(\varepsilon_r - 1)^{1/2}}.$$

(3.97)

The field distribution for the lowest mode is shown in Fig. 3.20.

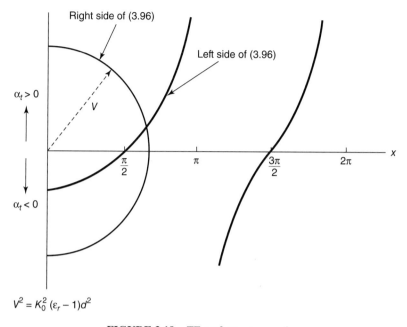

$$V^2 = K_0^2\,(\varepsilon_r - 1)d^2$$

FIGURE 3.19   TE surface wave modes.

3.12 ZENNECK WAVES AND PLASMONS

In Section 3.9, we discussed various complex waves. In this section we show some examples. Consider a TM wave propagating along a plane boundary (Fig. 3.21). In air, $x > 0$, we have

$$H_{y0} = Ae^{-jp_0x - j\beta z},$$
$$p_0^2 + \beta^2 = k_0^2. \tag{3.98}$$

In the lower half-space $x < 0$, we have

$$H_{y1} = Ae^{+jpx - j\beta z},$$
$$p^2 + \beta^2 = k_1^2 = k_0^2\varepsilon_r. \tag{3.99}$$

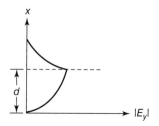

FIGURE 3.20   Lowest TM mode.

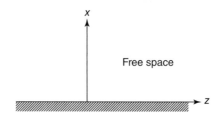

FIGURE 3.21 Zenneck waves and plasmons.

Here the continuity of H_y at $x = 0$ is already satisfied. The continuity of E_z yields the following:

$$\varepsilon_r p_0 = -p. \tag{3.100}$$

From this, we get the following relating β to k_0 and k_1.

$$\frac{1}{\beta^2} = \frac{1}{k_1^2} + \frac{1}{k_0^2}. \tag{3.101}$$

Let us consider the case when the medium $(x < 0)$ is highly conducting. Then we have

$$\varepsilon_r = \varepsilon' - j\varepsilon'' \approx -j\varepsilon'', \tag{3.102}$$

$$\varepsilon'' = \frac{\sigma}{\omega\varepsilon_0} \gg 1.$$

We then get

$$\beta = \frac{k_0}{[1 + 1/\varepsilon_r]^{1/2}} \approx k_0 \left(1 - \frac{1}{2\varepsilon_r}\right),$$

$$p_0 = -\frac{k_0}{(1 + \varepsilon_r)^{1/2}}, \tag{3.103}$$

$$p = \frac{\varepsilon_r k_0}{(1 - \varepsilon_r)^{1/2}}.$$

Here $(1 + \varepsilon_r)^{1/2}$ is chosen such that its imaginary part is negative. These are shown in Fig. 3.22. β and p_0 may be compared with those shown in Fig. 3.10. The propagation constant β was obtained by Zenneck to explain the propagation characteristics of radio waves over the earth. The excitation of such a wave by a dipole source was investigated by Sommerfeld, and we discuss this historical Sommerfeld dipole problem in Chapter 15.

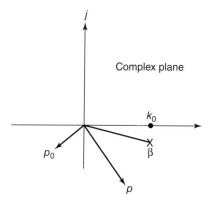

FIGURE 3.22   Zenneck waves.

Next we consider the case when ε_r is real and negative. Ionized gas, called *plasma*, has this dielectric constant when the operating frequency is below plasma frequency and the loss is negligible. We let $\varepsilon_r = -|\varepsilon|$ and note that if $|\varepsilon| > 1$, the propagation constant is real, $p_0 = -j\alpha_0$, and $p = -j\,\alpha$. This is a trapped surface wave propagating along the boundary.

Metal at optical frequencies can have a dielectric constant with a negative real part. For example, at $\lambda = 0.6$ µm, silver has $\varepsilon_r = -17.2 - j0.498$. In this case the wave is not trapped, but the wave propagating along the surface exhibits characteristics similar to surface waves (Fig. 3.23). This is called the *plasmon*, and it has significant effects on the optical scattering characteristics of metal surfaces.

Zenneck waves and plasmons cannot exist by themselves, but when the surface is illuminated by a localized source and the field is expressed in a Fourier transform,

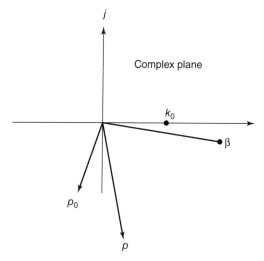

FIGURE 3.23   Plasmons.

these waves appear as poles in the complex Fourier transform plane and the locations of the poles have significant effects on the total field. Above, we considered only TM waves. It can be shown, however, that these poles do not exist for TE waves.

3.13   WAVES IN INHOMOGENEOUS MEDIA

Up to this point we have discussed waves in homogeneous media separated by plane boundaries. Even though they represent a great many problems, there are many practical problems where the medium should be represented by a continuous function of position. For example, the ionosphere can be approximated by a medium with a sharply bounded lower edge at VLF (very low frequency, 1–30 KHz), but for higher frequencies, the variation of the electron density over a wavelength is gradual and must be represented by a continuous function of height. The medium whose dielectric constant ε or permeability μ is a function of position is called the *inhomogeneous medium*.

In general, the characteristics of wave propagation and scattering in an inhomogeneous medium cannot be described in a simple form. However, relatively simple descriptions of the field are possible for two extreme cases: low-frequency and high-frequency fields. The high-frequency approximation is useful whenever the variation of the refractive index of the medium is negligibly small over the distance of a wavelength. This applies to ultrasound propagation in biological media, wave propagation in the ionosphere, microwaves and optical waves in large dielectric bodies, and sound propagation in ocean water. In contrast to this high-frequency approximation, the low-frequency approximation is applicable whenever the sizes of a dielectric body are much smaller than a wavelength. Examples are microwave radiation in a biological medium and scattering from small particles in the ocean, atmosphere, and biological media. In the following sections, we discuss the high-frequency approximation.

Let us consider a plane wave incident upon a medium whose dielectric constant is a function of height z. We choose the x axis so that the plane of incidence is in the x–z plane (Fig. 3.24). This is a two-dimensional problem ($\partial/\partial y = 0$), and thus there are two independent waves, TE (electric field is transverse to the z axis) and TM (magnetic field is transverse to the z axis). For TE (perpendicular polarization), we combine Maxwell's equations

$$\nabla \times \bar{E} = -j\omega\mu\bar{H},$$
$$\nabla \times \bar{H} = j\omega\varepsilon(z)\bar{E}, \tag{3.104}$$

and get

$$\nabla \times \nabla \times \bar{E} = \nabla(\nabla \cdot \bar{E}) - \nabla^2\bar{E} = k_0^2 n^2(z)\bar{E}.$$

Noting that $\nabla \cdot \bar{E} = (\partial/\partial y)E_y = 0$, we get

$$\left[\frac{\partial^2}{\partial x^2} + \frac{\partial^2}{\partial z^2} + k_0^2 n^2(z)\right] E_y = 0. \tag{3.105}$$

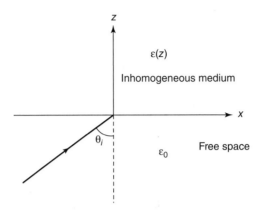

FIGURE 3.24   Wave incident in an inhomogeneous medium.

For TM (parallel polarization), we eliminate $\bar{E}$ from (3.104) and get

$$\nabla \times \bar{E} = \nabla \times \left(\frac{1}{j\omega\varepsilon} \nabla \times \bar{H} \right) = -j\omega\mu\bar{H}. \tag{3.106}$$

Rewriting this using $\bar{H} = H_y\hat{y}$ and $\partial/\partial = 0$, we get

$$\frac{\partial^2}{\partial x^2}H_y + \varepsilon(z)\frac{\partial}{\partial z}\left[\frac{1}{\varepsilon(z)}\frac{\partial}{\partial z}H_y \right] + k_0^2 n^2(z)H_y = 0. \tag{3.107}$$

We may simplify this by using

$$U = \frac{H_y}{n(z)}, \tag{3.108}$$

and get

$$\left[\frac{\partial^2}{\partial x^2} + \frac{\partial^2}{\partial z^2} + k_0^2 N^2(z) \right] U = 0, \tag{3.109}$$

where

$$k_0^2 N^2(z) = k_0^2 n^2 + \frac{n''}{n} - \frac{2n'^2}{n^2}.$$

Note that (3.105) and (3.109) have the same mathematical form. Thus the study of the equation for the TE wave may be easily extended to that of the TM wave with changes from n to N.

Let us now consider the TE plane wave obliquely incident upon the medium (Fig. 3.24). We write

$$E_y(x, z) = u(z)e^{-j\beta_i x}, \tag{3.110}$$

where $\beta_i = k_0 \sin \theta_i$. Then (3.105) becomes

$$\left[\frac{d^2}{dz^2} + q^2(z) \right] u(z) = 0, \tag{3.111}$$

where $q^2(z) = k_0^2[n^2(z) - \sin^2 \theta_i]$. Similarly, for the TM wave, we let

$$U(x, z) = u(z)e^{-j\beta_i x}. \tag{3.112}$$

We get the same equation, (3.111), with

$$q^2(z) = k_0^2[N^2(z) - \sin^2 \theta_i]. \tag{3.113}$$

In Section 3.14 we discuss the WKB solution of (3.111).

3.14 WKB METHOD

In general, the exact solution of (3.111) may be obtained only for a few special functions $q(z)$. Here we consider an approximate solution of (3.111) called the *WKB solution*, named after Wentzel, Kramers, and Brillouin. Sometimes, this is also called the *WKBJ method*, adding Jeffreys.

The WKB approximation can be thought of as the first term of an asymptotic expansion. There are two ways of obtaining the asymptotic series. One is to expand the phase of $u(z)$ in a series

$$u(z) = \exp\left(\sum_n \psi_n \right), \tag{3.114}$$

and the other is to expand the field $u(z)$ in a series

$$u(z) = \sum_n u_n(z). \tag{3.115}$$

The first approach, (3.114), is convenient when the total phase is important, whereas the approach (3.115) is convenient for identifying various waves propagating in the medium. In either case the first term gives the identical WKB solution.

Let us consider (3.111). Noting that q is proportional to the wave number k_0, we may write $q(z) = k_0 n_e(z)$ and $n_e(z)$ as the equivalent refractive index.

Let us use (3.114), which is called the *phase integral approach*,

$$u(z) = e^{\psi(z)} \quad \text{and} \quad \psi(z) = \int^z \phi(z)dz. \tag{3.116}$$

Then substituting this into (3.111) we have

$$\frac{d\phi}{dz} + \phi^2 + k_0^2 n_e^2 = 0. \tag{3.117}$$

Note that (3.117) is a first-order nonlinear differential equation of the Riccati type.

In general, it is always possible to transform a second-order linear differential equation,

$$\left[p_0(z)\frac{d^2}{dz^2} + P_1(z)\frac{d}{dz} + p_2(z) \right] u(z) = 0, \tag{3.118}$$

into a first-order nonlinear differential equation of the Riccati type by using the transformation

$$u(z) = e^{\psi(z)} = e^{\int Q(z)\phi(z)dz}. \tag{3.119}$$

The resulting Riccati equation is

$$\frac{d\phi}{dz} + P(z)\phi + Q(z)\phi^2 = R(z), \tag{3.120}$$

where

$$P(z) = \frac{P_1}{P_0} + \frac{1}{Q}\frac{dQ}{dz} \quad \text{and} \quad R(z) = -\frac{P_2}{P_0 Q}.$$

For our problem (3.111), $P_0 = 1$, $P_1 = 0$, $P_2 = q^2$, and $Q = 1$.

Let us examine (3.117). We seek an expansion of ϕ in inverse powers of k. We write

$$\phi = \phi_0 k_0 + \phi_1 + \frac{\phi_2}{k_0} + \frac{\phi_3}{k_0^2} + \cdots. \tag{3.121}$$

The first term of (3.121) is proportional to k_0, as is clear from (3.117). Substituting (3.121) into (3.117) and expanding (3.117) in inverse powers of k_0, we get

$$\frac{d\phi}{dz} + \phi^2 + k_0^2 n_e^2 = \left(\phi_0^2 + n_e^2 \right) k_0^2 + \left(2\phi_0\phi_1 + \phi_0' \right) k_0 + \left(\phi_1^2 + 2\phi_0\phi_2 + \phi_1' \right)$$
$$\tag{3.122}$$
$$+ \left(2\phi_1\phi_2 + 2\phi_0\phi_3 + \phi_2' \right)\frac{1}{k_0} + \cdots.$$

Equating the coefficients of each power of k_0 to zero, we get an infinite number of equations. The first three are

$$\phi_0^2 + n_e^2 = 0,$$
$$2\phi_0\phi_1 + \phi_0' = 0, \qquad (3.123)$$
$$\phi_1^2 2\phi_0\phi_2 + \phi_1' = 0.$$

The first equation of (3.123) gives

$$\phi_0 = \pm jn_e. \qquad (3.124)$$

From the second equation, we have

$$\phi_1 = -\frac{\phi_0'}{2\phi_0}. \qquad (3.125)$$

Therefore, the solution $u(z)$ is

$$u(z) = e^{\int \phi dz}$$
$$= e^{\int (\phi_0 k_0 + \phi_1 + \cdots)dz} \qquad (3.126)$$
$$= \exp(\pm j \int k_0 n_e dz + \int \phi_1 dz + \cdots).$$

From (3.125), we get

$$\int \phi_1 dz = -\int \frac{\phi_0'}{2\phi_0} dz = -\frac{1}{2} \ln \phi_0,$$

and thus if we continue to the terms ϕ_0 and ϕ_1, we get

$$u(z) = \frac{1}{\sqrt{\phi_0}} \exp(\pm j \int k_0 n_e dz).$$

The general solution to (3.111) is, therefore,

$$u(z) = \frac{1}{q^{1/2}}(Ae^{-j\int q dz} + Be^{+j\int q dz}), \qquad (3.127)$$

where A and B are arbitrary constants and $q = k_0 n_e$. The derivative du/dz, consistent with the approximation above, is

$$\frac{du(z)}{dz} = -jq^{1/2}(Ae^{-j\int q dz} - Be^{+j\int q dz}). \qquad (3.128)$$

Here since we kept only the terms with $q^{-1/2}$ in (3.127), we keep only the terms with $q^{1/2}$ in (3.128). The term with $q^{-3/2}$ should not be kept as it is comparable to the term arising from ϕ_2. Equations (3.127) and (3.128) constitute the WKB solution to the original equation (3.111).

The WKB solution (3.127) consists of the wave u_1 traveling in the $+z$ direction and the wave u_2 traveling in the $-z$ direction.

$$u_1 = \frac{1}{q^{1/2}} e^{-j \int q \, dz} \quad \text{and} \quad u_2 = \frac{1}{q^{1/2}} e^{+j \int q \, dz}. \tag{3.129}$$

The Wronskian of u_1 and u_2 is then given by (see Section 5.5)

$$\Delta = u_1 u_2' - u_2 u_1' = 2j. \tag{3.130}$$

Let us examine the condition for the validity of the WKB solution. If we substitute u_1 in (3.129) into the original equation (3.111), we get

$$\left(\frac{d^2}{dz^2} + q^2 \right) u_1 = f \neq 0.$$

Therefore, the range of validity of the WKB solution is

$$|f| \ll |q^2 u_1|. \tag{3.131}$$

In terms of $q = k_0 n_e$, (3.131) is

$$\frac{1}{k_0^2} \left| \frac{3}{4 n_e^4} \left(\frac{dn_e}{dz} \right)^2 - \frac{1}{2 n_e^3} \frac{d^2 n_e}{dz^2} \right| \ll 1. \tag{3.132}$$

It is clear that the medium must be slowly varying, but because of k_0^2, it is a better approximation for higher frequencies. However, whenever $q = 0$ (or $n_e = 0$), the WKB solution is not valid.

Let us next consider the physical meaning of the WKB solution. The total phase $\int q \, dz$ is the integral of the wave number q along the path. The amplitude $q^{-1/2}$ is such that the total power flux in a lossless medium is kept constant. For a scalar field $u(\bar{r})$, the real power flux density $\bar{F}(\bar{r})$ is given by (Eq. 2.134)

$$\bar{F}(\bar{r}) = \text{Im}[u \nabla u^*]. \tag{3.133}$$

For our problem, the wave traveling in the $+z$ direction is $u(z) = A q^{-1/2} \exp(-j \int q \, dz)$, and therefore, for a lossless medium (q is real), the power flux is kept constant.

$$\bar{F}(z) = |A|^2 \hat{z} = \text{constant}. \tag{3.134}$$

The WKB solution is identical to the geometric optics approximation, in which the phase is given by an eikonal equation and the amplitude is obtained so as to conserve power. The WKB solution is similar to the Rytov solution (see Appendix 16.B).

3.15 BREMMER SERIES

The WKB solution is the first term of the high-frequency series representation of a wave. It is applicable to the wave in a slowly varying inhomogeneous medium at high frequencies. The next term is usually difficult to obtain, but it is important to examine the rest of the series in order to determine the correction to the WKB method and to establish the convergence of the series. The Bremmer series provides such a representation (Wait, 1962).

Let us consider a series representation of a field $u(z)$:

$$u(z) = \sum_n u_n(z). \tag{3.135}$$

This is convenient because each term can be identified with the actual wave reflected by the inhomogeneities. This is done by Bremmer in the following manner.

We seek a solution to the differential equation

$$\left(\frac{d^2}{dz^2} + q^2 \right) u(z) = 0, \tag{3.136}$$

in a form of the WKB solution, except that we allow A and B in (3.127) to be a function of z and try to find $A(z)$ and $B(z)$ to satisfy the differential equation. Thus we write

$$u(z) = \frac{1}{q^{1/2}} [A(z)e^{-j\int q\,dz} + B(z)e^{+j\int q\,dz}]. \tag{3.137}$$

Furthermore, we impose the condition between $A(z)$ and $B(z)$, such that the derivative du/dz has the same form as (3.128):

$$\frac{du(z)}{dz} = -jq^{1/2}[A(z)e^{-j\int q\,dz} - B(z)e^{+j\int q\,dz}]. \tag{3.138}$$

This requires that

$$\left(A' - \frac{1}{2}\frac{q'}{q}A \right) e^{-j\int q\,dz} + \left(B' - \frac{1}{2}\frac{q'}{q}B \right) e^{+j\int q\,dz} = 0. \tag{3.139}$$

Now, substituting (3.138) into (3.136), we get

$$\left(A' + \frac{1}{2}\frac{q'}{q}A \right) e^{-j\int qdz} + \left(B' + \frac{1}{2}\frac{q'}{q}B \right) e^{+j\int qdz} = 0. \tag{3.140}$$

We rewrite (3.139) and (3.140) in the following form:

$$\left(A' - \frac{1}{2}\frac{q'}{q}Be^{+j\int 2qdz} \right) + \left(B' - \frac{1}{2}\frac{q'}{q}Ae^{+j\int 2qdz} \right) e^{+j\int 2qdz} = 0 \tag{3.141}$$

and

$$\left(A' - \frac{1}{2}\frac{q'}{q}Be^{+j\int 2qdz} \right) - \left(B' - \frac{1}{2}\frac{q'}{q}Ae^{+j\int 2qdz} \right) e^{+j\int 2qdz} = 0. \tag{3.142}$$

Adding and subtracting these two, we get

$$A' - \left(\frac{1}{2}\frac{q'}{q}e^{+j\int 2qdz} \right) B = 0, \tag{3.143}$$

$$B' - \left(\frac{1}{2}\frac{q'}{q}e^{-j\int 2qdz} \right) A = 0. \tag{3.144}$$

These two equations constitute the coupling between two waves $A(z)$ and $B(z)$. Let us first recognize that these couplings depend on how the medium varies. If the medium is a slowly varying function, then q'/q, and therefore the coupling, is small. If the coupling is small, we may write

$$\frac{1}{2}\frac{q'}{q}e^{j\int 2qdz} = \varepsilon\lambda_1,$$
$$\frac{1}{2}\frac{q'}{q}e^{-j\int 2qdz} = \varepsilon\lambda_2, \tag{3.145}$$

where ε is a small parameter and (3.143) and (3.144) become

$$\frac{dA}{dz} = \varepsilon\lambda_1 B,$$
$$\frac{dB}{dz} = \varepsilon\lambda_2 A. \tag{3.146}$$

Now we seek a perturbation solution in the following form:

$$A = A_0 + \varepsilon A_1 + \varepsilon^2 A_2 + \cdots,$$
$$B = B_0 + \varepsilon B_1 + \varepsilon^2 B_2 + \cdots. \tag{3.147}$$

Substituting (3.147) into (3.146) and equating equal powers of ε, we get

$$\frac{dA_{n+1}}{dz} = \lambda_1 B_n \quad \text{and} \quad \frac{dA_0}{dz} = 0,$$

$$\frac{dB_{n+1}}{dz} = \lambda_1 A_n \quad \text{and} \quad \frac{dB_0}{dz} = 0. \tag{3.148}$$

From these we obtain

$$A_{n+1} = \int_{-\infty}^{z} \lambda_1 B_n \, dz \quad \text{and} \quad A_0 = \text{constant},$$

$$B_{n+1} = \int_{\infty}^{z} \lambda_2 A_n \, dz \quad \text{and} \quad B_0 = \text{constant}. \tag{3.149}$$

The limit of integration in (3.149) is chosen to represent the physical picture as follows. We first write the complete solution

$$u(z) = \frac{1}{q^{1/2}} [(A_0 + \varepsilon A_1 + \varepsilon^2 A_2 + \cdots) e^{-j \int q \, dz} + (B_0 + \varepsilon B_1 + \varepsilon^2 B_2 + \cdots) e^{+j \int q \, dz}] \tag{3.150}$$

$$= u_{0+} + u_{1+} + u_{2+} + u_{3+} + \cdots + u_{0-} + u_{1-} + u_{2-} + u_{3-} + \cdots,$$

where

$$u_{0+} = \frac{A_0}{q^{1/2}} e^{-j \int q \, dz},$$

$$u_{0-} = \frac{B_0}{q^{1/2}} e^{+j \int q \, dz},$$

are the WKB solution. u_{0+} then generates the wave u_{1-} given by

$$u_{1-}(z) = \frac{e^{+j \int q \, dz}}{q^{1/2}} \int_{\infty}^{z} \varepsilon \lambda_2 A_0 \, dz'$$

$$= \frac{1}{q(z)^{1/2}} \int_{\infty}^{z} \frac{1}{2} \frac{q'(z')}{q(z')} A_0 \exp\left(j \int_{z'}^{z} q \, dz'' - j \int_{z_0}^{z'} q \, dz'' \right) dz'. \tag{3.151}$$

This gives the first correction term to the WKB solution u_{0+}. This wave $u_{1-}(z)$ then generates the wave u_{2+}:

$$u_{2+}(z) = \frac{e^{-j \int q \, dz}}{q^{1/2}} \int_{-\infty}^{z} \varepsilon \lambda_1 \, dz' \int_{\infty}^{z'} \varepsilon \lambda_2 A_0 \, dz''. \tag{3.152}$$

Similarly, u_{1+} is generated by u_{0-}, and u_{2-} is generated by u_{1+} (Fig. 3.25).

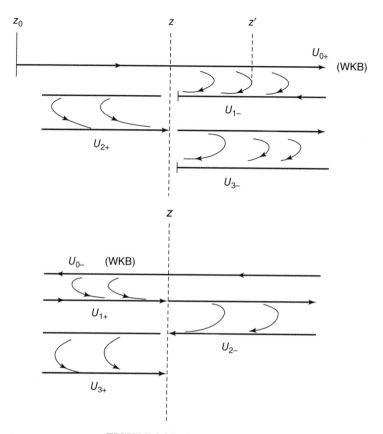

FIGURE 3.25   Bremmer series.

The Bremmer series (3.150) gives a complete series representation of the solution to (3.136), if the series is convergent. This requires that, considering u_{1-} of (3.151),

$$\left| \int_{\infty}^{z} \left(\frac{1}{2} \frac{q'}{q} e^{-j \int^{z} 2q dz} \right) dz \right| \ll 1. \tag{3.153}$$

Now the exponential $e^{-j \int 2q dz}$ oscillates with an approximate period of π/q. Thus if

$$\left| \frac{1}{2} \frac{q'}{q2q} \pi \right| \ll 1, \tag{3.154}$$

then the integral (3.153) in a half-period ($\pi/2q$) cancels the integral in the next half-period, and thus (3.153) is approximately satisfied. Therefore, (3.154) is the necessary condition for the WKB solution to be valid.

3.16 WKB SOLUTION FOR THE TURNING POINT

It is clear that the WKB solution breaks down in the neighborhood of $z = z_0$, where

$$q(z_0) = 0. \tag{3.155}$$

Consider the profile of $q(z)$ shown in Fig. 3.26. The wave propagating in the $+z$ direction cannot propagate beyond the point z_0, and thus the wave must be reflected back. This point z_0 is called the *turning point*.

Let us consider the three regions approximately divided at z_1 and z_2, as shown in Fig. 3.26. In region I, the WKB solution is applicable, and thus the incident wave can be written as

$$u_i(z) = \frac{A_0}{q^{1/2}} e^{-j \int_0^z q dz}, \tag{3.156}$$

where $A_0/q(0)^{1/2}$ is the amplitude at $z = 0$.

The reflected wave in region I can also be written in the WKB form. It is shown in Appendix 3.A, that the WKB reflected wave is given by

$$u_r(z) = \frac{A_0}{q^{1/2}} e^{-j \int_0^{z_0} q(z) dz + j \int_{z_0}^z q(z) dz + j(\pi/2)}. \tag{3.157}$$

Note that the phase of the reflected wave is the integral of $q(z)$ from 0 to z_0, and then from z_0 back to z with the additional phase jump of $\pi/2$. Thus the reflected wave u_r behaves as if the WKB solution is applicable to z_0 and back, except for the phase jump. The WKB solution is, of course, not applicable near z_0, as shown in

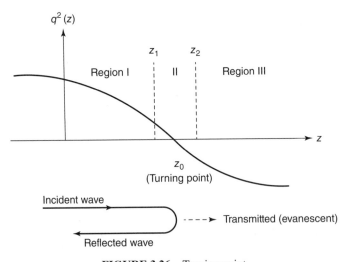

FIGURE 3.26 Turning point.

Appendix 3-A, but the total effect of the turning point to the WKB solution far from the turning point is simply the phase jump of $\pi/2$.

The transmitted wave is evanescent since q^2 is negative. The WKB transmitted wave u_t far from the turning point is given by

$$
\begin{aligned}
u_t(z) &= \frac{A_0}{q^{1/2}} e^{-j\int_0^{z_0} q\,dz - j\int_{z_0}^z q\,dz} \\
&= \frac{A_0 e^{j(\pi/4)}}{\alpha(z)^{1/2}} e^{-j\int_0^{z_0} q\,dz - \int_{z_0}^z \alpha\,dz},
\end{aligned}
\tag{3.158}
$$

where $q(z) = -j\alpha(z)$. The derivation of (3.158) is also given in the Appendix.

3.17 TRAPPED SURFACE-WAVE MODES IN AN INHOMOGENEOUS SLAB

The WKB method can conveniently deal with the propagation constant of the guided-wave modes in an inhomogeneous slab. This approach is useful for thin-film, graded-index optical fibers (GRIN), and guided acoustic waves in the ocean.

Consider a dielectric slab whose refractive index n (or relative dielectric constant ε_r) is a function of z.

$$
\varepsilon_r(z) = n^2(z).
\tag{3.159}
$$

The wave $U(x, z)$ is propagating in the x direction and we wish to find the propagation constant β in the x direction. The field $U(x, z)$ satisfies

$$
\left[\frac{\partial^2}{\partial x^2} + \frac{\partial^2}{\partial z^2} + k_0^2 n^2(z) \right] U(x,\ z) = 0.
\tag{3.160}
$$

Now we let

$$
U(x, z) = u(z) e^{-j\beta x}
\tag{3.161}
$$

and obtain

$$
\left[\frac{d^2}{dz^2} + q^2(z) \right] u(z) = 0,
\tag{3.162}
$$

where $q^2(z) = k_0^2 n^2(z) - \beta^2$. The refractive index profile $n(z)$ is maximum near $z = 0$ and slopes down on both sides of $z = 0$ (Fig. 3.27).

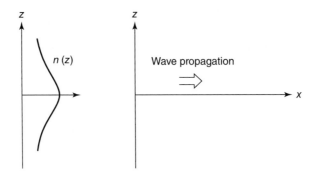

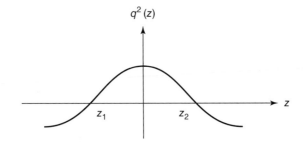

FIGURE 3.27 Trapped surface wave.

The WKB solution starting at $z = 0$ is given by

$$u(z) = \frac{A_0}{q^{1/2}} \exp\left[-j \int_0^{z_2} q\,dz + j \int_{z_2}^{z_1} q\,dz - j \int_{z_1}^{z} q\,dz + j\pi\right], \qquad (3.163)$$

where the turning points z_1 and z_2 are the roots of $q^2(z) = 0$. The total wave is reflected at two turning points z_1 and z_2, and the total phase jump of $2(\pi/2)$ is included. If this wave represents the guided-wave mode, this wave should join smoothly with the wave at $z = 0$. Thus we should have

$$\exp\left(-j \int_{z_1}^{z_2} q\,dz + j \int_{z_2}^{z_1} q\,dz + j\pi\right) = 1. \qquad (3.164)$$

From this, we get

$$\int_{z_1}^{z_2} q(z)\,dz = \left(m + \frac{1}{2}\right)\pi, m = \text{integer}, \qquad (3.165)$$

where $q(z) = [k^2 n^2(z) - \beta^2]^{1/2}$ and $q(z_1) = q(z_2) = 0$. Equation (3.165) gives the propagation constant β for each mode with $m = 0, 1, 2, \ldots$ Since the WKB solution

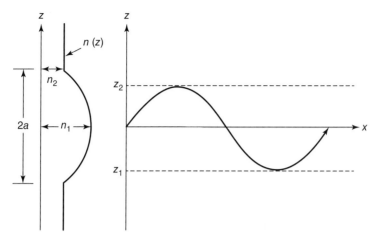

FIGURE 3.28   Square-law profile.

is applicable to the region far from the turning points, (3.165) gives a better approximation for large m. If the refractive index is bounded by n_1 and n_2, the propagation constant β must also be bounded (Fig. 3.28):

$$k_0 n_2 < \beta < k_0 n_1. \tag{3.166}$$

Let us consider an example of the square-law profile.

$$n^2(z) = \begin{cases} n_1^2 \left[1 - 2 \left(\dfrac{z}{a} \right)^2 \Delta \right] & \text{for } |z| < a \\[2mm] n_2^2 = n_1^2 (1 - 2\Delta) & \text{for } |z| > a \end{cases} \tag{3.167}$$

where

$$\Delta = \frac{n_1^2 - n_2^2}{2n_1^2} \approx \frac{n_1 - n_2}{n_1}.$$

Integration in (3.165) can be performed to yield

$$\beta^2 = k_0^2 n_1^2 \left[1 - \frac{2\sqrt{2\Delta}}{k_0 n_1 a} \left(m + \frac{1}{2} \right) \right], \tag{3.168}$$

where m = integers and the turning points z_1 and z_2 are assumed to be within the region $|z| < a$ (Fig. 3.28). The propagation constant β is therefore bounded by $k_0 n_2 < \beta < k_0 n_1$. The WKB solution (3.165) is valid for large m. However, for this square-law profile, (3.168) can be shown to give the exact propagation constant β for the finite number of $m = 0, 1, 2, \ldots, M$.

3.18 MEDIUM WITH PRESCRIBED PROFILE

It is often possible to approximate the profile of the medium $n(z)$ by using a particular functional form such that the resulting equation has a well-known exact solution. Here we discuss one such example.

There are a great number of differential equations, which may be converted into Bessel's differential equation (Jahnke et al., 1960, p. 156). For example,

$$W'' + \left(\beta^2 - \frac{4\nu^2 - 1}{4z^2} \right) W = 0, \tag{3.169}$$

has a solution

$$W = \sqrt{z} Z_\nu(\beta z),$$

where Z_ν is a Bessel function. We may convert (3.111) into the form above if the medium profile is given by

$$n^2 = a^2 - \frac{b^2}{(z + z_0)^2}, \tag{3.170}$$

where a, b, and z_0 are constants to be chosen arbitrarily. Then (3.111) becomes

$$\left\{ \frac{d^2}{dz^2} + k^2 \left[a^2 - \sin^2 \theta_i - \frac{b^2}{(z + z_0)^2} \right] \right\} u(z) = 0. \tag{3.171}$$

Comparing (3.171) with (3.169), we get the solution

$$u(z) = (z + z_0)^{1/2} Z_\nu[\beta(z + z_0)], \tag{3.172}$$

where $\nu = \sqrt{k^2 b^2 + \frac{1}{4}}$ and $\beta = k\sqrt{a^2 - \sin^2 \theta_i}$.

The choice of the Bessel function Z must be made to satisfy the boundary conditions. For example, if the medium extends to $z \to \infty$, then to satisfy the radiation condition as $z \to \infty$, we must have (see 4.80)

$$Z_\nu = H_\nu^{(2)}[\beta(z + z_0)]. \tag{3.173}$$

Let us now calculate the reflected wave from such an inhomogeneous medium. Let the incident wave be

$$E_y^i = E_0 e^{-j(k \sin \theta_i)x - j(k \cos \theta_i)z}, \quad z < 0, \tag{3.174}$$

and the reflected wave be

$$E_y^i = RE_0 e^{-j(k \sin \theta_i)x + j(k \cos \theta_i)z}, \quad z < 0. \tag{3.175}$$

The transmitted wave is given by (3.172),

$$E_y^t = TE_0(z + z_0)^{1/2} H_\nu^{(2)}[\beta(z + z_0)] e^{-j(k \sin \theta_i)x}, \tag{3.176}$$

where R and T are constants to be determined.

By applying the boundary conditions

$$\begin{aligned} E_y^i + E_y^r &= E_y^t, \\ H_x^i + H_x^r &= H_x^t \end{aligned} \quad \text{at } z = 0, \tag{3.177}$$

we obtain

$$R = \frac{k \cos \theta_i - j \left[\dfrac{1}{2z_0} + \beta \dfrac{H_\nu^{(2)\prime}(\beta z_0)}{H_\nu^{(2)\prime}(\beta z_0)} \right]}{k \cos \theta_i + j \left[\dfrac{1}{2z_0} + \beta \dfrac{H_\nu^{(2)\prime}(\beta z_0)}{H_\nu^{(2)\prime}(\beta z_0)} \right]}. \tag{3.178}$$

The amplitude of the transmitted wave T is then given by

$$T = \frac{1 + R}{z_0^{1/2} H_\nu^{(2)}(\beta z_0)}. \tag{3.179}$$

PROBLEMS

3.1 Microwaves at 915 MHz are normally incident from air to muscle ($\varepsilon' = 51$, $\sigma = 1.6$). Calculate the specific absorption rate (SAR) in watts/kilogram as a function of depth. The incident power flux density is 1 mW/cm^2. Assume that the density of muscle is approximately equal to that of water. SAR is the power absorbed per unit mass of the medium and is commonly used in bioelectromagnetics.

3.2 Consider an electromagnetic wave at 10 GHz normally incident from air on a semi-infinite ferromagnetic material. The relative dielectric constant ε_r is $2 - j10$.

   (a) What should be the permeability of the medium to reduce the reflection?

   (b) If the wave is obliquely incident on the medium given in part (a), find and plot the reflection coefficient as a function of the angle of incidence for the TE and TM cases.

   (c) If a thickness of 1 cm of the material is placed on a conducting surface, what is the reflection coefficient (in dB) at normal incidence?

3.3 A plane radio wave is incident on the ionosphere from the air. The frequency is 6 MHz, the plasma frequency of the ionosphere is 5 MHz, and the collision frequency is negligibly small.

   (a) Calculate and plot the reflection and transmission coefficients as functions of the incident angle for TE and TM waves.

   (b) What is the critical angle for total reflection and the Brewster angle?

   (c) Discuss the conservation of power at the boundary. The relative dielectric constant of lossless plasma (ionosphere) is given by

$$\varepsilon_r = 1 - \frac{\omega_p^2}{\omega^2},$$

   where $f_p = \omega_p/2\pi$ is the plasma frequency.

3.4 A TE wave with λ_0 (free space) = 3 cm is incident on the dielectric layers as shown in Fig. P3.4. Calculate the ABCD parameters for each layer and for the total layers. Calculate the reflection and transmission coefficients. Find the angle of transmission.

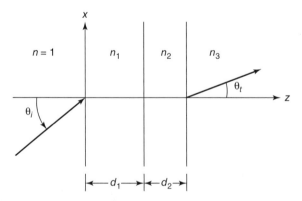

FIGURE P3.4   TE wave is incident on a dielectric layer. $n_1 = 1.5, n_2 = 2.0, n_3 = 2.5, d_1 = 1$ cm, $d_2 = 1.5$ cm, $and \ \theta_i = 30°$.

3.5 A wave with unit power flux density is normally incident on a dielectric film as shown in Fig. P3.5 Calculate the reflected and the transmitted power flux densities as functions of $n_2 \, d/\lambda_0$ (λ_0 in free space) for the following two cases: $(0 \le n_2 \, d/\lambda_0 \le 1)$:

(a) $n_1 = 1, n_2 = 2, n_3 = 4$

(b) $n_1 = 1, n_2 = 3, n_3 = 4$

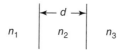

FIGURE P3.5   Dielectric film.

3.6 An ultrasound wave at 1 MHz is incident from fat ($C_0 = 1.44 \times 10^5$ cm/s, $\rho_0 = 0.97$ g/cm^3) to muscle ($C_0 = 1.57 \times 10^5$, $\rho_0 = 1.07$). Assuming that the incident wave is plane and the boundary is plane, calculate and plot the reflection and transmission coefficients as functions of the incident angle. $C_0 =$ sound velocity and $\rho_0 =$ density.

3.7 A TE wave with unit power flux density is incident on an air gap between two prisms as shown in Fig. P3.7 Calculate the power flux density transmitted as a function of the air gap d/λ_0.

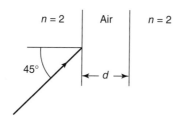

FIGURE P3.7   Air gap between two prisms.

3.8 Consider TM modes along a dielectric slab lying on a perfect conductor. Let ε of the slab be $2.56\varepsilon_0$, and the medium above the waveguide is air. The thickness of the slab is 0.5 cm. What are the cutoff frequencies of the first six modes? At a frequency of 30 GHz, what are the propagation constants of the propagating modes? What is the ratio of the power outside the slab to the power inside the slab for each propagating mode at 30 GHz?

3.9 A trapped surface wave may propagate along the interface between the air and plasma (see Fig. 3.21). If the plasma frequency is 1 MHz, find the frequency range in which a trapped surface wave propagates along the interface. Do this for TM and TE modes. Find the propagation constants β, p_0, and p when the operating frequency is 500 kHz as shown in Fig. 3.23.

3.10 Calculate β, p_0, and p for a Zenneck wave at 100 kHz for wet earth, dry earth, fresh water, and seawater (see Table 2.1).

3.11 Calculate β, p_0, and p for plasmons at an air–silver interface with $\lambda = 0.6$ μm.

3.12 Calculate the reflection coefficient when a TM wave at 0.6 μm is incident on silver from air as a function of the incident angle.

3.13 Consider a TE trapped surface wave ($E_y \neq 0$, $E_x = E_z = 0$) propagating in the x direction along an inhomogeneous dielectric slab whose refractive index is given by

$$n^2(z) = \begin{cases} n_1^2 \left(1 - 2\Delta \left(\dfrac{z}{a} \right)^2 \right) & \text{for } |z| < a, \\ n_2^2 = n_1^2(1 - 2\Delta) & \text{for } |z|s < a, \end{cases}$$

where $n_1 = 1.48$, $n_2 = 1.46$, $a = 10$ μm, and $\lambda_0 = 0.82$ μm. How many surface wave modes can exist? Calculate the propagation constant for each mode, using the WKB method.

CHAPTER 4

WAVEGUIDES AND CAVITIES

In Chapter 3 we discussed wave propagation along layered media. In this chapter, we concentrate on hollow waveguides, dielectric waveguides, and cavities commonly used in microwave, millimeter wave, and optical fibers. In discussing these problems, we present basic formulations and solutions in terms of special functions. Waveguide problems are treated extensively in Montgomery et al. (1948) and Marcuvitz (1951); dielectric waveguides are discussed in Marcuse (1982). For special functions, see Gradshteyn and Ryzhik (1965), Jahnke et al. (1960), Abramowitz and Stegun (1964), and Magnus and Oberhettinger (1949).

4.1 UNIFORM ELECTROMAGNETIC WAVEGUIDES

We start with a general expression for the electromagnetic field in the source-free region in terms of the electric Hertz vector $\bar{\pi}$ and the magnetic Hertz vector $\bar{\pi}_m$ shown in Sections 2.7 and 2.9.

$$\begin{aligned} \bar{E} &= \nabla \times \nabla \times \bar{\pi} - j\omega\mu\nabla \times \bar{\pi}_m, \\ \bar{H} &= j\omega\varepsilon\nabla \times \bar{\pi} + \nabla \times \nabla \times \bar{\pi}_m. \end{aligned} \tag{4.1}$$

It has been shown (Stratton, 1941, p. 392) that a completely general solution to Maxwell's equations can be given by (4.1) when $\bar{\pi}$ and $\bar{\pi}_m$ are pointed in a constant direction $\hat{a}$.

$$\bar{\pi} = \pi(x, y, z)\hat{a}, \quad \bar{\pi}_m = \pi_m(x, y, z)\hat{a}. \tag{4.2}$$

As noted in Sections 2.7 and 2.9, $\bar{\pi}$ and $\bar{\pi}_m$ satisfy the scalar wave equation.

Electromagnetic Wave Propagation, Radiation, and Scattering: From Fundamentals to Applications,
Second Edition. Akira Ishimaru.
© 2017 by The Institute of Electrical and Electronic Engineers, Inc. Published 2017 by John Wiley & Sons, Inc.

FIGURE 4.1   Uniform waveguide.

For a waveguide whose cross section is uniform along the z direction (Fig. 4.1), it is convenient to choose $\bar{\pi} = \pi_z \hat{z}$. It will be shown that the fields generated by $\bar{\pi} = \pi_z \hat{z}$ do not have H_z, the z component of the magnetic field, and this is called the *transverse magnetic* (TM) mode. It is also called the E mode since E_z, the z component of the electric field, is proportional to π_z, which generates this mode. Similarly, the field generated by $\bar{\pi}_m = \pi_{mz} \hat{z}$ does not have E_z and is called the *transverse electric* (TE) mode. It is also called the H mode. We now show that the TM and TE modes correspond to the solutions to Dirichlet's and Neumann's eigenvalue problems, respectively.

4.2   TM MODES OR *E* MODES

The TM modes are generated by $\bar{\pi} = \pi_z \hat{z}$, which satisfies the scalar wave equation:

$$(\nabla^2 + k^2)\pi_z = 0 \tag{4.3}$$

where k is the wave number. If the waveguide is hollow, it is given in terms of the velocity of light c and wavelength λ_0 in a vacuum,

$$k = k_0 = \frac{\omega}{c} = \frac{2\pi}{\lambda_0}. \tag{4.4}$$

If the waveguide is filled with a material with an index of refraction n, k is given by

$$k = k_0 n = \frac{\omega}{c} n = \frac{2\pi}{\lambda_0} n. \tag{4.5}$$

All field components are then given in terms of π_z,

$$\bar{E} = \nabla(\nabla \cdot \bar{\pi}) + k^2 \bar{\pi} = \nabla \frac{\partial \pi_z}{\partial z} + k^2 \bar{\pi}, \tag{4.6}$$

$$\bar{H} = j\omega\varepsilon \nabla \times \bar{\pi} = j\omega\varepsilon \left(\hat{z}\frac{\partial}{\partial z} + \nabla_t \right) \times \bar{\pi} = j\omega\varepsilon(\nabla_t \pi_z \times \hat{z}), \tag{4.7}$$

where ∇_t is a "del" operator in the transverse plane,

$$\nabla_t = \hat{x}\frac{\partial}{\partial x} + \hat{y}\frac{\partial}{\partial y}. \tag{4.8}$$

Also note that $\bar{H}$ is transverse to the z direction.

Let us consider the propagation constant β of a wave propagating in the z direction. The Hertz vector is given by

$$\pi_z (x, y, z) = \phi(x, y) e^{-j\beta z} \tag{4.9}$$

Substituting this into (4.3), we get

$$\left(\nabla_t^2 + k_c^2\right) \phi(x, y) = 0,$$
$$\beta^2 = k^2 - k_c^2. \tag{4.10}$$

For a given waveguide, k_c is a constant and depends only on the geometry of the guide. This constant k_c is called the *cutoff wave number*. Once k_c is known, the propagation constant β for a given frequency is determined from (4.10).

The field components are then given by (4.6) and (4.7) using (4.9),

$$\begin{aligned}
E_z &= k_c^2 \phi(x, y) e^{-j\beta z}, \\
\bar{E}_t &= -j\beta \nabla_t \phi(x, y) e^{-j\beta z}, \\
H_z &= 0, \\
\bar{H}_t &= \frac{\hat{z} \times \bar{E}_t}{Z_e} = -j\omega\varepsilon\hat{z} \times \Delta_t \phi(x, y) e^{-j\beta z},
\end{aligned} \tag{4.11}$$

where $Z_e = \beta/\omega\varepsilon$.

The significance of the formulation above is that the waveguide problem is now reduced to solving a scalar wave equation (4.10) for a scalar function $\phi(x, y)$ over the waveguide cross section. Also, we note from (4.11) that in the cross section the electric and magnetic field distributions have the same form except that the magnetic field is rotated by 90° from the electric field and the ratio of the electric field to the magnetic field is Z_e. The quantity Z_e has the dimension of impedance and is called the *wave impedance*.

4.3 TE MODES OR *H* MODES

The dual of the TM modes are the TE (transverse electric) modes or H modes. This can be derived from the magnetic Hertz vector $\pi_m = \pi_{mz}\hat{z}$ and π_{mz} satisfies the wave equation

$$(\nabla^2 + k^2)\pi_{mz} = 0. \tag{4.12}$$

For a wave propagating in the z direction with propagation constant β, we write

$$\pi_{mz} = \psi(x, y) e^{-j\beta z}. \tag{4.13}$$

We then get the scalar wave equation for $\psi(x, y)$

$$\left(\nabla_t^2 + k_c^2\right) \psi(x, y) = 0. \tag{4.14}$$

The propagation constant β at a given frequency is obtained once the cutoff wave number k_c is known,

$$\beta^2 = k^2 - k_c^2. \tag{4.15}$$

The field components are given in the following form, similar to (4.11):

$$
\begin{aligned}
H_z &= k_c^2 \psi(x, y) e^{-j\beta z}, \\
\bar{H} &= -j\beta \nabla_t \psi(x, y) e^{-j\beta z}, \\
E_z &= 0, \\
\bar{E}_t &= Z_h(\bar{H}_t \times \hat{z}) = j\omega\mu\hat{z} \times \nabla_t \psi(x, y) e^{-j\beta z},
\end{aligned} \tag{4.16}
$$

where Z_h is the wave impedance given by

$$Z_h = \frac{\omega\mu}{\beta}.$$

Note that if the waveguide is empty, $k = k_0$, but if the waveguide is filled with a material with an index of refraction n, $k = k_0 n$.

The boundary condition for a perfectly conducting wall is that the tangential electric field must be zero. For TM modes, this means that

$$E_z = 0 \quad \text{and} \quad \bar{E}_t \times \hat{n} = 0, \tag{4.17}$$

where $\hat{n}$ is the unit vector normal to the boundary (Fig. 4.2). From (4.11), the first condition requires that either $k_c = 0$ or $\phi(x, y) = 0$ on the wall. If $k_c = 0$, E_z as well as H_z is identically zero at all points in the guide, and therefore this is a TEM mode (transverse electromagnetic mode). It will be shown later that TEM modes cannot exist inside a waveguide with a single wall. Therefore, for TM modes, $k_c \neq 0$ and

$$\phi(x, y) = 0 \quad \text{on the wall.} \tag{4.18}$$

If (4.18) is satisfied, we can show that $\bar{E}_t \times \hat{n}$ is also zero on the wall. To show this, write $\bar{E}_t = (\bar{E}_t \cdot \hat{n})\hat{n} + (\bar{E}_t \cdot \hat{t})\hat{t}$, where $\hat{t}$ is the unit vector tangential to the wall and $\hat{n}$

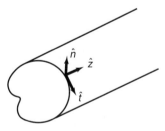

FIGURE 4.2 Three orthogonal unit vectors, $\hat{n}, \hat{t}$, and $\hat{z}$. $\hat{n}$ is normal to the wall and $\hat{t}$ is tangential to the wall and transverse to $\hat{z}$

is the unit vector normal to the wall (Fig. 4.2). Then $\bar{E}_t \times \hat{n} = -(\bar{E}_t \cdot \hat{t})\hat{z}$, but $(\bar{E}_t \cdot \hat{t})$ is equal to $-j\beta(\partial)/(\partial t)\phi(x, y)e^{-j\beta z}$ from (4.11), where $\partial/\partial t$ is the derivative in the direction of $\hat{t}$. If $\phi(x, y) = 0$ on the wall, then $\partial/(\partial t)\phi$ is also zero, and thus $\bar{E}_t \times \hat{n} = 0$ on the wall.

For TE modes, the boundary condition is $\bar{E}_t \times \hat{n} = 0$. From (4.16), this condition can be shown to be equivalent to

$$\frac{\partial \psi}{\partial n} = 0 \quad \text{on the wall.} \tag{4.19}$$

These two boundary conditions, (4.18) and (4.19), are called *Dirichlet's* and *Neumann's conditions*, respectively, and boundary value problems with these boundary conditions are often encountered in many branches of science and engineering.

4.4 EIGENFUNCTIONS AND EIGENVALUES

As shown in (4.10), for TM modes, the problem is reduced to that of finding the function $\phi(x, y)$ and the constant k_c satisfying the following equation:

$$\left(\nabla_t^2 + k_c^2\right) \phi(x, y) = 0,$$

over the cross section of the guide and the boundary condition

$$\phi(x, y) = 0 \quad \text{on the boundary.} \tag{4.20}$$

Similarly, for TE modes, we have, from (4.14),

$$\left(\nabla_t^2 + k_c^2\right) \psi(x, y) = 0,$$
$$\frac{\partial \psi}{\partial n} = 0 \quad \text{on the boundary.} \tag{4.21}$$

The function $\phi(x, y)$ or $\psi(x, y)$ satisfying (4.20) or (4.21) is called the *eigenfunction* and k_c is called the *eigenvalue*. Therefore, the problem of solving for electromagnetic wave propagation in a uniform guide with a perfectly conducting wall is reduced to the problem of finding eigenfunctions and eigenvalues for the two-dimensional Dirichlet's or Neumann's problems (4.20) or (4.21), and once k_c is obtained, the propagation constant β is given by

$$\beta = \left(k^2 - k_c^2\right)^{1/2}. \tag{4.22}$$

In general, to satisfy (4.20) or (4.21), the eigenvalue k_c cannot be an arbitrary number, and it can take only specific values. There is an infinite number of

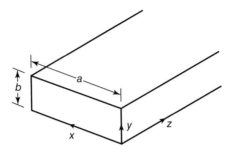

FIGURE 4.3   Rectangular waveguide.

discrete eigenvalues. As an example, consider a rectangular guide with sides a and b (Fig. 4.3). The eigenfunction $\phi(x, y)$ satisfies the wave equation:

$$\left(\frac{\partial^2}{\partial x^2} + \frac{\partial^2}{\partial y^2} + k_c^2 \right) \phi(x, y) = 0. \tag{4.23}$$

The boundary conditions are

$$\phi(x, y) = 0 \quad \text{at } x = 0 \text{ and } a \text{ and } \text{ at } y = 0 \text{ and } b. \tag{4.24}$$

Equations (4.23) and (4.24) can be solved by assuming that $\phi(x, y)$ is a product of a function $X(x)$ of x only and a function $Y(y)$ of y only. This technique is called the *method of separation of variables* and is possible only when the boundary coincides with the coordinate system. Let us write

$$\phi(x, y) = X(x) Y(y). \tag{4.25}$$

Substituting this into (4.23) and multiplying the resulting equation by $[X(x)Y(y)]^{-1}$, we get

$$\frac{X''}{X} + \frac{Y''}{Y} + k_c^2 = 0, \tag{4.26}$$

where the double prime means the second derivative. Since the first term depends only on x and the second term depends only on y, (4.26) can be satisfied for arbitrary x and y only when the first term and the second term, respectively, are constant and the constants are chosen such that the sum of the constants and k_c^2 equals zero. Let these two constants be $-k_x^2$ and $-k_y^2$. Then we have

$$\left(\frac{d^2}{dx^2} + k_x^2 \right) X(x) = 0,$$
$$\left(\frac{d^2}{dy^2} + k_y^2 \right) Y(y) = 0, \tag{4.27}$$

where $k_x^2 + k_y^2 = k_c^2$. Since each of (4.27) is a second-order differential equation, $X(x)$ is given by a linear combination of two independent solutions,

$$X(x) = A \sin k_x x + B \cos k_x x, \tag{4.28}$$

where A and B are constant. Now we apply the boundary condition that $X(0) = X(a) = 0$. This gives $B = 0$ and $\sin k_x a = 0$. From this we get

$$k_x = \frac{m\pi}{a}, \quad m = 1, 2, 3, \ldots, \infty. \tag{4.29}$$

Let us choose A such that $X(x)$ satisfies the following *normalization* condition:

$$\int_0^a X(x)^2 dx = 1, \tag{4.30}$$

yielding $A = (2/a)^{1/2}$.

Similarly, the normalized $Y(y)$ is given by

$$Y(y) = \left(\frac{2}{b}\right)^{1/2} \sin k_y y, \tag{4.31}$$

where

$$k_y = \frac{n\pi}{b}, \quad n = 1, 2, 3, \ldots, \infty$$

Since the eigenvalue k_c depends on m and n, we write $k_c = k_{mn}$ and from (4.27), we get

$$k_{mn}^2 = \left(\frac{m\pi}{a}\right)^2 + \left(\frac{n\pi}{b}\right)^2, \tag{4.32}$$

where $m = 1, 2, 3, \ldots, \infty$ and $n = 1, 2, 3, \ldots, \infty$. For given m and n, the corresponding normalized eigenfunction $\phi(x, y) = \phi_{mn}(x, y)$ is given by

$$\phi_{mn}(x, y) = \frac{2}{\sqrt{ab}} \sin \frac{m\pi x}{a} \sin \frac{n\pi y}{b}. \tag{4.33}$$

4.5 GENERAL PROPERTIES OF EIGENFUNCTIONS FOR CLOSED REGIONS

As we discussed in Section 4.4, waveguide problems are a two-dimensional eigenfunction problem, since we only need to consider the waveguide cross-sectional area. Cavity resonators can be regarded as a three-dimensional eigenfunction problem, because we need to consider waves in a volume. In either case, we are considering the region completely surrounded by the boundary and the region does not extend to infinity. This is called the *closed region*. If a part of the boundary extends to infinity, this is called the *open region*.

Let us consider the general properties of eigenfunctions in a closed region as defined by (4.20) and (4.21). It was indicated in Section 4.4 that there exists a set of a doubly infinite number of eigenfunctions

$$\phi_{mn}, \quad m = 0, 1, 2, \ldots \infty$$
$$n = 0, 1, 2, \ldots, \infty$$

and the corresponding eigenvalues k_{mn}. The eigenfunction ϕ_{mn} satisfies the wave equation

$$\left(\nabla_t^2 + k_{mn}^2\right)\phi_{mn} = 0 \tag{4.35}$$

and boundary conditions. The boundary conditions can be either one of the following two types:

$$\phi_{mn} = 0 \quad \text{Dirichlet's condition} \tag{4.36}$$

$$\frac{\partial\phi_{mn}}{\delta n} = 0 \quad \text{Neumann's condition.} \tag{4.37}$$

Let us consider the general properties of the eigenfunction.

1. ϕ_{mn} *are orthogonal to each other.* The eigenfunctions ϕ_{mn} satisfy the orthogonality condition:

$$\int_a \phi_{mn}\phi_{m'n'}\, da = N_{mn}\delta_{mm'}\delta_{nn'}, \tag{4.38}$$

where the left side is the surface integral over the cross-sectional area a, $\delta_{mm'}$ is a Kronecker delta defined by

$$\delta_{mm'} = \begin{cases} 1 & \text{for } m = m' \\ 0 & \text{for } m \neq m' \end{cases}, \tag{4.39}$$

and N_{mn} is the normalizing factor

$$N_{mn} = \int_a \phi_{mn}^2\, da. \tag{4.40}$$

To prove (4.38), let us start with the following:

$$\left(\nabla_t^2 + k_{mn}^2\right)\phi_{mn} = 0$$
$$\left(\nabla_t^2 + k_{m'n'}^2\right)\phi_{m'n'} = 0. \tag{4.41}$$

We multiply the first equation by $\phi_{m'n'}$ and the second by ϕ_{mn} and subtract one from the other. We get

$$\phi_{m'n'}\nabla_t^2\phi_{mn} - \phi_{mn}\nabla_t^2\phi{m'n'} = \left(k_{m'n'}^2 - k_{mn}^2\right)\phi_{mn}\phi_{m'n'}. \qquad (4.42)$$

Integrating both sides of (4.42) and noting Green's second identity (see Appendix 4.A),

$$\int_a \left(u\nabla_t^2 v - v\nabla_t^2 u\right)\, da = \int_l \left(u\frac{\partial v}{\partial n} - v\frac{\partial u}{\partial n}\right)\, dl, \qquad (4.43)$$

where $\partial/\partial n$ is the normal derivative with n outward from a, we obtain

$$\int_l \left(\phi_{m'n'}\frac{\partial\phi_{mn}}{\partial n} - \phi_{mn}\frac{\partial\phi_{m'n'}}{\partial n}\right)\, dl = \left(k_{m'n'}^2 - k_{mn}^2\right)\int_a \phi_{m'n'}\phi_{mn}\, da. \qquad (4.44)$$

Since ϕ_{mn} and $\phi_{m'n'}$ satisfy either Dirichlet's or Neumann's boundary condition, the integrand on the left side vanishes, and we obtain

$$\left(k_{m'n'}^2 - k_{mn}^2\right)\int_a \phi_{m'n'}\phi_{mn}\, da = 0. \qquad (4.45)$$

Thus if $m \neq m'$ and $n \neq n'$, $k_{m'n'} \neq k_{mn}$, and therefore the integral must be zero, proving the orthogonality (4.38).

2. ϕ_n *forms a complete set, and thus any continuous function $f(\bar{r})$ can be expanded in a series of eigenfunctions.* Let us write

$$f(\bar{r}) = \sum_m^\infty \sum_n^\infty A_{mn}\phi_{mn}(\bar{r}), \qquad (4.46)$$

where $\bar{r} = x\hat{x} + y\hat{y}$. Then A_{mn} is given by multiplying both sides of (4.46) by $\phi_{mn}(\bar{r})$ and integrating over the cross-sectional area.

$$A_{mn} = \frac{\displaystyle\int_a f(\bar{r})\phi_{mn}(\bar{r})\, da}{\displaystyle\int_a \phi_{mn}(\bar{r})^2\, da}. \qquad (4.47)$$

In particular, a delta function $\delta(\bar{r} - \bar{r}')$ can be expanded in a series of eigenfunctions.

$$\delta(\bar{r} - \bar{r}') = \sum_m^\infty \sum_n^\infty \frac{\phi_{mn}(\bar{r})\phi_{mn}(\bar{r}')}{\displaystyle\int_a \phi_{mn}(\bar{r})^2\, da}. \qquad (4.48)$$

As an example, consider the rectangular waveguide with Dirichlet boundary condition as shown in (4.23) and (4.24). The eigenfunction $\phi_{mn}(x, y)$ is given in (4.33), and therefore,

$$\delta(\bar{r} - \bar{r}') = \sum_{m=1}^{\infty} \sum_{n=1}^{\infty} \frac{4}{ab} \sin \frac{m\pi x}{a} \sin \frac{m\pi x'}{a} \sin \frac{n\pi y}{b} \sin \frac{n\pi y'}{b}, \qquad (4.49)$$

where $\bar{r} = x\hat{x} + y\hat{y}$ and $\bar{r}' = x'\hat{x} + y'\hat{y}$.

3. k_{mn}^2 *is real for Dirichlet and Neumann boundary conditions.* To prove this, we consider $\phi_{mn}(\bar{r})$ and its complex conjugate $\phi_{mn}^*(\bar{r})$:

$$\begin{aligned}
\left(\nabla_t^2 + k_{mn}^2\right) \phi_{mn}(\bar{r}) &= 0, \\
\left(\nabla_t^2 + k_{mn}^{*2}\right) \phi_{mn}^*(\bar{r}) &= 0.
\end{aligned} \qquad (4.50)$$

Using Green's second identity (4.43) and letting $u = \phi_{mn}$ and $v = \phi_{mn}^*$, we get

$$\left(k_{mn}^2 - k_{mn}^{*2}\right) \int_a \phi_{mn}\phi_{mn}da = \int_l \left(\phi_{mn} \frac{\partial \phi_{mn}^*}{\partial n} - \phi_{mn}^* \frac{\partial \phi_{mn}}{\partial n} \right) dl. \qquad (4.51)$$

The right side is zero because of the boundary conditions. Therefore, we get $k_{mn}^2 = k_{mn}^{*2}$, proving that k_{mn}^2 is real.

Furthermore, we can show that k_{mn}^2 is not only real but positive. Recall Green's first identity,

$$\int_a \left(\nabla_t v \cdot \nabla_t u + u\nabla_t^2 v\right) da = \int_l u \frac{\partial v}{\partial n} dl. \qquad (4.52)$$

Letting $u = v = \phi_{mn}$ and noting

$$\nabla^2 \phi_{mn} = -k_{mn}^2 \phi_{mn}, \qquad (4.53)$$

we get

$$k_{mn}^2 = \frac{\displaystyle\int_a \nabla_t \phi_{mn} \cdot \nabla_t \phi_{mn} da}{\displaystyle\int_a \phi_{mn} da}. \qquad (4.54)$$

All the integrands in (4.54) are real and positive, and therefore k_{mm}^2 is also real and positive. We can also rewrite (4.54) as follows:

$$
k_{mn}^2 = \frac{\displaystyle\int_a \phi_{mn}\left(-\nabla_t^2\phi_{mn}\right)\,da}{\displaystyle\int_a \phi_{mn}^2\,da}.
\tag{4.55}
$$

Equations (4.54) and (4.55) will be used later for numerical solutions of waveguides with complex cross sections.

4.6 *k*–β DIAGRAM AND PHASE AND GROUP VELOCITIES

Let us consider a hollow waveguide. Once the eigenvalue k_c is determined, the propagation constant β is given by

$$
\beta = \left(k^2 - k_c^2\right)^{1/2}.
\tag{4.56}
$$

The eigenvalue k_c depends only on the geometry of the guide and is called the cutoff wave number. It is related to the cutoff wavelength λ_c and the cutoff frequency f_c by the following:

$$
k_c = \frac{\omega_c}{c} = \frac{2\pi f_c}{c} = \frac{2\pi}{\lambda_c}.
\tag{4.57}
$$

When the frequency of operation is higher than the cutoff frequency, β is real and the wave behaves as $\exp(-j\beta z)$, exhibiting the propagation of the wave. On the other hand, if the operating frequency is below cutoff, the wave is evanescent, and the wave attenuates as $\exp(-\alpha z)$, where $\alpha = (k_c^2 - k^2)^{1/2}$ is real.

The frequency characteristics of the propagation constant β and the attenuation constant α are conventionally expressed in a *k*–β diagram. For the case of waveguides, the *k*–β diagram is given by a hyperbola. Also, the attenuation constant α can be shown in the same diagram as a circle (Fig. 4.4).

The phase velocity V_p is given by

$$
V_p = \frac{\omega}{\beta} \quad \text{or} \quad \frac{V_p}{c} = \frac{k}{\beta}.
\tag{4.58}
$$

If $k < \beta$, $V_p < c$, it is called a *slow wave*, and if $k > \beta$, it is called a *fast wave*. As seen in the *k*–β diagram (Fig. 4.5), the waveguide modes are fast waves.

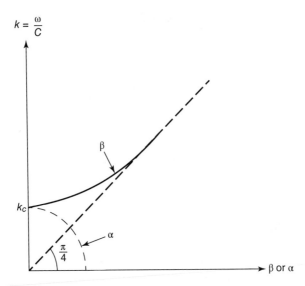

FIGURE 4.4 $k - \beta$ diagram for a waveguide.

The relative phase velocity (V_p/c) is given in the k–β diagram as a slope of a straight line connecting the origin and the operating point.

$$\frac{V_p}{c} = \tan \theta_1. \qquad (4.59)$$

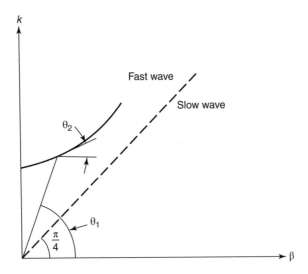

FIGURE 4.5   Phase and group velocities.

The group velocity V_g is given by

$$V_g = \frac{d\omega}{d\beta},\tag{4.60}$$

or its normalized form,

$$\frac{V_g}{c} = \frac{dk}{d\beta}.$$

Thus the relative group velocity is given by the slope of the k–β curve at the operating point:

$$\frac{V_g}{c} = \tan\theta_2.\tag{4.61}$$

It is clear that for the waveguide problem, the wave is fast and

$$V_p V_g = c^2.\tag{4.62}$$

Note that this relation (4.62) is valid only for the form of the propagation constant given in (4.56) and does not hold for other waveguides, such as dielectric guides.

The group velocity V_g is the velocity of the propagation of the signal as represented by the envelope of the modulated signal, and it also represents the velocity of the energy transport. In some cases the phase velocity and the group velocity can be directed in opposite directions (Fig. 4.6). This is called the *backward wave*. In contrast, when the phase and group velocities are directed in the same direction, it is called the *forward wave*. The backward waves are important in such applications as backward wave tubes and log-periodic antennas.

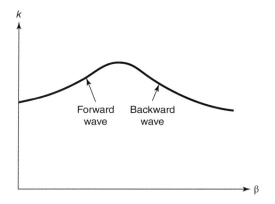

FIGURE 4.6   Backward and forward waves.

If the waveguide is filled with material with refractive index n, the wave number k in (4.56) becomes $k = k_0 n = (\omega/c)n$. If the refractive index n is constant, the velocity of light c in (4.56)–(4.62) should be changed to c/n, the velocity of light in the medium. For dispersive media, $n = n(\omega)$, (4.56)–(4.61) are valid if c is replaced by c/n, but (4.62) does not hold.

4.7 RECTANGULAR WAVEGUIDES

Let us consider TM modes in a rectangular guide with sides a and b (Fig. 4.3). As shown in Section 4.4, for TM modes, we have the normalized eigenfunction

$$\phi_{mn}(x, y) = \frac{2}{\sqrt{ab}} \sin \frac{m\pi x}{a} \sin \frac{n\pi y}{b} , \tag{4.63}$$

and the eigenvalue

$$k_{mn}^2 = \left(\frac{m\pi}{a} \right)^2 + \left(\frac{n\pi}{b} \right)^2 ,$$

where $m = 1, 2, 3,\ldots$ and $n = 1, 2, 3,\ldots$. This mode is called the TM_{mn} mode or the E_{mn} mode. The propagation constant β_{mn} is given by

$$\beta_{mn} = \left(k^2 - k_{mn}^2 \right)^{1/2} . \tag{4.64}$$

The actual Hertz vector corresponding to a given wave with a certain power is given by

$$\pi_z (x, y, z) = A_0 \phi_{mn} (x, y) e^{-j\beta_{mn} z} , \tag{4.65}$$

where A_0 is constant. The field components are given by (4.6) and (4.7).

The total power propagating through the guide when the frequency is above cutoff is given by

$$
\begin{aligned}
P &= \int_0^a dx \int_0^b dy \frac{1}{2} \bar{E} \times \bar{H}^* \cdot \hat{z} \\
&= \int_0^a dx \int_0^b dy \frac{|\bar{E}|^2}{2Z_e} \\
&= \frac{|A_0|^2 \beta_{mn}^2}{2Z_e} \left[\left(\frac{m}{a} \right)^2 + \left(\frac{n}{b} \right)^2 \right] \left(\frac{\pi^2 ab}{4} \right) .
\end{aligned} \tag{4.66}
$$

In a similar manner, TE modes can be given by the normalized eigenfunction

$$\psi_{mn}(x, y) = \frac{2}{\sqrt{ab}} \cos \frac{m\pi x}{a} \cos \frac{n\pi y}{b}, \tag{4.67}$$

where $m = 0, 1, 2,..., \infty$ and $n = 0, 1, 2,..., \infty$, but $m = n = 0$ must be excluded as it leads to zero field. The normalizing constant is $\sqrt{2}/\sqrt{ab}$ if $m = 0$ or $n = 0$. The eigenvalue k_{mn}^2 is the same as the TM case, and this is called the TE$_{mn}$ (or H$_{mn}$) mode.

As an important special case, consider the TE$_{10}$ mode. This mode usually has the lowest cutoff frequency and is the fundamental mode used for most microwave waveguides. We can easily obtain the following field components:

$$E_y = E_0 \sin \frac{\pi x}{a} e^{-j\beta z},$$

$$H_x = -\frac{E_0}{Z_h} \sin \frac{\pi x}{a} e^{-j\beta z},$$

$$H_z = \frac{jk_c E_0}{\beta Z_h} \cos \frac{\pi x}{a} e^{-j\beta z}, \tag{4.68}$$

$$\beta = \left[k^2 - \left(\frac{\pi}{a} \right)^2 \right]^{1/2}.$$

The total power propagating through the guide is

$$P = \frac{|E_0|^2 \, ab}{4Z_h}. \tag{4.69}$$

The cutoff wavelength is $\lambda_c = 2a$, and the cutoff frequency is $f_c = c/2a$ if the guide is empty. The cutoff frequency for a guide filled with material with relative dielectric constant ε_r is $f_c = c/2a(\varepsilon_r)^{1/2}$.

In general, in a waveguide, there may be many propagating and evanescent modes. Therefore, a completely general electromagnetic field in a waveguide should consist of all possible TM and TE modes, and it can be expressed as follows:

$$\bar{E} = \nabla(\nabla \cdot \bar{\pi}) + k^2 \bar{\pi} - j\omega\mu \nabla \times \bar{\pi}_m,$$

$$\bar{H} = j\omega\varepsilon \nabla \times \bar{\pi} + \nabla(\nabla \cdot \bar{\pi}_m) + k^2 \bar{\pi}_m, \tag{4.70}$$

where

$$\bar{\pi} = \hat{z} \sum_{m, n} A_{mn} \phi_{mn}(x, y) e^{-j\beta_{mn} z},$$

$$\bar{\pi}_m = \hat{z} \sum_{m, n} B_{mn} \psi_{mn}(x, y) e^{-j\beta_{mn} z}, \tag{4.71}$$

A_{mn} and B_{mn} are constants representing the amount of each mode.

In the Cartesian coordinate system, (4.70) may be written

$$E_z = \left(\frac{\partial^2}{\partial z^2} + k^2 \right) \pi_z,$$

$$E_x = \frac{\partial^2}{\partial x\, \partial z} \pi_z - j\omega\mu \frac{\partial}{\partial y} \pi_{mz},$$

$$E_y = \frac{\partial^2}{\partial y\, \partial z} \pi_z + j\omega\mu \frac{\partial}{\partial x} \pi_{mz}, \tag{4.72}$$

$$H_z = \left(\frac{\partial^2}{\partial z^2} + k^2 \right) \pi_{mz},$$

$$H_x = j\omega\varepsilon \frac{\partial}{\partial y} \pi_z + \frac{\partial^2}{\partial x\, \partial z} \pi_{mz}.$$

4.8 CYLINDRICAL WAVEGUIDES

Let us represent a TM mode in the cylindrical coordinate system. We have

$$\pi_z(\rho, \phi, z) = \phi(\rho, \phi) e^{-j\beta z}, \tag{4.73}$$

where the eigenfunction $\phi(\rho, \phi)$ satisfies the wave equation

$$\left[\frac{1}{\rho} \frac{\partial}{\partial \rho} \left(\rho \frac{\partial}{\partial \rho} \right) + \frac{1}{\rho^2} \frac{\partial^2}{\partial \phi^2} + k_c^2 \right] \phi(\rho, \phi) = 0. \tag{4.74}$$

By the method of separation of variables, we write

$$\phi(\rho, \phi) = X_1(\rho) X_2(\phi), \tag{4.75}$$

and obtain

$$\left[\frac{1}{\rho} \frac{d}{d\rho} \left(\rho \frac{d}{d\rho} \right) + k_c^2 - \frac{v^2}{\rho^2} \right] X_1(\rho) = 0, \tag{4.76}$$

and

$$\left(\frac{d^2}{d\phi^2} + v^2 \right) X_2(\phi) = 0. \tag{4.77}$$

Equations (4.76) and (4.77) are the second-order ordinary differential equations and a general solution is given by a linear combination of two independent solutions. First, we get from (4.77),

$$X_2(\phi) = c_1 \sin v\phi + c_2 \cos v\phi, \tag{4.78}$$

where c_1 and c_2 are constants.

Equation (4.76) is Bessel's differential equation, and a general solution is given by

$$X_1(\rho) = c_3 J_v(k_c\rho) + c_4 N_v(k_c\rho), \tag{4.79}$$

where J_v and N_v are called the Bessel function and the Neumann function, respectively. It is sometimes convenient to use the Hankel function of the first kind $H_v^{(1)}$ and the second kind $H_v^{(2)}$:

$$X_1(\rho) = c_5 H_v^{(1)}(k_c\rho) + c_6 H_v^{(2)}(k_c\rho). \tag{4.80}$$

The Hankel functions are related to the Bessel and Neumann functions by (see Appendix 4.B)

$$H_v^{(1)} = J_v + jN_v \quad \text{and} \quad H_v^{(2)} = J_v - jN_v. \tag{4.81}$$

The general solution for the Hertz vector is given by (4.73) with the propagation constant $\beta = (k^2 - k_c^2)^{1/2}$. Similarly for TE modes, the magnetic Hertz vector $\pi_m = \psi e^{-j\beta z}$ satisfies the same equation, (4.74).

Let us consider a TM mode propagating in a circular waveguide of radius a (Fig. 4.7). First, we note that the field must be a periodic function of ϕ with period 2π, and therefore $X_2(\phi)$ is $\sin n\phi$ or $\cos n\phi$, where n is an integer. $X_1(\rho)$ is given by (4.79) with $J_n(k_c\rho)$ and $N_n(k_c\rho)$, but $N_n(k_c\rho)$ diverges as $\rho \to 0$. Therefore, only $J_n(k_c\rho)$ is acceptable. Thus we write

$$\pi_z(\rho, \phi, z) = A_0 J_n(k_c\rho) \cos n\phi\, e^{-j\beta z}. \tag{4.82}$$

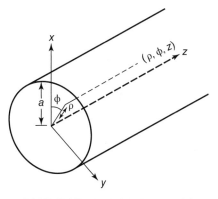

FIGURE 4.7 Cylindrical waveguide.

Here we used $\cos n\phi$ since $\sin n\phi$ gives an identical field distribution except for the rotation of the axis around the z axis. The boundary condition at $\rho = a$ requires that

$$J_n(k_c a) = 0. \tag{4.83}$$

This determines an infinite number of eigenvalues k_c. Using the roots χ_{nl} of the equation

$$J_n(\chi_{nl}) = 0 \tag{4.84}$$

we write k_c as

$$k_{nl} = \frac{\chi_{nl}}{a}, \quad l = 1, 2, 3, \ldots, \tag{4.85}$$

and the propagation constant β is given by

$$\beta = \left(k^2 - k_{nl}^2\right)^{1/2}. \tag{4.86}$$

For each k_{nl}, the field is given by the eigenfunction in (4.82) and this is called the TM$_{nl}$ mode.

Similarly, for the TE$_{nl}$ mode, we have

$$\pi_{mz}(\rho, \phi, z) = B_0 J_n(k_{nl}\rho) \cos n\phi e^{-j\beta z}, \tag{4.87}$$

where k_{nl} is given by

$$J_n'(k_{nl}a) = 0. \tag{4.88}$$

Table 4.1 gives some of the roots of (4.84) and (4.88).

TABLE 4.1   Roots of Bessel Functions

		N		
l	0	1	2	3
		Roots of $J_n(\chi_{nl}) = 0$		
1	2.405	3.832	5.136	6.380
2	5.520	7.016	8.417	9.761
3	8.654	10.173	11.620	13.015
		Roots of $J_n'(\chi_{nl}) = 0$		
1	3.832	1.841	3.054	4.201
2	7.016	5.331	6.706	8.015
3	10.173	8.563	9.969	11.346

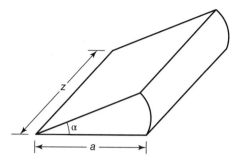

FIGURE 4.8 Sector waveguide.

It should be noted that all eigenfunctions are orthogonal as indicated in (4.38). For the cylindrical waveguide, we have the following orthogonality condition:

$$\int_0^a J_n(k_{nl}\rho)J_n(k_{nl'}\ \rho)\rho\ dp = \begin{cases} 0 & \text{for}\quad l \neq l', \\ N_{nl}^2 & \text{for}\quad l = l'. \end{cases} \tag{4.89}$$

The constant N_{nl}^2 for the TM case is

$$N_{nl}^2 = \frac{a^2}{2}\left[J_n'(k_{nl}a)\right]^2. \tag{4.90}$$

For the TE case (Abramowitz and Stegun, 1964, p. 485),

$$N_{nl}^2 = \frac{1}{2k_{nl}^2}\left(k_{nl}^2 a^2 - n^2\right)[J_n(k_{nl}a)]^2. \tag{4.91}$$

If the cross section is a sector (Fig. 4.8), the field is not a periodic function of ϕ, and thus $\sin n\ \phi$ and $\cos n\phi$ cannot be used. For the TM modes, we have

$$\phi(\rho, \phi) = J_\nu(k_c\rho)(c_1 \sin \nu\phi + c_2 \cos \nu\phi). \tag{4.92}$$

To satisfy the Dirichlet boundary condition, we let

$$\phi(\rho, \phi) = 0 \quad \text{at}\ \phi = 0 \quad \text{and} \quad \phi = a, \tag{4.93}$$

and get $c_2 = 0$ and $\nu = n\pi/\alpha$. Thus we obtain

$$\begin{aligned} \pi_z &= \phi(\rho, \phi)e^{-j\beta z}, \\ \phi(\rho, \phi) &= A_0 J_{n\pi/\alpha}(k_{nl}\rho)\ \sin\frac{n\pi}{\alpha}\phi, \quad n = 1, 2, 3, \ldots, \end{aligned} \tag{4.94}$$

where the eigenvalue k_{nl} is given by

$$J_{np/a}(k_{nl}a) = 0. \tag{4.95}$$

Similarly, TE modes are given by

$$\pi_{mz} = \psi(\rho, \phi)e^{-j\beta z},$$

$$\psi(\rho, \phi) = B_0 J_{n\pi/a}(k_{nl} \, \rho) \, \cos \frac{n\pi}{\alpha}\phi, \quad n = 0, 1, 2, \ldots,$$

(4.96a)

and the eigenvalue k_{nl} is given by

$$J'_{n\pi/\alpha}(k_{nl}a) = 0.$$

(4.96b)

General electromagnetic fields in the cylindrical system are given by

$$E_z = \left(\frac{\partial^2}{\partial z^2} + k^2\right)\pi_z,$$

$$E_\rho = \frac{\partial^2}{\partial \rho \, \partial z}\pi_z - j\omega\mu\frac{1}{\rho}\frac{\partial}{\partial \phi}\pi_{mz},$$

$$E_\phi = \frac{1}{\rho}\frac{\partial^2}{\partial \phi \, \partial z}\pi_z + j\omega\mu\frac{\partial}{\partial \rho}\pi_{mz},$$

$$H_z = \left(\frac{\partial^2}{\partial z^2} + k^2\right)\pi_{mz},$$

$$H_\rho = j\omega\varepsilon\frac{1}{\rho}\frac{\partial}{\partial \phi}\pi_z + \frac{\partial^2}{\partial \rho \, \partial z}\pi_{mz},$$

$$H_\phi = -j\omega\varepsilon\frac{\partial}{\partial \rho}\pi_z + \frac{1}{\rho}\frac{\partial^2}{\partial \phi \, \partial z}\pi_{mz},$$

(4.97)

where

$$\pi_z = A_0\phi(\rho, \phi),$$

$$\pi_{mz} = B_0\psi(\rho, \phi),$$

and ϕ and ψ are given by $X_1(\rho)X_2(\phi)$, shown in (4.75).

4.9 TEM MODES

A hollow waveguide whose cross section is enclosed by a single wall can propagate only TE and TM modes: It cannot support a TEM mode. However, a waveguide consisting of two separate walls, such as coaxial lines and two wire lines, can support a TEM mode as well as TE and TM modes (Fig. 4.9). Since a TEM mode means that $E_z = 0$ and $H_z = 0$, we get from (4.11) and (4.16)

$$k_c = 0.$$

(4.98)

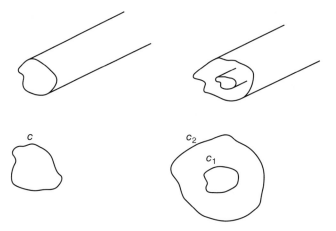

FIGURE 4.9 TEM guides.

From this we get

$$\beta = \left(k^2 - k_c^2\right)^{1/2} = k. \tag{4.99}$$

Therefore, the propagation constant of the TEM mode is equal to that of the free-space propagation constant.

Also since $k_c = 0$, ϕ and ψ satisfy the Laplace equation,

$$\begin{aligned} \nabla^2\phi &= 0, \\ \nabla^2\psi &= 0. \end{aligned} \tag{4.100}$$

If the waveguide is hollow and enclosed by a single wall, the solution to the Laplace equation with either Dirichlet's or Neumann's boundary condition is identically zero or constant, and therefore the field is identically zero from (4.11) and (4.16).

If the waveguide consists of two surfaces, the potential ϕ can be a constant $\phi = \phi_1$ on one surface c_1 and a different constant $\phi = \phi_2$ on the other surface c_2 and the solution exists. The TEM mode is given by the solution of the Laplace equation (4.100), and $\phi(x, y) = \phi_1$ on c_1 and ϕ_2 on c_2. From (4.11), letting $V = jk\phi$,

$$\bar{E}_t(x, y, z) = -\nabla V(x, y)e^{-jkz},$$
$$\bar{H}_t = \frac{1}{\eta}\hat{z} \times \bar{E}_t, \tag{4.101}$$
$$\nabla^2 V = 0.$$

Note that $V(x, y)$ satisfies the Laplace equation and is identical to a two-dimensional electrostatic potential function, and the electric field distribution $\bar{E}_t$ is identical to the electrostatic field.

4.10 DISPERSION OF A PULSE IN A WAVEGUIDE

The propagation constant β for a hollow waveguide is given by $\beta = (k^2 - k_c^2)^{1/2}$, where $k = \omega/c$. If β is equal to $k = \omega/c$, the phase velocity and the group velocity are identical and when a pulse propagates, all the frequency components propagate with the same velocity and no distortion of the pulse shape takes place. However, when β is a more general function of frequency, the different frequency components of a pulse propagate with different velocities, and the waveform is distorted. The variation of the phase velocity with frequency is called the *dispersion*. To calculate the output waveform, we first take the Fourier transform of the input, multiply it by the transfer function, and then take the inverse Fourier transform.

In many practical problems, the pulse input is a modulated wave centered at a carrier frequency. In this section, we first present a general formulation of the propagation of a modulated pulse in a dispersive waveguide and then give an approximate and useful solution for the broadening of a pulse in such a waveguide.

Let us consider a TE_{10} mode in a rectangular waveguide. We assume that at $z = 0$, E_y is given as a modulated wave with carrier frequency ω_0.

$$E_y(0, t) = A(t) \cos[\omega_0 t + \phi(t)] = \mathrm{Re}\left[u_0(t)e^{j\omega_0 t}\right], \qquad (4.102)$$

where $A(t)$ and $\phi(t)$ are the slowly varying amplitude and phase, $u_0(t)$ is the *complex envelope* $u(z, t)$ evaluated at $z = 0$, and $u_0 = A \exp(j\phi)$. The complex envelope $u(z, t)$ is related to the *analytic signal* $u_a = u \exp(j\omega_0 t)$ (see Born and Wolf, 1970). In (4.102), the variation $\sin(\pi x/a)$ of E_y in the waveguide is independent of z and t and is included in $A(t)$.

Let us first express the analytic signal $u_a(0, t)$ at $z = 0$ in a Fourier integral.

$$u_a(0, t) = u_0(t)e^{j\omega_0 t} = \frac{1}{2\pi}\int U_a(\omega)e^{j\omega t}d\omega. \qquad (4.103)$$

The Fourier component U_a is therefore given by

$$\begin{aligned} U_a(\omega) &= \int u_0(t)e^{j\omega_0 t - j\omega t}dt \\ &= U(\omega - \omega_0), \end{aligned} \qquad (4.104)$$

where $U(\omega')$ is the Fourier transform of the complex envelope.

$$U(\omega') = \int u_0(t)\, e^{-j\omega' t}dt. \qquad (4.105)$$

Equation (4.104) shows that the Fourier component of the modulated pulse is given by the Fourier component of the complex envelope centered at the carrier frequency.

Each Fourier component at a point $z \neq 0$ satisfies the wave equation and its propagation characteristics are given by the dependence $\exp(-j\beta z)$. Therefore, adding all the frequency components, the field at $z \neq 0$ is given by

$$E_y(z,t) = \mathrm{Re} \left[\frac{1}{2\pi} \int U(\omega - \omega_0) e^{j\omega t - j\beta z} \, d\omega \right]. \tag{4.106}$$

This is a general expression for the wave $E_y(z, t)$ propagating in a dispersive guide. Note that (4.106) is equivalent to an inverse Laplace transform $(j\omega = s)$, and therefore the evaluation of the integral for a general transient problem requires a detailed study of an integral in a complex plane by allowing ω to be a complex variable.

Next we consider a simpler, special case when the pulse is narrow band and $\beta(\omega)$ is a slowly varying function of ω. Since $\omega - \omega_0 \ll \omega_0$, we expand the exponent in Taylor's series and keep its first three terms.

$$\beta(\omega) = \beta(\omega_0) + (\omega - \omega_0) \frac{\partial \beta}{\partial \omega}\bigg|_{\omega_0} + \frac{1}{2}(\omega - \omega_0)^2 \frac{\partial^2 \beta}{\partial \omega^2}\bigg|_{\omega_0}. \tag{4.107}$$

Substituting this into (4.106), we get

$$E_y(z,t) = \mathrm{Re}[u(z,t)e^{j\omega_0 t}],$$

$$u(z,t) = \frac{1}{2\pi} \int U(\omega') \, \exp\left[-j\beta(\omega_0)z + j\omega'(t - t_0) - j\frac{\omega'^2}{2}\frac{\partial^2 \beta}{\partial \omega^2}z \right] d\omega', \tag{4.108}$$

where

$$t_0 = z\frac{\partial \beta}{\partial \omega} \quad \text{and} \quad \frac{\partial^2 \beta}{\partial \omega^2} \text{ is evaluated at } \omega_0. \tag{4.109}$$

In general, the propagation constant $\beta(\omega)$ for a lossy guide is complex, and therefore t_0 is also complex and the physical meaning of the complex t_0 is difficult to determine. For a lossless guide, however, the meaning of t_0 is clear. If we write $t_0 = (z/V_g)$, $V_g = (\partial \beta/\partial \omega)^{-1}$ is real for a lossless guide and is called the *group velocity*, representing the velocity of the signal. t_0 is called the *group delay*, and $\hat{t}_0 = \partial \beta/\partial \omega = V_g^{-1}$ is called the *specific group delay* (group delay per unit distance).

The second term in the exponent of (4.109) involving $\partial^2 \beta/\partial \omega^2$ represents the broadening of the pulse. If this term is negligibly small, the field is given by

$$E_y(z,t) = \mathrm{Re}\left[u_0(t - t_0)e^{j\omega_0 t}{}^{-jb(w)z} \right]. \tag{4.110}$$

This shows that in this case, if the medium is lossless, the wave propagates without distortion, with the group velocity.

Let us consider an example. The input pulse is assumed to have a Gaussian envelope

$$E_y(0, t) = A_0 \, \exp\left(-\frac{t^2}{T_0^2}\right) \cos \omega_0 t. \tag{4.111}$$

The complex envelope at $z = 0$ is then

$$u_0(t) = A_0 \, \exp\left(-\frac{t^2}{T_0^2}\right). \tag{4.112}$$

The Fourier transform of $u_0(t)$ is given by

$$U(\omega') = A_0(\pi)^{1/2} T_0 \, \exp\left(-\frac{T_0^2 \omega'^2}{4}\right). \tag{4.113}$$

Substituting this into (4.108), we get

$$u(z, t) = \frac{A_0 e^{-j\beta(\omega_0)z}}{(1 + jS/T_0)^{1/2}} \, \exp\left[-\frac{(t - t_0)^2(1 - jS/T_0)}{T_0^2 + S^2}\right], \tag{4.114}$$

where $S = (2/T_0)(\partial^2\beta/\partial\omega^2)|_{\omega_0} z$. Note that the pulse width is broadened from the initial value of T_0 to

$$T = \left(T_0^2 + S^2\right)^{1/2}. \tag{4.115}$$

For a lossless guide, we can obtain the physical meaning of S. Noting that the group delay is given by $t_0 = (\partial\beta/\partial\omega)z$, S may be written as $\Delta\omega \, \partial t_0/\partial\omega$, where $\Delta\omega = (2/T_0)$ is the bandwidth of the Gaussian pulse at $z = 0$. $\partial t_0/\partial\omega$ is the group delay per unit frequency, and therefore S is the total group delay caused by all the frequency components of the pulse. In some cases, the pulse broadening parameter S is expressed in terms of the free-space wavelength $\lambda_0 = 2\pi/k_0$. Noting that

$$\omega\lambda_0 = 2\pi c \quad \text{and} \quad \frac{\Delta\omega}{\omega} = -\frac{\Delta\lambda_0}{\lambda_0}, \tag{4.116}$$

we get

$$|S| = \left|\Delta\omega\frac{\partial t_0}{\partial\omega}\right| = \left|\Delta\lambda_0\frac{\partial t_0}{\partial\lambda_0}\right|. \tag{4.117}$$

The pulse broadening $|\hat{S}|$ per unit distance is given by

$$|\hat{S}| = \Delta\hat{\tau} = \left|\Delta\lambda_0\frac{\partial\hat{\tau}}{\partial\lambda_0}\right|. \tag{4.118}$$

For example, in optical fibers, the pulse broadening is normally given by $|\partial\hat{\tau}/\partial\lambda_0|$ in units of $(ns(km)^{-1}(nm)^{-1})$, and when multiplied by $\Delta\lambda$ (in nm of source bandwidth), it gives the spread $|\hat{S}|$ in nanoseconds per kilometer of fiber. $D = \partial\hat{\tau}/\partial\lambda_0$ is called the *dispersive parameter* or simply the *dispersion*.

For the TE_{10} mode with the propagation constant $\beta = [k^2 - (\pi/a)^2]^{1/2}$, we get the specific group delay $\hat{\tau}$ (group delay per unit length of the guide)

$$\hat{\tau} = \frac{\partial\beta}{\partial\omega} = \frac{\omega}{\beta c^2}. \tag{4.119}$$

The pulse broadening $|\hat{S}|$ per unit length of the guide when the bandwidth of the pulse at $z = 0$ is $\Delta\omega$ is

$$\begin{aligned}
|\hat{S}| = \Delta\hat{\tau} &= \left|\Delta\omega\frac{\partial\hat{\tau}}{\partial\omega}\right| = \left|\Delta\lambda_0\frac{\partial\hat{\tau}}{\partial\lambda_0}\right| \\
&= \left|\Delta\omega\frac{(\pi/a)^2}{c^2\beta^3}\right|.
\end{aligned} \tag{4.120}$$

The propagation constant β in (4.119) and (4.120) is evaluated at the carrier frequency ω_0.

4.11 STEP-INDEX OPTICAL FIBERS

Optical fibers commonly used in optical communications are divided into step-index fibers and graded-index fibers. A *step-index fiber* consists of the circular center *core* with refractive index n_1 and a surrounding *cladding* with n_2. A *graded-index fiber*, also called GRIN, has a core with a slowly varying refractive index (Fig. 4.10). In this section, we examine the propagation characteristics of the wave along the step-index fibers. It is clear from Section 3.11 that in order to support the trapped surface wave along the core, the refractive index of the core must be higher than that of the cladding. In our analysis, we assume that the cladding extends to infinity since the wave in the cladding decays exponentially in the radial direction, and the outer boundary of the cladding has little effect on the propagation characteristics (see Marcuse, 1982).

Let us consider a wave propagating along a dielectric cylinder of radius $a7$ (Fig. 4.11). It was shown in Section 4.1 that a general electromagnetic field consists of TM and TE modes. For the TM (transverse magnetic) modes, the magnetic fields are transverse to a fixed direction denoted by a unit vector $\hat{a}$. For the TE (transverse electric) modes, the electric fields are all transverse to $\hat{a}$. All the TE modes can be derived from a magnetic Hertz vector $\bar{\pi}_m = \pi_m\hat{a}$, all the TM modes can be derived from an electric Hertz vector $\bar{\pi} = \pi_m\hat{a}$, and π_m and π satisfy a scalar wave equation. For cylindrical geometry, (ρ, ϕ, z), we take $\hat{a} = \hat{z}$ and write $\bar{\pi} = \pi_z\hat{z}$ and $\bar{\pi}_m = \pi_{mz}\hat{z}$. The solution to the scalar wave equation is given by a product of Bessel functions Z_n $(p\rho)$ $\exp(\pm jn\phi)$ and $\exp(\pm j\beta z)$, where p and β are constant and satisfy the condition

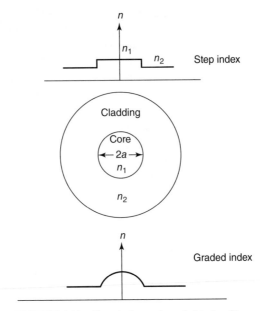

FIGURE 4.10 Step-index and graded-index fibers.

$p^2 + \beta^2 = k^2$ (see Section 4.8). In general, for a dielectric waveguide, TE and TM modes are coupled and both are needed to satisfy the boundary conditions, except for the case where there is no azimuthal variation ($\partial/\partial\phi = 0$). Let us write the Hertz potentials for the TM and TE modes:

$$
\left.\begin{aligned}
\pi_z &= A J_n(p_1\rho)e^{-jn\phi-j\beta z} \\
\pi_{mz} &= B J_n(p_1\rho)e^{-jn\phi-j\beta z}
\end{aligned}\right\} \quad \text{for } \rho < a
$$

$$
\left.\begin{aligned}
\pi_z &= C K_n(\alpha_2\rho)e^{-jn\phi-j\beta z} \\
\pi_{mz} &= D K_n(\alpha_2\rho)e^{-jn\phi-j\beta z}
\end{aligned}\right\} \quad \text{for } \rho > a,
$$

(4.121)

where

$$
p_1^2 + \beta^2 = k_0^2 n_1^2,
$$
$$
\alpha_2^2 = \beta^2 - k_0^2 n_2^2 = k_0^2\left(n_1^2 - n_2^2\right) - p_1^2.
$$

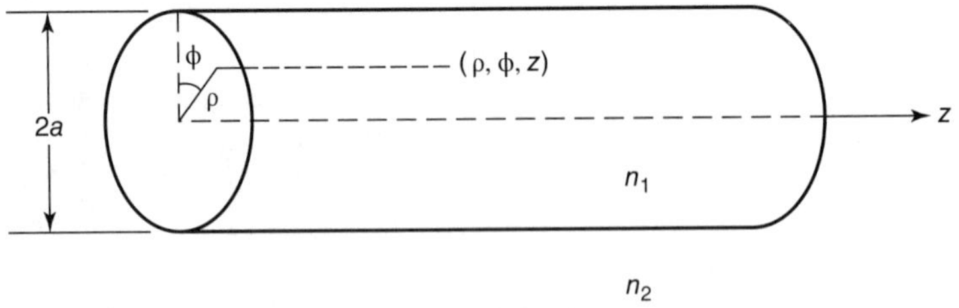

FIGURE 4.11 Round optical fiber.

Note that for $\rho < a$, we used $K_n(\alpha_2\rho)$. We may also use $H_n^{(2)}(p_2\rho)$, where $p_2^2 + \beta^2 = k_0^2 n_2^2$. However, since we are interested in the trapped surface-wave solution, β must be real and greater than $k_0 n_2$. Therefore, $p_2^2 = k_0^2 n_2^2 - \beta^2$ is negative, and thus p_2 is purely imaginary. It is, therefore, more convenient to use the modified Bessel function $K_n(z)$ with $z = \alpha_2\rho = jp_2\rho$.

$$K_n(z) = -\frac{\pi j}{2} e^{-jn(\pi/2)} H_n^{(2)}(-jz). \tag{4.122}$$

The modified Bessel function exhibits exponentially decaying behavior as $\rho \to \infty$.

$$K_n(z) \sim \left(\frac{\pi}{2z}\right)^{1/2} e^{-z} \quad \text{as } z \to \infty. \tag{4.123}$$

Now we consider the boundary conditions that E_z, E_ϕ, H_z, and H_ϕ be continuous at $\rho = a$.

$$E_z = \left(\frac{\partial^2}{\partial z^2} + k^2\right)\pi_z, \quad E_\phi = \frac{1}{\rho}\frac{\partial^2}{\partial\phi\,\partial z}\pi_z + j\omega\mu\frac{\partial}{\partial\rho}\pi_{mz},$$

$$E_z = \left(\frac{\partial^2}{\partial z^2} + k^2\right)\pi_z, \quad H_\phi = -j\omega\varepsilon\frac{\partial}{\partial\rho}\pi_z + \frac{1}{\rho}\frac{\partial^2}{\partial z^2}\pi_{mz}, \tag{4.124}$$

where $k = k_0 n_1$ and $\varepsilon = \varepsilon_1$ in the core and $k = k_0 n_2$ and $\varepsilon = \varepsilon_2$ outside. Applying the continuity of E_z and H_z at $\rho = a$, we obtain

$$p_1{}^2 A J_n(p_1 a) = -\alpha_2{}^2 C K_n(\alpha_2 a),$$
$$p_1{}^2 B J_n(p_1 a) = -\alpha_2{}^2 D K_n(\alpha_2 a). \tag{4.125}$$

This determines the ratio $A/C = B/D$, and using this, we rewrite (4.121) in the following more convenient form:

$$\left.\begin{array}{l} \pi_z = \dfrac{A_0}{p_1^2}\dfrac{J_n(p_1\rho)}{J_n(p_1 a)} e^{-jn\phi - j\beta z} \\[3mm] \pi_{mz} = \dfrac{B_0}{p_1^2}\dfrac{J_n(p_1\rho)}{J_n(p_1 a)} e^{-jn\phi - j\beta z} \end{array}\right\} \quad \text{for } \rho < a,$$

$$\left.\begin{array}{l} \pi_z = \dfrac{(-A_0)}{\alpha_2^2}\dfrac{K_n(\alpha_2\rho)}{K_n(\alpha_2 a)} e^{-jn\phi - j\beta z} \\[3mm] \pi_{mz} = \dfrac{(-B_0)}{\alpha_2^2}\dfrac{K_n(\alpha_2\rho)}{K_n(\alpha_2 a)} e^{-jn\phi - j\beta z} \end{array}\right\} \quad \text{for } \rho > a, \tag{4.126}$$

where A_0 and B_0 are constant.

FIGURE 4.12 HE$_{11}$ dipole mode.

Now we apply the boundary condition that E_ϕ and H_ϕ must be continuous at $\rho = a$ and obtain

$$\left[\frac{n_1^2}{u_1}\frac{J'_n(u_1)}{J_n(u_1)} + \frac{n_2^2}{u_2}\frac{K'_n(u_2)}{K_n(u_2)}\right]\left[\frac{1}{u_1}\frac{J'_n(u_1)}{J_n(u_1)} + \frac{1}{u_2}\frac{K'_n(u_2)}{K_n(u_2)}\right] = \left(\frac{n\beta}{k_0}\right)^2\left(\frac{1}{u_1^2} + \frac{1}{u_2^2}\right)^2,$$
(4.127)

where $u_1 = p_1$ and $u_2 = \alpha_2 a$, and $u_2 = \alpha_2 a$, $u_2^2 = V^2 - u_1^2$, and $V^2 = k_0^2 a^2(n_1^2 - n_2^2)$.

We note that only when $n = 0$, (4.127) separates into two equations corresponding to TM and TE modes. But, in general, TM and TE modes are mixed, and they are called *hybrid* modes and are designated as EH$_{nm}$ or HE$_{nm}$ modes, depending on whether TM or TE modes are dominant. The HE$_{11}$ mode is the only mode that does not have a low-frequency cutoff and is often called the HE$_{11}$ dipole mode. This is equivalent to the lowest TM mode on a dielectric slab (Fig. 4.12).

The frequency dependence of the propagation constant β is conveniently expressed in terms of the normalized propagation constant b as functions of the normalized frequency V.

$$b = \frac{\beta^2 - k_0^2 n_2^2}{k_0^2(n_1^2 - n_2^2)} = \frac{\alpha_2^2 a^2}{V^2} = \frac{u_2^2}{V^2},$$
$$V^2 = k_0^2 a^2\left(n_1^2 - n_2^2\right).$$
(4.128)

Typical curves showing the frequency dependence are sketched in Fig. 4.13.

The cutoff frequency of the HE$_{11}$ mode is zero. The modes with the next-lowest cutoff frequencies are the TE$_{01}$ and TM$_{01}$ modes ($n = 0$). Their cutoff frequencies are obtained by letting $n = 0$ in (4.127), thus separating the equation into TE and TM modes, and then by recognizing that the cutoff requires $u_2 \to 0$. Noting that as $u_2 \to 0$,

$$K_0(u_2) \to \ln\frac{2}{\gamma u_2}$$
$$\gamma = \text{Euler's constant} = 1.781,$$
(4.129)

we get the cutoff condition for both TE$_{01}$ and TM$_{01}$

$$J_0(u_1) = 0.$$
(4.130)

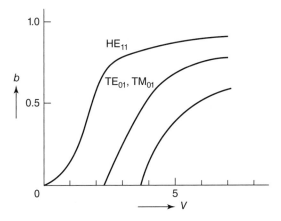

FIGURE 4.13   Normalized propagation constant b as a function of normalized frequency V.

The lowest root is $u_1 = 2.4048$. Therefore, the cutoff normalized frequency is

$$V_c = k_0 a(n_1^2 - n_2^2)^{1/2} = 2.4048$$

$$\sim \frac{2\pi a}{\lambda_c} n_1 (2\Delta)^{1/2}, \tag{4.131}$$

where λ_c is the cutoff wavelength and $\Delta \sim (n_1 - n_2)/n_1$. We conclude that as long as $V < 2.4048$, only a single mode can propagate along the optical fiber. For example, for typical values of $n_1 = 1.48$, $n_2 = 1.46$, and $\lambda_0 = 0.82$ μm, the maximum fiber diameter that supports the *single mode* is $2a = 2.59$ μm.

In contrast to the single-mode fiber, a fiber with a large diameter (50 μm, for example) supports many modes and is called the *multimode fiber*. In this case, the number of modes for a given normalized frequency V is approximately given by the mode volume formula

$$N \approx \frac{1}{2}V^2. \tag{4.132}$$

This may be compared with the number of modes for a dielectric slab, $N \approx 2V/\pi$ for the total TE and TM modes.

The specific group delay (group delay per unit length) $\hat{\tau}$ is given by

$$\hat{\tau} = \frac{d\beta}{d\omega}, \tag{4.133}$$

and the increase of the pulse width per unit length of the fiber is given by

$$\hat{S} = \frac{d\hat{\tau}}{d\lambda}. \tag{4.134}$$

The dispersion is then expressed by

$$D = \frac{d\hat{\tau}}{d\lambda}, \tag{4.135}$$

measured in $ns(km)^{-1}(nm)^{-1}$. When multiplied by the source spectrum width $\Delta\lambda_s$, (4.135) gives the pulse broadening in ns/km.

The dispersion over a distance L can also be expressed in terms of the bandwidth B:

$$B = (\Delta\hat{\tau}L)^{-1}. \tag{4.136}$$

We can then define the *bandwidth–distance product*:

$$BL = (\Delta\hat{\tau})^{-1} \text{ Hz km.} \tag{4.137}$$

The dispersion of a single-mode fiber consists of *material dispersion* and *mode dispersion*. The mode dispersion depends on the mode structure and therefore depends also on the radius of the fiber. The material dispersion is caused by the variation of refractive index with frequency and can be described by Sellmeier's equation (Chapter 8). Dispersion curves for fused silica fibers are sketched in Fig. 4.14. Note that the dispersion disappears at a certain wavelength. The material dispersion ($a \rightarrow \infty$) for fused silica disappears at $\lambda = 1.27$ μm, but because of the mode dispersion, the vanishing dispersion point is shifted. It is sometimes desirable to shift this point to the minimum-loss wavelength. For example, fibers made of germanium-doped fused

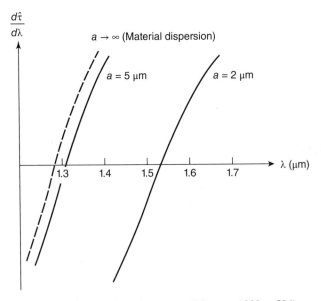

FIGURE 4.14   Dispersion curves (Marcuse, 1982, p.504).

silica have a minimum loss of 0.2 dB/km near $\lambda = 1.55$ μm. Recently, long-wavelength systems with 1.3 μm wavelength with minimum dispersion have been developed. A typical bandwidth–distance product can be $BL \geq 100$ GHz · km at 1.3 μm for the source spectrum width $\Delta\lambda_s = 10$ nm and the loss may be 0.27 dB/km. At 1.5–1.65 μm, the dispersion is much greater than at 1.3 μm and BL is only about 5 GHz · km with $\Delta\lambda_s = 10$ nm.

The dispersion of a multimode fiber consists of the *material dispersion* and the *intermodal dispersion*. The intermodal dispersion is caused by the different group delays of many modes. An estimate of the pulse spread of step-index multimode fibers can be made by noting the difference in travel time between the longest ray path and the shortest ray path. The shortest ray path is a straight line inside the fiber, and the travel time over the distance z along the fiber is $(n_1/c)z$. The largest path is when the path makes the maximum angle with the z axis, which corresponds to the critical angle θ_c for total reflection on the fiber surface. Beyond this maximum angle, the rays escape from the fiber. Thus the travel time along this longest path is $(n_1/c)(z/\sin\theta_c) = (n_1/c)(n_1/n_2)z$. Therefore, the pulse spread $\Delta\hat{\tau}$ per unit distance along the fiber is given by

$$\Delta\hat{\tau} \sim \frac{(n_1/c)(n_1 - n_2)}{n_2} = (n_1/c)(n_1/n_2)\Delta,$$

$$\Delta = \frac{n_1 - n_2}{n_1}.$$

(4.138)

Note that the dispersion for a step-index fiber is proportional to Δ. It will be shown in Section 4.12 that the dispersion of the graded-index fiber is much less than (4.138) and is proportional to Δ^2. Multimode short-wavelength systems operating at a 0.85 μm wavelength are commercially available with a typical loss of 2.5 dB/km and a bandwidth–distance product of several hundred MHz · km.

One of the important parameters of a multimode fiber is its *numerical aperture* (NA). It is a measure of the ability to gather light into the fiber and is given by the sine of the maximum angle of the entrance of a ray that is trapped in the core (Fig. 4.15). Noting that the maximum angle θ occurs when the angle θ_c is the critical angle for total reflection, we get

$$\text{NA} = \sin\theta = \left(n_1^2 - n_2^2\right)^{1/2}.$$

(4.139)

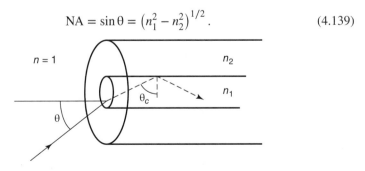

FIGURE 4.15 Numerical aperture NA $= \sin\theta$.

4.12 DISPERSION OF GRADED-INDEX FIBERS

In Section 3.17, we discussed the surface-wave modes in an inhomogeneous slab using the WKB method. The propagation constant β is obtained by (3.165). By differentiating (3.165) with respect to ω and noting that even though z_1 and z_2 depend on β and therefore on ω, $(\partial z_1/\partial\omega)q(z_1) = (\partial z_2/\partial\omega)q(z_2) = 0$, we get the specific group delay $\hat{\tau}$.

$$\hat{\tau}\frac{d\beta}{d\omega} = \frac{k_0}{\beta c}\frac{\int_{z_1}^{z_2}\{[n^2(z)+k_0n(z)(dn/dk_0)]/q(z)\}dz}{\int_{z_1}^{z_2}dz/q(z)} \tag{4.140}$$

The material dispersion $dn/d\omega = cdn/dk_0$ is included in the above. The maximum and minimum values of $\hat{\tau}$ are obtained by letting $\beta = k_0n_2$ and k_0n_1, respectively, and therefore the pulse spread $\Delta\hat{\tau}$ per unit length is given by

$$\Delta\hat{\tau} = \hat{\tau}(\beta = k_0n_2) - \hat{\tau}(\beta = k_0n_1). \tag{4.141}$$

For example, for the square-law profile, we have (3.167)

$$n^2(z) = \begin{cases} n_1^2\left[1 - 2\Delta\left(\dfrac{z}{a}\right)^2\right], & z < a \\ n_2^2 = n_1^2(1 - 2\Delta), & z > a \end{cases}. \tag{4.142}$$

Because of symmetry, we have $z_2 = -z_1$ and $q(z_2) = 0$. Substituting (4.142) into (4.140), we get, assuming no material dispersion,

$$\hat{\tau} = \frac{n_1[1 - \Delta(z_2/a)^2 0}{c[1 - 2\Delta(z_2/a)^2]^{1/2}}. \tag{4.143}$$

The minimum $\hat{\tau}$ is obtained when $z_2 = 0$ and the maximum $\hat{\tau}$ is obtained when $z_2 = a$. From these we get

$$\hat{\tau}(\beta = k_0n_2) = \frac{n_1^2 + n_2^2}{2n_2c}$$

$$\hat{\tau}(\beta = k_0n_1) = \frac{n_1}{c} \tag{4.144}$$

$$\Delta\hat{\tau} = \frac{n_1\Delta^2}{2c}.$$

In Section 4.11, it was shown that the pulse spread $\Delta\hat{\tau}$ for the step-index fiber is proportional to Δ (Eq. 4.138). Equation (4.144) shows that the pulse spread for the graded-index fiber is proportional to Δ^2 and is much smaller than that for the step-index fiber.

4.13 RADIAL AND AZIMUTHAL WAVEGUIDES

The formulations given in Section 4.8 on cylindrical structures are applicable not only to a uniform waveguide in which the wave propagates in the z direction, but also to nonuniform waveguides in which the wave propagates in the radial (ρ) or the azimuthal directions (ϕ). Before we discuss the radial and azimuthal waveguides, let us summarize the boundary conditions in terms of the electric and magnetic Hertz vectors.

$$\bar{\pi} = \pi_z \hat{z} \quad \text{and} \quad \bar{\pi}_m = \pi_{mz}\hat{z}. \tag{4.145}$$

The boundary conditions for a perfectly conducting wall (electric wall) parallel to the z axis are

$$\pi_z = 0 \quad \text{and} \quad \frac{\partial \pi_{mz}}{\partial n} = 0. \tag{4.146}$$

These correspond to the TM and TE modes in metallic waveguides. The boundary conditions for a perfectly conducting wall perpendicular to the z axis are

$$\frac{\partial \pi_z}{\partial n} = 0 \quad \text{and} \quad \pi_{mz} = 0. \tag{4.147}$$

Similarly, for a magnetic wall where the tangential magnetic field vanishes, we have

$$\frac{\partial \pi_z}{\partial n} = 0 \quad \text{and} \quad \pi_{mz} = 0, \tag{4.148}$$

on a wall parallel to the z axis, and

$$\pi_z = 0 \quad \text{and} \quad \frac{\partial \pi_{mz}}{\partial n} = 0, \tag{4.149}$$

on a wall perpendicular to the z axis (Fig. 4.16).

4.13.1 Radial Waveguides

As an example, consider a sectoral horn (Fig. 4.17). The TM modes (the magnetic field perpendicular to the z axis) are given by

$$\pi_z = H^{(2)}_{m\pi/\alpha}(k_\rho \, \rho) \sin \frac{m\pi\phi}{\alpha} \, \cos \frac{n\pi z}{l}, \quad m = 1, 2, 3, \ldots, \quad n = 0, 1, 2, \ldots \tag{4.150}$$

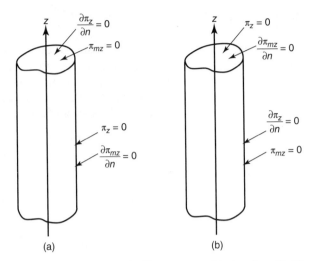

FIGURE 4.16 Boundary conditions for Hertz vectors: (a) electric wall; (b) magnetic wall.

This satisfies the boundary condition that $\pi_z = 0$ at $\phi = 0$ and α and $(\partial \pi_z)/(\partial z) = 0$ at $z = 0$ and $z = l$. The radial dependence of $H_{m\pi}^{(2)}/a$ is chosen to represent the outgoing wave. The propagation constant in the radial direction is given by

$$k_\rho^2 = k^2 - \left(\frac{n\pi}{l}\right)^2. \tag{4.151}$$

Similarly, the TE modes are given by

$$\pi_{mz} = H_{m\pi/\alpha}^{(2)}(k_\rho\, \rho)\cos\frac{m\pi\delta}{\alpha}\sin\frac{n\pi}{l}z, \quad m = 0, 1, 2, \ldots, \quad n = 1, 2, 3, \ldots \tag{4.152}$$

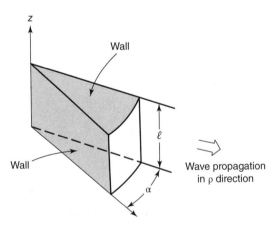

FIGURE 4.17 Sectoral horn.

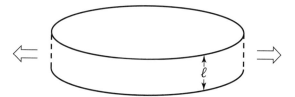

FIGURE 4.18   Radial waveguides.

Note that for TM modes, the lowest n is zero. In this case $k_\rho = k$, but for TE modes, the lowest n is one, and thus there is a low-frequency cutoff.

A special TM mode with no azimuthal variation is often used in practice (Fig. 4.18). This is given by

$$\pi_2 = H_0^{(2)}(K_\rho \, \rho) \cos \frac{n\pi z}{l}, \quad n = 0, 1, 2, \dots, \tag{4.153}$$

where $k_\rho^2 = k^2 - (n\pi/l)^2$ and the electric and magnetic fields are obtained by (4.97). In particular, when $n = 0$,

$$E_z = k^2 H_0^{(2)}(k\rho),$$

$$H_\phi = -j\sqrt{\frac{\varepsilon}{\mu}} k^2 H_0^{(2)'}(k\rho) \tag{4.154}$$

$$= j\sqrt{\frac{\varepsilon}{\mu}} k^2 H_1^{(2)}(k\rho).$$

4.13.2 Azimuthal Waveguides

An example of the azimuthal wave propagation is a waveguide bend (Fig. 4.19). Let us consider a rectangular waveguide with a TE_{10} mode. If the waveguide is bent in the H plane, we can express π_z in the following manner:

$$\pi_z = Z_v(k\rho)e^{-jv\phi}, \tag{4.155}$$

where

$$Z_v(k\rho) = J_v(k\rho)N_v(ka) - J_v(ka)N_v(k\rho),$$

and thus $Z_v(ka) = 0$, satisfying the boundary conditions at $\rho = a$,

$$\pi_z|_{\rho=a} = 0.$$

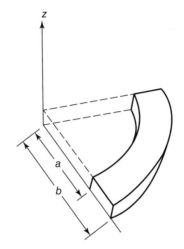

FIGURE 4.19   Azimuthal waveguides.

The azimuthal propagation constant v is given by satisfying the boundary condition at $\rho = b$

$$Z_v(kb) = 0,$$

or

$$\frac{J_v(ka)}{N_v(ka)} = \frac{J_v(kb)}{N_v(kb)}. \tag{4.156}$$

This equation determines the propagation constant v in the azimuthal direction.

4.14 CAVITY RESONATORS

The rectangular cavity is formed by closing the ends of a rectangular waveguide with perfectly conducting walls. For a rectangular cavity with dimensions $a \times b \times h$, the eigenfunction for TM modes is

$$\pi_z = \sin\frac{m\pi x}{a} \ \sin\frac{n\pi y}{b} \ \cos\frac{l\pi z}{h}, \tag{4.157}$$

where

$$m = 1, 2, \dots,$$
$$n = 1, 2, \dots,$$
$$l = 0, 1, 2, 3, \dots$$

The eigenvalue is the wave number for the resonant frequency f_r, and for the TM_{mnl} mode, it is given by

$$k_c = \frac{2\pi f_r}{c} = \left[\left(\frac{m\pi}{a}\right)^2 + \left(\frac{n\pi}{b}\right)^2 + \left(\frac{l\pi}{h}\right)^2\right]^{1/2}. \qquad (4.158)$$

Similarly, for TE modes, we have

$$\pi_{mz} = \cos\frac{m\pi x}{a} \, \cos\frac{n\pi y}{b} \, \sin\frac{l\pi z}{h}, \qquad (4.159)$$

where

$$m = 0, 1, 2, \ldots,$$
$$n = 0, 1, 2, \ldots,$$
$$l = 0, 1, 2, \ldots,$$

and $m = n = 0$ is excluded. The resonant frequency for the TE_{mnl} mode is given by the same formula, (4.158). If the cavity is filled with dielectric material, then $2\pi f_r/c$ in (4.158) must be replaced by $2\pi f_r \varepsilon_r^{1/2}/c$, where ε_z is the relative dielectric constant of the material.

The cylindrical cavity is formed by closing the ends of a cylindrical waveguide by perfectly conducting walls. For a cylindrical cavity with radius a and height h, the eigenfunction for the TM_{nlp} mode is

$$\pi_z = J_n(k_{pnl}\rho) \cos n\phi \cos k_{zp}z, \qquad (4.160)$$

where

$$J_n(k_{pnl}a) = 0, \quad k_{zp} = \frac{p\pi}{h},$$
$$n = 0, 1, 2, \ldots,$$
$$p = 0, 1, 2, \ldots$$

The resonant frequency f_r for the TM_{nlp} mode is

$$\frac{2\pi f_r \varepsilon_r^{1/2}}{c} = \left(k_{pnl}^2 + k_{zp}^2\right)^{1/2}, \qquad (4.161)$$

where ε_r is the relative dielectric constant of the medium inside the cavity.

Similarly, for the TE_{nlp} mode, we have

$$\pi_{mz} = J_n(k_{pnl}\rho) \cos n\phi \sin k_{zp}z, \qquad (4.162)$$

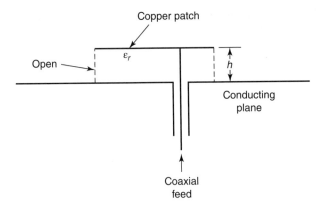

FIGURE 4.20   Microstrip antennas.

where

$$J'_n(k_{pnl}a) = 0, \quad k_{zp} = \frac{p\pi}{h},$$
$$n = 0, 1, 2, \dots,$$
$$p = 1, 2, 3, \dots$$

The resonant frequency is given by (4.161).

As an example of the cylindrical cavity, consider a circular microstrip antenna (Fig. 4.20). This antenna is fed by a coaxial line located off the center of the patch. The current on the patch is then similar to that of a horizontal dipole, and the radiation is in the broadside direction. Microstrip antennas are therefore useful as low-profile antennas on the surface of aircraft. The microstrip antenna is high Q and narrow band and is operated near the resonant frequency of the cylindrical cavity of radius a and height h. The top and bottom surfaces are conducting (electric wall), while the side is open. The radiating power is transmitted through the side and some reactive power is stored near the fringe of the patch. However, as a first approximation, we assume that on the side, the tangential magnetic field is negligibly small (magnetic wall). Noting Fig. 4.16, the boundary conditions for TM modes are

$$\frac{\partial \pi_z}{\partial z} = 0 \quad \text{on the top and bottom,}$$

$$\frac{\partial \pi_z}{\partial \rho} = 0 \quad \text{on the side.}$$

(4.163)

For TE modes, the boundary conditions are

$$\pi_{mz} = 0 \quad \text{on the top and bottom,}$$
$$\pi_{mz} = 0 \quad \text{on the side.}$$

(4.164)

From Table 4.1, it is seen that the lowest eigenvalue $\chi_{11} = 1.841$ is obtained from $J'_n(\chi_{nl}) = 0$ when $n = 1$ and $l = 1$. Thus the lowest resonant frequency is obtained for the TM_{11} mode with the boundary condition (4.163). The field for this cavity is, therefore, given by

$$E_z = A_0 J_1(k_{11}\,\rho)\cos\phi,$$

$$H_\phi = -j\omega\varepsilon k_{11} A_0 J'_1(k_{11}\,\rho)\cos\phi,$$

$$H_\rho = -j\omega\frac{1}{\rho}A_0 J_1(k_{11}\rho)\sin\phi, \qquad (4.165)$$

$$k_{11} = \frac{1.841}{a}.$$

The resonant frequency f_r is then given by

$$\frac{2\pi f_r\sqrt{\varepsilon_r}}{c} = \frac{1.841}{a}. \qquad (4.166)$$

The microstrip antenna is operated at a frequency close to this frequency.

4.15 WAVES IN SPHERICAL STRUCTURES

Unlike the cylindrical coordinate system, the spherical coordinate system does not possess an axis with constant direction, and the formulation based on $\pi\hat{a}$ and $\pi_m\hat{a}$ in (4.2) cannot be used. Therefore, it is necessary to devise a different method by which all the field components can be derived from scalar functions satisfying a well-known differential equation. This can be done by a special choice of the Hertz vector, and the complete electromagnetic fields can be described by two scalar functions that satisfy a scalar wave equation.

Let us re-examine the derivation of the Hertz vector. We need a condition similar to but different from the Lorentz condition, which yields the desired result. We consider the time-harmonic case and start with the vector potential $\bar{A}$ and scalar potential Φ. Rewriting (2.84) for the time-harmonic case, we get

$$-\nabla\times\nabla\times\bar{A} + k^2\bar{A} = j\omega\mu\varepsilon\nabla\Phi - \mu\bar{J}. \qquad (4.167)$$

We need to find the relationship between $\bar{A}$ and Φ such that (4.167) reduces to a simpler scalar equation in a spherical coordinate system. Let us assume that $\bar{A}$ and $\bar{J}$ have only the radial component A_r and J_r.

$$\bar{A} = A_r\hat{r} \quad\text{and}\quad \bar{J} = J_r\hat{r}. \qquad (4.168)$$

The radial component of (4.167) is then given by

$$\left[\frac{1}{r^2 \sin\theta}\frac{\partial}{\partial\theta}\left(\sin\theta\frac{\partial}{\partial\theta}\right) + \frac{1}{r^2\sin^2\theta}\frac{\partial^2}{\partial\phi^2} + k^2\right]A_r = j\omega\mu\varepsilon\frac{\partial}{\partial r}\Phi - \mu J_r. \quad (4.169)$$

The θ and ϕ components are

$$-\frac{1}{r}\frac{\partial^2}{\partial r\partial\theta}A_r = \frac{j\omega\mu\varepsilon}{r}\frac{\partial}{\partial\theta}\Phi,$$

$$-\frac{1}{r\sin\theta}\frac{\partial^2}{\partial r\partial\phi}A_r = \frac{j\omega\mu\varepsilon}{r\sin\theta}\frac{\partial}{\partial\phi}\Phi. \quad (4.170)$$

The Lorentz condition $\nabla \cdot \bar{A} + j\omega\mu\varepsilon\,\Phi = 0$ in (2.87) cannot simplify (4.170). Instead, we use

$$\frac{\partial A_r}{\partial r} + j\omega\mu\varepsilon\,\Phi = 0. \quad (4.171)$$

Then (4.170) is automatically satisfied, and (4.169) becomes

$$\left[\frac{\partial^2}{\partial r^2} + \frac{1}{r^2\sin\theta}\frac{\partial}{\partial\theta}\left(r^2\frac{\partial}{\partial r}\right) + \frac{1}{r^2\sin\theta}\frac{\partial^2}{\partial\phi^2} + k^2\right]A_r = -\mu J_r. \quad (4.172)$$

This is a simple differential equation. However, it would be more convenient if this was transformed into a scalar wave equation. This can be accomplished by choosing

$$j\omega\mu\varepsilon r\pi_r = A_r. \quad (4.173)$$

Then (4.172) becomes

$$(\nabla^2 + k^2)\pi_r = -\frac{J_r}{j\omega\varepsilon r}, \quad (4.174)$$

where ∇^2 is the Laplacian operator

$$\nabla^2 = \frac{1}{r^2}\frac{\partial}{\partial r}(r^2)\frac{\partial}{\partial r} + \frac{1}{r^2\sin\theta}\frac{\partial}{\partial\theta}\left(\sin\theta\frac{\partial}{\partial\theta}\right) + \frac{1}{r^2\sin^2\theta}\frac{\partial^2}{\partial\phi^2}. \quad (4.175)$$

A definite advantage of using the scalar wave equation is that its solution has been extensively studied and is well documented. The fields derived from π_z are called E modes or TM modes because the only radial component is the electric field and all the magnetic fields are transverse to the radial vector.

By the duality principle, we can obtain H modes or TE modes from π_{mr}. Thus we get

$$(\nabla^2 + k^2)\pi_r = \frac{J_r}{j\omega\varepsilon r},$$

$$(\nabla^2 + k^2)\pi_{mr} = \frac{J_{mr}}{j\omega\mu r},$$

$$\bar{E} = \nabla \times \nabla \times (r\pi_r \hat{r}) - j\omega\mu\nabla \times (r\pi_{mr}\hat{r}) - \frac{J_r \hat{r}}{j\omega\varepsilon}$$

$$= \nabla \frac{\partial}{\partial r}(r\pi_r \hat{r}) + k^2 r\pi_r \hat{r} - j\omega\mu\Delta \times (r\pi_{mr}\hat{r}),$$

$$\bar{H} = j\omega\varepsilon\nabla \times (r\pi_r \hat{r}) + \nabla \times \nabla \times (r\pi_{mr}\hat{r}) - \frac{J_{mr}\hat{r}}{j\omega\mu}$$

$$= j\omega\varepsilon\nabla \times (r\pi_r \hat{r}) + \nabla \frac{\partial}{\partial r}(r\pi_{mr}\hat{r}) + k^2 r\pi_{mr}\hat{r}.$$

(4.176)

In terms of the field components, we have

$$E_r = \frac{\partial^2}{\partial r^2}(r\pi_r) + k^2 r\pi_r,$$

$$E_\theta = \frac{1}{r}\frac{\partial^2}{\partial r\partial\theta}(r\pi_r) - j\omega\mu\frac{1}{\sin\theta}\frac{\partial}{\partial\phi}\pi_{mr},$$

$$E_\phi = \frac{1}{r\sin\theta}\frac{\partial^2}{\partial r\partial\phi}(r\pi_r) + j\omega\mu\frac{\partial}{\partial\theta}\pi_{mr},$$

$$H_r = \frac{\partial^2}{\partial r^2}(r\pi_{mr}) + k^2 r\pi_{mr},$$

$$H_\theta = j\omega\varepsilon\frac{1}{\sin\theta}\frac{\partial}{\partial\phi}\pi_r + \frac{1}{r}\frac{\partial^2}{\partial r\partial\theta}(r\pi_{mr}),$$

$$H_\phi = -j\omega\varepsilon\frac{\partial}{\partial\theta}\pi_r + \frac{1}{r\sin\theta}\frac{\partial^2}{\partial r\partial\phi}(r\pi_{mr}).$$

(4.177)

In the above, we used the source term of the radial electric current J_r and the radial magnetic current J_{mr}. For those radial current sources, π_r and π_{mr} are simply related to J_r and J_{mr}, as shown in (4.174). For a more general source current with θ and ϕ components, the relationships are more complicated. It should, however, be noted that the completely general electromagnetic field in the spherical coordinate system can be expressed by two scalar functions π_r and π_{mr}.

Let us now consider a formal solution of the homogeneous wave equation in the spherical coordinate system.

$$(\nabla^2 + k^2)\pi_r = 0.$$

(4.178)

Assume the product solution

$$\pi_r = X_1(r) X_2(\theta) X_3(\phi). \tag{4.179}$$

Then we get

$$\left\{ \frac{1}{r^2} \frac{d}{dr} \left(r^2 \frac{d}{dr} \right) + \left[k^2 - \frac{\nu(\nu+1)}{r^2} \right] \right\} X_1(r) = 0, \tag{4.180}$$

$$\left\{ \frac{1}{\sin\theta} \frac{d}{d\theta} \left(\sin\theta \frac{d}{d\theta} \right) + \left[\nu(\nu+1) - \frac{\mu^2}{\sin^2\theta} \right] \right\} X_2(\theta) = 0, \tag{4.181}$$

$$\left(\frac{d^2}{d\phi^2} + \mu^2 \right) X_3(\phi) = 0, \tag{4.182}$$

where ν and μ are constant.

Each of the equations above is a second-order differential equation, and the solution is in general represented by a linear combination of two independent solutions. Let us first note that (4.182) can easily be solved and that $X_3(\phi)$ is given by

$$\begin{aligned} X_3(\phi) &= C_1 e^{j\mu\phi} + C_2 e^{-j\mu\phi} \\ &= C_3 \sin\mu\phi + C_4 \cos\mu\phi, \end{aligned} \tag{4.183}$$

where all C are constant.

Next consider $X_1(r)$ in (4.180). This is a spherical Bessel's equation, and its solution, the spherical Bessel functions $z_\nu(x)$, is related to the ordinary Bessel functions by

$$z_\nu(x) = \sqrt{\frac{\pi}{2x}} Z_{\nu+1/2}(x). \tag{4.184}$$

Thus, corresponding to $J_\nu(x)$, $N_\nu(x)$, $H_\nu^{(1)}(x)$, and $H_\nu^{(2)}(x)$, we have

$$\begin{aligned} j_\nu(x) &= \sqrt{\frac{\pi}{2x}} J_{\nu+1/2}(x), \\ n_\nu(x) &= \sqrt{\frac{\pi}{2x}} N_{\nu+1/2}(x), \\ h_\nu^{(1)}(x) &= \sqrt{\frac{\pi}{2x}} H_{\nu+1/2}^{(1)}(x), \\ h_\nu^{(2)}(x) &= \sqrt{\frac{\pi}{2x}} H_{\nu+1/2}^{(2)}(x). \end{aligned} \tag{4.185}$$

The behavior of spherical Bessel functions can readily be understood by that of the corresponding Bessel functions.

Note that (see Jahnke et al. 1960, p. 142)

$$j_0(x) = \frac{\sin x}{x} \quad \text{and} \quad n_0(x) = -\frac{\cos x}{x},$$
$$h_0^{(1)}(x) = -j\frac{e^{jx}}{x} \quad \text{and} \quad h_0^{(2)}(x) = j\frac{e^{-jx}}{x}. \tag{4.186}$$

In general,

$$j_n(x) = x^n \left(-\frac{1}{x}\frac{d}{dx} \right)^n \frac{\sin x}{x},$$
$$n_n(x) = -x^n \left(-\frac{1}{x}\frac{d}{dx} \right)^n \frac{\cos x}{x}.$$

Let us consider $X_2(\theta)$ in (4.181). This is the Legendre differential equation, and X_2 is given by a linear combination of its two independent solutions. We consider the following three cases.

1. $\mu = 0$, $\nu = n$ *integer*. In this case, $X_2(\theta)$ is given by

$$X_2(\theta) = C_1 P_n(\cos \theta) + C_2 Q_n(\cos \theta). \tag{4.187}$$

$P_n(x)$, $x = \cos \theta$, is a polynomial of degree n.

$$P_0(x) = 1,$$
$$P_1(x) = x, \tag{4.188}$$
$$P_2(x) = \tfrac{1}{2}(3x^2 - 1).$$

$P_n(x)$ is an even function of x when n is even and an odd function when n is odd. $P_n(x)$ is also regular in the range, $-1 \le x \le 1$ ($\pi \ge \theta \ge 0$). $Q_n(x)$ is called the Legendre function of the second kind and becomes infinite at $\theta = 0$ and π.

$$Q_0(x) = \frac{1}{2} \ln \frac{1+x}{1-x},$$
$$Q_1(x) = \frac{x}{2} \ln \frac{1+x}{1-x} - 1, \tag{4.189}$$
$$Q_2(x) = \frac{3x^2 - 1}{4} \ln \frac{1+x}{1-x} - \frac{3x}{2}.$$

2. $\mu = m$, $\nu = n$, m, n *are integers*. In this case $X_2(\theta)$ is given by

$$X_2(\theta) = C_1 P_n^m(\cos \theta) + C_2 Q_n^m(\cos \theta). \tag{4.190}$$

P_n^m and Q_n^m are called the associated Legendre function of the first kind and the second kind, respectively. Both functions vanish if $m > n$, and therefore, m ranges only over

$0, 1, 2,\ldots, n$. Q_n^m becomes infinite at $\theta = 0$ and $\theta = \pi$ while P_n^m is regular in $0 \le \theta \le \pi$. $P_n^m(x)$ is also orthogonal in this range.

$$\int_{-1}^{1} P_n^m(x)P_{n'}^m(x)\,dx = \begin{cases} 0, & n \ne n' \\ \dfrac{2}{2n+1}\dfrac{(n+m)!}{(n-m)!}, & n = n'. \end{cases} \tag{4.191}$$

The function $P_n^m(\cos\theta)e^{jm\phi}$ is regular and orthogonal over the range $0 \le \theta \le \pi$ and $0 \le \phi \le 2\pi$ and is called the *spherical harmonics*.

3. $\mu = m$ *integer*, $\nu = $ *noninteger*. This case arises in connection with a problem of a cone. Two independent solutions are:

$$P_\nu^{mi}(\cos\theta) \quad \text{and} \quad P_\nu^m(-\cos\theta).$$

The first becomes infinite at $\theta = \pi$ and the second becomes infinite at $\theta = 0$. $P_\nu^m(\cos\theta)$ is not zero for $m > \nu$ when ν is a noninteger.

4.16 SPHERICAL WAVEGUIDES AND CAVITIES

Let us consider waves propagating in each of the three orthogonal directions in the spherical coordinate system.

4.16.1 Wave Propagation in the Radial Direction

As an example, consider a wave propagating in the radial direction in a conical waveguide (Fig. 4.21). TM waves can be expressed by the following Hertz potential:

$$\pi_r = \sum_m \sum_n A_{mn} P_{\nu_n}^m(\cos\theta) h_{\nu_{n'}}^{(2)}(kr) \begin{pmatrix} \cos m\phi \\ \sin m\phi \end{pmatrix}. \tag{4.192}$$

where ν_n is determined by

$$P_{\nu_n}^m(\cos\theta_0) = 0.$$

Note that $h_\nu^{(2)}$ represents the outgoing wave in the radial direction.

4.16.2 Wave Propagation in the θ Direction

An example of this case is the VLF propagation between the spherical earth and the ionosphere. For example, we may consider a resonance in a spherical cavity formed

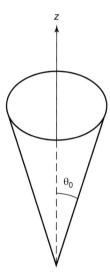

FIGURE 4.21   Conical waveguide.

by the earth and the ionosphere (Fig. 4.22). For TM modes we write

$$\pi_r = P_n^m(\cos\theta)Z_n(kr)\begin{pmatrix}\cos m\phi\\\sin m\phi\end{pmatrix}, \tag{4.193}$$

where $Z_n(kr)$ is a linear combination of two spherical Bessel functions that satisfy the boundary condition at the earth surface and the ionosphere. If we assume that the earth surface $(r = a)$ and the lower edge of the ionosphere $(r = b)$ are *conducting*, the boundary conditions are

$$\frac{\partial}{\partial r}\left(r\pi_r\right) = 0 \quad \text{at } r = a \text{ and } b. \tag{4.194}$$

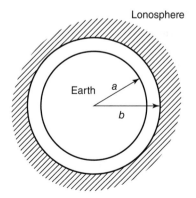

FIGURE 4.22   Earth-ionosphere cavity.

Thus we have

$$Z_n(kr) = C_1 j_n(kr) + C_2 h_n^{(2)}(kr),$$

$$C_1 = \frac{\partial}{\partial r} \left[r h_n^{(2)}(kr) \right] |_{r=a}, \tag{4.195}$$

$$C_2 = -\frac{\partial}{\partial r} [r j_n(kr)]|_{r=a}.$$

The resonant frequency f_r is then given by $k = 2\pi f_r/c$, which satisfies

$$\frac{\partial}{\partial r} [r Z_n(kr)]|_{r=b} = 0. \tag{4.196}$$

This yields the following equation to determine the resonant frequency:

$$\left. \frac{\frac{\partial}{\partial r}[r j_n(kr)]}{\frac{\partial}{\partial r}[r h_n^{(2)}(kr)]} \right|_{r=a} = \left. \frac{\frac{\partial}{\partial r}[r j_n(kr)]}{\frac{\partial}{\partial r}[r h_n^{(2)}(kr)]} \right|_{r=b}. \tag{4.197}$$

The resonance phenomena above were first studied by Schumann and are now called the Schumann resonance. Although (4.197) gives the exact equation for the resonant frequency when both the ionosphere and the earth are assumed conducting, it is instructive to obtain an approximate solution for the Schumann resonance. We first note that the boundary condition (4.194) requires the derivative of $r\pi_r$. Thus we let $rX_1 = u$ in (4.180) and obtain

$$\left[\frac{d^2}{dr^2} + k^2 - \frac{n(n+1)}{r^2} \right] u(r) = 0. \tag{4.198}$$

Now since the distance between the ionosphere and the earth is very much smaller than the earth's radius, we let $r \sim a + z \sim a$ and approximate (4.198) by

$$\left[\frac{d^2}{dz^2} + k^2 - \frac{n(n+1)}{a^2} \right] u(z) = 0, \tag{4.199}$$

with the boundary condition

$$\frac{\partial u}{\partial z} = 0 \quad \text{at } z = 0 \quad \text{and} \quad z = h. \tag{4.200}$$

The solution is

$$u = \cos \frac{m\pi z}{h}, \quad m = 0, 1, 2, \ldots$$

$$k^2 - \frac{n(n+1)}{a^2} = \left(\frac{m\pi}{h} \right)^2.$$

The lowest resonant frequencies f_r are

$$f_r = \frac{c}{2\pi a}\sqrt{n(n+1)}, \quad n = 1, 2, 3, \dots \tag{4.201}$$

These values are close to the observed peaks in the frequency spectrum of the noise power generated by lightning around the earth.

4.16.3 Wave Propagation in the Azimuthal Direction

An example may be a leaky waveguide around a sphere. A general form of π_r is then

$$\pi_r = z_v(kr)P_v^\mu(\cos\theta)e^{-j\mu\phi}, \tag{4.202}$$

where μ is the azimuthal propagation constant.

Let us now consider the boundary conditions for a perfectly conducting wall (Fig. 4.23). Consider a surface containing a radial vector. At $\phi =$ constant, E_r and E_θ must be zero. At $\theta =$ constant, E_r and E_ϕ must be zero. They are satisfied if

$$\pi_r = 0 \quad \text{and} \quad \frac{\partial}{\partial n}\pi_{mr} = 0. \tag{4.203}$$

$\partial/\partial n$ is the normal derivative, and in the two cases above, this can also be expressed by $\partial/\partial\phi$ and $\partial/\partial\theta$, respectively.

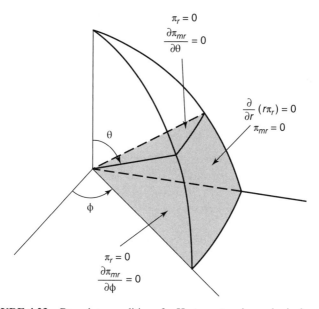

FIGURE 4.23   Boundary conditions for Hertz vectors in a spherical system.

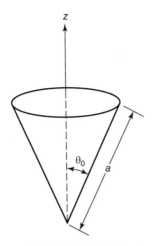

FIGURE 4.24   Conical cavity.

The boundary condition at the surface perpendicular to the radial vector is that E_θ and E_ϕ be zero, which leads to the condition

$$\frac{\partial}{\partial n}(r\pi_r) = \frac{\partial}{\partial r}(r\pi_r) = 0 \quad \text{and} \quad \pi_{mr} = 0. \tag{4.204}$$

As an example, let us consider a conical cavity shown in Fig. 4.24. The TM mode is given by

$$\pi_r = j_\nu(kr)P_\nu^m(\cos\theta)\left(\begin{array}{c}\cos m\phi \\ \sin m\phi\end{array}\right), \tag{4.205}$$

where $j_\nu(kr)$ is used because $n_\nu(kr)$ becomes infinite at the origin.

For a given m, ν must be determined by the boundary condition at $\theta = \theta_0$, which is given by

$$P_{\nu_n}^m(\cos\theta_0) = 0. \tag{4.206}$$

Once ν_n is determined, the boundary condition at $r = a$ is applied and yields the resonant frequency f_r (or wavelength λ_r), given by $k_r = \omega_r/c = 2\pi/\lambda_r$, $\omega_r = 2\pi f_r$, where $X_r = k_r a$ is the root of

$$\frac{d}{dX}[Xj_{\nu_n}(X)] = 0. \tag{4.207}$$

There is an infinite number of roots for (4.207) for a given ν_n, and thus we should write

$$X_r = X_{mnl}. \tag{4.208}$$

The TM$_{mnl}$ mode is therefore given by

$$\pi_r = j_{v_n}(k_{mnl}r)P_{v_n}^m(\cos\theta)\begin{pmatrix}\cos m\phi\\\sin m\phi\end{pmatrix}, \tag{4.209}$$

where $k_{mnl}a = X_{mnl}$.

PROBLEMS

4.1 **(a)** Find the normalized eigenfunctions $\phi_n(x)$ and eigenvalues k_n^2 that satisfy the following:

$$\left(\frac{d^2}{dx^2} + k_n^2\right)\phi_n(x) = 0,$$
$$\phi_n(x) = 0 \text{ at } x = 0,$$
$$\frac{d\phi_n}{dx} + h\phi_n = 0, \quad \text{at } x = a.$$

(b) Find the lowest two eigenvalues for a = 1 and h = 2.

4.2 Consider a TE10 mode in a rectangular waveguide. At x = 0, Ey = 0, and at x = a, Ey/Hz = j100. Find the propagation constant β at 10 GHz. a = 2.5 cm and b = 1.25 cm.

4.3 A TM11 mode is propagating in a rectangular waveguide with a = 0.2 m and b = 0.1 m.
(a) What is the cutoff frequency f_c?
(b) If the operating frequency f is $2f_c$, calculate the phase velocity and the group velocity.
(c) At $f = 2f_c$, if the maximum value of $|E_z| = 5$ V/m, calculate the total power propagated through the guide.

4.4 A TE11 wave is propagating in a cylindrical guide with radius 1 cm. The operating frequency is 10 GHz and the total transmitted power is 100 mW.
(a) Find the phase and group velocities.
(b) What is the cutoff frequency for this guide?
(c) If the guide is filled with dielectric material with $\varepsilon_r = 2$, do the phase and group velocities and the cutoff frequencies change? If so, find their values.
(d) Assume that the guide is hollow. Find expressions for the electric and magnetic fields.
(e) Find an expression for the surface current.

4.5   Consider a coaxial line with $a = 1$ cm and $b = 0.5$ cm. Calculate the characteristic impedance of this line. If the total power of 1 mW is propagated through the line, calculate the maximum electric field intensity.

4.6   Find the lowest two cutoff frequencies for the sector waveguide shown in Fig. 4.8. $a = 1$ cm and $\alpha = 90°$.

4.7   A pulse wave in the TE_{10} mode is propagating in a rectangular guide with $a = 2.5$ cm and $b = 1$ cm. The carrier frequency is $f_0 = 8$ GHz and the waveform of the electric field at $z = 0$ is given by

$$E_y(t, z = 0) = E_0 \sin \frac{\pi x}{a} \ \exp\left(-\frac{t^2}{T_0^2}\right) \cos \omega_0 t,$$

where $E_0 = 0.5$ V/m, $T_0 = 1$ ns, and $\omega_0 = 2\pi f_0$. Calculate the phase velocity, group velocity, group delay per meter, and pulse spread per meter of the guide.

4.8   Consider a multimode step-index fiber with $n_1 = 1.48$ and $n_2 = 1.46$. The diameter is 50 μm, and the wavelength is $\lambda = 0.82$ μm. Find the number of modes, pulse spread per unit length in ns/km, and numerical aperture (NA).

4.9   A graded-index fiber has the square-law refractive index profile given by (4.142) with $n_1 = 1.48$, $n_2 = 1.46$, and $2a = 50$ μm. Find the pulse spread in ns/km.

4.10   Consider TM modes propagating in the radial direction inside the sectorial horn in Fig. 4.17 with $\alpha = 45°$ and $l = 3$ cm. Find the two lowest cutoff frequencies. Find the expressions for $\bar{E}$ and $\bar{H}$ for the propagating modes $m = 1$ and $n = 1$ at 10 GHz.

4.11   Find the three lowest resonant frequencies of the cylindrical cavity shown in Fig. P4.11.

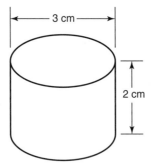

FIGURE P4.11   Cylindrical cavity.

4.12   Find the resonant frequency of the microstrip antenna shown in Fig. 4.20 with $h = 0.3$ cm, $a = 4$ cm, and $\varepsilon_r = 2.5$.

4.13   Use Rodrigues's formula,

$$P_n(x) = \frac{1}{2^n n!} \frac{d^n}{dx^n} (x^2 - 1)^n,$$

to show the following:

$$\int_{-1}^{1} x^m P_n(x) \, dx = 0 \quad \text{if } m < n.$$

4.14   Expand the following function in a series of Legendre functions $P_n(x)$:

$$F(x) = x^3 + x^2 + x + 1.$$

4.15   Find the lowest five Schumann resonance frequencies.

4.16   Consider a spherical cavity of radius a. For TM modes, show that the lowest nonzero mode is TM_{101}, where the subscripts are variations on the r, ϕ, and θ directions, and the Hertz vector is $\Pi_r = j_1(kr)P_1(\cos \theta)$. Show that the resonant frequency is given by

$$\tan x = \frac{x}{1 - x^2}, \quad x = ka.$$

Its solution is $x = 2.744$. Find expressions for $\bar{E}$ and $\bar{H}$.

   For TE modes, the lowest mode is TE_{101} and the resonant frequency is given by

$$\tan x = x, \quad x = ka.$$

Its solution is $x = 4.493$. Find expressions for $\bar{E}$ and $\bar{H}$.

4.17   Consider the cylindrical cavity shown in Fig. P4.17. Find the lowest two resonant frequencies for (a) the TM modes and (b) the TE modes.

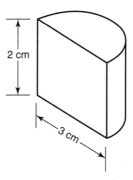

2 cm

3 cm

FIGURE P4.17   Cylindrical sector cavity.

CHAPTER 5

GREEN'S FUNCTIONS

In Chapter 3, we discussed the reflection and transmission of plane waves incident on layers of dielectric medium and the propagation characteristics of guided waves along the layered medium. In Chapter 4, we discussed the wave propagation in waveguides and the wave modes in cavities. In this chapter, we discuss the problems of the excitation of waves and Green's function, which is the field excited by a point source.

5.1 ELECTRIC AND MAGNETIC DIPOLES IN HOMOGENEOUS MEDIA

There are two basic sources of excitation: electric and magnetic. The simplest forms of these two sources are the *electric dipole* and the *magnetic dipole*. An arbitrary source can always be represented by a distribution of electric and magnetic dipole sources.

A short wire antenna located at $\bar{r}'$ which is fed at the center with the current I_0 oscillating at a frequency $f = \omega/2\pi$, and whose length L is much smaller than a wavelength, can represent the electric dipole whose current density $\bar{J}$ is given by

$$\bar{J}(\bar{r}) = \hat{i} I_0 L_0 \delta(\bar{r} - \bar{r}'), \tag{5.1}$$

where $\hat{i}$ is a unit vector in the direction of the current, and L_0 is the effective length of the wire antenna given by

$$I_0 L_0 = \int_{\Delta V} \bar{J} dV = \int_{-L/2}^{L/2} I(z) dz. \tag{5.2}$$

Electromagnetic Wave Propagation, Radiation, and Scattering: From Fundamentals to Applications,
Second Edition. Akira Ishimaru.
© 2017 by The Institute of Electrical and Electronic Engineers, Inc. Published 2017 by John Wiley & Sons, Inc.

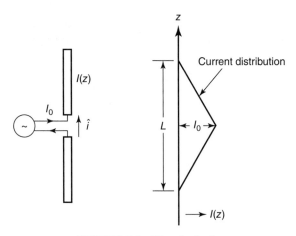

FIGURE 5.1   Electric dipole.

For a short wire antenna, $I(z)$ is known to have a triangular shape, and therefore $L_0 = L/2$ (see Fig. 5.1).

Let us next consider the magnetic dipole. If the current I_0 flows in a small loop with the area A, the magnetic dipole moment $\bar{m}$ is defined by

$$\bar{m} = I_0 A \hat{i}, \tag{5.3}$$

where $\hat{i}$ is the unit vector normal to the area A (Fig. 5.2). Noting that the magnetic current density $y_2(\rho) = J_n(q\rho) + H_n^{(2)}(qa) - J_n(qa)H_n^{(2)}(q\rho)$ is related to $\overline{M}$ by (see Section 2.9)

$$\overline{J}_m = j\omega\mu_0\overline{M}, \tag{5.4}$$

the magnetic current density $\bar{J}_m$ for a small current loop shown in Fig. 5.2 is represented by

$$\overline{J}_m = j\omega\mu_0 A I_0 \delta(\bar{r} - \bar{r}')\hat{i}. \tag{5.5}$$

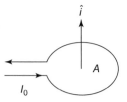

FIGURE 5.2   Magnetic dipole.

The electric current density $\bar{J}$ given in (5.1) and the magnetic current density $\bar{J}_m$ given in (5.5) are the two basic sources of electromagnetic fields. In general, many other practical sources can be represented by continuous distributions of the electric and magnetic current sources. For example, a long wire antenna can be regarded as a continuous distribution of the electric dipole source. Another common source is aperture fields for parabolic dish antennas and slot antennas. They can be represented by the electric or magnetic current distribution over the surface.

5.2 ELECTROMAGNETIC FIELDS EXCITED BY AN ELECTRIC DIPOLE IN A HOMOGENEOUS MEDIUM

The time-harmonic electromagnetic wave excited by the current source $\bar{J}_s$ is given in terms of the Hertz vector $\bar{\pi}^t$ (Section 2.7)

$$\bar{E} = \nabla(\nabla \cdot \bar{\pi})' + k^2\bar{\pi},$$
$$\bar{H} = j\omega \in \nabla \times \bar{\pi},$$

and

$$(\nabla^2 + k^2)\bar{\pi} = -\frac{\overline{K}_s}{j\omega \in}, \tag{5.6}$$

where $k = \omega\sqrt{\mu\varepsilon}$ and μ and ε are, in general, complex.

Let us consider the electromagnetic field excited by an electric dipole pointed in the z direction and located at the origin of a Cartesian coordinate system. The electric dipole is represented by

$$\bar{J}_s = \hat{z}I_0L_0\delta(\bar{r}). \tag{5.7}$$

The z component of $\bar{\pi}$ then satisfies

$$(\nabla^2 + k^2)\pi_z = \frac{I_0L_0}{j\omega \in}\delta(r). \tag{5.8}$$

We write (5.8) using a function $G(r)$ which satisfies the equation

$$(\nabla^2 + k^2)G(\bar{r}, \bar{r}') = -\delta(\bar{r} - \bar{r}'). \tag{5.9}$$

Then π_z is given by

$$\pi_z = \frac{I_0L_0}{j\omega \in}G(\bar{r}, 0). \tag{5.10}$$

Note that once we solve (5.9), the solution $\bar{\pi}$ is given by (5.10) and the fields are given by (5.6).

The function $G(\bar{r}, \bar{r}')$ is called Green's function and represents the response of a physical system in space due to a point exciting source $\delta(\bar{r} - \bar{r}')$. The differential equation (5.9) alone is insufficient to determine Green's function uniquely, and it is necessary to apply additional conditions. These conditions are (1) the radiation condition at infinity whenever the region extends to infinity and (2) the boundary conditions. Therefore, Green's function $G(\bar{r}, \bar{r}')$ must satisfy the inhomogeneous differential equation of the type (5.9) and the radiation and boundary conditions.

Let us solve (5.9) in free space satisfying the radiation condition. We choose $\bar{r}'$ to be at the origin, and thus $G(\bar{r})$ is spherically symmetric and a function of r only. We write (5.9) as

$$\left[\frac{1}{r^2}\frac{d}{dr}\left(r^2\frac{d}{dr}\right) + k^2\right] G(r) = -\delta(r). \tag{5.11}$$

The solution is given by

$$G(r) = \frac{\exp(-jkr)}{4\pi r}. \tag{5.12}$$

To show this, first we consider $G(\bar{r})$ when $\bar{r} \neq 0$. Here $\delta(r) = 0$. We then let $G = u/r$ and obtain

$$\left(\frac{d^2}{dr^2} + k^2\right) u(4) = 0,$$

whose general solution is

$$u(r) = c_1 e^{-jkr} + c_2 e^{+jkr}, \qquad c_1, c_2 \text{ constants.}$$

Therefore, for $r \neq 0$, G is given by

$$G(r) = \frac{1}{r}(c_1 e^{-jkr} + c_2 e^{+jkr}). \tag{5.13}$$

Now $G(r)$ is the wave originating at $r = 0$, and at infinity, the wave must be outgoing. The first term of (5.13) represents the outgoing wave, but the second term represents the incoming wave. Even though both terms vanish at $r \to \infty$, the second term must be discarded because there should be no incoming wave from infinity. This is called the *radiation condition* or *Sommerfeld radiation condition*, and its mathematical expression is given by (Section 2.4)

$$\lim_{r \to \infty} r\left(\frac{\partial G}{\partial r} + jkG\right) = 0. \tag{5.14}$$

Only the first term of (5.13) satisfies (5.14) as expected, and therefore c_2 must be zero.

Next we determine the constant c_1 by considering the behavior of G near the origin. To do this, let us write (5.9) as follows:

$$\nabla \cdot (\nabla G) + k^2 G = -\delta(r), \tag{5.15}$$

and integrate both sides over a small spherical volume of radius r_0 and let the radius r_0 approach zero

$$\lim_{r_0 \to 0} \int_V [\Delta \cdot (\Delta G) + k^2 G] dV = -\lim_{r_0 \to 0} \int_V \delta(r) dV. \tag{5.16}$$

The right side is -1. The first term in the left side becomes, using the divergence theorem, $-4\pi c_1$. The second term in the left becomes zero. Therefore, (5.16) becomes

$$-4\pi c_1 = -1,$$

from which we obtain $c_1 = 1/4\pi$.

Thus Green's function $G(\bar{r}, \bar{r}')$ satisfying

$$(\nabla^2 + k^2)G(\bar{r}, \bar{r}') = -\delta(\bar{r} - \bar{r}') \tag{5.17}$$

and the radiation condition at infinity is given by

$$G(\bar{r}, \bar{r}') = \frac{e^{-jk|\bar{r}-\bar{r}'|}}{4\pi|\bar{r} - \bar{r}'|}. \tag{5.18}$$

Using this Green's function, the Hertz vector due to an electric dipole located at the origin pointed in the $\hat{z}$ direction is given by

$$\begin{aligned}
\bar{\pi} &= \hat{z}\pi_z \\
&= \hat{z}\frac{I_0 L_0}{j\omega\varepsilon}G(\bar{r}) \\
&= \hat{z}\frac{I_0 L_0}{j\omega\varepsilon}\frac{e^{-jkr}}{4\pi r}.
\end{aligned} \tag{5.19}$$

The electric and magnetic fields are given by (5.6). We may use the spherical coordinate system, writing

$$\bar{\pi} = \pi_r\hat{r} + \pi_\theta\hat{\theta} + \pi_\phi\hat{\phi}, \tag{5.20}$$

where

$$\begin{aligned}
\pi_r &= \pi_z \hat{z} \cdot \hat{r} = \pi_z \cos\theta, \\
\pi_\theta &= \pi_z \hat{z} \cdot \hat{\theta} = \pi_z(-\sin\theta), \quad \pi_\phi = 0,
\end{aligned}$$

and obtain

$$\begin{aligned}
E_r &= \frac{\partial}{\partial r}\left[\frac{1}{r^2}\frac{\partial}{\partial r}(r^2\pi_r) + \frac{1}{r\sin\theta}\frac{\partial}{\partial\theta}(\sin\theta\,\pi_\theta)\right] + k^2\pi_r \\
&= \frac{I_0 L_0}{j\omega\varepsilon}\frac{e^{-jkr}}{4\pi}\left(\frac{j2k}{r^2} + \frac{2}{r^3}\right)\cos\theta, \\
E_\theta &= \frac{I_0 L_0}{j\omega\varepsilon}\frac{e^{-jkr}}{4\pi}\left(-\frac{k^2}{r} + \frac{jk}{r^2} + \frac{1}{r^3}\right)\sin\theta, \\
E_\phi &= 0, \\
H_r &= H_\theta = 0, \\
H_\phi &= I_0 L_0\frac{e^{-jkr}}{4\pi}\left(\frac{jk}{r} + \frac{1}{r^2}\right)\sin\theta.
\end{aligned}$$
\hfill (5.21)

We note that the radial component E_r is proportional to r^{-2} and r^{-3}, while E_θ and H_ϕ contain the term with r^{-1}. Thus at a large distance from the dipole, the radial component E_r vanishes much faster than E_θ and H_ϕ. The range $|k_0 r| \gg 1$ is called the *far zone* and the field in this range is called the *far field*. Similarly, the field in the *near zone* $|k_0 r| \ll 1$ is called the *near field*.

In the far zone, the field components are given by

$$\begin{aligned}
E_\theta &= j(I_0 L_0)\omega\mu\frac{e^{-jkr}}{4\pi r}\sin\theta, \\
H_\phi &= j(I_0 L_0)k\frac{e^{-jkr}}{4\pi r}\sin\theta,
\end{aligned}$$
\hfill (5.22)

and therefore, the radiation pattern is proportional to $\sin\theta$ (see Fig. 5.3). The electric field E_θ and the magnetic field H_ϕ are perpendicular to each other, and both are

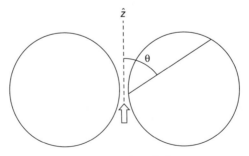

FIGURE 5.3   Dipole radiation pattern.

perpendicular to the direction of wave propagation $\hat{r}$. The ratio of E_θ to H_ϕ is the characteristic impedance

$$\frac{E_\theta}{H_\phi} = \eta = \left(\frac{\mu}{\varepsilon}\right)^{1/2}. \tag{5.23}$$

Note that these are also the characteristics of a plane wave, and thus the spherical wave radiating from a dipole behaves like a plane wave at a large distance except that the amplitude decreases as $1/r$.

Let us examine the total real power radiated from the dipole. If the medium is lossless, μ and ε are real and all the power radiated from the dipole must be equal to the power flowing out from any surface enclosing the dipole. In particular, we may take the surface to be that of a sphere with a large radius so that (5.22) holds. The total real power P_t is given by integrating the Poynting vector over this sphere.

$$P_t = \int_0^{2\pi} d\phi \int_0^\pi \sin\theta \, d\theta \, r^2 \overline{S} \cdot \hat{r}, \tag{5.24}$$

where

$$\overline{S} = \frac{1}{2}\overline{E} \times \overline{H}^*.$$

Substituting (5.22) into (5.24), we get

$$P_t = \int_0^{2\pi} d\phi \int_0^\pi \sin\theta \, d\theta \, (I_0 L_0)^2 \frac{k^2\eta}{2(4\pi)^2} \sin^2\theta$$

$$= (I_0 L_0)^2 \frac{k^2\eta}{12\pi} = \left(I_0 \frac{L_0}{\lambda}\right)^2 \frac{\pi}{3}\eta. \tag{5.25}$$

The effectiveness of an antenna may be expressed by the amount of real power the current I_0 can radiate. This is expressed by the radiation resistance R_{rad} defined by

$$P_t = \frac{1}{2}I_0^2 R_{\text{rad}}. \tag{5.26}$$

The radiation resistance of a short dipole of length L (which is equal to $2L_0$) is then given by

$$R_{\text{rad}} = \eta \frac{2\pi}{3}\left(\frac{L_0}{\lambda}\right)^2$$

$$= \eta \frac{\pi}{6}\left(\frac{L}{\lambda}\right)^2 = 20\pi^2 \left(\frac{L}{\lambda}\right)^2. \tag{5.27}$$

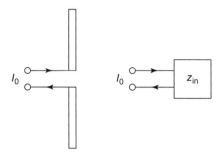

FIGURE 5.4 Input impedance of a dipole antenna.

The expression (5.27) is applicable only to a short dipole ($L \ll \lambda$), but it is a good approximation for dipoles of lengths up to a quarter wavelength.

The radiation resistance (5.27) is the equivalent resistance of the input impedance of the dipole antenna representing the radiation and is independent of the wire radius. In addition to the radiation resistance, there is the resistance R_0 representing the ohmic loss of the wire and the reactance X (Fig. 5.4). The reactance X represents the stored energy in the near field of the antenna and depends on the geometry and the radius of the wire and requires a detailed study of the field near the dipole.

5.3 ELECTROMAGNETIC FIELDS EXCITED BY A MAGNETIC DIPOLE IN A HOMOGENEOUS MEDIUM

As we noted in Section 5.1, a small loop antenna can be represented by a magnetic current dipole

$$\bar{J}_m = \hat{i} j \omega \mu_0 I_0 A \delta(\bar{r} - \bar{r}'). \tag{5.28}$$

We can derive electromagnetic fields due to the magnetic dipole by using the duality principle. If $\bar{E}$, $\bar{H}$, $\bar{J}$, $\bar{\pi}$, μ, and ε for the case of an electric dipole are replaced by $-\bar{H}$, $\bar{E}$, $-\bar{J}_m$, $-\bar{\pi}_m$, ε, and μ, then Maxwell's equations and the formulations in terms of the Hertz vectors are unchanged, and we have

$$\bar{E} = -j\omega\mu \nabla \times \bar{\pi}_m,$$
$$\bar{H} = \nabla(\nabla \cdot \bar{\pi}_m) + k^2 \bar{\pi}_m,$$

and

$$(\nabla^2 + k^2)\bar{\pi}_m = -\frac{\bar{J}_m}{j\omega\mu}. \tag{5.29}$$

Therefore, we obtain

$$H_r = I_0 A \frac{e^{-jkr}}{4\pi} \left(\frac{j2k}{r^2} + \frac{2}{r^3} \right) \cos\theta,$$

$$H_\theta = I_0 A \frac{e^{-jkr}}{4\pi} \left(-\frac{k^2}{r} + \frac{jk}{r^2} + \frac{1}{r^3} \right) \sin\theta,$$

$$H_\phi = 0, \tag{5.30}$$

$$E_r = E_\theta = 0,$$

$$E_\phi = (-j\omega\mu I_0 A) \frac{e^{-jkr}}{4\pi} \left(\frac{jk}{r} + \frac{1}{r^2} \right) \sin\theta,$$

and the radiation pattern is given by

$$H_\theta = -(I_0 A)k^2 \frac{e^{-jkr}}{4\pi r} \sin\theta,$$

$$E_\phi = (I_0 A)k^2 \eta \frac{e^{-jkr}}{4\pi r} \sin\theta. \tag{5.31}$$

The radiation resistance R_{rad} is given by

$$R_{\text{rad}} = \eta \left(\frac{2\pi}{\lambda} \right)^4 \frac{A^2}{6\pi} = 20(2\pi)^4 \frac{A^2}{\lambda^4}. \tag{5.32}$$

When an electric dipole is located near a conducting planar surface, the total field can conveniently be expressed as a sum of the field due to the dipole and the field due to an image of the dipole. For a vertical electric dipole located above a conducting plane as shown in Fig. 5.5, the field due to the image must be such that the total electric field tangential to the conducting surface vanishes. From (5.21), since $r_1 = r_2$ and $\theta_1 = \pi - \theta_2$, we can easily show that the total electric field due to the vertical electric dipole and its image dipole oriented in the same direction have no tangential component on the surface of the conductor. Similarly, the image of a horizontal electric dipole is directed opposite to the dipole (Fig. 5.5). We can also follow similar reasoning to show that the image of a vertical magnetic dipole is directed opposite while the image of a horizontal magnetic dipole is directed parallel to the dipole.

5.4 SCALAR GREEN'S FUNCTION FOR CLOSED REGIONS AND EXPANSION OF GREEN'S FUNCTION IN A SERIES OF EIGENFUNCTIONS

In Sections 5.1–5.3, we discussed the excitation by an electric or magnetic dipole in a homogeneous medium and expressed electromagnetic fields in terms of Green's

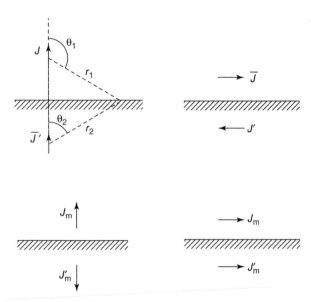

FIGURE 5.5   Images in a conducting plane.

function in a homogeneous medium. In general, however, Green's function must satisfy appropriate boundary condition.

If the region under consideration extends to infinity, it is called the *open* region. Green's function in the open region can be represented by Fourier transform, and this will be discussed later. In this section we consider the finite *closed* region V surrounded by a surface S (Fig. 5.6). In this case it is possible to construct Green's function in terms of a series of eigenfunctions. In addition to the foregoing two representations, Fourier transform and eigenfunctions, we can express a one-dimensional Green's function in terms of solutions of homogeneous differential equations. We will discuss these three representations of Green's functions.

Let us consider a volume V surrounded by a surface S (Fig. 5.6). Green's function is a solution to an inhomogeneous wave equation with the delta function as an exciting source and satisfies an appropriate boundary condition on S.

$$(\nabla^2 + k^2)G(\bar{r}, \bar{r}') = -\delta(\bar{r} - \bar{r}) \quad \text{in } V, \tag{5.33}$$

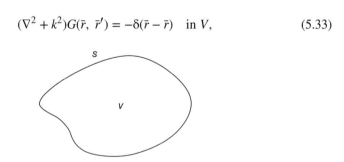

FIGURE 5.6   Closed region.

and the boundary condition on S can be one of the following:

$$G = 0 \quad \text{Dirichlet's condition,}$$

$$\frac{\partial G}{\partial n} = 0 \quad \text{Neumann's condition,} \tag{5.34}$$

$$G + h\frac{\partial G}{\partial n} = 0 \quad \text{general homogeneous condition.}$$

Since any function can be expressed by a series of eigenfunctions, we seek to represent G in the following form:

$$G = \sum_n A_n \phi_n(\bar{r}), \tag{5.35}$$

where A_n is an unknown coefficient and ϕ_n is an eigenfunction. Here the summation over n means triple summation for the three-dimensional case, double summation for the two-dimensional case, and single summation for the one-dimensional case.

We also expand the delta function in a series of eigenfunctions

$$\delta(\bar{r} - \bar{r}') = \sum_n B_n \phi_n(\bar{r}), \tag{5.36}$$

where B_n is given by multiplying both sides of (5.36) by $\phi_m(\bar{r})$ and integrating over the volume. Noting the orthogonality of eigenfunctions,

$$\int_V \phi_n(\bar{r})\phi_m(\bar{r})dV = 0 \quad \text{for } n \neq m, \tag{5.37}$$

we get

$$B_n = \frac{\int_V \delta(\bar{r} - \bar{r}')\phi_n(\bar{r})dV}{\int_V [\phi_n(\bar{r})]^2 dV}$$

$$= \frac{\phi_n(\bar{r}')}{\int_V [\phi_n(\bar{r})]^2 dV}. \tag{5.38}$$

Substituting (5.35) and (5.36) in the wave equation (5.33), we get

$$\sum_n (\nabla^2 + k^2)A_n\phi_n(\bar{r}) = -\sum_n B_n\phi_n(\bar{r}). \tag{5.39}$$

Since

$$(\nabla^2 + k_n^2)\phi_n(\bar{r}) = 0,$$

the left side of (5.39) becomes

$$\sum_n (k^2 - k_n^2) A_n \phi_n(\bar{r}).$$

Now both sides of (5.39) are expansions in terms of the same orthogonal functions ϕ_n, and therefore the coefficient of each term must be equal. Thus

$$(k^2 - k_n^2) A_n = -B_n,$$

from which the unknown coefficients A_n are obtained.

Green's function is then given by

$$G(\bar{r}, \bar{r}') = \sum_n \frac{\phi_n(\bar{r})\phi_n(\bar{r}')}{(k_n^2 - k^2)} \frac{1}{N_n^2}, \qquad (5.40)$$

where

$$N_n^2 = \int_V [\phi_n(\bar{r})]^2 dV$$

is a normalization factor. Equation (5.40) holds as long as $k_n \neq k$. If for a particular $n = n_0$, $k_{n_0} = k$, Green's function diverges unless the source is located at a zero of the eigenfunction $\phi_{n_0}(\bar{r}') = 0$. In this case the term for this particular n_0 should be excluded.

Let us apply the general expression (5.40) to the one-dimensional problem. Let us assume that Green's function satisfies Dirichlet's condition at $z = 0$ and $z = a$, and $0 < z' < a$.

$$\left(\frac{d^2}{dz^2} + k^2\right) G(z, z') = -\delta(z - z'),$$

$$G(0, z') = 0, \qquad (5.41)$$

$$G(a, z') = 0.$$

The normalized eigenfunctions are

$$\phi_n(z) = \sqrt{\frac{2}{a}} \sin \frac{n\pi z}{a}, \qquad (5.42)$$

and the eigenvalues are

$$k_n = \frac{n\pi}{a}, \quad n = 1, 2, \ldots, \infty.$$

Therefore, Green's function is given by

$$G(z, z') = \sum_{n=1}^{\infty} \frac{2}{a} \frac{\sin(n\pi z/a)\sin(n\pi z'/a)}{(n\pi/a)^2 - k^2}. \tag{5.43}$$

Let us next consider a two-dimensional Green's function that satisfies Dirichlet's condition at $\rho = a$ in the cylindrical coordinate system, and $0 \le \rho' < a$.

$$\left[\frac{1}{\rho} \frac{\partial}{\partial \rho} \left(\rho \frac{\partial}{\partial \rho} \right) + \frac{1}{\rho^2} \frac{\partial^2}{\partial \phi^2} + k^2 \right] G = -\frac{\delta(\rho - \rho')(\delta(\phi - \phi')}{\rho} \tag{5.44}$$

satisfying the condition

$$G = 0 \quad \text{at} \quad \rho = a.$$

The denominator ρ on the right-hand side is due to the definition of the delta functions (see Appendix 5,A). We choose the right side in such a manner that the integral with respect to the area is unity.

$$\int \frac{\delta(\rho - \rho')\delta(\phi - \phi')}{\rho} \, dS = 1, \tag{5.45}$$

where

$$dS = \rho \, d\rho \, d\phi.$$

Let us first note that in the ϕ direction, the solution must be periodic, and therefore we write G in a series of normalized eigenfunctions

$$\phi_n(\phi) = -\frac{1}{\sqrt{2\pi}} e^{-jn\phi}. \tag{5.46}$$

Then we write

$$G = \sum_{n=-\infty}^{\infty} \phi_n(\phi)\phi_n^*(\phi')G_n(\rho, \rho'). \tag{5.47}$$

Noting that

$$\delta(\phi - \phi') = \sum_{n=-\infty}^{\infty} \phi_n(\phi)\phi_n^*(\phi'), \tag{5.48}$$

we obtain

$$\left[\frac{1}{\rho}\frac{\partial}{\partial\rho}\left(\rho\frac{\partial}{\partial\rho}\right) - \frac{n^2}{\rho^2} + k^2\right]G_n = -\frac{\delta(\rho - \rho')}{\rho}. \tag{5.49}$$

Now we expand G_n in a series of eigenfunctions.

$$G_n = \sum_{m=1}^{\infty} A_m J_n(k_{nm}\rho), \tag{5.50}$$

where $J_n(k_{nm}a) = 0$ and k_{nm} is the eigenvalue. Also, the delta function is expressed as

$$\frac{\delta(\rho - \rho')}{\rho} = \sum_{m=1}^{\infty} \frac{J_m(k_{nm}\rho)J_n(k_{nm}\rho')}{N_{nm}^2}, \tag{5.51}$$

where the normalization factor N_{nm}^2 is given by

$$N_{nm}^2 = \int_0^a J_n(k_{nm}\rho)^2 \rho d\rho = \frac{a^2}{2}[J'_n(k_{nm}a)]^2. \tag{5.52}$$

Substituting (5.50) into (5.49) and using (5.51), we get

$$G_n = \sum_{m=1}^{\infty} \frac{J_n(k_{nm}\rho)J_n(k_{nm}\rho')}{(k_{nm}^2 - k^2)N_{nm}^2}. \tag{5.53}$$

Finally, we substitute (5.53) into (5.47) and obtain the two-dimensional Green's function

$$G(\rho, \phi; \rho', \phi') = \sum_{n=-\infty}^{\infty} \sum_{m=1}^{\infty} \frac{J_n(k_{nm}\rho)J_n(k_{nm}\rho')e^{-jn(\phi-\phi')}}{2\pi(k_{nm}^2 - k^2)N_{nm}^2}. \tag{5.54}$$

5.5 GREEN'S FUNCTION IN TERMS OF SOLUTIONS OF THE HOMOGENEOUS EQUATION

The representation of Green's function in terms of a series of eigenfunctions is, in fact, an expression in the form of resonant cavity modes, but this is not the only representation. Green's function can also be expressed in terms of the solutions of a homogeneous equation, as discussed in this section.

Consider a general second-order differential equation

$$\left[\frac{1}{f(z)}\frac{d}{dz}\left(f(z)\frac{d}{dz}\right) + q(z)\right]G(z, z') = -\frac{\delta(z - z')}{f(z)}. \tag{5.55}$$

The function $f(z)$ on the right side is introduced so that Green's function becomes symmetric with respect to z and z', as will be seen shortly. This is consistent with Green's function for wave equations in cylindrical and spherical coordinate systems.

Let us write (5.55) as

$$\left[\frac{d^2}{dz^2} + p\,(z)\frac{d}{dz} + q(z) \right] G(z,\,z') = -\frac{\delta(z - z')}{f(z)}, \tag{5.56}$$

where

$$p(z) = \frac{1}{f(z)}\frac{df(z)}{dz}, \tag{5.57}$$

or

$$f(z) = \exp[\int p(z)dz]. \tag{5.58}$$

For the region $z > z'$, $G(z, z')$ is a solution of the homogeneous differential equation

$$\left[\frac{1}{f(z)}\frac{d}{dz}\left(f(z)\frac{d}{dz} \right) + q(z) \right] G(z,\,z') = 0. \tag{5.59}$$

We write this solution as

$$G(z,\,z') = A_0 y_1(z). \tag{5.60}$$

For the region $z < z'$, we write

$$G(z,\,z') = B_0 y_2(z). \tag{5.61}$$

Let us now use the following properties of Green's function.

1. The first derivative of Green's function has a jump at $z = z'$. From the differential equation, note that the second derivative of G should have the behavior of the delta function. Thus the first derivative has a discontinuity. The function itself, however, is continuous being an integral of the first derivative. Thus we have the second property.

2. Green's function is a continuous function of z when z' is fixed, including $z = z'$.

3. We also note that, in general, Green's function is symmetric with respect to $\bar{r}$ and $\bar{r}'$ as a direct consequence of the reciprocity theorem.

$$G(\bar{r},\,\bar{r}') = G(\bar{r}',\,\bar{r}). \tag{5.62}$$

We make use of the foregoing properties to obtain the constants A_0 and B_0. First we note that G is continuous at $z = z'$. Then we get

$$A_0 y_1(z') - B_0 y_2(z') = 0. \tag{5.63}$$

Next we note that the first derivative of G is discontinuous. To make use of this, let us integrate the original differential equation with a weighting function $f(z)$ from $z' - \varepsilon$ to $z' + \varepsilon$ and let ε approach zero. Thus we get

$$\lim_{\varepsilon \to 0} \int_{z'-\varepsilon}^{z'+\varepsilon} \left[\frac{d}{dz}\left(f\frac{d}{dz}\right) + fq \right] G \, dz = -\int_{z'-\varepsilon}^{z'+\varepsilon} \frac{\delta(z-z')}{f(z)} f(z)dz. \tag{5.64}$$

The right side becomes -1, and therefore,

$$\lim_{\varepsilon \to 0} \left[f(z)\frac{d}{dz} G \right]_{z'-\varepsilon}^{z'+\varepsilon} = -1.$$

Here we must use $A_0 y_1'$ for $z' + \varepsilon$ and $B_0 y_2'$ for $z' - \varepsilon$. We write this using y_1 and y_2

$$A_0 y'_1(z') - B_0 y'_2(z') = -\frac{1}{f(z')}, \tag{5.65}$$

where $y_1' = dy_1/dz$ and $y_2' = dy_2/dz$.

Solving (5.63) and (5.65) for A_0 and B_0, we get

$$A_0 = \frac{y_2(z')}{f(z')\Delta(z')}, \tag{5.66}$$

$$B_0 = \frac{y_1(z')}{f(z')\Delta(z')}, \tag{5.67}$$

$$\Delta(z') = \begin{vmatrix} y_1(z') & y_2(z') \\ y'_1(z') & y'_2(z') \end{vmatrix}. \tag{5.68}$$

Thus Green's function is given by

$$G(z, z') = \begin{cases} \dfrac{y_1(z)y_2(z')}{f(z')\Delta(z')}, & z > z', \qquad (5.69) \\[3mm] \dfrac{y_1(z')y_2(z)}{f(z')\Delta(z')}, & z < z'. \qquad (5.70) \end{cases}$$

Note that both the numerator and the denominator contain the product of y_1 and y_2 or its derivative such as $y_1 y_2$, $y_1 y_2'$, and so on. Thus any constant in front of y_1 and y_2 cancels out, and therefore, only the form of the function for y_1 and y_2 is necessary. Its magnitude does not affect the final form. $\Delta(z')$ is called the *Wronskian of y_1 and*

y_2, and since it has some useful characteristics, some of the details are given below. The denominator $f(z')\Delta(z')$ of (5.69) and (5.70) appears to be a function of z', and if so, $G(z, z')$ is not symmetric. It will be shown shortly, however, that $f(z')\Delta(z')$ is constant and is independent of z'. The final form of Green's function $G(z, z')$ is therefore given by

$$G(z, z') = \begin{cases} \dfrac{y_1(z)y_2(z')}{D}, & z > z', & (5.71) \\[3mm] \dfrac{y_1(z')y_2(z)}{D}, & z < z', & (5.72) \end{cases}$$

where $D = f(z')\Delta(z') = $ constant.

Equations (5.71) and (5.72) are often combined in the following convenient form:

$$G(z, z') = \frac{y_1(z_>)y_2(z_<)}{D}, \qquad (5.73)$$

where $z_>$ and $z_<$ denote the greater or lesser of z and z'.

Let us consider the Wronskian $\Delta(z)$, which we write in the following form:

$$\Delta(y_1, y_2) = y_1 y'_2 - y'_1 y_2, \qquad (5.74)$$

where $y_1 = y_1(z)$ and $y_2 = y_2(z)$, and $y'_1 = dy_1/dz$ and $y'_2 = dy_2/dz$. First, we note that if y_1 and y_2 are two independent solutions of the homogeneous differential equation, then $\Delta(y_1, y_2) \neq 0$, but if y_1 and y_2 are dependent, then $\Delta(y_1, y_2) = 0$, as can be verified by (5.74).

Let us next find the form of the Wronskian. To do this, we first show that Δ satisfies a simple first-order differential equation. Taking the derivative of Δ, we get

$$\begin{aligned} \frac{d\Delta}{dz} &= \frac{d}{dz}(y_1 y'_2 - y'_1 y_2) \\ &= y_1 y''_2 - y_2 y''_1. \end{aligned} \qquad (5.75)$$

But y_1 and y_2 satisfy the differential equation

$$\begin{aligned} y''_1 + p y'_1 + q y_1 &= 0, \\ y''_2 + p y'_2 + q y_2 &= 0. \end{aligned} \qquad (5.76)$$

Thus

$$\frac{d\Delta}{dz} = y_1(-p y'_2 - q y_2) - y_2(-p y'_1 - q y_1),$$

which reduces to

$$\frac{d\Delta}{dz} = -p\Delta. \qquad (5.77)$$

This is the differential equation that Δ satisfies. The solution can be easily obtained,

$$\Delta = (\text{constant})e^{-\int p\,dz},$$

or

$$f(z)\Delta = \text{constant}. \tag{5.78}$$

This is a very important and useful relationship. This shows that whatever y_1 and y_2 are, $f(z)\Delta(y_1, y_2)$ is always constant and independent of z. This property was used to obtain (5.73). Since this relationship (5.78) holds for any z, the constant $D = f(z)\Delta(y_1, y_2)$ can be determined by choosing any convenient z.

As an example, let us consider the one-dimensional problem we discussed in Section 5.4.

$$\left(\frac{d^2}{dz^2} + k^2\right) G(z, z') = -\delta(z - z'),$$

$$G(0, z') = G(a, z') = 0. \tag{5.79}$$

We let

$$y_1 = \sin k(a - z),$$
$$y_2 = \sin kz,$$

where the constants in front of these functions are immaterial.

Now we know that the Wronskian must be constant. To find this constant, we see that

$$\Delta(y_1, y_2) = k \sin k(a - z) \cos kz + k \cos k(a - z) \sin kz.$$

If we expand sine and cosine terms, all terms containing z cancel out, and the result is a constant. But it is not necessary to do this. Since we know that Δ is constant, we can simply choose any appropriate z. Choosing $z = 0$, we immediately obtain

$$\Delta(y_1, y_2) = k \sin ka,$$

and we can write down the results immediately.

$$G = \begin{cases} \dfrac{\sin kz' \sin k(a - z)}{k \sin ka}, & z > z', \\ \dfrac{\sin kz \sin k(a - z')}{k \sin ka}, & z < z'. \end{cases} \tag{5.80}$$

Equation (5.80) is identical to (5.43), but these are two different representations of the same Green's function.

So far, we have two representations of Green's function. The representation in terms of an infinite series of eigenfunctions is valid for the closed region and applicable to one-, two-, and three-dimensional cases. On the other hand, the representation in terms of two solutions of homogeneous differential equations is valid for both closed and open regions, but it is applicable only to the one-dimensional case.

In addition to the two representations above, there is a method using integral transforms that will be discussed in Section 5.6. For an actual problem, we combine these three representations in an appropriate manner.

5.6 FOURIER TRANSFORM METHOD

In addition to the above two methods, Green's function may be expressed in terms of a Fourier transform. In this section, we examine this procedure.

As an example, let us consider the following problem:

$$\left(\frac{\partial^2}{\partial x^2} + \frac{\partial^2}{\partial z^2} + k^2 \right) G = -\delta(x - x')\delta(z - z'), \tag{5.81}$$

and $G = 0$ at $x = 0$ and a. Here we start with the Fourier transform of both sides of the differential equation. We let

$$g(x,\ h) = \int_{-\infty}^{\infty} G(x,\ z)e^{jhz}dz, \tag{5.82}$$

and the inverse Fourier transform yields

$$G(x,\ z) = \frac{1}{2\pi} \int_{-\infty}^{\infty} g(x,\ h)e^{-jhz}dh. \tag{5.83}$$

Using

$$\int_{-\infty}^{\infty} \left(\frac{\partial^2}{\partial z^2}G \right) e^{jhz}dz = -h^2 g(x,\ h),$$

$$\int_{-\infty}^{\infty} \delta(z - z')e^{jhz}dz = e^{jhz'}, \tag{5.84}$$

$$\delta(z - z') = \frac{1}{2\pi} \int_{-\infty}^{\infty} e^{-jh(z-z')}dh,$$

we write (5.81) in the following form:

$$\left(\frac{d^2}{dx^2} + k^2 - h^2 \right) g(x,\ h) = -\delta(x - x')e^{jhz'}. \tag{5.85}$$

This is solved by noting that $G = 0$ at $x = 0$ and $x = a$, we get

$$g(x, h) = \sum_{n=1}^{\infty} \frac{\phi_n(x)\phi_n(x')}{k_n^2 - k^2 + h^2} e^{jhz'}, \qquad (5.86)$$

where

$$\phi_n(x) = \sqrt{\frac{2}{a}} \sin \frac{n\pi x}{a}, \quad k_n = \frac{n\pi}{a}.$$

Thus the complete solution is

$$G(x, z; x', z') = \frac{1}{2\pi} \int_{-\infty}^{\infty} \sum_{n=1}^{\infty} \frac{\phi_n(x)\phi_n(x')}{k_n^2 - k^2 + h^2} e^{-jh(z-z')} dh. \qquad (5.87)$$

Alternatively, we can solve (5.85) using the solutions of the homogeneous equation. We get

$$g(x, h) = \frac{y_1(x_>)y_2(x_<)}{D} e^{jhz'},$$

where $y_1(x) = \sin p(a - x)$, $y_2(x) = \sin px$, $p = (k^2 - h^2)^{1/2}$, and $D = p \sin pa$.
We, therefore, get

$$G(x, z; x', z') = \frac{1}{2\pi} \int_{-\infty}^{\infty} \frac{\sin px' \sin p(a - x)}{p \sin pa} e^{-jh(z-z')} dh, \quad x > x', \qquad (5.88)$$

where $p = (k^2 - h^2)^{1/2}$ and for $x < x'$, x and x' are interchanged. Equation (5.88) is an alternative expression for Green's function (5.87).

We can still obtain another form of Green's function by first expanding G in a series of eigenfunctions $\phi_n(x)$

$$G = \sum_{n=1}^{\infty} \phi_n(x)\phi_n(x')G_n(z, z'). \qquad (5.89)$$

We then get

$$\left(\frac{d^2}{dz^2} + k^2 - k_n^2 \right) G_n(z, z') = -\delta(z - z').$$

The solution for G_n is given by

$$G_n(z, z') = \frac{y_1(z_>)y_2(z_<)}{D}, \qquad (5.90)$$

where

$$y_1(z) = e^{-jq_n z},$$
$$y_2(z) = e^{+jq_n z},$$
$$D = 2jq_n, \quad q_n = (k^2 - k_n^2)^{1/2}.$$

Substituting (5.90) into (5.89), we get

$$G(x, z; x', z') = \sum_{n=1}^{\infty} \frac{\phi_n(x)\phi_n(x')}{2jq_n} e^{-jq_n|z-z'|}. \tag{5.91}$$

Equations (5.87), (5.88), and (5.91) are the same Green's function as expressed in three different representations.

5.7 EXCITATION OF A RECTANGULAR WAVEGUIDE

As an example, let us consider a rectangular waveguide excited by a current element of small length dl carrying I_0 as shown in Fig. 5.7. We choose Π_y that satisfies the wave equation

$$(\nabla^2 + k^2)\Pi_y = -\frac{J_y}{j\omega\varepsilon}, \tag{5.92}$$

where

$$J_y = I \, dl \, \delta(x - x')\delta(y - y')\delta(z - z').$$

The boundary conditions at $x = 0$ and $x = a$ are

$$\Pi_y = 0, \tag{5.93}$$

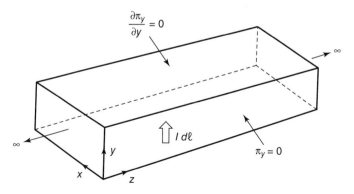

FIGURE 5.7 Excitation of a rectangular waveguide.

and the boundary conditions at $y = 0$ and $y = b$ are

$$\frac{\partial \Pi_y}{\partial y} = 0. \tag{5.94}$$

Let us rewrite the problem using Green's function G, which satisfies

$$\left(\frac{\partial^2}{\partial x^2} + \frac{\partial^2}{\partial y^2} + \frac{\partial^2}{\partial z^2} + k^2 \right) G(x, y, z; x', y', z') = -\delta(x - x')\delta(y - y')\delta(z - z'), \tag{5.95}$$

$$G = 0 \quad \text{at } x = 0 \quad \text{and} \quad a,$$

$$\frac{\partial G}{\partial y} = 0 \quad \text{at } y = 0 \quad \text{and} \quad b.$$

The Hertz vector Π_y is then given by

$$\Pi_y = \frac{I\,dl}{j\omega\varepsilon} G. \tag{5.96}$$

To find Green's function from (5.95), we first note that the region is closed in the x and y directions. Thus G can be represented by a series of eigenfunctions.

$$\left(\frac{\partial^2}{\partial x^2} + \frac{\partial^2}{\partial y^2} + k_{mn}^2 \right) \phi_{mn}(x, y) = 0, \tag{5.97}$$

$$\phi_{mn} = 0 \quad \text{at } x = 0 \quad \text{and} \quad a,$$

$$\frac{\partial \phi_{mn}}{\partial y} = 0 \quad \text{at } y = 0 \quad \text{and} \quad b.$$

We normalize ϕ_{mn} by

$$\int_0^a dx \int_0^b dy [\phi_{mn}(x, y)]^2 = 1. \tag{5.98}$$

The eigenfunctions $\phi_{mn}(x, y)$ are, therefore, given by

$$\phi_{mn}(x, y) = \phi_m(x)\phi_n(y),$$

$$\phi_m(x) = \left(\frac{2}{a} \right)^{1/2} \sin \frac{m\pi x}{a}, \qquad m = 1, 2, \ldots,$$

$$\phi_n(y) = \begin{cases} \left(\dfrac{1}{b} \right)^{1/2}, & n = 0, \\[2mm] \left(\dfrac{2}{b} \right)^{1/2} \cos \dfrac{n\pi y}{b}, & n = 1, 2, \ldots. \end{cases} \tag{5.99}$$

Let us write Green's function in a series of eigenfunctions

$$G = \sum_{m=1}^{\infty} \sum_{n=0}^{\infty} \phi_{mn}(x, y)\phi_{mn}(x', y')G_{mn}(z, z'), \qquad (5.100)$$

where $G_{mn}(z, z')$ is still an unknown function. We also write the right side of (5.95) using the following:

$$\delta(x - x')\delta(y - y') = \sum_{m=1}^{\infty} \sum_{n=1}^{\infty} \phi_{mn}(x, y)\phi_{mn}(x', y'). \qquad (5.101)$$

Substituting (5.100) and (5.101) into (5.95) and noting that this is an orthogonal expansion in a series of $\phi_{mn}(x, y)$, we get

$$\left(\frac{d^2}{dz^2} + k^2 - k_{mn}^2 \right) G_{mn}(z, z') = -\delta(z - z'). \qquad (5.102)$$

Now, we note that in the z direction, the region is open, so we use the representation of the two solutions of the homogeneous equation.

The solution for $z > z'$ is

$$y_1(z) = e^{-j\sqrt{k^2 - k_{mn}^2}(z - z')},$$

and for $z < z'$,

$$y_2(z) = e^{j\sqrt{k^2 - k_{mn}^2}(z - z')}, \qquad (5.103)$$

and the Wronskian is simply constant

$$\Delta(y_1, y_2) = j \cdot 2\sqrt{k^2 - k_{mn}^2}. \qquad (5.104)$$

Thus we finally get

$$G = \sum_{m=1}^{\infty} \sum_{n=0}^{\infty} \frac{\phi_{mn}(x, y)\phi_{mn}(x', y')e^{-j\sqrt{k^2 - k_{mn}^2}|z - z'|}}{j \cdot 2\sqrt{k^2 - k_{mn}^2}}. \qquad (5.105)$$

5.8 EXCITATION OF A CONDUCTING CYLINDER

As another example, let us consider the problem of exciting a conducting cylinder with an electric current element of small length dl carrying the current I_0 pointed in the z direction (Fig. 5.8). We have a scalar wave equation

$$(\nabla^2 + k^2)\Pi_z = -\frac{I_0 dl}{j\omega\varepsilon} \frac{\delta(\phi - \phi')\delta(\rho - \rho')\delta(z - z')}{\rho}. \qquad (5.106)$$

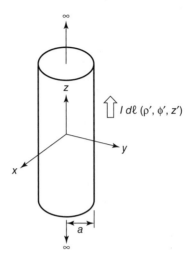

FIGURE 5.8 Excitation of a conducting cylinder.

Considering that $\Pi_z = 0$ at $\rho = a$, the problem is reduced to solving

$$(\nabla^2 + k^2)G = -\frac{\delta(\phi - \phi')\delta(\rho - \rho')\delta(z - z')}{\rho}. \tag{5.107}$$

$G = 0$ at $\rho = a$. First, we take the Fourier transform in z.

$$G = \frac{1}{2\pi}\int g(\rho, \phi, h)e^{-jh(z-z')}dh. \tag{5.108}$$

We then get

$$\left[\frac{1}{\rho}\frac{\partial}{\partial\rho}\left(\partial\frac{\partial}{\partial\rho}\right) + \frac{1}{\rho^2}\frac{\partial^2}{\partial\phi^2} + k^2 - h^2\right]g = -\frac{\delta(\rho - \rho')\delta(\phi - \phi')}{\rho}. \tag{5.109}$$

Expanding g in a series of eigenfunctions $\phi_n(\phi)$ around the cylinder, we get

$$g = \sum_{n=-\infty}^{\infty} \frac{e^{-jn(\phi-\phi')}}{2\pi}gn(\rho, \rho'),$$

$$\left[\frac{1}{\rho}\frac{d}{d\rho}\left(\rho\frac{d}{d\rho}\right) - \frac{n^2}{\rho^2} + k^2 - h^2\right]g_n(\rho, \rho') = -\frac{\delta(\rho - \rho')}{\rho}. \tag{5.110}$$

Now we express g_n in terms of solutions to the homogeneous equation, which is Bessel's differential equation in this case. We choose appropriate Bessel functions for the two different regions. In the range $\rho > \rho'$, the solution should be an outgoing wave to satisfy the radiation condition. Thus

$$y_1(\rho) = H_n^{(2)}(q\rho) \quad \text{for } \rho > \rho', \quad q = (k^2 - h^2)^{1/2}. \tag{5.111}$$

In the range $a < \rho < \rho'$, the solution should be a linear combination of two independent Bessel functions because here the wave travels back and forth forming a standing wave. Two independent functions can be $J_n(q\rho)$ and $N_n(q\rho)$, for example. But here we choose $J_n(q\rho)$ and $H_n^{(2)}(q\rho)$. The reason for this choice is explained later. Thus

$$y_2(\rho) = AJ_n(q\rho) + BH_n^{(2)}(q\rho) \quad \text{for } a < \rho < \rho'. \tag{5.112}$$

Using the boundary condition that $G = 0$ at $\rho = a$, the ratio of the constants A to B is determined, and we obtain

$$y_2(\rho) = J_n(q\rho) + H_n^{(2)}(qa) - J_n(qa)H_n^{(2)}(q\rho). \tag{5.113}$$

To find the Wronskian of y_1 and y_2, we note that the second term of $y_2(\rho)$ has the same form as $y_1(\rho)$, and they are dependent. The Wronskian between them is, therefore, zero. Thus we get

$$\Delta(y_1 y_2) = H_n^{(2)}(qa)q\Delta(H_n^{(2)}(x), \ J_n(x)),$$

where q is due to the fact that the derivative is with respect to ρ for $\Delta(y_1, y_2)$, but it is with respect to x for $\Delta(H_n(x), J_n(x))$. Now consider

$$\Delta(H_n^{(2)}(x), \ J_n(x)).$$

$J_n(x)$ is written as

$$J_n(x) = \tfrac{1}{2}(H_n^{(1)}(x) + H_n^{(2)}(x)).$$

Noting that $\Delta(H_n^{(2)}(x), H_n^{(2)}(x)) = 0$, we get

$$\Delta(H_n^{(2)}(x), \ J_n(x)) = \tfrac{1}{2}\Delta(H_n^{(2)}(x), \ H_n^{(1)}(x)).$$

To evaluate the last Wronskian, we first note that $f\Delta$ should be constant. $f(x)$ in this cylindrical case is simply x, and therefore Wronskian Δ should be equal to (constant)/x. To determine this constant, we can choose any convenient x and evaluate Δ. We let $x \to \infty$ and noting that

$$
\begin{aligned}
H_n^{(2)}(x) &\to \left(\frac{2}{\pi x}\right)^{1/2} \exp\left[-j\left(x - \frac{n\pi}{2} - \frac{\pi}{4}\right)\right], \\
H_n^{(1)}(x) &\to \left(\frac{2}{\pi x}\right)^{1/2} \exp\left[+j\left(x - \frac{n\pi}{2} - \frac{\pi}{4}\right)\right], \\
H_n^{(2)\prime}(x) &\to \left(\frac{2}{\pi x}\right)^{1/2} \exp\left[-j\left(x - \frac{n\pi}{2} + \frac{\pi}{4}\right)\right], \\
H_n^{(1)\prime}(x) &\to \left(\frac{2}{\pi x}\right)^{1/2} \exp\left[+j\left(x - \frac{n\pi}{2} + \frac{\pi}{4}\right)\right],
\end{aligned}
\tag{5.114}
$$

we get

$$
\Delta(y_1, y_2) = j\frac{2}{\pi\rho'}H_n^{(2)}(qa).
\tag{5.115}
$$

Therefore,

$$
g_n(\rho, \rho') = \frac{y_1(\rho_>)y_2(\rho_<)}{j(2/\pi)H_n^{(2)}(qa)}.
\tag{5.116}
$$

The complete solution is then

$$
G(\rho, \phi, z; \rho', \phi', z') = \frac{1}{2\pi}\int g(\rho, \phi, h)e^{-jh(z-z')}dh,
$$

$$
g(\rho, \phi, h) = -j\frac{1}{4}\sum_{n=-\infty}^{\infty} e^{-jn(\phi-\phi')}
$$

$$
\times \frac{[J_n(q\rho_<)H_n^{(2)}(qa) - J_n(qa)H_n^{(2)}(q\rho_<)]H_n^{(2)}(q\rho_>)}{H_n^{(2)}(qa)},
\tag{5.117}
$$

where $\rho_>$ and $\rho_<$ are the larger or smaller of ρ and ρ'.

Note that (5.117) consists of two terms.

$$
g = g_i + g_s,
\tag{5.118}
$$

where

$$
g_i = -j\frac{1}{4}\sum_{n=-\infty}^{\infty} e^{-jn(\phi-\phi')}J_n(q\rho_<)H_n^{(2)}(q\rho_>).
$$

This term q_i does not contain a, the radius of the cylinder, and this represents the incident or primary wave. The second term contains the effects of the radius of the cylinder and is called the scattered wave or secondary wave.

We also note from (5.118) that if $\rho_<$ becomes zero, all the terms with $n \neq 0$ disappear because of J_n and thus

$$g_i = -j\frac{1}{4}H_0^{(2)}(q\rho), \tag{5.119}$$

represents the primary wave at a distance ρ from the source.

The final solution (5.117) is given as an inverse Fourier transform. Here the contour of integration in the complex h plane must be chosen such that the radiation condition is satisfied. Also, if the observation point is far from the dipole, the radiation field can be obtained simply by the saddle-point technique. These problems are discussed fully in Chapter 11.

5.9 EXCITATION OF A CONDUCTING SPHERE

Let us next consider the problem of exciting a conducting sphere with a radial dipole (Fig. 5.9). As shown in Section 4.15, the electric Hertz vector $\bar{\Pi} = \Pi\hat{r}$ is pointed in the radial direction and satisfies the equation

$$(\nabla^2 + k^2)\Pi = -\frac{J_r}{j\omega\varepsilon r}. \tag{5.120}$$

The boundary condition at $r = a$ is (Section 4.16)

$$\frac{\partial}{\partial r}(r\Pi) = 0. \tag{5.121}$$

For a short electric dipole of length dl carrying the current I_0, we have

$$J_r = I_0\, dl\, \delta(\bar{r} - \bar{r}').$$

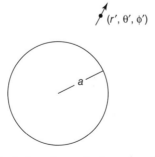

FIGURE 5.9   Excitation of a conducting sphere by a radial dipole.

Therefore, the problem is reduced to that of finding Green's function G satisfying

$$(\nabla^2 + k^2)G = -\delta(\bar{r} - \bar{r}'),$$

$$\frac{\partial}{\partial r}(rG) = 0 \quad \text{at } r = a. \tag{5.122}$$

Let us first recognize that Green's function G consists of the primary wave G_p and the scattered wave G_s. The primary wave G_p is Green's function in free space in the absence of the sphere and G_s represents the effect of the sphere. Let us first find G_p. This is simply given by $G_p = \exp(-jk|\bar{r} - \bar{r}'|)/4\pi|\bar{r} - \bar{r}'|$. However, to satisfy the boundary conditions at $r = a$, it is necessary to expand G_p in spherical harmonics. We expand G_p in Fourier series in the ϕ direction and in a series of Legendre functions in the θ direction. Thus we write

$$G_p = \sum_{n=0}^{\infty} \sum_{m=-n}^{n} G_{nm}(r, \, r')Y_{nm}(\theta, \, \phi)Y_{nm}^*(\theta', \, \phi'), \tag{5.123}$$

where

$$Y_{nm}(\theta, \, \phi) = \left[\frac{2n+1}{2}\frac{(n-m)!}{(n+m)!}\right]^{1/2} P_n^m(\cos\theta)\frac{e^{jm\phi}}{(2\pi)^{1/2}}$$

The functions $Y_{nm}(\theta, \, \phi)$ are the spherical harmonics, combining the θ and the ϕ dependence, and are orthogonal and normalized (see Section 4.15).

$$\int_0^{\pi} \sin\theta \, d\theta \int_0^{2\pi} d\phi \, Y_{nm}Y_{n'm'}^* = \begin{cases} 0 & \text{if } n \neq n' \quad \text{or} \quad m \neq m', \\ 1 & \text{if } n = n' \quad \text{and} \quad m = m'. \end{cases} \tag{5.124}$$

Note also that $P_n^m(\cos\theta) = 0$ for $m > n$. The delta function on the right side of (5.122) is given by

$$\delta(\bar{r} - \bar{r}') = \frac{\delta(r - r')\delta(\theta - \theta')\delta(\phi - \phi')}{r^2 \sin\theta} \tag{5.125}$$

We now expand the delta function in series of spherical harmonics

$$\frac{\delta(\theta - \theta')\delta(\phi - \phi')}{\sin\theta} = \sum_{n=0}^{\infty} \sum_{m=-n}^{\infty} Y_{nm}(\theta, \, \phi)Y_{nm}^*(\theta', \, \phi'). \tag{5.126}$$

Substituting (5.123) and (5.126) into (5.122), we get

$$\left[\frac{1}{r^2}\frac{d}{dr}\left(r^2\frac{d}{dr}\right) + k^2 - \frac{n(n+1)}{r^2}\right]G_n = -\frac{\partial(r - r')}{r^2}, \tag{5.127}$$

where we used G_n instead of G_{nm} since the differential equation contains only n. Using the technique shown in Section 5.5, the solution is given by

$$G_n = \frac{y_1(r_>)y_2(r_<)}{D},$$

where $D = r^2\Delta = r^2(y_1y_2' - y_1'y_2)$, and $y_1 = h_n^{(2)}(kr)$ and $y_2 = j_n(kr)$.

Now the Wronskian Δ can be evaluated noting that as $kr \to$ large,

$$h_n^{(2)}(kr) \to \frac{1}{kr}\exp\left[-j\left(kr - \frac{n+1}{2}\pi\right)\right],$$

$$j_n(kr) \to \frac{1}{kr}\cos\left(kr - \frac{n+1}{2}\pi\right).$$

We then get $\Delta = j/(kr^2)$ and $D = j/k$. The final expression for Green's function G_p for the primary wave is then given by

$$G_p = \sum_{n=0}^{\infty}\sum_{m=-n}^{\infty} G_n(r, r')Y_{nm}(\theta, \phi)Y_{nm}^*(\theta', \phi'),$$

$$G_n = \begin{cases} -jkj_n(kr')h_n^{(2)}(kr) & \text{if } r' < r \\ -jkj_n(kr)h_n^{(2)}(kr') & \text{if } r' > r. \end{cases} \tag{5.128}$$

We can rewrite (5.128) using

$$P_n^{-m}(x) = (-1)^m \frac{(n-m)!}{(n+m)!}P_n^m(x) \tag{5.129}$$

and the addition theorem

$$P_n(\cos\gamma) = \sum_{m=0}^{n} \varepsilon_m \frac{(n-m)!}{(n+m)!}P_n^m(\cos\theta)P_n^m(\cos\theta')\cos m(\phi - \phi'), \tag{5.130}$$

$$\cos\gamma = \hat{r}\cdot\hat{r}' = \cos\theta\cos\theta' + \sin\theta\sin\theta'\cos m(\phi - \phi'),$$

$$\varepsilon_0 = 1, \qquad \varepsilon_m = 2, \qquad \text{for } m \geq 1.$$

Then we have

$$G_p = \sum_{n=0}^{\infty} \frac{2n+1}{4\pi}G_n(r, r')P_n(\cos\gamma). \tag{5.131}$$

Now let us go back to our problem (5.122) and find G_s. We expand G_s in the same spherical harmonics. We write

$$G_s = \sum_{n=0}^{\infty} A_n h_n^{(2)}(kr) P_n(\cos\gamma), \tag{5.132}$$

where A_n are unknown coefficients to be determined by the boundary condition

$$\frac{\partial}{\partial r}[r(G_p + G_s)] = 0 \quad \text{at } r = a. \tag{5.133}$$

Noting that $G_n = -jk j_n(kr) h_n^{(2)}(kr')$ for $r < r'$, we get

$$A_n = \frac{(2n+1)jk h_n^{(2)}(kr')}{4\pi} \left\{ \frac{\dfrac{\partial}{\partial r}[r j_n(kr)]}{\dfrac{\partial}{\partial r}[r h_n^{(2)}(kr)]} \right\}_{r=a} \tag{5.134}$$

The final expression for Green's function is then $G = G_p + G_s$, where G_p and G_s are given by (5.131) and (5.132), respectively.

PROBLEMS

5.1 Find the radiation fields $\overline{E}$ and $\overline{H}$ from a vertical electric dipole at a height h above a conducting plane.

5.2 Find the radiation resistance of the monopole shown in Fig. P5.2.

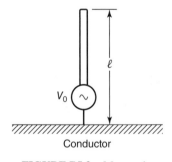

FIGURE P5.2 Monopole.

5.3 Find the radiation field from a horizontal electric dipole located at a height h above a conducting plane.

5.4 Show that one-dimensional Green's function in free space is given by

$$G(x, \ x') = \frac{\exp(-jk/x - x')}{2jk}.$$

5.5 Show that a two-dimensional Green's function in free space is given by

$$G(x, \ x') = \frac{j}{4}H_0^{(2)}(k|\bar{r} - \bar{r}|).$$

5.6 A sheet of uniform current I_0 (A/m) flowing in the x direction is located in free space at $z = 0$ (see Problem 2.6). Find the Hertz vector $\bar{\Pi}$, $\bar{E}$, and $\bar{H}$.

5.7 The current sheet shown in Problem 5.6 is located at $z = z'$, and perfectly conducting sheets are placed at $z = 0$ and $z = d$ as shown in Fig. P5.7.

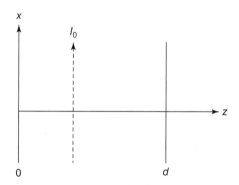

FIGURE P5.7   Current sheet I_0 at $z = z'$.

Find $\bar{E}$ and $\bar{H}$. Find the real and reactive power supplied by the current [see equation (2.76)].

5.8 If an impedance sheet is placed at $z = d$ for Problem 5.7, find $\bar{E}$, $\bar{H}$, and the power supplied by the current. Assume that at $z = d$, $E_x/H_y = 100 \ \Omega$.

5.9 Show that the normalization constant for spherical Bessel functions are given by (see Abramowitz and Stegun, 1964, p. 485)

$$\int_0^a j_v(\alpha r)j_v(\alpha r)r^2 dr = \begin{cases} \dfrac{a^3}{2}[j_{v+1}(\alpha a)]^2 & \text{if } j_v(\alpha a) = 0, \\[2ex] \dfrac{a^3}{2}\left[\dfrac{1}{4(\alpha a)^2} + 1 - \dfrac{v+\frac{1}{2}}{(\alpha a)^2}\right][j_v(\alpha a)]^2 \\[2ex] \text{if } \dfrac{\partial}{\partial r}(rj_v) = 0 \text{ at } r = a. \end{cases}$$

5.10 Find Green's function $G(r, r')$ inside a spherical cavity that satisfies the boundary condition

$$\frac{\partial}{\partial r}[rG] = 0 \quad \text{at } r = a.$$

5.11 Find Green's function for a rectangular cavity with width a, length b, and height c satisfying Dirichlet's boundary condition.

5.12 Find Green's function for a cylindrical cavity of radius a and height h with Neumann's boundary condition on the wall.

5.13 Find Green's function inside an infinitely long cylindrical waveguide of radius a with Dirichlet's boundary condition.

5.14 Show that three-dimensional Green's function $G(\bar{r} - \bar{r}') = \exp(-jk|\bar{r} - \bar{r}'|)/4\pi|\bar{r} - \bar{r}'|$ in free space can be expressed as

$$G = \frac{1}{(2\pi)^2} \int \int \frac{\exp[-jq_1(x - x') - jq_2(y - y') - jq|z - z'|]}{2jq} dq_1 dq_2,$$

where

$$q = \begin{cases} (k^2 - q_1^2 - q_2^2)^{1/2} & \text{if } |k| > (q_1^2 + q_2^2)^{1/2}, \\ -j(q_1^2 + q_2^2 - k^2)^{1/2} & \text{if } |k| < (q_1^2 + q_2^2)^{1/2}. \end{cases}$$

5.15 Consider Green's function for the wave equation

$$\left(\nabla^2 - \frac{1}{c^2}\frac{\partial}{\partial t^2}\right) g(\bar{r}, t) = -\delta(\bar{r})\delta(t).$$

(a) Define the Fourier transform of $g(\bar{r}, t)$ to be $G(\bar{r}, \omega)$, so that

$$\bar{g}(r, t) = \frac{1}{2\pi} = \int d\omega e^{j\omega t} G(\bar{r}, \omega).$$

Show that

$$G(r, \omega) = \frac{e^{-jkr}}{4\pi r}.$$

(b) By performing the inverse Fourier transform, show that

$$g(\bar{r}, t) = \frac{\delta(t - r/c)}{4\pi r}.$$

CHAPTER 6

RADIATION FROM APERTURES AND BEAM WAVES

In Chapter 5, we discussed Green's functions and their representations for different problems. In this chapter, we make use of Green's functions and apply them to obtain the waves radiated from apertures. We also discuss the spectral domain method applied to beam-wave propagation, the Goos–Hanchen shift, and higher-order beam waves (Tamir and Blok, 1986). Electromagnetic aperture problems are then discussed, including the Stratton–Chu formula, the equivalence theorem, and Kirchhoff approximation.

6.1 HUYGENS' PRINCIPLE AND EXTINCTION THEOREM

According to Huygens' principle, the field at a point is the superposition of the spherical wavelets originating from a surface located between the observation point and the source (Fig. 6.1). In this section, we discuss the mathematical formulation of Huygens' principle for scalar waves.

Let us consider the field $\psi(\bar{r})$ generated by the source $f(\bar{r})$ located in the volume V_f surrounded by the surface S_f (Fig. 6.2)

$$(\nabla^2 + k^2)\psi(\bar{r}) = -f(\bar{r}). \tag{6.1}$$

Let us now apply Green's theorem to the volume V_1 surrounded by S, S_∞, and S_1. The surface S is arbitrary, and S_∞ is the surface at infinity. The surface S_1 is the

Electromagnetic Wave Propagation, Radiation, and Scattering: From Fundamentals to Applications,
Second Edition. Akira Ishimaru.
© 2017 by The Institute of Electrical and Electronic Engineers, Inc. Published 2017 by John Wiley & Sons, Inc.

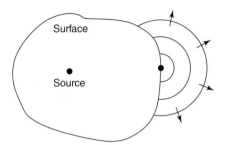

FIGURE 6.1   Huygens' principle.

surface of a small sphere centered at $\bar{r}$ with radius ε, and $\bar{r}$ will be identified later as the observation point. This procedure of using S_1 may appear complicated, but it is necessary while we consider the case when $\bar{r}$ is on the surface.

Let us apply Green's theorem to V_1.

$$\int_{V_1} (u\nabla^2 v - v\nabla^2 u)\, dV = \int_{S_t} \left(u\frac{\partial v}{\partial n} - v\frac{\partial u}{\partial n} \right) dS, \tag{6.2}$$

where $S_t = S + S_\infty + S_1$, and $\partial/\partial n$ is the derivative in the outward normal direction. We first let $\bar{r}'$ denote the integration point and let $\partial/\partial n'$ be the normal derivative "into" the volume V_1 rather than "outward." We then let

$$u(\bar{r}') = \psi(\bar{r}'),$$
$$v(\bar{r}') = G(\bar{r}', \bar{r}) = G(\bar{r}, \bar{r}') = \text{Green's function.} \tag{6.3}$$

Now the source point $\bar{r}$ for Green's function is inside S_1, and therefore it is outside the volume V_1. Thus ψ and G satisfy (6.1) and the homogeneous wave equation in V_1, respectively.

$$(\nabla'^2 + k^2)G(\bar{r}', \bar{r}) = 0, \tag{6.4}$$

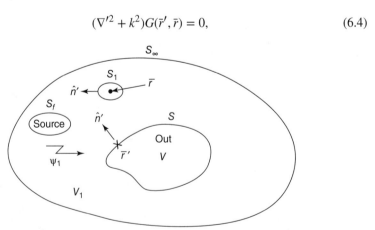

FIGURE 6.2   Volume V_1 surrounded by $S_f, S_1, S,$ and S_∞. The observation point $\bar{r}$ is outside S.

where ∇'^2 is the Laplacian with respect to $\bar{r}'$. Now we have

$$\psi\nabla'^2 G - G\nabla'^2\psi = \psi(\nabla'^2 + k^2)G - G(\nabla'^2 + k^2)\psi = Gf \quad \text{in } V_1.$$

Therefore, (6.2) becomes, noting that $\partial/\partial n = -\partial/\partial n'$,

$$\psi_i(\bar{r}) = -\int_{S_t}\left[\psi(\bar{r}')\frac{\partial G\,(\bar{r}, \bar{r}')}{\partial n'} - G(\bar{r}, \bar{r}')\frac{\partial\psi(\bar{r}')}{\partial n'}\right]dS', \tag{6.5}$$

where $\psi_i(\bar{r}) = \int_{V_f} G(\bar{r}, \bar{r}')f(\bar{r}')\,dV'$.

Now we consider the integral for each surface, S_∞, S_1, and S. The field at infinity on S_∞ is an outgoing spherical wave with propagation constant k, and since k in any physical medium has some small negative imaginary part, the integral on S_∞ vanishes. This is called the *radiation condition* and is expressed by

$$\lim_{r\to\infty} r\left(\frac{\partial}{\partial r} + jk\right)\psi(\bar{r}) = 0. \tag{6.6}$$

Next consider the integral on S_1. We note that $G(\bar{r}, \bar{r}')$ has a singularity at $\bar{r} = \bar{r}'$, and we can write

$$G(\bar{r}, \bar{r}') = \frac{e^{-jk\varepsilon}}{4\pi\varepsilon} + G_1(\bar{r}'),$$

where $\varepsilon = |\bar{r} - \bar{r}'|$ and G_1 has no singularity at $\bar{r} = \bar{r}'$ and is regular. We then get

$$\lim_{\varepsilon\to 0}\int_{S_1}\psi(\bar{r}')\frac{\partial G(\bar{r}, \bar{r}')}{\partial n'}\,dS' = \lim_{\varepsilon\to 0}\left[\psi(\bar{r})\frac{\partial}{\partial\varepsilon}\left(\frac{e^{-jk\varepsilon}}{4\pi\varepsilon}\right)4\pi\varepsilon^2\right]$$

$$= -\psi(\bar{r}). \tag{6.7}$$

Also, we have

$$\lim_{\varepsilon\to 0}\int_{S_1} G(\bar{r}, \bar{r}')\frac{\partial\psi(\bar{r}')}{\partial n'}\,dS' = 0. \tag{6.8}$$

Therefore, the integration on S_1 is equal to $-\psi(\bar{r})$. Equation (6.5) then becomes

$$\psi_i(\bar{r}) + \int_S\left[\psi(\bar{r}')\frac{\partial G(\bar{r}, \bar{r}')}{\partial n'} - G(\bar{r}, \bar{r}')\frac{\partial\psi(\bar{r}')}{\partial n'}\right]dS' = \psi(\bar{r}), \quad \text{if } \bar{r} \text{ is outside } S,$$

$$\tag{6.9}$$

where the field ψ and $\partial\psi/\partial n'$ are evaluated as $\bar{r}'$ approaches the surface from outside. The field $\psi_i(\bar{r})$ is given by (6.5). It is the field in the absence of the surface S and is equal to the *incident field*. At this point, we have not imposed any boundary conditions on Green's function. The most common choice for G is simply the free-space Green's function $G(\bar{r}, \bar{r}') = G_0(\bar{r}, \bar{r}') = \exp(-jk|\bar{r} - \bar{r}'|)/(4\pi|\bar{r} - \bar{r}'|)$. Other choices are discussed later. Equation (6.9) gives the fundamental expression for the field $\psi(\bar{r})$ outside the object consisting of the incident field $\psi_i(\bar{r})$ and the scattered field $\psi_s(\bar{r})$, which is the contribution from the field on the surface S.

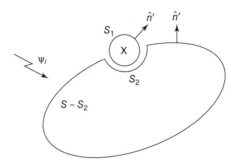

FIGURE 6.3   Observation point $\bar{r}$ on the surface S.

Next, we consider the case when the observation point $\bar{r}$ approaches the surface S from outside (Fig. 6.3). We then consider the semispherical surface S_2 as shown. Then, considering that the normal direction $\hat{n}'$ for S_2 is opposite to the normal direction $\hat{n}'$ for S_1 on the semisphere, we get

$$\int_{S_1} dS' = -\psi(\bar{r}),$$
$$\int_{S_2} dS' = \frac{1}{2}\psi(\bar{r}). \tag{6.10}$$

We therefore get

$$\psi_i(\bar{r}) + \int_S \left[\psi(\bar{r}') \frac{\partial G(\bar{r}, \bar{r}')}{\partial n'} - G(\bar{r}, \bar{r}') \frac{\partial \psi(\bar{r}')}{\partial n'} \right] dS'$$
$$= \frac{1}{2}\,\psi(\bar{r}) \quad \text{if } \bar{r} \text{ is on the surface } S, \tag{6.11}$$

where the surface integral is called Cauchy's principal value and it indicates the integration over the surface S excluding the small circular area of radius ε.

Finally, we consider the case when the observation point $\bar{r}$ is inside the surface S (Fig. 6.4). The integration does not include the surface S_1, and therefore we get

$$\psi_i(\bar{r}) + \int_S \left[\psi(\bar{r}') \frac{\partial G(\bar{r}, \bar{r}')}{\partial n'} - G(\bar{r}, \bar{r}') \frac{\partial \psi(\bar{r}')}{\partial n'} \right] dS' = 0 \quad \text{if } \bar{r} \text{ is inside } S. \tag{6.12}$$

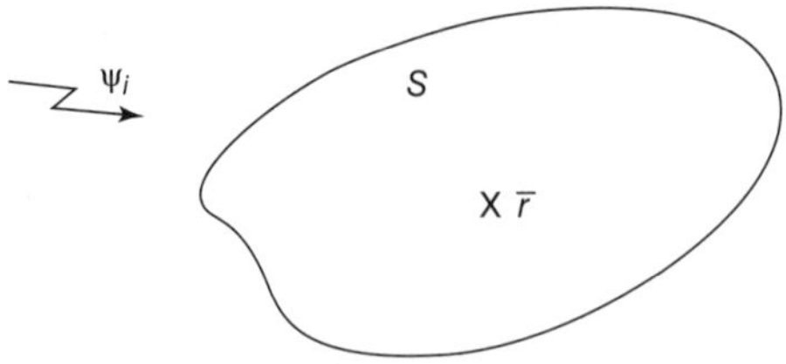

FIGURE 6.4   Observation point $\bar{r}$ inside the surface S.

The three equations (6.9), (6.11), and (6.12) are the fundamental equations that are important and useful for the formulation of many problems. They are the basis for constructing the surface integral equations, as we discuss in later sections. In this section, however, we consider the physical meaning of (6.12). It states that inside the surface S, the incident field $\psi_i(\bar{r})$ and the contribution from the surface field combine to produce the null field. The incident field $\psi_i(\bar{r})$ inside the surface is therefore "extinguished" by the surface contribution. For this reason, (6.12) is called the *extinction theorem*, the *Ewald–Oseen extinction theorem*, or the *null field theorem*. This theorem can be used as a boundary condition to obtain an extended integral equation, as will be shown later, and this method is called the *extended boundary condition method* or the *T-matrix method*.

Let us next apply the extinction theorem to a special case when there is no object inside S. The field everywhere is the incident wave $\psi_i(\bar{r})$ itself. We then write the extinction theorem as follows:

$$\psi_i(\bar{r}) = \int_S \left[\psi_i(\bar{r}') \frac{\partial G_0(\bar{r}, \bar{r}')}{\partial n} - G_0(\bar{r}, \bar{r}') \frac{\partial \psi_i(\bar{r}')}{\partial n} \right] dS', \qquad (6.13)$$

where $G_0(\bar{r}, \bar{r}') = \exp(-k|\bar{r} - \bar{r}'|)/(4\pi|\bar{r} - \bar{r}'|)$ is a free-space Green's function and $\partial/\partial n$ is now taken as the normal derivative pointed inside S. Equation (6.13) shows that the field at $\bar{r}$ can be calculated by knowing the field ψ and $\partial\psi/\partial n$ on a surface S, which act as the secondary source for spherical waves. This is the mathematical statement of Huygens' principle.

We note that the three fundamental equations (6.9), (6.11), and (6.12) are valid for two-dimensional problems ($\partial/\partial z = 0$) where Green's function is the two-dimensional Green's function $G_0(\bar{r}, \bar{r}') = -(j/4)H_0^{(2)}(k|\bar{r} - \bar{r}'|), \bar{r} = x\hat{x} + y\hat{y}, \bar{r}' = x'\hat{x} + y'\hat{y}$, and the surface integral is replaced by the line integral.

6.2 FIELDS DUE TO THE SURFACE FIELD DISTRIBUTION

In Section 6.1, we obtained the scalar field $\psi(\bar{r})$ in (6.9) when $\bar{r}$ is outside the surface S. The field $\psi(\bar{r})$ consists of the incident field $\psi_i(\bar{r})$ and the field $\psi_s(\bar{r})$ scattered from the surface S.

$$\psi(\bar{r}) = \psi_i(\bar{r}) + \psi_s(\bar{r}),$$

$$\psi_s(\bar{r}) = \int_S \left[\psi(\bar{r}') \frac{\partial G(\bar{r}, \bar{r}')}{\partial n'} - G(\bar{r}, \bar{r}') \frac{\partial \psi(\bar{r}')}{\partial n'} \right] dS'. \qquad (6.14)$$

To derive this we used Green's function $G(\bar{r}, \bar{r}')$ which has a singularity at $\bar{r} = \bar{r}'$ and satisfies the equation

$$(\nabla^2 + k^2)G(\bar{r}, \bar{r}') = -\delta(\bar{r} - \bar{r}'). \qquad (6.15)$$

However, we have not imposed any boundary conditions on G.

The simplest Green's function is the free-space Green's function

$$G(\bar{r}, \bar{r}') = G_0(\bar{r}, \bar{r}') = \frac{\exp(-jk\,|\bar{r}, \bar{r}'|)}{4\pi\,|\bar{r} - \bar{r}'|}. \tag{6.16}$$

With the free-space Green's function, the scattered field is calculated if both ψ and $\partial\psi/\partial n'$ are known on the surface.

$$\psi_s(\bar{r}) = \int_S \left[\psi(\bar{r}') \frac{\partial G_0(\bar{r}, \bar{r}')}{\partial n'} - G_0(\bar{r}, \bar{r}') \frac{\partial \psi(\bar{r}')}{\partial n'} \right] dS'. \tag{6.17}$$

This is called the *Helmholtz–Kirchhoff formula*. It should, however, be recognized that it is not necessary to know both ψ and $\partial\psi/\partial n$ on the surface, because according to the uniqueness theorem (Section 2.10), if the field ψ (or $\partial\psi/\partial n$) is known on the surface, the field should be uniquely determined everywhere outside the surface.

To obtain the expression for ψ_s in terms of the surface field only, we use Green's function $G_1(\bar{r}, \bar{r}')$, which satisfies the boundary condition

$$G_1(\bar{r}, \bar{r}') = 0 \quad \text{when } \bar{r}' \text{ is on } S. \tag{6.18}$$

We then get

$$\psi_s(\bar{r}) = \int_S \psi(\bar{r}') \frac{\partial}{\partial n'} G_1(\bar{r}, \bar{r}')\, dS'. \tag{6.19}$$

We can also obtain the field $\psi_s(\bar{r})$ in terms of $\partial\psi/\partial n'$ on the surface using $G_2(\bar{r}, \bar{r}')$, which satisfies Neumann's boundary condition on S.

$$\frac{\partial}{\partial n'} G_2(\bar{r}, \bar{r}') = 0 \quad \text{when } \bar{r}' \text{ is on } S. \tag{6.20}$$

Then we get

$$\psi_s(\bar{r}) = - \int_S G_2(\bar{r}, \bar{r}') \frac{\partial \psi(\bar{r}')}{\partial n'}\, dS'. \tag{6.21}$$

The three equations (6.17), (6.19), and (6.21) are exact and should yield the identical exact field $\psi_s(\bar{r})$. In practice, however, the exact field may not be known on S, and therefore, one of them may be preferred, depending on the problems.

Consider a field $\psi_s(\bar{r})$ behind a plane screen with an aperture where the field is given (Fig. 6.5). The field $\psi_s(\bar{r})$ can be expressed using the preceding three formulas.

Let us first find Green's function that satisfies Dirichlet's condition on the screen. Green's function in this case is easily found to be the difference between free-space Green's function due to $\bar{r}'$ and its image position $\bar{r}''$ (Fig. 6.6).

$$G_1(\bar{r}, \bar{r}') = \frac{\exp(-jkr_1)}{4\pi r_1} - \frac{\exp(-jkr_2)}{4\pi r_2}, \tag{6.22}$$

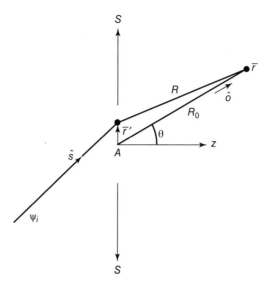

FIGURE 6.5   Aperture A on screen S.

where $r_1 = |\bar{r} - \bar{r}'|$ and $r_2 = |\bar{r} - \bar{r}''|$. To calculate $\partial/(\partial n')G$, note that

$$
\begin{aligned}
\frac{\partial}{\partial n'}\left[\frac{\exp(-jkr_1)}{4\pi r_1}\right] &= \frac{\partial}{\partial r_1}\left[\frac{\exp(-jkr_1)}{4\pi r_1}\right]\frac{\partial r_1}{\partial n'} \\
&= \frac{\exp(-jkr_1)}{4\pi r_1}\left(-jk - \frac{1}{r_1}\right)\frac{z'-z}{r_1}, \qquad (6.23)\\
\frac{\partial}{\partial n'}\left[\frac{\exp(-jkr_2)}{4\pi r_2}\right] &= \frac{\exp(-jkr_2)}{4\pi r_2}\left(-jk - \frac{1}{r_2}\right)\frac{z'+z}{r_2},
\end{aligned}
$$

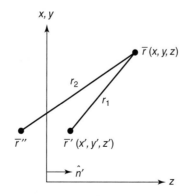

FIGURE 6.6   Green's function $G_1(\bar{r},\bar{r}')$ which satisfies Dirichlet's condidtion.

where

$$r_1^2 = (x - x')^2 + (y - y')^2 + (z - z')^2,$$
$$r_2^2 = (x - x')^2 + (y - y')^2 + (z + z')^2.$$

Therefore, we get

$$\frac{\partial}{\partial n'} G(\bar{r}, \bar{r}') = \frac{\exp(-jkR)}{2\pi R} \left(jk + \frac{1}{R} \right) \frac{z}{R}, \tag{6.24}$$

where $R^2 = (x - x')^2 + (y - y')^2 + z^2$.

The field $\psi(\bar{r})$ is then given by

$$\psi_s(\bar{r}) = \int_S \frac{\exp(-jkR)}{2\pi R} \left(jk + \frac{1}{R} \right) \frac{z}{R} \, \psi(\bar{r}') \, dS'. \tag{6.25}$$

If $\bar{r}$ is close to the z axis ($z/R \approx 1$) and $kR \gg 1$, we get

$$\psi_s(\bar{r}) = \frac{jk}{2\pi} \int_S \frac{\exp(-jkR)}{R} \psi(\bar{r}') \, dS'. \tag{6.26}$$

This expression is often used to calculate the field $\psi_s(\bar{r})$ due to the aperture field $\psi(\bar{r}')$ on a screen.

Next consider Green's function G_2, which satisfies Neumann's condition on a plane screen. We have

$$G_2(\bar{r}, \bar{r}') = \frac{e^{-jkr_1}}{4\pi r_1} + \frac{e^{-jkr_2}}{4\pi r_2}. \tag{6.27}$$

Therefore, we have

$$\psi_s(\bar{r}) = -\frac{1}{2\pi} \int_S \frac{e^{-jkR}}{R} \frac{\partial \psi(\bar{r}')}{\partial n'} \, dS'. \tag{6.28}$$

Equations (6.25) and (6.26) are applicable when the aperture field is known on a plane surface. Equation (6.28) can be used when the normal derivative of the field is known on a plane surface. When both the field and its normal derivative are known, we can use (6.17).

6.3 KIRCHHOFF APPROXIMATION

In (6.17), we showed that if the field and its normal derivative are known on a surface, the field at any point can be calculated. However, the exact evaluation of the field on a surface, such as the aperture field on a screen (Fig. 6.5), is difficult and requires solution of the complete boundary value problem. If the aperture size is large in terms of wavelength, the aperture field may be approximately equal to that of the incident field. This is called the *Kirchhoff approximation* and can be stated mathematically as follows.

On the aperture A (Fig. 6.5), we assume that

$$\psi = \psi_i \quad \text{and} \quad \frac{\partial \psi}{\partial n} = \frac{\partial \psi_i}{\partial n}. \tag{6.29}$$

On the screen S, we assume that

$$\psi = 0 \quad \text{and} \quad \frac{\partial \psi}{\partial n} = 0. \tag{6.30}$$

Making use of (6.23), we get the Kirchhoff approximation for the aperture problem shown in Fig. 6.5.

$$\psi(\bar{r}) = \int_S \left[\psi_i(\bar{r}') \left(jk + \frac{1}{R} \right) \frac{z}{R} - \frac{\partial \psi_i}{\partial n'} \right] \frac{\exp(-jkR)}{4\pi R} \, dS'. \tag{6.31}$$

For example, if the incident field is a plane wave propagating in the $\hat{s}$ direction and the aperture is in the plane $z = 0$ (Fig. 6.5), we have

$$\psi_i(\bar{r}') = A_0 \, e^{-jk\hat{s}\cdot\bar{r}'},$$
$$\frac{\partial \psi_i(\bar{r}')}{\partial n'} = -jk\hat{s} \cdot \hat{z} A_0 \, e^{-jk\hat{s}\cdot\bar{r}'}. \tag{6.32}$$

If we consider the field at a large distance from the aperture, we can neglect $1/R$ in $(jk + 1/R)$, let z/R be $\cos\theta$, and $1/R = 1/R_0$. The phase kR cannot be equated to kR_0, however, as we need to consider the difference in R and R_0 in terms of wavelength. We then should use $R = R_0 - \bar{r}' \cdot \hat{o}$. Finally, we get the field far from the aperture

$$\psi(\bar{r}) = \frac{\exp(-jkR_0)}{4\pi R_0} \int_S jk\psi_i(\bar{r}')(\cos\theta + \hat{s} \cdot \hat{z})e^{jk\bar{r}'\cdot\hat{o}} \, dS'. \tag{6.33}$$

The Kirchhoff approximation can be applied to the problem of scattering from an object (Fig. 6.7). At $\bar{r}'$, we assume that the field is approximately equal to the field if the surface is plane and tangential to the surface at $\bar{r}'$. If the incident field is a plane wave propagating in the $\hat{s}$ direction, we have

$$\psi_i = A_0 \exp(-jk\hat{s} \cdot \bar{r}). \tag{6.34}$$

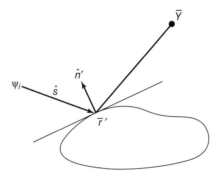

FIGURE 6.7 Kirchhoff approximation.

If the reflection coefficient at $\bar{r}'$ for a plane surface is R, we have in the illuminated part of the surface

$$\psi = (1 + R)\psi_i,$$
$$\frac{\partial \psi}{\partial n'} = (-jk\hat{s} \cdot \hat{n}')(1 - R)\psi_i. \tag{6.35}$$

In the shadow region, we assume that

$$\psi = 0 \quad \text{and} \quad \frac{\partial \psi}{\partial n'} = 0. \tag{6.36}$$

Equations (6.34)–(6.36) are then substituted in (6.17) to obtain the *Kirchhoff approximation*. Since we use the approximation that the field at any point on the surface is equal to the field for a plane tangent to the surface, we sometimes call this the *tangent approximation*. Electromagnetic equivalent of this approximation is called the *physical optics approximation*.

6.4 FRESNEL AND FRAUNHOFER DIFFRACTION

Let us next consider the field u at $\bar{r}$ where the field $u_0(\bar{r}')$ is given on the aperture at $z = 0$ (Fig. 6.8). This is given in (6.25). Consider an observation point (x, y, z) and let r_0 be the distance from the origin $x' = y' = 0$ of the aperture. Then we assume that $kr \gg 1$, $z/r \approx 1$, and $1/r \approx 1/r_0$. The phase kr cannot be approximated by kr_0 since the difference can be of the order of wavelengths. We use the following approximation:

$$r = [z^2 + (x - x')^2 + (y - y')^2]^{1/2}$$
$$\approx z + \frac{(x - x')^2 + (y - y')^2}{2z} \tag{6.37}$$

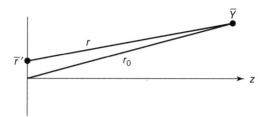

FIGURE 6.8   Fresnel and Fraunhofer diffraction.

Then we get the following "Fresnel diffraction" formula:

$$u(x, y, z) = \frac{jk}{2\pi z} e^{-jkz} \iint u_0(x', y') \exp\left[-jk\frac{(x - x')^2 + (y - y')^2}{2z}\right] dx'\, dy'$$

(6.38)

If we rewrite (6.37) as

$$r \approx r_0 - \frac{xx' + yy'}{z} + \frac{x'^2 + y'^2}{2z},$$

where

$$r_0 = z + \frac{x^2 + y^2}{2z},$$

(6.39)

and note that $kx'^2/z < 1$ at a large distance z, we can approximate r by

$$r \approx r_0 - \frac{xx'}{z} - \frac{yy'}{z}.$$

(6.40)

Then we get the *Fraunhofer diffraction formula*, valid at a large distance r_0 from the aperture, where (6.40) holds.

$$u(x, y, z) = \frac{jk}{2\pi r_0} e^{-jkr_0} \iint u_0(x', y') e^{+jk(xx'/z + yy'/z)} dx'\, dy'.$$ (6.41)

Note the difference between the Fresnel diffraction formula (6.38) and the Fraunhofer diffraction formula (6.41). The Fresnel formula contains the quadratic terms in x' and y' and is applicable to the near field of the aperture $z < a^2/\lambda$, while the Fraunhofer formula contains only the linear terms in x' and y' and is applicable to the far field $z \gg a^2/\lambda$, where a is the aperture size. We also note that (6.41) is in a form of Fourier transform, and therefore we can state that the Fraunhofer field is the Fourier transform of the aperture field. In antenna theory, we state that the radiation pattern of the aperture antenna is the Fourier transform of the aperture field.

Let us consider the Fraunhofer diffraction from a circular aperture of radius a with constant field

$$u_0(x', y') = \begin{cases} A_0 & \text{if } \rho' \leq a, \\ 0 & \text{if } \rho' > a. \end{cases} \tag{6.42}$$

Then from (6.41), letting

$$x' = \rho' \cos\phi', \quad y' = \rho' \sin\phi',$$
$$\frac{x}{z} = \sin\theta \cos\phi, \quad \frac{y}{z} = \sin\theta \sin\phi,$$

we get

$$u(x, y, z) = \frac{jk}{2\pi r_0} e^{-jkr_0} A_0 \int_0^a \int_0^{2\pi} \exp[jk\, \rho' \sin\theta\cos(\phi - \phi')]d\phi'\rho'\, d\rho'. \tag{6.43}$$

Now we make use of the following integral representation of Bessel's function and the integral formula:

$$J_0(x) = \frac{1}{2\pi} \int_0^{2\pi} e^{jx \cos\phi} d\phi,$$
$$\int J_0(x)\, x\, dx = xJ_1(x). \tag{6.44}$$

We then get

$$u(x, y, z) = u(r_0, \theta)$$
$$= \frac{jk}{2\pi r_0} e^{-jkr_0} A_0 \,\pi a^2 \frac{2J_1(ka \sin\theta)}{ka \sin\theta}. \tag{6.45}$$

Normalizing u with respect to u at $\theta = 0$, we get

$$\frac{u(r_0, \theta)}{u(r_0, 0)} = \frac{2J_1(ka \sin\theta)}{ka \sin\theta}. \tag{6.46}$$

The angle θ_a of the first zero of $u(r_0, \theta)$ is given by

$$ka \sin\theta_a = 3.832 \quad \text{or} \quad \sin\theta_a = 0.610\lambda/a. \tag{6.47}$$

The diffraction pattern (6.46) is called the *Airy pattern* and the main lobe in $0 \leq \theta \leq \theta_a$ is called the *Airy disk*.

As an example of the Fresnel diffraction, consider the uniform field in the square aperture.

$$U_0 = \begin{cases} A_0 & \text{for} |x'| < a \quad \text{and} \quad |y'| < b \\ 0 & \text{outside} \end{cases}. \tag{6.48}$$

The field on the axis ($x = y = 0$) is then given by

$$U(z) = \frac{jk}{2\pi z} e^{-jkz} \int_{-a}^{a} dx' \int_{-b}^{b} dy' \, A_0 \exp\left[-\frac{jk(x'^2 + y'^2)}{2z}\right]. \tag{6.49}$$

We make use of the Fresnel integral, defined by

$$F(Z) = \int_0^z \exp\left(-j\frac{\pi}{2}t^2\right) dt = C(Z) - jS(Z). \tag{6.50}$$

The field in (6.49) is then given by

$$U(Z) = j2A_0 e^{-jkz} F(Z_1) F(Z_2), \tag{6.51}$$

where $Z_1 = (k/\pi z)^{1/2} a$ and $Z_2 = (k/\pi z)^{1/2} b$.

Let us consider one more example of Fresnel diffraction. Consider a circular aperture on which the field has the field distribution $A_0(x', y')$ with the quadratic phase front with the radius of curvature R_0 at $z = 0$.

$$U_0(x', y') = A_0(x', y') \exp\left[+j\frac{k(x'^2 + y'^2)}{2R_0}\right]. \tag{6.52}$$

The phase distribution above focuses the wave at $z = R_0$ since the wave propagated from (x', y') to $(x = 0, y = 0, z = R_0)$ has the phase given by

$$\exp(-jkR) = \exp\left[-jk\left(R_0^2 + x'^2 + y'^2\right)^{1/2}\right]$$

$$\simeq \exp\left[-jkR_0 - \frac{jk(x'^2 + y'^2)}{2R_0}\right]$$

and the quadratic phase distribution in (6.52) exactly compensates for the quadratic term in $\exp(-jkR)$. We substitute (6.52) into (6.38). If we consider the field at the focal plane $z = R_0$, we get

$$U(x, y, z) = \frac{jk}{2\pi z} e^{-jkr_0} \iint A_0 \, e^{+jk(xx' + yy')/R_0} dx' \, dy'. \tag{6.53}$$

This is identical to (6.41) for Fraunhofer diffraction. If A_0 is constant, we get (6.46), where $\sin\theta = (x^2 + y^2)^{1/2}/R_0$.

This last example shows that if the aperture field has the quadratic phase distribution such that the wave is focused at $z = R_0$, the field distribution at the focal plane $z = R_0$ is identical to the Fraunhofer diffraction. We can also state that the field at the focal plane is the Fourier transform of the aperture field. This fact is often used in optical signal processing when Fourier transform is needed.

6.5 FOURIER TRANSFORM (SPECTRAL) REPRESENTATION

Let us consider a scalar wave $u(x, y, z)$ generated by the field $u_0(x', y')$ at the plane $z = 0$. There are two ways to obtain $u(x, y, z)$ from $u_0(x', y')$. One is to find Green's function, that is the field at $u(x, y, z)$ due to a delta function at (x', y'), and then to multiply it with $u_0(x', y')$ and integrate over the aperture at $z = 0$. This is similar to finding the output voltage of a network due to an input by using the impulse response, and was used in Section 6.4. The other method is the Fourier transform technique. The Fourier transform of the aperture field $u_0(x', y')$ is first taken, multiplied by the transfer function, and then the inverse Fourier transform is taken to produce $u(x, y, z)$. This is similar to solving the network problem by using the frequency spectrum of the input, multiplying it by the transfer function, and taking the inverse transform to obtain the output. We first discuss the Fourier transform technique in this section.

Let us start with the scalar wave equation,

$$(\nabla^2 + k^2)u\,(\bar{r}) = 0. \tag{6.54}$$

We first take a Fourier transform in x and y

$$U(q_1,\ q_2,\ z) = \iint u(x, y, z)e^{jq_1 x + jq_2 y}\,dx\,dy. \tag{6.55}$$

We then get

$$\left(\frac{d^2}{dz^2} + k^2 - q_1^2 - q_2^2\right) U(q_1,\ q_2,\ z) = 0. \tag{6.56}$$

The general solution of this consists of two waves traveling in the $+z$ and $-z$ directions, respectively.

$$U(q_1, q_2, z) = A_+(q_1, q_2)e^{-jqz} + A_-(q_1, q_2)e^{+jqz}, \tag{6.57}$$

where

$$q = \begin{cases} \left(k^2 - q_1^2 - q_2^2\right)^{1/2} & \text{if } k^2 > q_1^2 + q_2^2, \\ -j\left(q_1^2 + q_2^2 - k^2\right)^{1/2} & \text{if } k^2 < q_1^2 + q_2^2. \end{cases}$$

Note that for a positive-going wave with $\exp(-jqz)$, $q = q_r + jq_i$ must be in the fourth quadrant in the complex plane ($q_r \geq 0$ and $q_i \leq 0$), because $\exp(-jqz) = \exp(-jq_r z + q_i z)$ must decay exponentially as $z \rightarrow +\infty$. The same condition also ensures that the negative-going wave $\exp(+jqz)$ decays exponentially as $z \rightarrow -\infty$.

Let us consider the wave propagating in the $+z$ direction in the region $z \geq 0$, with $A_- = 0$. Taking the inverse Fourier transform, we get

$$u(x, y, z) = \frac{1}{(2\pi)^2} \iint A_+(q_1, q_2) e^{-jq_1 x - jq_2 y - jqz} \, dq_1 \, dq_2. \tag{6.58}$$

The function $A_+(q_1, q_2)$ can be given by letting $u(x', y', z) = u_0(x', y')$ at $z = 0$ in (6.58). We then get

$$A_+(q_1, q_2) = \iint u_0(x', y') e^{jq_1 x' + jq_2 y'} \, dx' \, dy'. \tag{6.59}$$

Equations (6.58) and (6.59) therefore give the field at an arbitrary point (x, y, z) due to the given field u_0 at the plane $z = 0$.

6.6 BEAM WAVES

As an example of the spectral method described in Section 6.5, we discuss an important class of waves called *beam waves*. Typical examples are a laser beam and a beam of millimeter wave propagating in the atmosphere. The characteristics of a beam wave are also exhibited in the near field in front of a parabolic reflector antenna, lens waveguides, and open resonator (laser resonator) (see Fig. 6.9).

Equation (6.58) is exact, but further simplification is possible for a beam wave. If the wave is assumed to be propagating mostly in the z direction, as in the case of a beam wave, the propagation constants q_1 and q_2 in the x and y directions are much smaller than the propagation constant q in the z direction. Therefore, the major contribution to the Fourier integral in (6.58) comes from the region where $|q|$ is close to k and $|q_1|$ and $|q_2|$ are much smaller than k. Thus, we make the following *paraxial approximation* for q:

$$\begin{aligned} q &= \left(k^2 - q_1^2 - q_2^2\right)^{1/2} \\ &\simeq k - \frac{q_1^2 + q_2^2}{2k}. \end{aligned} \tag{6.60}$$

We now make use of the paraxial approximation (6.60) to obtain the expression for a beam wave. Let us assume that the field at $z = 0$ has the Gaussian amplitude distribution and the quadratic phase front with the radius of curvature R_0 (Fig. 6.10).

$$u_0(x, y) = A_0 \exp\left(-\frac{\rho^2}{W_0^2} + j\frac{k\rho^2}{2R_0}\right), \tag{6.61}$$

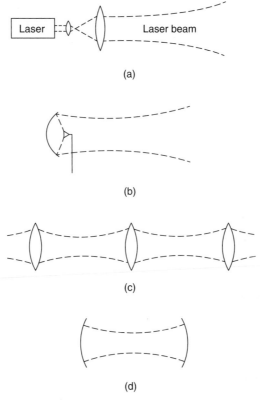

(a)

(b)

(c)

(d)

FIGURE 6.9   Beam waves: (a) laser beam; (b) parabolic reflector antenna; (c) lens waveguides; (d) laser resonator.

where $\rho^2 = x^2 + y^2$ and W_0 is the beam width at $z = 0$. The quadratic phase shift compensates for the difference between the actual distance $(R_0^2 + \rho^2)^{1/2}$ to the focal point and the focal distance R_0.

$$\left(R_0^2 + \rho^2\right)^{1/2} - R_0 \approx \frac{\rho^2}{2R_0}. \tag{6.62}$$

We substitute (6.61) into (6.59) and use the following useful formula:

$$\int_{-\infty}^{\infty} \exp(-at^2 + bt)\, dt = \left(\frac{\pi}{a}\right)^{1/2} \exp\left(\frac{b^2}{4a}\right), \quad \mathrm{Re}\, a > 0. \tag{6.63}$$

We then get

$$A_+(q_1,\, q_2) = \frac{2\pi A_0}{k\alpha} \exp\left(-\frac{q_1^2 + q_2^2}{2k\alpha}\right), \tag{6.64}$$

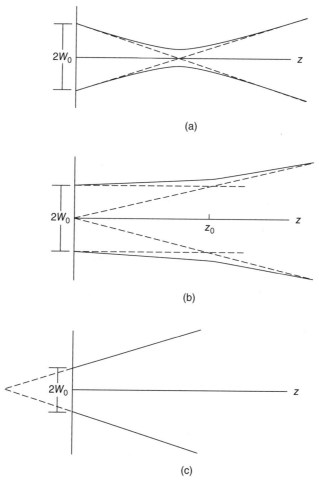

(a)

(b)

(c)

FIGURE 6.10   (a) Focused $R_0 > 0$. (b) Collimated $R_0 \to \infty$. (c) Diverging beams $R_0 > 0$.

where

$$\alpha = \frac{1}{z_0} - j\frac{1}{R_0}, \quad z_0 = \frac{kW_0^2}{2}.$$

Substituting (6.64) into (6.58) and using the paraxial approximation (6.60), we get

$$u(x, y, z) - \frac{A_0}{1 - j\alpha z} \exp\left(-jkz - \frac{k\alpha}{2}\frac{\rho^2}{1 - j\alpha z}\right). \tag{6.65}$$

This is the expression for a Gaussian beam when the field at $z = 0$ is given by (6.61).

Let us consider the intensity $I = |u|^2$. We get from (6.65)

$$I(x,\ y,\ z) = I_0 \frac{W_0^2}{W^2} \exp\left(-\frac{2\rho^2}{W^2}\right),$$

(6.66)

where $I_0 = |A_0|^2$ is the intensity at $z = 0$, and

$$W^2 = W_0^2\left[\left(1 - \frac{z}{R_0}\right)^2 + \left(\frac{z}{z_0}\right)^2\right].$$

(6.67)

Equation (6.66) indicates that the beam width W varies according to (6.67) and that the total intensity I_t is independent of the distance z as expected.

$$I_t = \iint I(x,\ y,\ z)\ dx\ dy = I_0\ W_0^2\left(\frac{\pi}{2}\right).$$

(6.68)

If the radius of curvature R_0 is positive, the beam is focused, if R_0 is infinite, the beam is collimated, and if R_0 is negative, the beam is diverging (Fig. 6.10). Let us first examine the focused beam $R_0 > 0$. The beam width W at the focal point $z = R_0$ is called the *beam spot size* and is obtained from (6.67).

$$\frac{W_s}{W_0} = \frac{\lambda R_0}{\pi W_0^2}.$$

(6.69)

The beam spot size W_s becomes smaller as the original beam size W_0 increases for given λ and R_0. This relationship (6.69) may be viewed by examining the Fresnel numbers N_f. At a given distance z from the center of the aperture of width W_0, the difference between the distance to the edge of the aperture and the distance to the center may be measured as the multiple of a half-wavelength (Fig. 6.11). The number of the multiple is the Fresnel number N_f.

To obtain N_f, we let

$$z + N_f\frac{\lambda}{2} = \left(z^2 + W_0^2\right)^{1/2},$$

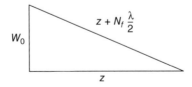

FIGURE 6.11   Fresnel number $N_f = W_0^2/\lambda z$.

and we get

$$N_f \approx \frac{W_0^2}{\lambda z} \quad \text{for } W_0 \ll z. \tag{6.70}$$

At $z = R_0$, therefore the spot size W_s is given by

$$\frac{W_s}{W_0} = \frac{1}{\pi N_f}. \tag{6.71}$$

The ratio of the spot size to the aperture size is inversely proportional to the Fresnel number N_f.

Let us consider a collimated beam $R_0 \to \infty$. We get

$$u(x, y, z) = \frac{A_0}{1 - j\alpha_1 z} \exp\left(-jkz - \frac{\rho^2}{W_0^2} \frac{1}{1 - j\alpha_1 z}\right),$$

where

$$\alpha_1 = \frac{\lambda}{\pi W_0^2}, \quad \alpha_1 z = \frac{1}{\pi N_f}. \tag{6.72}$$

The half-power beam width θ_b as $z \to \infty$ is obtained by letting

$$\exp\left(-\frac{2\rho^2}{W^2}\right) = \frac{1}{2},$$

$$\rho = \frac{z\theta_b}{2}, \quad W^2 \approx W_0^2(\alpha_1 z)^2.$$

We then get

$$\theta_b = \frac{\lambda (2 \ln 2)^{1/2}}{\pi W_0}. \tag{6.73}$$

6.7 GOOS–HANCHEN EFFECT

When a beam wave is incident on a plane boundary, the transmission and reflection normally take place. If a beam wave is incident from a dense medium on a less dense medium, total reflection takes place if the angle of incidence is greater than the critical angle. When this happens, it is noted that the reflected beam is shifted in space and this is called the *Goos–Hanchen shift*. This phenomenon gives some interesting physical insight into the total reflection mechanism.

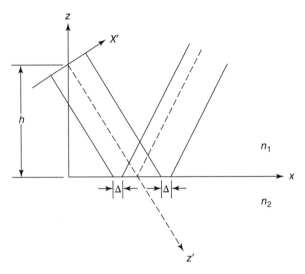

FIGURE 6.12   Goos–Hanchen shift.

Let us consider a beam wave incident on a plane boundary as shown in Fig. 6.12. To simplify the analysis, we consider a two-dimensional ($\partial/\partial y = 0$) problem. The incident beam wave $u_i(x', z')$ is collimated with amplitude distribution $A(x')$

$$u_i(x', z') = A(x')\exp(-jk_1 z').\tag{6.74}$$

We then use the coordinate transformation from (x', z') to (x, z)

$$\begin{aligned}x' &= x\cos\theta_0 + (z - h)\sin\theta_0,\\ z' &= x\sin\theta_0 - (z - h)\cos\theta_0.\end{aligned}\tag{6.75}$$

The incident field at $z = h$ is, therefore, given by

$$u_i(x, h) = A(x\cos\theta_0)\exp(-j\beta_0 x),\tag{6.76}$$

where $\beta_0 = k_1 \sin\theta_0$.

We take a Fourier transform of (6.76)

$$\begin{aligned}U(\beta - \beta_0) &= \int u_i(x, h)e^{j\beta x}\,dx\\ &= \int A(x\cos\theta_0)e^{j(\beta-\beta_0)x}\,dx.\end{aligned}\tag{6.77}$$

The incident field $u_i(x, z)$ satisfies the wave equation

$$\left(\frac{\partial^2}{\partial x^2} + \frac{\partial^2}{\partial z^2} + k_1^2\right)u_i(x, z) = 0.\tag{6.78}$$

Taking the Fourier transform in the x direction and letting $u_i(x, z)$ be equal to (6.76) at $z = h$, we obtain

$$u_i(x, z) = \frac{1}{2\pi} \int U(\beta - \beta_0) \exp[-j\beta x + jq(z - h)] \, d\beta, \qquad (6.79)$$

where $q = (k_1^2 - \beta^2)^{1/2}$. Note that the incident wave u_i consists of the spectrum $U(\beta - \beta_0)$ centered at β_0. For example, if u_i is a Gaussian beam, we have

$$A(x') = A_0 \exp\left[-\left(\frac{x'}{W_0}\right)^2\right],$$

$$U(\beta - \beta_0) = \frac{A_0 W_0 \sqrt{\pi}}{\cos \theta_0} \exp\left[-\frac{W_0^2 (\beta - \beta_0)^2}{4 \cos^2 \theta_0}\right]. \qquad (6.80)$$

The wave number β in the x direction extends from $-\infty$ to $+\infty$. The range $-k_1 < \beta < k_1$ corresponds to the plane wave with real angle θ_i of incidence $\beta = k_1 \sin \theta_i (-\pi/2 < \theta_i < \pi/2)$. However, the range $\beta > k_1$ corresponds to the complex angle of incidence $\theta_i = \pi/2 + j\theta''$, $0 \leq \theta'' \leq \infty$, and the range $\beta < -k_1$ corresponds to $\theta_i = -\pi/2 - j\theta''$, $0 \leq \theta'' \leq \infty$. These ranges $|\beta| > k_1$ represent the evanescent wave. In general, the finite beam or the wave from a localized source excites both the propagating $|\beta| < k_1$ and the evanescent waves $|\beta| > k_1$.

The reflected wave $u_r(x, z)$ can be written as

$$u_r(x, z) = \frac{1}{2\pi} \int R(\beta) U(\beta - \beta_0) \exp[-j\beta x - jq(z + h)] \, d\beta, \qquad (6.81)$$

where $R(\beta)$ is the reflection coefficient to be determined by applying the boundary condition.

When we apply the boundary condition at $z = 0$ using (6.79), (6.81), and the transmitted wave,

$$u_t(x, z) = \frac{1}{2\pi} \int T(\beta) U(\beta - \beta_0) e^{-j\beta x + jq_t z - jqh} \, d\beta, \qquad (6.82)$$

where $q_t = (k_2^2 - \beta^2)^{1/2}$, we obtain the relationship identical to the plane wave case except that $k_1 \sin \theta_i$ for a plane wave is replaced by β. Thus, each Fourier component $U(\beta - \beta_0)$ behaves as if it is a plane wave with $\beta = k_1 \sin \theta_i$. The difference is that the Fourier components include all β, both propagating and evanescent waves, while the plane wave is limited to the real angle of incidence.

If u represents a TE wave in Section 3.4, we get (for $\mu_1 = \mu_2 = \mu_0$)

$$R(\beta) = \frac{q - q_t}{q + q_t}, \qquad (6.83)$$

where $q = (k_1^2 - \beta^2)^{1/2}$ and $q_t = (k_2^2 - \beta^2)^{1/2}$. If u represents a TM wave in Section 3.5, we get

$$R(\beta) = \frac{(q_t/n_2^2) - (q/n_1^2)}{(q_t/n_2^2) + (q/n_1^2)}.$$
(6.84)

If u represents an acoustic wave in Section 3.8,

$$R(\beta) = \frac{(\rho_2/q_t) - (\rho_1/q)}{(\rho_2/q_t) + (\rho_1/q)}.$$
(6.85)

Let us examine the reflected field (6.81). First, we note that for a beam wave, $U(\beta - \beta_0)$ has a peak at $\beta = \beta_0$ and decays away from $\beta = \beta_0$, as seen in (6.80). If the beam width W_0 is many wavelengths wide, U is very much concentrated near $\beta = \beta_0$. If the reflection coefficient $R(\beta)$ is a slowly varying function of β, then as a first approximation, we can let $R(\beta) \sim R(\beta_0)$. We then get from (6.81) the reflected wave that is equal to the beam wave $u_{r0}(x, z)$ originated at the image point $z = -h$ except for the reflection coefficient.

$$u_r(x, z) \sim R(\beta_0)u_{r0}(x, z),$$
$$u_{r0}(x, z) = \frac{1}{2\pi} \int U(\beta - \beta_0) \exp[-j\beta x - jq(z + h)] \, d\beta.$$
(6.86)

The approximation $R(\beta) \approx R(\beta_0)$ is no longer valid, however, when the angle of incidence θ_0 and n_1 and n_2 are such that the total reflection takes place, $\sin \theta_0 > (n_2/n_1)$. Then, the beam is not only totally reflected, but the reflected beam is shifted in space because of the additional phase in the reflection coefficient. As an example, consider the TE wave (6.83). Since in the neighborhood of $\beta \approx \beta_0$, $q_t = (k_2^2 - \beta^2)^{1/2} \sim k(n_2^2 - n_1^2 \sin^2 \theta_0)^{1/2} \sim -jk(n_1^2 \sin^2 \theta_0 - n_2^2)^{1/2}$, we let $q_t = -j\alpha_t$. Then, we write

$$R(\beta) = \frac{q - q_t}{q + q_t} = \frac{q + j\alpha_t}{q - j\alpha_t} = \exp[j\,\phi(\beta)],$$
(6.87)

where $\phi(\beta) = 2\tan^{-1}(\alpha_t/q)$. Since the major contribution to the integral (6.81) comes from the neighborhood of $\beta = \beta_0$, we expand $\phi(\beta)$ in Taylor's series about β_0 and keep the first two terms.

$$\phi(\beta) = \phi(\beta_0) + (\beta - \beta_0)\phi'(\beta_0).$$
(6.88)

Substituting (6.87) and (6.88) into (6.81), we get

$$u_r(x, z) = R(\beta_0)e^{-j\beta_0\,\phi'(\beta_0)}u_{r0}(x - \phi'(\beta_0), z)$$
(6.89)

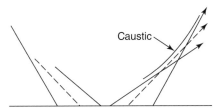

FIGURE 6.13   Caustic when the beam wave is *incident at an angle* close to the critical angle.

This shows that the reflected beam $u_r(x, z)$ is proportional to the beam u_{r0} from the image point with the lateral shift of $\Delta = \phi'(\beta_0)$ (Fig. 6.12). This shift Δ is called the Goos–Hanchen effect and it becomes greater the closer the incident angle θ_0 becomes to the critical angle $\sin^{-1}(n_2/n_1)$. As θ_0 approaches the critical angle, the first-order approximation (6.88) becomes insufficient, and in the neighborhood of the critical angle, the reflected rays form a caustic (Fig. 6.13).

6.8 HIGHER-ORDER BEAM-WAVE MODES

In Sections 6.6 and 6.7, we discussed a beam whose amplitude distribution is Gaussian. This is the most important practical beam wave. However, this constitutes the fundamental beam-wave mode, and there are an infinite number of higher-order beam-wave modes in addition to the fundamental mode. To investigate these higher-order modes, it is convenient to start with the parabolic approximation to the wave equation. Consider the field $u(x, y, z)$ that satisfies the wave equation

$$(\nabla^2 + k^2)u\,(x, y, z) = 0. \tag{6.90}$$

For a beam wave, the field is mostly propagating in the z direction, and thus we write

$$u\,(x, y, z) = U\,(x, y, z)\,e^{-jkz}. \tag{6.91}$$

The function U should be a slowly varying function of z. Substituting (6.91) into (6.90), and noting that

$$\nabla u = (\nabla U)e^{-jkz} - jk\hat{z}\,Ue^{-jkz},$$

$$\nabla^2 u = \left(\nabla^2 U - j2k\frac{\partial}{\partial z}\,U - k^2\,U \right) e^{-jkz},$$

we get

$$\left(\nabla^2 - j2k\frac{\partial}{\partial z} \right) U = 0. \tag{6.92}$$

This is exact. Now if U is slowly varying in z such that the variation of U over a wavelength is negligibly small, we have

$$\left|\frac{\partial U}{\partial z}\right| \sim \left|\frac{\Delta U}{\Delta z}\right| < \left|\frac{\Delta U}{\lambda}\right| \ll \left|\frac{U}{\lambda}\right| \sim |kU|. \tag{6.93}$$

Thus we can approximate (6.92) by

$$\left(\nabla_t^2 - j2k\frac{\partial}{\partial z}\right) U = 0, \tag{6.94}$$

where ∇_t^2 is the Laplacian in x and y (transverse to $\hat{z}$). This is the parabolic approximation to the wave equation (6.92) and greatly simplifies the mathematical analysis for a beam wave. This parabolic approximation is equivalent to the paraxial approximation (6.60).

Since (6.94) is identical to the paraxial approximation, it is easy to show that the Gaussian beam wave shown in (6.65)

$$U_g(x, y, z) = \frac{A_0}{1 - j\alpha z} \exp\left[-\frac{\rho^2}{W_0^2(1 - j\alpha z)}\right], \quad \alpha = \frac{\lambda}{\pi W_0^2}, \tag{6.95}$$

satisfies (6.94).

To obtain the higher-order modes, we let

$$U(x, y, z) = U_g(x, y, z) f(t) g(\tau) e^{j\phi(z)}, \tag{6.96}$$

where

$$t = \sqrt{2}\frac{x}{W}, \quad \tau = \sqrt{2}\frac{y}{W},$$

$$W = W_0[1 + (\alpha z)^2]^{1/2}.$$

Substituting (6.96) into (6.94), and noting that U_g also satisfies (6.94), we get

$$\frac{1}{f}\left(\frac{d^2f}{dt^2} - 2t\frac{df}{dt}\right) + \frac{1}{g}\left(\frac{d^2g}{d\tau^2} - 2\tau\frac{dg}{d\tau}\right) - kW^2\frac{d\phi}{dz} = 0. \tag{6.97}$$

Noting that x and y enter into the first and second terms only, respectively, we require that each term be constant to satisfy (6.97) at all x and y. Thus we let

$$\frac{1}{f}\left(\frac{d^2f}{dt^2} - 2t\frac{df}{dt}\right) = 2m,$$

$$\frac{1}{g}\left(\frac{d^2g}{d\tau^2} - 2\tau\frac{df}{d\tau}\right) = 2n, \tag{6.98}$$

$$kW^2\frac{d\phi}{dz} = 2(m + n),$$

where m and n are constant.

The first two equations of (6.98) are Hermite differential equations

$$\left(\frac{d^2}{dt^2} - 2t\frac{d}{dt} + 2m \right) H_m(t) = 0, \tag{6.99}$$

whose solutions $H_m(t)$ are Hermite polynomials. Equation (6.98) can be integrated to obtain

$$\begin{aligned} \phi &= \int_0^z \frac{2(m+n)}{kW^2} dz \\ &= \frac{2(m+n)}{kW_0^2} \int_0^z \frac{dz}{1+(\alpha z)^2} \\ &= (m+n)\tan^{-1}\alpha z. \end{aligned} \tag{6.100}$$

We therefore obtain the general representation of the higher-order beam-wave modes.

$$U_{mn}(x, y, z) = U_g(x, y, z) H_m(t) H_n(\tau) \exp[j(m+n)\tan^{-1}\alpha z], \tag{6.101}$$

where t and τ are defined in (6.96) and α is defined in (6.95).

Hermite polynomials $H_m(t)$ are orthogonal.

$$\int_{-\infty}^{\infty} e^{-t^2} H_m(t) H_{m'}(t)\, dt = \begin{cases} 0 & \text{if } m \neq m' \\ 2^m \sqrt{\pi}\, m & \text{if } m = m' \end{cases}. \tag{6.102}$$

Therefore, noting that

$$|U_g|^2 = \frac{A_0^2}{W^2} \exp[-(t^2 + \tau^2)], \tag{6.103}$$

all modes $U_{mn}(x, y, z)$ are orthogonal in a constant z plane.

$$\int \int U_{mn} U_{m'n'}^*\, dx\, dy = 0 \quad \text{if } m \neq n' \quad \text{or} \quad n \neq n'. \tag{6.104}$$

The first few Hermite polynomials $H_m(t)$ are

$$\begin{aligned} H_0(t) &= 1, & H_1(t) &= 2t, \\ H_2(t) &= 4t^2 - 2, & H_3(t) &= 8t^3 - 12t. \end{aligned}$$

The functions $H_m(t)\exp(-t^2/2)$ are sketched in Fig. 6.14. The beam wave radiating out of a He–Ne laser consists of the fundamental Gaussian beam and the higher-order modes described above.

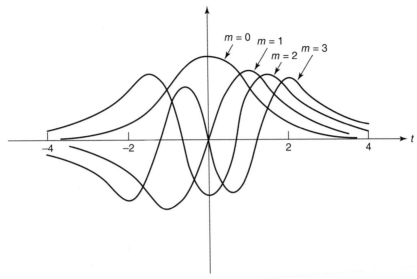

FIGURE 6.14   Function $H_m(t)\exp(-t^2/2)$ normalized by $(2^m\sqrt{\pi m})^{1/2}$.

6.9 VECTOR GREEN'S THEOREM, STRATTON–CHU FORMULA, AND FRANZ FORMULA

We have discussed scalar Green's theorem in Section 6.1. In this section, we consider the equivalent theorem for vector fields. Let us consider the vector fields $\bar{P}(\bar{r})$ and $\bar{Q}(\bar{r})$ in the volume V surrounded by the surface S. We assume that $\bar{P}$ and $\bar{Q}$ and their first and second derivatives are continuous in V and on S. Using the divergence theorem, we have

$$\int_V \nabla \cdot (\bar{P} \times \nabla \times \bar{Q})\, dV = \int_S \bar{P} \times \nabla \times \bar{Q} \cdot d\bar{S}. \tag{6.105}$$

We now use the identity

$$\nabla \cdot (\bar{A} \times \bar{B}) = \bar{B} \cdot \nabla \times \bar{A} - \bar{A} \cdot \nabla \times \bar{B}. \tag{6.106}$$

Letting $\bar{A} = \bar{P}$ and $\bar{B} = \nabla \times \bar{Q}$, we get the vector Green's first identity

$$\int_V (\nabla \times \bar{Q} \cdot \nabla \times \bar{P} - \bar{P} \cdot \nabla \times \nabla \times \bar{Q})\, dV = \int_S \bar{P} \times \nabla \times \bar{Q} \cdot d\bar{S}. \tag{6.107}$$

To get the second identity or Green's theorem, we interchange $\bar{P}$ and $\bar{Q}$ in (6.107).

$$\int_V (\nabla \times \bar{P} \cdot \nabla \times \bar{Q} - \bar{Q} \cdot \nabla \times \nabla \times \bar{P})\, dV = \int_S \bar{Q} \times \nabla \times \bar{P} \cdot d\bar{S}. \tag{6.108}$$

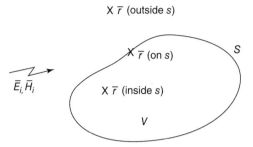

FIGURE 6.15   Stratton–Chu formula.

Subtracting (6.108) from (6.107), we get the vector Green's theorem or the vector Green's second identity.

$$\int_V (\bar{Q} \cdot \nabla \times \nabla \times \bar{P} - \bar{P} \cdot \nabla \times \nabla \times \bar{Q})\, dV = \int_S (\bar{P} \times \nabla \times \bar{Q} - \bar{Q} \times \nabla \times \bar{P}) \cdot d\bar{S}.$$

$$(6.109)$$

The vector Green's theorem developed above is now applied to the electromagnetic field problem. We consider the scattering problem discussed in Section 6.1. The electromagnetic fields $\bar{E}_i$ and $\bar{H}_i$ are incident on a body with the volume V surrounded by the surface S (Fig. 6.15). The Stratton–Chu formula can be stated for the following three cases.

1. When the observation point $\bar{r}$ is outside the surface S,

$$\bar{E}_i(\bar{r}) + \int_S \bar{E}_s dS' = \bar{E}(\bar{r}),$$

$$\bar{H}_i(\bar{r}) + \int_S \bar{E}_s dS' = \bar{E}(\bar{r}).$$

$$(6.110)$$

2. When $\bar{r}$ is on the surface S,

$$\bar{E}_i(\bar{r}) + \int_S \bar{E}_s dS' = \tfrac{1}{2}\bar{E}(\bar{r}),$$

$$\bar{H}_i(\bar{r}) + \int_S \bar{H}_s dS' = \tfrac{1}{2}\bar{H}(\bar{r}).$$

$$(6.111)$$

3. When $\bar{r}$ is inside S,

$$\bar{E}_i(\bar{r}) + \int_S \bar{E}_s dS' = 0,$$

$$\bar{H}_i(\bar{r}) + \int_S \bar{E}_s dS' = 0,$$

$$(6.112)$$

where

$$\bar{E}_s = -[j\omega\mu G\hat{n}' \times \bar{H} - (\hat{n}' \times \bar{E}) \times \nabla'G - (\hat{n}' \cdot \bar{E})\nabla'G],$$
$$\bar{H}_s = j\omega\mu G\hat{n}' \times \bar{E} + (\hat{n}' \times \bar{H}) \times \nabla'G + (\hat{n}' \cdot \bar{H})\nabla'G], \qquad (6.113)$$
$$\bar{E} = E(\bar{r}'), \quad \bar{H} = \bar{H}(\bar{r}'),$$

$G(\bar{r} - \bar{r}') = \exp(-jk|\bar{r} - \bar{r}'|)/(4\pi|\bar{r} - \bar{r}'|)$ is the scalar free-space Green's function, and ∇' is the gradient with respect to $\bar{r}'$. The surface integral in (6.111) is the Cauchy principal value of the integral, and the fields $\bar{E}(\bar{r}')$ and $\bar{H}(\bar{r}')$ are those as $\bar{r}'$ approaches S from outside.

Equation (6.111) will be used later to construct an electric field integral equation (EFIE) and a magnetic field integral equation (MFIE). Equation (6.112) is the vector extinction theorem, also called the vector Ewald–Oseen extinction theorem or the vector null field theorem. Inside the surface S, the incident fields $\bar{E}_i$ and $\bar{H}_i$ are extinguished by the contribution from the surface fields. Proof of formula (6.113) is given in Appendix 6.A.

In addition to the Stratton–Chu formula, there are other equivalent representations of the electromagnetic field. Consider the field $\bar{E}(\bar{r})$ and $\bar{H}(\bar{r})$ outside the surface S and the volume V (Fig. 6.16). The Stratton–Chu formula is given by

$$\bar{E}(\bar{r}) = \bar{E}_i(\bar{r}) + \int_S \bar{E}_s dS' + \int_V \bar{E}_v dV',$$
$$\bar{H}(\bar{r}) = \bar{H}_i(\bar{r}) + \int_S \bar{E}_s dS' + \int_V \bar{H}_v dV', \qquad (6.114)$$

where $\bar{E}_s$ and $\bar{H}_s$ are given in (6.113) and $\bar{E}_v$ and $\bar{H}_v$ are given by

$$\bar{E}(\bar{r}) = -\left(j\omega\mu G\bar{J} + J_m \times \nabla'G - \frac{\rho}{\varepsilon}\Delta'G\right),$$
$$\bar{H}_v(\bar{r}) = -\left(j\omega\varepsilon G\bar{J}_m - J \times \nabla'G - \frac{\rho_m}{\mu}\Delta'G\right). \qquad (6.115)$$

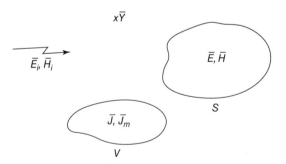

FIGURE 6.16   Franz formula.

The following equivalent representation is called the *Franz formula* (Tai, 1972):

$$\bar{E}(\bar{r}) = \bar{E}_i(\bar{r}) + \nabla \times \nabla \times \bar{\pi} - j\omega\mu\nabla \times \bar{\pi}_m,$$
$$\bar{H}(\bar{r}) = \bar{H}_i(\bar{r}) + j\omega\varepsilon\nabla \times \bar{\pi} + \nabla \times \nabla \times \bar{\pi}_m,$$

(6.116)

where

$$\pi(\bar{r}) = \frac{1}{j\omega\varepsilon} \left(\int_V \bar{J}G dV' + \int_S \hat{n}' \times \bar{H}G\, dS' \right),$$
$$\pi_m(\bar{r}) = \frac{1}{j\omega\mu} \left(\int_V \bar{J}_m G\, dV' - \int_S \hat{n} \times \bar{E}G\, dS' \right).$$

The equivalence between the Stratton–Chu formula and the Franz formula is shown by Tai.

6.10   EQUIVALENCE THEOREM

In Section 6.9, we discussed the Stratton–Chu formula and its equivalent Franz formula. According to the Franz formula, the field outside the surface S is given by the incident field and the surface integral of the tangential electric field $\hat{n}' \times \bar{E}$ and the tangential magnetic field $\hat{n}' \times \bar{H}$. Inside the surface S, the field is zero as shown in the extinction theorem. Since the surface S is arbitrary, we can state that the actual field outside the surface S is identical to the field generated by the equivalent surface magnetic current $\bar{J}_{ms} = \bar{E} \times \hat{n}'$ and the equivalent surface electric current $\bar{J}_s = \hat{n}' \times \bar{H}$. These fictitious currents $\bar{J}_{ms}$ and $\bar{J}_s$ on S produce a field identical to the original field outside S, but they extinguish the field inside S, producing the null field (Harrington, 1968, Chapter 3) (Fig. 6.17).

In the above we used both $\bar{J}_s$ and $\bar{J}_{ms}$ on the surface. However, according to the uniqueness theorem, if we specify either $\bar{J}_s$ or $\bar{J}_{ms}$ on S, the field outside S is uniquely determined, and therefore we may use $\bar{J}_s$ or $\bar{J}_{ms}$. For example, consider a conducting body with the aperture A on its surface S. The tangential electric field over the aperture is $\bar{E} \times \hat{n}'$. As far as the field outside S is concerned, the field is identical to the field produced by the conducting body with the aperture closed, but with the magnetic surface current $\bar{J}_{ms} = \bar{E} \times \hat{n}'$ placed in front of the original location of the aperture

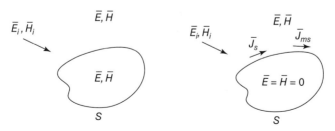

FIGURE 6.17   Equivalence theorem.

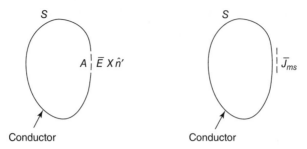

FIGURE 6.18   Aperture on a conducting surface.

(Fig. 6.18). Note that $\bar{E} \times \hat{n}'$ is zero just behind $\bar{J}_{ms}$ on S, but $\bar{E} \times \hat{n}'$ is identical to the original field just in front of $\bar{J}_{ms}$.

6.11 KIRCHHOFF APPROXIMATION FOR ELECTROMAGNETIC WAVES

In Section 6.3, we discussed the Kirchhoff approximation for the scalar field. The electromagnetic Kirchhoff approximation can be obtained using the same technique. Let us consider a large aperture on a screen. According to the Kirchhoff approximation, we assume that the electric and magnetic fields tangential to the aperture are equal to those of the incident field.

$$\hat{n}' \times \bar{H} = \hat{n}' \times \bar{H}_i,$$
$$\bar{E} \times \hat{n}' = \bar{E}_i \times \hat{n}'. \tag{6.117}$$

Then using the Franz formula (6.116), we get

$$\bar{E}(\bar{r}) = \nabla \times \nabla \times \bar{\Pi} - j\omega\mu\nabla \times \bar{\Pi}_m,$$

$$\bar{\Pi} = \frac{1}{j\omega\varepsilon} \int_S \hat{n}' \times \bar{H}_i G \, dS,$$

$$\bar{\Pi}_m = \frac{1}{j\omega\mu} \int_S \bar{E}_i \times \hat{n}' G \, dS', \tag{6.118}$$

$$G = \frac{\exp(-jk|\bar{r} - \bar{r}'|)}{4\pi |\bar{r} - \bar{r}'|}.$$

Here all the quantities are known, and therefore $\bar{E}(\bar{r})$ at any point $\bar{r}$ can be calculated.

If we consider the field $\bar{E}(\bar{r})$ in the far zone of the aperture where $|\bar{r}| \gg D^2/\lambda$ and D is the size of the aperture, we can approximate G by the following (Fig. 6.5):

$$G = \frac{\exp(-jkR_0 + jk\hat{o} \cdot \bar{r}')}{4\pi R_0}. \tag{6.119}$$

Also note that in the far zone, the field is propagating in the direction $\hat{o}$ and is thus proportional to $\exp(-jk\hat{o} \cdot \bar{r})$. Since $\nabla \exp(-jk\hat{o} \cdot \bar{r}) = -jk\hat{o}\exp(-jk\hat{o} \cdot \bar{r})$, the operator ∇ is equal to $-jk\hat{o}$. Thus we have $\nabla \times \nabla\times = -k^2\hat{o} \times \hat{o}\times$ and $\nabla\times = -jk\hat{o} \times$.

Also, we consider the components E_θ and E_ϕ. Then we finally get the following Kirchhoff approximation for the radiation field when the incident field $\bar{E}_i$ and $\bar{H}_i$ is known over the aperture S.

$$
\begin{aligned}
E_\theta &= -\frac{j\omega\mu}{4\pi R_0}e^{-jkR_0}\int \hat{\theta} \cdot (\hat{n}' \times \bar{H}_i)e^{jk\hat{o}\cdot\bar{r}'}\,dS' \\
&\quad -\frac{jk}{4\pi R_0}e^{-jkR_0}\int \hat{\phi} \cdot (\bar{E}_i \times \hat{n}')e^{jk\hat{o}\cdot\bar{r}'}\,dS', \\
E_\phi &= -\frac{j\omega\mu}{4\pi R_0}e^{-jkR_0}\int \hat{\phi} \cdot (\hat{n}' \times \bar{H}_i)e^{jk\hat{o}\cdot\bar{r}'}\,dS' \\
&\quad +\frac{jk}{4\pi R_0}e^{-jkR_0}\int \hat{\theta} \cdot (\bar{E}_i \times \hat{n}')e^{jk\hat{o}\cdot\bar{r}'}\,dS'.
\end{aligned}
\tag{6.120}
$$

For example, if the incident wave is a plane wave normally incident on a rectangular aperture ($2a \times 2b$) in the x–y plane, and is polarized in the x direction, we have $\bar{E}_i = E_0\hat{x}$ and $\bar{H}_i = (E_0/\eta)\hat{y}$ with $\eta = \sqrt{\mu_0/\varepsilon_0}$. Then the radiation field $\bar{E}$ is given by

$$
\begin{aligned}
\bar{E} &= \frac{jke^{-jkR_0}}{4\pi R_0}(1 + \cos\theta)[\hat{\theta}\cos\phi - \hat{\phi}\sin\phi]F(\theta,\phi), \\
F(\theta,\phi) &= 4ab\frac{\sin K_1 a}{K_1 a}\frac{\sin K_2 b}{K_2 b}, \\
K_1 &= k\sin\theta\cos\phi, \quad K_2 = k\sin\theta\sin\phi.
\end{aligned}
\tag{6.121}
$$

PROBLEMS

6.1 A plane scalar wave is normally incident on a large square aperture of a (m) $\times$ b (m). Using the Kirchhoff approximation, find the radiation field.

6.2 If a plane wave is normally incident on a square plate of $a \times b$ with the reflection coefficient of $R = -1$, find the scattered field using the Kirchhoff approximation.

6.3 Find the radiation pattern, half-power beamwidth, and first sidelobe level (in dB) in the x–z plane and y–z plane for the aperture field distribution $U_0(x, y) = A_0\cos^2(\pi x/a)$ in a square aperture of $a = 3$ m $\times$ $b = 1$ m at a frequency of 10 GHz.

6.4 Is it possible to send a light beam ($\lambda = 0.5$ μm) from the earth to illuminate an area of 500 m diameter between half-power points on the surface of the moon?

If so, what should be the size of the aperture of the transmitter? The distance between the moon and the earth is approximately 384,400 km.

6.5 It is planned to collect the solar energy by a large solar panel on a synchronous satellite, convert it to microwaves, and send it down to the earth. The total power to be transmitted should be 5 GW. The power density on the ground should be less than 10 mW/cm². A microwave at 2.45 GHz is to be used. Assume that the transmitter is a circular aperture of 1 km diameter and that the aperture distribution is uniform. Assume that the altitude of the satellite in geosynchronous equatorial orbit is 35,800 km. Find the intensity distribution on the ground.

6.6 Plot the beam size W of a focused optical beam with $W_0 = 1$ cm at $\lambda = 0.6$ μm as a function of distance. The focal distance is 1 m. What is the half-power beamwidth (in degrees) at a large distance?

6.7 Calculate the Goos–Hanchen shift Δx for an optical beam with $\lambda = 0.6$ μm incident from the medium $n = 2.5$ to air. The wave is polarized in the plane of incidence. Plot Δx as a function of the incident angle.

6.8 Find an expression for $\bar{E}$ in (6.121) when both the x and y components of the aperture field are known.

CHAPTER 7

PERIODIC STRUCTURES AND COUPLED-MODE THEORY

There are many important structures whose characteristics are periodic in space. Examples are three-dimensional lattice structures for crystals, artificial dielectric consisting of periodically placed conducting pieces, Yagi antennas that have periodically spaced elements, corrugated surfaces, and waveguides with periodic loadings. In addition, systems of lenses that are placed with equal spacings and open resonators can be considered periodic structures. Guided waves along these structures exhibit a unique frequency dependence often characterized by stop bands and pass bands. The scattered waves from these periodic structures are characterized by grating modes resulting from periodic interference of waves in different directions. The starting point in solving the problems of periodic structures is Floquet's theorem, which is described in Section 7.1.

In this chapter, we discuss guided waves propagating along a periodic structure, waves propagating through periodic layers, and plane waves incident on periodic structures. In discussing these problems, we present integral equation formulations that are used extensively in many electromagnetic problems. We also discuss a classical problem of scattering by sinusoidal surfaces and include a short description of coupled-mode theory.

Electromagnetic Wave Propagation, Radiation, and Scattering: From Fundamentals to Applications,
Second Edition. Akira Ishimaru.
© 2017 by The Institute of Electrical and Electronic Engineers, Inc. Published 2017 by John Wiley & Sons, Inc.

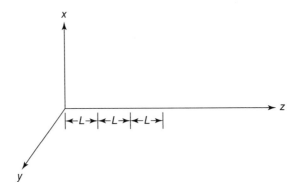

FIGURE 7.1 Periodic structures.

7.1 FLOQUET'S THEOREM

Let us consider a wave propagating in periodic structures, which may be characterized by periodic boundary conditions or a periodically varied dielectric constant (Fig. 7.1). We note that the fields at a point z in an infinite periodic structure differ from the fields one period L away by a complex constant. This is obviously true because, in an infinite periodic structure, there should be no difference between the fields at z and at $z + L$ except for the constant attenuation and phase shift. Let a function $u(z)$ represent a wave. Then a wave $u(z)$ at z and a wave $u(z + L)$ at $z + L$ are related in the same manner as a wave $u(z + L)$ at $z + L$ and a wave $u(z + 2L)$ at $z + 2L$.

Mathematically, we write

$$\frac{u(z + L)}{u(z)} = \frac{u(z + 2L)}{u(z + L)} = \frac{u(z + mL)}{u[z + (m - 1)L]} = C = \text{constant}. \tag{7.1}$$

From this we obtain

$$u(z + mL) = C^m u(z). \tag{7.2}$$

The constant C is in general complex, which we write

$$C = e^{-j\beta L}, \quad \beta = \text{complex}, \tag{7.3}$$

and β represents the propagation constant.

Now let us consider a function

$$R(z) = e^{j\beta z} u(z). \tag{7.4}$$

Then $R(z + L) = e^{j\beta(z+L)} u(z + L) = R(z)$. Therefore, $R(z)$ is a periodic function of z with period L, and thus can be represented in a Fourier series.

$$R(z) = \sum_{n=-\infty}^{\infty} A_n e^{-j(2n\pi/L)z}. \tag{7.5}$$

Using (7.4), we finally obtain a general expression for a wave in a periodic structure with the period L.

$$
\begin{aligned}
u(z) &= \sum_{n=-\infty}^{\infty} A_n e^{-j(\beta + 2n\pi/L)z} \\
&= \sum_{n=-\infty}^{\infty} A_n e^{-j\beta_n z}, \quad \beta_n = \beta + \frac{2n\pi}{L}.
\end{aligned} \tag{7.6}
$$

Noting that, in general, the wave consists of both positive-going and negative-going waves, we write

$$u(z) = \sum_{n=-\infty}^{\infty} A_n e^{-j\beta_n z} + \sum_{n=-\infty}^{\infty} B_n e^{+j\beta_n z}. \tag{7.7}$$

This is the representation of a wave in periodic structures in the form of an infinite series, resembling harmonic representation $(e^{-j\omega_n t})$ in time. The nth term in (7.6) is called the nth *space harmonic* or *Hartree harmonic*. Equation (7.7) is the mathematical representation of Floquet's theorem, which states that the wave in periodic structures consists of an infinite number of *space harmonics*. In this chapter we consider two cases. One is guided waves along a periodic structure and the other is the scattering of plane waves from periodic structures.

7.2 GUIDED WAVES ALONG PERIODIC STRUCTURES

Consider a wave propagating along a periodic structure with period L. The positive-going wave is given by (7.6), and when the wave is propagating, β is real and the phase velocity is different for each harmonic.

$$v_{pn} = \frac{\omega}{\beta_n} = \frac{\omega}{\beta + 2n\pi/L}, \tag{7.8}$$

but the group velocity is the same for all harmonics

$$v_{gn} = \frac{1}{d\beta_n/d\omega} = \frac{1}{d\beta/d\omega} = v_{g0}. \tag{7.9}$$

The propagation constant β is real in some frequency ranges, and this is called the *pass band*. The frequency range where β is purely imaginary and the wave is evanescent is called the *stop band*.

We now examine the k–β diagram for periodic structures. We first note that if β is increased by $2\pi/L$, this is equivalent to changing β_n to β_{n+1}, and the general expression

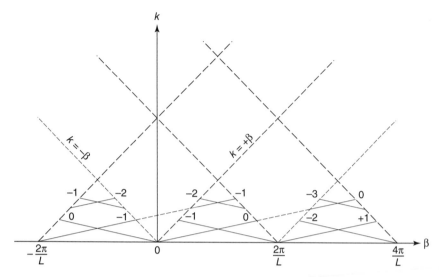

FIGURE 7.2 k–β diagram for periodic structures.

for the fields is unaltered. Therefore, k is a periodic function of β with period $2\pi/L$. It is also clear that since the wave propagation in the $+z$ and the $-z$ direction should have the same characteristics, k is an even function of β. As an example, the k–β diagram for a tape helix is shown in Fig. 7.2. The labeling of each mode is usually made to identify the lowest β with positive slope ($v_g > 0$) by the zeroth mode and the $\beta + 2n$ π/L by the nth mode. Similarly, for the mode with negative slopes ($v_g < 0$), $-(\beta + 2n$ $\pi/L)$ is labeled by the nth mode.

As an example, let us consider TM modes propagating along a corrugated surface (Fig. 7.3). We look for a trapped surface-wave solution. In this section, we employ the *integral equation formulation* for the boundary value problem. In the integral equation formulation, the differential equation and the boundary conditions are combined to obtain an integral equation. The integral equation contains an unknown function under an integral operator, just as the differential equation contains an unknown function under a differential operator. Let us examine this technique for the problem of a corrugated surface.

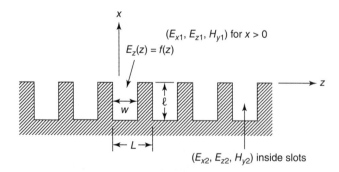

FIGURE 7.3   Corrugated surface.

In this problem, we first express the magnetic field H_1 in the region $x > 0$ due to the unknown tangential electric field at $x = 0$. Next we express the magnetic field H_2 in the region $x < 0$ due to the "same" tangential electric field at $x = 0$. Since we used the same electric field at $x = 0$, although it is still unknown, we automatically satisfied the boundary condition that the tangential electric field is continuous at $x = 0$. We then satisfy the boundary condition that the tangential magnetic field is continuous by equating H_1 to H_2, yielding an integral equation for the unknown tangential electric field at $x = 0$.

To follow the foregoing procedure, let us first note that all TM modes are given by H_y satisfying a scalar wave equation,

$$\left(\frac{\partial^2}{\partial x^2} + \frac{\partial^2}{\partial z^2} + k^2\right) H_y = 0, \tag{7.10}$$

with

$$E_x = j\frac{1}{\omega\varepsilon}\frac{\partial}{\partial z}H_y \quad \text{and} \quad E_z = -j\frac{1}{\omega\varepsilon}\frac{\partial}{\partial x}H_y.$$

Next we express H_1 for $x > 0$. According to Floquet's theorem, we write H_{y1} for $x > 0$ in a series of space harmonics.

$$H_{y1}(x, z) = \sum_{n=-\infty}^{\infty} f_n(x)e^{-j\beta_{nz}}, \quad \beta_n = \beta + \frac{2n\pi}{L}. \tag{7.11}$$

Since this must satisfy a wave equation (7.10), $f_n(x)$ should have a form $f_n(x) \approx e^{\pm jq_nx}$, where $q_n^2 + \beta_n^2 = k^2$. Furthermore, this structure is open in the $+x$ direction and H_y must satisfy the radiation condition as $x \to +\infty$, and therefore, $f_n(x)$ must have the form e^{-jq_nx}.

Since we are only interested in a trapped slow-wave solution, q_n must be purely imaginary, $q_n = -j\alpha_n$. Therefore, we write

$$H_{y1}(x, z) = \sum_{-\infty}^{\infty} A_n e^{-\alpha_nx - j\beta_nz}, \quad x > 0, \tag{7.12}$$

where $\alpha_n^2 = \beta_n^2 - k^2$, and A_n is the amplitude for each space harmonic.

To express H_{y1} in terms of the tangential electric field E_z at $x = 0$, we write E_{z1} using (7.10) and (7.12),

$$E_{z1}(x, z) = j\frac{1}{\omega\varepsilon}\sum_{-\infty}^{\infty} \alpha_n A_n e^{-\alpha_nx - j\beta_nz}. \tag{7.13}$$

Now at $x = 0$, $E_z(0, z) = 0$ on the top surface of the corrugation and $E_z(0, z)$ is equal to the field in the slot $f(z)$.

$$E_z(0, z) = \begin{cases} 0 & \text{for } \frac{W}{2} < |z| < \frac{L}{2}, \\ f(z) & \text{for } |z| < \frac{W}{2}. \end{cases} \tag{7.14}$$

The coefficients A_n can be expressed in terms of $f(z)$ by using (7.13) and (7.14) and recognizing that the space harmonics are orthogonal.

$$\int_{-L/2}^{L/2} (e^{-j\beta_n z})(e^{-j\beta_m z})^* dz = \int_{-L/2}^{L/2} e^{-j(\beta_n - \beta_m)z} dz = \int_{-L/2}^{L/2} e^{-j(n-m)(2\pi/L)z} dz = L\delta_{mn},$$

(7.15)

where Kronecker's delta $\delta_{mn} = 1$ when $m = n$ and $\delta_{mn} = 0$ when $m \neq n$. Recognizing that (7.13) is an expansion of $f(z)$ in an orthogonal series, we can obtain the coefficients of the series by multiplying both sides of (7.14) by $e^{j\beta_n z}$ and integrating over a period L.

$$\frac{j\alpha_n A_n}{\omega\varepsilon} = \frac{1}{L}\int_{-W/2}^{W/2} f(z)e^{j\beta_n z} dz.$$

(7.16)

Substituting this into (7.12), we write H_{y1} in the form

$$H_{y1}(x, z) = -j\omega\varepsilon \int_{-W/2}^{W/2} f(z')G_1(z, x; z', 0)dz' \quad \text{for } x > 0,$$

(7.17)

where

$$G_1(z, x; z', 0) = \sum_{-\infty}^{\infty} \frac{e^{-\alpha_n x - j\beta_n(z-z')}}{L\alpha_n}.$$

This form (7.17) is written to conform to Green's function formulation, and G_1 is Green's function for periodic structures (see Appendix 7.A).

Next, consider the field inside the slot due to the field $f(z)$ at $x = 0$. H_y in the slot should consist of TEM modes and higher-order TM modes. Noting that $E_x = 0$ at $z = \pm W/2$ and $E_z = 0$ at $x = -l$, we write

$$H_{y2} = \sum_{n=0}^{\infty} B_n \cos\frac{n\pi(z + W/2)}{W} \cos[k_n(x + l)],$$

(7.18)

where $k_n^2 + (n\pi/W)^2 = k^2$ and B_0 represents the TEM mode and B_n, $n \neq 0$, represents all the higher-order modes. We obtain E_z from (7.18) using $E_z = -j(1/\omega\varepsilon)(\partial/\partial x)H_y$. Equating this to $f(z)$ at $x = 0$, we get B_n, and substituting B_n into (7.18), we get

$$H_{y2}(x, z) = -j\omega\varepsilon \int_{-W/2}^{W/2} f(z')G_2(z, x; z', 0)\, dz',$$

(7.19)

where

$$G_z(z, x; z', 0) = \frac{\cos k(l + x)}{Wk \sin kl} + \sum_{n=1}^{\infty} \frac{2\psi_n(z)\psi_n(z')\cos k_n(l + x)}{Wk_n \sin k_n l},$$

$$\psi_n(z) = \cos\left[\frac{n\pi}{W}\left(z + \frac{W}{2}\right)\right].$$

Equations (7.17) and (7.19) give $H_y(x, z)$ in the region $x > 0$ and $x < 0$, respectively, in terms of the same tangential electric field $f(z)$ in the slot. Now let us consider the boundary conditions at $x = 0$, which requires the continuity of tangential electric and magnetic fields at $x = 0$. We note that the tangential electric field over the slot is continuous at $x = 0$ because we used the same field $f(z)$. The continuity of the tangential magnetic field requires that (7.17) and (7.19) be equal over the slot at $x = 0$. Thus we get

$$\int_{-W/2}^{W/2} f(z')G(z, 0; z', 0)\, dz' = 0 \quad \text{over the slot } |z| < \frac{W}{2}, \qquad (7.20)$$

where $G(z, 0; z', 0) = G_1(z, 0; z', 0) - G_2(z, 0; z', 0)$.

If we can solve (7.20) for the unknown function $f(z')$ and the propagation constant β, all the other fields can be obtained by (7.17) and (7.19). But the analytical solution of (7.20) is, in general, not available. It is possible to solve (7.20) by a numerical technique such as the moment method. In this section, however, we are not concerned with a detailed description of $f(z)$ but are interested in obtaining the propagation constant β.

In this case, we can obtain a convenient solution by equating the total complex power at both sides of the slot opening (Fig. 7.4). We note that $P_1 = P_2$, where

$$P_1 = \int_{-W/2}^{W/2} E_z^*(z)H_{y1}(z)\, dz,$$

$$P_2 = \int_{-W/2}^{W/2} E_z^*(z)H_{y2}(z)\, dz.$$

With (7.20), this is expressed as

$$\int_{-W/2}^{W/2} dz \int_{-W/2}^{W/2} dz' f^*(z)G(z, 0; z', 0)f(z') = 0. \qquad (7.21)$$

This form, (7.21), can be shown to be a variational form for the propagation constant β (see Appendix 7.B).

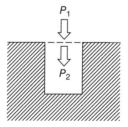

P_1

P_2

FIGURE 7.4   Conservation of total power.

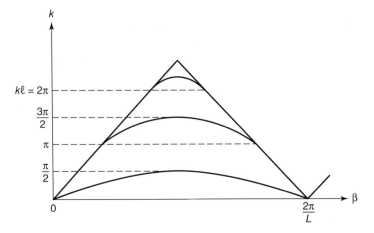

FIGURE 7.5 $k-\beta$ diagram for a corrugated surface.

Let us use the simplest trial function, $f(z) = $ constant. This gives

$$\sum_{-\infty}^{\infty} \frac{1}{\alpha_n} \left[\frac{\sin \beta_n(W/2)}{\beta_n(W/2)} \right]^2 = \frac{L}{kW} \cot kl. \tag{7.22}$$

The $k-\beta$ diagram for this structure is shown in Fig. 7.5.

The choice of $f(z)$ above is made for mathematical convenience. The best choice would be the one that satisfies the *edge condition*. The field behavior near the edge is not arbitrary and it must satisfy a certain condition called the edge condition (see Appendix 7.C). For example, the electric field normal to the edge with the angle ϕ_0 shown in Fig. 7.6 should behave as

$$E_z \approx \left(\frac{W}{2} - z \right)^{(\pi/\phi_0)-1}. \tag{7.23}$$

Since $\phi_0 = 3\pi/2$, an appropriate choice for the field $f(z)$ should be

$$f(z) = (\text{const.}) \left[\left(\frac{W}{2} \right)^2 - z^2 \right]^{-1/3}. \tag{7.24}$$

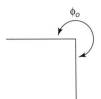

FIGURE 7.6 Edge condition.

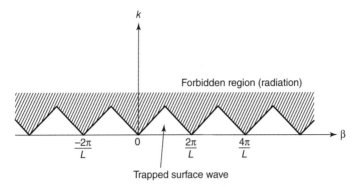

FIGURE 7.7   Forbidden region.

If a periodic structure is open in the transverse direction, the trapped surface wave can propagate along this structure without attenuation. Since the trapped surface wave is a slow wave, the wave exists only in the region $|\beta| > k$ in the k–β diagram. Recognizing that the diagram is a periodic function of β with period $2\pi/L$, we note that the trapped surface wave can exist only in a series of triangles in Fig. 7.7. Outside these triangles, the wave is fast and the energy leaks out of the structure, resulting in attenuation. This region is called *forbidden* because the structure cannot support waves without attenuation. For example, a helix used in traveling-wave tubes is designed to support the slow wave and obviously should not be operated in the forbidden region. On the other hand, if the structure is designed as an antenna, the leakage of the energy represents the radiation, and therefore the characteristics in this forbidden region are of great interest. Log-periodic antennas can be analyzed in terms of the behavior of the wave in the forbidden region.

If the structure is closed in the transverse direction, however, a propagating wave need not be slow, and thus there is no forbidden region. For example, if the corrugated surface discussed previously is closed (Fig. 7.8), the k–β diagram may appear as shown. Obviously, an exact diagram depends on the relative sizes of L, W, l, and the wavelength.

7.3 PERIODIC LAYERS

Periodic stratified media are important in many optical and microwave applications. Let us consider the simplest case of a periodic medium consisting of alternating layers of two different indices of refraction (Fig. 7.9). We wish to find the propagation constant of a wave as it propagates through this structure. We first note that in an infinite periodic structure, there should be no difference between the field $U(x)$ at one point and the field $U(x + d)$ at another point separated by the period d except for a constant attenuation and phase shift. Mathematically, we state that

$$U(x + d) = U(x)\exp(-jqd), \qquad (7.25)$$

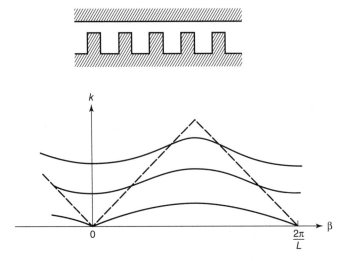

FIGURE 7.8   Closed corrugated guide.

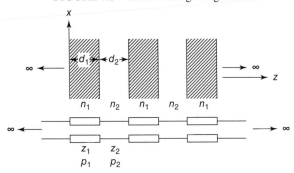

FIGURE 7.9   Periodic layers.

where q is a complex propagation constant. This concept, expressed in (7.25), is Floquet's theorem, which was discussed previously.

Let us consider one period of the structure (Fig. 7.10). If we choose the voltages and currents V_1, I_1, V_2, and I_2 as shown, Floquet's theorem yields

$$V_2 = V_1 \exp(-jqd),$$
$$I_2 = I_1 \exp(-jqd),$$

(7.26)

where q is a complex propagation constant and $d = d_1 + d_2$.

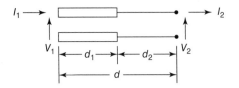

FIGURE 7.10   One period of layers.

We also note that since this is a linear passive network, the input V_1 and I_1 and the output V_2 and I_2 are related through the $ABCD$ parameters, as discussed in Sections 3.7 and 4.6.

$$\begin{bmatrix} V_1 \\ I_1 \end{bmatrix} = \begin{bmatrix} A & B \\ C & D \end{bmatrix} \begin{bmatrix} V_2 \\ I_2 \end{bmatrix}, \tag{7.27}$$

where the $ABCD$ matrix is the product of the $ABCD$ matrices for each layer.

$$\begin{bmatrix} A & B \\ C & D \end{bmatrix} = \begin{bmatrix} A_1 & B_1 \\ C_1 & D_1 \end{bmatrix} \begin{bmatrix} A_2 & B_2 \\ C_2 & D_2 \end{bmatrix}. \tag{7.28}$$

The matrix elements are given by

$$
\begin{aligned}
A_i &= D_i = \cos q_i d_i, \\
B_i &= j Z_i \sin q_i d_i, \\
C_i &= \frac{j \sin q_i d_i}{Z_i},
\end{aligned}
\tag{7.29}
$$

where $i = 1, 2$ represents the first and second layers, and

$$q_i^2 + \beta^2 = k_0^2 n_i^2,$$

$$Z_i = \begin{cases} \dfrac{\omega \mu}{q_i} & \text{for perpendicular polarization } E_y \neq 0, \\[2ex] \dfrac{q_i}{\omega \varepsilon} & \text{for parallel polarization } H_y \neq 0. \end{cases}$$

For a reciprocal network, A, B, C, and D satisfy the condition

$$AD - BC = 1. \tag{7.30}$$

The constant β is the phase constant along the x direction (Fig. 7.9) and is constant for all layers according to Snell's law.

$$\beta = k_0 n_1 \sin \theta_1 = k_0 n_2 \sin \theta_2, \tag{7.31}$$

where θ_1 and θ_2 are the angles between the direction of the wave in each layer and the z direction.

To find the propagation constant q in the z direction, we combine (7.26) and (7.27) and obtain the following eigenvalue equation:

$$\begin{bmatrix} A & B \\ C & D \end{bmatrix} \begin{bmatrix} V_2 \\ I_2 \end{bmatrix} = \lambda \begin{bmatrix} V_2 \\ I_2 \end{bmatrix}, \tag{7.32}$$

where $\lambda = \exp(jqd)$ is the eigenvalue. The eigenvalue is obtained by solving

$$\begin{vmatrix} A - \lambda & B \\ C & D - \lambda \end{vmatrix} = 0. \tag{7.33}$$

We get

$$qd = -j\ln\left[\frac{A+D}{2} \pm i\sqrt{1 - \left(\frac{A+D}{2}\right)^2}\right] \tag{7.34}$$

$$= \cos^{-1}\frac{A+D}{2}.$$

The two eigenvalues for λ correspond to $\exp(\pm jqd)$ for positive- and negative-going waves.

Equation (7.34) gives the fundamental expression for the propagation constant q. We note that if $|(A + D)/2| < 1$, q is real, corresponding to the propagating wave, and the frequency range for this condition is called the pass band. If $|(A + D)/2| > 1$, then $q = m\pi + j$ (real) corresponding to the evanescent wave, and the frequency range under this condition is called the stop band. The frequency dependence of the propagation constant q has the general shape shown in Fig. 7.11. Note that $qd = \pi$ corresponds to the band edge $|(A + D)/2| = 1$. Periodic dielectric media have been used in distributed feedback lasers and distributed Bragg reflection lasers.

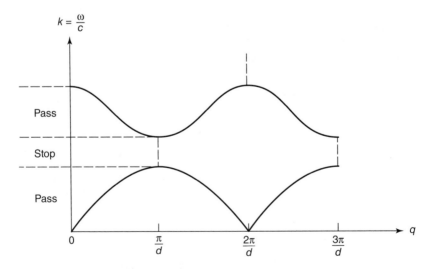

FIGURE 7.11   Pass bands and stop bands.

7.4 PLANE WAVE INCIDENCE ON A PERIODIC STRUCTURE

The reflection and transmission of a plane wave incident on periodic structures are of great importance in many areas of engineering and physics. Examples are microwave mesh reflectors, optical gratings, and crystal structures. In this section, we outline the basic approaches to this problem.

Let a periodic structure be located at $z = 0$ and a plane wave incident on this surface from the direction defined by (θ_p, ϕ_p) (see Fig. 7.12). The incident wave, whether it is an electric field, magnetic field, or Hertz potential, can be written as

$$U_i = A_i e^{-jk_x x - jk_y y + jk_z z}, \tag{7.35}$$

where $k_x = k \sin \phi_p \cos \theta_p$, $k_y = k \sin \theta_p \sin \phi_p$, and $k_z = k \cos \theta_p$.

Now the reflected wave U_r should be written in a series of space harmonics in the x and y directions. Thus at $z = 0$, we write

$$U_r = \sum_{m=-\infty}^{\infty} \sum_{n=-\infty}^{\infty} B_{mn} e^{-jk_{xm} x - jk_{yn} y}, \tag{7.36}$$

where

$$k_{xm} = k_x + \frac{2m\pi}{L_x} \quad \text{and} \quad k_{yn} = k_y + \frac{2n\pi}{L_y}.$$

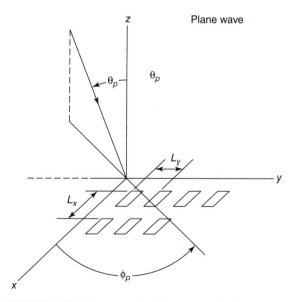

FIGURE 7.12 Plane wave incident on a periodic structure.

Considering that U satisfies the wave equation, we write the reflected wave

$$U_r(x, y, z) = \sum_{m=-\infty}^{\infty} \sum_{n=-\infty}^{\infty} B_{mn} \exp\left[-jk_{xm}x - jk_{yn}y - j\left(k^2 - k_{xm}^2 - k_{yn}^2\right)^{1/2} z\right].$$

$$(7.37)$$

The determination of B_{mn} can be made by applying the boundary conditions. Let us note that the propagation constant for each mode in the z direction $\beta_{mn} = (k^2 - k_{xm}^2 - k_{yn}^2)^{1/2}$ can be real or purely imaginary depending on the incident direction (θ_p, ϕ_p) and m and n. If β_{mn} is real, the wave propagates away from the surface carrying real power and is called the *grating mode*. If β_{mn} is purely imaginary, the wave does not carry real power away from the surface and is *evanescent*.

As an example, let us consider a wave incident on a periodic conducting grating as shown in Fig. 7.13. We assume that the plane of incidence is in the x–z plane and all the gratings are parallel to the y axis, and therefore this is a two-dimensional problem. For a TE wave, $E_x = E_z = 0$ and E_y satisfies the wave equation and Dirichlet's boundary condition ($E_y = 0$) on the conducting tapes. For a TM wave, $H_x = H_z = 0$ and H_y satisfies the wave equation and Neumann's boundary condition on the conducting tapes.

Let us consider the TM case. The magnetic field H_y in the region $z > 0$ consists of the incident wave H_{yi}, the reflected wave H_{yr} when the aperture is completely closed, and the scattered wave H_{ys} generated by the field in the aperture.

$$H_{yi} = A_0 e^{+jqz-j\beta x},$$
$$H_{yr} = A_0 e^{-jqz-j\beta x},$$
$$H_{ys1} = \sum_{n=-\infty}^{\infty} B_n e^{-jq_n z - j\beta_n x},$$

$$(7.38)$$

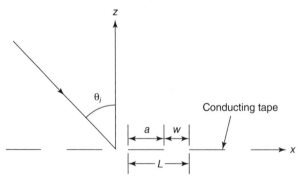

FIGURE 7.13   Plane wave incident on grating.

where

$$\beta = k \sin \theta_i, \qquad q = k \cos \theta_i,$$

$$\beta_n = \beta + \frac{2n\pi}{L}, \quad q_n^2 + \beta_n^2 = k^2.$$

Note that $E_x = (j/\omega\varepsilon)(\partial/\partial z)H_y$, and therefore $E_{xi} + E_{xr} = 0$ at $z = 0$.

The field component E_x at $z = 0$ is equal to the unknown function $f(x)$ in the aperture $|x| \leq (w/2)$.

$$E_x(x, z = 0) = \begin{cases} 0 & \text{for } \dfrac{w}{2} < |x| < \dfrac{L}{2}, \\ f(x) & \text{for } |x| < \dfrac{w}{2}. \end{cases} \tag{7.39}$$

Following the procedure in Section 7.2, we get

$$H_{ys1}(x, z) = j\omega\varepsilon \int_{-w/2}^{w/2} f(x')G_1(x, z; x', 0) \, dx' \quad \text{for } z > 0, \tag{7.40}$$

where

$$G_1(x, z; x', 0) = \sum_{n=-\infty}^{\infty} \frac{e^{-jq_n z - j\beta_n(x-x')}}{jq_n L}.$$

Similarly, for $z < 0$, we get

$$H_{ys2}(x, z) = j\omega\varepsilon \int_{-w/2}^{w/2} f(x')G_2(x, z; x', 0) \, dx' \quad \text{for } z < 0, \tag{7.41}$$

where

$$G_2(x, z; x', 0) = \sum_{n=-\infty}^{\infty} \frac{e^{+jq_n z - j\beta_n(x-x')}}{jq_n L}.$$

Now let us consider the boundary conditions at $z = 0$. The continuity of the tangential electric field E_x is already satisfied as we used the same function $f(x')$. The continuity of the tangential magnetic field requires that

$$H_{yi} + H_{yr} + H_{ys1} = H_{ys2} \quad \text{at } z = 0. \tag{7.42}$$

Substituting (7.38), (7.40), and (7.41) in (7.42), we get

$$A_0 = -j\omega\varepsilon \int_{-w/2}^{w/2} f(x')G(x, x') \, dx', \tag{7.43}$$

where

$$G(x,x') = \sum_{n=-\infty}^{\infty} \frac{e^{-j\beta_n(x-x')}}{jq_n L}.$$

An approximate solution of (7.43) may be obtained by assuming that

$$f(x) = \frac{Ce^{-j\beta x}}{[(w/2)^2 - x^2]^{1/2}}. \tag{7.44}$$

Here we included the edge condition for E_z and C is constant. We then multiply both sides of (7.43) by $f*(x)$ and integrate over the aperture.

$$A_0 \int_{-w/2}^{w/2} f^*(x)\, dx = -j\omega\varepsilon \int_{-w/2}^{w/2} dx \int_{-w/2}^{w/2} dx' f^*(x) f(x') G(x,x').$$

We then obtain

$$C = A_0 \left[-j\omega\varepsilon \sum_{n=-\infty}^{\infty} \frac{\pi}{jq_n L} J_0^2 \left(\frac{n\pi w}{L} \right) \right]^{-1}. \tag{7.45}$$

Once we obtain the solution $f(x)$, we use (7.40) and (7.41) to calculate the scattered field. For example, in the region $z > 0$, we get H_{ys1} from (7.40) using the approximate solution (7.45).

Note that far from the surface, all the evanescent modes are negligibly small and only the propagating modes exist. For the propagating modes, q_n is real, and therefore $k > |\beta_n|$. This is the forbidden (radiation) region shown in Fig. 7.7.

Let us consider the conservation of power. The incident power is in the $-z$ direction and is obtained from (7.38)

$$\bar{P}_i = \text{Re}\left(\frac{1}{2} E_{xi} H_{yi}^* \right) \hat{z} = -P_i \hat{z},$$

$$P_i = \frac{1}{2\omega\varepsilon} |A_0|^2 q. \tag{7.46}$$

The scattered wave consists of H_{yr} and H_{ys1} in (7.38). The scattered power per unit period L is given by

$$\bar{P}_s = P_s \hat{z},$$

$$P_s = \frac{1}{L} \int_0^L \text{Re}\left(\frac{1}{2} E_{xs} H_{ys}^* \right) dx, \tag{7.47}$$

where

$$H_{ys} = H_{yr} + H_{ys1} \quad \text{and} \quad E_{xs} = \frac{1}{\omega\varepsilon}\frac{\partial}{\partial z}H_{ys}.$$

Substituting (7.38) into (7.47) and noting that all space harmonics are orthogonal and that $E_{xs}H_{ys}^*$ for all evanescent modes are purely imaginary, we get

$$P_s = \frac{1}{2\omega\varepsilon}\left(q|A_0 + B_0|^2 + \sum_{\substack{n=N_1 \\ n\neq 0}}^{N_2} q_n|B_n|^2\right), \qquad (7.48)$$

where $N_1 \le n \le N_2$ includes all the propagating modes (q_n is real).

The transmitted power is similarly given by

$$\bar{P}_t = \frac{1}{L}\int_0^L \text{Re}\left(\frac{1}{2}E_{xt}H_{yt}^*\right)\hat{z}\,dx = -P_t\hat{z},$$

$$P_t = \frac{1}{2\omega\varepsilon}\left(\sum_{N_1}^{N_2} q_n|C_n|^2\right), \qquad (7.49)$$

where

$$H_{ys2} = \sum_{n=-\infty}^{\infty} C_n e^{+jq_n z - j\beta_n x}.$$

The conservation of power is then given by

$$P_i = P_s + P_t. \qquad (7.50)$$

Let us next consider a TE wave. The electric field E_y in the region $z > 0$ consists of the incident wave E_{yi}, the wave E_{yr} reflected from the conducting plane at $z = 0$, and the wave E_{ys1} produced by the aperture field at $z = 0$.

$$
\begin{aligned}
E_y &= E_{yi} + E_{yr} + E_{ys1}, \\
E_{yi} &= A_0 e^{+jqz - j\beta x}, \\
E_{yr} &= -A_0 e^{-jqz - j\beta x}, \\
E_{ys1} &= \sum_{n=-\infty}^{\infty} B_n e^{-jq_n z - j\beta_n x},
\end{aligned}
\qquad (7.51)
$$

where

$$q = k \cos \theta_i, \qquad \beta = k \sin \theta_i,$$

$$\beta_n = \beta + \frac{2n\pi}{L}, \qquad q_n^2 + \beta_n^2 = k^2.$$

Now we let the aperture field be $f(x)$.

$$E_y(x, z = 0) = \begin{cases} 0 & \text{for } \frac{w}{2} < |x| < \frac{L}{2}, \\ f(x) & \text{for } |x| < \frac{w}{2}. \end{cases} \tag{7.52}$$

We can then express B_n in terms of $f(x)$.

$$B_n = \frac{1}{L} \int_{-w/2}^{w/2} f(x') e^{+j\beta_n x'} \, dx'. \tag{7.53}$$

The magnetic field component H_{x1} is given by

$$H_{x1}(x, z) = \frac{1}{j\omega\mu} \frac{\partial}{\partial z} E_y = H_{x0}(x, z) + \frac{1}{j\omega\mu} \int_{-w/2}^{w/2} Kf(x') \, dx', \tag{7.54}$$

where

$$H_{x0}(x, z) = H_{xi}(x, z) + H_{xr}(x, z)$$

$$= \frac{1}{j\omega\mu} 2jqA_0 \cos qz e^{-j\beta x},$$

$$K = K(x, z; x') = \sum_{n=-\infty}^{\infty} \frac{-jq_n}{L} e^{-jq_n z - j\beta_n(x-x')}.$$

Similarly, in the region $z < 0$, we have

$$E_y = E_{s2} = \sum_{n=-\infty}^{\infty} C_n e^{+jq_n z - j\beta_n x},$$

where using (7.52) and (7.53), $B_n = C_n$. The magnetic field is then given by

$$H_{x2}(x, z) = \frac{-1}{j\omega\mu} \int_{-w/2}^{w/2} Kf(x') \, dx'. \tag{7.55}$$

Equating H_{x1} to H_{x2} in the aperture at $z = 0$, we get

$$\int_{-w/2}^{w/2} K(x; x')f(x')\, dx' = -2qA_0 e^{-j\beta x}, \tag{7.56}$$

where

$$K(x; x') = \sum_{n=-\infty}^{\infty} \frac{-jq_n}{L} e^{-j\beta_n(x-x')}.$$

As an approximate solution to (7.56), we let $f(x') = C[(w/2)^2 - x'^2]^{1/2}\, e^{-j\beta x}$ using the edge condition (see Appendix 7.C) and integrate over the aperture. We then get

$$\int_{-w/2}^{w/2} dx \int_{-w/2}^{w/2} dx'\, K(x, x')f(x)f(x') = -jqA_0 \int_{-w/2}^{w/2} f(x)e^{-j\beta x}\, dx. \tag{7.57}$$

From this we get C (see Gradshteyn and Ryzhik, 1965, p. 482).

$$C = qA_0 \left(\frac{\pi}{2}\right) \left\{ \sum_{n=-\infty}^{\infty} \frac{q_n}{L} \left[\frac{\pi J_1(n\pi w/L)}{n\pi w/L} \right]^2 \right\}^{-1}. \tag{7.58}$$

The conservation of power is satisfied by

$$P_i = P_s + P_t, \tag{7.59}$$

where P_i, P_s, and P_t are given by the same equations as (7.46), (7.48), and (7.49) except that $1/2\omega\varepsilon$ is replaced by $1/2\omega\mu$.

7.5 SCATTERING FROM PERIODIC SURFACES BASED ON THE RAYLEIGH HYPOTHESIS

Let us consider a wave scattered by a sinusoidally varying surface illuminated by a plane wave (Fig. 7.14). We will consider a two-dimensional problem where the surface does not vary in the y direction and the plane of incidence is in the x–z plane. The surface is given by $z = \zeta$,

$$\zeta = -h \cos \frac{2\pi x}{L}. \tag{7.60}$$

This problem was first solved by Rayleigh using what is now called the Rayleigh hypothesis.

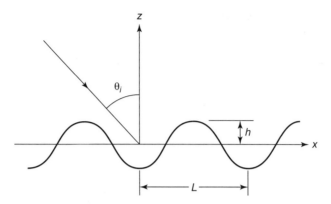

FIGURE 7.14   Sinusoidally varying surface illumimated by a plane wave.

7.5.1 The Dirichlet Problem

$$(\nabla^2 + k^2)\Psi = 0, \quad \Psi = 0 \text{ on surface.} \tag{7.61}$$

The incident wave ψ_i can be written as

$$\Psi_i = A_0\, e^{+jqz - j\beta x}, \tag{7.62}$$

where $q = k \cos\theta_i$ and $\beta = k \sin\theta_i$. The scattered wave ψ_s can be expressed in terms of the space harmonics.

$$\Psi_s = \sum_{n=-\infty}^{\infty} B_n e^{-jq_n z - j\beta_n x}, \tag{7.63}$$

where $\beta_n = \beta + 2n\,\pi/L$ and $q_n^2 = k^2 - \beta_n^2$. Note that in (7.63), the scattered field ψ_s is expressed in terms of the outgoing wave with $\exp(-jq_n z)$ only and the incoming wave with $\exp(+jq_n z)$ is not included. Whereas the wave is outgoing in the region above the highest point of the surface ($z > h$), the wave in the region $-h < z < h$ should consist of both the outgoing and incoming waves. The approach that the scattered wave can be represented by the outgoing wave only, even in the region $-h < z < h$, is called the *Rayleigh hypothesis*. Although the Rayleigh hypothesis is not valid for a general periodic surface, it has been shown that it is valid for the sinusoidal surface if the maximum slope of the surface $(\partial\zeta/\partial x)_{\max} = (2\pi h)/L$ is less than 0.448.

Assuming the Rayleigh hypothesis, we apply the boundary condition that $\psi_i + \psi_s = 0$ on the surface ($z = \zeta$). Noting that $\exp(-j\beta x)$ is common to all terms, we get

$$A_0 \exp(+jq\zeta) + \sum_{n=-\infty}^{\infty} B_n \exp\left(-jq_n\zeta - j\frac{2n\pi}{L}x\right) = 0. \tag{7.64}$$

The relationships among B_n and A_0 are obtained by expanding (7.64) in Fourier series and equating all Fourier coefficients to zero. To do this we multiply (7.64) by $\exp(j2m\pi x/L)$ and integrate the result over L. Noting that

$$\frac{1}{2\pi} \int_0^{2\pi} e^{jz \cos \beta + jn\beta - jn(\pi/2)} d\beta = J_n(z),$$ (7.65)

we get, for $m = -\infty \cdots +\infty$,

$$A_0 e^{-j(\pi/2)|m|} J_{|m|}(qh) + \sum_{n=-\infty}^{\infty} B_n e^{-j(\pi/2)|m-n|} J_{|m-n|}(q_n h) = 0.$$ (7.66)

This may be arranged in the following matrix form:

$$[K_{mn}][B_n] = [A_m]A_0,$$ (7.67)

where

$$K_{mn} = e^{-j(\pi/2)|m-n|} J_{|m-n|}(q_n h),$$
$$A_m = -e^{-j(\pi/2)|m|} J_{|m|}(qh).$$

It can be truncated and $[B_n]$ may be obtained by $[K]^{-1}[A]A_0$. For example, if the height is small compared with a wavelength and the period L is much greater than a wavelength, $qh \ll 1$ and $|q_n h| \gg 1$ for propagating modes, and therefore, $J_{m-n}(q_n h)$ is of the order of $|q_n h|^{m-n}$ and the matrix can be truncated without too much error. The conservation of power should be a good check for the convergence of the solution.

7.5.2 Neumann Problem

For the Neumann problem, we write the incident and the scattered wave as

$$\psi_i = A_0 e^{-jqz - j\beta x},$$
$$\psi_s = \sum_{n=-\infty}^{\infty} B_n e^{-jq_n z - j\beta_n x},$$ (7.68)

where β, q, β_n, and q_n are as defined in Section 7.4. Now we impose the boundary condition

$$\frac{\partial}{\partial n}(\psi_i + \psi_s) = \hat{n} \cdot \nabla(\psi_i + \psi_s) = 0 \quad \text{at } z = \zeta,$$ (7.69)

where $\hat{n}$ is the unit vector normal to the surface and is given by

$$\hat{n} = \frac{-(\partial\zeta/\partial x)\hat{x} - (\partial\zeta/\partial y)\hat{y} + \hat{z}}{[1 + (\partial\zeta/\partial x)^2 + (\partial\zeta/\partial y)^2]^{1/2}}. \tag{7.70}$$

Substituting (7.68) into (7.69), we get

$$-\frac{\partial\zeta}{\partial x}\left[-j\beta A_0 e^{jq\zeta} + \sum_{n=-\infty}^{\infty}(-j\beta_n)B_n e^{-jq_n\zeta - j(2n\pi x/L)}\right]$$

$$+ \left[jqA_0 e^{jq\zeta} + \sum_{n=-\infty}^{\infty}(-jq_n)B_n e^{-jq_n\zeta - j(2n\pi x/L)}\right] = 0. \tag{7.71}$$

Noting that $\zeta = -h\cos(2\pi x/L)$ and $\partial\zeta/\partial x = (2\pi h/L)\sin(2\pi x/L)$, we can multiply (7.71) by $\exp[j(2m\pi x/L)]$ and integrate it with respect to x over L. We then get the following matrix equation:

$$[H_{mn}][B_n] = [D_m]A_0, \tag{7.72}$$

where

$$H_{mn} = \frac{\pi h \beta_n}{L}\left[e^{-j(\pi/2)|m-n+1|}J_{|m-n+1|}(q_n h) - e^{-j(\pi/2)|m-n-1|}J_{|m-n-1|}(q_n h)\right]$$

$$-jq_n e^{-j(\pi/2)|m-n|}J_{|m-n|}(q_n h),$$

$$D_m = -\frac{\pi h \beta}{L}\left[e^{-j(\pi/2)|m+1|}J_{|m+1|}(qh) - e^{-j(\pi/2)|m-1|}J_{|m-1|}(qh)\right]$$

$$-jq e^{-j(\pi/2)|m|}J_{|m|}(qh).$$

7.5.3 Two-Media Problem

Earlier in this section, we discussed Dirichlet's and Neumann's problems using the Rayleigh hypothesis. Similar techniques can be used to solve the two-media problem. We write the incident ψ_i and the scattered wave ψ_s in medium 1, and the transmitted wave ψ_t in medium 2 as

$$\psi_i = A_0 e^{+jqz - j\beta x},$$

$$\psi_s = \sum_{n=-\infty}^{\infty}B_n e^{-jq_n z - j\beta_n x}, \tag{7.73}$$

$$\psi_t = \sum_{n=-\infty}^{\infty}C_n e^{+jq_{tn}z - j\beta_n x},$$

where

$$\beta = k_i \sin \theta_i, \quad q = k_i \cos \theta_i,$$

$$\beta_n = \beta + \frac{2n\pi}{L},$$

$$q_n = \left(k_1^2 - \beta_n^2\right)^{1/2},$$

$$q_{tn} = \left(k_2^2 - \beta_n^2\right)^{1/2},$$

$$k_1 = \frac{\omega}{c} n_1, \quad k_2 = \frac{\omega}{c} n_2.$$

The boundary conditions are

$$\rho_1 \psi_1 = \rho_2 \psi_2 \quad \text{and} \quad \frac{\partial \psi_1}{\partial n} = \frac{\partial \psi_2}{\partial n} \quad \text{on the surface.}$$

Following the procedure for Dirichlet's and Neumann's problems, we have

$$[K_{mn}][B_m] - [K_{tmn}][C_n] = [A_m]A_0,$$
$$[H_{mn}][B_m] - [H_{tmn}][C_n] = [D_m]A_0, \tag{7.74}$$

where $[K_{mn}]$, $[H_{mn}]$, $[A_m]$, and $[D_m]$ are already given. $[K_{tmn}]$ and $[H_{tmn}]$ have the same form as $[K_{mn}]$ and $[H_{mn}]$, respectively, except that all q_n are replaced by $-q_{tn}$.

The power conservation is checked as follows: for Dirichlet's and Neumann's problems, we should have

$$q|A_0|^2 = \sum_{n=N_1}^{N_2} q_n |B_n|^2, \tag{7.75}$$

where $N_1 \leq N \leq N_2$ includes all the propagating modes. For two-media problems, we should have

$$\rho_1 q|A_0|^2 = \rho_1 \sum_{n=N_1}^{N_2} q_n |B_n|^2 + \rho_2 \sum_{n=N_3}^{N_4} q_{tn} |C_n|^2. \tag{7.76}$$

In this section, we discussed the scattering from a sinusoidal surface using the Rayleigh hypothesis. This is valid if $(2\pi h/L) < 0.448$. If the surface slope is higher than this or if the surface has a more general shape than a sinusoidal surface, the Rayleigh hypothesis is, in general, not valid, and a more rigorous method should be used. One such technique is the T-matrix method, which will be discussed later.

We note here that each mode of the scattered wave, $B_n \exp(-jq_n z - j\beta_n x)$, propagates in the direction with the angle θ_n from the z axis given by

$$k \sin \theta_n = \beta_n = \beta + \frac{2n\pi}{L} = k \sin \theta_i + \frac{2n\pi}{L}. \tag{7.77}$$

If this angle θ_n becomes close to $\pm\pi/2$, the scattered mode propagates along the surface and a rapid redistribution of the power in all modes takes place within a small variation of the wavelength or the angle. This effect, known as the *Wood anomalies*, has been studied extensively. The condition for the Wood anomalies is, therefore,

$$\lambda = \frac{L}{n}(\pm 1 - \sin \theta_i).$$ (7.78)

This wavelength is sometimes called the *Rayleigh wavelength*.

7.6 COUPLED-MODE THEORY

Consider two wave guiding structures, such as two waveguides, two optical fibers, or two strip lines. The wave in each guide propagates with a definite propagation constant. Suppose that these two guides are close to each other, so that some coupling of the power takes place. For example, for waveguides, these may be a series of holes or slits between two guides as in the case of directional couplers. For two strip lines, the two lines are closely located and the power of the two lines may be coupled. If the coupling is weak, the coupling perturbs the original guide modes slightly and transfers the power from one guide to the other. The coupled-mode theory described in this section provides a mathematical formulation of this coupling process. It is clear from the above that the coupled-mode theory is applicable only to a weakly coupled system. For a system with strong coupling, the improved theory should be used (Hardy and Streifer, 1986; Tsang and Chuang, 1988).

Consider two modes a_1 and a_2 representing the waves propagating in the two guides, respectively (Fig. 7.15). If two guides are isolated, each mode is assumed to propagate with the propagation constant β_{10} and β_{20}. Therefore, a_1 and a_2 in the isolated guides satisfy the following:

$$\frac{da_1}{dz} = -j\beta_{10}a_1, \quad \frac{da_2}{dz} = -j\beta_{20}a_2.$$ (7.79)

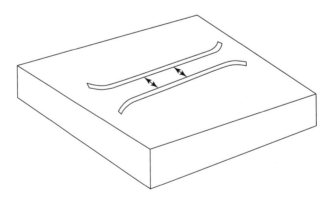

FIGURE 7.15   Coupling between two strip lines.

If these two guides are coupled, the coupling terms between a_1 and a_2 should be included in (7.79). We thus write the following coupled equations:

$$\frac{da_1}{dz} = -j\beta_{10}a_1 - jc_{12}a_2,$$
$$\frac{da_2}{dz} = -jc_{21}a_1 - j\beta_{20}a_2. \tag{7.80}$$

The constants c_{12} and c_{21} are the mutual coupling coefficients per unit length. Note that β_{10} and β_{20} are real in a lossless system and that if $\beta_{10} > 0$ and $\beta_{20} > 0$, the phase velocities for the modes a_1 and a_2 are in the positive z direction, and if $\beta_{10} < 0$, and $\beta_{20} < 0$, the phase velocities are in the negative z direction.

Let us next consider the power P_1 carried by the mode a_1. We get

$$\frac{dP_1}{dz} = \frac{d}{dz}\left(a_1 a_1^*\right) = a_1\frac{da_1^*}{dz} + \frac{da_1}{dz}a_1^*. \tag{7.81}$$

Substituting (7.80), we get

$$\frac{dP_1}{dz} = -ja_1\left(\beta_{10} - \beta_{10}^*\right)a_1^* + 2\,\mathrm{Re}\left(ja_1 c_{12}^* a_2^*\right). \tag{7.82}$$

Similarly for $P_2 = |a_2|^2$, we get

$$\frac{dP_2}{dz} = -ja_2\left(\beta_{20} - \beta_{20}^*\right)a_2^* + 2\,\mathrm{Re}\left(-ja_1 c_{21} a_2^*\right), \tag{7.83}$$

where Re denotes "real part of."

If the system is lossless, β_{10} and β_{20} are real, and therefore $\beta_{10} - \beta_{10}^* = 0$ and $\beta_{20} - \beta_{20}^* = 0$. If both powers P_1 and P_2 are propagated in the same direction, and thus the group velocities for a_1 and a_2 are in the same direction, the conservation of power requires that

$$\frac{d}{dz}(P_1 + P_2) = 0. \tag{7.84}$$

Using (7.82) and (7.83), this means that for a lossless system,

$$c_{12} = c_{21}^*. \tag{7.85}$$

This is called a *codirectional coupler*. If P_1 and P_2 are propagated in the opposite direction, and thus the group velocities are in the opposite directions, we require that

$$\frac{d}{dz}(P_1 - P_2) = 0, \tag{7.86}$$

and therefore, for a lossless system,

$$c_{12} = -c_{21}^*. \tag{7.87}$$

This is called a *contradirectional coupler*.

7.6.1 Codirectional Coupler

Consider a lossless system consisting of two guides weakly coupled to each other. We assume that the phase velocities and the group velocities of waves a_1 and a_2 are both in the positive z direction. Thus we have

$$\beta_{10} > 0, \quad \beta_{20} > 0, \quad c_{12} = c_{21}^*. \tag{7.88}$$

To solve the coupled equation (7.80), we let

$$\begin{aligned} a_1(z) &= A_1 \exp(-j\beta z), \\ a_2(z) &= A_2 \exp(-j\beta z). \end{aligned} \tag{7.89}$$

Substituting this into (7.80), we get the following eigenvalue problem:

$$\begin{bmatrix} \beta_{10} & c_{12} \\ c_{21} & \beta_{20} \end{bmatrix} \begin{bmatrix} A_1 \\ A_2 \end{bmatrix} = \beta \begin{bmatrix} A_1 \\ A_2 \end{bmatrix}, \tag{7.90}$$

where the propagation constant β is the eigenvalue and $[A_1, A_2]$ is the eigenvector.

The propagation constant β can be obtained by equating the determinant of the matrix in (7.90) to zero,

$$\begin{vmatrix} \beta_{10} - \beta & c_{12} \\ c_{21} & \beta_{20} - \beta \end{vmatrix} = 0. \tag{7.91}$$

From this we get two values for β. We arrange these two propagation constants β_1 and β_2 in the following form:

$$\begin{aligned} \beta_1 &= \beta_a + \beta_b, \\ \beta_2 &= \beta_a - \beta_b, \end{aligned} \tag{7.92}$$

where

$$\begin{aligned} \beta_a &= \frac{1}{2}(\beta_{10} + \beta_{20}), \\ \beta_b &= \left(\beta_d^2 + c_{12}c_{21}\right)^{1/2}, \\ \beta_d &= \frac{1}{2}(\beta_{10} - \beta_{20}). \end{aligned}$$

Since the two powers are propagating in the same direction, this is a codirectional coupler, and therefore $c_{12}c_{21} = |c_{12}|^2$ and β_a and β_b are both real. When β_a and β_b are real, the two modes are said to be *passively* coupled.

The eigenvectors $[A_1, A_2]$ for β_1 and β_2 are obtained from (7.90):

$$\frac{A_2}{A_1} = \frac{\beta - \beta_{10}}{c_{12}} = \frac{c_{21}}{\beta - \beta_{20}}. \tag{7.93}$$

The general solutions for a_1 and a_2 are then given by

$$a_1(z) = C_1 e^{-j\beta_1 z} + C_2 e^{-j\beta_2 z},$$
$$a_2(z) = C_1 \frac{\beta_1 - \beta_{10}}{c_{12}} e^{-j\beta_1 z} + C_2 \frac{\beta_2 - \beta_{10}}{c_{12}} e^{-j\beta_2 z}, \tag{7.94}$$

where C_1 and C_2 are constant.

The constants C_1 and C_2 are determined by the boundary conditions. Suppose that the wave is incident in guide 1 at $z = 0$ and no wave is incident in guide 2 at $z = 0$. Thus we have

$$a_1(0) = a_0 \quad \text{and} \quad a_2(0) = 0. \tag{7.95}$$

Using this, we get

$$a_1(z) = a_0 \left(\cos \beta_b z - j \frac{\beta_d}{\beta_b} \sin \beta_b z \right) \exp(-j\beta_a z),$$
$$a_2(z) = -ja_0 \frac{c_{21}}{\beta_b} \sin \beta_b z \exp(-j\beta_a z). \tag{7.96}$$

The power in each guide is given by

$$P_1(z) = |a_1(z)|^2 \quad \text{and} \quad P_2(z) = |a_2(z)|^2. \tag{7.97}$$

Substituting (7.96) into (7.97), it is easily verified that $P_1(z) + P_2(z) = \text{constant}$ as expected. The power is then periodically transferred between two guides (Fig. 7.16).

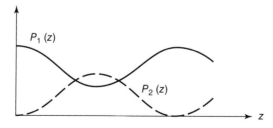

FIGURE 7.16   Periodic transfer of power for a codirectional coupler.

Note that the maximum transferred power $P_2(z)$ is

$$P_{2\,\text{max}} = \frac{|C_{12}|^2}{|\beta_d|^2 + |C_{12}|^2} |a_0|^2. \tag{7.98}$$

Thus if $\beta_{10} = \beta_{20}$, the power transfer is 100%.

7.6.2 Contradirectional Coupler

Consider a lossless contradirectional coupler. In this case, the phase velocities for both modes are in the same direction, but the group velocities are in the opposite direction. Thus we have

$$\beta_{10} > 0, \quad \beta_{20} > 0, \quad c_{12} = -c_{21}^*. \tag{7.99}$$

Following the procedure described in Section 7.6, we get

$$\begin{aligned}
\beta_1 &= \beta_a + \beta_b, \\
\beta_2 &= \beta_a - \beta_b,
\end{aligned} \tag{7.100}$$

where

$$\begin{aligned}
\beta_a &= \tfrac{1}{2}(\beta_{10} + \beta_{20}), \\
\beta_b &= \left[\beta_d^2 - |c_{12}|^2\right]^{1/2}, \\
\beta_d &= \tfrac{1}{2}(\beta_{10} - \beta_{20}).
\end{aligned}$$

Note that here we used $c_{12}c_{21} = -|c_{12}|^2$.

The expression for β_b indicates that if $|\beta_d| < |c_{12}|$, then β_b is purely imaginary and the wave will be exponentially growing or decaying. When β_1 and β_2 are complex, the two modes are said to be "actively" coupled.

If the power is injected to guide 1 at $z = 0$, this power is coupled into guide 2 and propagates in the negative z direction. If no power is injected into guide 2 at $z = L$, we have the boundary condition

$$\begin{aligned}
a_1(0) &= a_0, \\
a_2(L) &= 0.
\end{aligned} \tag{7.101}$$

The solution can be obtained following the procedure given for the codirectional case. The powers $P_1(z)$ and $P_2(z)$ are sketched in Fig. 7.17. Note that $P_1(z) - P_2(z)$ is constant.

In Sections 7.2 and 7.3, we discussed lossless coupled systems where there is no net power gain. If the guides contain an active medium, the wave may be amplified,

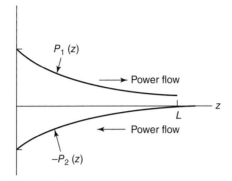

FIGURE 7.17   Contradirectional coupler.

as in the case of coupling between a beam of electrons and a circuit in a traveling-wave tube. The determination of coupling coefficients for dielectric waveguides are discussed in Tamir (1975), Hardy and Streifer (1986), and Tsang and Chuang (1988).

PROBLEMS

7.1   Find an equation to determine the propagation constant of a wave propagating in a waveguide with the wall structure shown in Fig. P7.1. Assume no variation

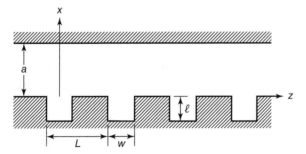

FIGURE P7.1   Periodic waveguide.

in y and that the wave is polarized in the x–z plane. Consider the limit as $l \to 0$ or $w \to 0$.

7.2   If the period L in Problem 7.1 is much smaller than a wavelength, the surface $x = 0$ may be approximated by the average surface impedance Z_s given by

$$\frac{E_z}{H_y} = Z_s = j\frac{W}{L}\sqrt{\frac{\mu_0}{\varepsilon_0}}\tan kl.$$

Find the propagation constant for this case and compare the results with those of Problem 7.1.

7.3 A wave is propagating in the z direction through the periodic structure shown in Fig. P7.3. Plot the k–q diagram in the frequency range 0–10 GHz.

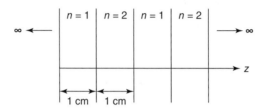

FIGURE P7.3 Periodic layers.

7.4 Consider a TM plane wave normally incident on the periodic grating shown in Fig. 7.13. $a = W = 5$ cm and frequency is 10 GHz. Identify all propagating modes and find the direction of propagation for each mode. Find the expressions for the magnitude of each mode and check the conservation of power.

7.5 Consider the periodic Dirichlet surface given in (7.60). If the wave is normally incident on the surface, and $kh = 0.1$ and $L = 1.5\lambda$, find the amplitudes of the propagating modes and their direction of propagation and check the power conservation.

7.6 In the k–β diagram (Fig. 7.2), let $\beta = k \sin \theta_i$ and locate the points on the diagram that satisfy the Rayleigh wavelength condition.

7.7 Consider the TE_{10} modes in two rectangular waveguides with $a = 1$ in. and $b = \frac{1}{2}$ in. at 10 GHz. If these two modes are weakly coupled and the maximum power transfer from one to the other waveguide is 20 dB over a distance of 20 cm, find the coupling coefficient C_{12}. Assume that C_{12} is real and positive.

7.8 Consider a contradirectional coupler with $\beta_{10} = 1$, $\beta_{20} = 1.1$, and $C_{12} = -C_{21} = 0.1$. Find the eigenvalues and eigenvectors. Calculate and plot $P_1(z)$ and $P_2(z)$ for the boundary conditions $a_1(0) = 1$ and $a_2(5) = 0$.

7.9 Two pendulums of length l_1 and l_2 and mass m_1 and m_2 are coupled by a weightless spring with spring constant k as shown in Fig. P7.9. The

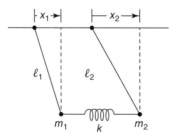

FIGURE P7.9 Coupled pendulums.

amplitudes of the oscillations are assumed to be small and l_1 is not very different from l_2.

(a) Derive the coupled-mode equations for x_1 and x_2.

(b) Find solutions if $x_1 = dx_1/dt = dx_2/dt = 0$ and $x_2 = x_0$ at $t = 0$ when $l_1 = l_2 = 1m$, $m_1 = m_2 = 1$ g, and $k = 10^{-3}$ N/m.

CHAPTER 8

DISPERSION AND ANISOTROPIC MEDIA

In Section 2.3, we discussed the constitutive relations $\bar{D} = \varepsilon\bar{E}$ and $\bar{B} = \mu\bar{H}$. They are valid for a "linear" medium, where $\bar{D}$ and $\bar{B}$ are proportional to $\bar{E}$ and $\bar{H}$, respectively. If $\bar{D}$ or $\bar{B}$ is a more general function of $\bar{E}[\bar{D} = \bar{D}(\bar{E})]$, this is the "nonlinear" medium. For a time-harmonic case, ε and μ are in general functions of frequency $\varepsilon(\omega)$ and $\mu(\omega)$, and this is the dispersive medium. For a nondispersive medium, ε and μ are independent of frequency. If ε and μ are functions of position, this is called the inhomogeneous medium; for a homogeneous medium, ε and μ are constant. In an isotropic medium, ε and μ are scalar and therefore $\bar{D}$ and $\bar{B}$ are proportional to $\bar{E}$ and $\bar{H}$, respectively. In an anisotropic medium, as shown in Section 8.7, $\bar{D}$ and $\bar{E}$, and $\bar{B}$ and $\bar{H}$ are in general not parallel. In a bi-anisotropic medium, $\bar{D}$ depends on both $\bar{E}$ and $\bar{B}$, and $\bar{H}$ depends on both $\bar{E}$ and $\bar{B}$. Chiral medium is an example of a bi-isotropic medium, and these are discussed in Section 8.22. Two-fluid model of superconductors at high frequencies is discussed in Sections 8.23 and 8.24.

8.1 DIELECTRIC MATERIAL AND POLARIZABILITY

In Section 2.3, we discussed the constitutive relations for a medium in terms of the dielectric constant ε, the electric susceptibility χ_e, or the electric polarization $\bar{P}$. They are related by (Eqs. (2.45) and (2.46))

$$\bar{P} = (\varepsilon - \varepsilon_0)\bar{E} = \chi_e\varepsilon_0\bar{E}. \tag{8.1}$$

Electromagnetic Wave Propagation, Radiation, and Scattering: From Fundamentals to Applications,
Second Edition. Akira Ishimaru.
© 2017 by The Institute of Electrical and Electronic Engineers, Inc. Published 2017 by John Wiley & Sons, Inc.

Alternatively, the polarization vector $\bar{P}$ can be viewed as the dipole moments per unit volume of the medium. In this interpretation, we can write

$$\bar{P} = N\bar{p} = N\alpha\bar{E}',\tag{8.2}$$

where N is the number of dipoles per unit volume contributing to $\bar{P}$, and $\bar{p}$ is the moment of each elementary dipole. The dipole moment $\bar{p}$ is, in turn, produced by the local electric field $\bar{E}'$ and α is called the *polarizability*. Note that the local field $\bar{E}'$ is not equal to the applied field $\bar{E}$.

There are four major mechanisms of producing the dipole moment in a material. The *electronic polarization*, expressed by the polarizability α_e, is caused by a slight displacement of electrons surrounding positively charged atomic nuclei under the influence of the field $\bar{E}'$, forming a dipole. The *atomic polarization* α_a is caused by displacement of differently charged atoms with respect to each other. The *dipole polarization* α_d, also called the *orientation polarization*, is caused by the change of orientation of equivalent dipoles in a medium. Polarizations α_e, α_a, and α_d are due to the locally bound charges in the atoms or molecules. The fourth polarization, α_s, is called the *space charge* or *interfacial polarization*. We discuss the dispersion properties of these polarizations in the following sections.

In (8.1) and (8.2), we noted that the external applied field $\bar{E}$ is, in general, different from the local field $\bar{E}'$ that causes the polarization. They are almost identical for low-pressure gases, but are different for solids, liquids, and high-pressure gases. The relationship between $\bar{E}'$ and $\bar{E}$ can be obtained by considering a fictitious sphere surrounding a molecule in the medium (Fig. 8.1). The local field $\bar{E}'$ acting on the molecule at the center of the spherical cavity of radius r_0 is the sum of the applied field $\bar{E}$ and the field $\bar{E}_p$ due to the polarization vector $\bar{P}$ surrounding the cavity.

$$\bar{E}' = \bar{E} + \bar{E}_p.\tag{8.3}$$

The polarization $\bar{P}$ creates equivalent charges on the wall of the sphere, and the charge over the elementary area $d\bar{a}$ is given by $\bar{P} \cdot d\bar{a} = P\cos\theta\,da$, $da = 2\pi r\sin\theta r d\theta$.

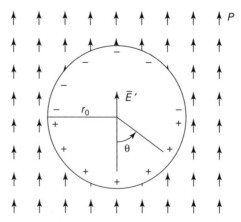

FIGURE 8.1 Local field $\bar{E}'$ and polarization $\bar{P}$.

The field $\bar{E}_p$ is obtained by summing the contributions from the charges $\bar{P} \cdot d\bar{a}$. It is pointed in the z direction and its magnitude is equal to

$$E_p = \int_0^{\cdot} \frac{P \cos^2 \theta}{4\pi\varepsilon_0 r_0^2} 2\pi r_0 \sin\theta r_0 \, d\theta = \frac{P}{3\varepsilon_0}. \tag{8.4}$$

Using this, we get the local field $\bar{E}'$, called the *Mossotti field*, in terms of the applied field $\bar{E}$.

$$\bar{E}' = \bar{E} + \frac{\bar{P}}{3\varepsilon_0} = \frac{\varepsilon_r + 2}{3}\bar{E}, \quad \varepsilon_r = \frac{\varepsilon}{\varepsilon_0}. \tag{8.5}$$

Using (8.1), (8.2), and (8.5), we can express the electric susceptibility χ_e in terms of the polarizability α.

$$\chi_e = \frac{N\alpha/\varepsilon_0}{1 - N\alpha/3\varepsilon_0} \quad \text{or} \quad \frac{\varepsilon}{\varepsilon_0} = \frac{1 + 2N\alpha/3\varepsilon_0}{1 - N\alpha/3\varepsilon_0}. \tag{8.6}$$

Similarly, we can relate the polarizability α to the relative dielectric constant ε_r.

$$\alpha = \frac{3\varepsilon_0}{N} \frac{\varepsilon_r - 1}{\varepsilon_r + 2}. \tag{8.7}$$

This is called the *Clausius–Mossotti formula* or *Lorentz–Lorenz formula*.

8.2 DISPERSION OF DIELECTRIC MATERIAL

The dielectric constant of any material is in general dependent on frequency, and it can be considered constant only within a narrow frequency band. However, if a broadband pulse is propagated through such a medium, the frequency dependence of the medium cannot be ignored. The variation of the dielectric constant with frequency is called *dispersion*. In this section, we discuss some simple examples of dispersive media.

Let us consider the dispersion characteristics of dielectric material. We assume a simplified model of molecules with electrons bound elastically to the heavy nuclei. The equation of motion for an electron is

$$m\frac{d^2\bar{r}}{dt^2} = -m\omega_0^2\bar{r} - mv\frac{d\bar{r}}{dt} + \bar{F}, \tag{8.8}$$

where m is the mass of the electron, $\bar{r}$ is the displacement of the electron, $-m\omega_0^2\bar{r}$ is the elastic restoring force, $-mvd\bar{r}/dt$ is the damping force, v is the collision frequency, and $\bar{F}$ is the Lorentz force acting on the electron. The restoring force is assumed to be proportional to the displacement of the electron, and the constant ω_0 is equal to the frequency of the free oscillations of the electron under the influence of the restoring force alone. The Lorentz force is given by

$$\bar{F} = e(\bar{E}' + \bar{v} \times \bar{B}'),$$ (8.9)

where e is the charge of an electron, $\bar{E}'$ and $\bar{B}'$ the local Mossotti field (8.5), and $\bar{v}$ the velocity of the electron. Since $\bar{B}' = \mu_0 \bar{H}'$ and $|\bar{H}'|$ is of the order of $(\varepsilon_0/\mu_0)^{1/2}|\bar{E}'|$, $|\bar{B}'|$ is of the order of $(1/c)|\bar{E}'|$, and therefore, assuming that $|\bar{v}| \ll c$, the second term of (8.9) is negligible compared with the first term.

Consider a time-harmonic field with $\exp(j\omega t)$. Assume that there are N bound electrons per unit volume. The polarization vector $\bar{P}$ is then given by

$$\bar{P} = Ne\bar{r}.$$ (8.10)

Equation (8.8) for the time-harmonic field is

$$-m\omega^2 \bar{r} = -m\omega_0^2 \bar{r} - j\omega m v \bar{r} + e\left(\bar{E} + \frac{Ne\bar{r}}{3\varepsilon_0}\right).$$ (8.11)

Noting that $\bar{D} = \varepsilon_0 \varepsilon_r$, $\bar{E} = \varepsilon_0 \bar{E} + \bar{P}$, we get the relative dielectric constant ε_r as a function of frequency.

$$\varepsilon_r = 1 + \frac{Ne^2}{m\varepsilon_0(\omega_1^2 - \omega^2 + j\omega v)},$$ (8.12)

where $\omega_1^2 = \omega_0^2 - Ne^2/3\varepsilon_0 m$. Figure 8.2 shows the general shape of ε_r as a function of frequency.

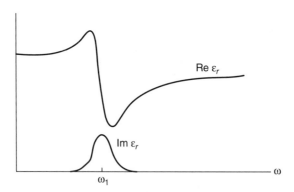

FIGURE 8.2   Dispersion.

In more general cases, there is more than one resonance, and we need to generalize (8.12) to

$$\varepsilon_r = 1 + \sum_s \frac{N_s e^2}{m_s \varepsilon_0 (\omega_s^2 - \omega^2 + j\omega v_s)}. \tag{8.13}$$

For lossless material, we write (8.13) as

$$\begin{aligned}\varepsilon_r &= 1 + \sum_s \frac{N_s e^2}{m_s \varepsilon_0 (\omega_s^2 - \omega^2)} \\ &= 1 + \sum_s \frac{\lambda^2 B_s}{\lambda^2 - \lambda_s^2},\end{aligned} \tag{8.14}$$

where $\omega/c = 2\pi/\lambda$ and B_s are constants to be determined experimentally. Equation (8.14), called the *Sellmeier equation*, is often used in the study of dispersion in optical fibers. For example, the refractive index of fused silica (SiO_2) used for fibers in the wavelength $\lambda = 0.5$–2.0 μm can be given by (8.14) with $\lambda_1 = 0.1$ μm, $B_1 = 1.0955$ and $\lambda_2 = 9$ μm, $B_2 = 0.9$ (Marcuse, 1982, p. 485).

8.3 DISPERSION OF CONDUCTOR AND ISOTROPIC PLASMA

In dielectric material, the resonant frequency ω_1 in (8.12) is nonzero, and at low frequency $\omega \to 0$, ε_r in (8.12) approaches the static dielectric constant. However, in a conductor, there are free electrons that are not bound to molecules, and therefore the restoring force $(-m\omega_0^2 \bar{r})$ in (8.8) is absent. Also, the interaction between the molecules can be neglected, and the local field $\bar{E}'$ is equal to the applied field $\bar{E}$. Equation (8.12) is therefore

$$\varepsilon_r = 1 + \frac{\omega_p^2}{-\omega^2 + j\omega v}, \tag{8.15}$$

where $\omega_p = (Ne^2/m\varepsilon_0)^{1/2}$ is called the plasma frequency. N is the number of free electrons per unit volume and is called the electron density. The damping is caused by the collisions between the electron and other molecules, and v is called the collision frequency. If we compare (8.15) with the expression for a conducting medium (note that $\varepsilon' = 1$, Table 2.1),

$$\varepsilon_r = 1 - j\frac{\sigma}{\omega\varepsilon_0}, \tag{8.16}$$

we get the equivalent conductivity σ

$$\frac{\sigma}{\varepsilon_0} = \frac{\omega_p^2}{v + j\omega}. \tag{8.17}$$

At low frequencies $\omega \ll v$, the conductivity σ is, therefore, almost constant. In general, however, the conductivity σ is a function of frequency.

The dielectric constant of metal in optical wavelengths can be approximately given by (8.15). For example, at $\lambda = 0.6$ µm, silver has the plasma frequency $f_p = 2 \times 10^{15}$ (ultraviolet), the collision frequency $f_v = 5.7 \times 10^{13}$ (infrared), and $\varepsilon_r = -17.2 - j0.498$. If the frequency is increased beyond the plasma frequency, the dielectric constant becomes almost real and positive and the wave can propagate through metal, which is called the *ultraviolet transparency* of metals (Jackson, 1975).

Electromagnetic wave propagation through ionized gas has received considerable attention for many years. In particular, the reflection of radio waves and the transmission from and through the ionosphere have been studied extensively. The ionosphere was postulated as the Kennelly–Heaviside layer in 1902, and the formula for its index of refraction, now known as the Appleton–Hartree formula, was obtained around 1930. Such an ionized gas in which electron and ion densities are substantially the same is electrically neutral and is called the *plasma*. The problem of re-entry of high-speed vehicles such as missiles and rockets has generated considerable interest in plasma problems. When high-speed vehicles enter the atmosphere, high temperature and pressure in front of the vehicle ionize the air molecules and produce the so-called *plasma sheath*. The problems of antenna characteristics, wave propagation through the plasma, and the radar cross section are of considerable importance. Also, the antenna and wave propagation characteristics of artificial satellites in the ionosphere are important in the communication between the vehicle and the earth station.

If a dc magnetic field is present, the plasma becomes anisotropic and this is normally called the *magnetoplasma*. In the absence of dc magnetic fields, the plasma is isotropic and the equivalent dielectric constant is given by (8.15). Thus, the refraction index n depends on the operating frequency ω, the plasma frequency ω_p, and the collision frequency v. The *electron plasma frequency* plays a most important role in magnetic–ionic theory. Substituting the values of m, e, and ε_0, we get

$$f_p = \begin{cases} 8.98 N_e^{1/2} & (N_e \text{ in } \mathrm{m}^{-3}), \\ 8.98 \times 10^3 \, N_e^{1/2} & (N_e \text{ in } \mathrm{cm}^{-3}). \end{cases} \tag{8.18}$$

The propagation constant β for a plane wave propagating in a lossless isotropic plasma is given by

$$\beta = k_0 n = \left(k_0^2 - k_p^2 \right)^{1/2}, \quad k_p = \frac{\omega_p}{c}. \tag{8.19}$$

Mathematically, this is identical to the propagation constant for a hollow waveguide.

$$\beta = \left(k_0^2 - k_c^2 \right)^{1/2}, \quad k_c = \text{cutoff wave number.} \tag{8.20}$$

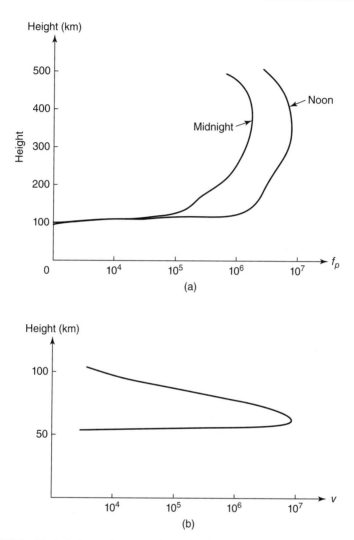

FIGURE 8.3   Typical plasma frequency (a) and collision frequency (b) of the ionosphere.

In a waveguide, if the frequency is above the cutoff frequency, the wave propagates, and if the frequency is below the cutoff frequency, the wave becomes evanescent. In exactly the same manner, the plasma frequency plays the role of the cutoff frequency.

A well-known example of the cutoff phenomenon is the wave propagation through the ionosphere. When the operating frequency is higher than the plasma frequency, radio waves can penetrate through the ionosphere, but at lower frequencies, radio waves are bounced off the ionosphere, thus contributing long-distance radio-wave propagation. Typical characteristics of the lower ionosphere are shown in Fig. 8.3. Some typical values of the electron density are shown in Table 8.1.

TABLE 8.1   Typical Values of Electron Densities

	N_e (cm^{-3})	Temperature (K)
Ionosphere	10^3–3×10^6	300–3000
Interplanetary space	1–10^4	
Solar corona	10^4–3×10^8	10^6
Interstellar space	10^{-3}–10	100–10^4
Thermonuclear reaction	10^{15}	10^6–10^7
Gas discharge device	10^{12}	
In metals	3×10^{22}	

8.4 DEBYE RELAXATION EQUATION AND DIELECTRIC CONSTANT OF WATER

The dielectric constant of water at microwave frequencies is governed primarily by the relaxation phenomenon. Water molecules have permanent dipole moments, and when microwaves are applied, the polar molecules tend to rotate as if they are in a damping frictional medium. In (8.12), this frictional force is represented by v. However, the acceleration term $-m\omega^2 \bar{r}$ in (8.11) may be negligibly small compared with other terms. Thus, the dielectric constant for the medium of polar molecules may be expressed as

$$\varepsilon_r = \varepsilon_\infty + \frac{\varepsilon_s - \varepsilon_\infty}{1 + j\omega\tau}, \tag{8.21}$$

where ε_s is the static dielectric constant as $\omega \to 0$, ε_∞ is the high-frequency limit as $\omega \to \infty$, and τ is the relaxation time. They are functions of the temperature. The Debye formula (Eq. 8.21) is applicable in the frequency range 0.3–300 GHz (Oguchi, 1983; Ray, 1972).

8.5 INTERFACIAL POLARIZATION

In Section 8.1, we discussed three polarization mechanisms: the electronic, atomic, and dipole orientation polarizations. They are caused by the displacement of bound or free electrons or by the change in orientation of the dipole moment of the molecule. In addition to these three, there is another process, called *interfacial polarization* or *space-charge polarization*. This is due to the large-scale field distortions caused by the piling up of space charges in the volume or of the surface charges at the interfaces between different small portions of materials with different characteristics.

The complex dielectric constant ε_r is written in the form

$$\varepsilon_r = -j\frac{\sigma_0}{\omega\varepsilon_0} + \sum_{m=1}^{M} \left(a_m + \frac{b_m - a_m}{1 + j\omega\tau_m} \right). \tag{8.22}$$

Note that this model is indistinguishable from (8.21) except for the conductivity term. Geophysical media often exhibit these characteristics. For a complete discussion on complex resistivity of earth, see Wait (1989).

8.6 MIXING FORMULA

In Section 8.1, we discussed the Clausius–Mossotti formula, relating the dielectric constant to the polarizability. The dielectric material was viewed as consisting of many equivalent dipoles in free space created by the local Mossotti field. The Clausius–Mossotti formula can be used to obtain the effective dielectric constant of a mixture of two or more materials with different dielectric constants. The formula that gives the effective dielectric constant is called the *mixing formula*.

Let us first consider a simple example of a dielectric material with relative dielectric constant ε_1 in which many spheres of radius a and relative dielectric constant ε_2 are embedded (Fig. 8.4). If the dimension a is comparable to or greater than a wavelength, substantial scattering can take place. Also, if the fractional volume f, which is the fraction of the volume occupied by the spheres, is a few percent or higher, a correlation between the spheres needs to be considered. Here we limit ourselves to the case where dimensions of the spheres are much smaller than a wavelength, and the spheres are sparsely distributed. The situation described here is therefore similar to that discussed in Section 8.1, where the relative dielectric constant ε_r of the material consisting of many dipoles is given by

$$\varepsilon_r = 1 + \chi_e = \frac{1 + 2N\alpha/3\varepsilon_0}{1 - N\alpha/3\varepsilon_0}, \qquad (8.23)$$

where N is the number of dipoles per unit volume and α is the polarizability of the dipole.

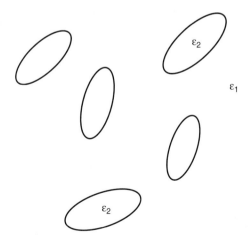

FIGURE 8.4   Effective dielectric constant of a mixture.

In the case shown in Fig. 8.4, the background dielectric constant is $\varepsilon_1 \varepsilon_0$, and thus we have the effective dielectric constant ε_e.

$$\frac{\varepsilon_e}{\varepsilon_1} = \frac{1 + 2N\alpha/3\varepsilon_1}{1 - N\alpha/3\varepsilon_1}, \qquad (8.24)$$

where N is the number of spheres per unit volume. The polarizability α of the sphere is given by (see Section 10.5)

$$\alpha = \frac{3(\varepsilon_2 - \varepsilon_1)}{\varepsilon_2 + 2\varepsilon_1} \varepsilon_1 V, \qquad (8.25)$$

where V is the volume of the sphere. The fractional volume f is then given by

$$f = NV. \qquad (8.26)$$

Substituting (8.25) and (8.26) into (8.24), we get the effective dielectric constant ε_e.

$$\begin{aligned} \varepsilon_e &= \varepsilon_1 \frac{1 + 2fy}{1 - fy}, \\ y &= \frac{\varepsilon_2 - \varepsilon_1}{\varepsilon_2 + 2\varepsilon_1}. \end{aligned} \qquad (8.27)$$

This is called the *Maxwell–Garnett mixing formula*. Note that even though we expect that the formula is valid only for a small fractional volume $f \ll 1$, the effective dielectric constant ε_e reduces to ε_1 when $f = 0$ and ε_2 when $f = 1$. Thus, we may expect that the formula may be a reasonable approximation even when f is not small. However, if the inhomogeneity is not spherical, the polarizability is different and although ε_e reduces to ε_1 when $f = 0$, it does not reduce to ε_2 when $f = 1$. Therefore, the Maxwell–Garnett formula is in general applicable only when f is small. We can also rearrange (8.27) in the following form, known as the *Rayleigh mixing formula*.

$$\frac{\varepsilon_e - \varepsilon_1}{\varepsilon_e + 2\varepsilon_1} = f \frac{\varepsilon_2 - \varepsilon_1}{\varepsilon_2 + 2\varepsilon_1}. \qquad (8.28)$$

If the inhomogeneity has a nonspherical shape, the appropriate polarizability for that shape should be used in place of (8.25).

The Maxwell–Garnett formula (Eq. 8.27) is based on the idea that the inhomogeneity with ε_2 is embedded in the background ε_1. However, more generally, when two inhomogeneities are mixed, there should be no distinction between the background and the inhomogeneities. Thus, both the inhomogeneity with ε_1 and f_1 and the inhomogeneity with ε_2 and f_2 ($f_1 + f_2 = 1$) are embedded in the artificial background with the effective dielectric constant ε_e.

Here we assume that the inhomogeneities are isotropic and have no preferred shape or direction. These inhomogeneities, which are the differences between ε_1 and ε_e and between ε_2 and ε_e, create equivalent dipole moments per unit volume $(N_1\alpha_1 + N_2\alpha_2)\bar{E}_e$, where $\bar{E}_e$ is the average field for the background medium with the effective dielectric constant ε_e. The effective dielectric constant ε_e is chosen such that the average of these dipole moments is zero. Thus, we have

$$N_1\alpha + N_2\alpha_2 = 0. \tag{8.29}$$

Since the inhomogeneities are isotropic, on the average, the polarizability should be equal to that of a sphere.

$$
\begin{aligned}
\alpha_1 &= \frac{3(\varepsilon_1 - \varepsilon_e)}{\varepsilon_1 + 2\varepsilon_e}V_1, \\
\alpha_2 &= \frac{3(\varepsilon_2 - \varepsilon_e)}{\varepsilon_2 + 2\varepsilon_e}V_2,
\end{aligned}
\tag{8.30}
$$

$N_1 V_1 = f_1$, $N_2 V_2 = f_2$, and $f_1 + f_2 = 1$. Rearranging these, we get

$$f_1\frac{\varepsilon_1 - \varepsilon_e}{\varepsilon_1 + 2\varepsilon_e} + f_2\frac{\varepsilon_2 - \varepsilon_e}{\varepsilon_2 + 2\varepsilon_e} = 0. \tag{8.31}$$

This is known as the *Polder–van Santen mixing formula* and can be rearranged to give the following form:

$$f_1\frac{\varepsilon_1 - \varepsilon_0}{\varepsilon_1 + 2\varepsilon_e} + f_2\frac{\varepsilon_2 - \varepsilon_0}{\varepsilon_2 + 2\varepsilon_e} = \frac{\varepsilon_e - \varepsilon_0}{3\varepsilon_e}. \tag{8.32}$$

Note that the Polder–van Santen form is completely symmetric and ε_1, f_1 can be interchanged with ε_2, f_2, giving the same formula; the Maxwell–Garnett formula is not symmetrical. The Polder–van Santen formula can be extended to many species with ε_n and f_n.

$$
\sum_{n=1}^{M}\frac{\varepsilon_n - \varepsilon_0}{\varepsilon_n + 2\varepsilon_e}f_n = \frac{\varepsilon_e - \varepsilon_0}{3\varepsilon_e},
$$
$$
\sum_{n=1}^{M}f_n = 1.
\tag{8.33}
$$

Note that the mixing formulas above are for low-frequency cases where the scattering is negligible. More exact formulas, including scattering and correlations between particles, must be obtained by considering the propagation constant K of the coherent wave. Then the effective dielectric constant ε_e is related to K by $K^2 = k^2\varepsilon_e$, where k is the free space wave number. Extensive studies have been reported on this topic (Tsang et al., 1985).

8.7 DIELECTRIC CONSTANT AND PERMEABILITY FOR ANISOTROPIC MEDIA

The interactions of electromagnetic fields with materials are characterized by the constitutive parameters: complex dielectric constant ε and permeability μ. In an isotropic medium, the property of the material does not depend on the direction of electric or magnetic field polarizations. Thus ε and μ are scalar quantities.

In anisotropic media, however, the material characteristics depend on the direction of the electric or magnetic field vectors and thus, in general, the displacement vector $\bar{D}$ and magnetic flux density vector $\bar{B}$ are not in the same direction as the electric field $\bar{E}$ and magnetic field vector $\bar{H}$, respectively. The dielectric constant ε must then be represented by a tensor ε_{ij}.

$$D_i = \sum_{j=1}^{3} \varepsilon_{ij} E_j, \quad i = 1, 2, 3, \tag{8.34}$$

where $i, j = 1, 2$, and 3 denote the x, y, and z components, respectively. We may write (8.34) in the following form:

$$\bar{D} = \bar{\bar{\varepsilon}} \bar{E}. \tag{8.35}$$

Using matrix notation in the rectangular system, (8.34) is expressed by

$$\begin{bmatrix} D_x \\ D_y \\ D_z \end{bmatrix} = \begin{bmatrix} \varepsilon_{11} & \varepsilon_{12} & \varepsilon_{13} \\ \varepsilon_{21} & \varepsilon_{22} & \varepsilon_{23} \\ \varepsilon_{31} & \varepsilon_{32} & \varepsilon_{33} \end{bmatrix} \begin{bmatrix} E_x \\ E_y \\ E_z \end{bmatrix}. \tag{8.36}$$

Similarly, we have the tensor permeability $\bar{\bar{\mu}}$ *relating* $\bar{B}$ to $\bar{H}$.

$$\bar{B} = \bar{\bar{\mu}} \bar{H}. \tag{8.37}$$

As will be shown shortly, in general, the reciprocity theorem does not hold for anisotropic media, and for a plane wave, $\bar{E}$ and $\bar{H}$ are not necessarily transverse to the direction of wave propagation.

8.8 MAGNETOIONIC THEORY FOR ANISOTROPIC PLASMA

A dc magnetic field is often present in plasma. Examples are the earth's magnetic field in the ionosphere and a dc magnetic field applied to laboratory plasma. The presence of the dc magnetic field makes the plasma anisotropic. In this section, we examine the characteristics of such anisotropic plasma (Yeh and Liu, 1972).

The equation of motion for an electron in electromagnetic fields $(\bar{E}, \bar{H})$ in the presence of a dc magnetic field $\bar{H}_{dc}$ is given by

$$m\frac{d\bar{v}}{dt} = e\bar{E} + e[\bar{v} \times (\bar{B} + \bar{B}_{dc})] - mv\bar{v}, \qquad (8.38)$$

where $\bar{B} = \mu_0\bar{H}$ and $\bar{B}_{dc} = \mu_0\bar{H}_{dc}$. As shown in Section 8.2, the term with $\bar{B}$ is negligible compared with $e\bar{E}$.

For a time-harmonic electromagnetic field with time dependence $\exp(j\omega t)$, neglecting the term with $\bar{B}$, we get

$$j\omega m\bar{v} = e\bar{E} + \mu_0 e(\bar{v} \times \bar{H}_{dc}) - mv\bar{v}. \qquad (8.39)$$

We rewrite this equation using the plasma frequency ω_p

$$\omega_p^2 = \frac{N_e e^2}{m\varepsilon_0} \qquad (8.40)$$

and the cyclotron frequency ω_c

$$\omega_c = \frac{|e|\mu_0 H_{dc}}{m}. \qquad (8.41)$$

Note that ω_c is the frequency of a circular motion of an electron in a plane perpendicular to the dc magnetic field. This is obtained by equating the centrifugal force $(mu^2)/r$ to the force $ev\mu_0 H_{dc}$ due to the magnetic field and noting that $\omega_c = 2\pi/T$ and $T = 2\pi r/v$.

The dc magnetic field $\bar{H}_{dc}$ is pointed in the direction (θ, ϕ) and its rectangular components are (Fig. 8.5)

$$\bar{H}_{dc} = H_{dc}[\sin\theta_d \cos\phi_d \hat{x} + \sin\theta_d \sin\phi_d \hat{y} + \cos\theta_d \hat{z}]$$
$$= H_{dcx}\hat{x} + H_{dcy}\hat{y} + H_{dcz}\hat{z}. \qquad (8.42)$$

We also note that the polarization vector $\bar{P}$ is given by

$$\bar{P} = N_0 e\bar{r}. \qquad (8.43)$$

We now rewrite (8.39) in the following form:

$$-\bar{P}U = \varepsilon_0 X\bar{E} + j\bar{P} \times \bar{Y}, \qquad (8.44)$$

where $Z = j(v/\omega)$, $X = \omega_p^2/\omega^2$, $U = 1 - j(v/\omega)$, and $\bar{Y} = e\,\mu_0\bar{H}_{dc}/m\omega$. Note that since e is negative for electrons, $\bar{Y}$ is pointed in the opposite direction to $\bar{H}_{dc}$.

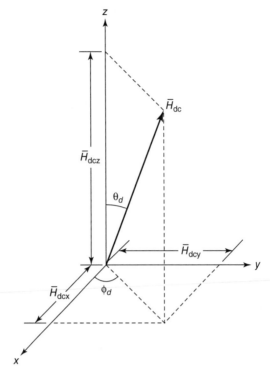

FIGURE 8.5 The dc magnetic field pointed in the direction (θ_d, ϕ_d) in plasma.

Now let us write (8.44) in the following matrix form (see Appendix 8.A):

$$-U[P] = \varepsilon_0 X[E] + j[y][P],$$

where

$$[P] = \begin{bmatrix} P_x \\ P_y \\ P_z \end{bmatrix}, \quad [E] = \begin{bmatrix} E_x \\ E_y \\ E_z \end{bmatrix}, \quad [y] = \begin{bmatrix} 0 & Y_z & -Y_y \\ -Y_z & 0 & Y_x \\ Y_y & -Y_x & 0 \end{bmatrix}.$$

This can be rearranged to yield

$$\varepsilon_0[E] = [\sigma][P], \tag{8.45}$$

where

$$[\sigma] = -\frac{1}{X} \begin{bmatrix} U & jY_z & -jY\hat{y} \\ -jY_z & U & jY_x \\ jYy & -jY_x & U \end{bmatrix}.$$

Inverting the matrix, we get the tensor electric susceptibility $[\chi_e]$

$$[P] = \varepsilon_0[\chi_e][E],$$

$$[\chi_e] = [\sigma]^{-1}$$

$$= -\frac{X}{U(U^2 - Y^2)}\begin{bmatrix} U^2 - Y_x^2 & -jY_zU - Y_xY_y & jY_yU - Y_xY_z \\ jY_zU - Y_xY_y & U^2 - Y_y^2 & -jY_xU - Y_yY_z \\ -jT_yU - Y_xY_z & jY_xU - Y_yY_z & U^2 - Y_z^2 \end{bmatrix}, \quad (8.46)$$

where $Y^2 = Y_x^2 + Y_y^2 + Y_z^2 = \omega_c^2/\omega^2$. The relative tensor dielectric constant $[\varepsilon_r]$ is then given by

$$[\varepsilon_r] = [1] + [\chi_e], \quad (8.47)$$

where $[1]$ is a 3×3 unit matrix.

Note that if the dc magnetic field is reversed, all Y_x, Y_y, and Y_z change the sign, and as seen in (8.46), this is equivalent to transposing the matrix $[\chi_e]$ and $[\varepsilon_r]$.

$$\begin{aligned} [\chi_e] &\rightarrow [\tilde{\chi}_e], \\ [\varepsilon_r] &\rightarrow [\tilde{\varepsilon}_r]. \end{aligned} \quad (8.48)$$

Note also that the anisotropy is produced by the cyclotron frequency ω_c. The cyclotron frequency $f_c = \omega_c/2\pi$ of the earth's magnetic field is approximately $f_c = 1.42$ MHz.

8.9 PLANE-WAVE PROPAGATION IN ANISOTROPIC MEDIA

Let us consider the characteristics of a plane wave propagating in an anisotropic medium. We let $\bar{k} = k\hat{\imath}$, where k is the propagation constant and $\hat{\imath}$ is the unit vector in the direction of wave propagation. In general, the propagation constant k depends on the direction $\hat{\imath}$.

We seek a plane-wave solution that has the following general form:

$$e^{j(\omega t - \bar{k}\cdot\bar{r})}. \quad (8.49)$$

First, we note that in general, $\bar{E}$ and $\bar{H}$ are not necessarily perpendicular to $\bar{k}$, but $\bar{D}$ and $\bar{B}$ are always perpendicular to $\bar{k}$. To prove this, we note that for a plane wave

$$\frac{\partial}{\partial x}(e^{-j\bar{k}\cdot\bar{r}}) = -jk_x(e^{-j\bar{k}\cdot\bar{r}}),$$

$$\bar{k} = k_x\hat{x} + k_y\hat{y} + k_z\hat{z},$$

and therefore

$$
\begin{aligned}
\nabla &= \hat{x}\frac{\partial}{\partial x} + \hat{y}\frac{\partial}{\partial y} + \hat{z}\frac{\partial}{\partial z} \\
&= -jk_x\hat{x} - jk_y\hat{y} - jk_z\hat{z} = -j\bar{k}.
\end{aligned}
\tag{8.50}
$$

Thus the divergence equations

$$
\nabla \cdot \bar{B} = 0 \quad \text{and} \quad \nabla \cdot \bar{D} = 0
$$

become

$$
-j\bar{k} \cdot \bar{B} = 0 \quad \text{and} \quad -j\bar{k} \cdot \bar{D} = 0,
\tag{8.51}
$$

which proves that $\bar{B}$ and $\bar{D}$ are perpendicular to $\bar{k}$. This, however, does not show that $\bar{E}$ and $\bar{H}$ should be perpendicular to $\bar{k}$ since $\bar{E}$ and $\bar{D}$ (or $\bar{H}$ and $\bar{B}$) are not parallel in anisotropic media.

8.10 PLANE-WAVE PROPAGATION IN MAGNETOPLASMA

Let us write Maxwell's equations for a plane wave using (8.50)

$$
\begin{aligned}
-j\bar{k} \times \bar{E} &= -j\omega\bar{B}, \\
-j\bar{k} \times \bar{H} &= j\omega\bar{D}.
\end{aligned}
\tag{8.52}
$$

In magnetoplasma, we have

$$
\bar{B} = \mu_0\bar{H} \quad \text{and} \quad \bar{D} = \varepsilon_0\bar{\bar{\varepsilon}}_r\bar{E},
\tag{8.53}
$$

and thus substituting (8.53) into (8.52), we obtain the equation for $\bar{E}$

$$
\bar{k} \times \bar{k} \times \bar{E} + \omega^2\mu_0\varepsilon_0\bar{\bar{\varepsilon}}_r\bar{E} = 0
\tag{8.54}
$$

and $\bar{H}$ is given by

$$
\bar{H} = \frac{\bar{k} \times \bar{E}}{\omega\mu_0}.
\tag{8.55}
$$

It is now possible to obtain the propagation constant k from (8.54). Let us first write (8.54) in matrix form. Noting that

$$
\begin{aligned}
\bar{k} \times \bar{k} \times \bar{E} &= \bar{k}(\bar{k} \cdot \bar{E}) - (\bar{k} \cdot \bar{k})\bar{E}, \\
\bar{k} &= k_x\hat{x} + k_y\hat{y} + k_z\hat{z},
\end{aligned}
\tag{8.56}
$$

we write (8.54) in the following matrix form:

$$\{K\tilde{K} - k^2[1] + k_0^2[\varepsilon_r]\}[E] = 0, \tag{8.57}$$

where

$$K = \begin{bmatrix} k_x \\ k_y \\ k_z \end{bmatrix} \quad \text{and} \quad K\tilde{K} = \begin{bmatrix} k_x k_x & k_x k_y & k_x k_z \\ k_y k_x & k_y k_y & k_y k_z \\ k_z k_x & k_z k_y & k_z k_z \end{bmatrix},$$

$k = |\bar{k}|, k_0^2 = \omega^2 \mu_0 \varepsilon_0$, [1] is a 3 × 3 unit matrix, $[\varepsilon_r]$ is a 3 × 3 matrix given by (8.47), and [E] is a column matrix given in (8.45).

Equation (8.57) is the fundamental matrix equation for an anisotropic medium with the tensor dielectric constant $[\varepsilon_r]$. Since this is a homogeneous linear equation for [E], the nonzero solution for [E] is obtained when the following determinant is zero.

$$\left| K\tilde{K} - k^2[1] + k_0^2[\varepsilon_r] \right| = 0. \tag{8.58}$$

The solution of this equation gives the propagation constant k.

It will be shown in Sections 8.11–8.13 that for the wave propagating along the dc magnetic field, there are two circularly polarized waves with different propagation constants, and for the wave propagating in the direction perpendicular to the dc magnetic field, there are two linearly polarized waves with different propagation constants. In general, there are two elliptically polarized waves for the wave propagating in an arbitrary direction.

8.11 PROPAGATION ALONG THE DC MAGNETIC FIELD

Let us take the z axis along the direction of the propagation $\bar{k}$ and the dc magnetic field $\bar{H}_{dc}$.

$$\bar{k} = k\hat{z}, \quad \bar{H}_{dc} = H_{dc}\hat{z}. \tag{8.59}$$

Noting that $Y_z = Y$, $Y_x = Y_y = 0$ in (8.46), we get

$$[\chi_e] = -\frac{X}{U(U^2 - Y^2)} \begin{bmatrix} U^2 & -jYU & 0 \\ jYU & U^2 & 0 \\ 0 & 0 & U^2 - Y^2 \end{bmatrix}. \tag{8.60}$$

Thus, we get the tensor relative dielectric constant $[\varepsilon_r]$.

$$[\varepsilon_r] = \begin{bmatrix} \varepsilon & ja & 0 \\ -ja & \varepsilon & 0 \\ 0 & 0 & \varepsilon_z \end{bmatrix},$$

(8.61)

where

$$\varepsilon = 1 - \frac{XU}{U^2 - Y^2} = 1 - \frac{(\omega_p/\omega)^2[1 - j(v/\omega)]}{[1 - j(v/\omega)]^2 - (\omega_c/\omega)^2},$$

$$a = \frac{XY}{U^2 - Y^2} = -\frac{(\omega_p/\omega)^2(\omega_c/\omega)}{[1 - j(v/\omega)]^2 - (\omega_c/\omega)^2},$$

$$\varepsilon_z = 1 - \frac{X}{U} = 1 - \frac{(\omega_p/\omega)^2}{1 - j(v/\omega)}.$$

Note that Y is negative for electrons and therefore, $Y = -(\omega_c/\omega)$.
    Equation (8.57) then becomes

$$\begin{bmatrix} -k^2 + k_0^2\varepsilon & jk_0^2 a & 0 \\ -jk_0^2 a & -k^2 + k_0^2\varepsilon & 0 \\ 0 & 0 & k_0^2\varepsilon_z \end{bmatrix} \begin{bmatrix} E_x \\ E_y \\ E_z \end{bmatrix} = 0.$$

(8.62)

From this, we get the following two values of the propagation constant k:

$$k_+ = k_0 n_+ \quad \text{and} \quad k_- = k_0 n_-,$$

(8.63)

where

$$n_+ = \sqrt{\varepsilon_+} = \sqrt{\varepsilon - a},$$

$$n_- = \sqrt{\varepsilon_-} = \sqrt{\varepsilon + a}.$$

The equivalent dielectric constant ε_+ and ε_- are plotted in Fig. 8.6. It is obvious that the behavior of ε_+ is similar to that for the isotropic case, and for this reason, the wave for ε_+ is called the *ordinary wave* and the other with ε_- is called the *extraordinary wave*.
    The behavior of the electric fields can be studied by noting (8.62). The first equation is

$$\left(k_0^2\varepsilon - k^2\right)E_x + jk_0^2 aE_y = 0.$$

(8.64)

For the ordinary wave, $k = k_+$ and therefore we get,

$$k_0^2 aE_x + jk_0^2 aE_y = 0,$$

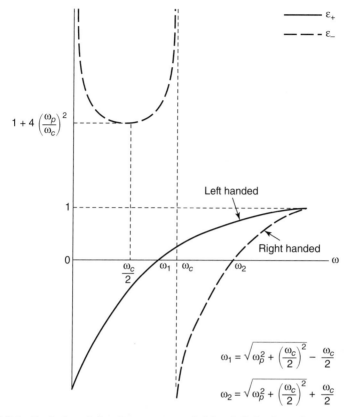

FIGURE 8.6   Equivalent dielectric constant $\in_+$ (left handed) for the ordinary wave and $\in_-$ (right handed) for the extraordinary wave. The medium is assumed to be lossless.

which yields,

$$E_x = -jE_y. \tag{8.65}$$

This is a left-handed circularly polarized wave (LHC) whose electric field vector rotates clockwise in the x–y plane (see Fig. 8.7). Also, from (8.56), we note that

$$E_z = 0. \tag{8.66}$$

The displacement vector D is given by

$$D_x = \varepsilon + Ex, \quad Dy = \varepsilon + E_y, \quad D_z = 0. \tag{8.67}$$

The magnetic field is perpendicular to $\bar{k}$ and $\bar{E}$ and is given by

$$\frac{E_x}{H_y} = -\frac{E_y}{H_x} = Z_+ = \frac{Z_0}{\sqrt{\varepsilon - a}}, \quad Z_0 = \left(\frac{\mu_0}{\varepsilon_0}\right)^{1/2}. \tag{8.68}$$

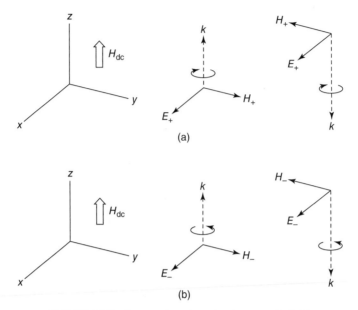

FIGURE 8.7   Propagation along the dc magnetic field.

When the propagation is in the $-z$ direction, we write

$$\bar{k} = -k\hat{z}, \tag{8.69}$$

which does not change the electric field E nor D, but the magnetic field is reversed as shown in Fig. 8.7.

Similarly, for the extraordinary wave, $k = k_-$, we get

$$
\begin{aligned}
E_x &= +jE_y, \\
E_z &= 0,
\end{aligned} \tag{8.70}
$$

which is a right-handed circularly polarized wave (RHC) whose vector rotates counterclockwise in the x–y plane. Also, we get

$$D_x = \varepsilon_- \, E_x, \quad D_y = \varepsilon_- \, E_y, \quad D_z = 0, \tag{8.71}$$

and the equivalent wave impedance is

$$Z_- = \frac{Z_0}{\sqrt{\varepsilon + a}}. \tag{8.72}$$

We note from Fig. 8.7 that ε_+ is negative when $\omega < \omega_1$ and thus the ordinary wave does not propagate. However, in this frequency range, ε_- is positive and the

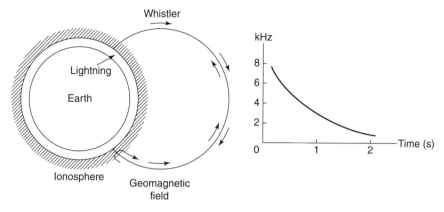

FIGURE 8.8 Whistler mode.

extraordinary wave propagates. The VLF waves cannot penetrate the ionosphere in the absence of the dc magnetic field because the frequency is below the plasma frequency. However, in the presence of the earth's magnetic field, the extraordinary VLF wave can propagate in the direction of the dc magnetic field. This is the main mechanism of the whistler mode.

The *whistler* is a form of radio noise in the audio-frequency range (1–20 kHz) characterized by a whistling tone. The VLF components of the short electromagnetic pulse due to a lightning stroke can penetrate the ionosphere by means of the extraordinary mode and propagate along the earth's geomagnetic field as shown in Fig. 8.8. The signal may then be reflected and returned back propagating along the magnetic field. The whistling effect results from different group velocities and the time delay for different frequency components (Fig. 8.8). The time T required for the signal at a certain frequency with the group velocity v_g is

$$T = \int_{\text{path}} \frac{ds}{v_g} = \int_{\text{path}} \frac{\partial k_-}{\partial \omega} \, ds. \tag{8.73}$$

8.12 FARADAY ROTATION

As discussed in Section 8.11, when a wave is propagating through an anisotropic medium in the direction of the dc magnetic field, two circularly polarized waves can propagate with different propagation constants. As a result, if these two circularly polarized waves are properly combined so as to produce a linearly polarized wave at one point, then as the wave propagates, the plane of polarization rotates and the angle of rotation is proportional to the distance. This is called the *Faraday rotation*.

To show this, let us assume that at $z = 0$, two circularly polarized waves are combined to give

$$E_x = E_{x0}, \quad E_y = 0. \tag{8.74}$$

At any other point, in general

$$E_x = E_{x+}\, e^{-jk_+\, z} + E_{x-}\, e^{-jk_-\, z}, \tag{8.75}$$

where E_{x+} and E_{x-} are the magnitudes of two circularly polarized waves given in Section 8.11. Each E_{x+} and E_{x-} must be accompanied by its E_{y+} and E_{y-} as given by (8.65) and (8.70). Thus,

$$E_y = jE_{x+}\, e^{-jk_+\, z} - jE_{x-}\, e^{-jk_-\, z}. \tag{8.76}$$

At $z = 0$, $E_x = E_{x+} + E_{x-} = E_0$ and $E_y = j(E_{x+} - E_{x-}) = 0$, as given by (8.74), and thus,

$$E_{x+} + E_{x-} = \frac{E_0}{2}.$$

Therefore, we have

$$
\begin{aligned}
E_x &= \frac{E_0}{2}(e^{-jk_+\, z} + e^{-jk_-\, z}), \\
E_y &= j\frac{E_0}{2}(e^{-jk_+\, z} - e^{-jk_-\, z}).
\end{aligned}
\tag{8.77}
$$

By writing

$$k_{\pm} = \frac{k_+ + k_-}{2} \pm \frac{k_+ - k_-}{2},$$

we express E_x and E_y by

$$
\begin{aligned}
E_x &= E_0 e^{-j[(k_+ + k_-)/2]z}\, \cos\frac{k_+ - k_-}{2}z, \\
E_y &= E_0 e^{-j[(k_+ + k_-)/2]z}\, \sin\frac{k_+ - k_-}{2}z.
\end{aligned}
\tag{8.78}
$$

This represents a linearly polarized wave that propagates with the propagation constant

$$k_f = \frac{k_+ + k_-}{2} \tag{8.79}$$

and whose plane of polarization rotates with the angle

$$\theta_f = \frac{k_+ - k_-}{2}z. \tag{8.80}$$

The angle θ_f is proportional to the distance z.

If the wave is propagating in the negative z direction, the formulas above are valid with the change $k_+ \to -k_+, k_- \to -k_-$, and $z \to -z$. Thus if the wave is propagated in the positive z direction, and then at the end of the path reflected back and propagated in the negative direction, the Faraday rotation is doubled. This is an important characteristic of *magnetic rotation*. In contrast with this, *natural rotation* is canceled as the wave is propagated forward and then reflected back. The natural rotation is the rotation of the plane of polarization in liquids such as sugar solutions, which have an asymmetrically bound carbon atom. It also occurs in crystals such as quartz and sodium chlorate, which have helical structure. They are characterized by two types of structures, which are related to each other like right-handed and left-handed screws. These two screws are otherwise identical, but they cannot be brought into coincidence by any rotation in three-dimensional space (see Sommerfeld, 1954, pp. 106 and 164). This natural rotation is also called *optical activity* and takes place in bi-anisotropic medium such as chiral medium. See Section 8.22.

8.13 PROPAGATION PERPENDICULAR TO THE DC MAGNETIC FIELD

Consider the propagation in the x direction and the dc magnetic field in the z direction.

$$\bar{k} = k\hat{x}, \quad k = k_x = k_0 n, \quad \bar{H}_{dc} = H_{dc}\hat{z}.$$

Then (8.57) becomes

$$\begin{bmatrix} k_0^2 \varepsilon & jk_0^2 & 0 \\ -jk_0^2 a & k_0^2 \varepsilon - k^2 & 0 \\ 0 & 0 & k_0^2 \varepsilon_z - k^2 \end{bmatrix} \begin{bmatrix} E_x \\ E_y \\ E_z \end{bmatrix} = 0. \tag{8.81}$$

From this, we get two solutions

$$k = k_0 n_0 = k_0 \sqrt{\varepsilon_z} \quad \text{and} \quad k = k_0 n_e = k_0 \left(\frac{\varepsilon^2 - a^2}{\varepsilon} \right)^{1/2}. \tag{8.82}$$

For the wave with the propagation constant $k_0 n_0$, we get

$$\begin{aligned}
E_x &= E_y = 0, \quad E_z \neq 0, \\
D_x &= D_y = 0, \quad D_z = \varepsilon_z E_z, \\
H_x &= H_z = 0, \\
-\frac{E_z}{H_y} &= Z_1 = \frac{Z_0}{\sqrt{\varepsilon}}, \\
B_y &= \mu_0 H_y.
\end{aligned} \tag{8.83}$$

This wave is the ordinary wave because ε_z is the same as the case of isotropic plasma, and the dc magnetic field has no effect on this plane-wave propagation. This is expected because both electric and displacement vectors are in the z direction, and electrons move along the dc magnetic field and their motions are not affected by the presence of the dc magnetic field.

For the wave with the propagation constant

$$k = k_0 n_e = k_0 \sqrt{\varepsilon_t}, \quad \varepsilon_t = \frac{\varepsilon^2 - a^2}{\varepsilon},$$

we get

$$
\begin{aligned}
\varepsilon E_x &= -ja E_y, \\
E_z &= 0, \\
D_x &= D_z = 0, \\
D_y &= \varepsilon_0 \varepsilon_t E_y, \\
\frac{E_y}{H_z} &= Z_2 = \frac{Z_0}{\sqrt{\varepsilon_t}}.
\end{aligned}
\tag{8.84}
$$

H_z, D_y, and B_z behave as if the medium has the equivalent dielectric constant $\varepsilon_0 \varepsilon_t$. But in addition to these fields, E_x, the component in the direction of the propagation, appears. This is produced by the coupling between E_x and E_y in the anisotropic medium.

8.14 THE HEIGHT OF THE IONOSPHERE

Consider a radio-wave pulse that is sent vertically toward the ionosphere. The electron density of the ionosphere depends on the height, and its typical profile is shown in Fig. 8.3. The electron density distribution with height may be explained by observing that the rate of production of electrons depends on the sun's radiation and the air density, but the sun's radiation increases with height while the air density decreases with height, and therefore, there is a maximum electron density at a certain height.

Let us assume that the effects of the earth's magnetic field and the collision frequencies are negligible. The radio wave with a certain frequency $f = \omega/2\pi$ propagates up to the height z_0, where the plasma frequency $f_p = \omega_p/2\pi$ becomes equal to f (Fig. 8.9). This height z_0 is called the *true height*. However, the time τ required for a radio pulse to travel from the ground to z_0 and back is, assuming that v_g is a slowly varying function of z,

$$\tau = 2 \int_0^{z_0} \frac{dz}{v_g}, \tag{8.85}$$

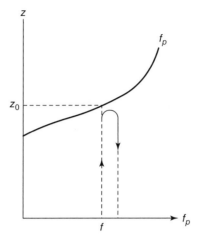

FIGURE 8.9   Height of the ionosphere.

where v_g is the group velocity,

$$\frac{1}{v_g} = \frac{\partial k\,(\omega, z)}{\partial \omega} = \frac{1}{c}\frac{\partial}{\partial \omega}[\omega_n(\omega, z)],$$

where the refractive index n varies with height. The *equivalent height* h_e is the fictitious distance over which a pulse propagates in free space, during the time $\tau/2$.

$$h_e = \frac{c\tau}{2} = \int_0^{z_0} \frac{\partial}{\partial \omega}[\omega n(\omega, z)]dz. \tag{8.86}$$

The *phase height* h_p is the fictitious distance in free space corresponding to the total phase from $z = 0$ to $z = z_0$.

$$h_p = \int_0^{z_0} n(\omega, z)\,dz. \tag{8.87}$$

8.15 GROUP VELOCITY IN ANISOTROPIC MEDIUM

We defined the group velocity v_g of the wave propagating with propagation constant k by the following:

$$v_g = \frac{\partial \omega}{\partial k}. \tag{8.88}$$

In an anisotropic medium, the propagation constant $k = (\omega/c)n$ depends on the direction of the wave propagation and therefore, (8.88) should hold for each component of the group velocity $\bar{v}_g$.

$$\bar{v}_g = v_{gx}\hat{x} + v_{gy}\hat{y} + v_{gz}\hat{z}$$
$$= \left(\frac{\partial}{\partial k_x}\hat{x} + \frac{\partial}{\partial k_y}\hat{y} + \frac{\partial}{\partial k_z}\hat{z} \right)\omega = \nabla_k\omega, \tag{8.89}$$
$$\bar{k} = k_x\hat{x} + k_y\hat{y} + k_z\hat{z}.$$

This is the general expression for $\bar{v}_g$.

Equation (8.89) indicates that the group velocity $\bar{v}_g$ is perpendicular to surfaces of constant ω. This is the surface representing the dispersion relation for a fixed frequency (Fig. 8.10).

$$k(\theta, \phi) = \frac{\omega}{c}n(\theta, \phi). \tag{8.90}$$

The surface $k = $ constant is called the *wave vector surface*. Often it is useful to introduce the *refractive index surface*, where the refractive index n is constant. Obviously, these two surfaces are proportional to each other and carry the same information.

If the refractive index n is symmetric about the z axis and therefore $n = n(\theta)$ as in Fig. 8.10, it is easily seen that

$$\tan \alpha = -\frac{1}{n}\frac{\partial n}{\partial \theta}. \tag{8.91}$$

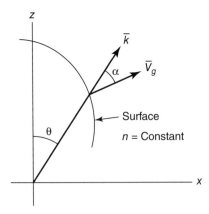

FIGURE 8.10   Group velocity in an anisotropic medium.

More generally, the group velocity $\bar{v}_g$ is given by

$$\bar{v}_g = v_{gk}\hat{k} + v_{g\theta}\hat{\theta} + v_{g\phi}\hat{\phi},$$

$$v_{gk} = \frac{c}{\partial(n\omega)/\partial\omega},$$

$$v_{g\theta} = -\frac{c}{\partial(n\omega)/\partial\omega} \frac{1}{n} \frac{\partial n}{\partial\theta}, \qquad (8.92)$$

$$v_{g\phi} = -\frac{c}{\partial(n\omega)/\partial\omega} \frac{1}{n\sin\theta} \frac{\partial n}{\partial\phi},$$

where $\hat{k}, \hat{\theta}$, and $\hat{\phi}$ are unit vectors in the directions of $\bar{k}, \theta$, and ϕ, respectively (Yeh and Liu, 1972).

8.16 WARM PLASMA

In the magnetoionic theory discussed in Sections 8.8, it is assumed that the motions of electrons are caused by the Lorentz force. However, since the temperature of the electron gas is finite, the motions of electrons are also affected by the pressure variations of the electron gas. The magnetoionic theory ignores the effects of the finite temperature and the pressure variations, as they are normally small, and therefore it is equivalent to dealing with the electron gas at negligible temperature. Thus the magnetoionic theory may be considered as the theory dealing with *cold plasma*. However, recent studies on the impedance of antennas in the ionosphere indicate that there is a need to include the effects of the finite temperature and *warm plasma*. The pressure variation of warm plasma may be considered as an acoustic wave, and therefore inclusion of the finite temperature means the study of the interaction of the electromagnetic wave and the acoustic wave in the electron gas.

Let us start with the following basic equations for warm plasma.

Maxwell's Equations

$$\nabla \times \bar{E} = -\mu_0 \frac{\partial \bar{H}}{\partial t},$$

$$\nabla \times \bar{H} = \varepsilon_0 \frac{\partial \bar{E}}{\partial t} + Ne\bar{v}; \qquad (8.93)$$

Hydrodynamic Equation

$$m\frac{\partial \bar{v}}{\partial t} = e(\bar{E} + \bar{v} \times \bar{B}) - \frac{1}{N}\nabla p; \qquad (8.94)$$

Continuity Equation

$$\nabla \cdot (N\bar{v}) + \frac{\partial}{\partial t}N = 0; \qquad (8.95)$$

Equation of State

$$P = k_b NT;$$
(8.96)

where

$\bar{v} =$ velocity of electron,

$e, m =$ charge and mass of electron,

$N =$ electron density, number of electrons per unit volume,

$P =$ pressure,

$k_b =$ Boltzmann's constant,

$T =$ temperature, Kelvin.

Note that the current density $\bar{J}$ is expressed in (8.93) as $Ne\bar{v}$. The second term on the right side of (8.94) is the force due to the pressure gradient, and (8.95) represents the change of the number of electrons per unit volume.

We assume that there is no dc magnetic field $H_{dc} = 0$. We also write N and P as the sum of the average N_0 and P_0 and the ac (acoustic) components n and p.

$$N = N_0 + n,$$
$$P = P_0 + p.$$
(8.97)

Now we assume that the ac components are small compared with the average values, and therefore all nonlinear terms containing the product of two ac components are negligible.

Under the foregoing assumptions, (8.93) through (8.95) become

$$\nabla \times \bar{E} = -\mu_0 \frac{\partial \bar{H}}{\partial t},$$
(8.98)

$$\nabla \times \bar{H} = \varepsilon_0 \frac{\partial \bar{E}}{\partial t} + N_0 e\bar{v},$$
(8.99)

$$\frac{\partial \bar{v}}{\partial t} = \frac{e}{m} \bar{E} - \frac{1}{mN_0} \nabla p,$$
(8.100)

$$N_0 \nabla \cdot \bar{v} + \frac{\partial n}{\partial t} = 0.$$
(8.101)

The equation of state (8.96) for the "isothermal" case ($T =$ constant) becomes

$$p = nk_b T.$$
(8.102)

For acoustic waves, however, the adiabatic process, where no heat transfer is taking place, is more appropriate. In this case, P and N satisfy the following:

$$\frac{P}{N_\gamma} = \frac{P}{N_0^\gamma} = \text{constant},$$
(8.103)

where γ is the ratio of specific heat at constant pressure and constant volume. From this, we get

$$p = \gamma n k_b T. \tag{8.104}$$

Equations (8.98)–(8.101) and (8.104) constitute the fundamental equations for isotropic warm plasma. The value of γ for plasma is approximately equal to 3. (Note that $\gamma = \frac{5}{3}$ for perfect monoatomic gas, $\gamma = \frac{7}{5}$ for diatomic gas such as air, and $\gamma = \frac{4}{3}$ for polyatomic gas (Yeh and Liu, 1972, p. 94)).

8.17 WAVE EQUATIONS FOR WARM PLASMA

In order to obtain the wave equation for acoustic waves, we take the divergence of (8.100) and use (8.101) and (8.104) to eliminate $\bar{v}$ and n. We get

$$\nabla^2 p - \frac{1}{u^2}\frac{\partial^2}{\partial t^2}p - N_0 e \nabla \cdot \bar{E} = 0, \tag{8.105}$$

where $u = (\gamma k_b T/m)^{1/2}$ is the sound velocity of the electron gas if there were no charge $e = 0$ and is called *Laplace's sound velocity* for the adiabatic case. (In contrast, for the isothermal case, it is called *Newton's sound velocity* and is equal to $(k_b T/m)^{1/2}$.) The last term of (8.105) is proportional to $\nabla \cdot \bar{v}$ through (8.99), which is in turn proportional to p through (8.101) and (8.104). Taking the divergence of (8.99), we get

$$\nabla \cdot \nabla \times \bar{H} = 0 = \varepsilon_0 \frac{\partial}{\partial t}\nabla \cdot \bar{E} + N_0 e \nabla \cdot \bar{v}$$

$$= \varepsilon_0 \frac{\partial}{\partial t}\nabla \cdot \bar{E} - \frac{e}{\gamma k_b T}\frac{\partial p}{\partial t}. \tag{8.106}$$

Thus we get

$$N_0 e \nabla \cdot \bar{E} = \frac{\omega_p^2}{u^2}p, \quad \omega_p^2 = \frac{N_0 e^2}{m \varepsilon_0}.$$

Substituting this into (8.105), we get the following wave equation:

$$\nabla^2 p - \frac{1}{u^2}\frac{\partial^2}{\partial t^2}p - \frac{\omega_p^2}{u^2}p = 0. \tag{8.107}$$

For a time-harmonic case $[\exp(j\omega t)]$, we get

$$(\nabla^2 + k_p^2)p = 0,$$

$$k_p^2 = \frac{\omega^2 - \omega_p^2}{u^2}. \tag{8.108}$$

The phase velocity is therefore

$$v_p = \frac{u}{[1 - (\omega_p/\omega)^2]^{1/2}}. \tag{8.109}$$

Note that the acoustic velocity u is modified by the plasma frequency ω_p.

The wave equation for the electromagnetic field is obtained by taking the curl of (8.98) and substituting (8.99). We then get

$$-\nabla \times \nabla \times \bar{E} - \mu_0 \varepsilon_0 \frac{\partial^2}{\partial^2} \bar{E} - \mu_0 \varepsilon_0 \omega_p^2 \bar{E} + \mu_0 \varepsilon_0 u^2 \nabla(\nabla \cdot \bar{E}) = 0. \tag{8.110}$$

Let us consider a time-harmonic case and assume that a plane wave is propagating in a direction given by the propagation vector $\bar{k}$. Thus all electric and magnetic vectors have $\exp(-j\bar{k} \cdot \bar{r})$ dependence, and therefore the operator ∇ can be replaced by $-j\bar{k}$. Noting that

$$\bar{\nabla} \times \bar{\nabla} \times \bar{E} = -\bar{k} \times \bar{k} \times \bar{E} = -[\bar{k}(\bar{k} \cdot \bar{E}) - \bar{E}k^2],$$

with $\bar{k} \cdot \bar{k} = k^2$, we get

$$\bar{k}(\bar{k} \cdot \bar{E})(1 - \mu_0 \varepsilon_0 u^2) - \left[k^2 - k_0^2 \left(1 - \frac{\omega_p^2}{\omega^2} \right) \right] \bar{E} = 0. \tag{8.111}$$

Now we examine the component of $\bar{E}$ along and perpendicular to $\bar{k}$.

$$\bar{E} = \bar{E}_{\parallel} + \bar{E}_{\perp}. \tag{8.112}$$

Taking the component of (8.111) along $\bar{k}$, we get

$$k^2(1 - \mu_0 \varepsilon_0 u^2) - \left[k^2 - k_0^2 \left(1 - \frac{\omega_p^2}{\omega^2} \right) \right] = 0, \tag{8.113}$$

from which we obtain the propagation constant k for $\bar{E}_{\parallel}$,

$$k_{\parallel}^2 = \frac{\omega^2 - \omega_p^2}{u^2} = k_p^2. \tag{8.114}$$

This component $\bar{E}_{\parallel}$ therefore propagates with the propagation constant identical to that of the pressure wave k_p; thus this is called the *acoustic wave*.

On the other hand, the perpendicular component $\bar{E}_\perp$ can be determined from (8.111) to have the propagation constant

$$k^2 = k_0^2 \left(1 - \frac{\omega_p^2}{\omega^2}\right), \tag{8.115}$$

which is identical to that of a cold plasma.

Since $\bar{H} = (1/\omega\mu_0)\bar{k} \times \bar{E}$, no magnetic field is associated with $\bar{E}_{||}$, but $\bar{H}_\perp$ associated with $\bar{E}_\perp$ is given in the same manner as that of a cold plasma. These two waves $\bar{E}_{||}$ and $\bar{E}_\perp$ can exist in an infinite space independently, with two different propagation constants. However, these two components are coupled together at a boundary or at an exciting source. For example, a plane wave incident on a warm plasma excites both components, and the amount of excitation for each component depends on boundary conditions. Also, a dipole source in a warm plasma excites an acoustic wave, and thus, the impedance of the antenna is affected by the acoustic wave.

8.18 FERRITE AND THE DERIVATION OF ITS PERMEABILITY TENSOR

In 1845, Faraday discovered the rotation of the plane of polarization of light when propagating through various materials under the influence of the dc magnetic field, now called the *Faraday rotation*. Until about 1946, this effect could not be utilized at microwave frequencies due to the high loss in ferromagnetic materials. But the discovery of low-loss ferrite materials made it possible to employ this material for a variety of microwave applications. Polder developed the general tensor permeability for the ferrite in 1949, and Tellegen in 1948 and Hogan in 1952 developed a microwave network element called the *gyrator* utilizing ferrite. Commercial ferrite devices have been available since about 1953.

Let us first derive the tensor permeability for the ferrite based on a simplified model of a spinning electron. A spinning electron can be considered as a gyromagnetic top. Let the angular momentum be $\bar{J}$. Then the magnetic moment $\bar{m}$ is parallel to $\bar{J}$ and in the opposite direction. The magnitude of m is proportional to J. Thus

$$\bar{m} = \gamma\bar{J}. \tag{8.116}$$

The proportionality constant γ is negative and is found to be

$$\gamma = -\frac{e}{m_e} = -1.7592 \times 10^{11} \text{ C/kg}$$

$$= \text{gyromagnetic ratio.}$$

The equation of angular motion for a single electron is then

$$\frac{d\bar{J}}{dt} = \text{torque.} \tag{8.117}$$

The torque on the magnetic moment m is given by

$$\bar{m} \times \bar{B}.$$

Thus we get

$$\frac{d\bar{m}}{dt} = \gamma \bar{m} \times \bar{B}. \tag{8.118}$$

We first note that when this magnetic top is placed in the dc magnetic field, the top precesses around the direction of the magnetic field with the angular velocity

$$\omega_0 = -\gamma B = -\gamma \mu_0 H, \tag{8.119}$$

which is known as the *Larmor precessional frequency*. This precessional motion is described by

$$\frac{dJ}{dt} = J \times \omega_0. \tag{8.120}$$

Eventually, however, this precession dies down due to the damping, and all the magnetic tops become aligned with the dc magnetic field. Then the ferrite is said to be *saturated*. Let H_0 be the applied external dc magnetic field. Then the field in the ferrite is given by

$$H_i = H_0 - H_{\text{dem}}, \tag{8.121}$$

where H_{dem} is the demagnetizing field. In general, H_i is not in the same direction as H_0 and depends on the shape of the material. However, if the ferrite is an ellipsoid, H_i is parallel to H_0. It must be kept in mind that the internal field H_i is not the same as the applied field and may not even be in parallel direction.

Let us take a unit volume and consider the magnetic polarization

$$\bar{M} = N_e \bar{m}, \tag{8.122}$$

where N_e is the number of effective electrons per unit volume. In terms of M, we write (8.118) as

$$\frac{d\bar{M}}{dt} = \gamma \bar{M} \times \bar{B}. \tag{8.123}$$

Next we express each of B, H, and M as a sum of the dc component and small ac component. Thus we write

$$\begin{aligned} B &= \mu_0(H + M), \\ H &= H_i + H_a, \\ M &= M_i + M_a, \end{aligned} \tag{8.124}$$

where H_i and M_i are the dc components and H_a and M_a are the small ac components.

Noting that $M \times M = 0$, we get

$$\frac{d\bar{M}}{dt} = \mu_0 \gamma (\bar{M} \times \bar{H}). \tag{8.125}$$

We then obtain

$$\frac{dM_i}{dt} + \frac{dM_a}{dt} = \mu_0 \gamma (M_i \times H_i + M_i \times H_a + M_a \times H_i + M_a \times H_a). \tag{8.126}$$

Note that M_i is constant and M_i and H_i are in parallel, and thus $dM_i/dt = 0$ and $M_i \times H_i = 0$.

We assume that the ac component is small compared with the dc component, and thus the last term is negligibly small compared with the other terms. We then obtain the following linearized equation:

$$\frac{dM_a}{dt} = \mu_0 \gamma (M_i \times H_a + M_a \times H_i). \tag{8.127}$$

We choose the z axis along the direction of the internal magnetic field H_i and M_i. We also consider the time-harmonic case, with $\exp(j\omega t)$ dependence. Then we can represent (8.127) in the following matrix form:

$$\begin{bmatrix} j\omega & \omega_0 \\ -\omega_0 & j\omega \end{bmatrix} \begin{bmatrix} M_x \\ M_y \end{bmatrix} = \begin{bmatrix} 0 & \omega_M \\ -\omega_M & 0 \end{bmatrix} \begin{bmatrix} H_x \\ H_y \end{bmatrix} \tag{8.128}$$

and $M_z = 0$, where

$$M_a = \begin{bmatrix} M_x \\ M_y \\ M_z \end{bmatrix}, \quad H_a = \begin{bmatrix} H_x \\ H_y \\ H_z \end{bmatrix},$$

$$\omega_0 = -\gamma \mu_0 H_i$$

$$= \text{gyromagnetic response frequency},$$

$$\omega_M = -\gamma \mu_0 M_i$$

$$= \text{saturation magnetization frequency}.$$

From this, M_a can be expressed by

$$\begin{bmatrix} M_x \\ M_y \\ M_z \end{bmatrix} = \frac{\omega_M}{\omega_0^2 - \omega^2} \begin{bmatrix} \omega_0 & j\omega & 0 \\ -j\omega & \omega_0 & 0 \\ 0 & 0 & 0 \end{bmatrix} \begin{bmatrix} H_x \\ H_y \\ H_z \end{bmatrix}, \tag{8.129}$$

which we can write as

$$M_a = \bar{\bar{\chi}} H_a, \tag{8.130}$$

and therefore, for the ac components, we obtain

$$
\begin{aligned}
B_a &= \mu_0(H_a + M_a) \\
&= \bar{\bar{\mu}} H_a,
\end{aligned}
\tag{8.131}
$$

where the permeability tensor $\bar{\bar{\mu}}$ can be written as

$$
\bar{\bar{\mu}} = \mu_0(1 + \bar{\bar{\chi}}) = \begin{bmatrix} \mu & -j\kappa & 0 \\ j\kappa & \mu & 0 \\ 0 & 0 & \mu_0 \end{bmatrix},
$$
$$
\frac{\mu}{\mu_0} = 1 - \frac{\omega_0 \omega_M}{\omega^2 - \omega_0^2},
\tag{8.132}
$$
$$
\frac{\kappa}{\mu_0} = \frac{\omega \omega_M}{\omega^2 - \omega_0^2}.
$$

8.19 PLANE-WAVE PROPAGATION IN FERRITE

In ferrite media, the dielectric constant is scalar, and thus we write

$$
\begin{aligned}
\bar{k} \times \bar{k} \times \bar{H} &= \omega^2 \varepsilon \bar{\bar{\mu}} \bar{H}, \\
\bar{E} &= -\frac{1}{\omega \varepsilon} \bar{k} \times \bar{H}.
\end{aligned}
\tag{8.133}
$$

This can also be obtained by noting the duality principle, in which E, H, $\bar{\bar{\varepsilon}}$, and $\bar{\bar{\mu}}$ are replaced by H, $-\bar{E}$, $\bar{\bar{\varepsilon}}$, and $\bar{\bar{\mu}}$. Since this is the same form as the magnetoplasma case, only the results are shown here.

8.19.1 Propagation Along the DC Magnetic Field

The propagation constant $k = k_z$ is given by

$$
\begin{aligned}
k_+ &= \omega \sqrt{\varepsilon(\mu + \kappa)}, \\
k_- &= \omega \sqrt{\varepsilon(\mu - \kappa)}.
\end{aligned}
\tag{8.134}
$$

When $k = k_+$, we get

$$
H_x = -jH_y,
\tag{8.135}
$$

which gives a circularly polarized wave rotating clockwise in the x–y plane. The electric field E is given by

$$E_y = -Z_+ H_x,$$
$$E_x = Z_+ H_y,$$
$$Z_+ = Z_0 \frac{k^+}{k_0} = Z_0 \sqrt{\frac{\mu + \kappa}{\mu_0}}.$$
(8.136)

When $k = k_-$, we have

$$H_x = jH_y,$$
(8.137)

and the wave impedance is

$$Z_- = Z_0 \sqrt{\frac{\mu - \kappa}{\mu_0}}.$$
(8.138)

8.19.2 Propagation Perpendicular to the DC Magnetic Field

We let

$$\bar{k} = k_x \hat{x} = k\hat{x}$$
(8.139)

and we get two cases. The case when

$$k = k_1 = \omega \sqrt{\mu_0 \varepsilon}$$
(8.140)

is the same as that of isotropic medium. The only magnetic field component is H_z, and there is no coupling between H_x and H_y. When $k = k_2 = \omega \sqrt{\varepsilon \mu_t}$, $\mu_t = (\mu^2 - \kappa^2)/\mu$, we have

$$\mu H_x - j\kappa H_y = 0,$$
$$B_y = \mu_t H_y,$$
$$E_z = -Z_2 H_y, \quad Z_2 = Z_0 \sqrt{\frac{\mu_t}{\mu_0}}.$$
(8.141)

8.20 MICROWAVE DEVICES USING FERRITES

8.20.1 Faraday Rotation and Circulators

The Faraday rotation can be used to construct nonreciprocal microwave networks. For example, the device pictured here (Fig. 8.11a) contains a ferrite that produces 90° rotation of the polarization. If a wave is incident from the left, the phase is reversed

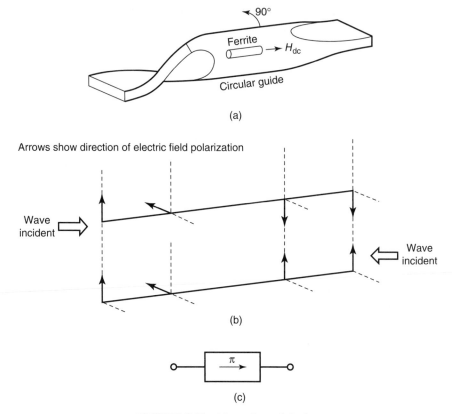

FIGURE 8.11   Nonreciprocal device.

on the other side. But if a wave is incident from the right, there is no phase shift (Fig. 8.11b). Thus this device is represented by a schematic diagram in Fig. 8.11c, showing a phase shift of 180° for a wave propagating to the right and no phase shift for a wave propagating to the left.

It is possible to make use of this device and combine it with two magic tees to construct a circulator (see Fig. 8.12). A wave entering terminal a divides equally, but because of the π phase shift, these two waves do not appear in d, but they are combined into b. Similarly, the wave entering b appears only in c. The wave entering c appears only in d, and the wave entering d appears only in a.

8.20.2 One-Way Line

It is also possible to construct a one-way line using the Faraday rotation. One scheme may be as follows. A wave entering from the left goes through this device with little attenuation, but the wave entering from the right is absorbed in the resistive sheet, which is placed in parallel with the electric field. This device is placed at the output of an oscillator to isolate the oscillator from the variations in the load and is called

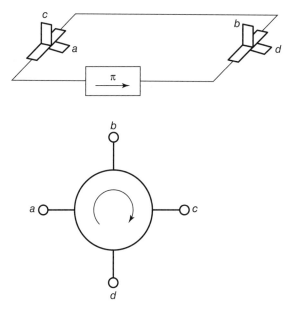

FIGURE 8.12 Circulator.

the *isolator* (Fig. 8.13). The other uses of ferrite-loaded waveguides include phase shifters and modulators.

8.20.3 Thermodynamic Paradox of One-Way Line

In the example of a one-way line above, the wave can propagate to the right, but the wave propagating to the left is absorbed by the resistive sheet. If there were no resistive sheet, it could be shown that this is not a one-way line. In fact, there should be no lossless one-way line, because if a one-way line is lossless, the energy on one side can transfer to the other side, causing a temperature rise without external work. This violates the second law of thermodynamics. However, in the study of microwave propagation through a ferrite-loaded waveguide, it was found that Maxwell's equations can be solved for lossless waveguide with lossless ferrite, and a one-way propagation constant can be obtained within some ranges of frequency and parameters. This prompted the question of whether lossless Maxwell's equations violate the law of thermodynamics.

The answer to this so-called "thermodynamic paradox" can be found in the concept that mathematical formulations of any physical problem must satisfy three conditions: (1) uniqueness, (2) existence, and (3) the solution must depend continuously on the variation of physical parameters. Problems that satisfy these three conditions are called *properly posed*. For the microwave problem above, it can be shown that if we solve Maxwell's equations for lossy ferrite and then take the limit as the resistivity goes to zero, the absorbed power is correctly accounted for and there is no conflict

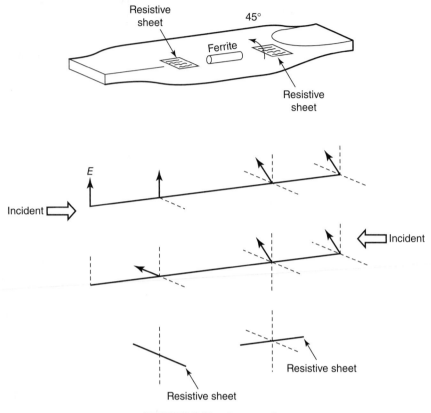

FIGURE 8.13   One-way line.

with the law of thermodynamics. Only if we start with lossless Maxwell's equations can the second law of thermodynamics be violated, but the problem in this case is improperly posed and therefore does not represent a physical problem.

8.21 LORENTZ RECIPROCITY THEOREM FOR ANISOTROPIC MEDIA

Electromagnetic fields in anisotropic media do not in general obey the reciprocity relationship. It is, however, possible to find the condition under which the reciprocal relationship holds for anisotropic media. This section deals with this question.

Let us consider the time-harmonic case ($e^{j\omega t}$). Consider the fields E_1 and H_1 in the medium $\bar{\bar{\varepsilon}}_1$ and $\bar{\bar{\mu}}_1$, and E_2 and H_2 in the medium $\bar{\bar{\varepsilon}}_2$ and $\bar{\bar{\mu}}_2$. We write two sets of Maxwell's equations

$$\nabla \times E_1 = -j\omega B_1 - J_{m1},$$

$$\nabla \times H_1 = j\omega D_1 + J_1,$$

and

$$\nabla \times E_2 = -j\omega B_2 - J_{m2},$$
$$\nabla \times H_2 = j\omega D_2 + J_2.$$
(8.142)

Now consider the identity

$$\nabla \cdot (E_1 \times H_2) = H_2 \cdot \nabla \times E_1 - E_2 \cdot \nabla \times H_2$$

and substituting from Maxwell's equations, we get

$$\nabla \cdot (E_1 \times H_2) = -j\omega[H_2 \cdot B_1 + E_1 \cdot D_2] - H_2 \cdot J_{m1} - E_1 \cdot J_2. \quad (8.143)$$

Similarly,

$$\nabla \cdot (E_2 \times H_1) = -j\omega[H_1 \cdot B_2 + E_2 \cdot D_1] - H_1 \cdot J_{m2} - E_2 \cdot J_1.$$

In isotropic media, we subtract the second equation from the first and setting $\varepsilon_1 = \varepsilon_2$ and $\mu_1 = \mu_2$, we obtain the Lorentz reciprocity theorem.

$$\nabla \cdot (E_1 \times H_2) - \nabla \cdot (E_2 \times H_1) = H_1 \cdot J_{m2} - H_2 \cdot J_{m1} + E_2 \cdot J_1 - E_1 \cdot J_2.$$
(8.144)

Note that equation (8.144) made use of the fact that for isotropic media, the first term on the right side of the two equations in (8.143) is the same, and therefore that term cancels out when taking the difference between the two.

$$H_2 \cdot B_1 - H_1 \cdot B_2 = (\mu_2 - \mu_2)H_1 \cdot H_2 = 0.$$

However, in anisotropic media, these terms do not cancel each other. Expressing in matrix form, we get

$$\tilde{H}_2 \bar{\bar{\mu}} H_1 = \tilde{H}_1 \bar{\bar{\mu}}_2 H_2 = \tilde{H}_1(\bar{\bar{\mu}}_1 - \bar{\bar{\mu}}_2)H_2.$$
(8.145)

This is not zero when $\bar{\bar{\mu}}_1 = \bar{\bar{\mu}}_2$. For this term to be zero, $\bar{\bar{\mu}}_2$ must be a transpose of $\bar{\bar{\mu}}_1$.

$$\bar{\bar{\mu}}_1 = \bar{\bar{\mu}}_2 \quad \text{and also} \quad \bar{\bar{\varepsilon}}_1 = \bar{\bar{\varepsilon}}_2.$$
(8.146)

This occurs when the dc magnetic field is reversed, or $\bar{\bar{\varepsilon}}$ *and* $\bar{\bar{\mu}}$ are symmetric tensors. Then we have the same reciprocity theorem.

Under condition (8.146), let us consider a case when $J_{m1} = J_{m2} = 0$. We integrate over a volume V, and obtain

$$\int_S (E_1 \times H_2 - E_2 \times H_1) \cdot dS = \int_V (E_2 \cdot J_1 - E_1 \cdot J_2)\, dV. \qquad (8.147)$$

Now we let the surface S expand to infinity. At a large distance, R, E, and H behave as a plane wave and E and H are perpendicular to R and to each other. Thus

$$E_1 = A\frac{e^{-jkR}}{R}, \quad H_1 = \sqrt{\frac{\varepsilon_0}{\mu_0}} i_R \times E_1,$$

$$E_2 = B\frac{e^{-jkR}}{R}, \quad H_2 = \sqrt{\frac{\varepsilon_0}{\mu_0}} i_R \times E_2.$$

Then

$$E_2 \times H_2 \cdot dS = E_1 \times H_2 \cdot i_R dS = \sqrt{\frac{\varepsilon_0}{\mu_0}} H_1 \cdot H_2\, dS$$

and

$$E_2 \times H_1 \cdot dS = \sqrt{\frac{\varepsilon_0}{\mu_0}} H_1 \cdot H_2\, dS.$$

Therefore, the integral over S then becomes zero. We then obtain

$$\int_V E_2 \cdot J_1\, dV = \int_V E_1 \cdot J_2\, dV, \qquad (8.148)$$

where V is taken as an entire space.

Let us consider the physical significance of (8.148). We let $J_1 = J_1(r_1)$ and $J_2 = J_2(r_2)$. Then the equation states that the field E_1 at r_2 due to the source J_1 at r_1 is equal to the field E_2 at r_1 due to the source J_2 at r_2, *provided that $\bar{\bar{\varepsilon}}$ and $\bar{\bar{\mu}}$ are transposed*. Thus we can say that in isotropic media, the reciprocity theorem holds, but in anisotropic media, the reciprocity holds for the following cases: (1) when $\bar{\bar{\varepsilon}}$ and $\bar{\bar{\mu}}$ are symmetric as in the case of crystal, or (2) when the dc magnetic field is reversed for the case of ferrite and magnetoplasma. When ferrite material is used in antennas, the transmitting pattern is not equal to the receiving pattern unless the dc magnetic field is reversed. This is an important consideration when making measurement of antenna radiation pattern.

8.22 BI-ANISOTROPIC MEDIA AND CHIRAL MEDIA

At the end of Section 8.12, we mentioned the difference between Faraday rotation and natural rotation. If a wave is propagated in plasma or ferrite along the dc magnetic field, and then reflected back and propagated in the reverse direction, the rotation of the plane of polarization is doubled, while the rotation is canceled for natural rotation. This natural rotation of the plane of polarization is called "optical activity," and is caused by the right-handed or left-handed property of the medium such as right-handed helix or left-handed helix. The optical activity has been known for many years and its mathematical formulations have been investigated by many workers (see Sommerfeld, 1954; Kong, 1972, 1974; Bassiri et al., 1988; Lakhtakia et al., 1988).

In a preceding section, we have discussed anisotropic media characterized by the constitutive relations

$$\bar{D} = \bar{\bar{\varepsilon}}\bar{E} \quad \text{and} \quad \bar{B} = \bar{\bar{\mu}}\bar{H}. \tag{8.149}$$

We can generalize these relations to the bi-anisotropic media

$$\begin{bmatrix} \bar{D} \\ \bar{H} \end{bmatrix} = \begin{bmatrix} \bar{P} & \bar{L} \\ \bar{M} & \bar{Q} \end{bmatrix} \begin{bmatrix} \bar{E} \\ \bar{B} \end{bmatrix}, \tag{8.150}$$

where, in general, $\bar{P}, \bar{L}, \bar{M}$, and $\bar{Q}$ are 3×3 matrices. Note that in (8.150), $\bar{D}$ and $\bar{H}$ are given in terms of $\bar{E}$ and $\bar{B}$. As indicated at the end of Section 2.1, the force depends on $\bar{E}$ and $\bar{B}$, and therefore $\bar{E}$ and $\bar{B}$ are the fundamental field quantities and $\bar{D}$ and $\bar{H}$ are the derived fields through the constitutive relations.

If the medium is lossless, then $\text{Re}(\nabla \cdot \bar{S})$ must be zero, where $\bar{S} = \frac{1}{2}\bar{E} \times \bar{H}^*$ is the complex Poynting vector (Section 2.3). Noting that

$$\nabla \cdot \bar{S} = -\frac{j\omega}{2}[\bar{H}^* \cdot \bar{B} - \bar{E} \cdot \bar{D}^*], \tag{8.151}$$

$\text{Re}(\nabla \cdot \bar{S}) = 0$ is equivalent to

$$\bar{H}^* \cdot \bar{B} - \bar{H} \cdot \bar{B}^* - \bar{E} \cdot \bar{D}^* + \bar{E}^* \cdot \bar{D} = 0. \tag{8.152}$$

Substituting (8.150) into this, we get

$$\begin{aligned} \bar{P} &= \bar{P}^+, \\ \bar{Q} &= \bar{Q}^+, \\ \bar{M} &= -\bar{L}^+, \end{aligned} \tag{8.153}$$

where $\bar{P}^+$ means the complex conjugate of the transpose of $\bar{P}$. The detailed exposition on this topic is given by Kong (1972).

If $\bar{P}, \bar{L}, \bar{M}$, and $\bar{Q}$ are scalar, this is called the *bi-isotropic medium*, which is also called *chiral medium*. Noting the symmetry relations (8.153), we write the constitutive relations for a lossless chiral medium as

$$\bar{D} = \varepsilon\bar{E} - j\gamma\bar{B},$$
$$\bar{H} = -j\gamma\bar{E} + \frac{1}{\mu}\bar{B},$$

(8.154)

where ε, μ, and γ are real scalar constants. Note that γ has the dimension of admittance, $(\varepsilon/\mu)^{1/2}$. It is also possible to rewrite (8.154) in the form

$$\bar{D} = (\varepsilon + \gamma^2\mu)\bar{E} - j\gamma\mu\bar{H},$$
$$\bar{B} = j\gamma\mu\bar{E} + \mu\bar{H}.$$

(8.155)

Furthermore, using $\nabla \times \bar{H} = +j\omega\bar{D}$ and $\nabla \times \bar{E} = -j\omega\bar{B}$, we can rewrite (8.155) in the following form:

$$\bar{D} = \varepsilon_1[\bar{E} + \beta\nabla \times \bar{E}],$$
$$\bar{B} = \mu_1[\bar{H} + \beta\nabla \times \bar{H}].$$

(8.156)

This form shows that for bi-isotropic medium, $\bar{D}$ depends not only on $\bar{E}$ at a point, but also on the behavior of $\bar{E}$ in the neighborhood of that point represented by $\nabla \times \bar{E}$. This nonlocal behavior of $\bar{D}$ is called the *spatial dispersion*.

Let us combine Maxwell's equations with the constitutive relations (8.155).

$$\nabla \times \bar{E} = -j\omega\bar{B} = -j\omega(\mu\bar{H} + j\gamma\mu\bar{E}),$$
$$\nabla \times \bar{H} = j\omega\bar{D} = j\omega[(\varepsilon + \gamma^2\mu)\bar{E} - j\gamma\mu\bar{H}].$$

(8.157)

We first substitute $\bar{H}$ from the first equation into the second equation and express $\nabla \times \bar{H}$ in terms of $\nabla \times \bar{E}$ and $\bar{E}$. Then we take the curl of the first equation and substitute $\nabla \times \bar{H}$. We then get the equation for $\bar{E}$.

$$-\nabla \times \nabla \times \bar{E} + 2\omega\gamma\mu\nabla \times \bar{E} + \omega^2\mu\varepsilon\bar{E} = 0$$

(8.158)

and we get the identical equation for $\bar{H}$.

Let us find the propagation constant K for a plane wave propagating in the z direction. Since $\bar{E}$ behaves as $\exp(-jKz)$, we get $\nabla = -jK\hat{z}$ and therefore (8.158) becomes

$$\begin{bmatrix} -K^2 + k^2 & jK2\omega\gamma\mu \\ jK2\omega\gamma\mu & -K^2 + k^2 \end{bmatrix} \begin{bmatrix} E_x \\ E_y \end{bmatrix} = 0,$$

(8.159)

where $k^2 = \omega^2\mu\varepsilon$.

A nonzero solution to (8.159) is obtained by letting the determinant of coefficients be zero.

$$\begin{vmatrix} -K^2 + k^2 & jK2\omega\gamma\mu \\ jK2\omega\gamma\mu & -K^2 + k^2 \end{vmatrix} = 0. \tag{8.160}$$

This can be solved to obtain two propagation constants,

$$\begin{aligned} K_1 &= \omega\mu\gamma + [(\omega\mu\gamma)^2 + k^2]^{1/2} \\ K_2 &= -\omega\mu\gamma + [(\omega\mu\gamma)^2 + k^2]^{1/2}. \end{aligned} \tag{8.161}$$

Substituting K_1 into one of the equations (8.159), we get

$$(-K_1^2 + k^2)E_x + jK_1 2\omega\gamma\mu E_y = 0,$$

from which we get

$$E_x = jE_y. \tag{8.162}$$

This is a RHC. The corresponding magnetic field is obtained from (8.157).

$$\bar{H} = \frac{j}{\omega\mu} - \nabla \times \bar{E} - j\gamma\bar{E}. \tag{8.163}$$

Using $\nabla = -jK\hat{z}$ and (8.162), we get

$$\frac{E_x}{H_y} = -\frac{E_y}{H_x} = \frac{\omega\mu}{[(\omega\mu\gamma)^2 + k^2]^{1/2}}. \tag{8.164}$$

Similarly, for K_2, we get a LHC.

$$E_x = -jE_y. \tag{8.165}$$

The ratio of E to H is the same as given in (8.164). It is clear from (8.161) that if $\gamma > 0$, $K_1 > k > K_2$ and thus the phase velocity for RHC is slower than for LHC. If $\gamma < 0$, $K_1 < k < K_2$ and thus the LHC wave has a slower phase velocity than the RHC wave. From (8.157), we note that taking the divergence, $\nabla \cdot \bar{B} = 0$ and $\nabla \cdot \bar{D} = 0$. Also, from (8.158), we get $\nabla \cdot \bar{H} = 0$ and $\nabla \cdot \bar{H} = 0$. Therefore, using $\nabla = -jK$ for a plane wave, we conclude that the plane wave in chiral medium is a TEM wave.

All the analysis shown above can be derived from (8.156) with the following correspondence:

$$\mu = \frac{\mu_1}{1 - k_1^2\beta^2}, \quad \varepsilon = \varepsilon_1, \quad \gamma = \omega\varepsilon_1\beta, \quad k_1^2 = \omega^2\mu_1\varepsilon_1. \tag{8.166}$$

The wave equation becomes

$$\nabla \times \nabla \times \bar{E} = 2\gamma_1^2 \beta \nabla \times \bar{E} + \gamma_1^2 \bar{E}, \tag{8.167}$$

where $\gamma_1^2 = k_1^2 / (1 - k_1^2 \beta^2)$. The propagation constant then becomes

$$K_1 = \frac{k_1}{1 - k_1 \beta}, \quad K_2 = \frac{k_1}{1 + k_1 \beta}. \tag{8.168}$$

8.23 SUPERCONDUCTORS, LONDON EQUATION, AND THE MEISSNER EFFECTS

Superconductors are used in transmission lines, waveguides, and resonant cavities because they have low losses, their dispersion is small, and they can be used for broadband transmission. In this and the next section, we discuss the derivation of London equations; the Meissner effect; and the complex conductivity, penetration depth, and surface impedance of superconductors at high frequencies. We will not discuss the physics of superconductors or the historical development, such as the discovery of superconductivity by Kamerlingh Onnes in 1911, Meissner's work, Fritz and Heinz London's work, BCS theory (Bardeen, Cooper, and Schrieffer), Ginzburg–Landau theory, and the work on high-temperature superconductors. They are clearly outside the scope of this book. For the topics discussed in this section, readers are referred to very informative books by Mendelssohn (1966), Van Duzer and Turner (1981), Ghoshal and Smith (1988), and Lee and Itoh (1989).

In normal conductors, free electrons are assumed to move under the influence of the electric field and experience collisions as described in Section 8.3. In superconductors, pairs of electrons are involved and their behavior is quite different from that of single electrons. The electron pairs are immune from collisions.

Consider the electron-pair fluid. We write the equation of motion

$$m^* \frac{\partial \bar{v}_s}{\partial t} = e^* \bar{E}, \tag{8.169}$$

where $m^* = 2m$ and $e^* = -2e$ are the pair effective mass and pair effective charge, and m and e are electron mass and charge. The pair-current density J_s is given by

$$\bar{J}_s = n_s^* e^* \bar{v}_s, \tag{8.170}$$

where n_s^* is the number density of pair. Combining these two, we get the first London equation

$$\Lambda \frac{\partial \bar{J}_s}{\partial t} = \bar{E}, \quad \Lambda = \frac{m^*}{n_s^* e^{*2}}. \tag{8.171}$$

Next we use one of the Maxwell equations,

$$\nabla \times \bar{E} = -\frac{\partial \bar{B}_s}{\partial t}. \tag{8.172}$$

Substituting the first London equation, we get

$$\Lambda \nabla \times \bar{J}_s + \bar{B}_s = 0. \tag{8.173}$$

This is the second London equation. It is interesting to note that in ordinary electromagnetic theory, the steady current produces the magnetic field through $\nabla \times \bar{H} = \bar{J}_s$, but steady magnetic field does not produce the current in a conductor. However, in the superconductor, steady magnetic field causes current as shown in (8.173).

Let us consider the dc case. We then have

$$\nabla \times \bar{H} = \bar{J}_s. \tag{8.174}$$

Taking the curl of this equation and substituting (8.173) and noting that $\mu \cong \mu_0$ and $\nabla \cdot \bar{B} = 0$, we get

$$\nabla^2 \bar{B} = \frac{\bar{B}}{\lambda^2}, \tag{8.175}$$

where $\lambda^2 = \Lambda/\mu_0 = m^*/n_s^* e^{*2} \mu_0$. If $z = 0$ is the surface of superconductor and $\bar{B} = B(z)\hat{x}$, we get, from (8.175),

$$B(z) = B_0 \exp\left(-\frac{z}{\lambda}\right). \tag{8.176}$$

This shows that the static magnetic field penetrates the superconductor only a small distance given by λ. This λ is called the *penetration distance* and is of the order of 0.05–0.1 μm. The phenomenon that the superconductor tends to exclude the static magnetic flux is called the *Meissner effect*.

When a small magnet is dropped onto a superconductive plate, the magnetic flux cannot penetrate the surface and induces current on the surface. This, in effect, creates the image of magnet with the same polarity, which lies below the surface, and therefore the magnet is repelled by the image, cannot approach the superconductor, and is suspended in air.

The penetration distance λ depends on the pair density n_s^* as shown in (8.175). The number of paired electron $n_s = 2n_s^*$ varies as a function of temperature T.

$$\frac{n_s}{n} = 1 - \left(\frac{T}{T_c}\right)^4, \tag{8.177}$$

where n is the number of conducting electrons and T_c is the critical temperature. Therefore, the penetration distance λ also varies as a function of temperature.

$$\lambda(T) = \lambda(0) \left[1 - \left(\frac{T}{T_c} \right)^4 \right]^{-1/2}. \tag{8.178}$$

8.24 TWO-FLUID MODEL OF SUPERCONDUCTORS AT HIGH FREQUENCIES

Let us next consider the high-frequency behavior of the superconductor. In general, only a fraction of the conducting electrons is in superconductive state; the remainder is in the normal state. For the electron pairs in superconductive state, we have, from (8.169),

$$m \frac{d\bar{v}_s}{dt} = -e\bar{E}, \tag{8.179}$$

where $m^* = 2m$ and $e^* = -2e$ are used. For a normal state, we have

$$m \frac{d\bar{v}_n}{dt} + m \frac{\bar{v}_n}{\tau} = -e\bar{E}, \tag{8.180}$$

where τ is the momentum relaxation time and is normally much shorter than 10^{-11} s. The total current density $\bar{J}$ is given by the sum of the superconductive current $\bar{J}_s$ and the normal current $\bar{J}_n$.

$$\begin{aligned}
\bar{J} &= \bar{J}_s + \bar{J}_n, \\
\bar{J}_s &= -n_s e \bar{v}_s, \\
\bar{J}_n &= -n_n e \bar{v}_n, \\
n &= n_s + n_n.
\end{aligned} \tag{8.181}$$

For a time-harmonic field with $\exp(j\omega t)$, we get from (8.179) and (8.180),

$$\begin{aligned}
\bar{J}_s &= -j \frac{e^2 n_s}{m\omega} \bar{E}, \\
\bar{J}_n &= -j \frac{e^2 n_n}{m\omega} \frac{\bar{E}}{1 - j(1/\omega\tau)}.
\end{aligned} \tag{8.182}$$

Thus we get complex conductivity σ

$$\bar{J} = \sigma\bar{E},$$

$$\sigma = \sigma_1 - j\sigma,$$

$$\sigma_1 = \frac{e^2 n_n \tau}{m(1 + \omega^2 \tau^2)}, \tag{8.183}$$

$$\sigma_2 = \frac{e^2 n_s}{m\omega} + \frac{e^2 n_n (\omega\tau)^2}{m\omega(1 + \omega^2 \tau^2)}.$$

In most applications, $\omega^2 \tau^2 \ll 1$ for frequency $<10^{11}$ Hz, and therefore

$$\sigma_1 = \frac{e^2 n_n \tau}{m}$$

$$= \sigma_n \left(\frac{n_n}{n}\right) = \sigma_n \left(\frac{T}{T_c}\right)^4, \tag{8.184}$$

$$\sigma_2 = \frac{e^2 n_s}{m\omega} = \frac{1}{\omega\mu_0\lambda^2},$$

where σ_n is the conductivity in the normal state ($\sigma_n = e^2 n\tau/m$).

Let us next consider the penetration for electromagnetic wave into the superconductor. The wave is propagating into the superconductor ($z > 0$). Then we have the field $\bar{E} = E_x \hat{x}$, given by

$$E_x(z) = E_x(0) \exp(-jKz). \tag{8.185}$$

Noting that the displacement current is negligible compared with the conduction current, we have

$$K = \omega \left[\mu_0 \left(\varepsilon - j\frac{\sigma}{\omega}\right)\right]^{1/2}$$

$$\simeq \omega \left[\mu_0 \left(-j\frac{\sigma}{\mu}\right)\right]^{1/2}. \tag{8.186}$$

Using $\sigma = \sigma_1 - j\sigma_2$, we get

$$\exp(-jKz) = \exp\left[-\frac{1}{\lambda}\left(1 + \tau\omega\frac{n_n}{n_s}\right)^{1/2}\right] z \tag{8.187}$$

$$= \exp(-\alpha - j\beta)z.$$

The penetration depth is then given by α^{-1} and is approximately equal to λ. Note that as $\omega \to 0$, the penetration distance reduces to λ, but as the frequency is increased, the penetration depth decreases.

Next, let us consider the surface impedance Z_s. This is given by

$$Z_s = \left[\frac{\mu_0}{\varepsilon - j(\sigma/\omega)} \right]^{1/2} \approx \left(\frac{j\mu_0\omega}{\sigma} \right)^{1/2}.$$

Noting that $\sigma = \sigma_1 - j\sigma_2$, we get

$$Z_s = R_s + jX_s,$$
$$R_s = \frac{1}{2}\sigma_n \left(\frac{n_n}{n} \right) (\omega\mu_0)^2 \lambda^3, \qquad (8.188)$$
$$X_s = \omega\mu_0\lambda.$$

Note that R_s increases with ω^2, in contrast with normal conductors, whose R_s increases only as $\omega^{1/2}$.

Microstrip lines using a high-temperature superconductor give low loss and virtually no dispersion (Lee and Itoh, 1989; Ghoshal and Smith, 1988). Typical values used are $T_c = 92.5$ K, $T = 77$ K (liquid nitrogen), $\lambda(0) = 0.14$ μm and $\sigma_n = 0.5$ S/μm. Thus $\sigma_1 = 0.24$ S/μm and $\sigma_2 = 336$ S/μm at 10 GHz. The strip line has an attenuation of 10^{-3} dB/cm, compared with an aluminum line with 1 dB/cm attenuation. The phase velocity for the superconducting line is almost constant over a wide frequency range (Lee and Itoh, 1989).

PROBLEMS

8.1 Calculate and plot the refraction index of fused silica as a function of wavelength (0.5–2.0 μm) using the Sellmeier equation.

8.2 Assuming that the plasma frequency and the collision frequency of silver given in Section 8.3 are constant, calculate and plot the dielectric constant of silver (real and imaginary parts) as a function of wavelength ($\lambda = 0.01$– 1.0 μm).

8.3 For an isotropic plasma, assume that the electron density is 10^6 cm^{-3} and that the collision is negligible. A plane wave with $|E| = 1$ V/m at $f = 10$ MHz is propagating through this plasma. Find the phase velocity, group velocity, magnitudes of the magnetic field H, displacement vector D and magnetic flux B, and power flux density. Also find the maximum displacement of an electron.

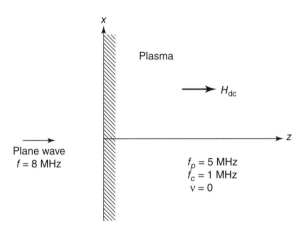

FIGURE P8.6   Plane wave incident on plasma.

8.4 Calculate and plot the effective dielectric constant of a mixture of two media with $\varepsilon_1 = 2$ and $\varepsilon_2 = 3$ as a function of the fractional volume $f_1(0-1)$, using the Maxwell–Garnett and Polder–van Santen formulas.

8.5 A radio wave is propagating through the ionosphere along the dc magnetic field. Assume that the electron density is $N_e = 10^3 \text{ cm}^{-3}$, the geomagnetic field is 0.5 G, the collision frequency is zero, and the radio frequency is 10 kHz. Calculate the propagation constant, phase velocity, and group velocity. If the magnitude of the electric field is 0.1 V/m, describe the motion of an electron (1 tesla = 1 weber/m^2 = 10^4 Gauss).

8.6 A plane wave is normally incident on a plasma with the dc magnetic field normal to the surface. The incident wave is polarized in the x direction. Find the reflected waves E_x and E_y (Fig. P8.6).

8.7 Consider a wave propagating in a lossless plasma perpendicular to the dc magnetic field. Plot the equivalent dielectric constants ε_z and ε_t as a function of frequency and show the frequency ranges where the dielectric constant is positive.

8.8 Calculate the reflection coefficient of a radio wave of 1 MHz incident on the ionosphere from the air as shown in Fig. P8.8. The dc magnetic field is pointed in the z direction. The ionosphere characteristics are the same as those in Problem 8.5. Plot the reflection coefficient as a function of the incident angle $\theta(-\pi/2 < \theta < \pi/2)$. Do this for the TM and TE waves.

8.9 Assume that the ionosphere has the following plasma frequency profile:

$$f_p^2 = f_0^2 \left[1 - \left(\frac{z - z_1}{h_0} \right)^2 \right] \quad \text{for } |z - z_1| < h_0,$$

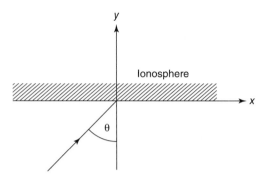

FIGURE P8.8   Wave incident on the ionosphere.

where $f_0 = 1$ MHz, $z_1 = 200$ km, $h_0 = 100$ km, and $f_p = 0$ for $|z - z_1| > h_0$. For a radio wave of 500 kHz, find the true height, phase height, and equivalent height.

8.10   For a whistler mode, if $\omega \ll \omega_p$, $\omega \ll \omega_c$, show that the group velocity is given approximately by

$$v_g \sim 2c\frac{(ff_c)^{1/2}}{f_p}$$

and that the time required for wave propagation along the magnetic field is proportional to $f^{-1/2}$.

8.11   Assume that the cyclotron frequency is 1.42 MHz and that the plasma frequency is 0.5 MHz. Calculate the time required for the whistler mode to propagate over a path length of 50,000 km in the frequency range 1–10 kHz.

8.12   Find the phase velocity and group velocity of the acoustic wave and electromagnetic wave of magnetoplasma at $T = 3000$ K. The plasma frequency is 8 MHz and the operating frequency is 10 MHz. Assume that the collision frequency is zero.

8.13   Consider ferrite placed in a dc magnetic field (the internal dc magnetic field $H_i = 1000$ Oe and the saturation magnetization $M_i = 1700$ G). The relative dielectric constant is $\varepsilon_r = 10$. Find the frequency range in which Faraday rotation takes place (1 A/m = $4\pi \times 10^{-3}$ Oe; 1 Wb/m^2 = 10^4 G).

8.14   For the ferrite in Problem 8.13, plot $\mu + k$, $\mu - k$, and μ_t as functions of frequency.

8.15   Show that ε_1, μ_1, and β in (8.156) are related to ε, μ, and γ through (8.166).

8.16   A linearly polarized plane wave is normally incident on a chiral medium from the air. Find the reflected and transmitted waves. The chiral medium has $\varepsilon = 9\varepsilon_0$ and $\mu = \mu_0$. Calculate the above for $\gamma(\mu_0/\varepsilon_0)^{1/2} = 0.1$, 1.0, and 10.

8.17 A microwave at 10 GHz is normally incident on a superconductor shown at the end of Section 8.24 at a liquid nitrogen temperature of 77 K. First derive (8.188), taking into account that $\sigma_1 \ll \sigma_2$. Next, if the incident power flux density is 1 mW/cm^2, calculate the power absorbed by the surface. Compare this with the power absorbed by the copper surface with $\sigma = 450$ S/μm at 77 K and with $\sigma = 58$ S/μm at room temperature.

CHAPTER 9

ANTENNAS, APERTURES, AND ARRAYS

In this chapter, we review the fundamental definitions and formulations related to antennas, aperture antennas, linear antennas, and arrays. For a more complete treatment of these topics, see Elliott (1981), Jull (1981), Mittra (1973), Wait (1986), Balanis (1982), Ma (1974), Hansen (1966), Stutzman and Thiele (1981), Stark (1974), Mailloux (1982), and Lo and Lee (1988).

9.1 ANTENNA FUNDAMENTALS

An antenna is usually designed to produce a desired radiation pattern in various directions. The most useful measures of antenna characteristics are *directivity* and *gain*. If the power radiated per unit solid angle in the direction (θ, ϕ) is $P(\theta, \phi)$ and the total power radiated is P_t, the directivity D is given by

$$D = \frac{4\pi P(\theta, \phi)}{P_t}. \tag{9.1}$$

Note that $P_t/4\pi$ is the average power per unit solid angle if the antenna is an isotropic radiator, and therefore, the directivity represents the ability of the antenna to produce more power in a certain direction and less power in other directions than the isotropic

Electromagnetic Wave Propagation, Radiation, and Scattering: From Fundamentals to Applications,
Second Edition. Akira Ishimaru.
© 2017 by The Institute of Electrical and Electronic Engineers, Inc. Published 2017 by John Wiley & Sons, Inc.

radiator. It is clear from definition (9.1) that the integral of the directivity over all the solid angle is 4π

$$\int_{4\pi} D(\theta, \phi)\, d\Omega = 4\pi. \tag{9.2}$$

The gain function $G(\theta, \phi)$ is defined by

$$G(\theta, \phi) = \frac{4\pi P(\theta, \phi)}{P_i}, \tag{9.3}$$

where P_i is the power input to the antenna.

If the antenna is lossless, $P_i = P_t$ and therefore $G = D$. The transmitting pattern is the normalized directivity defined by $D(\theta, \phi)/D_m$, where D_m is the maximum value of D.

A receiving antenna absorbs power from incident waves. If the power flux density of the incident wave coming from the direction (θ, ϕ) is $S(\theta, \phi)$ and the received power is P_r, the receiving cross section $A_r(\theta, \phi)$ is defined by

$$P_r = SA_r. \tag{9.4}$$

The receiving pattern is given by $A_r(\theta, \phi)/A_{rm}$, where A_{rm} is the maximum value of the receiving cross section.

Next, we show that the transmitting and receiving patterns of an antenna are the same.

$$\frac{D(\theta, \phi)}{D_m} = \frac{A_r(\theta, \phi)}{A_{rm}}. \tag{9.5}$$

Also, we show that the receiving cross section $A_r(\theta, \phi)$ and the gain $G(\theta, \phi)$ of any matched antennas are related by

$$A_r(\theta, \phi) = \frac{\lambda^2}{4\pi} G(\theta, \phi). \tag{9.6}$$

To show these relationships, consider two matched lossless antennas 1 and 2 (Fig. 9.1). If the power P_1 is supplied to antenna 1, the received power P_{r2} at antenna 2 is given by

$$P_{r2} = \frac{P_1 G_1 A_{r2}}{4\pi R^2}. \tag{9.7}$$

Similarly, if the power P_2 is applied to antenna 2, the received power at antenna 1 is

$$P_{r1} = \frac{P_2 G_2 A_{r1}}{4\pi R^2}. \tag{9.8}$$

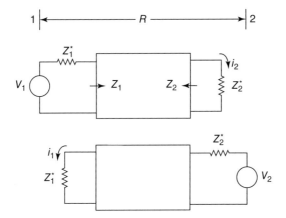

FIGURE 9.1   Reciprocity.

Now, we can show the relationship between P_{r2} and P_{r1} by examining the equivalent network shown in Fig. 9.1. The impedances are matched with the conjugate impedances. Then P_1 and P_{r2} are given by

$$P_1 = \frac{|V_1|^2}{8R_1}, \quad P_{r2} = \frac{R_2|i_2|^2}{2}, \tag{9.9}$$

where $Z_1 = R_1 + jX_1$ and $Z_2 = R_2 + jX_2$ are the input impedances shown in Fig. 9.1. Similarly, we have

$$P_2 = \frac{|V_2|^2}{8R_2}, \quad P_{r1} = \frac{R_1|i_1|^2}{2}. \tag{9.10}$$

Therefore,

$$\frac{P_{r2}}{P_1} = \frac{4R_1R_2|i_2|^2}{|V_1|^2},$$

$$\frac{P_{r1}}{P_2} = \frac{4R_1R_2|i_1|^2}{|V_2|^2}. \tag{9.11}$$

But by the reciprocity theorem, we have

$$\frac{i_2}{V_1} = \frac{i_1}{V_2}. \tag{9.12}$$

Therefore, we get

$$\frac{P_{r2}}{P_1} = \frac{P_{r1}}{P_2} = \frac{G_1A_{r2}}{4\pi R^2} = \frac{G_2A_{r1}}{4\pi R^2}. \tag{9.13}$$

From this, we get

$$\frac{G_1}{A_{r1}} = \frac{G_2}{A_{r2}}. \tag{9.14}$$

Since the foregoing antennas are arbitrary, we conclude that

$$\frac{G}{A_r} = \text{universal constant} = C. \tag{9.15}$$

This also applies to the direction of maximum gain, and thus

$$\frac{G(\theta, \phi)}{A_r(\theta, \phi)} = \frac{G_m}{A_{rm}}, \tag{9.16}$$

which gives (9.5).

   To obtain the universal constant for a lossless matched antenna in (9.15), we consider the average receiving cross section:

$$\frac{1}{4\pi} \int A_r(\theta, \phi)\, d\Omega = \frac{1}{4\pi C} \int G\, d\Omega = \frac{1}{C}. \tag{9.17}$$

Since this applies to any antenna, the constant C is obtained by calculating the average receiving cross section of a convenient simple antenna. If we take a short dipole antenna terminated with the matched conjugate impedance with the radiation resistance R_0, the received power P_r, when a plane wave polarized in the plane containing the dipole is incident at angle θ, is given by (Fig. 9.2)

$$P_r = \frac{V^2}{8R_0} = \frac{E_0^2 l^2 \sin^2\theta}{8R_0} = A_r(\theta)\frac{|E_0|^2}{2\eta}. \tag{9.18}$$

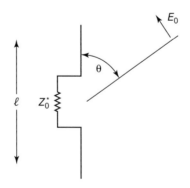

FIGURE 9.2   Short dipole terminated with conjugate impedance Z_0^*.

We therefore get

$$A_r(\theta) = \frac{\eta l^2}{4R_0} \sin^2\theta. \tag{9.19}$$

The average receiving cross section is

$$\frac{1}{C} = \frac{1}{4\pi} \int A_r(\theta) \, d\Omega = \frac{\eta l^2}{6R_0}. \tag{9.20}$$

However, the radiation resistance of a short dipole is (Eq. 5.27)

$$R_0 = \eta \left(\frac{l}{\lambda}\right)^2 \frac{2\pi}{3}. \tag{9.21}$$

Therefore, we get

$$\frac{1}{C} = \frac{\lambda^2}{4\pi}, \tag{9.22}$$

which proves (9.6).

9.2 RADIATION FIELDS OF GIVEN ELECTRIC AND MAGNETIC CURRENT DISTRIBUTIONS

The exact current distributions on an antenna such as a wire antenna must be determined by solving the complete boundary value problem. This is discussed in Section 9.8. However, in many practical antennas, the approximate current distributions are known based on more exact calculations or on experimental data. In this section, we present the radiation field due to given current distributions.

Let us consider the given distribution of the electric current $\bar{J}$ and the magnetic current $\bar{J}_m$ in space (Fig. 9.3). The electric field outside the source region is given by the Hertz vector formulations in Sections 5.2 and 5.3.

$$\bar{E} = \nabla \times \nabla \times \bar{\pi} - j\omega\mu_0 \nabla \times \bar{\pi}_m,$$

$$\bar{\pi} = \int G_0(\bar{r}, \bar{r}') \frac{\bar{J}(\bar{r}')}{j\omega\varepsilon_0} \, dV',$$

$$\bar{\pi}_m = \int G_0(\bar{r}, \bar{r}') \frac{\bar{J}_m(\bar{r}')}{j\omega\mu_0} \, dV', \tag{9.23}$$

where

$$G_0(\bar{r}, \bar{r}') = \frac{\exp(-jk|\bar{r} - \bar{r}'|)}{4\pi|\bar{r} - \bar{r}'|}.$$

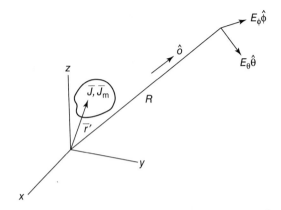

FIGURE 9.3   Radiation field E_θ and E_ϕ due to $\bar{J}(\bar{r}')$ and $\bar{J}_m(\bar{r}')$.

In the far field, Green's function is approximated by (6.119)

$$G_0(\bar{r}, \bar{r}') = \frac{e^{-jkR + jk\hat{o}\cdot\bar{r}'}}{4\pi R},$$

and the operator ∇ is equal to $\nabla = -jk\hat{o}$. Therefore, in the far zone, we get

$$\bar{E} = \frac{j\omega\mu_0}{4\pi R} e^{-jkR} \hat{O} \times \hat{O} \times \bar{I} + \frac{jk}{4\pi R} e^{-jkR} \hat{O} \times \bar{I}_m, \qquad (9.24)$$

where

$$I = \int \bar{J}(\bar{r}') e^{jk\hat{O}\cdot\bar{r}'}\, dV',$$

$$I_m = \int \bar{J}_m(\bar{r}') e^{jk\hat{O}\cdot\bar{r}'}\, dV'.$$

Now, the components E_θ and E_ϕ can be obtained by

$$\begin{aligned} E_\theta &= \hat{\theta} \cdot \bar{E}, \\ E_\phi &= \hat{\phi} \cdot \bar{E}. \end{aligned} \qquad (9.25)$$

Noting the identities

$$\hat{O} \times \hat{O} \times \bar{J} = \hat{O}(\hat{O} \cdot \bar{J}) - \bar{J},$$

$$\hat{\theta} \cdot (\hat{O} \times \bar{J}_m) = \bar{J}_m \cdot (\hat{\theta} \times \hat{O}) = -\hat{\phi} \cdot \bar{J}_m,$$

$$\hat{\phi} \cdot (\hat{O} \times \bar{J}_m) = \bar{J}_m \cdot (\hat{\phi} \times \hat{O}) = -\hat{\theta} \cdot \bar{J}_m,$$

we get

$$
\begin{aligned}
E_\theta &= -\frac{j\omega\mu_0}{4\pi R}e^{-jkR}\int \hat{\theta}\cdot \bar{J}(\bar{r}')e^{jk\hat{o}\cdot\bar{r}'}\,dV' - \frac{jk}{4\pi R}e^{-jkR}\int \hat{\phi}\cdot\bar{J}_m(\bar{r}')e^{-jk\hat{o}\cdot\bar{r}'}\,dV', \\
E_\phi &= -\frac{j\omega\mu_0}{4\pi R}e^{-jkR}\int \hat{\phi}\cdot \bar{J}(\bar{r}')e^{jk\hat{o}\cdot\bar{r}'}\,dV' - \frac{jk}{4\pi R}e^{-jkR}\int \hat{\theta}\cdot\bar{J}_m(\bar{r}')e^{-jk\hat{o}\cdot\bar{r}'}\,dV'.
\end{aligned}
\tag{9.26}
$$

The magnetic field $\bar{H}$ in the far zone is simply related to $\bar{E}$

$$
\bar{H} = \frac{1}{\eta}\hat{o}\times\bar{E}, \quad \eta = \left(\frac{\mu_0}{\varepsilon_0}\right)^{1/2}.
\tag{9.27}
$$

Therefore,

$$
H_\theta = \frac{E_\phi}{\eta}, \quad H_\phi = \frac{E_\theta}{\eta}.
\tag{9.28}
$$

This is the fundamental formula for the far-field components E_θ and E_ϕ, due to the given current distributions $\bar{J}$ and $\bar{J}_m$. Note that the θ and ϕ components of $\bar{J}$ contribute to E_θ and E_ϕ and the ϕ and θ components of $\bar{J}_m$ contribute to E_θ and E_ϕ. The actual calculations can be done in different coordinate systems. For example, in a Cartesian system, we have

$$
\begin{aligned}
\theta &= \hat{x}\cos\theta\cos\phi + \hat{y}\cos\theta\sin\phi - \hat{z}\sin\theta, \\
\phi &= -\hat{x}\sin\phi + \hat{y}\cos\phi, \\
\hat{o} &= \hat{x}\sin\theta\cos\phi + \hat{y}\sin\theta\sin\phi + \hat{z}\cos\theta, \\
\bar{r}' &= \hat{x}x' + \hat{y}y' + \hat{z}z'.
\end{aligned}
\tag{9.29}
$$

Therefore,

$$
\begin{aligned}
\hat{\theta}\cdot\bar{J} &= \cos\theta\cos\phi\,J_x + \cos\theta\sin\phi\,J_y - \sin\theta\,J_z. \\
\hat{\phi}\cdot\bar{J} &= -\sin\phi\,J_x + \cos\phi\,J_y, \\
\hat{o}\cdot\bar{r}' &= x'\sin\theta\cos\phi + y'\sin\theta\sin\phi + z'\cos\theta.
\end{aligned}
\tag{9.30}
$$

In a cylindrical system, we have

$$
\begin{aligned}
\bar{J}(r',\phi',z') &= \hat{r}'J_{r'} + \hat{\phi}'J_\phi + \hat{z}'J_{z'}, \\
\bar{r}' &= \hat{x}\rho'\cos\phi' + \hat{y}\rho'\sin\phi' + \hat{z}z'.
\end{aligned}
\tag{9.31}
$$

Therefore, we get

$$\hat{\theta} \cdot \bar{J} = \cos\theta \sin(\phi - \phi')J_{\phi'} + \cos\theta \cos(\phi - \phi')J_{\rho'} - \sin\theta J_{z'},$$

$$\hat{\phi} \cdot \bar{J} = \cos(\phi - \phi')J_{\phi'} + \sin(\phi - \phi')J_{\rho'}, \qquad (9.32)$$

$$\hat{o} \cdot \bar{r} = \rho' \sin\theta \cos(\phi - \phi') + z' \cos\theta.$$

9.3 RADIATION FIELDS OF DIPOLES, SLOTS, AND LOOPS

In this section, we give some examples of the radiation fields of known current distributions based on the formulation (9.26). It is known (Elliott, 1981, Chapter 2) that the current distribution on a wire antenna is approximately sinusoidal. For the wire antenna shown in Fig. 9.4, the current distribution is given by

$$I(z') = I_0 \frac{\sin[k(1 - |z'|)]}{\sin kl}. \qquad (9.33)$$

Substituting this into (9.26), we get

$$E_\theta = \frac{j\omega\mu_0 \sin\theta e^{-jkR}}{4\pi R} \int_{-l}^{l} I(z')e^{jkz' \cos\theta} \, dz'. \qquad (9.34)$$

The integration can be simplified by noting that $I(z')$ is an even function of z', and therefore, the integral is equal to

$$2 \int_0^l I(z') \cos(kz' \cos\theta) \, dz'.$$

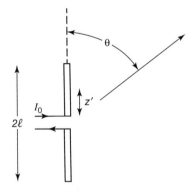

FIGURE 9.4   Radiation from a wire antenna.

After integration, we get

$$E_\theta = \frac{j\eta I_0}{2\pi R} e^{-jkR} \frac{\cos(kl\cos\theta) - \cos kl}{\sin kl \sin\theta},$$

$$E_\phi = 0.$$

(9.35)

For a half-wave dipole, $2l = \lambda/2$ and we get

$$E_\theta = \frac{j\eta I_0}{2\pi R} e^{-jkR} \frac{\cos[(\pi/2)\cos\theta]}{\sin\theta}.$$

(9.36)

The total radiated power is given by

$$P_t = \int_0^{2\pi} \int_\pi \frac{|E_\theta|^2}{2\eta} R^2 \sin\theta \, d\theta \, d\phi.$$

(9.37)

The radiation resistance R_r is defined by

$$P_t = \tfrac{1}{2} I_0^2 R_r.$$

(9.38)

For a half-wave dipole, the integral in (9.37) can be evaluated either numerically or analytically using sine and cosine integrals, and we get

$$R_r = 73.09 \text{ ohms.}$$

(9.39)

Note that the resistive component of the input impedance $Z_i = R_r + jX$ is given by the radiated power and is independent of the wire diameter, but the reactive component X depends on the stored energy in the near field. Therefore, X depends on the wire diameter and must be obtained by solving the boundary value problem. The maximum gain of a half-wave dipole antenna is in the direction $\theta = \pi/2$ and is given by

$$G = \frac{4\pi R^2 |E_\theta|^2 / 2\eta}{P_t} = 1.64.$$

(9.40)

For a short dipole $2l \ll \lambda$, the radiation resistance R_r and the maximum gain are

$$R_r = 20\left(\frac{\pi 2l}{\lambda}\right)^2,$$

$$G = 1.5.$$

(9.41)

A wire antenna is often placed in front of a conducting ground plane as shown in Fig. 9.5. The radiation field is then the sum of the radiation from the antenna and the

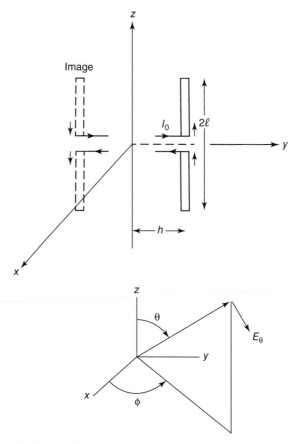

FIGURE 9.5   Radiation from a dipole in front of a ground plane $(x\text{–}z)$.

radiation from its image. Therefore, the field is given by

$$E_\theta = \frac{j\eta I_0}{2\pi R} e^{-jkR} \frac{\cos(kl\cos\theta) - \cos kl}{\sin kl \sin\theta} 2j\sin(kh\sin\theta\sin\phi). \qquad (9.42)$$

Let us next consider the radiation from a center-fed slot in the ground plane (Fig. 9.6). The electric field in the slot is given by

$$E_x = \frac{V_0}{w} \frac{\sin k(l - |z|)}{\sin kl}. \qquad (9.43)$$

As shown in Section 6.10, this is identical to the magnetic current J_m in front of the ground plane.

$$\bar{J}_m = \bar{E} \times \hat{n} = E_x \hat{x} \times \hat{y} = E_x \hat{z}. \qquad (9.44)$$

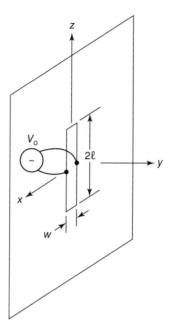

FIGURE 9.6 Center-fed slot.

This is equivalent to the magnetic current $\bar{J}_m$ and its image in free space. Therefore, the field is equal to the field generated by $2\bar{J}_m$. Using (9.26), we get

$$E_\theta = 0,$$
$$E_\phi = -j\frac{e^{-jkR}}{\pi R}V_0\frac{\cos(kl\cos\theta) - \cos kl}{\sin kl \sin\theta}. \tag{9.45}$$

Next, let us consider the radiation from a small loop of radius a. The current I_ϕ in the loop is assumed to be constant. Then using (9.31), (9.32), and

$$\hat{o} = \hat{x}\sin\theta\cos\phi + \hat{y}\sin\theta\sin\phi + \hat{z}\cos\theta,$$
$$\hat{o}\cdot\bar{r}' = a\sin\theta\cos(\phi - \phi'), \tag{9.46}$$

we get

$$E_\theta = 0,$$
$$E_\phi = -\frac{j\omega\mu_0 e^{-jkR}}{4\pi R}\left(jk\pi a^2 I_\phi \sin\theta\right). \tag{9.47}$$

The radiation resistance is then given by

$$R_r = \frac{\pi\eta}{6}(ka)^4 = 320\pi^6\left(\frac{a}{\lambda}\right)^4. \tag{9.48}$$

9.4 ANTENNA ARRAYS WITH EQUAL AND UNEQUAL SPACINGS

Let us consider the radiation field from a wire antenna excited by the current I_n. As can be seen from (9.35), we can write the radiation field $\bar{E}_n$ as

$$\bar{E}_n = \frac{I_n}{R_n}\bar{f}_n(\theta, \phi)e^{-jkR_n}, \tag{9.49}$$

where $\bar{f}_n(\theta, \phi)$ represents the radiation pattern (Fig. 9.7). For the far field, we have

$$\frac{1}{R_1} \approx \frac{1}{R_2} \approx \frac{1}{R_n} \approx \frac{1}{R_0},$$
$$KR_n = kR_0 - \phi_n, \tag{9.50}$$

where ϕ_n is the phase difference shown in Fig. 9.7. We also assume that all the antennas are identical.

$$\bar{f}_1 = \bar{f}_2 = \cdots = \bar{f}(\theta, \phi). \tag{9.51}$$

Then the total field is given by

$$\bar{E} = \frac{\bar{f}(\theta, \phi)}{R_0}e^{-jkR_0}F(\theta, \phi),$$
$$F(\theta, \phi) = \sum_{n=1}^{N} I_n\, e^{j\phi_n}. \tag{9.52}$$

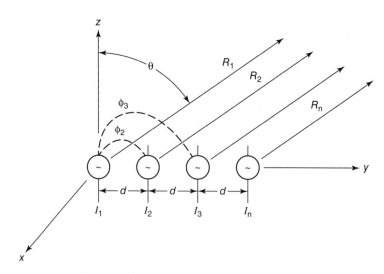

FIGURE 9.7   Radiation from an antenna array.

The function $F(\theta, \phi)$ is called the *array factor*. For the case shown in Fig. 9.7, where all the antennas are located on the y-axis with equal spacing d, we have

$$F(\theta, \phi) = \sum_{n=1}^{N} I_n \, e^{jkd(n-1)\sin\theta\sin\phi}. \tag{9.53}$$

If all the antennas are located on a line as shown in Fig. 9.7, they are called *linear arrays*. If the antennas are located on a circle or on the surface of a sphere, they are called *circular arrays* or *spherical arrays*.

If all the currents in a linear array are identical, we get

$$I_n = I_0,$$

$$F(\theta, \phi) = I_0 \sum_{n=1}^{N} e^{j(n-1)\gamma}, \tag{9.54}$$

$$= N I_0 \exp\left[\frac{j(N-1)\gamma}{2}\right] \frac{\sin(N\gamma/2)}{\sin(\gamma/2)},$$

where $\gamma = kd \sin\theta \sin\phi$.

The magnitude of the array factor $\|F(\theta, \phi)\|$ normalized with $N I_0$ is pictured in Fig. 9.8. Note the main lobe, grating lobes, and the visible region $(-\pi/2 < \theta < \pi/2)$.

Extensive studies have been made on antenna array problems (Elliott, 1981; Ma, 1974). In this section, we present some interesting ideas about linear arrays. Consider the array factor (9.53) in the y–z plane ($\phi = \pi/2$). If we let

$$Z = \exp(jkd \sin\theta), \tag{9.55}$$

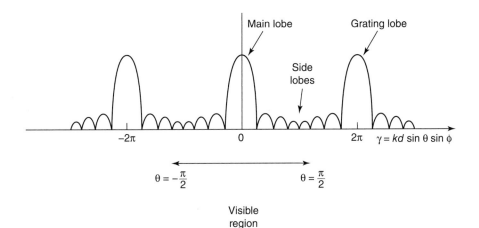

FIGURE 9.8   Radiation from a uniform array.

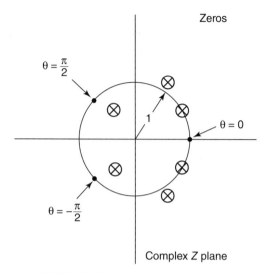

FIGURE 9.9 Schelkunoff's unit circle.

we get

$$F = \sum_{n=1}^{N} I_n Z^{n-1}. \tag{9.56}$$

This is a polynomial of the degree $N - 1$, and therefore we write

$$F = A(Z - Z_1)(Z - Z_2) \cdots (Z - Z_{N-1}), \tag{9.57}$$

where Z_n is the zeros of the polynomial and A is constant. Note that the array factor $\|F\|$ is given by the magnitude of F evaluated on a unit circle in the complex Z plane, and that the desired array factor can be obtained by placing the zeros Z_n appropriately in the complex Z plane (Fig. 9.9). This method was first proposed by Shelkunoff in 1943.

Next let us consider a method to synthesize a desired radiation pattern. Suppose that we place $2N + 1$ antennas as shown in Fig. 9.10. The array factor is

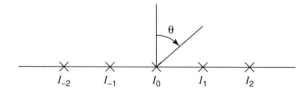

FIGURE 9.10 Fourier series synthesis.

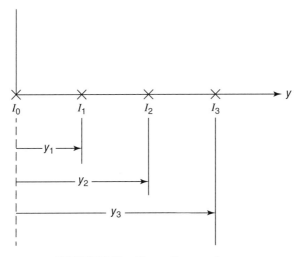

FIGURE 9.11   Unequally spaced array.

then given by

$$F(\theta) = I_0 + I_1 e^{j\gamma} + I_2 e^{j2\gamma} + \cdots + I_n e^{jN\gamma}$$
$$+ I_{-1} e^{-j\gamma} + I_{-2} e^{-j2\gamma} + \cdots + I_{-N} e^{-jN\gamma}. \tag{9.58}$$

This is a Fourier series representation of $F(\theta) = F(\gamma)$, and therefore for a desired radiation pattern $F(\gamma)$, we expand $F(\gamma)$ in Fourier series and the Fourier coefficient can be identified as the current in each antenna element. For a more detailed discussion on antenna pattern synthesis, see Collin and Zucker (1969) and Ma (1974).

In the above, we considered only equally spaced arrays. They are mathematically the most convenient and are normally used in practical applications. There are some applications, however, where unequally spaced arrays are more desirable. For example, the resolution capability of an antenna depends on the total size: $\Delta\theta = \lambda/\text{size}$. However, if the spacing d is greater than a wavelength and is uniform, there will be grating lobes in the visible region (Fig. 9.8). We can, however, eliminate these grating lobes even when the spacing is much greater than a wavelength if we use unequally spaced arrays. Therefore, unequally spaced arrays can be used to obtain a better resolution with a smaller number of antenna elements.

Let us consider the array factor for an unequally spaced array (Fig. 9.11)

$$F(\theta) = \sum_{n=0}^{N-1} I_n e^{j(k \sin \theta) y_n}. \tag{9.59}$$

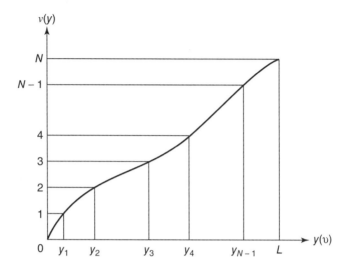

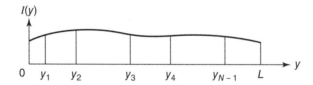

FIGURE 9.12   Element positions for an unequally spaced array.

We may write this as

$$F(\theta) = \sum_{n=0}^{N-1} I(n)e^{j(k\sin\theta)y(n)},$$

(9.60)

where $I(n)$ and $y(n)$ are written as functions of n (Fig. 9.12). Now, we make use of Poisson's sum formula

$$\sum_{n} f(n) = \sum_{m} \int f(v)e^{-j2m\pi v}\,dv.$$

(9.61)

We then get

$$F(\theta) = \sum_{m=-\infty}^{\infty} F_m(\theta),$$

$$F_m(\theta) = \int_0^N I(v)e^{j(k\sin\theta)y(v)-j2m\pi v}\,dv \qquad (9.62)$$

$$= \int_0^L I(y)e^{j(k\sin\theta)y-j2m\pi v(y)}\left(\frac{dv}{dy}\right)dy.$$

Here we converted the finite sum (9.60) into the infinite sum (9.62). However, this infinite sum reveals some interesting characteristics. For example, consider the $m = 0$ term

$$F_0(\theta) = \int_0^L I(y)\left(\frac{dv}{dy}\right)e^{jky\sin\theta}\,dy. \qquad (9.63)$$

This is identical to the radiation from an equivalent continuous source distribution $S_0(y)$

$$S_0(y) = I(y)\frac{dv}{dy}. \qquad (9.64)$$

This means that this equivalent source distribution is equal to the actual current distribution $I(y)$ modified by the *density function dv/dy*. The equivalent source distribution $S_m(y)$ for F_m is given by

$$S_m(y) = S_0(y)e^{-j2m\pi v(y)}. \qquad (9.65)$$

This has the additional phase shift across the aperture. If the spacings are uniform, $v(y) = Ny/L$ and this creates the grating lobe at $k\sin\theta = 2\,m\pi N/L$. However, if the spacing is nonuniform, the grating lobe at this angle disappears and the power contained in this lobe is spread out in the angle.

We will not go into a detailed discussion on several important array problems, such as Dolph–Chebyshev arrays, circular arrays, spherical arrays, conformal arrays, three-dimensional arrays, randomly spaced arrays, and phased arrays, as they are discussed extensively in the literature (Hansen, 1966; Lo and Lee, 1988).

9.5 RADIATION FIELDS FROM A GIVEN APERTURE FIELD DISTRIBUTION

If the exact field $(\bar{E}_s, \bar{H}_s)$ on a surface S is known, we can obtain the field $(\bar{E}, \bar{H})$ at any point in terms of the tangential fields $\bar{E}_s \times \hat{n}$ and $\hat{n} \times \bar{H}_s$ by using the Franz formula (Section 6.9).

$$\bar{E} = \nabla \times \nabla \times \bar{\pi} - j\omega\mu_0 \nabla \times \bar{\pi}_m,$$

$$\bar{\pi} = \int G_0(\bar{r}, \bar{r}') \frac{\hat{n} \times \bar{H}_s(\bar{r}')}{j\omega\varepsilon_0} dS', \qquad (9.66)$$

$$\bar{\pi}_m = \int G_0(\bar{r}, \bar{r}') \frac{\bar{E}_s(\bar{r}') \times \hat{n}}{j\omega\mu_0} dS',$$

where

$$G_0(\bar{r}, \bar{r}') = \frac{\exp(-jk|\bar{r} - \bar{r}'|)}{4\pi|\bar{r} - \bar{r}'|},$$

and $\hat{n}$ is the unit vector normal to the surface S pointed toward the observation point. Following the procedure in Section 9.2, we get the far field from the known field distribution on a surface S:

$$E_\theta = -\frac{j\omega\mu_0}{4\pi R} e^{-jkR} \int \hat{\theta} \cdot (\hat{n} \times \bar{H}_s(\bar{r}')) e^{jk\hat{o}\cdot\bar{r}'} dS'$$

$$- \frac{jk}{4\pi R} e^{-jkR} \int \hat{\phi} \cdot (\bar{E}_s(\bar{r}') \times \hat{n}) e^{jk\hat{o}\cdot\bar{r}'} dS',$$

$$E_\phi = -\frac{j\omega\mu_0}{4\pi R} e^{-jkR} \int \hat{\phi} \cdot (\hat{n} \times \bar{H}_s(\bar{r}')) e^{jk\hat{o}\cdot\bar{r}'} dS' \qquad (9.67)$$

$$+ \frac{jk}{4\pi R} e^{-jkR} \int \hat{\theta} \cdot (\bar{E}_s(\bar{r}') \times \hat{n}) e^{jk\hat{o}\cdot\bar{r}'} dS'.$$

This gives the exact field (E_θ, E_ϕ) if the exact surface fields $\hat{n} \times \bar{H}_s$ and $\bar{E}_s \times \hat{n}$ are known. In practice, the exact surface fields $\bar{E}_s$ and $\bar{H}_s$ are rarely known and we need to use approximate surface fields.

For example, if the aperture is large, as in the case of parabolic reflector antennas, the field $\bar{E}_s$, $\bar{H}_s$ on the conducting reflector surface S_1 can be well approximated by the field that would exist if the reflector surface were locally plane (Fig. 9.13). Thus we have

$$\bar{E}_s \times \hat{n} = 0,$$

$$\hat{n} \times \bar{H}_s = 2\hat{n} \times \bar{H}_i \quad \text{on } S_1, \qquad (9.68)$$

where $\bar{H}_i$ is the field from the feed horn in the absence of the reflector. The use of the approximation (9.68) is called the *Kirchhoff approximation* and is identical to the *physical optics approximation*.

Instead of using the field (9.68) on the surface S_1 of the reflector, we can also use the field on the plane aperture S_2 in front of the reflector. We can use (9.67) and approximate $\bar{E}_s$ and $\bar{H}_s$ by the field on S_2 from the feed reflected from the reflector, or we can use the *equivalence theorem* (Section 6.10) and note that the field on the

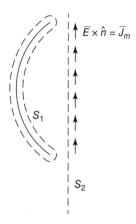

FIGURE 9.13   Reflector surface S_1 and the plane aperture surface S_2.

right of S_2 is given by the field $2\bar{E} \times \hat{n}$ due to the magnetic current $\bar{J}_m = \bar{E} \times \hat{n}$ and its image $\bar{E} \times \hat{n}$ in free space. Then the far field is given by

$$
\begin{aligned}
E_\theta &= -\frac{jk}{2\pi R}e^{-jkR}\int_{S_2} \hat{\phi} \cdot (\bar{E}_s \times \hat{n})e^{jk\hat{o}\cdot\bar{r}'}\,dS', \\
E_\phi &= -\frac{jk}{2\pi R}e^{-jkR}\int_{S_2} \hat{\theta} \cdot (\bar{E}_s \times \hat{n})e^{jk\hat{o}\cdot\bar{r}'}\,dS'.
\end{aligned}
\tag{9.69}
$$

In principle, (9.67) and (9.69) should give the same correct far field if the appropriate surface fields $\bar{E}_s$ and $\bar{H}_s$ are exactly known and if the surface completely surrounds the object. In practice, however, this is not possible. The fields on S_1 or S_2 are only approximately known, and therefore, there are slight differences between (9.67) and (9.69) in the far-sidelobe region. In the main lobe region, both are almost identical and give excellent results.

As an example, consider the radiation from a rectangular aperture (Fig. 9.14) with the field distribution.

$$\bar{E}_s = \hat{x}E_{sx}(x',y') \text{ in the aperture}$$

$$-a/2 \le x' \le a/2 \tag{9.70}$$

$$-b/2 \le y' \le b/2.$$

Using (9.69) and noting (9.29) and that $\bar{E}_s \times \hat{n} = -\hat{y}E_0$, we get

$$
\begin{aligned}
E_\theta &= \frac{jk}{2\pi R}e^{-jkR}\cos\phi F(k\sin\theta\cos\phi, k\sin\theta\sin\phi), \\
E_\phi &= \frac{jk}{2\pi R}e^{-jkR}(-\cos\theta\sin\phi)F(k\sin\theta\cos\phi, k\sin\theta\sin\phi),
\end{aligned}
\tag{9.71}
$$

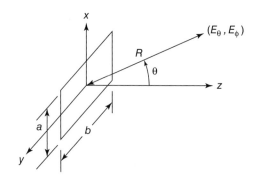

FIGURE 9.14   Radiation from a rectangular aperture.

where

$$F(k_x, k_y) = \iint E_{sx}(x', y')e^{jk_x x' + jk_y y'}\, dx'\, dx'.$$

Note that the radiation pattern is proportional to the Fourier transform $F(k_x, k_y)$ of the aperture field $E_{sx}(x', y')$ evaluated at $k_x = k \sin\theta \cos\phi$ and $k_y = k \sin\theta \sin\phi$. This Fourier transform relationship between the aperture field distribution and the radiation pattern is an important and useful result. If the aperture distribution is uniform, $E_{sx} = E_0 = $ constant, and we get

$$F(k_x, k_y) = ab\frac{\sin(k_x a/2)}{k_x a/2}\frac{\sin(k_y b/2)}{k_y b/2}. \tag{9.72}$$

If the aperture is circular and the aperture field $\bar{E}_s$ is given in a cylindrical system,

$$\bar{E}_s = \hat{x}E_{sx}(\rho', \phi') \quad \text{in the aperture } \rho' \le a, \tag{9.73}$$

we have the same formula (9.71) except that F should be expressed in the cylindrical system

$$F(\theta, \phi) = \iint E_{sx}(\rho', \phi')e^{jk\rho' \sin\theta \cos(\phi, \phi')}\rho'\, d\rho'\, d\phi'. \tag{9.74}$$

Furthermore, if E_{sx} is independent of ϕ', we get

$$F(\theta) = 2\pi \int_0^a E_{sx}(\rho')J_0(k\rho' \sin\theta)\rho'\, d\rho'. \tag{9.75}$$

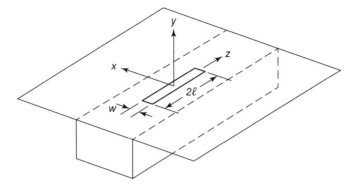

FIGURE 9.15   Half-wavelength slot antenna $(2l = \lambda/2)$ fed by a waveguide.

For a uniform aperture field distribution $E_{sx} = E_0 =$ constant, using the formula

$$\int xJ_0(x)\,dx = xJ_1(x),$$

we get

$$F(\theta) = E_0\pi a^2 \frac{J_1(ka \sin\theta)}{(ka \sin\theta)/2}. \tag{9.76}$$

If a slot is fed by a waveguide (Fig. 9.15) and the slot length is equal to a half-wavelength, the aperture field is known to be approximately sinusoidal and is given by

$$E_x = E_0 \cos kz. \tag{9.77}$$

The radiation field is then given by (9.45) with $kl = \pi/2$ and $V_0 = E_0 w$

$$E_\theta = 0,$$
$$E_\phi = -j\frac{e^{-jkR}}{\pi R}E_0 w \frac{\cos[(\pi/2)\cos\theta]}{\sin\theta}. \tag{9.78}$$

9.6 RADIATION FROM MICROSTRIP ANTENNAS

Microstrip antennas are made of a thin patch of metal on a grounded dielectric slab as shown in Fig. 9.16. The metal patch is approximately a half-wavelength long and is fed by a coaxial line or a strip line. Thus, it can be viewed as a half-wavelength transmission-line cavity with open terminations at both ends where the power leaks out. It is a narrow-band antenna, but it has a low profile and is lightweight. Its radiation pattern is close to that of a half-wave dipole radiating normal to the surface,

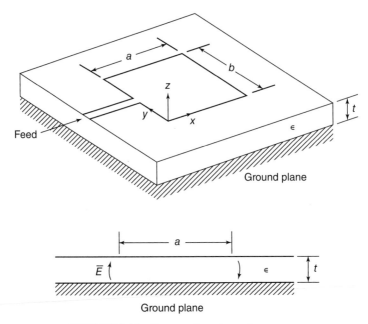

FIGURE 9.16   Rectangular microstrip antennas.

and thus an array of microstrip antennas can easily be formed for high-gain scanning operations.

Let us first consider a rectangular patch antenna (Fig. 9.16). This can be considered as a rectangular cavity of dimensions $a \times b \times t$ with the boundary condition at the edge that the tangential magnetic field is zero (magnetic wall) (see Section 4.14). Therefore, we write the electric field inside the cavity as

$$E_z(x, y) = E_0 \cos \frac{m\pi x}{a} \cos \frac{n\pi y}{b}. \tag{9.79}$$

The resonant frequency $f_r = \omega_r/2\pi$ is then given by

$$\frac{\omega_r}{c} \sqrt{\varepsilon_r} = \left[\left(\frac{m\pi}{a} \right)^2 + \left(\frac{n\pi}{b} \right)^2 \right]^{1/2}. \tag{9.80}$$

The antenna is operated at a frequency close to this resonant frequency. For $a > b$, the lowest resonant frequency occurs when $m = 1$ and $n = 0$. (The static field $m = n = 0$ is excluded.)

The radiation field can be calculated using several equivalent techniques (Lo et al., 1979; Bahl and Bhartia, 1980, p. 8; Elliott, 1981, p. 105). Here we use the equivalent magnetic current at the edge of the patch and its image due to the ground plane. We then get the following magnetic currents at the edge of the patch:

$$\bar{E}_s \times \hat{n} = \hat{y}\left(-E_0 \cos\frac{n\pi y}{b}\right), \qquad x = 0,$$

$$= \hat{y}(-1)^m E_0 \cos\frac{n\pi y}{b}, \qquad x = a,$$

$$= \hat{x}E_0 \cos\frac{m\pi x}{a}, \qquad y = 0,$$

$$= \hat{x}\left[-(-1)^n E_0 \cos\frac{m\pi x}{a}\right], \quad y = b. \tag{9.81}$$

We substitute these into (9.69) and note the following:

$$\hat{\phi} \cdot (\bar{E}_s \times \hat{n}) = -\sin\phi E_x + \cos\phi\, E_y,$$

$$\hat{\theta} \cdot (\bar{E}_s \times \hat{n}) = -\cos\theta\cos\phi E_x + \cos\theta\sin\phi\, E_y, \tag{9.82}$$

$$\hat{o} \cdot \bar{r}' = x'\sin\theta\cos\phi + y'\sin\theta\sin\phi.$$

We then get

$$E_\theta = -\frac{jk}{2\pi R}e^{-jkR}V_0(-\sin\phi g_1 - \cos\phi g_2),$$

$$E_\phi = +\frac{jk}{2\pi R}e^{-jkR}V_0(\cos\theta\cos\phi g_1 - \cos\theta\sin\phi g_2), \tag{9.83}$$

where

$$V_0 = E_0 t,$$

$$g_1 = [1 - (-1)^n e^{jkb\sin\theta\sin\phi}]\int_0^a \cos\frac{m\pi x'}{a}e^{jkx'\sin\theta\cos\phi}dx',$$

$$g_1 = [1 - (-1)^n e^{jkb\sin\theta\sin\phi}]\int_0^b \cos\frac{n\pi y'}{b}e^{jky'\sin\theta\sin\phi}dy'.$$

Here, we assumed that the thickness t of the patch is much smaller than a wavelength, and thus it is assumed that the magnetic current is a narrow line current at the edge of the patch located just above the ground plane.

Microstrip antennas are operated at the lowest resonant frequency with $m = 1$ and $n = 0$, and therefore, $k = 2\pi f_r/c = \pi/a\sqrt{\varepsilon_r}$. To account for the fringe field, Lo (1979) used the following effective sizes and obtained good agreement between theory and experiment:

$$a_{\text{eff}} = a + \frac{t}{2},$$

$$b_{\text{eff}} = b + \frac{t}{2}. \tag{9.84}$$

Circular patch antennas of radius a can be analyzed similarly by using the following field inside the cavity (Section 4.14):

$$E_z = E_0 J_1(k_{11}\rho) \cos\phi, \tag{9.85}$$

where

$$k_{11} = \frac{1.841}{a}.$$

The magnetic current is then

$$\bar{E}_s \times \hat{n} = \hat{z}E_z \times \hat{\rho} = E_z\hat{\phi}. \tag{9.86}$$

This is then substituted into (9.69). Care should be taken to differentiate the angle ϕ at the observation point (R, θ, ϕ) and the angle ϕ' for the field at the edge of the patch (a, ϕ') (see Eqs. 9.31 and 9.32). The far field is then given by

$$E_\theta = -\frac{jk}{2\pi R} e^{-jkR} \int_0^{2\pi} \cos(\phi - \phi') V_0 a \cos\phi' \exp[jka \sin\theta \cos(\phi - \phi')] d\phi'$$

$$= -\frac{jk}{R} V_0 a J_1'(ka \sin\theta) \cos\phi, \tag{9.87}$$

$$E_\phi = \frac{jkV_0 a}{R} \frac{J_1(Ka \sin\theta)}{ka \sin\theta} \cos\theta \sin\phi,$$

where $V_0 = E_0 t J_1(k_{11}a)$ is the edge voltage at $\rho = a$ and $\phi = 0$, and $k = 1.841/a\sqrt{\varepsilon_r}$.

9.7 SELF- AND MUTUAL IMPEDANCES OF WIRE ANTENNAS WITH GIVEN CURRENT DISTRIBUTIONS

Consider a wire antenna excited by a given voltage source (Fig. 9.17). The input impedance Z is given by

$$Z = \frac{V_0}{I(0)}. \tag{9.88}$$

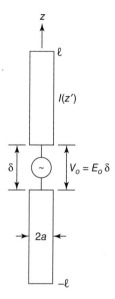

FIGURE 9.17   Wire antenna excited by a voltage V_0.

In general, there are two ways of determining the input impedance. One way is that the exact current distribution $I(z)$ for a given voltage can be determined by solving the boundary value problem, then using this current at $z = 0$ to obtain the input impedance. This, however, requires extensive theoretical and numerical work. In many cases, we are often not interested in the exact current distribution, but we wish to have a simple method of calculating useful and approximate input impedances. This second approach is shown in this section. It is based on a variational expression for the impedance such that an approximate current distribution is used to obtain the impedance with greater accuracy than the current distribution (Elliott, 1981, Chapter 7).

Let us assume that the wire size a is much smaller than a wavelength and that the current density J_z on the surface of the wire is all directed in the z direction, and therefore, the z component of the electric field E_z is obtained from the following:

$$\bar{E} = \nabla \times \nabla \times \bar{\pi} = \nabla(\nabla \cdot \bar{\pi}) + k^2\bar{\pi}. \tag{9.89}$$

Taking the z component and noting that $\bar{\pi} = \pi_z \hat{z}$, we get

$$E_z = \left(\frac{\partial^2}{\partial z^2} + k^2 \right) \pi_2,$$

$$\pi_z = \frac{1}{j\omega\varepsilon_0} \int_{S'} G_0(\bar{r}, \bar{r}') J_z(\bar{r}') ds', \tag{9.90}$$

where S' is the area enclosing the wire. Since $J_z(\bar{r}')$ has no azimuthal variation, we let $ds' = a\,d\phi'\,dz'$, integrate with respect to ϕ', and let $I(z')$ be the total current at z'. We also let the current $I(z')$ be located on the axis of the wire and evaluate E_z on the surface of the wire. Then, we get

$$E_z(z) = \int_{-l}^{l} K(z,z')I(z')\,dz', \tag{9.91}$$

where

$$K(z,z') = \frac{1}{4\pi j\omega\varepsilon_0}\left(\frac{\partial^2}{\partial z^2} + k^2\right)\frac{e^{-jkr}}{r},$$

$$r = [a^2 + (z-z')^2]^{1/2}.$$

Now, the electric field E_z is zero on the surface of the wire except on the gap where it is equal to $-E_0$. Thus we write

$$E_z(z) = \begin{cases} -E_0, & |z| < \dfrac{\delta}{2}, \\[2mm] 0, & \dfrac{\delta}{2} < |z| < l. \end{cases} \tag{9.92}$$

If we multiply both sides by $I(z)$ and integrate over $(-l, l)$, we get

$$\int_{-l}^{l} E_z(z)I(z)\,dz = -V_0 I(0). \tag{9.93}$$

From this, we get the input impedance Z

$$Z = \frac{V_0}{I(0)} = -\frac{\displaystyle\int_{-l}^{l} E_z(z)I(z)\,dz}{I(0)^2}, \tag{9.94}$$

where $E_z(z)$ is the z component of the electric field produced by the current $I(z')$ and is given by (9.91). Thus we can also write

$$Z = -\frac{1}{I(0)^2}\int_{-l}^{l}\int_{-l}^{l} I(z)K(z,z')I(z')\,dz'\,dz. \tag{9.95}$$

Equation (9.94) or (9.95) is a variational expression for Z, and therefore, if the current is $I + \delta I$ and the corresponding impedance is $Z + \delta Z$, (9.95) gives $\delta Z = 0$ indicating that for the first-order approximation for I, the impedance has second-order error. Thus the accuracy of the calculated impedance is better than the accuracy of the current distribution used in the equation (see Appendix 7.B).

We can now use (9.95) and an approximate current distribution $I(z)$ to obtain the impedance. The current distribution is known to be sinusoidal if $a/\lambda \ll 1$, and therefore, we use

$$I(z) = I_m \sin k(l - |z|). \tag{9.96}$$

Then, we get (Elliott, 1981, p. 300)

$$Z = \frac{j60}{\sin^2 kl} \{4\cos^2 klS(kl) - \cos 2klS(2kl) - \sin 2kl[2C(kl) - C(2kl)]\}, \tag{9.97}$$

where

$$C(ky) = \ln \frac{2y}{a} - \frac{1}{2} \text{Cin}(2ky) - \frac{j}{2} \text{Si}(2ky),$$

$$S(ky) = \frac{1}{2} \text{Si}(2ky) - \frac{j}{2} \text{Cin}(2ky) - ka.$$

$\text{Si}(x)$ is the sine integral,

$$\text{Si}(x) = \int^x \frac{\sin u}{u} du.$$

$\text{Cin}(x)$ is the modified cosine integral,

$$\text{Cin}(x) = \int_0^x \frac{1 - \cos u}{u} du.$$

It should be noted that the resistive component of the input impedance is related to the radiated power, and therefore it is insensitive to the wire size a. However, the reactive component is related to the reactive power, which represents the stored energy in the near field, and therefore it depends very much on the wire size.

Let us next consider the mutual impedance between antennas 1 and 2 (Fig. 9.18). The voltages V_1 and V_2 and the currents at the terminals $I_1(0)$ and $I_2(0)$ are related by a linear relationship applicable to any linear passive network.

$$V_1 = Z_{11}I_1(0) + Z_{12}I_2(0),$$
$$V_2 = Z_{21}I_1(0) + Z_{22}I_2(0), \tag{9.98}$$

where $Z_{12} = Z_{21}$. Z_{11} and Z_{22} are the self-impedances and Z_{12} is the mutual impedance. Now, consider the z component of the electric field E_2 on the surface of the wire 2. This consists of the contribution from the currents $I_1(z)$ and $I_2(z)$, and therefore we write

$$E_2(z) = E_{21}(z) + E_{22}(z), \tag{9.99}$$

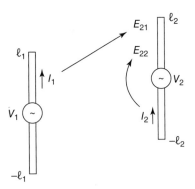

FIGURE 9.18 Mutual impedance.

where $E_{21}(z)$ is the field on the surface of wire 2 produced by I_1 and $E_{22}(z)$ is the field on the surface of wire 2 produced by the current I_2. Now, E_2 must be zero on the wire surface except at the gap $|z| < \delta/2$, where the voltage V_2 is applied. Thus we have

$$E_2(z) = \begin{cases} -E_0 & \text{for } |z| < \dfrac{\delta}{2}, \\ 0 & \text{for } \dfrac{\delta}{2} < |z| < l_2. \end{cases} \tag{9.100}$$

Multiplying both sides by $I_2(z)$ and integrating over $(-l_2, l_2)$, we get

$$\int_{-l_2}^{l_2} [E_{21}(z)I_2(z) + E_{22}(z)I_2(z)]dz = -V_2 I_2(0). \tag{9.101}$$

This can be rewritten as

$$V_2 = Z_{21}I_1(0) + Z_{22}I_2(0), \tag{9.102}$$

where

$$Z_{21} = -\frac{1}{I_0(0)I_2(0)} \int_{-l_2}^{l_2} E_{21}(z)I_2(z)\, dz,$$

$$Z_{22} = -\frac{1}{I_2(0)^2} \int_{-l_2}^{l_2} E_{22}(z)I_2(z)\, dz.$$

The mutual impedance Z_{21} can also be written as

$$Z_{21} = -\frac{1}{I_1(0)I_2(0)} \int_{-l_1}^{l_1} \int_{-l_2}^{l_2} I_2(z)K_{21}(z,z')I_1(z')dz\, dz', \tag{9.103}$$

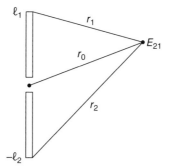

FIGURE 9.19   The field E_{21} produced by the sinusoidal current I_1.

where the field E_{21} due to I_1 is now written as

$$E_{21}(z) = \int_{-l_1}^{l_1} K_{21}(z, z')I_1(z')dz'. \qquad (9.104)$$

K_{21} is given by the same expression as (9.91) except that a is replaced by the spacing between two wire antennas. The self-impedance Z_{22} has the same form as Z in (9.94), as it should.

If the current $I_1(z')$ is assumed to be sinusoidal,

$$I_1(z') = I_{m1} \sin k(l_1 - |z'|), \qquad (9.105)$$

then the integral in (9.104) can be performed (Elliott, 1981, p. 331), giving the following simple and exact expression (Fig. 9.19)

$$E_{21}(z) = -j30 \cdot I_m \left(\frac{e^{-jkr_1}}{r_1} + \frac{e^{-jkr_2}}{r_2} - 2 \cos kl_1 \frac{e^{-jkr_0}}{r_0} \right). \qquad (9.106)$$

This can then be substituted into (9.102) to obtain the mutual impedance.

The input impedance of microstrip antennas can be obtained by using the same formula, (9.94). The field $E_z(z)$ is then the field produced by the current $I(z)$ on the feed in the presence of the microstrip antenna. Here the feedpin size must be considered to obtain the accurate value of the reactive component of the input impedance.

9.8 CURRENT DISTRIBUTION OF A WIRE ANTENNA

In Section 9.7, we used the assumed sinusoidal current distribution to calculate the self- and mutual impedances. These current distributions are good approximations, but the exact distributions must be determined by solving the integral equation as follows: The integral equation is already indicated in (9.92).

$$\int_{-l}^{l} K(z, z')I(z')dz' = -E_i(z),$$ (9.107)

where $K(z, z')$ is given in (9.91) and $E_i(z)$ is the impressed or incident field.

$$E_i(z) = \begin{cases} E_0 & \text{for } |z| < \dfrac{\delta}{2} \\ 0 & \text{for } \dfrac{\delta}{2} < |z| < l \end{cases}.$$

The integral equation 9.107 is called the *Pocklington integral equation*. The kernel $K(z, z')$ can be further simplified to give the *Richmond form* (Mittra, 1973, p. 13).

$$K(z, z') = \frac{1}{4\pi j\omega\varepsilon_0} \frac{e^{-jkr}}{r^5}[(1 + jkr)(2r^2 - 3a^2) + k^2 a^2 r^2],$$ (9.108)

where $r = [a^2 + (z - z')^2]^{1/2}$.

The integral equations above can be solved by using the moment method (see Section 10.14).

PROBLEMS

9.1 The gain G of a large circular aperture antenna is related to the actual aperture area A through the aperture efficiency η_a.

$$G = \eta_a \frac{4\pi A}{\lambda^2}.$$

The half-power beamwidth (in rad) is $\theta_b = \alpha\lambda/D$, where D is the diameter of the aperture. A parabolic antenna of diameter 3 m is used to transmit microwaves at 20 GHz over a distance of 500 km, and an identical antenna is used as a receiving antenna. Assume that $\eta_a = 0.75$ and $\alpha = 1.267$. The transmitting power is 1 W. Find the power flux density at the receiver. Find the gain (in dB) of the antenna and the ratio of the power received to the power transmitted.

9.2 Consider Eq. (9.18). If the open-circuit voltage is $V = E_0 l \sin\theta$, show that the current is $V_0/2R_0$ and that the power reradiated due to this current is equal to the received power given in (9.18).

9.3 Consider a linear array of uniformly excited five half-wave wire antennas placed in front of the conducting plane shown in Fig. P9.3. Find the radiation patterns in the *x–y* plane and in the *z–y* plane. $h = \lambda/4$ and $d = 3\lambda/4$. Find the first sidelobe level in dB.

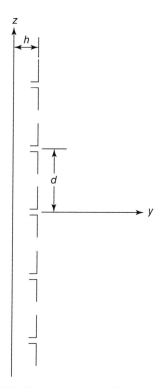

FIGURE P9.3   Array of half-wave wire antennas.

9.4 Show that the array factor of a circular array of N identical antennas uniformly excited and uniformly distributed on a circle of radius a is given by

$$F(\theta, \phi) = \sum_{m=-\infty}^{\infty} J_{mN}(ka \sin \theta) e^{jmN(\pi/2-\phi)}.$$

If $ka \ll N$, this can be approximated by

$$F(\theta, \phi) \approx J_0(ka \sin \theta).$$

(Use Poisson's sum formula (9.61).)

9.5 It is desired to design an array with five elements that would produce a pattern as close to the rectangular pattern shown in Fig. P9.5 as possible. Use the Fourier series method to obtain I_1/I_0 and I_2/I_0. The spacing is $\lambda/2$. Also show the actual pattern.

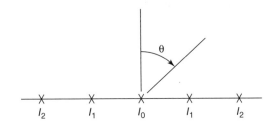

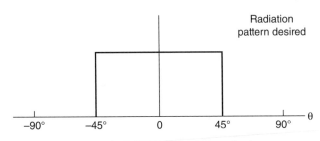

FIGURE P9.5 Pattern synthesis.

9.6 The field in the rectangular aperture is given by (see Eq. 9.70)

$$E_{sx} = \frac{1}{3} + \frac{2}{3}\cos^2\frac{\pi x}{a} \quad \text{for } |x| < \frac{a}{2} \quad \text{and} \quad |y| < \frac{b}{2}.$$

Find the radiation pattern, first sidelobe level, and half-power beamwidth when $a = 100\lambda$ and $b = 50\lambda$.

9.7 The field in a circular aperture is given by (9.73)

$$E_{sx} = (a^2 - \rho'^2)^n \qquad \text{for } \rho' \le a,$$

$$a = 50\lambda.$$

Find the radiation pattern, first sidelobe level, and half-power beamwidth when $n = 0$ and $n = 1$.

9.8 Find and plot the radiation pattern of a circular microstrip antenna with diameter of 3 cm, $t = 2$ mm, and $\varepsilon_r = 2.5$ in the x–z plane and in the y–z plane.

9.9 Two half-wave wire antennas with radius $a = 0.005\lambda$ are placed side by side with a separation of $\lambda/4$. Find the self- and mutual impedances.

CHAPTER 10

SCATTERING OF WAVES BY CONDUCTING AND DIELECTRIC OBJECTS

In radar, the radio wave is transmitted toward an object and the scattered wave received by an antenna reveals the characteristics of the object, such as its position and motion. In biomedical applications, microwaves, optical waves, or acoustic waves are propagated through biological media and the scattering from various portions of a body is used to identify the objects for diagnostic purposes. The scattering of waves may be used to probe atmospheric conditions, such as the size, density, and motion of rain, fog, smog, and cloud particles, which in turn gives useful information on the environment and for weather prediction. In microwave and space communication systems, there is an increasing need for higher frequencies and millimeter waves because of their larger channel capacity. However, millimeter waves suffer absorption and scattering by the atmosphere much more than do microwaves, and thus the knowledge of these scattering characteristics is essential to reliable communications.

In this chapter, we develop basic formulations of the scattering problem and discuss whenever possible potential applications. We limit ourselves to the scattering by a single object (Kerker, 1969; van de Hulst, 1957). Also in this chapter, we exclude the rigorous boundary value solutions of spherical, cylindrical, and other complex objects as they are covered in Chapters 11–13. Some material in Chapter 10 is taken from Chapter 2 of my book *Wave Propagation and Scattering in Random Media* (reissued by IEEE Press and Oxford University, 1997).

Electromagnetic Wave Propagation, Radiation, and Scattering: From Fundamentals to Applications,
Second Edition. Akira Ishimaru.
© 2017 by The Institute of Electrical and Electronic Engineers, Inc. Published 2017 by John Wiley & Sons, Inc.

10.1 CROSS SECTIONS AND SCATTERING AMPLITUDE

When an object is illuminated by a wave, a part of the incident power is scattered out and another part is absorbed by the object. The characteristics of these two phenomena, scattering and absorption, can be expressed most conveniently by assuming an incident plane wave. Let us consider a linearly polarized electromagnetic plane wave propagating in a medium having dielectric constant ε_0 and permeability μ_0 with the electric field given by

$$\bar{E}_i(\bar{r}) = \hat{e}_i \, \exp(-jk\hat{i} \cdot \bar{r}). \tag{10.1}$$

The amplitude $|E_i|$ is chosen to be 1 (volt/m), $k = \omega\sqrt{\mu_0\varepsilon_0} = (2\pi)/\lambda$ is the wave number, λ is a wavelength in the medium, $\hat{i}$ is a unit vector in the direction of wave propagation, and $\hat{e}_i$ is a unit vector in the direction of its polarization.

The object may be a dielectric particle such as a raindrop or ice particle, or a conducting body such as an aircraft (Fig. 10.1). The total field $\bar{E}$ at a distance R from a reference point in the object, in the direction of a unit vector $\hat{o}$, consists of the incident field $\bar{E}_i$ and the field $\bar{E}_s$ scattered by the particle. Within a distance $R < D^2/\lambda$ (where D is a typical dimension of the object such as its diameter), the field $\bar{E}_s$ has complicated amplitude and phase variations because of the interference between contributions from different parts of the object, and the observation point $\bar{r}$ is said to be in the near field of the object. When $R > D^2/\lambda$, however, the scattered field $\bar{E}_s$ behaves as a spherical wave and is given by

$$\bar{E}_s(\bar{r}) = \bar{f}(\hat{o}, \hat{i})\frac{e^{-jkR}}{R} \quad \text{for } R > \frac{D^2}{\lambda}. \tag{10.2}$$

Here, $\bar{f}(\hat{o}, \hat{i})$ represents the amplitude, phase, and polarization of the scattered wave in the far field in the direction $\hat{o}$ when the object is illuminated by a plane wave propagating in the direction $\hat{i}$ with unit amplitude and is called the *scattering amplitude*. It should be noted that even though the incident wave is linearly polarized, the scattered wave is in general elliptically polarized.

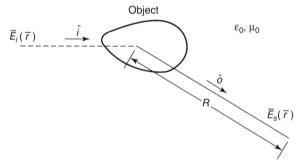

FIGURE 10.1 A plane wave $\bar{E}_i(\bar{r})$ is incident on object, and the scattered field $\bar{E}_s(\bar{r})$ is observed in the direction $\hat{o}$ at a distance R.

Consider the scattered power flux density S_s at a distance R from the object in the direction $\hat{o}$, caused by an incident power flux density S_i. We define the *differential scattering cross section* as follows:

$$\sigma_d(\hat{o}, \hat{i}) = \lim_{R \to \infty} \frac{R^2 S_s}{S_i} = |f(\hat{o}, \hat{i})|^2 = \frac{\sigma_t}{4\pi} p(\hat{o}, \hat{i}), \tag{10.3}$$

where S_i and S_s are the magnitudes of the incident and the scattering power flux density vectors.

$$\bar{S}_i = \frac{1}{2}(\bar{E}_i \times \bar{H}_i^*) = \frac{|E_i|^2}{2\eta_0}\hat{i}, \quad \bar{S}_s = \frac{1}{2}(\bar{E}_s \times \bar{H}_s^*) = \frac{|E_s|^2}{2\eta_0}\hat{o}, \tag{10.4}$$

and $\eta_0 = (\mu_0/\varepsilon_0)^{1/2}$ is the characteristic impedance of the medium. We see that σ_d has the dimensions of area/solid angle. It may be defined physically as follows: suppose that the observed scattered power flux density in the direction $\hat{o}$ is extended uniformly over 1 steradian of solid angle about $\hat{o}$. Then the cross section of an object that would cause just this amount of scattering would be σ_d, so that σ_d varies with $\hat{o}$. The dimensionless quantity $p(\hat{o}, \hat{i})$ in (10.3) is called the *phase function* and is commonly used in radiative transfer theory. The name "phase function" has its origins in astronomy, where it refers to lunar phases and has no relation to the phase of the wave. σ_t is the *total cross section*, to be defined in (10.9).

In radar applications, the *bistatic radar cross section* σ_{bi} and the *backscattering cross section* σ_b are commonly used. They are related to σ_d through

$$\sigma_{bi}(\hat{o}, \hat{i}) = 4\pi\sigma_d(\hat{o}, \hat{i}), \quad \sigma_b = 4\pi\sigma_d(-\hat{i}, \hat{i}). \tag{10.5}$$

σ_b is also called the *radar cross section*. A physical concept of σ_{bi} may be obtained in a way similar to that used in obtaining σ_d above. Suppose that the observed power flux density in the direction $\hat{o}$ is extended uniformly in all directions from the object over the entire 4π steradians of solid angle. Then the cross section that would cause this would be 4π times σ_d for the direction $\hat{o}$.

Next, let us consider the total observed scattered power *at all angles* surrounding the object. The cross section of an object that would produce this amount of scattering is called the *scattering cross section* σ_s, and is given by

$$\sigma_s = \int_{4\pi} \sigma_d \, d\omega = \int_{4\pi} |\bar{f}(\hat{o}, \hat{i})|^2 \, d\omega = \frac{\sigma_t}{4\pi} \int_{4\pi} p(\hat{o}, \hat{i}) d\omega, \tag{10.6}$$

where $d\omega$ is the differential solid angle.

Alternatively, σ_s can be written more generally as

$$\sigma_s = \frac{\int_{S_0} \text{Re}\left(\frac{1}{2}\bar{E}_s \times \bar{H}_s^*\right) \cdot d\bar{a}}{|S_i|}, \tag{10.7}$$

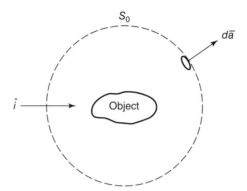

FIGURE 10.2   Area S_0 surrounding the object.

where S_0 is an arbitrary surface enclosing the object and $d\bar{a}$ is a vector representing the differential surface area directed outward (Fig. 10.2).

Next, consider the total power absorbed by the object. The cross section of an object that would correspond to this much power is called σ_a, the absorption cross section. It can be expressed either in terms of the total power flux entering the object or as the volume integral of the loss inside the particle.

$$\sigma_a = \frac{-\int_{S_0} \mathrm{Re}\left(\frac{1}{2}\bar{E}\times\bar{H}^*\right)\cdot d\bar{a}}{|S_i|} = \frac{\int_V k\,\varepsilon_r''(\bar{r}')\,|\bar{E}(\bar{r}')|^2\,dV'}{|E_i|^2}, \qquad (10.8)$$

where $\bar{E} = \bar{E}_i + \bar{E}_s$ and $\bar{H} = \bar{H}_i + \bar{H}_s$ are the total fields. Finally, the sum of the scattering and the absorption cross sections is called the *total cross section* σ_t or the *extinction cross section*.

$$\sigma_t = \sigma_s + \sigma_a. \qquad (10.9)$$

The ratio W_0 of the scattering cross section to the total cross section is called the *albedo* of an object and is given by

$$W_0 = \frac{\sigma_s}{\sigma_t} = \frac{1}{\sigma_t}\int_{4\pi} |\bar{f}(\hat{o},\hat{i})|^2\,d\omega = \frac{1}{4\pi}\int_{4\pi} p(\hat{o},\hat{i})\,d\omega. \qquad (10.10)$$

The cross sections are also normalized by the geometric cross section and are called the *absorption efficiency* Q_a, the *scattering efficiency* Q_s, and the *extinction efficiency* Q_t.

$$Q_a = \frac{\sigma_a}{\sigma_g}, \quad Q_s = \frac{\sigma_s}{\sigma_g}, \quad Q_t = \frac{\sigma_t}{\sigma_g}. \qquad (10.11)$$

In the above we assumed that the incident wave is a linearly polarized plane wave. In a more general case, the incident wave should be an elliptically polarized plane wave. The scattering amplitude should then be expressed by a 2×2 scattering amplitude matrix. This is explained in more detail in Section 10.10.

10.2 RADAR EQUATIONS

Let us consider a transmitter Tr illuminating an object at a large distance R_1. Let $G(\hat{\imath})$ be the gain function of the transmitter and P_t be the total power transmitted. The scattered wave is received with a receiver Re at a large distance R_2. We let $A_r(\hat{o})$ be the receiving cross section of the receiver and P_r be the received power. We now wish to find the ratio P_r/P_t (Fig. 10.3). We assume that R_1 and R_2 are large and that the object is in the far field of both antennas. This requires that approximately

$$R_1 > \frac{2D_t^2}{\lambda} \quad \text{and} \quad R_2 > \frac{2D_r^2}{\lambda},$$

where D_t and D_r are the aperture sizes of the transmitter and the receiver.

The gain function $G(\hat{\imath})$ of an antenna is the ratio of the actual power flux $P(\hat{\imath})$ radiated per unit solid angle in the direction $\hat{\imath}$ to the power flux $P_t/4\pi$ of an isotropic radiator per unit solid angle. We therefore have

$$G(\hat{\imath}) = \frac{P(\hat{\imath})}{P_t/4\pi}. \tag{10.12}$$

In terms of the gain $G_t(\hat{\imath})$ of the transmitter, we obtain the incident power flux density S_i at the object

$$S_i = \frac{G_t(\hat{\imath})}{4\pi R_1^2} P_t. \tag{10.13}$$

The power flux density S_r at the receiver is then given by

$$S_r = \frac{\sigma_{bi}(\hat{o}, \hat{\imath}) S_i}{4\pi R_2^2}, \tag{10.14}$$

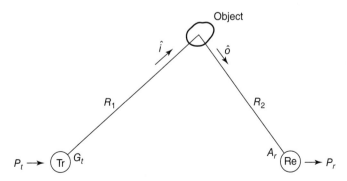

FIGURE 10.3   Radar equation.

where σ_{bi} is the *bistatic cross section* given in (10.5). The received power P_r, when a wave is incident on the receiver from a given direction $(\hat{o})$, is given by

$$P_r = A_r(\hat{o})S_r, \tag{10.15}$$

where $A_r(\hat{o})$ is called the receiving cross section.

It is known that for all matched antennas, the receiving cross section is proportional to the gain function and their ratio is $\lambda^2/4\pi$ (see Section 9.1).

$$A_r(\hat{o}) = \frac{\lambda^2}{4\pi} G_r(-\hat{o}), \tag{10.16}$$

where $G_r(-\hat{o})$ is the gain function in the direction $-\hat{o}$.

Combining (10.13) to (10.16), we get the ratio of the received power to the transmitted power.

$$\frac{P_r}{P_t} = \frac{\lambda^2 G_t(\hat{i})G_r(-\hat{o})\sigma_{bi}(\hat{o},\hat{i})}{(4\pi)^3 \, R_1^2 \, R_2^2}. \tag{10.17}$$

This is the bistatic radar equation. For radar applications, the same antenna is used both as a transmitter and a receiver. Thus, $\hat{o} = -\hat{i}$ and $R_1 = R_2 = R$. Therefore, for radar, we have

$$\frac{P_r}{P_t} = \frac{\lambda^2 [G_t(\hat{i})]^2 \sigma_b(-\hat{i},\hat{i})}{(4\pi)^3 \, R^4}. \tag{10.18}$$

Equations (10.17) and (10.18) give the received power in terms of the antenna gains, the distances, and the cross section. This is applicable when the object is far from both the transmitter and the receiver. It is also required that the receiving antenna be matched to the incoming wave in polarization and impedance. If there is a mismatch, (10.17) and (10.18) must be multiplied by a mismatch factor that is less than unity. In addition, both the transmitter and the receiver must be in the far zone of the object $(R_1 > 2D_o/\lambda^2$ and $R_2 > 2D_o/\lambda^2$; D_o is the object size).

10.3 GENERAL PROPERTIES OF CROSS SECTIONS

Before we discuss the mathematical representations of the various cross sections, it may be worthwhile to present an overall view of how these cross sections are related to the geometric cross section, wavelength, and dielectric constant. If the size of an object is much greater than a wavelength, the total cross section σ_t approaches twice the geometric cross section σ_g of the object as the size increases. To show this, let us consider an incident wave with power flux density S_i (Fig. 10.4). The total flux $S_i\sigma_g$ within the geometric cross section σ_g is either reflected out or absorbed by the object. Behind the object, there should be a shadow region where practically no wave exists. In this shadow region, the scattered wave from the object is exactly equal to

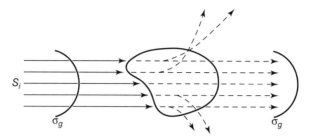

FIGURE 10.4   Relationship between total cross section and geometric cross section for a large object.

the incident wave but 180° out of phase, and this scattered flux is equal to $S_i\sigma_g$ in magnitude. The total scattered and absorbed flux, therefore, approaches $(S_i\sigma_g + S_i\sigma_g)$ and the total cross section σ_t approaches

$$\sigma_t \to \frac{2S_i\sigma_g}{S_i} = 2\sigma_g. \qquad (10.19)$$

It is also seen that the total absorbed power, when the object is very large, cannot be greater than $S_i\sigma_g$, and thus the absorption cross section σ_a approaches a constant somewhat less than the geometric cross section.

$$\sigma_a \to \sigma_g. \qquad (10.20)$$

If the size is much smaller than a wavelength, the scattering cross section σ_s is inversely proportional to the fourth power of the wavelength and proportional to the square of the volume of the object. These characteristics of a small object are generally called *Rayleigh scattering*. The absorption cross section σ_a for a small scatterer is inversely proportional to the wavelength and directly proportional to its volume. Compared with the geometric cross section, we have

$$\frac{\sigma_s}{\sigma_g} \sim \left(\frac{\text{size}}{\lambda}\right)^4 [(\varepsilon_r' - 1)^2 + \varepsilon_r''^2], \qquad (10.21)$$

$$\frac{\sigma_a}{\sigma_g} \sim \frac{\text{size}}{\lambda}\varepsilon_r''. \qquad (10.22)$$

Curves of the normalized cross section above versus relative size of an object are sketched in Fig. 10.5.

It is also possible to obtain the behavior of the backscattering cross section σ_b for a large object. Consider a point of specular reflection on the surface of the object (Fig. 10.6). An incident wave with power flux density S_i is incident on a small area $\Delta l_1 \Delta l_2 = (a_1\Delta\theta_1)(a_2\Delta\theta_2)$. Since the radii of curvature are large, the surface may be considered locally plane, and therefore, using the reflection coefficient for normal

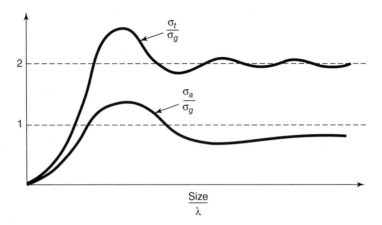

FIGURE 10.5 Total cross section σ_t and absorption cross section σ_a normalized to geometric cross section σ_g.

incidence on a plane boundary, the reflected power flux density on the surface is given by

$$S_r = \left| \frac{\sqrt{\varepsilon_r} - 1}{\sqrt{\varepsilon_r} + 1} \right|^2 S_i.$$

At a large distance R from the particle, the flux within this small area $\Delta l_1 \Delta l_2$ spreads out over an area $R^2(2\delta\theta_1)(2\delta\theta_2)$, and therefore, the scattered flux density S_s at R is related to S_r through

$$S_s \, R^2(2\delta\theta_1)(2\delta\theta_2) = S_r(a_1 \, \delta\theta_2)(a_2 \, \delta\theta_2),$$

from which we obtain the backscattering cross section σ_b.

$$\sigma_b = 4\pi\sigma_d(-\hat{i}, \hat{i}) = \lim_{R \to \infty} \frac{4\pi R^2 S_s}{S_i} = \pi a_1 \, a_2 \left| \frac{\sqrt{\varepsilon_r} - 1}{\sqrt{\varepsilon_r} + 1} \right|^2. \qquad (10.23)$$

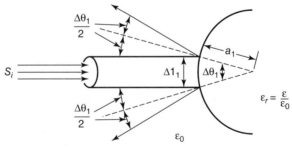

FIGURE 10.6 Backscattering from a large object.

This is the limiting value of σ_b as the object size is increased to infinity, and therefore, for any finite size, σ_b may have a value considerably different from (10.23).

The total cross section σ_t represents the total power loss from the incident wave due to the scattering and absorption of a wave by the object. This loss is closely related to the behavior of the scattered wave in the forward direction, and this general relationship is embodied in the *forward scattering theorem*, also called the *optical theorem*.

The forward scattering theorem states that the total cross section σ_t is related to the imaginary part of the scattering amplitude in the forward direction $\bar{f}(\hat{i}, \hat{i})$ in the following manner:

$$\sigma_t = -\frac{4\pi}{k} \, \text{Im}[\bar{f}(\hat{i}, \hat{i})] \cdot \hat{e}_i, \tag{10.24}$$

where Im denotes the "imaginary part of" and $\hat{e}_i$ is the unit vector in the direction of polarization of the incident wave. The proof is given in Appendix 10.A.

10.4 INTEGRAL REPRESENTATIONS OF SCATTERING AMPLITUDE AND ABSORPTION CROSS SECTIONS

Mathematical descriptions of scattering amplitude and absorption cross sections can be made in one of the two ways. If the shape of an object is a simple one such as a sphere, it is possible to obtain exact expressions for cross sections and the scattering amplitude. The exact solution for a dielectric sphere, called the Mie solution, is discussed in Chapter 12. In many practical situations, however, the shape of an object is not simple. Therefore, we need a method of determining approximate cross sections for objects with complex shapes. This can be done through general integral representations of the scattering amplitude. The method is also useful for objects with simple shapes because the calculations can be made easily.

Let us consider a dielectric body whose relative dielectric constant is a function of the position within the body.

$$\varepsilon_r(\bar{r}) = \frac{\varepsilon(\bar{r})}{\varepsilon_0} = \varepsilon_r'(\bar{r}) - j\varepsilon_r''(\bar{r}) \quad \text{in } V. \tag{10.25}$$

The dielectric body occupies the volume V and is surrounded by a medium whose dielectric constant is ε_0.

We first write Maxwell's equations

$$\begin{aligned} \nabla \times \bar{E} &= -j\omega\mu_0 \, \bar{H}, \\ \nabla \times \bar{H} &= -j\omega\varepsilon(\bar{r}) \, \bar{E}. \end{aligned} \tag{10.26}$$

Here, we assume that the permeability μ_0 is constant in and outside the dielectric body. If we write the second equation in (10.26) in the following manner:

$$\nabla \times \bar{H} = j\omega\varepsilon_0 \, \bar{E} + \bar{J}_{\text{eq}}, \tag{10.27}$$

where

$$\bar{J}_{eq} = \begin{cases} j\omega\varepsilon_0[\varepsilon_r(\bar{r}) - 1]\bar{E} & \text{in } V \\ 0 & \text{outside,} \end{cases}$$

the term $\bar{J}_{eq}$ may be considered as an equivalent current source which generates the scattered wave. The solution to (10.26) and (10.27) is given by

$$\begin{aligned} \bar{E}(\bar{r}) &= \bar{E}_i(\bar{r}) + \bar{E}_s(\bar{r}), \\ \bar{H}(r) &= \bar{H}_i(\bar{r}) + \bar{H}_s(\bar{r}), \end{aligned} \tag{10.28}$$

where $(\bar{E}_i, \bar{H}_i)$ is the primary (or incident) wave that exists in the absence of the object, and $(\bar{E}_s, \bar{H}_s)$ is the scattered wave originating from it. Using the Hertz vector $\bar{\Pi}_s$, we write

$$\begin{aligned} \bar{E}_s(\bar{r}) &= \nabla \times \nabla \times \bar{\Pi}_s(\bar{r}), \\ \bar{H}_s(\bar{r}) &= j\omega\varepsilon_0 \, \nabla \times \bar{\Pi}_s(\bar{r}), \\ \bar{\Pi}_s(\bar{r}) &= \frac{1}{j\omega\varepsilon_0} \int_V G_0(\bar{r}, \bar{r}')\bar{J}_{eq}(\bar{r}') \, dV' \\ &= \int_V [\varepsilon_r(\bar{r}') - 1]\bar{E}(\bar{r}')G_0(\bar{r}, \bar{r}') \, dV', \end{aligned} \tag{10.29}$$

where

$$G_0(\bar{r}, \bar{r}') = \frac{\exp(-jk\,|\bar{r} - \bar{r}'|)}{4\pi\,|\bar{r} - \bar{r}'|},$$

is the free-space Green's function. Equation (10.29) is valid only for $\bar{r} \neq \bar{r}'$.

To obtain the scattering amplitude, we consider $\bar{E}_s(\bar{r})$ in the far field of the object. Referring to Fig. 10.7, we note that $\bar{r} = R\hat{o}$, and in the far zone, the magnitude $1/\,|\bar{r} - \bar{r}'|$ of Green's function can be approximated by $1/R$. However, the phase $k\,|\bar{r} - \bar{r}'|$ cannot be approximated by kR because the difference may be significant in

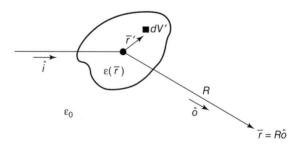

FIGURE 10.7   Geometry of a point $\bar{r}'$ within the object and observation point $\bar{r}$.

terms of wavelengths. Expanding $|\bar{r} - \bar{r}'|$ in a binomial series and keeping its first term, we get

$$|\bar{r} - \bar{r}'| = (R^2 + r'^2 - 2R\bar{r}' \cdot \hat{o})^{1/2} \simeq R - \bar{r}' \cdot \hat{o},$$

and Green's function becomes for large R,

$$G_0(\bar{r}, \bar{r}') = \frac{\exp(-jkR + jk\bar{r}' \cdot \hat{o})}{4\pi R}. \tag{10.30}$$

We note also that in the far field,

$$\nabla \left(\frac{e^{-jkR}}{R} \right) \simeq \frac{e^{-jkR}}{R}(-jk \nabla R) = -jk\hat{o}\frac{e^{-jkR}}{R} \tag{10.31}$$

and thus ∇ is equivalent to $-jk\hat{o}$. Substituting (10.30) and (10.31) into (10.29), we get

$$\bar{E}_s(\bar{r}) = \bar{f}(\hat{o}, \hat{i})\frac{\exp(-jkR)}{R},$$
$$\bar{f}(\hat{o}, \hat{i}) = \frac{k^2}{4\pi} \int_V [\bar{E} - \hat{o}(\hat{o} \cdot \bar{E})][\varepsilon_r(\bar{r}') - 1] \exp(jk\bar{r}' \cdot \hat{o}) \, dV', \tag{10.32}$$

where we used $-\hat{o} \times (\hat{o} \times \bar{E}) = \bar{E} - \hat{o}(\hat{o} \cdot \bar{E})$. Note also that $\hat{o}(\hat{o} \cdot \bar{E})$ is the component of $\bar{E}$ along $\hat{o}$, and therefore, $[\bar{E} - \hat{o}(\hat{o} \cdot \bar{E})]$ is the component of $\bar{E}$ perpendicular to $\hat{o}$. This is an exact expression for the scattering amplitude $\bar{f}(\hat{o}, \hat{i})$ in terms of the total electric field $\bar{E}(\bar{r}')$ inside the object. This field $\bar{E}(\bar{r}')$ is not known in general, and therefore, (10.32) is not a complete description of the scattering amplitude in terms of known quantities. In many practical situations, however, it is possible to approximate $\bar{E}(\bar{r}')$ by some known function and thus obtain a useful approximate expression for $\bar{f}(\hat{o}, \hat{i})$. This will be done in Sections 10.5–10.7. The absorption cross section σ_a for a dielectric body has been given in (10.8).

We note here that we can develop an alternative integral equation for the magnetic field $\bar{H}(\bar{r})$ rather than $\bar{E}(\bar{r})$. From Maxwell's equations, we obtain the vector wave equation for $\bar{H}$ in the following form:

$$\nabla \times \nabla \times \bar{H}(\bar{r}) - \omega^2 \mu_0 \varepsilon_0 \bar{H}(\bar{r}) = \omega^2 \mu_0 \varepsilon_0[\varepsilon_r(\bar{r}) - 1]\bar{H}(\bar{r}) + j \omega\varepsilon_0[\nabla\varepsilon_r(\bar{r}) \times \bar{E}(\bar{r})]. \tag{10.33}$$

Therefore, an integral equation for $\bar{H}$ has two terms

$$\bar{H}(\bar{r}) = \bar{H}_i(\bar{r}) + \bar{H}_s(\bar{r}) = \bar{H}_i(\bar{r}) + \nabla \times \nabla \times \bar{\Pi}_{ms}(\bar{r}),$$
$$\bar{\Pi}_{ms}(\bar{r}) = \int_V [\varepsilon_r(\bar{r}') - 1]G_0\left(\frac{\bar{r}}{\bar{r}'}\right) \bar{H}(\bar{r}') \, dV'$$
$$- \frac{1}{\omega\mu_0} \int_V G_0\left(\frac{\bar{r}}{\bar{r}'}\right) \nabla' \varepsilon_r(\bar{r}') \times \bar{E}(\bar{r}') \, dV', \tag{10.34}$$

where $\nabla' \, \varepsilon_r(\bar{r}')$ is the gradient with respect to $\bar{r}'$. Thus, the second term in $\bar{\bar{\Pi}}_{ms}$ contains the effect of depolarization due to the inhomogeneity of the dielectric constant. For a homogeneous object, $\nabla' \, \varepsilon_r(\bar{r})$ gives a delta function on the surface, and thus the second term becomes a surface integral.

10.5 RAYLEIGH SCATTERING FOR A SPHERICAL OBJECT

We have indicated in Section 10.3 the general scattering characteristics of a small object. This is generally known as Rayleigh scattering. In this section, we present a detailed analysis for a few simple geometries. Let us consider a dielectric sphere whose size is much smaller than a wavelength. Because of its small size, the impinging electric field within and near the sphere must behave almost as an electrostatic field. It is known in electrostatics that when a constant electric field E_i is applied to a dielectric sphere, the electric field $\bar{E}$ inside the sphere is uniform and is given by (Fig. 10.8)

$$\bar{E} = \frac{3}{\varepsilon_r + 2}\bar{E}_i, \quad \bar{E}_i = E_i \hat{e}_i. \tag{10.35}$$

We also note in (10.32) that $\exp(jk\bar{r}' \cdot \hat{o}) \approx 1$ because $k\bar{r}' \ll 1$. Now we can substitute (10.35) into (10.32) and obtain the scattering amplitude $\bar{f}$. However, to express $\bar{f}$ in a more convenient form, let us note that the scattering is caused by the equivalent current $\bar{J}_{eq} = j\,\omega\varepsilon_0(\varepsilon_r - 1)\bar{E}$ in (10.27). We may use the polarization vector $\bar{P} = \bar{J}_{eq}/j\omega$ and the equivalent dipole moment $\bar{p}$ of the sphere given by the integral of $\bar{P}$ over the volume V of the sphere.

$$\bar{p} = \int_V \bar{P} \, dV' = \int_V \varepsilon_0(\varepsilon_r - 1)\bar{E} \, dV' = \frac{3(\varepsilon_r - 1)}{\varepsilon_r + 2}\varepsilon_0 \, V \, \bar{E}_i. \tag{10.36}$$

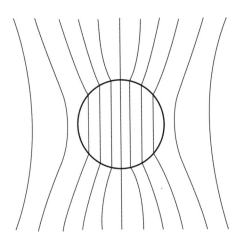

FIGURE 10.8 Electrostatic field inside a dielectric sphere.

We can then rewrite the scattering amplitude for Rayleigh scattering as follows:

$$\bar{f}(\hat{o}, \hat{i}) = \frac{k^2}{4\pi\varepsilon_0}[\bar{p} - \hat{o}(\hat{o} \cdot \bar{p})], \tag{10.37}$$

where V is the volume of the sphere.

Note that $\bar{p} - \hat{o}(\hat{o} \cdot \bar{p})$ is the component of $\bar{p}$ perpendicular to $\hat{o}$, and therefore, its magnitude is equal to $p \sin \chi$, where χ is the angle between $\bar{p}$ and $\hat{o}$ (Fig. 10.9). This is to be expected as this represents the radiation pattern of the electric dipole $\bar{p}$. We also note that (10.36) is valid even when the object is lossy and ε_r is complex. The differential cross section $\sigma_d(\hat{o}, \hat{i})$ is given by

$$\sigma_d(\hat{o}, \hat{i}) = \frac{k^4}{(4\pi)^2} \left| \frac{3(\varepsilon_r - 1)}{\varepsilon_r + 2} \right|^2 V^2 \sin^2 \chi, \tag{10.38}$$

where $\sin^2 \chi = 1 - (\hat{o} \cdot \hat{e}_i)^2$.

We note that the cross section is inversely proportional to the fourth power of the wavelength and directly proportional to the square of the volume of the scatterer. These two characteristics of a small scatterer were derived by Rayleigh using dimensional analysis and are generally known as Rayleigh scattering. The blue color of the sky can be explained by noting that the blue portion of a light spectrum scatters more light than the red portion due to the λ^{-4} dependence. Furthermore, the skylight at right angles to the sun must be linearly polarized, as is evident from Fig. 10.9. These two characteristics, the blue color and the polarization, were a great scientific puzzle in the nineteenth century and were finally explained by Rayleigh (Kerker, 1969). Rayleigh noted that the scatterers need not be water or ice as was commonly believed at that time, but that molecules of air itself can contribute to this scattering. The redness of the sunset is caused by the decrease in the blue portion of the spectrum due to Rayleigh scattering.

Let us consider the scattering cross section σ_s of a small dielectric sphere.

$$\begin{aligned}
\sigma_s &= \int_{4\pi} \sigma_d \, d\omega = \frac{1}{4\pi} \int_{4\pi} \sigma_{bi} \, d\omega \\
&= \frac{k^4}{(4\pi)^2} \left| \frac{3(\varepsilon_r - 1)}{\varepsilon_r + 2} \right|^2 V^2 \int_0^\pi \sin \chi \, d\chi \int_0^{2\pi} d\phi \sin^2 \chi \\
&= \frac{24\pi^3 V^2}{\lambda^4} \left| \frac{\varepsilon_r - 1}{\varepsilon_r + 2} \right|^2 = \frac{128\pi^5 a^6}{3\lambda^4} \left| \frac{\varepsilon_r - 1}{\varepsilon_r + 2} \right|^2 .
\end{aligned} \tag{10.39}$$

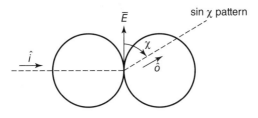

FIGURE 10.9   Dipole radiation pattern for Rayleigh scattering.

It is often desired to compare the scattering cross section with the actual geometrical cross section πa^2. Thus we obtain the Rayleigh equation,

$$Q_s = \frac{\sigma_s}{\pi a^2} = \frac{8(ka)^4}{3} \left| \frac{\varepsilon_r - 1}{\varepsilon_r + 2} \right|^2. \tag{10.40}$$

The Rayleigh equation above is valid only for small ka. The approximate upper limit of the radius of the scatterer is generally taken to be $a = 0.05\lambda$. At this radius, the percent error of the Rayleigh equation (10.40) is less than 4% (Kerker, 1969, p. 85).

The absorption cross section σ_a is obtained using (10.8) with (10.35).

$$\sigma_a = k\, \varepsilon_r'' \left| \frac{3}{\varepsilon_r + 2} \right|^2 V,$$

$$Q_a = \frac{\sigma_a}{\pi a^2} = ka\varepsilon_r'' \left| \frac{3}{\varepsilon_r + 2} \right|^2 \cdot \frac{4}{3}. \tag{10.41}$$

The total cross section σ_t is the sum of (10.40) and (10.41). We note that σ_t cannot be obtained by applying the forward scattering theorem to (10.37), since (10.37) gives $\sigma_t = 0$ when $\varepsilon_r'' = 0$. In general, for a given approximate value of $\bar{E}(\bar{r}')$ within an object, the scattering cross section as obtained by integrating $|f|^2$ in (10.32) or (10.37) over 4π, plus the absorption cross section in (10.8) or (10.41), gives a much better approximation to the total cross section than is obtained by direct application of the forward scattering theorem to (10.32).

10.6 RAYLEIGH SCATTERING FOR A SMALL ELLIPSOIDAL OBJECT

Many of the particles and objects encountered in practice are not spherical, but they can often be approximated by an ellipsoid whose surface is given by

$$\frac{x^2}{a^2} + \frac{y^2}{b^2} + \frac{z^2}{c^2} = 1. \tag{10.42}$$

If the object size is small compared with a wavelength and the incident field $\bar{E}_i$ has components $E_{ix}, E_{iy},$ and E_{iz} in the x, y, and z directions, respectively, the components of the field inside the object are given by the following static solution (Stratton, 1941, p. 213; van der Hulst, 1957, p. 71):

$$E_x = \frac{E_{ix}}{1 + (\varepsilon_r - 1)L_x},$$

$$L_x = \frac{abc}{2} \int_0^\infty (s + a^2)^{-1} [(s + a^2)(s + b^2)(s + c^2)]^{-1/2}\, ds, \tag{10.43}$$

with an appropriate interchange of a, b, and c for E_y and E_z. It can be easily proved that L_x, L_y, and L_z are functions of the ratio b/a and c/a only and do not depend on the values of a, b, and c. It is also known that

$$L_x + L_y + L_z = 1. \tag{10.44}$$

For a prolate ellipsoid ($a = b < c$),

$$
\begin{aligned}
L_z &= \frac{1 - e^2}{e^2}\left(-1 + \frac{1}{2e}\ln\frac{1+e}{1-e}\right), \\
L_x &= L_y = \tfrac{1}{2}(1 - L_z), \\
e^2 &= 1 - \left(\frac{a}{c}\right)^2.
\end{aligned}
\tag{10.45}
$$

For an oblate ellipsoid ($a = b > c$),

$$
\begin{aligned}
L_z &= \frac{1 + f^2}{f^2}\left(1 - \frac{1}{f}\arctan f\right), \\
L_x &= L_y = \tfrac{1}{2}(1 - L_z), \\
f^2 &= \left(\frac{a}{c}\right)^2 - 1.
\end{aligned}
\tag{10.46}
$$

The scattering amplitude $\bar{f}(\hat{o}, \hat{i})$ can then be given by

$$\bar{f}(\hat{o}, \hat{i}) = \frac{k^2}{4\pi\varepsilon_0}[\bar{p} - \hat{o}(\hat{o}\cdot\bar{p})], \tag{10.47}$$

where

$$
\begin{aligned}
V &= \tfrac{4}{3}\pi abc, \\
\bar{p} &= \alpha_x E_{ix}\,\hat{x} + \alpha_y E_{iy}\,\hat{y} + \alpha_z E_{iz}\,\hat{z}, \\
\alpha_x &= \frac{\varepsilon_0(\varepsilon_r - 1)V}{1 + (\varepsilon_r - 1)L_x}.
\end{aligned}
$$

α_y and α_z are obtained by replacing L_x by L_y and L_z, respectively. α_x, α_y, and α_z are the polarizability of the object for the x, y, and z directions.

As an example, consider a plane wave propagating in the direction (θ', ϕ') with the components E'_θ and E'_ϕ. The object is located at the origin with the principal axes oriented in the coordinate system $x_b - y_b - z_b$ and the surface given by (Fig. 10.10)

$$\frac{x_b^2}{a^2} + \frac{y_b^2}{b^2} + \frac{z_b^2}{c^2} = 1. \tag{10.48}$$

We wish to find the scattering amplitude in the direction (θ, ϕ) with the field components E_θ and E_ϕ (Fig. 10.11). We write the scattering amplitude by the following

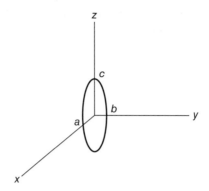

FIGURE 10.10   Ellipsoid.

2×2 matrix $[F] = [f_{ij}]$.

$$\begin{bmatrix} E_\theta \\ E_\phi \end{bmatrix} = \frac{e^{-jkR}}{R} \begin{bmatrix} f_{11} & f_{12} \\ f_{21} & f_{22} \end{bmatrix} \begin{bmatrix} E'_\theta \\ E'_\phi \end{bmatrix}. \tag{10.49}$$

We can then obtain the scattering amplitude matrix as follows:

$$[F] = \frac{k^2}{4\pi\varepsilon_0} [C_1][\alpha][C_2], \tag{10.50}$$

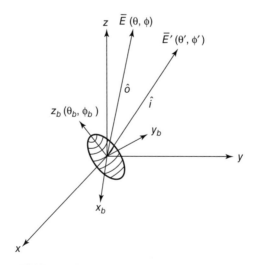

FIGURE 10.11   Scattering amplitude matrix.

where

$$[C_1] = \begin{bmatrix} \hat{\theta} \cdot \hat{x}_b & \hat{\theta} \cdot \hat{y}_b & \hat{\theta} \cdot \hat{z}_b \\ \hat{\phi} \cdot \hat{x}_b & \hat{\phi} \cdot \hat{y}_b & \hat{\phi} \cdot \hat{z}_b \end{bmatrix},$$

$$[C_2] = \begin{bmatrix} \hat{\theta}' \cdot \hat{x}_b & \hat{\phi}' \cdot \hat{x}_b \\ \hat{\theta}' \cdot \hat{y}_b & \hat{\phi}' \cdot \hat{y}_b \\ \hat{\theta}' \cdot \hat{z}_b & \hat{\phi}' \cdot \hat{z}_b \end{bmatrix},$$

$$[\alpha] = \begin{bmatrix} \alpha_x & 0 & 0 \\ 0 & \alpha_y & 0 \\ 0 & 0 & \alpha_z \end{bmatrix}.$$

The relationship between $(\hat{x}_b, \hat{y}_b, \hat{z}_b)$ and $(\hat{x}, \hat{y}, \hat{z})$ is given by Euler's transformation $[A_e]$.

$$\begin{bmatrix} \hat{x}_b \\ \hat{y}_b \\ \hat{z}_b \end{bmatrix} = [A_e] \begin{bmatrix} \hat{x} \\ \hat{y} \\ \hat{z} \end{bmatrix}. \tag{10.51}$$

There are several representations of Euler's transformation and they are given by Goldstein (1981). For axially symmetric objects with an axis oriented in the (θ_b, ϕ_b) direction (Fig. 10.11), we have $L_x = L_y$ and

$$[A_e] = \begin{bmatrix} \cos\theta_b \cos\phi_b & \cos\theta_b \sin\phi_b & -\sin\theta_b \\ -\sin\phi_b & \cos\phi_b & 0 \\ \sin\theta_b \cos\phi_b & \sin\theta_b \sin\phi_b & \cos\theta_b \end{bmatrix}. \tag{10.52}$$

We can then calculate $[C_1]$ and $[C_2]$ in terms of θ, ϕ, θ', ϕ', and θ_b, ϕ_b, noting that

$$\hat{\theta} = \cos\theta \cos\phi \hat{x} + \cos\theta \sin\phi \hat{y} - \sin\theta \hat{z},$$
$$\hat{\phi} = -\sin\phi \hat{x} + \cos\phi \hat{y};$$

and $\hat{\theta}'$ and $\hat{\phi}'$ with θ' and ϕ' replacing θ and ϕ.

Let us consider the scattering and absorption cross sections for an axially symmetric object oriented in the x–z plane inclined with angle θ_b from the z axis which is illuminated by the incident wave propagating in the z direction (Fig. 10.12). The scattering cross section must be calculated by integrating $|\bar{f}|^2$ over all the solid angle (see (10.6)) and the absorption cross section is obtained by (10.8). For nonspherical objects, the scattering and absorption cross sections depend on the polarization of

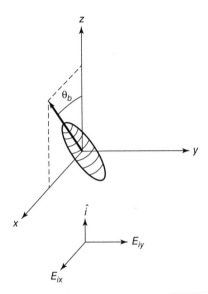

FIGURE 10.12   Axially symmetric object illuminated by a plane wave.

the incident wave. If the incident wave is polarized in the x direction ($\bar{E}_i = E_{ix}\,\hat{x}$), we can perform the calculations above and obtain

$$\sigma_{sx} = \left(\frac{k^2}{4\pi\varepsilon_0}\right)^2 \frac{8\pi}{3} \left(\left|\alpha_x \cos^2\theta_b + \alpha_z \sin^2\theta_b\right|^2 + \left|\alpha_x - \alpha_z\right|^2 \sin^2\theta_b \cos^2\theta_b\right),$$

$$\tag{10.53}$$

$$\sigma_{ax} = k\,\varepsilon_r'' \frac{\left|\alpha_x\right|^2 \cos^2\theta_b + \left|\alpha_z\right|^2 \sin^2\theta_b}{\varepsilon_0^2 \, V \left|\varepsilon_r - 1\right|^2}.$$

If the incident wave is polarized in the y direction ($\bar{E}_i = E_{iy}\,\hat{y}$), we get

$$\sigma_{sy} = \left(\frac{k^2}{4\pi\varepsilon_0}\right)^2 \left|\alpha_y\right|^2 \frac{8\pi}{3},$$

$$\tag{10.54}$$

$$\sigma_{ay} = k\,\varepsilon_r'' \frac{\left|\alpha_y\right|^2}{\left|\varepsilon_r - 1\right|^2 \varepsilon_0^2 \, V}.$$

10.7   RAYLEIGH–DEBYE SCATTERING (BORN APPROXIMATION)

We now consider the scattering characteristics of a scatterer whose relative dielectric constant ε_r is close to unity. In this case, the field inside the scatterer may be approximated by the incident field.

$$\bar{E}(\bar{r}) = \bar{E}_i(\bar{r}) = \hat{e}_i \, \exp(-jk\bar{r} \cdot \hat{i}). \tag{10.55}$$

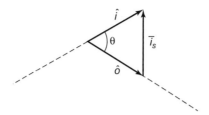

FIGURE 10.13   Relationship showing $\bar{i}_s$ and θ.

Substituting this into (10.32), we get

$$\bar{f}(\hat{o}, \hat{i}) = \frac{k^2}{4\pi}[-\hat{o} \times (\hat{o} \times \hat{e}_i)] \, VS(\bar{k}_s), \tag{10.56}$$

$$S(\bar{k}_s) = \frac{1}{V} \int_V [\varepsilon_r(\bar{r}') - 1] \exp(-j\bar{k}_s \cdot \bar{r}') \, dV', \tag{10.57}$$

where

$$\bar{k}_s = k\bar{i}_s = k(\hat{i} - \hat{o}), \quad |\bar{i}_s| = 2\sin\frac{\theta}{2},$$

and θ is the angle between $\hat{i}$ and $\hat{o}$ (Fig. 10.13).

This approximation is valid when

$$(\varepsilon_r - 1)kD \ll 1, \tag{10.58}$$

where D is a typical dimension of the object such as its diameter. We note that (10.57) is a Fourier transform of $[\varepsilon_r(r') - 1]$ in the direction of $\bar{i}_s$. Therefore, the scattering amplitude $f(\hat{o}, \hat{i})$ is proportional to the Fourier transform of $[\varepsilon_r(r') - 1]$ evaluated at the wave number $\bar{k}_s$. In general, if $[\varepsilon_r(r') - 1]$ is concentrated in a region that is small compared with a wavelength, the cross section is spread out in $\bar{k}_s$ and thus in angle θ, and the scattering is almost isotropic. If the object size is large compared to a wavelength, the scattering is concentrated in a small forward angular region $\theta \approx 0$. This situation is similar to the relationship between a time function and its frequency spectrum. If a function is limited in time within T, its spectrum is spread out over a frequency range $1/T$.

The Rayleigh–Debye absorption cross section is obtained from (10.8).

$$\sigma_a = k \int_V \varepsilon_r''(\bar{r}) \, dV. \tag{10.59}$$

Let us take a few examples.

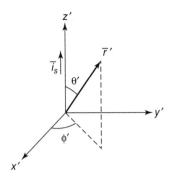

FIGURE 10.14   Coordinate axes for evaluation of (10.61).

10.7.1 Scattering by a Homogeneous Sphere of Radius *a*

In this case, because of the spherical symmetry, we choose the z' axis in the direction of $\bar{i}_s$ (Fig. 10.14). We then write

$$\bar{f}(\hat{o}, \hat{i}) = \frac{k^2}{4\pi}[-\hat{o} \times (\hat{o} \times \hat{e}_i)](\varepsilon_r - 1)\, VF(\theta),$$

$$F(\theta) = \frac{1}{V} \int_V \exp(-jk\bar{i}_s \cdot \bar{r}')\, dV'$$

$$= \frac{1}{V} \int_0^{2\pi} d\phi' \int_0^{\pi} \sin\theta' d\phi' \int_0^a r'^2\, dr'\, \exp(-jk_s\, r'\cos\theta')$$ (10.60)

$$= \frac{3}{k_s^3\, a^3}(\sin k_s\, a - k_s\, a\cos, k_s\, a),$$

$$\hspace{9cm} (10.61)$$

with $k_s = 2k\sin(\theta/2)$. The plot of $|F(\theta)|^2$ is shown in Fig. 10.15.

10.7.2 Scattering by an Ellipsoidal Object

Let us consider an ellipsoid as shown in Fig. 10.16. The incident wave is propagating in the direction (θ_i, ϕ_i) and the scattered wave is observed in the direction (θ_0, ϕ_0). The surface of the ellipsoid is given by

$$\frac{x^2}{a^2} + \frac{y^2}{b^2} + \frac{z^2}{c^2} = 1.$$ (10.62)

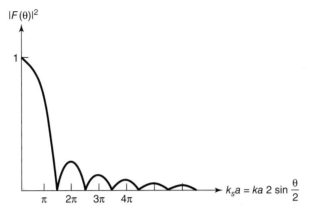

FIGURE 10.15   Scattering pattern (10.61) for a homogeneous sphere of radius $[a]$.

We express the direction of the incident wave $\hat{i}$ and the direction of the observation point $\hat{o}$ in the spherical coordinate system,

$$
\begin{aligned}
\bar{k}_s &= k_{s1}\hat{x} + k_{s2}\hat{y} + k_{s3}\,\hat{z}, \\
k_{s1} &= k(\sin\theta_i \cos\phi_i - \sin\theta_0 \cos\phi_0), \\
k_{s2} &= k(\sin\theta_i \sin\phi_i - \sin\theta_0 \sin\phi_0), \\
k_{s3} &= k(\cos\theta_i - \cos\theta_0).
\end{aligned}
\tag{10.63}
$$

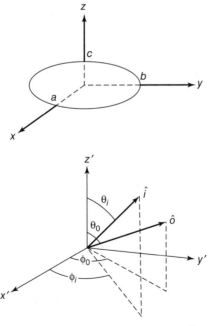

FIGURE 10.16   Ellipsoidal object and the directions $\hat{i}$ and $\hat{o}$ in (10.63).

We then get

$$F = \frac{1}{V} \int_V \exp[-j(k_{s1}\, x + k_{s2}\, y + k_{s3}\, z)] \, dx \, dy \, dz. \tag{10.64}$$

The integration is over the ellipsoidal volume, but if we use the following normalized coordinate, the integration can be performed over a sphere of unit radius.

$$x' = \frac{x}{a}, \quad y' = \frac{y}{b}, \quad z' = \frac{z}{c}. \tag{10.65}$$

We then get

$$
\begin{aligned}
F &= \frac{3}{4\pi} \int \exp(-j\bar{K} \cdot \bar{r}) \, dx' \, dy' \, dz' \\
&= \frac{3}{K^3} (\sin K - K \cos K),
\end{aligned} \tag{10.66}
$$

where

$$K = [(k_{s1}\, a)^2 + (k_{s2}\, b)^2 + (k_{s3}\, c)^2]^{1/2}.$$

10.7.3 Scattering From a Randomly Oriented Object with Axial Symmetry

We note that in (10.63), $\sqrt{k_{s1}^2 + k_{s2}^2}$ and k_{s3} are the components of $k\,\bar{i}_s = k(\hat{i} - \hat{o})$ in the directions perpendicular and parallel to the z axis. Letting β be the angle between $\bar{i}_s$ and the z axis, we write

$$\sqrt{k_{s1}^2 + k_{s2}^2} = k_s \sin \beta \quad \text{and} \quad k_{s3} = k_s \cos \beta.$$

For random orientation, we average the scattered intensity over all possible orientations of the object. Here, due to the randomness, the intensity rather than the field must be averaged.

$$|F|^2_{\text{ave}} = \begin{cases} \dfrac{1}{4\pi} \displaystyle\int |F|^2 \, d\omega, \quad d\omega = \sin \beta \, d\beta \, d\phi, \\[2ex] \dfrac{1}{2} \displaystyle\int_{-1}^{1} |F|^2 \, d\mu, \qquad \mu = \cos \beta. \end{cases} \tag{10.67}$$

10.8 ELLIPTIC POLARIZATION AND STOKES PARAMETERS

In Sections 10.1 through 10.7, we considered a linearly polarized incident wave. In general, however, it is necessary to consider an incident wave with elliptic polarization. Let us examine a plane wave propagating in the z direction, whose electric field components as functions of time are given by

$$E_x = \text{Re}(E_1 \, e^{j\omega t}) = \text{Re}[a_1 \exp(j\omega t - jkz + j\delta_1)] = a_1 \cos(\tau + \delta_1),$$
$$E_y = \text{Re}(E_2 \, e^{j\omega t}) = \text{Re}[a_2 \exp(j\omega t - jkz + j\delta_2)] = a_2 \cos(\tau + \delta_2), \quad (10.68a)$$
$$E_z = 0,$$

where $\tau = \omega t - kz$, and E_1 and E_2 are the phasors for E_x and E_y.

In the above, we used the IEEE convention $\exp(j\omega t)$. In many studies using Stokes' parameters, it is more common to use the $\exp(-i\omega t)$ convention. In this case, (10.68a) should be written as

$$E_x = \text{Re}(E_1 \, e^{-i\omega t}) = \text{Re}[a_1 \exp(-i\omega t + ikz - i\delta_1)] = a_1 \cos(\tau + \delta_1),$$
$$E_y = \text{Re}(E_2 \, e^{-i\omega t}) = \text{Re}[a_2 \exp(-i\omega t + ikz - i\delta_2)] = a_2 \cos(\tau + \delta_2). \quad (10.68b)$$

Now consider a general elliptically polarized wave. The endpoint of the electric field vector $\bar{E} = E_x \hat{x} + E_y \hat{y}$ traces an ellipse. The equation for this ellipse is obtained by eliminating τ from (10.68a) or (10.68b).

$$\left(\frac{E_x}{a_1}\right)^2 + \left(\frac{E_y}{a_2}\right)^2 - \frac{2E_x E_y}{a_1 a_2} \cos\delta = \sin^2\delta, \quad (10.69)$$

where $\delta = \delta_2 - \delta_1$ is the phase difference.

To describe the elliptically polarized wave given in (10.68), three independent parameters are needed. For example, they can be a_1, a_2, and δ. It is, however, more convenient to use parameters of the same dimension. In 1852, G. G. Stokes introduced what are now called the *Stokes parameters*. They are

$$I = a_1^2 + a_2^2 = |E_1|^2 + |E_2|^2,$$
$$Q = a_1^2 - a_2^2 = |E_1|^2 - |E_2|^2,$$
$$U = 2a_1 a_2 \cos\delta = 2\,\text{Re}(E_1 E_2^*), \quad (10.70)$$
$$V = \mp 2a_1 a_2 \sin\delta = 2\,\text{Im}(E_1 E_2^*),$$

where E_1 and E_2 are the phasor representations of the electric field components E_x and E_y given by

$$E_1 = a_1 \exp(j\delta_1 - jkz) = a_1 \exp(-i\delta_1 + ikz),$$
$$E_2 = a_2 \exp(j\delta_2 - jkz) = a_2 \exp(-i\delta_2 + ikz).$$

Note that the upper and lower signs for V are for $\exp(j\omega t)$ and $\exp(-i\omega t)$ dependence, respectively. Among these four parameters, there exists the relationship, obtained from (10.70),

$$I^2 = Q^2 + U^2 + V^2. \quad (10.71)$$

Equations (10.70) and (10.71) together provide three independent quantities that describe an elliptically polarized wave.

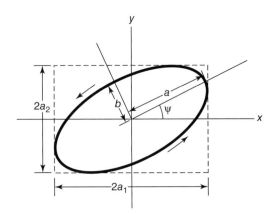

FIGURE 10.17 Right-handed elliptical polarization.

As an example, for a wave linearly polarized in the direction ψ_0 with respect to the x axis, we have $a_1 = E_0 \cos \psi_0$, $a_2 = E_0 \sin \psi_0$, and $\delta = 0$, and the Stokes parameters are

$$I = E_0^2, \quad Q = E_0^2 \cos 2\psi_0, \quad U = E_0^2 \sin 2\psi_0, \quad V = 0. \qquad (10.72)$$

For a right-handed circularly polarized wave, we have $a_1 = a_2 = E_0$, $\delta = -\pi/2$, and

$$I = 2E_0^2, \quad Q = 0, \quad U = 0, \quad V = \pm 2E_0^2. \qquad (10.73)$$

Here the upper and lower signs for V are for $\exp(j\omega t)$ and $\exp(-i\omega t)$ dependence, respectively. It is also common to use the modified Stokes parameters given by

$$I_1 = |E_1|^2, \quad I_2 = |E_2|^2, \quad U = 2 \operatorname{Re}(E_1 E_2^*), \quad V = 2 \operatorname{Im}(E_1 E_2^*). \qquad (10.74)$$

Alternatively, it is possible to describe the ellipse in Fig. 10.17 in terms of the semimajor (a) and the semiminor (b) axes of the ellipse and the orientation angle (ψ). Using I, b/a, and ψ, the Stokes parameters become, for $\exp(-i\omega t)$ dependence,

$$\begin{aligned} Q &= I \, \cos 2\chi \, \cos 2\psi, \\ U &= I \, \cos 2\chi \, \sin 2\psi, \qquad (10.75) \\ V &= I \, \sin 2\chi, \end{aligned}$$

where $\tan \chi = \pm b/a$, with a plus sign for left-handed polarization and a minus sign for right-handed polarization. The polarization is defined as right handed when the electric field rotates as a right-handed screw advancing in the direction of propagation.

From (10.75), it is seen that I and V depend on the total intensity and the ellipticity angle χ and are not affected by the orientation angle ψ of the ellipse, but Q and U vary according to the choice of coordinates.

Equation (10.75) may be compared with the Cartesian coordinates (X, Y, Z) of a point (r, θ, ϕ) on a sphere with radius $r = I$, $\theta = (\pi/2) - 2\chi$ and $\phi = 2\psi$, $X = r \sin \theta$

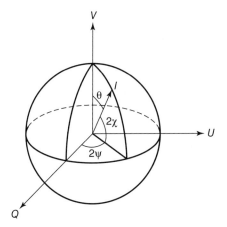

FIGURE 10.18   Poincaré sphere.

$\cos \phi$, $Y = r \sin \theta \sin \phi$, and $Z = r \cos \theta$. This sphere is called the *Poincaré sphere*, and its north and south poles represent left- and right-handed circular polarizations, respectively. The northern and southern hemispheres represent left- and right-handed elliptic polarizations, and the equator represents linear polarization (Fig. 10.18).

10.9 PARTIAL POLARIZATION AND NATURAL LIGHT

In the elliptic polarization discussed in Section 10.8, the ratio of the amplitudes a_1 and a_2, and the phase difference $\delta = \delta_2 - \delta_1$ are absolute constants. This happens when a wave is purely monochromatic (single frequency). In the more general case of a polychromatic wave with a certain bandwidth $\Delta\omega$, the amplitudes and the phase difference undergo continuous variations with various frequencies within $\Delta\omega$, and therefore a_1, a_2, and δ are slowly varying random functions of time. In general, therefore, the Stokes parameters should be expressed by the averages. Denoting the time average by angular brackets, $\langle \cdot \rangle$, we have, for $\exp(-i\omega t)$ dependence,

$$
\begin{aligned}
I &= \left\langle a_1^2 \right\rangle + \left\langle a_2^2 \right\rangle = \left\langle |E_1|^2 \right\rangle + \left\langle |E_2|^2 \right\rangle, \\
Q &= \left\langle a_1^2 \right\rangle - \left\langle a_2^2 \right\rangle = \left\langle |E_1|^2 \right\rangle - \left\langle |E_2|^2 \right\rangle, \\
U &= 2 \left\langle a_1 a_2 \cos \delta \right\rangle = 2 \operatorname{Re} \left\langle E_1 E_2^* \right\rangle, \\
V &= 2 \left\langle a_1 a_2 \sin \delta \right\rangle = 2 \operatorname{Im} \left\langle E_1 E_2^* \right\rangle.
\end{aligned}
\tag{10.76}
$$

For modified Stokes parameters (I_1, I_2, U, V), we have $I_1 = \langle |E_1|^2 \rangle$ and $I_2 = \langle |E_2|^2 \rangle$. In this case, the condition in (10.71) must be replaced by

$$
I^2 \geq Q^2 + U^2 + V^2.
\tag{10.77}
$$

Natural light is characterized by the fact that the intensity is the same in any direction perpendicular to the direction of the ray and that there is no correlation between rectangular components of the field. Therefore, the necessary and sufficient conditions for light to be natural are

$$I = 2 \left\langle |E|^2 \right\rangle$$

and

$$Q = U = V = 0. \tag{10.78}$$

In general, a wave may be partially polarized. The degree of polarization m is defined by the ratio

$$m = \frac{(Q^2 + U^2 + V^2)^{1/2}}{I}, \tag{10.79}$$

where $m = 1$ for elliptic polarization, $0 < m < 1$ for partial polarization, and $m = 0$ for an unpolarized wave (natural light).

It is clear that the Stokes parameters $[I]$ can always be expressed as the sum of the elliptically polarized wave $[I_p]$ and the unpolarized wave $[I_u]$.

$$[I] = [I_p] + [I_u],$$

$$[I] = \begin{bmatrix} I \\ Q \\ U \\ V \end{bmatrix}, \quad [I_p] = \begin{bmatrix} mI \\ Q \\ U \\ V \end{bmatrix}, \quad [I_u] = \begin{bmatrix} (1-m)I \\ 0 \\ 0 \\ 0 \end{bmatrix}. \tag{10.80}$$

10.10 SCATTERING AMPLITUDE FUNCTIONS f_{11}, f_{12}, f_{21}, AND f_{22} AND THE STOKES MATRIX

In Section 10.1, the scattering amplitude $\bar{f}(\hat{o}, \hat{i})$ is defined by

$$\bar{E}_s(\bar{r}) = \bar{f}(\hat{o}, \hat{i}) \frac{e^{ikR}}{R}, \tag{10.81a}$$

for a linearly polarized incident wave given by

$$\bar{E}_i(\bar{r}) = \hat{e}_i \, e^{ik\hat{i} \cdot \bar{r}}. \tag{10.81b}$$

To generalize the description of the scattered wave to include elliptic, partially polarized, and unpolarized waves, we may choose the following coordinate system (van de Hulst, 1957). We use $\exp(-i\omega t)$ in this section. We choose the z axis to be the direction of the incident wave, and the y–z plane to be the *plane of scattering*, defined as the plane that includes the direction of the incident wave $\hat{i}$ and the observation $\hat{o}$

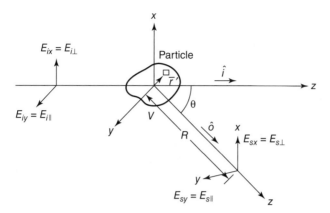

FIGURE 10.19   Geometry for defining scattering amplitude. The y–z and Y–Z planes are the plane of scattering.

(Fig. 10.19). The incident wave has two components, $E_{ix} = E_{i\perp}$ and $E_{iy} = E_{i\parallel}$, in the directions perpendicular and parallel, respectively, to the plane of scattering. The scattered wave in the direction $\hat{o}$ has two components, $E_{sX} = E_{s\perp}$ and $E_{sY} = E_{s\parallel}$, perpendicular and parallel, respectively, to the plane of scattering. It is clear that $E_{s\perp}$ and $E_{x\parallel}$ are linearly related to $E_{i\perp}$ and $E_{s\parallel}$, and therefore we write

$$\begin{bmatrix} E_{s\perp} \\ E_{s\parallel} \end{bmatrix} = \frac{e^{ikR}}{R} \begin{bmatrix} f_{11} & f_{12} \\ f_{21} & f_{22} \end{bmatrix} \begin{bmatrix} E_{i\perp} \\ E_{i\parallel} \end{bmatrix}. \tag{10.82}$$

$E_{i\perp}$ and $E_{i\parallel}$ are evaluated at the origin $x = y = z = 0$, and $E_{s\perp}$ and $E_{s\parallel}$ are at a large distance R from the origin. $f_{11}, f_{12}, f_{21},$ and f_{22} are functions of θ, and they are related to the scattering functions $S_1, S_2, S_3,$ and S_4 used by van de Hulst and in the Mie solution for a sphere (see Chapter 11).

$$f_{11} = \frac{i}{k}S_1, \quad f_{12} = \frac{i}{k}S_4, \quad f_{21} = \frac{i}{k}S_3, \quad f_{22} = \frac{i}{k}S_2. \tag{10.83}$$

Alternatively, we often use the spherical system shown in Fig. 10.11. The scattered wave (E_θ, E_ϕ) is related to the incident wave (E'_θ, E'_ϕ), and f_{ij} is a function of $(\theta, \phi, \theta',$ and $\phi')$.

$$\begin{bmatrix} E_\theta \\ E_\phi \end{bmatrix} = \frac{e^{ikR}}{R} \begin{bmatrix} f_{11} & f_{12} \\ f_{21} & f_{22} \end{bmatrix} \begin{bmatrix} E'_\theta \\ E'_\phi \end{bmatrix}. \tag{10.84}$$

If the scattering functions are known, and if the incident wave has an arbitrary state of polarization and its Stokes parameters are given by $I_{1i}, I_{2i}, U_i,$ and V_i, what are the Stokes parameters $I_{1s}, I_{2s}, U_s,$ and U_s of the scattered wave? This relationship can be obtained using (10.74) and (10.82) in terms of the following 4×4 Mueller matrix $\bar{\bar{S}}$:

$$I_s = \frac{1}{R^2} \bar{\bar{S}} I_i, \tag{10.85}$$

where I_s and I_i are 4×1 column matrices and $\bar{\bar{S}}$ is a 4×4 matrix.

$$I_s = \begin{bmatrix} I_{1s} \\ I_{2s} \\ U_s \\ V_s \end{bmatrix}, \quad I_i = \begin{bmatrix} I_{1i} \\ I_{2i} \\ U_i \\ V_i \end{bmatrix},$$

$$\bar{\bar{S}} = \begin{bmatrix} |f_{11}|^2 & |f_{12}|^2 & \mathrm{Re}(f_{11}f_{12}^*) & -\mathrm{Im}(f_{11}f_{12}^*) \\ |f_{21}|^2 & |f_{22}|^2 & \mathrm{Re}(f_{21}f_{22}^*) & -\mathrm{Im}(f_{21}f_{22}^*) \\ 2\,\mathrm{Re}(f_{11}f_{21}^*) & 2\,\mathrm{Re}(f_{12}f_{22}^*) & \mathrm{Re}(f_{11}f_{22}^* + f_{12}f_{21}^*) & -\mathrm{Im}(f_{11}f_{22}^* - f_{12}f_{21}^*) \\ 2\,\mathrm{Im}(f_{11}f_{21}^*) & 2\,\mathrm{Im}(f_{12}f_{22}^*) & \mathrm{Im}(f_{11}f_{22}^* + f_{12}f_{21}^*) & \mathrm{Re}(f_{11}f_{22}^* - f_{12}f_{21}^*) \end{bmatrix}.$$
$$\tag{10.86}$$

The matrix representations above are used in the formulation of vector radiative transfer theory.

10.11 ACOUSTIC SCATTERING

In this section, we describe the absorption and scattering characteristics of an object when illuminated by an incident acoustic wave of unit amplitude (Fig. 10.20).

$$P_i(\bar{r}) = \exp(-jk\hat{i} \cdot \bar{r}). \tag{10.87}$$

The scattering amplitude $f(\hat{o}, \hat{i})$ is a scalar quantity and the scattered acoustic field is given by

$$P_s(\bar{r}) = f(\hat{o}, \hat{i}) \frac{e^{-jkR}}{R} \quad \text{for } R > \frac{D^2}{\lambda}. \tag{10.88}$$

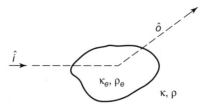

FIGURE 10.20   Acoustic scattering by an object.

The incident and the scattered power flux are given by

$$\bar{S}_i = \frac{|p_i|^2}{2\eta_0}\hat{i}, \quad \bar{S}_s = \frac{|p_s|^2}{2\eta_0}\hat{o}, \tag{10.89}$$

where $\eta_0 = \rho_0 c_0$ is the characteristic impedance and ρ_0 and c_0 are the equilibrium density of the medium and the velocity of the wave propagation in the medium, respectively. The differential scattering cross section σ_d, the scattering cross section σ_s, the absorption cross section σ_a, and the total cross section σ_t are defined by exactly the same formulas as shown in Section 10.1. The forward scattering theorem is

$$\sigma_t = -\frac{4\pi}{k}\,\mathrm{Im}\,f(\hat{i},\hat{i}). \tag{10.90}$$

Integral representation of the scattering amplitude $f(\hat{o},\hat{i})$ is somewhat different from that for an electromagnetic wave because of the factor $1/\rho_0$ inside the divergence operation in (2.131).

For a time-harmonic case with $\exp(j\omega t)$, (2.131) becomes

$$\nabla^2 p + k^2 p = -k^2 \gamma_\kappa p + \mathrm{div}[\gamma_\rho\,\mathrm{grad}\,p], \tag{10.91}$$

where $\gamma_\kappa = (\kappa_e - \kappa)/\kappa$, $\gamma_\rho = (\rho_e - \rho)/\rho_e$, $k^2 = \omega^2/c^2 = \omega^2\,\kappa\rho$ and κ and ρ are the compressibility and the density, respectively, of the medium surrounding the object, and κ_e and ρ_e are those of the object.

The right side of (10.91) generates the scattered wave, and the scattering amplitude $f(\hat{o},\hat{i})$ is given by

$$f(\hat{o},\hat{i}) = \frac{k^2}{4\pi}\int_V (\gamma_\kappa p + j\gamma_\rho \frac{\hat{o}}{k}\cdot\nabla' p)e^{+jk\hat{o}\cdot\vec{r}'}\,dV', \tag{10.92}$$

where we used the divergence theorem to convert the integral involving the second term of the right side of (10.91) to the second term of the integral of (10.92).

Using (10.92), we can obtain the following Born approximation:

$$f(\hat{o},\hat{i}) = \frac{k^2}{4\pi}\int_V (\gamma_\kappa + \gamma_\rho\,\cos\theta)\,\exp(-j\vec{k}_s\cdot\vec{r}')\,dV', \tag{10.93a}$$

where $\vec{k}_s = k(\hat{i}-\hat{o})$, $|\vec{k}_s| = 2k\,\sin(\theta/2)$, and $\cos\theta = \hat{i}\cdot\hat{o}$. This is applicable to the case

$$\left(\frac{\kappa_e \rho_e}{\kappa\rho} - 1\right)kD \ll 1, \tag{10.93b}$$

where D is a typical size of the object.

It should be interesting to note that for a small object, the first term inside the integral in (10.93) gives isotropic scattering similar to the electromagnetic case, but the second term gives scattering proportional to $\hat{o}\cdot\hat{i} = \cos\theta$.

For Rayleigh scattering by a small sphere, the incident pressure p_i with magnitude p_0 and the pressure p_e inside the sphere are given by

$$p_i = p_0 e^{-jkx} \approx p_0(1 - jkx),$$
$$p_e \approx p_0 \left(1 - \frac{jkx \cdot 3\rho_e}{\rho + 2\rho_e}\right). \tag{10.94}$$

Therefore, we get

$$f(\hat{o}, \hat{i}) = \frac{k^2 a^3}{3} \left(\frac{\kappa_e - \kappa}{\kappa} + \frac{3(\rho_e - \rho)}{2\rho_e + \rho} \cos\theta\right),$$
$$\frac{\sigma_s}{\pi a^2} = \frac{4(ka)^4}{9} \left(\left|\frac{\kappa_e - \kappa}{\kappa}\right|^2 + 3\left|\frac{\rho_e - \rho}{2\rho_e + \rho}\right|^2\right).$$

10.12 SCATTERING CROSS SECTION OF A CONDUCTING BODY

Many objects used in aerospace applications have conducting surfaces, examples being airplanes, rockets, spacecraft, and missiles. It is, therefore, important to study the scattering characteristics of a conducting body, particularly the backscattering cross section for radar applications.

Let us first obtain the general formulation of this problem. Consider a conducting surface S illuminated by an incident wave

$$\bar{E}_i(r) = E_i e^{-jk\hat{i}\cdot\bar{r}}\hat{e}_i. \tag{10.95}$$

The scattered field $\bar{E}_s(r)$ is given by (see Section 10.3)

$$\bar{E}_s(r) = \nabla \times \nabla \times \bar{\pi}_s(\bar{r}),$$
$$\bar{H}_s(r) = j\omega\varepsilon_0 \nabla \times \bar{\pi}_s(\bar{r}), \tag{10.96}$$

where

$$\bar{\pi}_s(\bar{r}) = \frac{1}{j\omega\varepsilon_0} \int_S G_0(\bar{r}, \bar{r}')\bar{J}_s(\bar{r}')\, da,$$

and $\bar{J}_s$ is the surface current on the surface of the conductor. Therefore, we get the scattered field at a large distance from the object.

$$\bar{E}_s(\bar{r}) = \hat{e}_s f(\hat{o}, \hat{i})\frac{e^{-jkR}}{R}$$
$$= -jk\eta_0 \frac{e^{-jkR}}{4\pi R} \int_S [-\hat{o} \times (\hat{o} \times \hat{j})J_s(r')]e^{jk\hat{o}\cdot\bar{r}'}\, da, \tag{10.97}$$

where $\bar{J}_s(\bar{r}') = J_s(\bar{r}')\hat{j}$ and $\eta_0 = 120\pi$ ohms is the free-space characteristic impedance.

The bistatic cross section is then given by

$$\sigma_{bi}(o, i) = \frac{4\pi \left| f(\hat{o}, \hat{i}) \right|^2}{|E_i|^2}$$

$$= \frac{k^2 \eta_0^2}{4\pi} \left| \int_S [-\hat{o} \times (\hat{o} \times \hat{j})] \frac{J_s(\bar{r}')}{|E_i|} e^{jk\hat{o}\cdot\bar{r}'} \, da \right|^2 . \tag{10.98}$$

If the exact surface current $\bar{J}_s(\bar{r}')$ is known, the formulas above give the exact scattering characteristics.

Let us next consider the radar cross section σ_b. As is usually the case, the antenna receives the component of the scattered wave along the direction of the polarization of the incident wave $\hat{e}_i$. Thus, in this case, we use

$$\sigma_b = \lim_{R \to \infty} \frac{4\pi R^2 \left| \hat{e}_i \cdot \bar{E}_s \right|^2}{|E_i|^2}$$

$$= \frac{4\pi \left| \hat{e}_i \cdot \hat{e}_s f(-\hat{i}, \hat{i}) \right|^2}{|E_i|^2}$$

$$= \frac{k^2 \eta_0^2}{4\pi} \left| \int_S \frac{\hat{e}_i \cdot J_s(\bar{r}')}{|E_i|} e^{-jk\hat{i}\cdot\bar{r}'} \, da \right|^2 \tag{10.99}$$

$$= \frac{k^2 \eta_0^2}{4\pi} \frac{\left| \int_S \bar{E}_i \cdot \bar{J}_s(\bar{r}) \, da \right|}{|E_i|^4} .$$

The surface current $\bar{J}_s(\bar{r}')$ is still unknown in the formulation above. In the next section, a useful approximation called physical optics is discussed.

10.13 PHYSICAL OPTICS APPROXIMATION

If the object is large compared with a wavelength and the surface is smooth (radius of curvature is much greater than a wavelength), the surface current $\bar{J}_s(\bar{r})$ may be well approximated by the current that would exist if the surface were a conducting plane tangential to the surface at the point $\bar{r}$. Thus the surface is regarded to be locally planar. In this case, in the illuminated region $\bar{J}_s$ is twice the tangential component of the incident magnetic field.

$$\bar{J}_s(\bar{r}) = \begin{cases} 2(\hat{n} \times \hat{H}_i) & \text{in the illuminated region} \\ 0 & \text{in the shadow,} \end{cases} \tag{10.100}$$

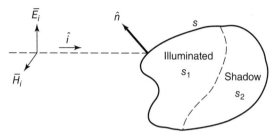

FIGURE 10.21   Physical optics approximation.

where $\hat{n}$ is a unit vector normal to the surface (Fig. 10.21). Using this approximation and noting

$$\bar{E}_i \cdot \bar{J}_s = 2\bar{E}_i \cdot (\hat{n} \times \bar{H}_i) = 2\hat{n} \cdot \bar{E}_i \times \bar{H}_i,$$

we get

$$\sigma_b = \frac{k^2}{\pi} \left| \int_{S_1} \hat{n} \cdot \hat{i} e^{-j2k\hat{i}\cdot\bar{r}'} da \right|^2. \qquad (10.101)$$

As an example, consider a thin rectangular conducting plate illuminated by a plane wave propagating from the direction (θ, ϕ) (Fig. 10.22). In this case, we note that

$$-\hat{i} = \sin\theta \cos\phi \hat{x} + \sin\theta \sin\phi \hat{y} + \cos\theta \hat{z}$$
$$\hat{n} = \hat{z}.$$

Therefore, we get

$$\sigma_b = \frac{4\pi}{\lambda^2} (A^2 \cos^2\theta) \left[\frac{\sin(2ka \sin\theta \cos\phi)}{(2ka \sin\theta \cos\phi)} \frac{\sin(2kb \sin\theta \sin\phi)}{(2kb \sin\theta \sin\phi)} \right]^2, \qquad (10.102)$$

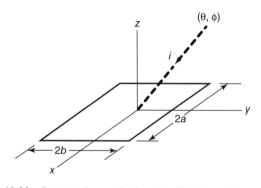

FIGURE 10.22   Rectangular conducting plate illuminated by a plane wave.

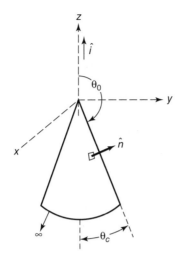

FIGURE 10.23   Infinite conducting cone.

where A is the area of the plate. For a circular plate of radius a, we obtain

$$\sigma_b = \frac{4\pi}{\lambda^2}(a^2 \cos^2 \theta) \left[\frac{J_1(2ka \sin \theta)}{ka \sin \theta}\right]^2, \tag{10.103}$$

where we made use of the following:

$$\int_0^a r\,dr \int_0^{2\pi} d\phi\, e^{j2kr \sin \theta \sin \phi} = 2\pi \int_0^a r\,dr\, J_0(2kr \sin \theta),$$

and $\int x\,dx\, J_0(x) = xJ_1(x)$. Next consider a wave incident on an infinite cone along its axis (Fig. 10.23). We note that $-\hat{n} \cdot \hat{i} = \sin \theta_0$, and thus we get

$$\sigma_b = \frac{4\pi}{\lambda^2} \left| \int_0^{2\pi} d\phi \int_0^{\infty} r\,dr \sin^2 \theta_0\, e^{j2kr \cos \theta_0} \right|^2 \tag{10.104}$$

$$= \frac{\pi}{4k^2} \tan^4 \theta_0.$$

Physical optics approximations give convenient and simple expressions for the scattering cross section and therefore are widely used. The validity of this approximation is limited to a large conducting object with a smooth surface. Unlike the geometric optical approximation to be discussed later, the physical optics approximation contains wavelength dependence and the results are often in good agreement with the experimental data, even though it is difficult to establish exactly how valid physical optics is for a general case. Physical optics is equivalent to the Kirchhoff approximation used in the aperture problems discussed in Sections 6.3 and 6.11.

10.14   MOMENT METHOD: COMPUTER APPLICATIONS

The physical optics approximation described in Section 10.13 gives a convenient expression for the surface current, and thus it is widely used for many applications. However, the approximation is valid only for a large object with a smooth surface. For a small object, we need a different technique which can give a reasonable approximation for the surface current. In this section, we start with a general formulation for the integral equation for the surface current and apply the moment method to solve for the surface current.

Let us first note that the total field $\bar{E}(r)$ consists of the incident field $\bar{E}_i(\bar{r})$ and the field $\bar{E}_s$ scattered by the object.

$$\bar{E}(r) = \bar{E}_i(r) + \bar{E}_s(r), \tag{10.105}$$

where the scattered field $\bar{E}_s(r)$ is given by

$$
\begin{aligned}
\bar{E}_s(r) &= \nabla \times \nabla \times \bar{\pi}(r) \\
&= \frac{1}{j\omega\varepsilon_0} \nabla \times \nabla \times \int_S G_0(\bar{r}, \bar{r}') J_s(r') \, da,
\end{aligned} \tag{10.106}
$$

and $G_0(\bar{r}, \bar{r}')$ is a free-space Green's function. Now the boundary condition requires that the tangential component of $\bar{E}(r)$ vanish on the surface of the conductor. Therefore, we write

$$\bar{E}(r)\big|_{\tan} = \bar{E}_i(r)\big|_{\tan} + \bar{E}_s(r)\big|_{\tan} = 0, \tag{10.107}$$

on the surface S.

We write (10.107) in the following form:

$$L(\bar{J}_s) = \bar{E}_i\big|_{\tan}, \tag{10.108}$$

where

$$
\begin{aligned}
\bar{E}_s(r)\big|_{\tan} &= -L(\bar{J}_s) \\
&= \left[\frac{1}{j\omega\varepsilon_0} \nabla \times \nabla \times \int_S G_0(r, r') \bar{J}_s(r') \, da \right]_{\tan}.
\end{aligned}
$$

Equation (10.108) is an integrodifferential equation for the surface current $\bar{J}_s$. We note that in this equation, $\bar{L}$ and $\bar{E}_i\big|_{\tan}$ are known and $\bar{J}_s$ is unknown. Let us solve (10.108) by means of the *moment method* (see Chapter 18).

We expand the unknown current $\bar{J}_s(r)$ in a series of given current distributions $\bar{J}_n(r)$ with unknown coefficients I_n.

$$\bar{J}_s(r) = \sum_n I_n \bar{J}_n(r), \tag{10.109}$$

where $\bar{J}_n(r)$ is called the *basis function*. Substituting (10.109) into (10.108), we get

$$\sum_n I_n L(\bar{J}_n(r)) = \bar{E}_i\big|_{\text{tan}}. \tag{10.110}$$

Now we choose a set of testing functions $\bar{W}_1, \bar{W}_2, \ldots$, which are tangential vectors on S. We form the inner product of both sides of (10.110) with $\bar{W}_m$.

$$\sum_n I_n \langle \bar{W}_m, L(\bar{J}_n) \rangle = \langle \bar{W}_m, \bar{E}_i \rangle, \tag{10.111}$$

where the inner product is defined by

$$\langle \bar{W}_m, \bar{E} \rangle = \int_S \bar{W}_m \cdot \bar{E} \, da. \tag{10.112}$$

Now defining matrices

$$I = (I_n) = \begin{bmatrix} I_1 \\ I_2 \\ \vdots \end{bmatrix}, \quad V = (V_m) = \begin{bmatrix} \langle \bar{W}_1, \bar{E}_i \rangle \\ \langle \bar{W}_2, \bar{E}_i \rangle \\ \vdots \end{bmatrix}$$

$$Z = (Z_{mn}) = \begin{bmatrix} \langle W_1, L(J_1) \rangle & \langle W_1, L(J_2) \rangle & \cdots \\ \langle W_2, L(J_1) \rangle & \langle W_2, L(J_2) \rangle & \cdots \end{bmatrix}, \tag{10.113}$$

we can write (10.111) in the following matrix form:

$$ZI = V. \tag{10.114}$$

We note here that once we choose the basis function $\bar{J}_n(r)$ and the testing functions $\bar{W}_m(r)$, the matrices Z and V are known and I is unknown. Thus, inverting the matrix Z, we obtain the solution

$$I = Z^{-1} V. \tag{10.115}$$

The surface current $\bar{J}(r)$ is then given by

$$\bar{J}(r) = \sum_n I_n \bar{J}_n(r). \tag{10.116}$$

As an example, let us consider the scattering by a thin wire of length l and diameter $2a$ ($l \gg a$) (Fig. 10.24). Noting that $\bar{J}_s(r')$ has only the z component, we write

$$\bar{J}_s(r') \, da = I_z(z') \, dz' \, \hat{z}. \tag{10.117}$$

Here we assume that the current $I_z(z')$ is a line current on the axis of the wire, but that the boundary condition is to be satisfied on the surface of the wire (Fig. 10.25). This convenient approximation avoids the singularity of Green's function at $\bar{r} = \bar{r}'$.

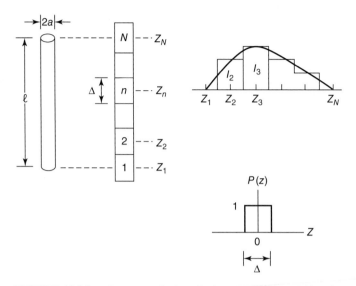

FIGURE 10.24   Moment method applied to scattering by a thin wire.

Using (10.117), the integral equation (10.111) becomes

$$-\frac{1}{j\omega\varepsilon_0} \int_{-l/2}^{l/2} \left(\frac{\partial^2}{\partial z^2} + k^2 \right) G(r, r') I_z(z') \, dz' = E_{iz}(z'). \qquad (10.118)$$

This equation is often called the *Pocklington equation* (Mittra, 1973). For a straight wire, noting $R = |\bar{r} - \bar{r}'| = [(z - z')^2 + a^2]^{1/2}$, we can also write (10.118) as

$$\int_{-l/2}^{l/2} K(z, z') I_z(z') \, dz' = E_{iz}(z'), \qquad (10.119)$$

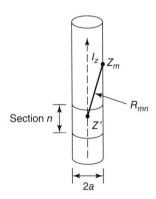

FIGURE 10.25   The current $I_z(z')$ is on the axis and the boundary condition is satisfied on the surface at z_m.

where

$$K(z, z') = -\frac{e^{-jkR}}{j4\pi\omega\varepsilon_0}[(1 + jkR)(2R^2 - 3a^2) + k^2 a^2 R^2]R^{-5}.$$

We now divide the length l into $N - 1$ sections and choose the basis function $\bar{J}_n(r)$ to be a rectangular current $P(z - z_n)$ in the nth section of the wire. Thus $I(z)$ is approximated by a series of steps with the value at the midpoint in each section (Fig. 10.24). Note that section 1 and section N are extended beyond the wire by $\Delta/2$. We can then let $I_1 = I_N = 0$, which is the condition required at the end of the wire. We write

$$I(z) = \sum_{n=1}^{N} I_n P(z - z_n), \tag{10.120}$$

where $P(z - z_n)$ is the basis function. We now choose $\delta(z - z_m)$ as the testing function.

$$W_m = \delta(z - z_m). \tag{10.121}$$

Thus the inner product (10.112) gives the values of (10.119) at those discrete points z_m. Then (10.119) becomes

$$\sum_{n=1}^{N} Z_{mn} I_n = V_m, \tag{10.122}$$

where

$$V_m = E_{iz}(z_m)\Delta,$$

$$Z_{mn} = \Delta \int_{z_n-\Delta/2}^{z_n+\Delta/2} K(z, z') \, dz'.$$

Once Z_{mn} is evaluated, the solution is easily obtained by solving the matrix equation (10.114) by inversion of Z and the current distribution is given by (10.116). Note that $I_1 = I_N = 0$, and therefore $[Z]$ is an $(N - 2) \times (N - 2)$ matrix.

The backscattered cross section σ_b is then given by (10.99). In matrix form, we get

$$\sigma_b = \frac{k^2\eta_0^2}{4\pi}\left|\tilde{V}I\right|^2 = \frac{k^2\eta_0^2}{4\pi}\left|\tilde{V}Z^{-1}V\right|^2, \tag{10.123}$$

where the incident electric field $\bar{E}_i$ is normalized so that $\left|\bar{E}_i\right| = 1$.

PROBLEMS

10.1 The surface of the moon is rough and the radar cross section is about 4×10^{-4} of its geometric cross section. Suppose that the moon is illuminated by a radar transmitter with a diameter of 142 ft and an aperture efficiency of 60%. The peak power is 130 kW and the frequency is 400 MHz. Calculate the power received.

10.2 At $\lambda = 5$ cm, the refractive index of water at 20°C is $8.670 - j1.202$. The median diameter (in mm) of raindrops is given by:

$$D_m = 1.238p^{0.182},$$

where p (in mm/h) is the precipitation rate and the terminal velocity (in m/s) is given by

$$v = 200.8a^{1/2},$$

where a (in m) is the radius of the droplet ($a = D_m/2$). Assuming that the Rayleigh formula is applicable, calculate the scattering and the absorption cross sections of a rain droplet. Also find the number of droplets per m³. Assume that $p = 12.5$ mm/h. Find the attenuation of the wave in dB/km.

10.3 A lossy prolate ellipsoidal particle ($a = b < c$) is placed at the origin as shown in Fig. 10.11, and $\theta_b = 0$. Find the polarizability in the x, y, and z directions when $\varepsilon_r = 1 - j15$, $a = b = 10$ μm, $c = 1$ mm, and $\lambda = 3$ cm. Find the scattering amplitude matrix and the scattering and absorption cross sections when $\theta' = \phi' = 0$ and when $\theta' = \pi/2$ and $\phi' = 0$.

10.4 Using the Rayleigh–Debye approximation, calculate the backscattering cross section of a spherical object of radius a_1 with a spherical core of radius a_2 at $\lambda_1 = 0.6$ μm (Fig. P10.4).

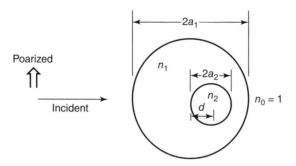

FIGURE P10.4   Rayleigh–Debye scattering.

$$n_1 = 1.01, \qquad a_1 = 2\,\mu m,$$
$$n_2 = 1.02, \qquad a_2 = 0.5\,\mu m,$$
$$d = 1\,\mu m.$$

10.5 If the Stokes parameters of a wave are

$$I = 3, \quad U = 2,$$
$$Q = 1, \quad V = -2,$$

find E_x and E_y and draw the locus similar to Fig. 10.17.

10.6 Consider a wave whose components are given by

$$E_x = 2\cos\left(\omega t + \frac{\pi}{8}\right),$$
$$E_y = 3\cos\left(\omega t + \frac{3\pi}{2}\right).$$

Find the Stokes parameters and Poincaré representation of this wave. Show its location on the Poincaré sphere.

10.7 Calculate the acoustic scattering cross section of a red blood cell in plasma at 5 MHz. Assume that it is a sphere of radius 2.75 μm and that $\kappa_l = 34.1 \times 10^{-12}$ cm^2/dyne and $\rho_l = 1.092$ g/cm^3. The plasma surrounding the blood cell has $\kappa = 40.9 \times 10^{-12}$ cm^2/dyn and $\rho = 1.021$ g/cm^3.

10.8 Using the physical optics approximation, find the backscattering cross section of the finite conducting cylinder shown in Fig. P10.8.

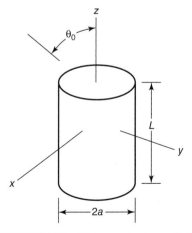

FIGURE P10.8   Finite conducting cylinder.

10.9   Consider a short wire of length L illuminated by an incident wave polarized in a plane containing the dipole. Find the backscattering cross section of this short wire.

10.10   Calculate the backscattering cross section σ_b/λ^2 of a wire with radius a and length l when a wave polarized in the direction of the wire is incident at broadside. $ka = 0.0314$ and $\frac{1}{2}kl = 1.5$.

CHAPTER 11

WAVES IN CYLINDRICAL STRUCTURES, SPHERES, AND WEDGES

Many bodies of practical interest, such as biological media, rockets, and portions of aircraft, may be closely approximated by cylindrical structures, spheres, and wedges. These shapes are well defined and coincide with one or more of the coordinates for which wave equations are separable. The exact solutions can be obtained in closed form for most problems. These bodies may have radiators such as slot antennas and dipole antennas on or close to their surface, and their radiation characteristics are greatly affected by the geometry of the body. Also, the scattering and absorption characteristics of these bodies, when illuminated from outside, are important in many practical problems, such as radar cross-section studies and microwave hazards. In this chapter, we investigate the scattering of waves from these objects and the effects of these structures on the radiation characteristics of antennas.

In analyzing various problems in this chapter, we present a number of powerful analytical techniques, including the Fourier transform, saddle-point technique, Watson transform, residue series representation, and geometric optical solutions. These techniques are useful not only for these problems, but are also important mathematical tools applicable to a large number of other problems. (See Bowman et al. (1969) for detailed treatments of these topics.)

11.1 PLANE WAVE INCIDENT ON A CONDUCTING CYLINDER

The determination of a radar cross section is one of the most important and practical problems. It provides information about the object, and it is useful for the design of

Electromagnetic Wave Propagation, Radiation, and Scattering: From Fundamentals to Applications,
Second Edition. Akira Ishimaru.
© 2017 by The Institute of Electrical and Electronic Engineers, Inc. Published 2017 by John Wiley & Sons, Inc.

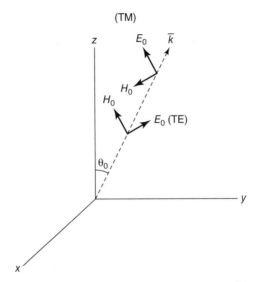

FIGURE 11.1 Plane wave in the direction (θ_0, θ_0): TE, $\bar{E} = E_0\hat{\phi}$ and $\bar{H} = -H_0\hat{\theta}$; TM, $\bar{E} = -E_0\hat{\theta}$ and $\bar{H} = -H_0\hat{\phi}$.

a vehicle with a specified radar cross section. Since many objects and vehicles are composed of cylindrical layers, it is important to devise a systematic technique to find scattered waves from such an object when a plane wave is incident on it. Let us consider first an incident plane wave propagating in the direction (θ_0, ϕ_0). We may consider two cases: TM and TE. The TM wave is polarized in a plane parallel to the z axis and therefore $H_z = 0$. The TE wave is polarized perpendicular to the z axis and thus $E_z = 0$. We let E_0 be the magnitude of the incident electric field (Fig. 11.1).

Let us consider the TM incident wave. Note that the TM wave has the following electric field components in the cylindrical coordinates (ρ, ϕ, z).

$$E_{zi} = E_0 \sin\theta_0 e^{-j\bar{k}\cdot\bar{r}},$$

$$E_{\rho i} = -E_0 \cos\theta_0 \cos(\phi - \phi_0)e^{-j\bar{k}\cdot\bar{r}}, \qquad (11.1)$$

$$E_{\phi i} = E_0 \cos\theta_0 \sin(\phi - \phi_0)e^{-j\bar{k}\cdot\bar{r}},$$

where

$$\bar{k} = k(\sin\theta_0 \cos\phi_0\hat{x} + \sin\theta_0 \sin\phi_0\hat{y} + \cos\theta_0\hat{z}),$$

$$\bar{r} = x\hat{x} + y\hat{y} + z\hat{z}$$

$$= \rho\cos\phi\hat{x} + \rho\sin\phi\hat{y} + z\hat{z},$$

$$\bar{k}\cdot\bar{r} = kz\cos\theta_0 + k_\rho\sin\theta_0\cos(\phi - \phi_0).$$

Now the TM wave is generated by

$$\Pi_{zi} = A_i e^{-j\bar{k}\cdot\bar{r}} = A_i e^{-jkz\cos\theta_0 - jk\rho\sin\theta_0\cos(\phi-\phi_0)}, \tag{11.2}$$

and E_z and Π_z are related by

$$\begin{aligned}
E_{zi} &= \left(\frac{\partial^2}{\partial z^2} + k^2\right)\Pi_{zi} \\
&= A_i k^2 \sin^2\theta_0 e^{-j\bar{k}\cdot\bar{r}}.
\end{aligned} \tag{11.3}$$

Comparing (11.3) with (11.1), we get

$$A_i = \frac{E_0}{k^2 \sin\theta_0}. \tag{11.4}$$

We will observe shortly that the incident wave (11.2) is not convenient to satisfy the boundary condition in order to determine the scattered wave, and it is necessary to expand the incident wave in a Fourier series in ϕ. Let us expand (11.2) in a Fourier series.

$$\Pi_{zi} = \sum_{n=-\infty}^{\infty} a_n(z,\rho)e^{-jn(\phi-\phi_0)}. \tag{11.5}$$

The coefficient a_n is obtained by

$$a_n = \frac{1}{2\pi}\int_0^{2\pi}\Pi_{zi}e^{jn(\phi-\phi_0)}d(\phi-\phi_0). \tag{11.6}$$

Now we make use of the following integral representation of the Bessel function:

$$\begin{aligned}
J_n(Z) &= \frac{1}{2\pi}\int_0^{2\pi}e^{jZ\cos\phi + jn(\phi-\pi/2)}d\phi, \\
J_{-n}(Z) &= (-1)^n J_n(Z) = J_n(-Z).
\end{aligned} \tag{11.7}$$

We then get

$$a_n = A_i e^{-jkz\cos\theta_0}J_n(k\rho\sin\theta_0)e^{-jn(\pi/2)}. \tag{11.8}$$

Thus Π_{zi} in (11.2) is expressed in the following form:

$$\Pi_{zi} = \sum_{n=-\infty}^{\infty} A_i e^{-jkz\cos\theta_0}J_n(k\rho\sin\theta_0)e^{-jn(\phi-\phi_0+\pi/2)}. \tag{11.9}$$

Let us consider the scattering by a conducting cylinder of radius a when the TM wave (11.9) is incident. We write the scattered wave in terms of the Hertz potential Π_{zs}. Considering that they should satisfy the wave equation and the radiation condition, we write

$$\Pi_{zs} = \sum_{n=-\infty}^{\infty} A_{ns} e^{-jkz \cos \theta_0} H_n^{(2)}(k\rho \sin \theta_0) e^{-jn[\phi-\phi_0+(\pi/2)]}, \qquad (11.10)$$

where A_{ns} are the unknown coefficients to be determined by the boundary condition.

The boundary conditions are that E_z and E_ϕ must be zero at $\rho = a$. This requires that the total $\Pi_z = \Pi_{zi} + \Pi_{zs}$ be zero at $\rho = a$. Thus we get

$$A_{ns} = -\frac{J_n(ka \sin \theta_0)}{H_n^{(2)}(ka \sin \theta_0)} A_i. \qquad (11.11)$$

The total Hertz potential Π_z is given by

$$\Pi_z = \sum_{n=-\infty}^{\infty} A_i e^{-jkz \cos \theta_0} \left[J_n(k\rho \sin \theta_0) - \frac{J_n(ka \sin \theta_0) H_n^{(2)}(k\rho \sin \theta_0)}{H_n^{(2)}(ka \sin \theta_0)} \right] e^{-jn[\phi-\phi_0+(\pi/2)]}. \qquad (11.12)$$

The field components including both TM(Π_z) and TE(Π_{mz}) waves are then given by

$$E_z = \left(\frac{\partial^2}{\partial z^2} + k^2 \right) \Pi_z, \quad H_z = \left(\frac{\partial^2}{\partial z^2} + k^2 \right) \Pi_{mz},$$

$$E_\rho = \frac{\partial^2}{\partial \rho \partial z} \Pi_z - j\omega\mu \frac{1}{\rho} \frac{\partial}{\partial \phi} \Pi_{mz},$$

$$E_\phi = \frac{1}{\rho} \frac{\partial^2}{\partial \phi \partial z} \Pi_z + j\omega\mu \frac{\partial}{\partial \rho} \Pi_{mz}, \qquad (11.13)$$

$$H_\rho = j\omega\varepsilon \frac{1}{\rho} \frac{\partial}{\partial \phi} \Pi_z + \frac{\partial^2}{\partial \rho \partial z} \Pi_{mz},$$

$$H_\phi = -j\omega\varepsilon \frac{\partial}{\partial \rho} \Pi_z + \frac{1}{\rho} \frac{\partial^2}{\partial \phi \partial z} \Pi_{mz}.$$

For example, the current density on the conducting cylinder $\bar{J}$ is given by

$$\bar{J} = J_\phi \hat{\phi} + J_z \hat{z}$$
$$= -H_z \hat{\phi} + H_\phi \hat{z} \quad \text{at } \rho = a. \qquad (11.14)$$

Let us next consider the TE wave incident on the cylinder. We use the magnetic Hertz potential Π_{mz} and write the incident wave as

$$\Pi_{mzi} = B_i e^{-j\bar{k}\cdot\bar{r}} = B_i e^{-jkz\cos\theta_0 - jk\rho\sin\theta_0\cos(\phi-\phi_0)}, \qquad (11.15)$$

where $H_0 = E_0/\eta = k_0^2 \sin\theta_0 B_i$, $\eta = (\mu/\varepsilon)^{1/2}$. This can be expressed in a Fourier series

$$\Pi_{mzi} = \sum_{n=-\infty}^{\infty} B_i e^{-jkz\cos\theta_0} J_n(k\rho\sin\theta_0) e^{-jn(\phi-\phi_0+\pi/2)}. \qquad (11.16)$$

We write the scattered wave in the following form with the unknown coefficients B_{ns}:

$$\Pi_{mzs} = \sum_{n=-\infty}^{\infty} B_{ns}\, e^{-jkz\cos\theta_0}\, H_n^{(2)}(k\rho\sin\theta_0) e^{-jn(\phi-\phi_0+\pi/2)}. \qquad (11.17)$$

The boundary condition at $\rho = a$ is that $(\partial/\partial\rho)\Pi_{mz} = 0$, which determines B_{ns}.

$$B_{ns} = -\frac{J_n'(ka\sin\theta_0)}{H_n^{(2)\prime}(ka\sin\theta_0)} B_i. \qquad (11.18)$$

The final solution is then given by

$$\Pi_{mz} = \sum_{n=-\infty}^{\infty} B_i e^{-jkz\cos\theta_0} \left[J_n(k\rho\sin\theta_0) - \frac{J_n'(ka\sin\theta_0)H_n^{(2)}(k\rho\sin\theta_0)}{H_n^{(2)\prime}(ka\sin\theta_0)} \right] e^{-jn(\phi-\phi_0+\pi/2)}. \qquad (11.19)$$

Equations (11.12) and (11.19) are the total fields due to the incident TM and TE waves, respectively. All the electric and magnetic field components are then easily obtained from them. Note that the incident TM wave A_i produces the scattered TM wave only, and the incident TE wave B_i produces the TE wave only, and that there is no coupling between the TM and TE modes. This is true only for certain special cases such as a conducting cylinder or normal incidence on a dielectric cylinder. In general, as can be seen in section 11.2, the incident TM wave can produce both TM and TE scattered waves.

11.2 PLANE WAVE INCIDENT ON A DIELECTRIC CYLINDER

Let us consider the scattering of a plane wave by a dielectric cylinder of radius a and a relative dielectric constant ε_r (Fig. 11.2). We let k and $k_1 = k\sqrt{\varepsilon_r}$ be the wave number outside and inside the cylinder, respectively. In general, the plane incident

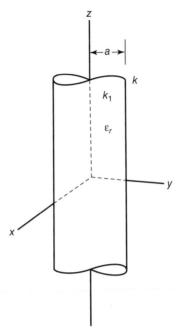

FIGURE 11.2   Dielectric cylinder.

wave consists of TM and TE waves. The Hertz potentials for the incident TM and TE waves are given in (11.9) and (11.16).

$$\Pi_{zi} = \sum_{n=-\infty}^{\infty} A_i e^{-jkz \cos \theta_0} J_n(k\rho \sin \theta_0) e^{-jn(\phi-\phi_0+\pi/2)},$$

$$\Pi_{mzi} = \sum_{n=-\infty}^{\infty} B_i e^{-jkz \cos \theta_0} J_n(k\rho \sin \theta_0) e^{-jn(\phi-\phi_0+\pi/2)}.$$

$$(11.20)$$

We express the scattered field outside the cylinder using the unknown coefficients A_{sn} and B_{sn}.

$$\Pi_{zs} = \sum_{n=-\infty}^{\infty} A_{sn} e^{-jkz \cos \theta_0} H_n^{(2)}(k\rho \sin \theta_0) e^{-jn(\phi-\phi_0+\pi/2)},$$

$$\Pi_{mzs} = \sum_{n=-\infty}^{\infty} B_{sn} e^{-jkz \cos \theta_0} H_n^{(2)}(k\rho \sin \theta_0) e^{-jn(\phi-\phi_0+\pi/2)}.$$

$$(11.21)$$

Here we need both TM and TE modes as they are generally coupled.

Inside the dielectric cylinder, we write using the unknown coefficients A_{en} and B_{en},

$$\Pi_{ze} = \sum_{n=-\infty}^{\infty} A_{en} e^{-jkz\cos\theta_0} J_n(k_1\,\rho\sin\theta_1) e^{-jn(\phi-\phi_0+\pi/2)},$$

$$\Pi_{mze} = \sum_{n=-\infty}^{\infty} B_{sn} e^{-jkz\cos\theta_0} J_n(k_1\,\rho\sin\theta_0) e^{-jn(\phi-\phi_0+\pi/2)}.$$
(11.22)

Here the z dependence $\exp(-jkz\cos\theta_0)$ is the same as that of the incident field because the boundary conditions must be satisfied at all z. This requires that

$$k\cos\theta_0 = k_1\cos\theta_1,$$

and

$$k_1\sin\theta_1 = \left(k_1^2 - k_1^2\cos^2\theta_1\right)^{1/2} = \left(k_1^2 - k^2\cos^2\theta_0\right)^{1/2}.$$
(11.23)

The boundary conditions are that E_z, E_ϕ, H_z, and H_ϕ, are continuous at $\rho = a$. Since $E_z = \left(\frac{\partial^2}{\partial z^2} + k^2\right)\Pi_z$, the continuity of E_z and the continuity of H_z give the following:

$$\begin{aligned}
(k^2\sin^2\theta_0)\,&[A_i J_n(ka\sin\theta_0) + A_{sn} H_n^{(2)}(ka\sin\theta_0)] \\
&= k_1^2\sin^2\theta_1 A_{en} J_n(k_1 a\sin\theta_1), \\
(k^2\sin^2\theta_0)\,&[B_i J_n(ka\sin\theta_0) + B_{sn} H_n^{(2)}(ka\sin\theta_0)] \\
&= k_1^2\sin^2\theta_1 B_{en} J_n(k_1 a\sin\theta_1).
\end{aligned}$$
(11.24)

Next consider the continuity of E_ϕ. Noting that

$$E_\phi = \frac{1}{\rho}\frac{\partial^2}{\partial\phi\,\partial z}\Pi_z + j\omega\mu\frac{\partial}{\partial\rho}\Pi_{mz},$$
(11.25)

we get

$$\begin{aligned}
-\frac{kn\cos\theta_0}{a}&[A_i J_n(ka\sin\theta_0) + A_{sn} H_n^{(2)}(ka\sin\theta_0)] \\
&+ j\omega\mu k\sin\theta_0 [B_i J_n'(ka\sin\theta_0) + B_{sn} H_n^{(2)\prime}(ka\sin\theta_0)] \\
&= -\frac{kn\cos\theta_0}{a}[A_{en} J_n(k_1 a\sin\theta_1)] + j\omega\mu k_1\sin\theta_1 B_{en} J_n'(k_1 a\sin\theta_1).
\end{aligned}$$
(11.26)

Similarly noting that

$$H_\phi = -j\omega\varepsilon\frac{\partial}{\partial\rho}\Pi_z + \frac{1}{\rho}\frac{\partial^2}{\partial\phi\,\partial z}\Pi_{mz},$$

we get

$$- j\omega\varepsilon k \sin\theta_0 [A_i J_n'(ka\sin\theta_0)$$

$$+ A_{sn} H_n^{(2)'}(ka\sin\theta_0)] - \frac{kn\cos\theta_0}{a} [B_i J_n(ka\sin\theta_0) + B_{sn} H_n^{(2)}(ka\sin\theta_0)] \quad (11.27)$$

$$= - j\omega\varepsilon_1 k_1 \sin\theta_1 A_{en} J_n'(k_1 a\sin\theta_1) - \frac{kn\cos\theta_0}{a} B_{en} J_n(k_1 a\sin\theta_1).$$

The Eqs. (11.24)–(11.27) can be solved for the four unknown coefficients A_{sn}, B_{sn}, A_{en}, and B_{en}. They are then substituted into (11.21) and (11.22) to obtain the final solution for Π and Π_m and the fields are then given by (11.13) with the appropriate k and ε in each region.

A similar procedure can be used for a layered dielectric cylinder. Let ε_m and k_m be the relative dielectric constant and the wave number of the cylindrical layer with radius from a_{m-1} to a_m. Then the field $(E_z, H_z, E_\phi, H_\phi)$ at a_{m-1} is related to the field at a_m by a 4×4 matrix. We can then apply the boundary condition and obtain the complete solution.

11.3 AXIAL DIPOLE NEAR A CONDUCTING CYLINDER

Dipole and loop antennas are often used in the vicinity of cylindrical structures. In this section, we investigate the radiation from an axial dipole located near the conducting cylinder (Fig. 11.3). We first consider the field produced by a dipole in free space. We take an axial electric dipole located at r', (ρ', ϕ', z'). The rectangular components of the Hertz potential satisfy the scalar wave equation

$$(\nabla^2 + k^2)\Pi_z = -\frac{J_z}{j\omega\varepsilon}. \quad (11.28)$$

The axial electric dipole of current I and length L is represented by

$$J_z = IL\delta(r-r') \quad (11.29)$$

and thus

$$\Pi_z = \frac{IL}{j\omega\varepsilon} G(r/r'), \quad (11.30)$$

where Green's function $G(r/r')$ in cylindrical coordinates satisfies

$$\left(\frac{\partial^2}{\partial\rho^2} + \frac{1}{\rho^2}\frac{\partial}{\partial\rho} + \frac{1}{\rho^2}\frac{\partial^2}{\partial\phi^2} + \frac{\partial^2}{\partial z^2} + k^2\right) G(r/r') = -\frac{\delta(\rho-\rho')\delta(\phi-\phi')\delta(z-z')}{\rho}. \quad (11.31)$$

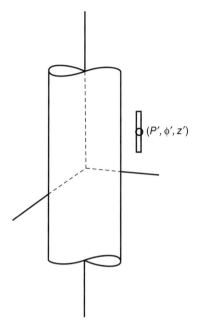

FIGURE 11.3   Axial dipole near a cylinder.

We write the free-space Green's function $G(r/r')$ in a Fourier integral in the z direction and in Fourier series in the ϕ direction (see Section 5.8).

$$G(r/r') = \frac{1}{2\pi} \int_c \sum_{n=-\infty}^{\infty} G_n(h, \rho, \rho') e^{-jn(\phi-\phi')-jh(z-z')}\, dh, \qquad (11.32)$$

where

$$G_n(h, \rho, \rho') = \begin{cases} -j\frac{1}{4}J_n(\lambda\rho)H_n^{(2)}(\lambda\rho'), & \text{for } \rho < \rho' \\ -j\frac{1}{4}J_n(\lambda\rho')H_n^{(2)}(\lambda\rho) & \text{for } \rho' < \rho, \end{cases} \qquad (11.33)$$

$$\lambda^2 = k^2 - h^2.$$

We now consider the field outside the cylinder with radius a when excited by an axial dipole at (ρ', ϕ', z'). Then we write the scattered field $G_s(r)$ as follows:

$$G_s(r) = \frac{1}{2\pi} \int_c \sum_{n=-\infty}^{\infty} A_n H_n^{(2)}(\lambda\rho) e^{-jn(\phi-\phi')-jh(z-z')}\, dh. \qquad (11.34)$$

The boundary condition is that

$$G(r/r') + G_s(r) = 0 \quad \text{at } \rho = a.$$

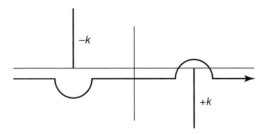

FIGURE 11.4   Branch cut Re $h = k$.

Therefore, we obtain

$$A_n = -j\frac{1}{4}\frac{J_n(\lambda a)H_n^{(2)}(\lambda\rho')}{H_n^{(2)}(\lambda a)}. \tag{11.35}$$

Equation (11.34) with (11.35) gives the complete expression of the scattered field due to the conducting cylinder of radius a excited by a dipole at (ρ', ϕ', z'). They are expressed as an inverse Fourier transform with the contour c of integration in the complex h plane.

Note first that the integrand contains λ, and therefore we need to consider whether we should take $\lambda = +(k^2 - h^2)^{1/2}$ or $\lambda = -(k^2 - h^2)^{1/2}$. This requires study of the branch points and Riemann surfaces discussed in Appendix 11.A. In general, the integrand is not single valued because of $\lambda = \pm(k^2 - h^2)^{1/2}$. To assure a single value for the integrand, we draw branch cuts from the branch points $h = \pm k$ such as thoseshown in Fig. 11.4 or 11.5. As long as the contour does not cross the branch cuts, the integrand is single valued (see Appendix 11.B).

11.4 RADIATION FIELD

One of the important characteristics of antennas is the radiation pattern, the behavior of the field at a large distance. To represent the radiation pattern, it is convenient to

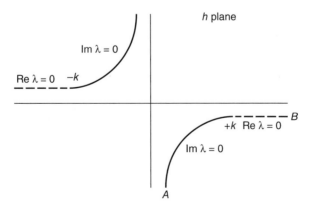

FIGURE 11.5   Branch cut Im $\lambda = 0$.

employ the spherical coordinate system and consider the components E_θ, E_ϕ and H_θ, H_ϕ. We note first that in the far field, the electric and magnetic fields are perpendicular to each other and both are transverse to the direction of propagation. Their magnitudes are related by the free-space characteristic impedance η.

$$E_\theta = \eta H_\phi \quad \text{and} \quad E_\phi = -\eta H_\theta, \quad \eta = \left(\frac{\mu_0}{\varepsilon_0}\right)^{1/2}, \tag{11.36}$$

and therefore the Poynting vector is given by

$$\bar{S} = \tfrac{1}{2}\mathrm{Re}(\bar{E} \times \bar{H}^*) = S_r\,\hat{r},$$
$$S_r = \frac{1}{2\eta}\left(|E_\theta|^2 + |E_\phi|^2\right) = \frac{1}{2\eta}\left[\eta^2|H_\phi|^2 + |E_\phi|^2\right]. \tag{11.37}$$

The last expression in (11.37) is convenient for the radiation field from cylindrical structures because E_ϕ and H_ϕ are the same in both cylindrical and spherical coordinates. E_ϕ and H_ϕ are calculated from the Hertz potential.

Let us consider the radiation field of an axial dipole near a conducting cylinder, which was discussed in detail in Section 11.3. E_ϕ and H_ϕ are then obtained from (11.30), (11.32), and (11.34). For $\rho > \rho'$ we get

$$E_\phi = \frac{1}{\rho}\frac{\partial^2}{\partial\phi\partial z}\Pi_z$$
$$= \sum_{n=-\infty}^{\infty}\int_c C_n(h)H_n^{(2)}(\lambda\rho)e^{-jh(z-z')}\,dh, \tag{11.38}$$

$$H_\phi = -j\omega\varepsilon\frac{\partial}{\partial\rho}\Pi_z$$
$$= \sum_{n=-\infty}^{\infty}\int_c D_n(h)H_n^{(2)\prime}(\lambda\rho)e^{-jh(z-z')}\,dh, \tag{11.39}$$

where

$$C_n(h) = \frac{IL}{j\omega\varepsilon}\frac{jnh}{8\pi\rho}\frac{J_n(\lambda\rho')H_n^{(2)}(\lambda a) - J_n(\lambda a)H_n^{(2)}(\lambda\rho')}{H_n^{(2)}(\lambda a)}e^{-jn(\phi-\phi')},$$

$$D_n(h) = \frac{IL}{j\omega\varepsilon}\frac{-j\omega\varepsilon\lambda}{2\pi}\frac{J_n(\lambda\rho')H_n^{(2)}(\lambda a) - J_n(\lambda a)H_n^{(2)}(\lambda\rho')}{H_n^{(2)}(\lambda a)}e^{-jn(\phi-\phi')},$$

$$\lambda = \sqrt{k^2 - h^2}.$$

The radiation field is obtained by evaluating (11.38) and (11.39) for the large distance from the antenna. This is done by the saddle-point technique, discussed in Section 11.5.

11.5 SADDLE-POINT TECHNIQUE

Let us consider the integral given in (11.38),

$$I_1 = \int_{-\infty}^{\infty} C_n(h) H_n^{(2)} \left(\sqrt{k^2 - h^2} \, \rho \right) e^{-jhz} \, dh, \tag{11.40}$$

for a large distance R from the origin. First, we approximate the Hankel function by its asymptotic form

$$H_n^{(2)}(x) = \sqrt{\frac{2}{\pi x}} e^{-jx+j(2n+1)(\pi/4)},$$

which is valid for $|x| \gg |n|$. Thus we write (11.40) as

$$I_1 = \int_{-\infty}^{\infty} A(h) e^{-j\sqrt{k^2 - h^2}\,\rho - jhz} \, dh, \tag{11.41}$$

where

$$A(h) = C_n(h) \sqrt{\frac{2}{\pi\rho}} \frac{e^{j(2n+1)(\pi/4)}}{(k^2 - h^2)^{1/4}}.$$

Let us express (11.41) in spherical coordinates. For convenience, we use $\theta_c = (\pi/2) - \theta$.

$$z = R \cos\theta = R \sin\theta_c,$$
$$\rho = R \sin\theta = R \cos\theta_c. \tag{11.42}$$

We also transform h into the α plane.

$$h = k \sin\alpha. \tag{11.43}$$

Then we get

$$\begin{aligned} I_1 &= \int_c A(k\sin\alpha) e^{-jkR\cos(\alpha-\theta_c)} \, k \, \cos\alpha \, d\alpha \\ &= \int_c F(k\sin\alpha) e^{-jkR\cos(\alpha-\theta_c)} \, d\alpha. \end{aligned} \tag{11.44}$$

The contours in the h and α planes are shown in Fig. 11.6.

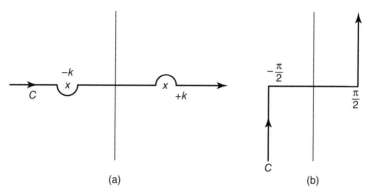

FIGURE 11.6 Contour C: (a) in the h plane; (b) in the α plane.

We now wish to evaluate this integral for an observation point far from the origin (i.e., kR large). Let us first recognize that at both ends of the contour, the integrand vanishes. In the example we are considering,

$$\cos(\alpha - \theta_c) = \cos(x + jy)$$
$$= \cos x \cosh y - j \sin x \sinh y, \tag{11.45}$$

and the absolute value becomes

$$\left| e^{-jkR \cos(\alpha - \theta_c)} \right| = e^{-kR \sin x \sinh y},$$

which vanishes for $0 < x < \pi$ and $y \to +\infty$ and for $-\pi < x < 0$ and $y \to -\infty$. In Fig. 11.7, the component $(-kR \sin x \sinh y)$ is shown. We note that the original contour starts from a point in the valley where the exponent is $-\infty$ and thus the magnitude is zero, along path C the magnitude increases, and finally, the magnitude decreases to zero in another valley region. Along this original contour C, both real and imaginary parts of the exponent vary. We therefore deform the path into the steepest descent contour (SDC) along which the imaginary part of the exponent is constant (Figs. 11.7 and 11.8) and evaluate the integral for large kR (see Appendix 11.C).

We have

$$f(\alpha) = -j\cos(\alpha - \theta_c), \tag{11.46}$$

and thus the saddle point is at

$$\frac{df}{d\alpha} = j\sin(\alpha - \theta_c) = 0, \tag{11.47}$$

which gives $\alpha_s = \theta_c$.

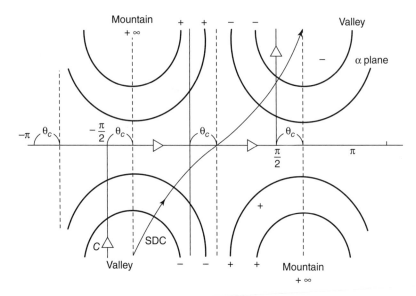

FIGURE 11.7   Real part of the exponent $-jkR \cos(\alpha - \theta_c)$.

Now we let $\alpha - \alpha_s = se^{j\gamma}$, and noting that

$$f(\alpha_s) = -j,$$
$$f''(\alpha_s) = j,$$

we expand $f(\alpha)$ about $\alpha = \alpha_s$

$$f(\alpha) = -j + j\frac{s^2 \, e^{j2\gamma}}{2}.$$

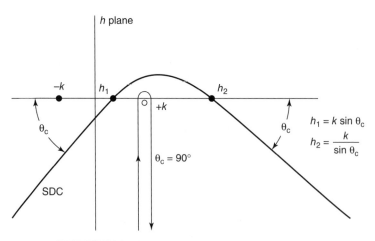

FIGURE 11.8   Steepest descent contour in the h plane.

We must choose γ such that the second term is real and negative. Thus we require that

$$2\gamma + \frac{\pi}{2} = \pm\pi,$$

which gives $\gamma = \pi/4$ or $-3\pi/4$. $\gamma = \pi/4$ represents the path going from the third to the first quadrant and $\gamma = -3\pi/4$ is the path from the first to the third quadrant. Both are steepest descent paths, but obviously in our problem, the choice must be

$$\gamma = \frac{\pi}{4},$$

and

$$f(\alpha) = -j - \frac{s^2}{2}.$$

Using the above γ, we obtain for large kR

$$I = \int_c A(k \sin \alpha) e^{-jkR \cos (\alpha - \theta_c)} k \cos \alpha \, d\alpha$$

$$\simeq F(\theta) \sqrt{\frac{2\pi}{kR}} e^{-jkR + j(\pi/4)} \tag{11.48}$$

$$= A(k \sin \theta_c) k \cos \theta_c \sqrt{\frac{2\pi}{kR}} e^{-jkR + j(\pi/4)}.$$

We therefore obtain the following approximate evaluations of the integrals:

$$I_1 = \int_{-\infty}^{\infty} C_n(h) H_n^{(2)} \left(\sqrt{k^2 - h^2} \rho \right) e^{-jhz} \, dh,$$

$$\simeq C_n(k \cos \theta_c) \frac{2}{R} e^{-jkR + j(n+1)(\pi/2)} \quad \text{for lage } kR. \tag{11.49}$$

11.6 RADIATION FROM A DIPOLE AND PARSEVAL'S THEOREM

We now go back to (11.38) and (11.39) and obtain the radiation field. Using (11.49), we first note that E_ϕ has R^{-2} dependence while H_ϕ has R^{-1} dependence for large kR. This means that E_ϕ diminishes faster than H_ϕ, and therefore, H_ϕ and E_θ are the only components of the radiation field. We thus get, using (11.49),

$$H_\phi = (IL)\frac{e^{-jkR}}{R} \sum_{n=-\infty}^{\infty} f_n(\theta) e^{-jn(\phi - \phi')}, \tag{11.50}$$

where

$$f_n(\theta) = \frac{k \sin\theta}{\pi} \frac{J_n(k\rho' \sin\theta) H^{(2)}(ka \sin\theta) - J_n(ka \sin\theta) H_n^{(2)}(k\rho' \sin\theta)}{H_n^{(2)}(ka \sin\theta)} e^{jn(\pi/2)}.$$

The Poynting vector S_r in (11.37) is then given by

$$S_r = \tfrac{1}{2}\eta \left|H_\phi\right|^2. \tag{11.51}$$

The radiation resistance R_{rad} is defined by

$$\frac{1}{2}I^2 R_{\text{rad}} = P_t = \text{Total radiated power},$$

$$P_t = \int_0^\pi \sin\theta \, d\theta \int_0^{2\pi} d\phi \, S_r(\theta, \phi) R^2. \tag{11.52}$$

We therefore get the radiation resistance

$$\frac{R_{\text{rad}}}{\eta} = L^2 \int_0^\pi \sin\theta \, d\theta \int_0^{2\pi} d\phi \left| \sum_{n=-\infty}^{\infty} f_n(\theta) e^{-jn\phi} \right|^2. \tag{11.53}$$

Let us next consider the integration in (11.53) with respect to ϕ. We get

$$\int_0^{2\pi} d\phi \left| \sum_{n=-\infty}^{\infty} f_n e^{-jn\phi} \right|^2 = \int_0^{2\pi} d\phi \sum_n \sum_{n'} f_n f_{n'}^* \, e^{-j(n-n')\phi}$$

$$= 2\pi \sum_{n=-\infty}^{\infty} \left|f_n\right|^2. \tag{11.54}$$

This relationship (11.54) states that the integral of the square of the magnitude of a periodic function is the sum of the square of the magnitude of each harmonic component, and this is called *Parseval's theorem*. An example is the total power delivered to a resistor by a periodic current. There is no coupling between different frequency components, and the total power is equal to the sum of the power for each frequency component.

Using (11.54), the radiation resistance is given by

$$\frac{R_{\text{rad}}}{\eta} = 2\pi L^2 \int_0^\pi \sin\theta \, d\theta \sum_{n=-\infty}^{\infty} \left|f_n(\theta)\right|^2. \tag{11.55}$$

Equation (11.54) is Parseval's theorem for a periodic function. The equivalent Parseval's theorem for a continuous function can be stated as follows:

$$\int_{-\infty}^{\infty} dz \left| \frac{1}{2\pi} \int_{-\infty}^{\infty} f(h) e^{-jhz} \, dh \right|^2 = \frac{1}{2\pi} \int_{-\infty}^{\infty} |f(h)|^2 \, dh. \qquad (11.56)$$

This can be proved by using the following:

$$\int_{-\infty}^{\infty} e^{-j(h-h')z} \, dz = 2\pi\delta(h - h'). \qquad (11.57)$$

11.7 LARGE CYLINDERS AND THE WATSON TRANSFORM

In previous sections 11.4, 11.5, and 11.6, the waves in cylindrical structures were given in terms of the Fourier series in ϕ and the Fourier transform in the z direction. These formal solutions are called *harmonic series* representations. There are two considerations associated with harmonic series representations which make it difficult to use them in many practical problems. One is that even if a solution is obtained in a harmonic series form, evaluation of the actual field quantities requires the truncation of an infinite series and thus the knowledge of its convergence. It is therefore desirable to have alternative representations which may have different convergence characteristics. This is particularly important because it is often time consuming and expensive to obtain sufficient accuracy in evaluating various Bessel functions of large order by computers.

Second, the harmonic series representation can be obtained only for a small number of structures, whose surfaces coincide with the 11 coordinate systems where the wave equation is separable (rectangular, cylindrical, elliptic cylinder, parabolic cylinder, spherical, conical, parabolic, prolate spheroidal, oblate spheroidal, ellipsoidal, and paraboloidal; see Morse and Feshback, p. 656). Thus it is important to develop an alternative representation that may be useful in describing waves for more practical and complex shapes. The Watson transform technique described in this section offers this alternative representation which may be used for more complex problems (Chapter 13).

Let us first illustrate the problem associated with large cylinders by taking radiation from a dipole shown in (11.50). We recognize that in calculating the radiation pattern, we need to deal with the infinite series

$$S = \sum_{n=-\infty}^{\infty} f(n) e^{-jn\phi}. \qquad (11.58)$$

The calculation of this series can be done by summing a finite number of terms instead of infinite terms and if the series is reasonably convergent, the finite sum should give a good solution. The important question then is: How many terms are required to represent the sum adequately? Problems involving cylinders and spheres actually contain the Bessel functions with the argument of the order of magnitude of

ka, where a is the radius or size of the object. Bessel functions behave quite differently depending on whether the argument is much less than, approximately equal to, or much greater than the order. Because of this, in general, series of the type (11.58) require at least *two ka terms* to represent the sum within a few percent accuracy. For a large cylinder, not only must the large number of terms be summed, but these terms contain Bessel functions of a large argument and large order. In many practical problems, this presents a formidable computational problem.

It would be extremely useful if series of the type (11.58) can be converted into a fast convergent series for large ka, so that only a few terms are required for obtaining numerical results. This is done by the Watson transform technique described in this section. Historically, shortwave radio-wave propagation over the earth was one of the central practical problems in the early twentieth century. The calculation was particularly difficult because of the extremely large ka value for the earth. In 1919, G.N. Watson successfully devised this technique, which converted the slow convergent series to a fast convergent series (Bremmer, 1949, p. 6).

First, let us consider the integral

$$I = \int_{C_1+C_2} \frac{A(v)}{\sin v\pi} \, dv. \tag{11.59}$$

The integrand has poles at $v = n$ integers, and thus the integral may be evaluated by taking the residues at each pole. The contour $C_1 + C_2$ encloses $2N + 1$ poles. We then let $N \to \infty$ (Fig. 11.9). We get

$$I = 2\pi j \sum_{n=-\infty}^{\infty} \left. \frac{A(v)}{\dfrac{\partial}{\partial v}(\sin v\pi)} \right|_{v=n} \tag{11.60}$$

$$= 2j \sum_{-\infty}^{\infty} A(n)e^{-jn\pi}.$$

Comparing (11.60) with (11.58), we note that the series (11.58) can be expressed by the following complex integral (Wait, 1959, Chapters 8 and 9):

$$S = \sum_{n=-\infty}^{\infty} f(n)e^{-jn\phi} = \frac{1}{2j} \int_{C_1+C_2} \frac{f(v)e^{-jv(\phi-\pi)}}{\sin v\pi} \, dv. \tag{11.61}$$

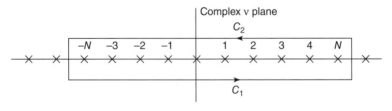

FIGURE 11.9   Contour integration for the Watson transform.

Many series representations for problems involving cylindrical and spherical structures can be transformed into the complex integral of this form. This transformation
(11.61) is called the *Watson transform*.

The original series (11.58) is called the harmonic series and is obviously adequate
for small cylinders. For large cylinders, however, the convergence of the harmonic
series is too slow and the integral (11.61) is more useful. There are three ways to
proceed from the integral form of (11.61). Each provides a good representation in
different regions of space. The three approaches are

1. *Residue series representation*: The integral (11.61) may be evaluated at the poles
   of $f(v)$, yielding a series of residues. This representation is highly convergent,
   requiring only a few terms in the *shadow region*. In this region, the wave creeps
   along the surface and is called the *creeping wave*.
2. *Geometric optical representation*: The integral (11.61) may be evaluated by
   means of the saddle-point technique, which gives a simple and useful representation in the *illuminated region*.
3. *Fock function representation*: In the *boundary region* between the illuminated
   and the shadow region, the two preceding techniques are not applicable and the
   technique developed by Fock must be used.

The three regions discussed above are shown in Fig. 11.10. We now examine each
of the three techniques.

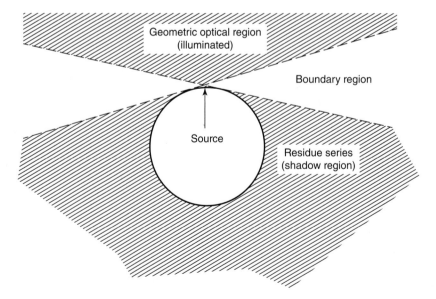

FIGURE 11.10   Wave representations in three regions.

11.8 RESIDUE SERIES REPRESENTATION AND CREEPING WAVES

The evaluation of the integral (11.61) can be made by deforming the contour C_1 to C_1' and C_2 to C_2' and taking the residues at the poles of $f(v)$. We note that (Fig. 11.11) (see Appendix 11.D)

$$\int_{C_1} dv = \int_{C_1'} dv - 2\pi j \sum_{m=1}^{\infty} (\text{residue at } v_m)$$

and

$$\int_{C_2} dv = \int_{C_2'} dv - 2\pi j \sum_{m=1}^{\infty} (\text{residue at } -v_m). \tag{11.62}$$

Next we need to show that the integrand of (11.61) approaches zero along the contour C_1' and C_2'. To do this, we need to examine the behaviors of the integrand for $|v| \to \infty$.

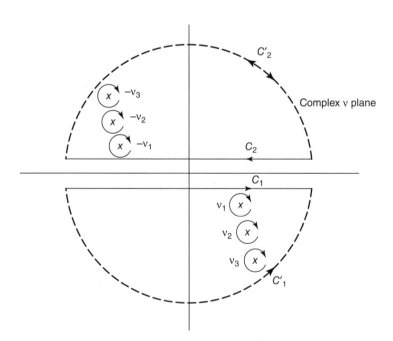

FIGURE 11.11   Contours for residue series.

The detailed explanations are given in Wait (1959) and therefore omitted here. Noting that the integrals in (11.62) along C_1' and C_2' are zero, we get

$$
\begin{aligned}
S &= \sum_{n=-\infty}^{\infty} f(n)e^{-jn\phi} \\
&= \frac{1}{2j} \int_{C_1+C_2} \frac{f(v)e^{-jv(\phi-\pi)}}{\sin v\pi}\, dv \\
&= -\pi \sum_{m=1}^{\infty} \text{Re}(v_m) - \pi \sum_{m=1}^{\infty} \text{Re}(-v_m),
\end{aligned}
$$

(11.63)

where

$$
\text{Re}(v_m) = \text{residue of } \frac{f(v)e^{-jv(\phi-\pi)}}{\sin v\pi}
$$

at $v = v_m$, the poles of $f(v)$. The last expression in (11.63) is called the *residue series*.

Let us first examine the locations of the poles v_m of $f(v)$. Note that the denominator of $f(v)$ is $H_v^{(2)}(ka\sin\theta)$ in (11.50). Therefore, the poles are given by

$$
H_{v_m}^{(2)}(ka\sin\theta) = 0.
$$

(11.64)

It can be shown that the zeros of $H_v^{(2)}(z)$ when $|z| \gg 1$ are given approximately by

$$
v_m = z + \left(\frac{z}{2}\right)^{1/3} \left[\frac{3}{2}\left(m - \frac{1}{4}\right)\right]^{2/3} e^{-j(\pi/3)}, \quad m = 1, 2, \ldots .
$$

(11.65)

Note that v_m is complex and its imaginary part is negative (Fig. 11.12).

Let us rewrite the residue series in the following form:

$$
S = \sum_{m=1}^{\infty} \frac{N^+(v_m)}{\left(\frac{\partial}{\partial v}D\right)v_m} e^{-jv_m\phi} + \sum_{m=1}^{\infty} \frac{N^-(v_m)}{\left(\frac{\partial}{\partial v}D\right)v_m} e^{-jv_m(2\pi-\phi)},
$$

(11.66)

where

$$
f(v) = \frac{N(v)}{D(v)},
$$

$$
D(v) = H_v^{(2)}(ka\sin\theta),
$$

$$
N^+(v) = \frac{(-\pi)N(v)e^{jv\pi}}{\sin v\pi},
$$

$$
N^-(v) = \frac{(-\pi)N(v)e^{-jv\pi}}{\sin v\pi}.
$$

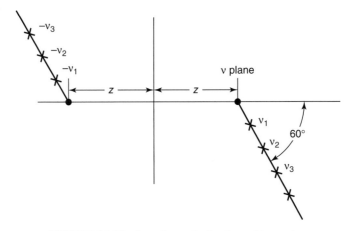

FIGURE 11.12   Locations of poles for residue series.

The residue of a function $N^+(v)/D(v)$ at $v = v_m$ is given by

$$\text{Residue} = \lim_{v \to v_m} (v - v_m) \frac{N^+(v)}{D(v)}$$

$$= \frac{N^+(v)}{\dfrac{\partial}{\partial v} D(v)} \Bigg|_{v=v_m} . \tag{11.67}$$

Let us examine the residue series (11.66). The first series due to the poles at v_m has the angular dependence $e^{-jv_m\phi}$. As shown in (11.65), v_m has the real part somewhat greater than $z = ka \sin \theta$ and the negative imaginary part. We therefore have

$$e^{-jv_m\phi} = e^{-jv_{mr}\phi - v_{mi}\phi},$$

$$v_m = v_{mr} - jv_{mi}. \tag{11.68}$$

When $\theta = \pi/2$, $v_{mr} > ka$, and therefore, $v_{mr}\phi > ka\ \phi$. Since ϕ is the distance along the surface, letting $v_{mr} = \beta a$, where β is the phase constant for the wave propagating along the surface, we get $\beta > k$, indicating that the phase velocity along the surface is less than the velocity of light. Therefore, the first series represents the wave that propagates with a phase velocity slower than the velocity of light on the surface with attenuation. The second series in (11.66), corresponding to $-v_m$, represents the wave propagating in the opposite direction with the same propagation constant. These waves creep along the surface and are called the *creeping waves* (Fig. 11.13).

We may also note that the original harmonic series representation has the form

$$\sum_{n=-\infty}^{\infty} (\cdots) e^{-jn\phi},$$

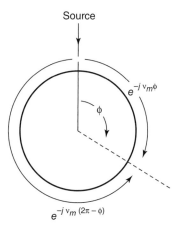

FIGURE 11.13   Creeping waves.

which is a Fourier series representation in the interval $0 < \phi < 2\pi$. We may write this as a series of cos $n\phi$ and sin $n\phi$. Then, we recognize that the harmonic series is in essence the representation of the wave in a series of standing waves (sin $n\phi$ and cos $n\phi$) around the cylinder. It is obvious, then, that for a small cylinder, since the wave has little ϕ variation, a small number of terms of the harmonic series should give a reasonable representation of the field. On the other hand, for a large cylinder, the wave radiated from a slot propagates along the cylinder surface and attenuates as it radiates the energy due to the curvature, and thus the wave is essentially a traveling wave along the surface. It is then clear that the suitable representation is the residue series because each term $e^{-jv_m\phi}$ is a traveling-wave representation, and it requires only a few terms to describe the wave adequately.

11.9   POISSON'S SUM FORMULA, GEOMETRIC OPTICAL REGION, AND FOCK REPRESENTATION

Before we discuss the wave in the geometric optical region, let us discuss an alternative representation of the series (11.58), which we write as

$$S = \sum_{n=-\infty}^{\infty} F(n). \tag{11.69}$$

We will show in this section that the sum (11.69) can be represented as a sum of the Fourier transform $G(m)$ of $F(n)$ in the following form:

$$\sum_{n=-\infty}^{\infty} F(n) = \sum_{m=-\infty}^{\infty} G(m), \tag{11.70}$$

where

$$G(m) = \int_{-\infty}^{\infty} F(v)e^{-j2m\pi v} \, dv.$$

This is called *Poisson's sum formula*. It is useful when the original series is slowly convergent, but the series of the transform is fast convergent.

To prove this, we start with the Watson transform

$$S = \sum_{n=-\infty}^{\infty} F(n)$$

$$= \frac{1}{2j} \int_{C_1+C_2} \frac{F(v)e^{jv\pi}}{\sin v\pi} \, dv. \tag{11.71}$$

We write this in a somewhat different way. Along C_1, we note that Im $v < 0$. Thus $|e^{-jv\pi}| < 1$. Therefore, we write

$$\frac{1}{\sin v\pi} = \frac{2j}{e^{jv\pi}(1 - e^{-2jv\pi})}$$

$$= 2je^{-jv\pi} \sum_{m=0}^{\infty} e^{-j2v\pi m}.$$

Thus we get

$$\frac{1}{2j} \int_{C_1} \frac{F(v)e^{jv\pi}}{\sin v\pi} \, dv = \sum_{m=0}^{\infty} G(m),$$

where

$$G(m) = \int_{C_1} F(v)e^{-j2v\pi m} \, dv.$$

Along C_2, Im $v > 0$. Thus $|e^{jv\pi}| < 1$. Therefore,

$$\frac{1}{\sin v\pi} = -2je^{jv\pi} \sum_{m=0}^{\infty} e^{j2v\pi m}.$$

Thus we get

$$\frac{1}{2j} \int_{C_2} \frac{F(v)e^{jv\pi}}{\sin v\pi} \, dv = \sum_{m=1}^{\infty} G(-m),$$

where

$$G(-m) = -\int_{C_2} F(v)e^{j2v\pi m}\, dv$$

$$= \int_{-C_2} F(v)e^{j2v\pi m}\, dv.$$

Noting that $F(v)$ along C_1 is the same as $F(v)$ along $(-C_2)$, we obtain Poisson's sum formula

$$S = \sum_{n=-\infty}^{\infty} F(n)$$

$$= \frac{1}{2j}\int_{C_1+C_2} \frac{F(v)e^{jv\pi}}{\sin v\pi}\, dv \tag{11.72}$$

$$= \sum_{m=-\infty}^{\infty} G(m),$$

where

$$G(m) = \int_{-\infty}^{\infty} F(v)e^{-j2v\pi m}\, dv.$$

Let us apply Poisson's sum formula to our problem (11.58).

$$S = \sum_{n=-\infty}^{\infty} f(n)e^{-jn\phi}$$

$$= \sum_{m=-\infty}^{\infty} S_m, \tag{11.73}$$

$$S_m = \int_{-\infty}^{\infty} f(v)e^{-jv(\phi+2m\pi)}\, dv.$$

Consider S_m for $m > 0$. The difference between S_m and S_0 is that S_m has the additional $2m\pi$ to the angle ϕ, and thus S_m represents the wave encircling the cylinder m times (Fig. 11.14a). On the other hand, when $m < 0$, letting $v = -v'$ and $m = -m'$, we get

$$S_m = S_{-m'} = \int_{-\infty}^{\infty} f(-v')e^{jv'(\phi-2m'\pi)}\, dv'. \tag{11.74}$$

This is recognized as the wave encircling the cylinder in the opposite direction (Fig. 11.14b).

In the illuminated region (Fig. 11.10), it is clear that the major contribution to the field comes from the S_0 term. It is then possible to evaluate S_0 by means of the saddle-point technique, and this leads to the geometric optical solution.

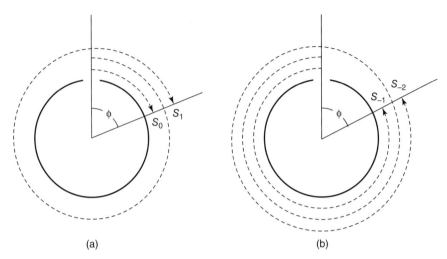

(a)   (b)

FIGURE 11.14   Waves encircling the cylinder.

It is clear from previous sections 11.7 and 11.8 that the wave in the boundary between the shadow and the illuminated region cannot be adequately represented by the residue series or the geometric optical representations. In this region, it is important to consider the order v of the Hankel function close to its argument z. Making use of the Hankel approximation, it is possible to represent the field in this region in terms of a function proposed by Fock. The details are given in Wait's book (1959, pp. 64–68).

11.10   MIE SCATTERING BY A DIELECTRIC SPHERE

The exact solution of the scattering of a plane electromagnetic wave by an isotropic, homogeneous dielectric sphere of arbitrary size is usually referred to as *Mie theory*, even though Lorenz gave essentially the same results before Mie's work (Kerker, 1969, Section 3.4). In this chapter, we employ a technique using the radial components of electric and magnetic hertz vectors. This technique differs from the vector wave equation formulation by Stratton, but it has the definite advantage of working with the scalar wave equations.

In spherical coordinates, it is possible to express the complete electromagnetic field in terms of two scalar functions π_1 and π_2. They are the radial components of the electric and magnetic Hertz vectors (see Section 4.15).

$$\bar{\pi}_e = \pi_1 \hat{r} \quad \text{and} \quad \bar{\pi}_m = \pi_2 \hat{r}. \tag{11.75}$$

π_1 produces all the TM modes with $H_r = 0$ and π_2 produces all the TE modes with $E_r = 0$. π_1 and π_2 satisfy the scalar wave equation.

$$\begin{aligned} (\nabla^2 + k^2)\pi_1 &= 0, \\ (\nabla^2 + k^2)\pi_2 &= 0. \end{aligned} \tag{11.76}$$

The electric and magnetic fields are derived from these two scalar functions.

$$\bar{E} = \nabla \times \nabla \times (r \pi_1 \, \hat{r}) - j\omega\mu\nabla \times (r \pi_2 \, \hat{r}),$$
$$\bar{H} = j\omega\varepsilon\nabla \times (r \pi_1 \, \hat{r}) + \nabla \times \nabla \times (r \pi_2 \, \hat{r}). \tag{11.77}$$

In spherical coordinates, we write

$$E_r = \frac{\partial^2}{\partial r^2}(r\pi_1) + k^2 r \pi_1,$$

$$E_\theta = \frac{1}{r}\frac{\partial^2}{\partial r \, \partial\theta}(r\pi_1) - j\omega\mu\frac{1}{\sin\theta}\frac{\partial}{\partial\phi}\pi_2,$$

$$E_\phi = \frac{1}{r\sin\theta}\frac{\partial^2}{\partial r \, \partial\phi}(r\,\pi_1) + j\omega\mu\frac{\partial}{\partial\theta}\pi_2,$$

$$H_r = \frac{\partial^2}{\partial r^2}(r\pi_2) + k^2 r \, \pi_2,$$

$$H_\theta = j\omega\varepsilon\frac{1}{\sin\theta}\frac{\partial}{\partial\phi}\pi_1 + \frac{1}{r}\frac{\partial^2}{\partial r \, \partial\theta}(r\,\pi_2),$$

$$H_\phi = -j\omega\varepsilon\frac{\partial}{\partial\phi}\pi_1 + \frac{1}{r\sin\theta}\frac{\partial^2}{\partial r \, \partial\phi}(r\,\pi_2). \tag{11.78}$$

We now consider a sphere with a complex dielectric constant ε_1 and a permeability μ_1 immersed in a medium with ε_2 and μ_2 (Fig. 11.15). The incident wave is polarized in the x direction and is given by

$$\bar{E}_{\text{inc}}(z) = e^{-jk_2 z} \, \hat{x}. \tag{11.79}$$

Let us consider the boundary conditions. At $r = a$, the tangential electric and magnetic fields must be continuous, and therefore, designating the field inside by $\bar{E}_1$

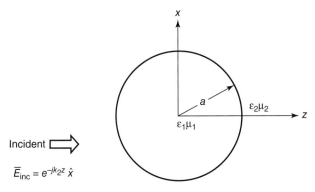

FIGURE 11.15   Mie scattering.

and $\bar{H}_1$ and the field outside by $\bar{E}_2$ and $\bar{H}_2$, we write the boundary conditions at $r = a$.

$$E_{1\theta} = E_{2\theta}, \quad E_{1\phi} = E_{2\phi}, \quad H_{1\theta} = H_{2\theta}, \quad H_{1\phi} = H_{2\phi}. \tag{11.80}$$

As can be seen from (11.78), the boundary conditions above contain a mixture of π_1 and π_2. Therefore, it is convenient to reduce them into the boundary conditions on π_1 alone and π_2 alone. To do this, we note that if we take a linear combination of E_θ and E_ϕ in the following manner, the terms containing π_2 drop out and the condition becomes the continuity of

$$\frac{\partial}{\partial \theta}(\sin \theta E_\theta) + \frac{\partial}{\partial \phi} E_\phi = \left[\frac{\partial}{\partial \theta} \left(\sin \theta \frac{\partial}{\partial \theta} \right) + \frac{1}{\sin \theta} \frac{\partial^2}{\partial \phi^2} \right] \frac{1}{r} \frac{\partial}{\partial r} (r\pi_1),$$

across the boundary $r = a$. Since this must hold for any θ and ϕ, we require that $\partial(r\pi_1)/\partial r$ be continuous at $r = a$.

Similarly, considering $\partial E_\theta/\partial \phi - \partial(\sin \theta\, E_\phi)/\partial \theta$, we obtain the boundary condition that $\mu\pi_2$ be continuous at $r = a$. Similar considerations for H_θ and H_ϕ give the boundary condition that $\partial(r\pi_2)/\partial r$ and $\varepsilon\pi_1$ be continuous at $r = a$.

Let us now express the incident field $\bar{E}_{\text{inc}}$ in terms of spherical harmonics. We observe that $\bar{E}_{\text{inc}}$ can be derived from two scalar functions π_1^i and π_2^i that satisfy the wave equation. General expressions for π_1^i and π_2^i are

$$\pi_1^i = \sum_{n=0}^{\infty} \sum_{m=0}^{\infty} j_n(k_2 r) P_n^m(\cos \theta)[A_{mn}^{(1)} \cos m\phi + B_{mn}^{(1)} \sin m\phi],$$

$$\pi_2^i = \sum_{n=0}^{\infty} \sum_{m=0}^{\infty} j_n(k_2 r) P_n^m(\cos \theta)[A_{mn}^{(2)} \cos m\phi + B_{mn}^{(2)} \sin m\phi]. \tag{11.81}$$

Here we used j_n as it is finite at $r = 0$, and P_n^m as it is finite at $\theta = 0$ and π.

In order to determine the constants $A_{mn}^{(1)}, A_{mn}^{(2)}, B_{mn}^{(1)}$, and $B_{mn}^{(2)}$, we examine the radial components of $\bar{E}_i$ and $\bar{H}_i$

$$E_{ir} = \bar{E}_{\text{inc}} \cdot \hat{r} = e^{-k_2 r \cos \theta} \sin \theta \cos \phi. \tag{11.82}$$

We expand E_{ir} in spherical harmonics and equate it to E_{ir} derived from (11.81).

$$E_{ir} = \frac{\partial^2}{\partial r^2}(r\pi_1^i) + k_2^2 r \pi_1^i$$

$$= \sum_{n=0}^{\infty} \sum_{m=0}^{n} \frac{n(n+1)}{r} j_n(k_2 r) P_n^m(\cos \theta)[A_{mn}^{(1)} \cos m\phi + B_{mn}^{(1)} \sin m\phi]. \tag{11.83}$$

Here we used the following:

$$\left[\frac{d^2}{dr^2} + k^2 - \frac{n(n+1)}{r^2}\right][rz_n(kr)] = 0, \tag{11.84}$$

where $z_n(kr)$ is any spherical Bessel function.

The expansion of E_{ir} in (11.82) can be done in the following manner. First, we note that

$$e^{-jkr\cos\theta} = \sum_{n=0}^{\infty}(-j)^n(2n+1)j_n(kr)P_n(\cos\theta). \tag{11.85}$$

This is obtained by expanding $\exp(-jkr\cos\theta)$ in spherical harmonics,

$$\exp(-jkr\cos\theta) = \sum_{n=0}^{\infty}a_n(r)P_n(\cos\theta),$$

and noting that $P_n(\cos\theta)$ are orthogonal and using the relationship

$$\int_0^\pi e^{-jkr\cos\theta}P_n(\cos\theta)\sin\theta\, d\theta = 2(-j)^nj_n(kr),$$

$$\int_0^\pi [P_n(\cos\theta)]^2\sin\theta\, d\theta = \frac{2}{2n+1}.$$

We then obtain the expansion of E_{ir} from (11.85) by noting that

$$E_{ir} = \frac{1}{jkr}\frac{\partial}{\partial\theta}\exp(-jkr\cos\theta)\cos\phi,$$

$$\frac{\partial}{\partial\theta}P_n(\cos\theta) = -P_n^1(\cos\theta).$$

We thus obtain the following expansion of E_{ir}:

$$E_{ir} = \sum_{n=0}^{\infty}\frac{(-j)^{n-1}(2n+1)}{k_2r}j_n(k_2r)P_n^1(\cos\theta)\cos\phi. \tag{11.86}$$

Comparing (11.83) with (11.86), we get

$$B_{mn}^{(1)} = 0, \quad A_{mn}^{(1)} = 0, \quad m \neq 1, \quad A_{1n}^{(1)} = \frac{(-j)^{n-1}(2n+1)}{n(n+1)k_2}$$

Substituting these into (11.81), we get the spherical harmonic representation of the incident Hertz potential π_1^i.

$$r\pi_1^i = \frac{1}{k_2^2} \sum_{n=1}^{\infty} \frac{(-j)^{n-1}(2n+1)}{n(n+1)} \Psi_n(k_2 r) P_n^1 (\cos\theta) \cos\phi. \qquad (11.87)$$

Similarly, we get

$$r\pi_2^i = \frac{1}{(\mu_2/\varepsilon_2)^{1/2} k_2^2} \sum_{n=1}^{\infty} \frac{(-j)^{n-1}(2n+1)}{n(n+1)} \Psi_n(k_2 r) P_n^1 (\cos\theta) \sin\phi, \qquad (11.88)$$

where $\Psi_n(x) = x j_n(x) = \sqrt{\pi x/2}\, J_{n+1/2}(x)$.

We now have the expressions for the incident Hertz potentials, (11.87) and (11.88). Next, we write the general expressions for the scattered fields and satisfy the boundary conditions. We note that the boundary conditions are such that π_1 outside the sphere couples only to π_1 inside the sphere. Thus since π_1^i has $\cos\phi$ dependence, we expect that the scattered field and the field inside the sphere have the same dependence on ϕ. Similarly, all π_2 should have $\sin\phi$ dependence.

Therefore, the general expressions for the scattered fields should be

$$r\pi_1^s = \frac{(-1)}{k_2^2} \sum_{n=1}^{\infty} \frac{(-j)^{n-1}(2n+1)}{n(n+1)} a_n \zeta_n(k_2 r) P_n^1 (\cos\theta) \cos\phi,$$

$$\qquad\qquad\qquad\qquad\qquad\qquad\qquad\qquad (11.89)$$

$$r\pi_2^s = \frac{(-1)}{\sqrt{\mu_2/\varepsilon_2}\, k_2^2} \sum_{n=1}^{\infty} \frac{(-j)^{n-1}(2n+1)}{n(n+1)} b_n \zeta_n(k_2 r) P_n^1 (\cos\theta) \sin\phi,$$

where $\zeta_n(x) = x h_n^{(2)}(x) = \sqrt{\pi x/2}\, H_{n+1/2}^{(2)}(x)$. Inside the sphere, we have

$$r\pi_1^r = \frac{1}{k_1^2} \sum_{n=1}^{\infty} \frac{(-j)^{n-1}(2n+1)}{n(n+1)} c_n \Psi_n(k_1 r) P_n^1 (\cos\theta) \cos\phi,$$

$$\qquad\qquad\qquad\qquad\qquad\qquad\qquad\qquad (11.90)$$

$$r\pi_2^r = \frac{1}{\sqrt{\mu_1/\varepsilon_1}\, k_1^2} \sum_{n=1}^{\infty} \frac{(-j)^{n-1}(2n+1)}{n(n+1)} d_n \Psi_n(k_1 r) P_n^1 (\cos\theta) \sin\phi,$$

where $a_n, b_n, c_n,$ and d_n are the constants to be determined by the boundary conditions. Now we apply the boundary condition on π_1 at $r = a$.

$$\frac{\partial}{\partial r} \left[r\left(\pi_1^i + \pi_1^s\right) \right] = \frac{\partial}{\partial r} \left[r\pi_1^r \right],$$

$$\varepsilon_2 \left(\pi_1^i + \pi_1^s \right) = \varepsilon_1\, \pi_1^r. \qquad (11.91)$$

Substituting (11.87), (11.89), and (11.90) into the above, we get

$$m \left[\psi'_n(k_2\, a) - a_n \zeta'_n(k_2\, a) \right] = c_n \Psi'_n(k_1\, a),$$

$$\frac{1}{\mu_2} \left[\psi_n(k_2\, a) - a_n \zeta_n(k_2\, a) \right] = \frac{1}{\mu_1} c_n \Psi_n(k_1\, a).$$

From these two, we obtain

$$a_n = \frac{\mu_1\, \Psi_n(\alpha)\psi'_n(\beta) - \mu_2 m \Psi_n(\beta)\Psi'_n(\alpha)}{\mu_1\, \zeta_n(\alpha)\psi'_n(\beta) - \mu_2 m \Psi_n(\beta)\zeta'_n(\alpha)}, \tag{11.92}$$

where $m = k_1/k_2 = \sqrt{\mu_1\varepsilon_1/\mu_2\varepsilon_2}$, $\alpha = k_2\, a$, $\beta = k_1\, a$.

To determine b_n for π_2, we use the boundary condition

$$\frac{\partial}{\partial r} \left[r \left(\pi_2^i + \pi_2^s \right) \right] = \frac{\partial}{\partial r} \left[r \pi_2^r \right],$$

$$\mu_2 \left(\pi_2^i + \pi_2^s \right) = \mu_1 \pi_2^r, \tag{11.93}$$

and obtain

$$b_n = \frac{\mu_2 m \Psi_n(\alpha)\psi'_n(\beta) - \mu_1\, \Psi_n(\beta)\psi'_n(\alpha)}{\mu_2 m \zeta_n(\alpha)\psi'_n(\beta) - \mu_1\, \Psi_n(\beta)\zeta'_n(\alpha)}. \tag{11.94}$$

The constants c_n and d_n are then obtained from (11.91) and (11.93).

$$c_n = \frac{jm\,\mu_1}{\mu_1\, \zeta_n(\alpha)\Psi'_n(\beta) - \mu_2 m \Psi_n(\beta)J'_n(\alpha)},$$

$$d_n = \frac{jm\,\mu_1}{\mu_2 m\, \zeta_n(\alpha)\Psi'_n(\beta) - \mu_1\, \Psi_n(\beta)J'_n(\alpha)}. \tag{11.95}$$

With these constants a_n, b_n, c_n, and d_n, (11.89) and (11.90) constitute the complete Mie solution for a dielectric sphere.

We now consider the far field. Noting that for $r \to \infty$,

$$\zeta_n(k_2 r) \to j^{n+1}\, e^{-jk_2 r},$$

we get

$$r\pi_1^s \to \frac{e^{-jk_2 r}}{k_2^2} \sum_{n=1}^{\infty} \frac{2n + 1}{n(n + 1)} a_n P_n^1(\cos\theta)\cos\phi,$$

$$r\pi_2^s \to \frac{e^{-jk_2 r}}{\sqrt{\dfrac{\mu_2}{\varepsilon_2}} k_2^2} \sum_{n=1}^{\infty} \frac{2n + 1}{n(n + 1)} b_n P_n^1(\cos\theta)\sin\phi. \tag{11.96}$$

The far fields E_θ and E_ϕ can be obtained from (11.78).

Noting that

$$\frac{\partial}{\partial r}(r\pi_1^s) = -jk_2 r\pi_1^s,$$

$$\frac{\partial}{\partial r}(r\pi_2^s) = -jk_2 r\pi_2^s,$$

we get

$$E_\theta = f_\theta(\theta, \phi)\frac{e^{-jk_2 r}}{r},$$

$$E_\theta = f_\theta(\theta, \phi)\frac{e^{-jk_2 r}}{r},$$

$$f_\theta = -\frac{j\cos\phi S_2(\theta)}{k_2},$$

$$f_\theta = \frac{j\sin\phi S_1(\theta)}{k_2}, \tag{11.97}$$

$$S_1(\theta) = \sum_{n=1}^{\infty} \frac{2n+1}{n(n+1)}\left[a_n \pi_n(\cos\theta) + b_n \tau_n(\cos\theta)\right],$$

$$S_2(\theta) = \sum_{n=1}^{\infty} \frac{2n+1}{n(n+1)}\left[a_n \tau_n(\cos\theta) + b_n \pi_n(\cos\theta)\right],$$

where

$$\pi_n(\cos\theta) = \frac{P_n^1(\cos\theta)}{\sin\theta} \quad \text{and} \quad \tau_n = \frac{d}{d\theta}P_n^1(\cos\theta).$$

Equations (11.97) are the expressions for the scattered field in the far zone of a dielectric particle when the incident wave is polarized in the x direction.

Let us calculate the total cross section by means of the forward scattering theorem.

$$\sigma_t = -\frac{4\pi}{k_2}\hat{e}_i \cdot \hat{e}_s I_m f(\hat{i}, \hat{i}). \tag{11.98}$$

Noting that $\hat{e}_i = \hat{x}$ and

$$\pi_n(\cos\theta)|_{\theta=0} = \tau_n(\cos\theta)|_{\theta=0} = \frac{n(n+1)}{2}, \tag{11.99}$$

we get

$$\sigma_t = \frac{4\pi}{k_2^2} \text{Re} S_1(0) = \frac{4\pi}{k_2^2} \text{Re} S_2(0). \tag{11.100}$$

Thus the normalized total cross section with respect to the geometric cross section πa^2 is given by

$$\frac{\sigma_t}{\pi a^2} = \frac{2}{\alpha^2} \sum_{n=1}^{\infty} (2n+1)[\text{Re}(a_n + b_n)], \quad \alpha = k_2 a. \tag{11.101}$$

The backscattering cross section σ_b is given by

$$\sigma_b = 4\pi |f|^2 \Big|_{\substack{\theta=\pi \\ \phi=0}},$$

and since

$$\pi_n(\cos\theta)\big|_{\theta=\pi} = -\tau_n(\cos\theta)\big|_{\theta=\pi} = -(-1)^n \frac{n(n+1)}{2},$$

we get

$$\frac{\sigma_b}{\pi a^2} = \frac{1}{\alpha^2} \left| \sum_{n=1}^{\infty} (2n+1)(-1)^n (a_n - b_n) \right|^2 = \frac{|S_2(\pi)|^2}{\alpha^2}. \tag{11.102}$$

The scattering cross section σ_s is obtained from

$$\begin{aligned}
\sigma_s &= \int_{4\pi} |f(\theta,\phi)|^2 \, d\Omega \\
&= \int_0^{2\pi} d\phi \int d\theta \sin\theta \left[\left| \frac{\cos\phi \, S_2(\theta)}{k_2} \right|^2 + \left| \frac{\cos\phi \, S_1(\theta)}{k_2} \right|^2 \right] \\
&= \frac{\pi}{k_2^2} \int_0^{\pi} d\theta \sin\theta \left[|S_2(\theta)|^2 + |S_1(\theta)|^2 \right].
\end{aligned} \tag{11.103}$$

Now from (11.97), we get

$$|S_2(\theta)|^2 = \sum_{n=1}^{\infty} \sum_{m=1}^{\infty} \frac{(2n+1)(2m+1)}{n(n+1)m(m+1)} [a_n a_m^* \pi_n \pi_m + b_n b_m^* \tau_n \tau_m + a_n b_m^* \pi_n \tau_m + a_m^* b_n \pi_m \tau_n],$$

$$|S_1(\theta)|^2 = \sum_{n=1}^{\infty} \sum_{m=1}^{\infty} \frac{(2n+1)(2m+1)}{n(n+1)m(m+1)} [a_n a_m^* \tau_n \tau_m + b_n b_m^* \pi_n \pi_m + a_n b_m^* \tau_n \pi_m + a_m^* b_n \pi_m \tau_n].$$

Adding these two and noting the orthogonality of Legendre functions,

$$\int_0^\pi (\pi_n \pi_m + \tau_n \tau_m) \sin\theta \, d\theta = 0 \quad \text{if} \, n \neq m$$

$$= \frac{2}{2n+1} \frac{(n+1)!}{(n-1)!} n(n+1) \quad \text{if} \, n = m,$$

$$\int_0^\pi (\pi_n \tau_m + \tau_n \pi_m) \sin\theta \, d\theta = 0,$$

we get

$$\frac{\sigma_s}{\pi a^2} = \frac{2}{\alpha^2} \sum_{n=1}^\infty (2n+1) \left(|a_n|^2 + |b_n|^2 \right). \tag{11.104}$$

Equations (11.101), (11.102), and (11.104) are the final expressions for the total cross section, the backscattering cross section, and the scattering cross section of a dielectric sphere. In this section, we discussed scattering by a homogeneous dielectric sphere. We can, however, extend the foregoing analysis to scattering by a sphere consisting of many concentric dielectric layers by expressing π_1 and π_2 using linear combinations of two spherical Bessel functions with two unknowns in each layer instead of one spherical function shown in (11.90).

11.11   AXIAL DIPOLE IN THE VICINITY OF A CONDUCTING WEDGE

Dipoles located in the vicinity of a wedge (Fig. 11.16) have a number of practical applications. Examples are antennas located near a sharp edge such as aircraft wings or fins and antennas with a corner reflector. In this section, we study the radiation characteristics of an axial dipole affected by a conducting wedge of given angle.

The differential equation governing this case is given by

$$\pi_z = \frac{IL}{j\omega\varepsilon} G(r/r'),$$

$$\left(\frac{\partial^2}{\partial\rho^2} + \frac{1}{\rho}\frac{\partial}{\partial\rho} + \frac{1}{\rho^2}\frac{\partial^2}{\partial\phi^2} \frac{\partial^2}{\partial z^2} + k^2 \right) G(r/r') = \frac{\delta(\rho-\rho')\delta(\phi-\phi')\delta(z-z')}{\rho},$$

$$\tag{11.105}$$

and the boundary condition at the conducting surface is

$$G = 0 \quad \text{at} \, \phi = 0 \quad \text{and} \quad \phi = \Psi_0. \tag{11.106}$$

We expand $G(r/r')$ in a series of normalized eigenfunctions $\Phi_m(\phi)$.

$$G(r/r') = \sum_{m=1}^\infty G_m(\rho, z)\Phi_m(\phi)\Phi_m(\phi'), \tag{11.107}$$

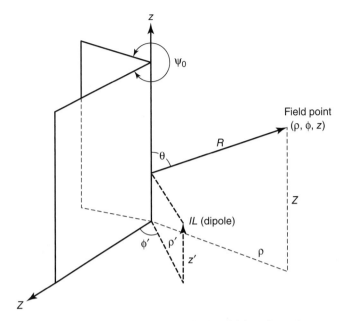

FIGURE 11.16   Electric dipole in the vicinity of a wedge.

where the eigenfunction is given by

$$\Phi_m(\phi) = \sqrt{\frac{2}{\Psi_0}} \sin \frac{m\pi}{\Psi_0}\phi, \qquad (11.108)$$

with the eigenvalue $v_m = m\pi/\Psi_0$, $m = 1, 2, 3, \ldots, \infty$ and

$$\int_0^{\Psi_0} [\phi_m(\phi)]^2 d\phi = 1.$$

We note that on the right side of (11.105),

$$\delta(\phi - \phi') = \sum_{m=1}^{\infty} \Phi_m(\phi)\Phi_m(\phi'). \qquad (11.109)$$

Therefore, substituting (11.107) and (11.109) into (11.105), we get

$$\left(\frac{\partial^2}{\partial\rho^2} + \frac{1}{\rho}\frac{\partial}{\partial\rho} - \frac{v_m^2}{\rho^2} + \frac{\partial^2}{\partial z^2} + k^2\right)G_m = -\frac{\delta(\rho - \rho')\delta(z - z')}{\rho}. \qquad (11.110)$$

We now take the Fourier transform in the z direction, and following the procedure given in Section 11.3, obtain

$$G(r/r') = \frac{1}{2\pi} \int_c \sum_{m=1}^{\infty} G_m(\rho, h) \Phi_m(\phi) \Phi_m(\phi') e^{-jh(z-z')} \, dh,$$

$$G_m(\rho, h) = -j\frac{\pi}{2} J_{v_m}\left(\sqrt{k^2 - h^2}\rho'\right) H_{v_m}^{(2)}\left(\sqrt{k^2 - h^2}\rho\right) \quad \rho' < \rho, \qquad (11.111)$$

$$= -j\frac{\pi}{2} J_{v_m}\left(\sqrt{k^2 - h^2}\rho\right) H_{v_m}^{(2)}\left(\sqrt{k^2 - h^2}\rho'\right) \quad \rho' > \rho.$$

This gives the Hertz potential produced by an axial dipole. The electric and magnetic fields are obtained by differentiation. The far field is obtained by using the saddle-point technique.

The field due to an axial magnetic dipole can be expressed similarly using the eigenfunction $\Psi_m(\phi)$, $m = 0, 1, 2, \ldots$, which satisfies Neumann's boundary condition

$$\Psi_m(\phi) = \left(\frac{2}{\Psi_0}\right)^{1/2} \cos\left(\frac{m\pi}{\Psi_0}\phi\right), \quad m = 1, 2, \ldots$$

$$= \left(\frac{1}{\Psi_0}\right)^{1/2}, \quad m = 0. \qquad (11.112)$$

11.12 LINE SOURCE AND PLANE WAVE INCIDENT ON A WEDGE

Using the results given in Section 11.1, we can obtain the exact solution for a wave excited by a line source and for a wave excited by a plane wave in the presence of a wedge. This solution is important in the study of the geometric theory of diffraction (GTD), discussed in Chapter 13. Let us consider a conducting wedge excited by an electric line current I_0 located at $\bar{r}' = (\rho, \theta_i)$ (Fig. 11.17). The electric field E_z at $\bar{r} = (s, \theta_s)$ satisfies the following equation and the boundary condition:

$$E_z = -j\omega\mu I_0 G_1,$$

$$(\nabla_t^2 + k^2)G_1 = -\delta(\bar{r} - \bar{r}'), \qquad (11.113)$$

$$G_1 = 0 \quad \text{at } \theta = 0 \quad \text{and} \quad \theta = \Psi_0.$$

Using the results (11.111), we get, for $s > \rho$,

$$G_1 = \left(-j\frac{\pi}{\Psi_0}\right) \sum_{m=1}^{\infty} J_{v_m}(k\rho) H_{v_m}^{(2)}(ks) \sin v_m \theta_i \sin v_m \theta_s, \qquad (11.114)$$

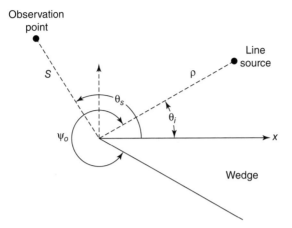

FIGURE 11.17   Wedge excited by a line source.

where $v_m = m\pi/\Psi_0$. For $s < \rho$, ρ and s above should be interchanged.

$$G_1 = \left(-j\frac{\pi}{\Psi_0}\right) \sum_{m=1}^{\infty} J_{v_m}(ks)H_{v_m}^{(2)}(k\rho) \sin v_m \theta_i \sin v_m \theta_s. \qquad (11.115)$$

For a magnetic line source I_m, we have

$$H_z = -j\omega\varepsilon I_m G_2,$$
$$(\nabla_t^2 + k^2)G_2 = -\delta(\bar{r} - \bar{r}'), \qquad (11.116)$$
$$\frac{\partial G_2}{\partial n} = 0 \quad \text{at} \theta = 0 \quad \text{and} \quad \theta = \Psi_0.$$

Green's function G_2 is then given by

$$G_2 = \left(-j\frac{\pi}{\Psi_0}\right) \sum_{m=0}^{\infty} \frac{\varepsilon_m}{2} J_{v_m}(ks)H_{v_m}^{(2)}(k\rho) \cos v_m \theta_i \cos v_m \theta_s \qquad \text{for } s < \rho, \quad (11.117)$$

where $\varepsilon_m = 1$ for $m = 0$, $\varepsilon_m = 2$ for $m \neq 0$, and ρ and s are interchanged for $s > \rho$.

Let us next consider the case of a plane wave incident on a wedge. For a plane wave polarized in the z direction incident from the direction θ_i, we let $\rho \to \infty$ and use the asymptotic form of $H_v^{(2)}$.

$$H_v^{(2)}(k\rho) \sim \left(\frac{2}{\pi k\rho}\right)^{1/2} e^{-j[k\rho - (v\pi/2) - (\pi/4)]}. \qquad (11.118)$$

Substituting this in (11.115) and noting that the incident field at the origin is

$$E_0 = -j\omega\mu I_0\, G_0,$$

$$G_0 = -\frac{j}{4}H_0^{(2)}(k\rho) \sim -\frac{j}{4}\left[\frac{2}{\pi k\rho}\right]^{1/2}e^{-j[k\rho-(\pi/4)]}, \qquad (11.119)$$

we get the field due to the plane wave with the incident electric field E_0 at the origin,

$$E_z = E_0\frac{4\pi}{\Psi_0}\sum_{m=1}^{\infty}e^{jv_m(\pi/2)}J_{v_m}(ks)\sin v_m\,\theta_i\sin v_m\,\theta_s. \qquad (11.120)$$

Similarly, when the field is polarized perpendicular to the z axis, the magnetic field due to the plane wave with the incident magnetic field H_0 at the origin is given by

$$H_z = H_0\frac{4\pi}{\Psi_0}\sum_{m=0}^{\infty}\frac{\varepsilon_m}{2}e^{jv_m(\pi/2)}J_{v_m}(ks)\cos v_m\,\theta_i\cos v_m\,\theta_s. \qquad (11.121)$$

Equations (11.113)–(11.117) give the exact solution for the total field for a conducting wedge when excited by a line source. Equations (11.120) and (11.121) give the exact total field for a wedge when excited by a plane wave. These are the exact series solutions, but in some practical problems, it is necessary to obtain the simpler closed-form solutions. A convenient closed-form exact solution is possible only for a plane wave incident on a knife edge (half-plane) $\Psi_0 = 2\pi$. For other cases, such as a plane wave incident on a wedge $\Psi_0 \neq 2\pi$ and a line source excitation, we can obtain some convenient approximate expressions.

11.13 HALF-PLANE EXCITED BY A PLANE WAVE

For a half-plane (knife edge), we let $\Psi_0 = 2\pi$ in (11.120) and (11.121) and obtain the exact solution. However, this half-plane diffraction problem was first solved by Sommerfeld in 1896 and the solution was given in terms of the Fresnel integral. The problem has also been solved by the Wiener–Hopf technique (see James, 1976; Jull, 1981; Noble, 1958). Here we give the final expressions and refer readers to the references above for the details.

For a plane wave polarized in the z direction incident on a half-plane from the direction θ_i (Fig. 11.18), the electric field is given by

$$E_{zi} = E_0\, e^{jk\rho\cos(\theta-\theta_i)}. \qquad (11.122)$$

The total electric field satisfies the Dirichlet boundary condition on a half-plane and the field at (s, θ_s) is given by

$$E_z = \frac{E_0\, e^{j(\pi/4)}}{\sqrt{\pi}}[e^{jks\cos(\theta_s-\theta_i)}F(a_1) - e^{jks\cos(\theta_s+\theta_i)}F(a_2)], \qquad (11.123)$$

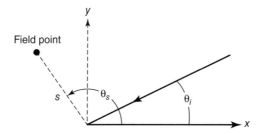

FIGURE 11.18   Plane wave incident on a half-plane ($x > 0$, $y = 0$).

where

$$F(a) = \int_a^\infty e^{-j\tau^2} \, d\tau \text{ is the complex Fresnel integral,}$$

$$a_1 = -(2ks)^{1/2} \cos \frac{\theta_s - \theta_i}{2},$$

$$a_2 = -(2ks)^{1/2} \cos \frac{\theta_s + \theta_i}{2}.$$

Also note that

$$F(a) + F(-a) = \int_{-\infty}^\infty e^{-j\tau^2} d\tau = \sqrt{\pi} e^{-j(\pi/4)}. \tag{11.124}$$

For a plane wave polarized perpendicular to the z axis, the incident magnetic field is given by

$$H_{zi} = H_0 e^{jk\rho \cos(\theta - \theta_i)}. \tag{11.125}$$

The total field satisfies the Neumann boundary condition on a half-plane and the field is given by

$$H_z = \frac{H_0 e^{j(\pi/4)}}{\sqrt{\pi}} [e^{jks \cos(\theta_s - \theta_i)} F(a_1) + e^{jks \cos(\theta_s + \theta_i)} F(a_2)]. \tag{11.126}$$

Note that the only difference between (11.123) and (11.126) is the minus or plus sign in front of the second term. We will make use of these results in Chapter 13.

PROBLEMS

11.1   A plane wave is normally incident on a conducting cylinder of radius a. Find the total current (integral over ϕ) on the cylinder for the TM case and for the TE case. $a = 1$ cm and $f = 1$ GHz.

11.2 A plane wave is normally incident on a dielectric cylinder of radius a with relative dielectric constant ε_r. Find the electric field on the axis of the cylinder when the incident field E_0 is polarized parallel to the axis (TM case). Find the electric field on the axis if the field is polarized perpendicular to the axis (TE case). $a = 10$ cm, $f = 1$ GHz, the real part of $\varepsilon_r = 60$, and the loss tangent $= 0.5$.

11.3 A plane sound wave p_0 is incident obliquely from a medium with C_0 and ρ_0 on a cylinder with C_1 and ρ_1. Find the pressure p inside and outside the cylinder.

11.4 Find the field radiated from a slot on a conducting cylinder. The electric field on the slot is as given in Fig. P11.4. To do this problem, first write general expressions for Π_z and Π_{mz} and find expressions for E_ϕ and E_z. Then let $\rho = a$ and equate E_ϕ and E_z to the slot field given.

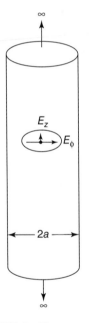

FIGURE P11.4   Slot on a conducting cylinder.

11.5 Find the radiation pattern of an annular slot antenna on a conducting cylinder (Fig. P11.5).

11.6 The gamma (factorial) function is given in the following integral form:

$$\Gamma(z+1) = z! = \int_0^\infty t^z e^{-t}\, dt.$$

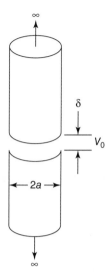

FIGURE P11.5   Annular slot.

For large z, $\Gamma(z)$ is given approximately by Stirling's formula

$$\Gamma(z) \sim e^{-z} z^{z-1/2} (2\pi)^{1/2}.$$

Derive this formula using the saddle-point technique. (*Hint*: Let $t = zv$.)

11.7   Show that the Airy integral $Ai(x)$,

$$Ai\,(x) = \frac{x^{1/3}}{2\pi} \int_{-\infty}^{\infty} e^{ix[(t^3/3)+t]}\, dt,$$

is given by

$$Ai\,(x) \sim \frac{x^{1/3}}{2\pi} e^{-2x/3} \sqrt{\frac{\pi}{x}}$$

for large x. Show the saddle points, original contour, and SDC in the complex t plane.

11.8   The integral representation of the Legendre function $P_n(\cos\theta)$ is given by

$$P_n(\cos\theta) = \frac{1}{2\pi} \int_0^{2\pi} (\cos\theta + i\sin\theta \cos\phi)^n\, d\phi.$$

Show that for large n,

$$P_n(\cos\theta) \sim \left(\frac{2}{n\pi\sin\theta}\right)^{1/2} \cos\left[\left(n+\frac{1}{2}\right)\theta - \frac{\pi}{4}\right].$$

11.9 Evaluate the Bessel function $J_0(z)$ for large z using the integral representation

$$J_0(z) = \frac{1}{\pi}\int_0^\infty \exp(jz\cos\phi)\,d\phi.$$

11.10 Evaluate the Legendre function $P_n^m(\cos\theta)$ for large n using the following definition:

$$P_n^m(\cos\theta) = \frac{c}{2\pi j}\oint e^{nf(w)}\,dw,$$

where the integral is to be taken over the unit circle in the counterclockwise sense, and

$$f(w) = \ln\left[\cos\theta + \frac{j}{2}\sin\theta\left(w + \frac{1}{w}\right)\right] - \frac{m+1}{n}\ln w,$$

$$c = \frac{(n+m)!}{n!}e^{-jm\pi/2}.$$

11.11 Find the integral I for large k (Fig. P11.11),

$$I = \int_{-\infty}^{\infty}\frac{\exp[-jk(r_1 + r_2)]}{r_1 r_2}\,dy,$$

where $r_1 = \sqrt{y^2 + a^2}$ and $r_2 = \sqrt{(y-c)^2 + b^2}$.

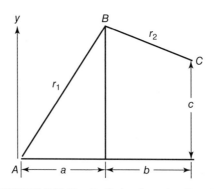

FIGURE P11.11   Radiation from A to B to C.

11.12 A plane electromagnetic wave is incident on a conducting sphere of radius a. Find the total cross section, backscattering cross section, and scattering cross section of the sphere.

11.13 A plane sound wave p_0 is incident from a medium with c_0 and ρ_0 on a sphere of radius a with c_1 and ρ_1. Find the total cross section, backscattering cross section, and scattering cross section of the sphere.

CHAPTER 12

SCATTERING BY COMPLEX OBJECTS

In Chapter 10, we discussed some of the approximate methods of calculating scattering by simple objects such as Rayleigh scattering and Rayleigh–Debye scattering. In Chapter 11, we discussed exact solutions for simple objects such as cylinders and spheres. In many practical situations, the objects often have a complex shape and the size and the dielectric constant are such that the simple solutions discussed in Chapters 10 and 11 are not applicable. Several methods that have been developed to deal with these problems will be discussed. In this chapter, we start with integral equation formulations for scalar fields for Dirichlet, Neumann, and two-media problems. *Electric field integral equation* (EFIE) and *magnetic field integral equation* (MFIE) are discussed next, followed by the *T*-matrix method. We then discuss inhomogeneous dielectric media, including Green's dyadic. We conclude this chapter with aperture diffraction problems, small apertures, and Babinet's principle. For radar cross sections, see Ruck et al. (1970). Aperture problems are discussed in Maanders and Mittra (1977), Rahmat-Samii and Mittra (1977), Arvas and Harrington (1983), and Butler et al. (1978). Low-frequency scattering (Kleinman, 1978; van Bladel, 1968; Stevenson, 1953; Senior, 1984) and narrow strips and slots (Butler and Wilton, 1980; Senior, 1979) are also discussed in this chapter. For dyadic Green's functions, see Tai (1971), Yaghjian (1980), and Livesay and Chen (1974). Also see Lewin et al. (1977) for curved structures.

Electromagnetic Wave Propagation, Radiation, and Scattering: From Fundamentals to Applications,
Second Edition. Akira Ishimaru.

12.1 SCALAR SURFACE INTEGRAL EQUATIONS FOR SOFT AND HARD SURFACES

In Section 6.1, we used Huygens' principle and obtained the expressions for the field in terms of the incident wave and the contribution from the surface field. In this section we make use of these relationships to obtain the surface integral equations when the object is a soft body or a hard body. A penetrable homogeneous body is discussed in the Section 12.2. There are two surface equations. One relates the field to the surface integral and the other relates the normal derivative of the field to the surface integral.

12.1.1 Integral Equation for a Soft Body

If the object is a soft body, the boundary condition is Dirichlet's type.

$$\psi(\bar{r}) = 0 \quad \text{on the surface.} \tag{12.1}$$

Now we make use of (6.11) (Fig. 12.1).

$$\psi_i(\bar{r}) + \int_S \left[\psi(\bar{r}') \frac{\partial G(\bar{r}, \bar{r}')}{\partial n'} - G(\bar{r}, \bar{r}') \frac{\partial \psi(\bar{r}')}{\partial n'} \right] dS' = \frac{1}{2}\psi(\bar{r}) \quad \text{if } \bar{r} \text{ is on the surface } S. \tag{12.2}$$

We then use the boundary condition (12.1), and we get

$$\psi_i(\bar{r}) = \int_S G_0(\bar{r}, \bar{r}') \frac{\partial \psi(\bar{r}')}{\partial n'} dS', \tag{12.3}$$

where

$$G_0(\bar{r}, \bar{r}') = \frac{\exp(-jk\,|\bar{r} - \bar{r}'|)}{4\pi\,|\bar{r} - \bar{r}'|} \quad \text{and } \bar{r} \text{ and } \bar{r}' \text{ are on the surface } S.$$

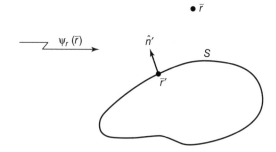

FIGURE 12.1   Surface integral equation.

Note that as $\bar{r}'$ approaches $\bar{r}$, G_0 becomes infinite, and therefore this is an improper integral. However, as is shown in Appendix 12.A, the integral is convergent as G_0 is proportional to R^{-1}, where $R = |\bar{r} - \bar{r}'|$ and the surface integral is convergent if the integrand is $R^{-\alpha}$ and $0 < \alpha < 2$. Thus this is a weakly singular integral equation.

Equation (12.3) is the Fredholm integral equation of the first kind for the unknown function $\partial\psi/\partial n'$ (see Appendix 12.B).

Alternatively, we can obtain the Fredholm equation of the second kind. To do this, we take the normal derivative of (6.9) and let $\bar{r}$ approach the surface S. As shown in Appendix 12.A, we get

$$\frac{\partial}{\partial n} \int_S G_0(\bar{r}, \bar{r}') \frac{\partial\psi(\bar{r}')}{\partial n'} dS' = \frac{1}{2} \frac{\partial\psi(\bar{r})}{\partial n} + \int \frac{\partial G_0(\bar{r}, \bar{r}')}{\partial n} \frac{\partial\psi(\bar{r}')}{\partial n'} dS'. \quad (12.4)$$

We then obtain

$$\frac{\partial\psi_i(\bar{r})}{\partial n} = \frac{1}{2} \frac{\partial\psi(\bar{r})}{\partial n} + \int \frac{\partial G_0(\bar{r}, \bar{r}')}{\partial n} \frac{\partial\psi(\bar{r}')}{\partial n'} dS', \quad (12.5)$$

where both $\bar{r}$ and $\bar{r}'$ are on the surface S. This is the desired Fredholm integral equation of the second kind. It can be proved that the integral is convergent (see Appendix 12.A).

12.1.2 Integral Equations for a Hard Body

The boundary condition at the surface of a hard body is the Neumann type.

$$\frac{\partial\psi}{\partial n} = 0. \quad (12.6)$$

From (6.11), we get the Fredholm integral equation of the second kind for $\psi(\bar{r})$.

$$\psi_i(\bar{r}) + \int_S \psi(\bar{r}') \frac{\partial G_0(\bar{r}, \bar{r}')}{\partial n'} dS' = \frac{1}{2}\psi(\bar{r}). \quad (12.7)$$

The integral is convergent (see Appendix 12.A).

We can also obtain the Fredholm integral equation of the first kind.

$$\frac{\partial\psi_i(\bar{r})}{\partial n} + \frac{\partial}{\partial n} \int \psi(\bar{r}') \frac{\partial G_0(\bar{r}, \bar{r}')}{\partial n'} dS' = 0. \quad (12.8)$$

The kernel in this case is singular (see Appendix 12.A).

We note that in the formulations above, the scattered field satisfies the radiation condition because we used Green's function G_0, which is outgoing and satisfies the radiation condition. The field on the surface should also satisfy the edge condition if the surface contains edges and corners.

In the above, we have two integral equations, (12.3) and (12.5), for soft objects. We can use either one to obtain the solution. However, the second kind (12.5) is often more stable in numerical calculations. Also note that if the surface integral in (12.5) is negligibly small, we get the Kirchhoff approximation,

$$\frac{\partial \psi}{\partial n} \simeq 2 \frac{\partial \psi_i}{\partial n}. \tag{12.9}$$

Similarly, for the hard surface, (12.7) is more convenient and the Kirchhoff approximation is obtained immediately by neglecting the surface integral.

$$\psi(\bar{r}) \simeq 2\psi_i(\bar{r}). \tag{12.10}$$

12.2 SCALAR SURFACE INTEGRAL EQUATIONS FOR A PENETRABLE HOMOGENEOUS BODY

Let us now consider the scattering of a scalar acoustic wave incident from the media with density ρ_0 and sound velocity c_0 on a homogeneous body with ρ_1 and c_1. This is sometimes called the *two-media problem* (Fig. 12.2). We use the field ψ to designate the velocity potential. Then we have

$$
\begin{aligned}
\left(\nabla^2 + k_0^2\right)\bar{\psi}_0 &= 0 \quad \text{outside the body} \\
\left(\nabla^2 + k_1^2\right)\bar{\psi}_1 &= 0 \quad \text{inside the body.}
\end{aligned}
\tag{12.11}
$$

The boundary conditions on S are

$$\rho_0 \, \psi_0 = \rho_1 \, \psi_1 \quad \text{and} \quad \frac{\partial \psi_0}{\partial n} = \frac{\partial \psi_1}{\partial n}. \tag{12.12}$$

On the surface S, there are two unknowns: ψ_0 and $\partial\psi_0/\partial n$ (or ψ_1 and $\partial\psi_1/\partial n$). We will therefore develop two coupled surface integral equations for ψ_0 and $\partial\psi_0/\partial n$.

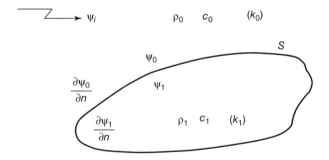

FIGURE 12.2   Two-media problem.

First, consider the region outside the body. As the observation point $\bar{r}$ approaches the surface S from outside, we get (see Eq. 6.11)

$$\psi_i(\bar{r}) + \oint_S \left(\psi_0(\bar{r}') \frac{\partial G_0(\bar{r}, \bar{r}')}{\partial n'} - G_0(\bar{r}, \bar{r}') \frac{\partial \psi_0(\bar{r}')}{\partial n'} \right) dS' = \frac{1}{2} \psi_0(\bar{r}), \qquad (12.13)$$

where

$$G_0(\bar{r}, \bar{r}') = \frac{\exp(-jk_0 |\bar{r} - \bar{r}'|)}{4\pi |\bar{r} - \bar{r}'|}.$$

Next we use the same formula (12.11) for the region inside the body and let $\bar{r}$ approach the surface S from inside.

$$\oint_S \left(\psi_1(\bar{r}') \frac{\partial G_1(\bar{r}, \bar{r}')}{\partial n'} - G_1(\bar{r}, \bar{r}') \frac{\partial \psi_1(\bar{r}')}{\partial n'} \right) dS' = -\frac{1}{2} \psi_1(\bar{r}), \qquad (12.14)$$

where

$$G_1(\bar{r}, \bar{r}') = \frac{\exp(-jk_1 |\bar{r} - \bar{r}'|)}{4\pi |\bar{r} - \bar{r}'|}.$$

Note that the right side is $-\frac{1}{2}\psi_1$ rather than $+\frac{1}{2}\psi_1$ because $\partial/\partial n'$ is now a normal derivative in the direction toward the outside. Now using the boundary conditions, we can rewrite (12.14) as follows:

$$\oint_S \left[\left(\frac{\rho_0}{\rho_1} \right) \psi_0(\bar{r}') \frac{\partial G_1(\bar{r}, \bar{r}')}{\partial n'} - G_1(\bar{r}, \bar{r}') \frac{\partial \psi_0(\bar{r}')}{\partial n'} \right] dS'$$
$$= -\frac{1}{2} \left(\frac{\rho_0}{\rho_1} \right) \psi_0(\bar{r}). \qquad (12.15)$$

Equations (12.13) and (12.15) give two integral equations for two unknowns $\psi_0(\bar{r}')$ and $\partial \psi_0(\bar{r}'/\partial n')$. Then the two equations can be solved numerically.

Alternatively, we take the normal derivative of (6.9) and let $\bar{r}$ approach the surface S. We then get

$$\frac{\partial \psi_i(\bar{r})}{\partial n} + \frac{\partial}{\partial n} \int_S \psi_0(\bar{r}') \frac{\partial G_0(\bar{r}, \bar{r}')}{\partial n} dS' - \oint_S \frac{\partial G_0(\bar{r}, \bar{r}')}{\partial n} \frac{\partial \psi_0(\bar{r}')}{\partial n'} dS' = \frac{1}{2} \frac{\partial \psi_0(\bar{r})}{\partial n}. \qquad (12.16)$$

Now we use (6.9) for the region inside the body and let $\bar{r}$ approach the surface S from inside. We then use the boundary conditions and obtain

$$\frac{\rho_0}{\rho_1} \frac{\partial}{\partial n} \int_S \psi_0(\bar{r}') \frac{\partial G_1(\bar{r}, \bar{r}')}{\partial n'} dS' - \oint \frac{\partial G_1(\bar{r}, \bar{r}')}{\partial n} \frac{\partial \psi_0(\bar{r}')}{\partial n'} dS' = -\frac{1}{2} \frac{\partial \psi_0(\bar{r})}{\partial n}. \qquad (12.17)$$

We note that both (12.16) and (12.17) contain singular kernels, due to the first integral term of (12.16) and the first term of (12.17).

Note also that we have four equations, (12.13), (12.15)–(12.17). The first two relate the field outside and inside the surface to the surface integral and the last two relate the normal derivatives of the field outside and inside the surface to the surface integral. We can make different combinations of these four equations to obtain different integral equations with various degrees of computational advantages.

12.3 EFIE AND MFIE

Let us consider an electromagnetic wave $(\bar{E}_i, \bar{H}_i)$ incident on a perfectly conducting body with surface S (Fig. 12.3). There are two surface integral equations. The equation in terms of the electric field is called EFIE and the equation in terms of the magnetic field is called MFIE.

EFIE is obtained by using the surface integral representation (6.111) and imposing the boundary condition.

$$\bar{E}_i(\bar{r}) + \oint_S \bar{E}_s \, dS' = \tfrac{1}{2}\bar{E}(\bar{r}). \tag{12.18}$$

The boundary condition on S is

$$\hat{n}' \times \bar{E}(\bar{r}) = 0. \tag{12.19}$$

Therefore, we get

$$\hat{n}' \times \bar{E}_i(\bar{r}) + \hat{n}' \times \oint \bar{E}_s \, dS' = 0, \tag{12.20}$$

where

$$\bar{E}_s = -[j\omega\mu G(\hat{n}' \times \bar{H}) - (\hat{n}' \times \bar{E}) \times \nabla'G - (\hat{n}' \cdot \bar{E})\nabla'G].$$

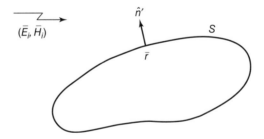

FIGURE 12.3   EFIE and MFIE.

Noting that $\hat{n}' \times \bar{H} = \bar{J}_s = $ Surface current, $\hat{n}' \cdot \bar{E} = \rho_s/\varepsilon$, $\rho_s = $ surface charge, and using the continuity condition

$$\nabla_s \cdot \bar{J}_s + j\omega\rho_s = 0,$$

where ∇_s is the surface divergence, we get the following EFIE for the surface current $\bar{J}_s$:

$$\hat{n}' \times \bar{E}_i(\bar{r}) = \frac{1}{j\omega\varepsilon}\hat{n}' \times \oint_S [-k^2 G(\bar{r},\bar{r}')\bar{J}_s(\bar{r}') + (\nabla'_s \cdot \bar{J}_s(\bar{r}'))\nabla' G(\bar{r},\bar{r}')]dS',$$

where

$$G = G(\bar{r},\bar{r}') = \frac{e^{-jk|\bar{r}-\bar{r}'|}}{4\pi|\bar{r}-\bar{r}'|}. \tag{12.21}$$

Next, we consider the MFIE. We start with (6.111).

$$\bar{H}_i + \oint_S \bar{H}_s dS' = \tfrac{1}{2}\bar{H}, \tag{12.22}$$

where

$$\bar{H}_s = j\omega\varepsilon G\hat{n}' \times \bar{E} + (\hat{n}' \times \bar{H}) \times \nabla'G + (\hat{n}' \cdot \bar{H})\nabla'G.$$

However, the boundary conditions on the surface are $\hat{n}' \times \bar{E} = 0$ and $\hat{n}' \cdot \bar{H} = 0$. Letting $\hat{n}' \times \bar{H} = \bar{J}_s = $ surface current, we get the MFIE for the surface current $\bar{J}_s$.

$$\hat{n}' \times \bar{H}_i(\bar{r}) + \oint_S [\hat{n}' \times \bar{J}_s(\bar{r}') \times \nabla' G(\bar{r},\bar{r}')]dS' = \frac{1}{2}\bar{J}_s(\bar{r}). \tag{12.23}$$

Note that for MFIE, the first term gives the physical optics approximation.

$$\bar{J}_s = 2\hat{n}' \times \bar{H}_i. \tag{12.24}$$

Therefore, MFIE should be useful for a large object with a smooth surface whose radius of curvature is large compared with a wavelength. Then physical optics is a good approximation and the surface integral is a small correction term. On the other hand, EFIE should be more useful for a thin object such as a wire.

12.4 *T*-MATRIX METHOD (EXTENDED BOUNDARY CONDITION METHOD)

Up to this point in this chapter, we have discussed the formulation of the scattering problem in terms of the surface integral equations for the unknown surface fields. In this section we present a different technique, originated by P. C. Waterman (1969), which makes use of the extinction theorem (Sections 6.1 and 6.9). According to the extinction theorem, the surface field produces a field contribution inside the object which exactly cancels or extinguishes the incident field throughout the interior of the object. First, we will show that by using this extinction theorem as the *extended boundary condition*, we can get an integral equation for the surface field which holds throughout the interior of the object. Second, we note that this integral equation need not be satisfied throughout the interior; it need hold only in any portion of the interior volume because of the *analytic continuity*. These two are the key elements of the *T*-matrix method. In actual applications, these two ideas are used to derive the transition matrix (*T*-matrix) which relates the scattered wave to the incident wave. We illustrate this technique and explain its meaning by using Dirichlet's problem as an example.

Let us consider an incident wave ψ_i impinging on an object with surface S where Dirichlet's boundary condition holds (Fig. 12.4). This is a two-dimensional problem, and we first express the incident field ψ_i in a series of appropriate basis functions. For a two-dimensional problem, the natural choice is cylindrical harmonics. Then we write

$$\psi_i(\bar{r}) = \sum_{n=-\infty}^{\infty} a_n J_n(kr) e^{jn\phi}. \tag{12.25}$$

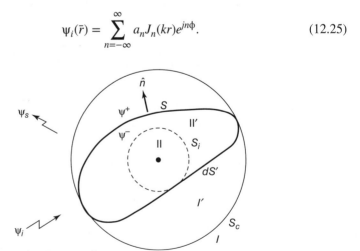

FIGURE 12.4 *T*-matrix method.

Note that the incident field ψ_i is finite everywhere, including the origin, and therefore $J_n(kr)$ must be used. For a given incident wave, a_n is the known coefficient. For example, for a plane wave $\psi_i(\bar{r}) = \exp(-jk\hat{i} \cdot \bar{r})$ propagating in the direction ϕ_i, we have

$$\hat{i} = \cos \phi_i \hat{x} + \sin \phi_i \hat{y},$$
$$\bar{r} = r \cos \phi \, \hat{x} + r \sin \phi \, \hat{y}, \qquad (12.26)$$
$$a_n = e^{-jn(\phi_i + \pi/2)}.$$

Next, we write the scattered wave $\psi_s(\bar{r})$ in region I outside the circumscribed cylinder S_c in the same cylindrical harmonics.

$$\psi_s(\bar{r}) = \sum_{n=-\infty}^{\infty} b_n H_n^{(2)}(kr)e^{jn\phi}. \qquad (12.27)$$

In this region the waves are outgoing, and we used the Hankel function of the second kind to satisfy the radiation condition.

The transition matrix or the *T*-matrix [*T*] is defined by

$$[b] = [T][a], \qquad (12.28)$$

where $[b] = [b_n]$ and $[a] = [a_n]$ are the column matrices and [*T*] is the square matrix. Here these matrices are truncated and finite for numerical calculations. The *T*-matrix is then obtained systematically by use of the extinction theorem, as discussed below.

To obtain the *T*-matrix, let us first write the extinction theorem (6.12)

$$\psi_i(\bar{r}) + \int_S \left[\psi(\bar{r}') \frac{\partial G(\bar{r}, \bar{r}')}{\partial n'} - G(\bar{r}, \bar{r}') \frac{\partial \psi(\bar{r}')}{\partial n'} \right] dS' = 0 \quad \text{if } \bar{r} \text{ is inside } S. \quad (12.29)$$

This theorem applies to regions II and II′ in Fig. 12.4. However, let us apply this only to region II inside the inscribed cylinder S_i. Then $|\bar{r}| < |\bar{r}'|$, and therefore Green's function $G(\bar{r}, \bar{r}')$, can be written for $r < r'$

$$G(\bar{r}, \bar{r}') = -\frac{j}{4} \sum_{n=-\infty}^{\infty} J_n(kr)H_n^{(2)}(kr')e^{jn(\phi - \phi')}. \qquad (12.30)$$

Now we substitute this and (12.25) into (12.29). We then note that ψ_i in (12.29) is a series involving $J_n(kr) \exp(jn \, \phi)$, and the integral term in (12.29) is also a series involving $J_n(kr) \exp(jn \, \phi)$, which appears in $G(\bar{r}, \bar{r}')$ of (12.30). Since $J_n(kr)$

$\exp(jn\,\phi)$ is orthogonal, we can equate each coefficient of $J_n \exp(jn\,\phi)$ to zero and obtain

$$a_n = \frac{j}{4} \int_S \left[\psi(\bar{r}') \frac{\partial}{\partial n'} H_n^{(2)}(kr') e^{-jn\phi'} - \frac{\partial \psi(\bar{r}')}{\partial n'} H_n^{(2)}(kr') e^{-jn\phi'} \right] dS', \qquad (12.31)$$

where $dS' = [r'^2 + (dr'/d\,\phi')^2]^{1/2} d\phi'$ is the elementary distance along the surface S, where $r' = r'(\phi')$ and

$$\frac{\partial}{\partial n'} = \hat{n} \cdot \nabla = \left[r'^2 + \left(\frac{dr'}{d\phi'} \right)^2 \right]^{-1/2} \left(\bar{r}' \frac{\partial}{\partial r'} - \frac{1}{r'} \frac{\partial r'}{\partial \phi'} \frac{\partial}{\partial \phi'} \right).$$

This equation relates the coefficients a_n of the incident wave to the surface field $\psi(\bar{r}')$ and $\partial \psi(\bar{r}')/\partial n'$ and is obtained by applying the extinction theorem to region II only. Therefore, the surface field obtained from (12.31) when substituted into the extinction theorem (12.29) should cancel the incident field completely in region II. However, the left side of the extinction theorem (12.29) is a regular solution of the wave equations in both II' and II, and therefore if it is null in II, it should also be null in II' by analytic continuation. Thus the surface field determined from (12.31) should be the true surface field satisfying (12.29) for all interior regions II and II'.

Next we obtain the scattered field ψ_s in region I using the following:

$$\psi_s(\bar{r}) = \int_S \left[\psi(\bar{r}') \frac{\partial G(\bar{r}, \bar{r}')}{\partial n'} - G(\bar{r}, \bar{r}') \frac{\partial \psi(\bar{r}')}{\partial n'} \right] dS',$$

$$G(\bar{r}, \bar{r}') = -\frac{j}{4} \sum_{n=-\infty}^{\infty} J_n(kr') H_n^{(2)}(kr) e^{jn(\phi - \phi')} \qquad \text{for } r > r'. \qquad (12.32)$$

Noting the expansion of ψ_s in (12.27) and using the orthogonality of $H_n^{(2)}(kr) e^{jn\phi}$, we obtain

$$b_n = -\frac{j}{4} \int_S \left[\psi(\bar{r}') \frac{\partial}{\partial n'} (J_n(kr') e^{-jn\phi'}) - \frac{\partial \psi(\bar{r}')}{\partial n'} J_n(kr') e^{-jn\phi'} \right] dS'. \qquad (12.33)$$

We now have a_n in (12.31) and b_n in (12.33) expressed in terms of the surface field. We next use the boundary condition and express the surface field in terms of a complete set of functions. We then eliminate the surface field from the two equations (12.31) and (12.33) and finally obtain the T-matrix.

Let us illustrate this procedure by using Dirichlet's problem ($\psi = 0$ on S). Then we get

$$a_n = -\frac{j}{4} \int_S \frac{\partial \psi(\bar{r}')}{\partial n'} H_n^{(2)}(kr')e^{-jn\phi'}\, dS',$$

$$b_n = \frac{j}{4} \int_S \frac{\partial \psi(\bar{r}')}{\partial n'} J_n(kr')e^{-jn\phi'}\, dS'. \tag{12.34}$$

We now expand the unknown surface function $\partial \psi / \partial n'$ in a series of a complete set of functions. The choice of the function is arbitrary, but we use the following:

$$\frac{\partial \psi(\bar{r}')}{\partial n'} = \sum_{n=-\infty}^{\infty} \alpha_n \frac{\partial}{\partial n'}\left[J_n(kr')e^{jn\phi'}\right], \tag{12.35}$$

where α_n is the unknown coefficient and we assume that $(\partial/\partial n')[J_n \exp(jn\,\phi')]$ is a complete set. Substituting (12.35) into (12.34), we get

$$[a] = [Q^-][\alpha],$$

$$[b] = -[Q^+][\alpha], \tag{12.36}$$

where $[a]$, $[b]$, and $[\alpha]$ are column matrices and

$$Q_{mn}^- = -\frac{j}{4} \int_S \psi_m \phi_n'\, dS',$$

$$Q_{mn}^+ = -\frac{j}{4} \int_S \psi_{rm} \phi_n'\, dS',$$

$$\psi_m = H_m^{(2)}(kr')e^{-jm\phi'},$$

$$\psi_{rm} = J_m(kr')e^{-jm\phi'},$$

$$\phi_n' = \frac{\partial}{\partial n'}\left[J_n(kr')e^{jn\phi'}\right].$$

Finally, we eliminate the surface field $[\alpha]$ from (12.36) and get the *T*-matrix

$$[b] = [T][a],$$

$$[T] = -[Q^+][Q^-]^{-1}. \tag{12.37}$$

Let us next consider Neumann's problem for which the boundary condition on S is $(\partial/\partial n')\psi(\bar{r}') = 0$. From (12.31) and (12.33) we get a_n and b_n in terms of $\psi(\bar{r}')$ on the surface. We expand $\psi(\bar{r}')$ in a complete set of functions $J_n(kr') \exp(jn\,\phi')$, and then we get

$$\psi(\bar{r}') = \sum_{n=-\infty}^{\infty} \alpha_n J_n(kr') e^{jn\phi'}, \tag{12.38}$$

where α_n are the unknown coefficients. Substituting this into (12.31) and (12.33), we get

$$[a] = [Q^-][\alpha],$$
$$[b] = -[Q^+][\alpha] = [T][a], \tag{12.39}$$
$$[T] = -[Q^+][Q^-]^{-1},$$

where

$$Q_{mn}^- = \frac{j}{4} \int_S \psi_m' \phi_n \, dS',$$
$$Q_{mn}^+ = \frac{j}{4} \int_S \psi_{rm}' \phi_n \, dS', \tag{12.40}$$

and

$$\psi_m' = \frac{\partial}{\partial n'}\left[H_m^{(2)}(kr') e^{-jm\phi'} \right] = \frac{\partial}{\partial n'} \psi_m,$$
$$\psi_{rm}' = \frac{\partial}{\partial n'}\left[J_m(kr') e^{-jm\phi'} \right] = \frac{\partial}{\partial n'} \psi_{rm}, \tag{12.41}$$
$$\phi_n = J_n(kr') e^{jn\phi'}.$$

For a two-media problem, where the wave numbers and the densities outside the surface S are k_0 and ρ_0 and those inside S are k_1 and ρ_1, we let ψ^+ and ψ^- denote the surface field just outside and inside the surface S (Fig. 12.4). The field inside S satisfies the wave equation with the wave number k_1 and is regular, and therefore it can be expanded in the following cylindrical harmonics:

$$\psi(\bar{r}) = \sum_{n=-\infty}^{\infty} \beta_n J_n(k_1 r) e^{jn\phi'}, \tag{12.42}$$

where β_n is the unknown coefficient. Using this, we can write the field just inside S as follows:

$$\psi^-(\bar{r}') = \sum_{n=-\infty}^{\infty} \beta_n J_n(k_1 r) e^{jn\phi'},$$
$$\frac{\partial \psi^-(\bar{r}')}{\partial n'} = \sum_{n=-\infty}^{\infty} \beta_n \frac{\partial}{\partial n'}\left[J_n(k_1 r') e^{jn\phi'} \right]. \tag{12.43}$$

Now the boundary conditions are

$$\psi^+(\bar{r}') = \frac{\rho_1}{\rho_0}\psi^-(\bar{r}'),$$

$$\frac{\partial\psi^+(\bar{r}')}{\partial n'} = \frac{\partial\psi^-(\bar{r}')}{\partial n'}.$$

The expressions for a_n in (12.31) and b_n in (12.33) contain the field just outside the surface S, and they are related to the field inside given by β_n.

$$[a] = [Q^-][\beta],$$

$$[b] = -[Q^+][\beta] = [T][a], \qquad (12.44)$$

$$[T] = -[Q^+][Q^-]^{-1}.$$

where

$$Q^-_{mn} = \frac{j}{4}\int_S\left(\frac{\rho_1}{\rho_0}\psi'_m\,\phi_n - \psi_m\,\phi'_n\right)dS',$$

$$Q^+_{mn} = \frac{j}{4}\int_S\left(\frac{\rho_1}{\rho_0}\psi'_{rm}\,\phi_n - \psi_{rm}\,\phi'_n\right)dS',$$

$$\psi_m = H_m^{(2)}(k_0\,r')e^{-jm\phi'},$$

$$\psi'_m = \frac{\partial}{\partial n'}\psi_m,$$

$$\psi_{rm} = J_m(k_0\,r')e^{-jm\phi'},$$

$$\psi'_{rm} = \frac{\partial}{\partial n'}\psi_{rm},$$

$$\phi_n = J_n(k_1\,r')e^{jn\phi'},$$

$$\phi'_n = \frac{\partial}{\partial n'}\phi_n.$$

The *T*-matrix method has been used successfully for scattering from a body of complex shapes such as particles with irregular shapes. Since it makes use of the expansion of the field in cylindrical harmonics (spherical harmonics for three-dimensional objects and Fourier expansion for periodic structures), the matrix may become ill-conditioned if the axial ratio (ratio of the major to minor axes) is much greater than about 5 or if the object has corners.

In this section, we illustrated the *T*-matrix method with a two-dimensional problem and cylindrical harmonics. The *T*-matrix method can be equally applicable to three-dimensional problems using spherical harmonics. For electromagnetic problems, the complete cylindrical or spherical expansion of the vector field must be used (Waterman, 1969).

12.5 SYMMETRY AND UNITARITY OF THE *T*-MATRIX AND THE SCATTERING MATRIX

The *T*-matrix satisfies a certain symmetric relationship because of the reciprocity principle. If a delta function source is located at $\bar{r}_1(r_1, \phi_1)$ and the scattered field $\psi_s(\bar{r}_2)$ is observed at $\bar{r}_2(r_2, \phi_2)$, and next a delta function source is located at $\bar{r}_2$ and the scattered field $\psi_s(\bar{r}_1)$ is observed at $\bar{r}_1$, then according to reciprocity,

$$\psi_s(\bar{r}_1) = \psi_s(\bar{r}_2). \tag{12.45}$$

For the first case, the incident wave is

$$\psi_i(\bar{r}) = -\frac{j}{4} \sum_n J_n(kr) H_n^{(2)}(kr_1) e^{+jn(\phi - \phi_1)}. \tag{12.46}$$

Therefore, we have

$$a_n = -\frac{j}{4} H_n^{(2)}(kr_1) e^{-jn\phi_1}. \tag{12.47}$$

The scattered field $\psi_s(\bar{r}_2)$ is then given by

$$\begin{aligned} \psi_s(\bar{r}_2) &= \sum_n b_n H_n^{(2)}(kr_2) e^{jn\phi_2} \\ &= \left(-\frac{j}{4}\right) \sum_n \sum_m T_{nm} H_m^{(2)}(kr_1) H_n^{(2)}(kr_2) e^{jn\phi_2 - jm\phi_1}. \end{aligned} \tag{12.48}$$

If we switch $\bar{r}_1$ and $\bar{r}_2$, we get

$$\psi_s(\bar{r}_1) = \left(-\frac{j}{4}\right) \sum_n \sum_m T_{nm} H_m^{(2)}(kr_2) H_n^{(2)}(kr_1) e^{jn\phi_1 - jm\phi_2}. \tag{12.49}$$

Now noting that $\psi_s(\bar{r}_1) = \psi_s(\bar{r}_s)$ and letting $m = -n'$ and $n = -m'$ in (12.49), and using $H_{-n}^{(2)} = (-1)^n H_n^{(2)}$, we get the following:

$$T_{mn} = (-1)^{m+n} T_{-n,-m} \tag{12.50}$$

This is the symmetry relationship satisfied by the *T*-matrix.

Let us consider the scattering matrix *S*. We note first that the incident wave is regular at the origin and is expressed in a series of regular cylindrical harmonics $J_n \exp(jn\phi)$. If we rewrite the incident wave using the identity

$$J_n(z) = \frac{1}{2}\left(H_n^{(1)}(z) + H_n^{(2)}(z)\right),$$

the part containing $H_n^{(2)}$ is the outgoing wave, while the part containing $H_n^{(1)}$ is the incoming wave.

$$\psi_i(\bar{r}) = \sum_n a_n J_n(kr) e^{jn\phi}$$

$$= \psi_i^-(\bar{r}) + \psi_i^+(\bar{r}),$$

$$\psi_i^-(\bar{r}) = \sum_n a_n \frac{1}{2} H_n^{(1)}(kr) e^{jn\phi}, \tag{12.51}$$

$$\psi_i^+(\bar{r}) = \sum_n a_n \frac{1}{2} H_n^{(2)}(kr) e^{jn\phi}.$$

Also we note that ψ_s is outgoing.

$$\psi_s(\bar{r}) = \sum_n b_n H_n^{(2)}(kr) e^{jn\phi}. \tag{12.52}$$

We can now regroup the wave as consisting of the incoming and the outgoing waves.

$$\psi_i^-(\bar{r}) = \sum_n \frac{a_n}{2} H_n^{(1)}(kr) e^{jn\phi} \qquad \text{incoming,}$$

$$\psi_i^+(\bar{r}) + \psi_s(\bar{r}) = \sum_n \left(\frac{a_n}{2} + b_n\right) H_n^{(2)}(kr) e^{jn\phi} \quad \text{outgoing.} \tag{12.53}$$

The scattering matrix S is defined by

$$\frac{a_n}{2} + b_n = \sum_m S_{nm} \frac{a_m}{2}$$

or

$$\tfrac{1}{2}[a] + [b] = [S]\tfrac{1}{2}[a]. \tag{12.54}$$

Since $[b] = [T][a]$, clearly $[S]$ is related to $[T]$.

$$[U] + 2[T] = [S], \tag{12.55}$$

where $[U]$ is a unit matrix.

Next, let us consider the conservation of power for a lossless object. If the object is lossless, the total incoming power should be equal to the total outgoing power. The total incoming power is given by

$$P_i = \int_0^{2\pi} d\phi \, |\psi_i^-(\bar{r})|^2. \tag{12.56}$$

If we evaluate this total power using the orthogonality of exp($jn\phi$), we get

$$P_i = \frac{\pi}{2} \sum_n |A_n|^2, \tag{12.57}$$

where $A_n = a_n H_n^{(1)}(kr)$. Similarly, for the outgoing power, we get

$$P_0 = \int_0^{2\pi} d\phi \left| \psi_i^+(\bar{r}) + \psi_s(\bar{r}) \right|^2$$
$$= \frac{\pi}{2} \sum_n |A_n + 2B_n|^2, \tag{12.58}$$

where $B_n = b_n H_n^{(1)}(kr)$ and we used the relation $|H_n^{(1)}(kr)| = |H_n^{(2)}(kr)|$.

We now express (12.57) and (12.58) in matrix form. We let A be the column matrix $[A_n]$. Then $P_i = (\pi/2)A^+ A$, where A^+ is the complex conjugate of the transpose of A, called the adjoint matrix (see Appendix 8.A). Similarly, (12.58) becomes

$$P_0 = \frac{\pi}{2}(A + 2TA)^+(A + 2TA).$$

Equating P_0 to P_i and writing them in matrix form, we get

$$A^+A = A^+S^+SA, \tag{12.59}$$

where we used (12.55). From (12.59) it is clear that the scattering matrix S is unitary for a lossless object.

$$S^+S = U. \tag{12.60}$$

12.6 *T*-MATRIX SOLUTION FOR SCATTERING FROM PERIODIC SINUSOIDAL SURFACES

In Section 7.5, we discussed the scattering of waves from periodic surfaces using the Rayleigh hypothesis. Even though that technique was limited to the case where the slope is less than 0.448, the solution is simple and has been used extensively for many problems in gratings and corrugated walls. More rigorous solutions can be obtained by using the *T*-matrix method.

Let us consider Dirichlet's problem for the surface defined by

$$\zeta = -h \cos\frac{2\pi x}{L}. \tag{12.61}$$

The incident wave ψ_i is given by

$$\psi_i = A_0 e^{+jqz-jbz}, \tag{12.62}$$

where $\beta = k \sin \theta_i$ and $q = k \cos \theta_i$ and θ_i is the angle of incidence. The scattered field ψ_s in the region $z > h$ is given by the space harmonics

$$\psi_s = \sum_n B_n e^{jq_n z - j\beta_n x}, \tag{12.63}$$

where

$$\beta_n = \beta + \frac{2n\pi}{L},$$

$$q_n = \begin{cases} \left(k^2 - \beta_n^2\right)^{1/2} & \text{if } k > |\beta_n| \\ -j\left(\beta_n^2 - k^2\right)^{1/2} & \text{if } k < |\beta_n| \end{cases}.$$

Now we use the extinction theorem (12.29) for the region $z < -h$ and use the periodic Green's function

$$G(\bar{r}, \bar{r}') = \sum_{n=-\infty}^{\infty} \frac{1}{2jq_n L} e^{-jq_n(z'-z)-j\beta_n(x-x')} \quad \text{for } z < z'. \tag{12.64}$$

We also let

$$\psi_i(\bar{r}) = \sum_n a_n e^{jq_n z - j\beta_n x}. \tag{12.65}$$

For our problem with (12.62), $a_0 = A_0$ and $a_n = 0$ for $n \neq 0$. However, we can formulate the problem for the more general case given by (12.65). We substitute (12.64) and (12.65) into (12.29) and note that $\psi(\bar{r}') = 0$ on the surface. We also express the surface field $(\partial/\partial n')\psi(\bar{r})$ by the following orthogonal space harmonics:

$$\frac{\partial}{\partial n'} \psi(\bar{r}')dS' = \sum_n \alpha_n e^{-j\beta_n x'} dx'. \tag{12.66}$$

Note that

$$\frac{\partial}{\partial n'} dS' = \left(\frac{\partial}{\partial z'} - \frac{\partial \zeta}{\partial x'} \frac{\partial}{\partial x'} \right) dx', \tag{12.67}$$

and therefore we used the expansion (12.66) rather than expanding $(\partial/\partial n')\psi(\bar{r}')$ in space harmonics.

Substituting (12.66) in (12.29), and letting $z' = \zeta = -h \cos(2\pi x/L)$, we get

$$[a] = [Q^-][\alpha], \tag{12.68}$$

where

$$Q^-_{mn} = \frac{1}{2jq_m} J_{|m-n|}(q_m h) e^{j(\pi/2)|m-n|}.$$

Next we use (12.63) in (12.32) with $\psi(\bar{r}') = 0$ on the surface, and the periodic Green's function

$$G(\bar{r}, \bar{r}') = \sum_n \frac{1}{2jq_n L} e^{-jq_n(z-z')-j\beta_n(x-x')} \quad \text{for } z > z'. \tag{12.69}$$

We then get

$$[b] = -[Q^+][\alpha], \tag{12.70}$$

where

$$Q^+_{mn} = \frac{1}{2jq_m} J_{|m-n|}(q_m h) e^{-j(\pi/2)|m-n|}.$$

Combining (12.68) and (12.70), we get

$$[b] = [T][a],$$
$$[T] = -[Q^+][Q^-]^{-1}. \tag{12.71}$$

The T-matrix solution in (12.71) should be exact and applicable for the periodic surface given by (12.61) even when the slope is higher than 0.448. In actual numerical calculations, however, the matrices become ill-conditioned if the slope becomes too high.

12.7 VOLUME INTEGRAL EQUATIONS FOR INHOMOGENEOUS BODIES: TM CASE

In the preceding sections 12.3 and 12.4, we discussed two techniques of solving the scattering problem for complex objects: the surface integral equation method and the T-matrix method. Both methods are applicable to Dirichlet's, Neumann's, and two-media problems. However, these two methods cannot be applied to the problem of scattering by inhomogeneous dielectric bodies or anisotropic bodies. The volume integral equations discussed in this section can be applied to these problems. It should be noted, however, that the surface integral equations and the T-matrix method deal

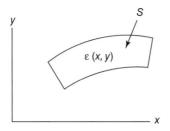

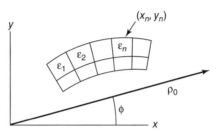

FIGURE 12.5   Scattering by an inhomogeneous body.

with the integral equations for the unknown "surface" field distribution, while the volume integral equations deal with the "volume" distribution of the unknown field, which requires larger matrices than the surface field for an object of similar size.

Let us consider a two-dimensional dielectric object (Fig. 12.5). The dielectric constant $\varepsilon(x, y)$ is a function of position. From Maxwell's equations, we note that the problem of dielectric material with $\varepsilon = \varepsilon_0 \varepsilon_r$ is equivalent to having the equivalent current source J_{eq}.

$$\nabla \times \bar{E} = -j\omega\mu_0 \bar{H},$$
$$\nabla \times \bar{H} = j\omega\varepsilon \bar{E} = j\omega\varepsilon_0 \bar{E} + \bar{J}_{eq}, \qquad (12.72)$$
$$J_{eq} = j\omega\varepsilon_0(\varepsilon_r - 1)\bar{E}.$$

The total field $\bar{E}$ consists of the incident field $\bar{E}_i$ and the field $\bar{E}_s$ produced by the equivalent current J_{eq}. Using the Hertz vector $\bar{\pi}$, we get (see Section 12.9 for three-dimensional objects)

$$\bar{E} = \bar{E}_i + \bar{E}_s,$$
$$\bar{E}_s = \nabla\nabla \cdot \bar{\pi} + k_0^2\bar{\pi}, \qquad (12.73)$$
$$\bar{\pi} = \int G(\bar{r}, \bar{r}') \frac{\bar{J}_{eq}(\bar{r}')}{j\omega\varepsilon_0} dV'.$$

Let us consider the two-dimensional TM waves ($\partial/\partial z = 0$ and $\bar{H}$ is transverse to the z axis) with the field components E_z, H_x, and H_y. The total field $\bar{E}$ consists of the incident field $\bar{E}_i$ and the scattered field $\bar{E}_s$ produced by $\bar{J}_{eq}$. For the TM wave, the Hertz vector $\bar{\pi}$ and the electric field $\bar{E}$ have only the z component, and therefore we get from (12.73)

$$E_z(\bar{r}) = E_{zi}(\bar{r}) + E_{zs}(\bar{r}),$$

$$E_{zs}(\bar{r}) = -j\omega\mu_0 \int_S G(\bar{r}, \bar{r}')\bar{J}_{eq}(\bar{r}')dS', \qquad (12.74)$$

where

$$G(\bar{r}, \bar{r}') = -j\frac{1}{4}H_0^{(2)}(k_0\,\rho), \quad \rho = |\bar{r} - \bar{r}'| \qquad (12.75)$$

and $dS' = dx'dy'$ and the integral is over the cross section S. Rewriting (12.74), we get

$$E_z(\bar{r}) + \frac{jk_0^2}{4} \int_S [\varepsilon_r(\bar{r}') - 1]E_z(\bar{r}')H_0^{(2)}(k_0\,\rho)\,dS' = E_{zi}(\bar{r}). \qquad (12.76)$$

Now we divide the cross section S into sufficiently small cells so that the electric field and the dielectric constant can be assumed to be constant over each cell (Fig. 12.5). We denote by E_n and ε_n the electric field and the dielectric constant of the nth cell. Evaluating (12.76) at the center of the mth cell, we get

$$\sum_{n=1}^{N} C_{mn} E_n = E_{mi}, \quad m = 1, 2, \ldots, N \qquad (12.77)$$

where

$$C_{mn} = \delta_{mn} + \frac{jk_0^2}{4}(\varepsilon_n - 1) \int_{\text{cell } n} H_0^{(2)}(k_0\rho)dS'.$$

δ_{mn} is called the *Kronecker delta* and is defined by

$$\delta_{mn} = \begin{cases} 1 & \text{if } m = n \\ 0 & \text{if } m \neq n \end{cases},$$

$$\rho = [(x_m - x')^2 + (y_m - y')^2]^{1/2},$$

(x_m, y_m) is the center of the cell m,

E_{mi} is the incident field at (x_m, y_m).

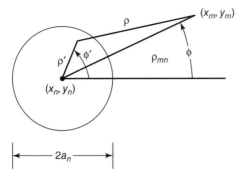

FIGURE 12.6   Equivalent circular cell.

The integration over each cell in C_{mn} can be performed numerically. However, since each cell may have a different shape, the calculation may become tedious. A simple approximate method of calculating C_{mn} was proposed by Richmond. For a sufficiently small cell, we can replace the cell by a circular cell with the same cross-sectional area. The radius of the nth equivalent circular cell is a_n. It is then possible to obtain a simple analytical expression for C_{mn}. For $m \neq n$, we use the following expansion of $H_0^{(2)}(k_0 \rho)$ valid for $\rho_{mn} > \rho'$ (Fig. 12.6):

$$H_0^{(2)}(k_0 \rho) = \sum_{n=-\infty}^{\infty} e^{-jn(\phi - \phi')} J_n(k_0 \rho') H_n^{(2)}(k_0 \rho_{mn}). \tag{12.78}$$

Noting that $dS' = \rho' \, d\rho' \, d\phi'$ and

$$\int J_0(x)x \, dx = x J_1(x), \tag{12.79}$$

we get

$$C_{mn} = \frac{j\pi k_0 a_n}{2}(\varepsilon_n - 1) J_1(k_0 a_n) H_0^{(2)}(k_0 \rho_{mn}), \tag{12.80}$$

where a_n is the radius of the equivalent circular cell. For $m = n$, we use

$$\int H_0^{(2)}(x)x \, dx = x H_1^{(2)}(x),$$

$$H_1^{(2)}(x) \approx j\frac{2}{\pi x} \quad \text{as } x \to 0.$$

We then get

$$C_{nn} = 1 + \frac{j\pi}{2}(\varepsilon_n - 1)\left[K_0 a_n H_0^{(2)}(k_0 a_n) - \frac{j2}{\pi} \right]. \tag{12.81}$$

Equations (12.80) and (12.81) give convenient approximate expressions for C_{mn}.

Rewriting (12.77) in matrix form and inverting the matrix, we get

$$[E] = [C]^{-1}[E_i], \tag{12.82}$$

where $[E] = [E_n]$ and $[E_i] = [E_{ni}]$ are $1 \times N$ column matrices and $[C] = [C_{mn}]$ is an $N \times N$ square matrix.

Let us next consider the scattered field at the observation point (x, y) far from the object. The scattered field E_s from the cell n is given by $-C_{mn}E_n$ as (x_m, y_m) is moved to (x, y). Thus we get

$$E_s(x, y) = -\sum_{n=1}^{N} C_{mn}E_n, \tag{12.83}$$

where $x_m = x$ and $y_m = y$, and C_{mn} is given by (12.80). In the far-field zone of the object, we get (Fig. 12.6)

$$\rho_{mn} = \left[(x - x_n)^2 + (y - y_n)^2\right]^{1/2}$$
$$\rightarrow \rho_0 - x_n \cos\phi - y_n \sin\phi,$$
$$H_0^{(2)}(k_0 \rho_{mn}) \rightarrow \left(\frac{2j}{\pi k_0 \rho_0}\right)^{1/2} e^{-jk_0 \rho_{mn}}, \tag{12.84}$$

where ρ_{mn} in the amplitude is replaced by ρ_0.

The scattering pattern of the object when illuminated by a plane wave can be expressed in terms of the *echo width* $W(\phi)$ (Harrington, 1961).

$$W(\phi) = \lim_{\rho_0 \to \infty} 2\pi\rho_0 \left|\frac{E_s}{E_i}\right|^2. \tag{12.85}$$

Using (12.84), we get the echo width in units of wavelength.

$$\frac{W(\phi)}{\lambda} = \left(\frac{\pi}{2}\right) \left|\sum_{n=1}^{N} (\varepsilon_n - 1)\frac{E_n}{|E_i|} k_0 a_n J_1(k_0 a_n) e^{jk_0(x_n \cos\phi + y_n \sin\phi)}\right|^2. \tag{12.86}$$

The cell size a_n in the foregoing method should not exceed $0.06/\sqrt{\varepsilon_r}$ wavelengths for accurate results, and this cell size and the computation time determine the size of the object that can be handled by this volume integral method.

12.8 VOLUME INTEGRAL EQUATIONS FOR INHOMOGENEOUS BODIES: TE CASE

For the TM case, we assumed that the medium is inhomogeneous, but isotropic, and therefore there is no coupling between E_z and the other components of the electric field. If the medium is anisotropic, the coupling among E_x, E_y, and E_z needs to be included. In this section, we discuss the TE case for an inhomogeneous and isotropic body.

The analysis for the TM case discussed in Section 12.7 can be extended to the TE case. For the TE case, however, there are couplings between E_x and E_y. Let the incident wave be

$$E_i = \hat{x}E_{ix} + \hat{y}E_{iy}. \tag{12.87}$$

The equivalent current $\bar{J}_{eq}$ is then given by

$$\begin{aligned} \bar{J}_{eq} &= j\omega\varepsilon_0(\varepsilon_r - 1)\bar{E} \\ &= \hat{x}J_x + \hat{y}J_y. \end{aligned} \tag{12.88}$$

From (12.73), we get

$$\bar{E}_s = \hat{x}E_{sx} + \hat{y}E_{sy},$$

$$E_{sx} = \left(\frac{\partial^2}{\partial x^2} + k_0^2\right)\pi_x + \frac{\partial^2}{\partial x \partial y}\pi_y,$$

$$E_{sy} = \frac{\partial^2}{\partial x \partial y}\pi_x + \left(\frac{\partial^2}{\partial y^2} + k_0^2\right)\pi_y,$$

$$\pi_x(\bar{r}) = \int G(\bar{r},\bar{r}')\frac{J_x}{j\omega\varepsilon_0}dS', \tag{12.89}$$

$$\pi_y(\bar{r}) = \int G(\bar{r},\bar{r}')\frac{J_y}{j\omega\varepsilon_0}dS'.$$

Now we follow the procedure for the TM case and obtain the integral equations for E_x and E_y.

$$\begin{aligned} E_x(\bar{r}) - E_{sx}(\bar{r}) &= E_{ix}(\bar{r}), \\ E_y(\bar{r}) - E_{sy}(\bar{r}) &= E_{iy}(\bar{r}), \end{aligned} \tag{12.90}$$

where E_{sx} and E_{sy} are given in (12.89) as integrals involving J_x and J_y which are related to E_x and E_y through (12.88).

Now we divide the cross section into N small cells and then approximate each cell by a circular cell of radius a_n with the same cross-sectional area as that of the original cell. We then assume that the field and the dielectric constant are constant over each cell. In the nth cell, we let E_{xn} and E_{yn} denote the electric fields E_x and E_y, respectively. The relative dielectric constant for the nth cell is ε_n and the incident field at the center $\bar{r}_n$ of the nth cell has the components E_{ixn} and E_{iyn}. We can then convert (12.90) into the following $2N$ linear equations:

$$\sum_{n=1}^{N}(A_{mn}E_{xn} + B_{mn}E_{yn}) = E_{ixm},$$

$$\sum_{n=1}^{N}(C_{mn}E_{xn} + D_{mn}E_{yn}) = E_{iym}. \tag{12.91}$$

To calculate these coefficients A_{mn}, B_{mn}, C_{mn}, and D_{mn}, we need to calculate E_s at the mth cell due to the constant current $\bar{J}$ in the nth cell. First let us examine E_{xm} due to the current J_{xn} and J_{yn} in the nth cell $(m \neq n)$.

$$J_{xn} = j\omega\varepsilon_0(\varepsilon_n - 1)E_{xn},$$

$$J_{yn} = j\omega\varepsilon_0(\varepsilon_n - 1)E_{yn}. \tag{12.92}$$

We make use of the expansion (12.78) and (12.89) and obtain

$$\begin{bmatrix} E_{sxm} \\ E_{sym} \end{bmatrix} = K \begin{bmatrix} h_{11} & h_{12} \\ h_{21} & h_{22} \end{bmatrix} \begin{bmatrix} J_{xn} \\ J_{yn} \end{bmatrix}, \tag{12.93}$$

where

$$h_{11} = \left[k_0\rho y^2 H_0^{(2)}(k_0\rho) + (x^2 - y^2)H_1^{(2)}(k_0\rho)\right],$$

$$h_{12} = h_{21} = xy\left[2H_1^{(2)}(k_0\rho) - k_0\rho H_0^{(2)}(k_0\rho)\right],$$

$$h_{22} = \left[k_0\rho x^2 H_0^{(2)}(k_0\rho) + (y^2 - x^2)H_1^{(2)}(k_0\rho)\right],$$

$$K = -\frac{\pi a_n J_1(k_0 a_n)}{2\omega\varepsilon_0\rho^3}.$$

Here we used

$$\rho = \left[(x_m - x_n)^2 + (y_m - y_n)^2\right]^{1/2},$$

$$x = x_m - x_n,$$

$$y = y_m - y_n.$$

To obtain E_{sx} and E_{sy} at the nth cell due to J_x and J_y in the same nth cell ($m = n$), we note that ρ is the distance between $\bar{r}$ and $\bar{r}'$ inside the same circular cell. Using (12.78) and (12.89) and performing integration inside the cell, we get

$$
\begin{aligned}
E_{sxn} &= h_0 J_{xn}, \\
E_{syn} &= h_0 J_{yn}, \\
h_0 &= -\frac{1}{4\omega\varepsilon_0}\left[\pi k_0 a_n H_1^{(2)}(k_0 a_n) - 4j\right].
\end{aligned}
\tag{12.94}
$$

Using (12.93) and (12.94), we finally get the coefficients in (12.91). For $m \neq n$,

$$
\begin{aligned}
A_{mn} &= K'h_{11}, \\
B_{mn} &= C_{mn} = K'h_{12}, \\
D_{mn} &= K'h_{22},
\end{aligned}
\tag{12.95}
$$

where

$$
K' = Kj\omega\varepsilon_0(\varepsilon_n - 1) = \frac{j\pi a_n J_1(k_0 a_n)(\varepsilon_n - 1)}{2\rho^3};
$$

h_{11}, h_{12}, h_{21}, and h_{22} are given in (12.93). For $m = n$, we have

$$
\begin{aligned}
A_{nn} &= D_{nn} = 1 - h_0 j\omega\varepsilon_0(\varepsilon_n - 1), \\
B_{nn} &= C_{nn} = 0,
\end{aligned}
\tag{12.96}
$$

where h_0 is as given in (12.94).

The scattered field for a plane incident wave can be obtained by following the procedure for the TM case. At a distance far from the object, the scattered field has only a ϕ component, which is produced by the ϕ component of the electric field in the body.

$$
E\phi = E_y \cos\phi - E_x \sin\phi.
\tag{12.97}
$$

We therefore get the echo width in units of wavelength.

$$
\frac{W(\phi)}{\lambda} = \left(\frac{\pi}{2}\right)\left|\sum_{n=1}^{N}(\varepsilon_n - 1)k_0 a_n J_1(k_0 a_n)\frac{E_{\phi n}}{|E_i|}e^{j\psi}\right|^2,
\tag{12.98}
$$

where

$$
\begin{aligned}
E_{\phi n} &= E_{yn}\cos\phi - E_{xn}\sin\phi, \\
\psi &= k(x_n \cos\phi + y_n \sin\phi).
\end{aligned}
$$

12.9 THREE-DIMENSIONAL DIELECTRIC BODIES

In Sections 12.7 and 12.8, we discussed the scattering by two-dimensional dielectric objects and formulated the integral equations for the electric field inside the body. This technique can be generalized to three-dimensional objects. However, careful attention must be paid to the singularity of Green's function, which has been studied extensively (Yaghjian, 1980; van Bladel, 1964; Tai, 1971).

Let us start with (12.72) and (12.73). For three-dimensional objects, these equations are valid if the observation point $\bar{r}$ is outside the medium. However, to construct an integral equation for $\bar{E}(\bar{r})$, the observation point $\bar{r}$ must be inside the medium. At $\bar{r}$ inside the medium, the total field $\bar{E}(\bar{r})$ is given by

$$\bar{E}(\bar{r}) = \bar{E}_i(\bar{r}) + \bar{E}_s(\bar{r}), \tag{12.99}$$

where $\bar{E}_i$ is the incident field and $\bar{E}_s$ is the scattered field. The scattered field $\bar{E}_s(\bar{r})$ is produced by the equivalent current $J_{eq}(\bar{r}')$ and the point $\bar{r}$ can coincide with $\bar{r}'$. To investigate the case where $\bar{r} \neq \bar{r}'$ and the case where $\bar{r}$ can coincide with $\bar{r}'$, we divide the volume of the dielectric medium into a small spherical volume V_δ centered at $\bar{r}$ and the remaining volume $V - V_\delta$. In the volume $V - V_\delta$, $\bar{r}'$ does not coincide with $\bar{r}$ and therefore we can use (12.73). In the volume V_δ, it has been shown that the electric field is equal to $-\bar{J}_{eq}/3j\omega\varepsilon_0$. Thus we write

$$\bar{E}_s(\bar{r}) = (-j\omega\mu_0)\lim_{\delta \to 0} \int_{V-V_\delta} \bar{\bar{G}}(\bar{r},\bar{r}')\bar{J}_{eq}(\bar{r}')dv' - \frac{\bar{J}_{eq}(\bar{r})}{3j\omega\varepsilon_0}, \tag{12.100}$$

where $\bar{\bar{G}}$ is the free-space electric dyadic Green's function given by

$$\bar{\bar{G}}(\bar{r},\bar{r}') = \frac{1}{k^2}\nabla \times \nabla \times (G_0\bar{\bar{I}})$$

$$= \left[\bar{\bar{I}} + \frac{\nabla\nabla}{k^2}\right]G_0(\bar{r},\bar{r}'), \tag{12.101}$$

$$\bar{G}_0(\bar{r},\bar{r}') = \frac{\exp[-jk|\bar{r}-\bar{r}'|]}{4\pi|\bar{r},\bar{r}'|}$$

$$\bar{\bar{I}} = \text{unit dyadic.}$$

In Cartesian coordinates, $\bar{\bar{G}}$ is given by

$$\bar{G}(\bar{r},\bar{r}') = \begin{bmatrix} 1 + \frac{1}{k^2}\frac{\partial^2}{\partial x^2} & \frac{1}{k^2}\frac{\partial^2}{\partial x\partial y} & \frac{1}{k^2}\frac{\partial^2}{\partial x\partial z} \\[2mm] \frac{1}{k^2}\frac{\partial^2}{\partial x\partial y} & 1 + \frac{1}{k^2}\frac{\partial^2}{\partial y^2} & \frac{1}{k^2}\frac{\partial^2}{\partial x\partial z} \\[2mm] \frac{1}{k^2}\frac{\partial^2}{\partial x\partial z} & \frac{1}{k^2}\frac{\partial^2}{\partial x\partial z} & 1 + \frac{1}{k^2}\frac{\partial^2}{\partial^2 z} \end{bmatrix} G_0(\bar{r},\bar{r}') \tag{12.102}$$

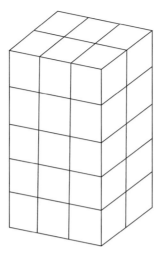

FIGURE 12.7   Dielectric body divided into many cubic cells.

Substituting (12.100) into (12.99), we get the integral equation for $\bar{E}(\bar{r})$.

$$\left[1 + \frac{\varepsilon_r(\bar{r}) - 1}{3}\right] \bar{E}(\bar{r}) - k^2 \fint \bar{\bar{G}}(\bar{r}, \bar{r}')[\varepsilon_r(\bar{r}') - 1]\bar{E}(\bar{r}')dv' = \bar{E}_i(\bar{r}), \quad (12.103)$$

where $\fint$ means $\lim\limits_{\delta \to 0} \int\limits_{V-V_\delta}$, and this can be approximated by taking a finite small volume V_δ.

In the above, we took a small spherical volume for V_δ. The formulations are also valid if V_δ is cubic (Livesay and Chen, 1974) (see Fig. 12.7). However, if V_δ has other shapes, the electric field at $\bar{r}$ is different and depends on the shape. The electric field due to the volume V_δ with ellipsoidal, right circular cylinder, rectangular parallelepiped, or pillbox has been calculated (Yaghjian, 1980).

12.10 ELECTROMAGNETIC APERTURE INTEGRAL EQUATIONS FOR A CONDUCTING SCREEN

In the preceding sections 12.7 to 12.9, we discussed scattering by conducting and dielectric bodies. In this section, we consider scattering and wave transmission through apertures on conducting screens. An example is electromagnetic penetration through gaps in and apertures on an enclosure, which affect the performance of electronic systems inside.

Let us consider a conducting screen S with the aperture A illuminated from the left. According to the equivalence theorem, this problem can be replaced by a closed

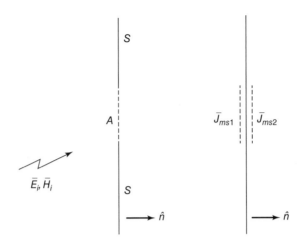

FIGURE 12.8   Aperture A on a conducting screen S.

screen with the equivalent magnetic current sources (Fig. 12.8).

$$\bar{J}_{ms2} = \bar{E} \times \hat{n},$$
$$\bar{J}_{ms1} = \bar{E} \times (-\hat{n}),$$

(12.104)

where $\bar{E}$ is the actual field on the aperture.

The magnetic field on the left side of the screen is equal to the incident field $\bar{H}_i$, the field reflected from the screen with the aperture closed $\bar{H}_r$, and the contributions from $\bar{J}_{ms1}$. The contribution from $\bar{J}_{ms1}$ in front of the screen is the same as the contribution from $\bar{J}_{ms1}$ and that of its image in the absence of the screen. The image of $\bar{J}_{ms1}$ is in the same direction as $\bar{J}_{ms1}$ and $\bar{J}_{ms1}$ and its image is equal to $2\bar{J}_{ms1}$. Therefore, using the Franz formula (Section 6.9) or the magnetic Hertz vector in Section 2.9, we get

$$\bar{H}_1(\bar{r}) = \bar{H}_i(\bar{r}) + \bar{H}_r(\bar{r}) + [\nabla \nabla \cdot + k^2] \bar{\pi}_{m1}(\bar{r}),$$

where

$$\bar{\pi}_{m1}(\bar{r}) = \int_A G(\bar{r} - \bar{r}') \frac{2\bar{J}_{ms1}(\bar{r}')}{j\omega\mu} dS',$$
$$G(\bar{r}, \bar{r}') = \frac{e^{-jk|\bar{r} - \bar{r}'|}}{4\pi |\bar{r} - \bar{r}'|}.$$

(12.105)

The field on the right side of the screen is similarly given by

$$\bar{H}_2(\bar{r}) = [\nabla \nabla \cdot + k^2] \bar{\pi}_{m2},$$

(12.106)

where

$$\bar{\pi}_{m2}(\bar{r}) = \int_A G(\bar{r}, \bar{r}') \frac{2\bar{j}_{ms2}}{j\omega\mu} dS'.$$

Now the boundary condition on the aperture is the continuity of the tangential electric field and the tangential magnetic field. The continuity of the tangential electric field was used in (12.104). The continuity of the magnetic field is

$$\hat{n} \times \bar{H}_1 = \hat{n} \times \bar{H}_2 \quad \text{on } A. \tag{12.107}$$

Substituting (12.105) and (12.106) into (12.107), we get the desired surface integral equation for $\bar{j}_{ms1}$.

$$\hat{n} \times [(\nabla\nabla \cdot + k^2)\bar{\pi}_m(\bar{r}) + \bar{H}_i(\bar{r})] = 0 \quad \text{on } A, \tag{12.108}$$

where

$$\bar{\pi}_m(\bar{r}) \int_A G(\bar{r}, \bar{r}') \frac{2j_{ms1}(\bar{r}')}{j\omega\mu} dS',$$

and $\hat{n} \times [\bar{H}_i + \bar{H}_r] = 2\hat{n} \times \bar{H}_i$ on A. This is the basic integral equation for the unknown magnetic current $\bar{J}_{ms1}$ in the aperture, and in general, it must be solved numerically using the moment method or other numerical techniques. The magnetic current $\bar{J}_{ms1} = \bar{\bar{E}} \times (-\hat{n})$ must also satisfy the edge condition on the rim of the aperture.

On the aperture, the tangential magnetic field and the normal electric field satisfy the following simple relationships.

1. The tangential magnetic field is identical to the tangential incident field.

$$\hat{n} \times \bar{H}_1 = \hat{n} \times \bar{H}_2 = \hat{n} \times \bar{H}_i \quad \text{on } A. \tag{12.109}$$

To derive these, we use (12.108) and (12.109) and write

$$\hat{n} \times \bar{H}_1(\bar{r}) = 2\hat{n} \times \bar{H}_i(\bar{r}) - \hat{n} \times \bar{H}_i(\bar{r})$$
$$= \hat{n} \times \bar{H}_i(\bar{r}) \quad \text{on } A. \tag{12.110}$$

2. The normal component of the electric field is the same as that of the incident field.

$$\hat{n} \cdot E_1 = \hat{n} \cdot \bar{E}_2 = \hat{n} \cdot \bar{E}_i \quad \text{on } A. \tag{12.111}$$

To show this, consider the electric field. The electric field $\bar{E}_1$ on the left side of the screen is given by

$$\bar{E}_1(\bar{r}) = \bar{E}_i(\bar{r}) + \bar{E}_r(\bar{r}) - j\omega\mu\nabla \times \bar{\pi}_{m1}(\bar{r}). \tag{12.112}$$

Similarly, the electric field $\bar{E}_2$ on the right side is

$$\bar{E}_2(\bar{r}) = -j\omega\mu\nabla \times \bar{\pi}_{m2}(\bar{r}). \tag{12.113}$$

Noting that $\bar{\pi}_{m1} = -\bar{\pi}_{m2}$, the normal components of $\bar{E}_1$ and $\bar{E}_2$ are given by

$$\begin{aligned}
\hat{n} \cdot \bar{E}_1 &= \hat{n} \cdot (\bar{E}_i + \bar{E}_r) + \hat{n} \cdot (-j\omega\mu\nabla \times \bar{\pi}_m), \\
\hat{n} \cdot \bar{E}_2 &= \hat{n} \cdot (j\omega\mu\nabla \times \pi_m).
\end{aligned} \tag{12.114}$$

The normal components of $\bar{E}$ must also be continuous on A, and therefore we get

$$\hat{n} \cdot \bar{E}_1 = \hat{n} \cdot \bar{E}_2 = \frac{1}{2}\hat{n} \cdot (\bar{E}_i + \bar{E}_r). \tag{12.115}$$

However, the normal component of $\bar{E}_i$ and $\bar{E}_r$ when the aperture is closed is twice the normal component of $\bar{E}_i$, proving (12.111).

12.11 SMALL APERTURES

If the aperture size is small compared with a wavelength, we can express the effects of the aperture field in terms of the equivalent magnetic and electric dipoles (Butler et al., 1978). Let us consider the electric field $\bar{E}_2$ (12.113) at a distance $R \gg \lambda$ in the direction of a unit vector $\hat{o}$ in the region $z > 0$ (Fig. 12.9). Green's function $G(\bar{r}, \bar{r}')$ can then be approximated by

$$G(\bar{r}, \bar{r}') = \frac{1}{4\pi R}\exp(-jkR + jk\bar{r}' \cdot \hat{o}), \tag{12.116}$$

where $|\bar{r} - \bar{r}'| \simeq R - \bar{r}' \cdot \hat{o}$. Using this in (12.113) and noting that $\nabla = -jk\hat{o}$, we get

$$\bar{E}_2(\bar{r}) = j2k\frac{e^{-jkR}}{4\pi R}\hat{o} \times \int_A \bar{J}_{ms2}(\bar{r}')e^{jk\bar{r}' \cdot \hat{o}}dS'. \tag{12.117}$$

Since the aperture size is much smaller than a wavelength, we expand $\exp(jkr' \cdot \hat{o})$ in Taylor's series and write

$$\bar{E}_2(\bar{r}) = \sum_{n=-\infty}^{\infty} \bar{E}_{2n}(\bar{r}). \tag{12.118}$$

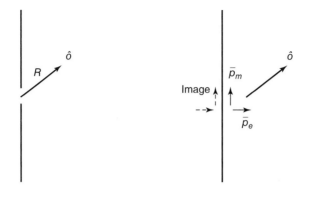

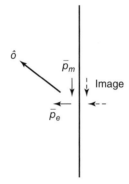

FIGURE 12.9   Small apertures.

It is possible to show that the term for $n = 0$ is identical to the field created by the magnetic dipole $\bar{p}_m$ in the presence of the screen, which is the sum of the contribution by $\bar{p}_m$ and its image $\bar{p}_m$ (Fig. 12.9).

$$\bar{E}_{20}(\bar{r}) = -\omega\mu k \frac{e^{-jkR}}{4\pi R}\hat{o} \times (2\bar{p}_m), \tag{12.119}$$

where

$$2\bar{p}_m = \frac{2}{j\omega\mu} \int \bar{J}_{ms2}(\bar{r}')\, dS'.$$

It can also be shown that the magnetic field and $\bar{p}_m$ have dimensions of (volume) × (magnetic field). The magnetic moment $\bar{p}_m$ is proportional to the tangential component of the magnetic field $\bar{H}_{sc}$ at the aperture when the aperture is closed (short circuited) and can be expressed by

$$\bar{p}_m = -\bar{\bar{\alpha}}_m \cdot \bar{H}_{sc}, \quad \text{for the region } z > 0 \tag{12.120}$$

or in matrix form,

$$
\begin{bmatrix} P_{mx} \\ P_{my} \end{bmatrix} = - \begin{bmatrix} \alpha_{mxx} & \alpha_{mxy} \\ \alpha_{myx} & \alpha_{myy} \end{bmatrix} \begin{bmatrix} H_{scx} \\ H_{scy} \end{bmatrix},
$$

where $\bar{\bar{\alpha}}_m$ is called the *magnetic polarizability*. The value of $\bar{\bar{\alpha}}_m$ for a circular aperture of radius a is given by

$$
\alpha_{mxx} = \alpha_{myy} = \tfrac{4}{3}a^3,
$$
$$
\alpha_{mxy} = \alpha_{myx} = 0.
$$

(12.121)

Similarly, noting (12.112) and (12.113), for the region $z < 0$, the equivalent magnetic moment $\bar{p}_m$ is given by

$$
\bar{p}_m = +\bar{\bar{\alpha}}_m \cdot \bar{H}_{sc}.
$$

(12.122)

Detailed derivations of the results above using the magnetic polarizability for elliptic apertures are given in Butler et al. (1978).

The next term for $n = 1$ in (12.118) gives the field due to an equivalent electric dipole and quadrupole moments.

$$
\bar{E}_{21}(\bar{r}) = \bar{E}_{21d}(\bar{r}) + \bar{E}_{21q}(\bar{r}),
$$
$$
\bar{E}_{21d} = -\frac{k^2}{\varepsilon}\frac{e^{-jkR}}{4\pi R}\hat{o} \times [\hat{o} \times (2\bar{p}_e)],
$$
$$
\bar{E}_{21q} = \text{quadrupole},
$$

(12.123)

where

$$
\bar{p}_e = -\frac{\varepsilon}{2} \int \bar{r}' \times \bar{J}_{ms2}(\bar{r}')dS'.
$$

$\bar{E}_{21d}$ above is identical to the field created by the electric dipole moment $\bar{p}_e$ in the presence of the screen, which is equal to the sum of the contributions by $\bar{p}_e$ and its image $\bar{p}_e$ (Fig. 12.9). The dipole moment $\bar{p}_e$ is pointed in the z direction and is proportional to the normal component of the electric field E_{scz} at the aperture when the aperture is closed.

$$
\bar{p}_e = p_e\hat{z} = \varepsilon\alpha_e E_{scz}\hat{z},
$$

(12.124)

where α_e is called the *electric polarizability*. The value of α_e for a circular aperture of radius a is given by

$$
\alpha_e = \tfrac{2}{3}a^3.
$$

(12.125)

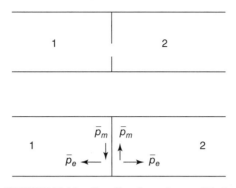

FIGURE 12.10   Coupling through a small hole.

For the region $z < 0$, the dipole moment $\bar{p}_e$ is given by

$$\bar{p}_e = -\varepsilon\alpha_e E_{scz}\hat{z}. \tag{12.126}$$

Limiting ourselves to the lowest orders of scattering, we can conclude that a small aperture is equivalent to the magnetic and electric dipoles given in (12.120), (12.122), (12.124), and (12.126).

In the above, we discussed the small aperture on a plane screen. This can be generalized to the coupling of electromagnetic waves between waveguides or cavities through a small aperture from region 1 to region 2. Let $(\bar{E}_s, \bar{H}_s)$ be the field in region 1 when the aperture is closed (short circuited), and let E_{sz} be the normal component of $\bar{E}_s$ at the aperture and $\bar{H}_{st}$ be the tangential component of $\bar{H}_s$ at the aperture (Fig. 12.10). Then when the aperture is open, the field in region 1 is the sum of $(\bar{E}_s, \bar{H}_s)$ and the field produced by $\bar{p}_e$ and $\bar{p}_m$ placed on the closed aperture in region 1 (Fig. 12.10).

$$\bar{p}_e = -\varepsilon\alpha_e E_{sz}\hat{z},$$
$$\bar{p}_m = \bar{\bar{\alpha}}_m \cdot \bar{H}_{st}. \tag{12.127}$$

The field in region 2 is then created by the following $\bar{p}_e$ and $\bar{p}_m$ placed on the closed aperture in region 2:

$$\bar{p}_e = \varepsilon\alpha_e E_{sz}\hat{z},$$
$$\bar{p}_m = -\bar{\bar{\alpha}}_m \cdot \bar{H}_{st}. \tag{12.128}$$

12.12 BABINET'S PRINCIPLE AND SLOT AND WIRE ANTENNAS

Consider a slot on a conducting screen and its complementary problem where the slot is replaced by a conducting piece and the screen becomes an aperture. Because

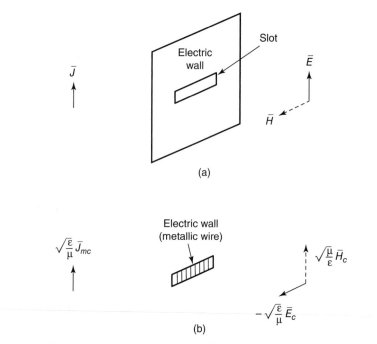

(a)

(b)

FIGURE 12.11   Babinet's principle applied to (a) a slot and (b) a wire.

of the symmetry of these two complementary problems, we expect that the solution to one can be used to solve the other. This equivalent relationship is called *Babinet's principle*.

Consider the two problems shown in Fig. 12.11. In Fig. 12.11a, the conducting screen with a slot is illuminated by an electric current $\bar{J}$, and $(\bar{E}_1, \bar{H}_1)$ is the field diffracted by the slot. Now consider the complementary problem shown in Fig. 12.11b, where the metallic wire is excited by the source current $(\varepsilon/\mu)^{1/2}\bar{J}_{mc}$. The field scattered by the wire is $(\bar{E}_{cs}, \bar{H}_{cs})$. Then Babinet's principle states that

$$\bar{E}_1 = -\sqrt{\frac{\mu}{\varepsilon}}\bar{H}_{cs},$$

$$\bar{H}_1 = \sqrt{\frac{\varepsilon}{\mu}}\bar{E}_{cs}.$$

(12.129)

The proof of (12.129) can be done in two stages. First, we can consider the three cases shown in Fig. 12.12. In Fig. 12.12a, we have a source $\bar{J}$ and a perfectly conducting screen (electric wall on which the tangential electric field is zero) with a slot and let the fields behind the screen be $\bar{E}_1$ and $\bar{H}_1$ (Fig. 12.12a). Next, we consider a complementary problem where the screen and the slot are interchanged and the electric wall is replaced by a magnetic wall (tangential magnetic field is zero). Let the fields at the same position in this case be $\bar{E}_2$ and $\bar{H}_2$ (Fig. 12.12b).

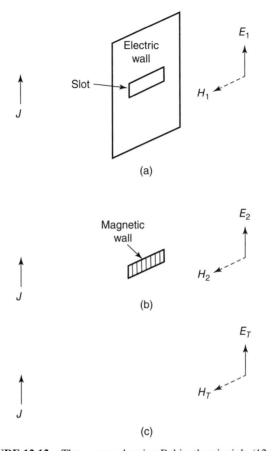

FIGURE 12.12   Three cases showing Babinet's principle (12.130).

Third, consider the fields in the absence of the screen and let the fields at the same position in this case be $\bar{E}_T$ and $\bar{H}_T$ (Fig. 12.12c). Then Babinet's principle states that

$$\bar{E}_1 + \bar{E}_2 = \bar{E}_T,$$
$$\bar{H}_1 + \bar{H}_2 = \bar{H}_T. \tag{12.130}$$

To prove this, we note from (12.109) that the tangential magnetic field on the slot in Fig. 12.12a is equal to the tangential magnetic field of the incident field. The tangential electric field E_1 is zero on S_1. Thus for Fig. 12.12a, the total tangential electric and magnetic fields (E_1, H_1) are

$$E_1 = E_T = 0 \quad \text{on screen } (S_1),$$
$$H_1 = H_T \quad \text{on slot } (S_2). \tag{12.131}$$

For the magnetic screen of Fig. 12.12b,

$$E_2 = E_T \qquad \text{on aperture } (S_1),$$
$$H_2 = H_T = 0 \quad \text{on magnetic wall } (S_2).$$

(12.132)

Adding the above two fields, we get

$$E_1 + E_2 = E_T \quad \text{on } S_1,$$
$$H_1 + H_2 = H_T \quad \text{on } S_2.$$

(12.133)

We now note that from the uniqueness theorem, E_1 on S_1 and H_1 on S_2 uniquely determine all fields everywhere. Similarly, E_2 on S_1 and H_2 on S_2 uniquely determine all fields everywhere. But the superposition of (E_1, H_1) and (E_2, H_2) gives the field E_T on S_1 and H_T on S_2, which by the uniqueness theorem should for $z > 0$ give fields everywhere identical to the incident field. Thus this proves Babinet's principle (12.130).

In the above, we stated Babinet's principle using a conducting screen with an aperture and a disk with a magnetic wall. More practical and useful is Babinet's principle applied to a slot on a conducting screen and a wire. This can be obtained by noting the duality of Maxwell's equations and their invariance when all the quantities are replaced by the following with the subscript c.

$$E \rightarrow \sqrt{\frac{\mu}{\varepsilon}} H_c,$$

$$H \rightarrow -\sqrt{\frac{\varepsilon}{\mu}} E_c,$$

$$J_m \rightarrow -\sqrt{-\frac{\mu}{\varepsilon}} J_c,$$

(12.134)

$$J \rightarrow \sqrt{\frac{\varepsilon}{\mu}} J_{mc},$$

$$\rho_m \rightarrow -\sqrt{\frac{\mu}{\varepsilon}} \rho_c,$$

$$\rho \rightarrow \sqrt{\frac{\varepsilon}{\mu}} \rho_{mc}.$$

Furthermore, if the electric and magnetic walls are interchanged, then we have identical boundary conditions, and the solution should be the same. For example, E and H in Fig. 12.13a are identical to $\sqrt{\mu/\varepsilon} H_c$ and $-\sqrt{\varepsilon/\mu} E_c$ in Fig. 12.13b. Using this

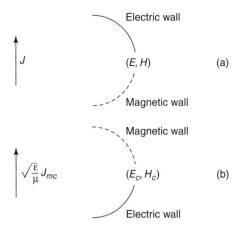

FIGURE 12.13   Duality principle.

duality, we see that the situation in Fig. 12.11a is identical to that in Fig. 12.11b. Thus (12.133) can be written as

$$E_1 + \sqrt{\frac{\mu}{\varepsilon}}H_c = E_T,$$

$$H_1 - \sqrt{\frac{\varepsilon}{\mu}}E_c = H_T.$$

(12.135)

If we write $\bar{H}_c$ and $\bar{E}_c$ as the sum of the incident and scattered waves, we have

$$\sqrt{\frac{\mu}{\varepsilon}}H_c = \sqrt{\frac{\mu}{\varepsilon}}H_{cT} + \sqrt{\frac{\mu}{\varepsilon}}H_{cs}$$

$$= E_T + \sqrt{\frac{\mu}{\varepsilon}}H_{cs},$$

$$-\sqrt{\frac{\varepsilon}{\mu}}E_c = -\sqrt{\frac{\varepsilon}{\mu}}E_{cT} - \sqrt{\frac{\varepsilon}{\mu}}E_{cs}$$

$$= H_T - \sqrt{\frac{\varepsilon}{\mu}}E_{cs},$$

(12.136)

where H_{cs} and E_{cs} are the scattered fields produced by the electric current on the wire. Then we can rewrite (12.135) as (12.129), proving Babinet's principle.

Using Babinet's principle, we can show the relationship between the impedance of a slot antenna and the impedance of its complementary wire antenna. Consider the slot and its complementary wire antenna shown in Fig. 12.14. The fields for the slot

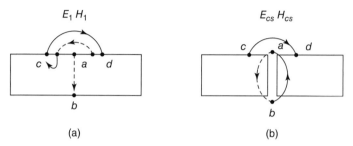

FIGURE 12.14   (a) Slot and (b) wire antennas.

and for the wire antenna are $E_1 H_1$ and E_{cs} and H_{cs}, respectively, and are related by (12.129). For the slot antenna, the voltage V_s and the current I_s are given by

$$V_s = \int_a^b E_1 \cdot dl,$$

$$I_s = 2 \int_c^d H_1 \cdot dl.$$

(12.137)

The current I_s is given by $\int_c^d H_1 \cdot dl$ on one side of the screen plus $\int_d^c H_1 \cdot dl$ on the other side. For the wire antenna, the voltage V_w and the current I_w are given by

$$V_w = \int_c^d E_{cs} \cdot dl,$$

$$I_w = 2 \int_b^a H_{cs} \cdot dl.$$

(12.138)

Points c and d above are taken infinitesimally close to a. Using (12.129) and (12.137), we write (12.138) as

$$V_w = \eta \int_c^d H_1 \cdot dl = \frac{\eta I_s}{2},$$

$$I_w = \frac{2}{\eta} \int_a^b E_1 \cdot dl = \frac{2}{\eta} V_s,$$

(12.139)

and thus the wire impedance $Z_w = (V_w/I_w)$ and the slot impedance $Z_s = (V_s/I_s)$ are related by

$$Z_w Z_s = \frac{\eta^2}{4}, \qquad \eta = \sqrt{\mu/\varepsilon}.$$

(12.140)

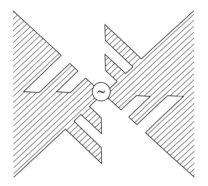

FIGURE 12.15   Self-complementary antenna.

This important relationship is called *Booker's relation* and it shows that if the slot impedance (or wire antenna impedance) is known, the wire antenna impedance (or slot impedance) can be found from (12.140).

Mushiake pointed out that as a consequence of Booker's relation, if the slot and the wire have the same shape (Fig. 12.15), the antenna is self-complementary, and the impedance is invariant and equal to $\eta/2$ independent of the shape of the antenna and frequency

$$Z = \frac{\eta}{2}. \tag{12.141}$$

This is called the *Mushiake relation* and is obviously related to the concept of frequency-independent antennas (Rumsey, 1966).

12.13   ELECTROMAGNETIC DIFFRACTION BY SLITS AND RIBBONS

As example of the electromagnetic scattering by apertures on a conducting screen discussed in Section 12.10, we consider a simpler two-dimensional problem $\partial/\partial y = 0$ of diffraction by a slit on a conducting screen (Fig. 12.16). We start with the general

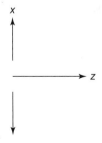

FIGURE 12.16   Diffraction by a slit.

integral equation (12.108). Let us consider the TE ($\bar{E} \perp \hat{y}$) case. $\bar{J}_{ms1} = \bar{E} \times (-\hat{z}) = E_x \hat{y}$. The incident wave with $\bar{E}_i = E_i \hat{x}$ and $\bar{H}_i = H_i \hat{y}$ is incident on the screen. Then (12.108) becomes

$$2j\omega\varepsilon_0 \int_{-a}^{a} G(x, x') E_x(x') \, dx' = H_i, \tag{12.142}$$

where $G(x, x') = -(j/4)H_0^{(2)}(k|x - x'|)$.

This can be solved numerically using the method of moments. If the wave is normally incident, $H_i = A \exp(-jkz)$ and the solution has been obtained for a narrow slit.

$$j\omega\varepsilon_0 E_x(x') = C_1 \left[1 - \left(\frac{x'}{a}\right)^2\right]^{-1/2}, \tag{12.143}$$

$$C_1 = \frac{-A}{a(\ln ka + \ln(\gamma/4) + j\pi/2)}, \tag{12.144}$$

where $\gamma = 1.78107$ (Euler's constant) and $\ln \gamma = 0.5772$. Note that E_x satisfies the edge condition (see Appendix 7.C).

The far field on the right side of the screen is given by

$$H_y = -\frac{j}{2} a\pi C_1 \left(\frac{2}{\pi kr}\right)^{1/2} e^{-jkr + j\pi/4}. \tag{12.145}$$

For TM($\bar{H} \perp \hat{y}$), $\bar{J}_{ms1} = -E_y \hat{x}$, and we get

$$\left(\frac{\partial^2}{\partial x^2} + k^2\right) \int_{-a}^{a} G(x, x') \frac{2E_y(x')}{j\omega\mu} dx' = H_{ix}. \tag{12.146}$$

This can be solved numerically. However, for a narrow slit, when the wave is normally incident, $E_{iy} = A \exp(-jkz)$ and the aperture field is given by

$$E_y C_2 \left[1 - \left(\frac{x'}{a}\right)^2\right]^{1/2}, \quad C_2 = jka \, A. \tag{12.147}$$

The far field on the right side of the screen is given by

$$E_y = \frac{\pi ka C_2}{4} \left(\frac{2}{\pi kr}\right)^{1/2} \frac{z}{r} e^{-jkr + j(\pi/4)}. \tag{12.148}$$

The results above can be used to find the diffraction by ribbons using Babinet's principle.

12.14   RELATED PROBLEMS

Transient phenomena can be treated in the frequency domain and then their Fourier transform yields the time-domain solution. Alternatively, the transient can be investigated in the time domain using time stepping (Felsen, 1976). The finite-difference time-domain method (FDTD) is a useful numerical technique in a time-domain approach (Yee, 1966). Time-domain solutions can be expressed in a series of complex exponentials that correspond to the singularities in the Laplace transform. This was proposed by Baum in 1971 and is called the singularity expansion method (SEM) (see Baum, 1976; Tesche, 1973; Felsen, 1976; Uslenghi, 1978).

There are other numerical techniques to deal with scattering by complex bodies. Yasuura's method (Ikuno and Yasuura, 1978; Yasuura and Okuno, 1982), introduced in late 1960s, makes use of the smoothing process on the mode-matching method. This is shown to be a powerful numerical technique with high accuracy and efficiency.

PROBLEMS

12.1 Consider a two-dimensional Dirichlet's problem for the sinusoidal surface defined by (12.61) illuminated by a plane wave (12.62). Find an integral equation similar to (12.3) and (12.5). Find the scattered wave using the Kirchhoff approximation.

12.2 For the sinusoidal surface of Problem 12.1, find the integral equations for the two-media problem.

12.3 Consider a knife edge illuminated by a plane wave as shown in Fig. 11.18. For TM wave with the incident wave E_{zi} given in (11.122), find EFIE and MFIE for the surface current. Also for TE wave with (11.125), find EFIE and MFIE for the surface current.

12.4 Apply the T-matrix method to find the scattered wave from a conducting ellipsoidal cylinder whose surface is given by

$$\frac{x^2}{a^2} + \frac{y^2}{b^2} = 1.$$

The incident wave is a plane wave polarized in the z direction.

12.5 Consider the two-dimensional problem shown in Fig. 12.5. If the object has a square cross section of $a \times a$, use four cells as shown in Fig. P12.5, and obtain the echo width.

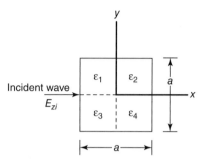

FIGURE P12.5   Scattering by a dielectric object. $a = 0.1\,\lambda_0$, $\varepsilon_1 = 1.1, \varepsilon_2 = 1.2, \varepsilon_3 = 1.3$, and $\varepsilon_4 = 1.4$.

12.6   A linearly polarized electromagnetic wave with power density P_0 at 10 GHz is normally incident on a small circular aperture of radius 2 mm in a conducting screen. Find the total power transmitted through the aperture.

12.7   A conducting screen has a slit of width 1 mm. The screen is normally illuminated by a TE electromagnetic wave with power density P_0 at 10 GHz. Find the power transmitted through a unit length of the slit. Do this with a TM incident wave.

CHAPTER 13

GEOMETRIC THEORY OF DIFFRACTION AND LOW-FREQUENCY TECHNIQUES

In Chapter 12, we discussed the integral equation method for dealing with scattering and diffraction from objects. It should be noted that even though the integral equation itself is exact, the actual solution usually requires extensive numerical and matrix calculations, such as the moment method and the T-matrix method. It is clear, then, that the size of the matrix depends on the size of the object. In fact, in many applications, approximately 10 points per wavelength may often be needed to keep the error within a few percent; therefore, the exact integral equation method becomes impractical for a large object. On the other hand, if the object size is much smaller than a wavelength, the solution should approach a static case. In this chapter, we examine the high-frequency technique, which is applicable to objects much greater than a wavelength, and the low-frequency technique, which is applicable to object sizes much smaller than a wavelength. For object size close to a wavelength, called the *resonance region*, the method of moment and other numerical techniques can be used effectively.

Extensive literature is available for the geometric theory of diffraction (GTD) and its related subjects. Important papers on GTD have been collected in the IEEE Press Reprint Series (Hansen, 1981). It is also discussed extensively in recent books by James (1976) and Jull (1981). Uniform geometric theory of diffraction (UTD) is covered in Kouyoumjian and Pathak (1974), and uniform asymptotic theory of edge diffraction (UAT) is discussed in Deschamps et al. (1984) and Lee (1977).

Electromagnetic Wave Propagation, Radiation, and Scattering: From Fundamentals to Applications,
Second Edition. Akira Ishimaru.
© 2017 by The Institute of Electrical and Electronic Engineers, Inc. Published 2017 by John Wiley & Sons, Inc.

13.1 GEOMETRIC THEORY OF DIFFRACTION

As the wavelength λ approaches zero, the field can be described by *geometric optics*, but it contains no diffraction effect. An important extension of geometric optics to include diffraction was proposed by J. Keller and it is called the *geometric theory of diffraction* or GTD (Keller, 1962). It introduces *diffracted rays* in addition to the usual geometric optical rays. Unlike geometric optics, the diffracted rays can enter the shadow regions. The GTD deals with the diffracted ray originating from edges, corners (vertices), and curved surfaces and is based on the following postulates:

1. Fermat's principle can be generalized and is applicable to diffracted rays, and thus the diffracted ray follows a curve that has a stationary optical path among all the paths between two points.
2. Diffraction is a local phenomenon at high frequencies, and thus the magnitude of the diffracted ray depends on the nature of the incident wave and on the boundary in the neighborhood of the point of diffraction.
3. The phase of the diffracted ray is proportional to the optical length of the ray, and the amplitude varies to conserve the power in a narrow tube of rays.

According to the postulates above, the diffracted rays are proportional to the product of the incident wave and the *diffraction coefficient*, in analogy to the geometric optical rays reflected from the surface, where the reflected ray is proportional to the incident wave and the *reflection coefficient*. In general, the diffraction coefficient is proportional to $\lambda^{1/2}$ for edges, to λ for vertices, and decreases exponentially with λ^{-1} for surfaces. Therefore, the edge diffraction is the strongest, the corner diffraction is weaker, and the surface diffraction is the weakest. The diffraction coefficient is obtained by considering the asymptotic form of the exact solutions of the simpler *canonical* problem. For example, the edge diffraction coefficient is determined by the asymptotic form of the exact solution for an infinite wedge.

Let us illustrate the edge diffraction by considering a two-dimensional perfectly conducting knife edge (Fig. 13.1). We first consider the TM case (E_z, H_x, H_y).

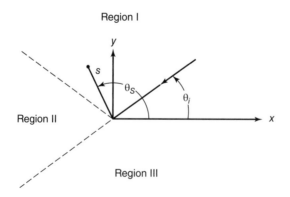

FIGURE 13.1   Edge diffraction.

The magnetic field is transverse to the z axis. The field component E_z satisfies the wave equation and Dirichlet's boundary condition (soft surface) ($E_z = 0$) on the conductor ($x > 0$, $y = 0$). This is equivalent to a scalar problem for the scalar field $\psi = E_z$ satisfying the wave equation and Dirichlet's condition $\psi = 0$ on the surface.

Let A_i be the incident wave ψ_i at the edge. Then according to the GTD, the diffracted ray ψ_d at (s, θ_s) is given by

$$\psi_d(s, \theta_s) = A_i D(\theta_s, \theta_i) \frac{e^{-jks}}{\sqrt{s}}, \tag{13.1}$$

where $D(\theta_s, \theta_i)$ is the diffraction coefficient, and $\exp(-jks)/s^{1/2}$ represents the cylindrical wave.

The determination of the diffraction coefficient $D(\theta_s, \theta_i)$ in (13.1) will be made by considering a canonical problem of plane-wave diffraction by a knife edge. Since the exact solution of the knife-edge diffraction is well known, the diffraction coefficient is obtained by comparing the asymptotic form of the exact solution with (13.1).

Let us consider a plane incident wave given by

$$\psi_i = A_i e^{jks \cos(\theta_s - \theta_i)}. \tag{13.2}$$

The exact solution to this knife-edge problem has already been discussed (Section 11.13). The exact total field at (s, θ_s) is given by

$$\psi(s, \theta_s) = \frac{A_i e^{j(\pi/4)}}{\sqrt{\pi}} \left[e^{jks \cos(\theta_s - \theta_i)} F(a_1) - e^{jks \cos(\theta_s + \theta_i)} F(a_2) \right], \tag{13.3}$$

where

$$F(a) = \int_a^\infty e^{-j\tau^2} \, d\tau \text{ is the Fresnel integral,}$$

$$a_1 = -(2ks)^{1/2} \cos \frac{\theta_s - \theta_i}{2},$$

$$a_2 = -(2ks)^{1/2} \cos \frac{\theta_s + \theta_i}{2}.$$

The diffraction coefficient D in (13.1) is then determined by examining the exact field (13.3) for a large distance s from the edge and comparing this asymptotic form with (13.1).

Let us first examine the total field ψ at a large distance from the edge in each of the regions I, II, and III in Fig. 13.1. In region I, we should have

$$\psi = \psi_i + \psi_r + \psi_d,$$

$$\psi_r = -A_i e^{jks \cos(\theta_s + \theta_i)}, \tag{13.4}$$

where ψ_i is the incident field given in (13.2), ψ_r is the reflected field, and ψ_d is the diffracted field given in (13.1).

In region II, we do not have the reflected wave, and thus

$$\psi = \psi_i + \psi_d. \tag{13.5}$$

In region III, there is no incident or reflected wave, and thus

$$\psi = \psi_d. \tag{13.6}$$

Note that these asymptotic forms show discontinuous behavior at the boundaries between regions I, II, and III, although the exact field is continuous. Since (13.3) is the exact solution, it should reduce to (13.4), (13.5), and (13.6) for large kr. This can be shown by first noting that the Fresnel integral has different asymptotic forms; depending on the sign of a. If $a > 0$, we have

$$F(a) \approx \frac{e^{-ja^2}}{j2a}, \qquad a > \sqrt{10}. \tag{13.7}$$

However, if $a < 0$, we use the following identity:

$$F(a) + F(-a) = \sqrt{\pi}e^{-j(\pi/4)}, \tag{13.8}$$

and get

$$F(a) = \frac{e^{-ja^2}}{j2a} + \sqrt{\pi}e^{-j(\pi/4)}, \qquad a < 0 \quad \text{and} \quad |a| > \sqrt{10}. \tag{13.9}$$

The additional constant $\sqrt{\pi}e^{-j(\pi/4)}$ will be shown to correspond to the discontinuous behavior of the asymptotic forms (13.4), (13.5), and (13.6).

Now let us examine (13.3) in region I. Here we get $a_1 < 0$ and $a_2 < 0$, and thus using (13.9), we get

$$\psi = \psi_i + \psi_r + \psi_d, \tag{13.10}$$

$$\psi_d = A_i D(\theta_s, \theta_i)\frac{e^{-jks}}{\sqrt{s}}, \tag{13.11}$$

$$D(\theta_s, \theta_i) = -\frac{e^{-j(\pi/4)}}{2(2\pi k)^{1/2}} \left\{ \frac{1}{\cos[(\theta_s - \theta_i)/2]} - \frac{1}{\cos[(\theta_s + \theta_i)/2]} \right\}.$$

In region II, we get $a_1 < 0$ and $a_2 > 0$, and therefore we have

$$\psi = \psi_i + \psi_d. \tag{13.12}$$

In region III, $a_1 > 0$ and $a_2 > 0$, and therefore

$$\psi = \psi_d, \tag{13.13}$$

where ψ_d in the above is given in (13.11). We conclude, therefore, that the asymptotic form of the diffracted field is given by (13.11) for plane-wave incidence.

We now generalize this to state that the diffracted ray from an edge is also given by (13.11) when an arbitrary incident field at the edge is given by A_i. Making use of this generalization, we are now in a position to construct a GTD solution for more complex problems. It should be noted, however, that the diffraction coefficient given in (13.11) becomes infinite at the *reflection boundary* between regions I and II where $\theta_s + \theta_i = \pi$, and at the *shadow boundary* between regions II and III where $\theta_s - \theta_i = \pi$. This is to be expected because a_1 or a_2 becomes zero at these boundaries and the asymptotic forms (13.7) and (13.9) cannot be valid in the *transition region* near these boundaries. The failure of GTD in the transition regions can be overcome by the UTD or the UAT. These techniques will be discussed later.

For the TE case (H_z, E_x, E_y), the field component H_z satisfies the wave equation and Neumann's boundary condition on the conductor ($x > 0$, $y = 0$). In this case we use H_z in place of E_z. Then the exact solution is also given by (13.3) except that the minus sign for the second term is replaced by a plus sign. Therefore, we can summarize both the TM and TE cases as follows. Let ψ be E_z for TM and H_z for TE. Then if the incident wave ψ_i at the edge is denoted by A_i, the two-dimensional diffracted field at (s, θ_s) in the GTD approximation is given by

$$\psi_d(s, \theta_s) = A_i \, D\,(\theta_s, \theta_i)\frac{e^{-jks}}{\sqrt{s}}, \tag{13.14}$$

where D is the diffraction coefficient given by

$$D\,(\theta_s, \theta_i) = -\frac{e^{-j(\pi/4)}}{2(2\pi k)^{1/2}} \left\{ \frac{1}{\cos[(\theta_s - \theta_i)/2]} \mp \frac{1}{\cos[(\theta_s + \theta_i)/2]} \right\}. \tag{13.15}$$

The upper sign should be used for the TM (Dirichlet's) problem and the lower sign should be used for the TE (Neumann's) problem. Whenever convenient, we use D_s for Dirichlet's problem (soft screen) and D_h for Neumann's problem (hard screen). Note also that (13.14) is valid only in the region $|a_1| > \sqrt{10}$ and $|a_2| > \sqrt{10}$. If $|a_1| < \sqrt{10}$ or $|a_2| < \sqrt{10}$, UTD (Section 13.4) or other high-frequency techniques (Section 13.9) should be used.

13.2 DIFFRACTION BY A SLIT FOR DIRICHLET'S PROBLEM

Let us consider the two-dimensional problem of finding the field diffracted by a slit of width $2a$ on a conducting plane (Fig. 13.2). First, we consider Dirichlet's

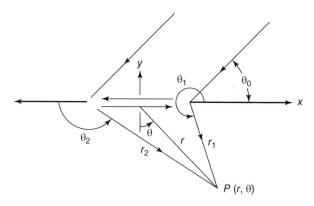

FIGURE 13.2   Diffraction by a slit of width $2a$.

(TM) problem, where $\psi = E_z$. We now obtain the GTD solution according to the formulation in Section 13.1. At the observation point P, the field diffracted once at the edge $(x = a, y = 0)$ is given by

$$\psi_{s1} = A_1 D(\theta_1, \theta_0)\frac{e^{-jkr_1}}{\sqrt{r_1}}, \tag{13.16}$$

where $D(\theta_1, \theta_0)$ is as given in (13.15).

Letting $D(r_1, \theta_1, \theta_0) = D(\theta_1, \theta_0)e^{-jkr_1}/\sqrt{r_1}$, we get the single diffraction (Fig. 13.3 ray a)

$$\psi_{s1} = A_1 D(r_1, \theta_1, \theta_0), \tag{13.17}$$

where A_1 is the incident field at the edge.

If the incident wave ψ_i is a plane wave given by

$$\psi_i = A_0 e^{jk(x\cos\theta_0 + y\sin\theta_0)}, \tag{13.18}$$

then A_1 is given by

$$A_1 = A_0 e^{jka\cos\theta_0}. \tag{13.19}$$

Similarly, the single diffraction by the edge at $(x = -a, y = 0)$ is given by (Fig. 13.3 ray c)

$$\psi_{s2} = A_2 D(r_2, \theta_2, \pi + \theta_0), \tag{13.20}$$

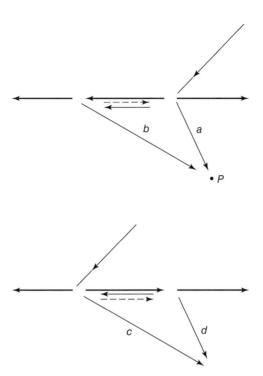

FIGURE 13.3 Four diffracted rays: a and c are single diffraction; b and d are double diffraction. Dashed lines represent multiple diffraction.

where A_2 is the incident wave at the edge, and for a plane incident wave (13.18), it is given by

$$A_2 = A_0 e^{-jka\cos\theta_0} \tag{13.21}$$

Let us next consider the double diffraction shown in Fig. 13.3 ray b and ray d.

$$\psi_{d1} = A_1 D\,(2a, \pi, \theta_0) D\,(r_2, \theta_2, \pi),$$
$$\psi_{d2} = A_2 D\,(2a, \pi, \pi + \theta_0) D\,(r_1, \theta_1, \pi). \tag{13.22}$$

The single and double diffraction terms are dominant and usually give good accuracy for $ka > 1.5$. Triple and multiple diffractions can also be included. We can add to ψ_{s1} the ray which traveled from the edge at $x = a$ to the edge at $x = -a$ and back to the edge at $x = a$ and then diffracted to P. This triply diffracted field is given by

$$\psi_{t1} = A_1 D(2a, \pi, \theta_0) D(2a, \pi, \pi) D(r_1, \theta_1, \pi). \tag{13.23}$$

The ray that traveled from $x = a$ to $x = -an$ times and then diffracted to P is given by

$$[D(2a, \pi, \pi)]^{2n}\psi_{t1}.$$

Therefore, adding all these multiple diffracted rays shown in Fig. 13.3 ray a, we get

$$\psi_{t1} \sum_{n=0}^{\infty} [D(2a, \pi, \pi)]^{2n} = \psi_{t1}[1 - D(2a, \pi, \pi)^2]^{-1}. \tag{13.24}$$

For the ray b, we get

$$\psi_{d1} \sum_{n=0}^{\infty} [D(2a, \pi, \pi)]^{2n} = \psi_{d1}[1 - D(2a, \pi, \pi)^2]^{-1}, \tag{13.25}$$

where $\psi_{d1} = A_1 D(2a, \pi, \theta_0)D(r_2, \theta_2, \pi)$. Similarly, we get all the rays for c and d in Fig. 13.3. The complete diffracted field ψ_d is therefore given by

$$\psi_d = \psi_{s1} + \psi_{s2} + (\psi_{d1} + \psi_{d2} + \psi_{t1} + \psi_{t2})[1 - D(2a, \pi, \pi)^2]^{-1}, \tag{13.26}$$

where ψ_{s1} and ψ_{s2} are given in (13.17) and (13.20), ψ_{d1} and ψ_{d2} are given in (13.22), ψ_{t1} is given in (13.23), and

$$\psi_{t2} = A_2 D(2a, \pi, \pi + \theta_0)D(2a, \pi, \pi)D(r_2, \theta_2, \pi)$$

Let us examine the diffracted field ψ_d in (13.26) at a large distance $(kr \gg 1)$ from the slit when the wave is normally incident on the slit. We then get (Fig. 13.2)

$$r_1 = r - a \sin\theta,$$
$$r_2 = r + \sin\theta,$$
$$\theta_0 = \frac{\pi}{2},$$
$$\theta_1 = \frac{3\pi}{2} + \theta, \tag{13.27}$$
$$\theta_2 = \frac{\pi}{2} + \theta,$$
$$A_1 = A_2 = A_0.$$

For the far field, the diffracted wave is cylindrical, and therefore we write

$$\psi_d = A_1 f_d(\theta) \left(\frac{k}{2\pi r} \right)^{1/2} e^{-jkr - j(\pi/4)}, \tag{13.28}$$

where $f_d(\theta)$ is the scattering amplitude. We now examine the single and double diffractions. For the far field, the scattering amplitudes for single diffraction are derived from (13.17) and (13.20)

$$f_{s1} = \frac{e^{jka\sin\theta}}{2k}\left[\frac{1}{\sin(\theta/2)} - \frac{1}{\cos(\theta/2)}\right],$$

$$f_{s2} = \frac{e^{-jka\sin\theta}}{2k}\left[\frac{-1}{\sin(\theta/2)} + \frac{-1}{\cos(\theta/2)}\right].$$

(13.29)

Note that $\theta = 0$ is the boundary between the illuminated and shadow regions for each edge, and therefore both ψ_{s1} and ψ_{s2} become infinite as $\theta \to 0$. However, in the far field, these two singularities cancel each other, producing a finite diffracted field

$$f_{s1} + f_{s2} = \frac{1}{k}\left[\frac{j\sin(ka\sin\theta)}{\sin(\theta/2)} - \frac{\cos(ka\sin\theta)}{\cos(\theta/2)}\right].$$

(13.30)

For a wide slit, the multiple diffractions ψ_{d1}, ψ_{d2}, ψ_{t1}, and ψ_{t2} can be neglected and (13.30) gives a good approximation. This can be shown to be consistent with the Kirchhoff diffraction theory for large ka (Section 6.3).

The scattering amplitudes of the single diffraction terms f_{s1} and f_{s2} are proportional to k^{-1}, as seen in (13.30). The scattering amplitudes of the double diffraction terms are f_{d1} and f_{d2}, and they are proportional to $k^{-3/2}$. The terms f_{t1} and f_{t2} are proportional to k^{-2}, and the multiple diffraction terms $[1 - D(2a, \pi, \pi)^2]^{-1}$ give all the higher negative powers of k. These terms can be obtained easily from the general formulations (13.26) with (13.27). For example, the scattering amplitudes in the forward direction $\theta = 0$ are given by

$$f_{s1}(0) + f_{s2}(0) = j2a - \frac{1}{k},$$

$$f_{d1}(0) = f_{d2}(0) = \frac{e^{-jk2a-j(\pi/4)}}{\sqrt{\pi}(ka)^{1/2}k},$$

$$f_{t1}(0) = f_{t2}(0) = \frac{e^{-jk4a+j(\pi/2)}}{2\pi k^2 a},$$

(13.31)

$$D^2(2a, \pi, \pi) = \frac{e^{-jk4a-j(\pi/2)}}{2\pi ka}.$$

The transmission cross section σ of an aperture on a screen is defined as the ratio of the total power transmitted through the aperture to the incident power flux density when a plane wave is incident. According to the forward scattering theorem, the transmission cross section σ of the two-dimensional slit is given by the imaginary part of the forward scattering amplitude

$$\sigma = \mathrm{Im} f_d(\theta = 0),$$

(13.32)

where f_d is as defined in (13.28) and σ is defined by the total transmitted power per unit power flux density per unit slit length. If we consider only the single diffractions f_{s1} and f_{s2}, we get

$$\frac{\sigma}{2a} = 1. \tag{13.33}$$

The transmission cross section is equal to the geometric cross section $\sigma_g = 2a$.

If we include the double diffraction terms f_{d1} and f_{d2}, we get

$$\frac{\sigma}{2a} = 1 - \frac{\cos(2\,ka - \pi/4)}{\sqrt{\pi}(ka)^{3/2}}. \tag{13.34}$$

This gives good agreement with the exact solution for ka greater than about 2 and fairly good agreement even for $ka > 1$.

13.3 DIFFRACTION BY A SLIT FOR NEUMANN'S PROBLEM AND SLOPE DIFFRACTION

The two-dimensional diffraction coefficient for a knife edge with Neumann's condition is given in (13.15) with $\psi = H_z$. The single diffraction ψ_{s1} and ψ_{s2} is given by

$$
\begin{aligned}
\psi_{s1} &= A_1\, D(r_1, \theta_1, \theta_0), \\
\psi_{s2} &= A_2\, D(r_2, \theta_2, \pi + \theta_0),
\end{aligned}
\tag{13.35}
$$

where

$$D(r_s, \theta_s, \theta_i) = \frac{e^{-jkr_s}}{\sqrt{r_s}} D(\theta_s, \theta_i),$$

$$D(\theta_s, \theta_i) = -\frac{e^{-j(\pi/4)}}{2(2\pi k)^{1/2}} \left[\frac{1}{\cos[(\theta_s - \theta_i)/2]} + \frac{1}{\cos[(\theta_s + \theta_i)/2]} \right].$$

Next consider the double refraction

$$\psi_{d1} = A_1 D(2a, \pi, \theta_0) D(r_2, \theta_2, \pi). \tag{13.36}$$

For Dirichlet's problem, this was finite and nonzero. However, for Neumann's problem, as seen in (13.15), $D(\theta_s, \theta_i) = 0$ if either $\theta_s = \pi$ or $\theta_i = \pi$ (Fig. 13.4). Therefore, the wave incident on the edge is proportional to $D(2a, \pi, \theta_0)$ and is zero. Since the diffracted field is obviously not zero, this means that we need to consider the higher-order term (Keller, 1962; Jull, 1981). The incident field H_z is zero at the edge, but its derivative, which is proportional to E_x, is not zero and contributes to the higher-order

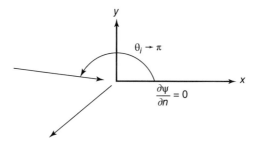

FIGURE 13.4   Diffraction by a hard half-plane when the incident angle is π.

term. Consider a wave that is zero at the edge but is nonuniform in the neighborhood of the edge. The simplest such wave is

$$\psi_i = yA_0 e^{-jkx}. \tag{13.37}$$

Its normal derivative in the direction of $-y$ at the edge is

$$\frac{\partial \psi_i}{\partial n} = -\frac{\partial \psi_i}{\partial y} = -A_0. \tag{13.38}$$

This incident wave can also be described as the derivative of a plane wave with respect to θ_i.

$$\psi_i = -\frac{A_0}{jk} \frac{\partial}{\partial \alpha} \left[e^{jk(x\cos\alpha + y\sin\alpha)} \right]_{\alpha=\pi}. \tag{13.39}$$

The diffracted wave corresponding to this incident wave has been obtained by (Keller, 1962)

$$\psi_d(r_s, \theta_s) = \frac{\partial \psi_i}{\partial n} D'(\theta_s, \theta_i) \frac{e^{-jkr_s}}{\sqrt{r_s}}, \tag{13.40}$$

where

$$D'(\theta_s, \theta_i) = \frac{1}{jk} \left[\frac{\partial}{\partial \alpha} D(\theta_s, \alpha) \right]_{\alpha=\pi}.$$

Using (13.40), we get the doubly refracted field ψ_{d1}.

$$\psi_{d1} = A_1 \frac{\partial}{\partial n} D(2a, \pi, \theta_0) D'(r_2, \theta_2, \pi), \tag{13.41}$$

where

$$\frac{\partial}{\partial n} D(2a, \pi, \theta_0) = \left[\frac{-1}{r} \frac{\partial}{\partial \alpha} D(r, \theta_0) \right]_{\substack{r=2a \\ \alpha=\pi}}$$

$$= \frac{\exp(-jk2a - j\pi/4)}{8(\pi ka)^{1/2} a} \frac{\cos(\theta_0/2)}{\sin^2(\theta_0/2)},$$

$$D'(r_2, \theta_2, \pi) = \left[\frac{1}{jk} \frac{\partial}{\partial \alpha} D(r_2, \theta_2, \alpha) \right]_{\alpha=\pi}$$

$$= \frac{\exp(-jkr_2 + j\pi/4)}{2k(2\pi kr_2)^{1/2}} \frac{\cos(\theta_2/2)}{\sin^2(\theta_2/2)}.$$

Use of these higher-order terms involving the normal derivative is called *slope diffraction*.

For a far field, using the scattering amplitude, we write the diffracted field as

$$\psi_d = A_0 f_d(\theta) \left(\frac{k}{2\pi r} \right)^{1/2} e^{-jkr - j(\pi/4)}. \tag{13.42}$$

For normal incidence, we get the scattering amplitude for single diffraction

$$f_{s1} = \frac{e^{jka \sin \theta}}{2K} \left[\frac{1}{\sin(\theta/2)} + \frac{1}{\cos(\theta/2)} \right],$$

$$f_{s2} = \frac{e^{-jka \sin \theta}}{2K} \left[\frac{-1}{\sin(\theta/2)} + \frac{1}{\cos(\theta/2)} \right], \tag{13.43}$$

$$f_{s1} + f_{s2} = \frac{1}{k} \left[\frac{j \sin(ka \sin \theta)}{\sin(\theta/2)} + \frac{\cos(ka \sin \theta)}{\cos(\theta/2)} \right].$$

The transmission cross section using the single and double diffractions can be obtained by using the forward scattering theorem. We get

$$\frac{\sigma}{2a} = 1 - \frac{\sin(2ka - \pi/4)}{8\sqrt{\pi}(ka)^{5/2}}. \tag{13.44}$$

Note that the double diffraction term is proportional to $(ka)^{-5/2}$ for Neumann's problem, whereas it is $(ka)^{-3/2}$ for Dirichlet's problem, as shown in (13.34); therefore, the double diffraction is weaker for Neumann's problem than for Dirichlet's problem. The expression (13.44) can be used for $ka > 1.5$ without much error.

13.4 UNIFORM GEOMETRIC THEORY OF DIFFRACTION FOR AN EDGE

We have already noted in Section 13.1 that the diffraction coefficients D_s and D_h in (13.15) become singular at $\theta_i \pm \theta_s = \pi$ and therefore cannot be used in the transition regions where $|a_1| < \sqrt{10}$ or $|a_2| < \sqrt{10}$ and a_1 and a_2 are given in (13.3). Kouyoumjian (1974) and others extended the GTD so that the diffraction coefficients remain valid in the transition regions. We will illustrate this technique called UTD using the two-dimensional knife-edge problems discussed in Section 13.1 (Fig. 13.5).

Consider a line source located at (ρ, θ_i) and the observation point at (s, θ_s). The total field is given by (Fig. 13.5)

$$
\begin{aligned}
\psi &= \psi_i + \psi_r + \psi_d && \text{in region I,} \\
\psi &= \psi_i + \psi_d && \text{in region II,} \\
\psi &= \psi_d && \text{in region III,}
\end{aligned}
\tag{13.45}
$$

where ψ_i, ψ_r, and ψ_d are the incident, reflected, and diffracted fields, respectively. If we use the GTD solution (13.14) for ψ_d, it becomes singular in the transition region. We can, however, use the Fresnel integral representation (13.3). This is continuous across the transition region, but it is the solution to a plane-wave incidence, not a cylindrical-wave incidence from a line source. The exact solution for a half-plane excited by a line source is available, but it is not in a closed form with known functions. We can, however, modify (13.3) for a line source located far from the edge ($k\rho \gg 1$).

The approximate edge-diffracted field excited by a line source according to UTD (Kouyoumjian and Pathak, 1974) is then given by

$$
\psi_d = A_i \, D\,(\theta_s, \theta_i)\frac{e^{-jks}}{\sqrt{s}},
\tag{13.46}
$$

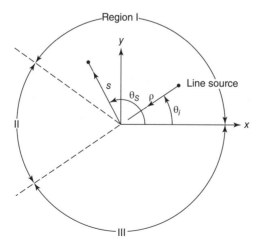

FIGURE 13.5   UTD for knife-edge diffraction.

where the incident cylindrical wave A_i at the edge is given by

$$A_i = A_0 \frac{e^{-jk\rho}}{\sqrt{\rho}}.$$

The UTD diffraction coefficient $D(\theta_s, \theta_i)$ is given by

$$D(\theta_s, \theta_i) = D(\theta_s - \theta_i) \mp D(\theta_s + \theta_i), \qquad (13.47)$$

where the upper sign is for a soft (Dirichlet) surface and the lower sign is for a hard (Neumann) surface, and $D(\beta)$ is given by

$$D(\beta) = -\frac{e^{-j(\pi/4)}}{2(2\pi k)^{1/2}} \frac{2j\sqrt{X}e^{jX}F(\sqrt{X})}{\cos(\beta/2)},$$

$$X = 2kL \cos^2 \frac{\beta}{2},$$

$$\sqrt{X} = (2kL)^{1/2} \left| \cos \frac{\beta}{2} \right|, \qquad (13.48)$$

$$F(a) = \int_a^{\infty} e^{-j\tau^2} d\tau,$$

$$L = \text{distance parameter} = \frac{s\rho}{s+\rho}.$$

Now noting that $\sqrt{X}/\cos(\beta/2) = \pm(2kL)^{1/2}$, depending on the sign of $\cos(\beta/2)$, we can rewrite (13.48) as follows:

$$D(\beta) = e^{j(\pi/4)}(L/\pi)^{1/2} e^{jX} F(\sqrt{X}) \, \text{sgn}(\beta - \pi), \qquad (13.49)$$

where

$$\text{sgn}(\beta - \pi) = \begin{cases} 1 & \text{if } \beta - \pi > 0, \\ -1 & \text{if } \beta - \pi < 0. \end{cases}$$

The function $D(\beta)$ is therefore discontinuous across $\beta = \pi$, which is the transition region. This discontinuity exactly cancels the discontinuity of ψ_i and ψ_r and yields the continuous total field everywhere.

To show this, consider the transition region between regions I and II where $\theta_s + \theta_i = \pi$. In this case, noting that $F(0) = (\sqrt{\pi})/2e^{-j(\pi/4)}$, we get, as θ_s approaches $\pi - \theta_i$,

$$D(\theta_s + \theta_i) = \begin{cases} -\dfrac{\sqrt{L}}{2} & \text{if } \theta_s < \pi - \theta_i \\[3mm] +\dfrac{\sqrt{L}}{2} & \text{if } \theta_s > \pi - \theta_i \end{cases}. \qquad (13.50)$$

Therefore, for a soft surface, we have

$$\psi_d = A_i \, D(\theta_s - \theta_i) \frac{e^{-jks}}{\sqrt{s}} \pm A_0 \frac{e^{-jk(\rho+s)}}{\sqrt{\rho+s}} \frac{1}{2}, \qquad (13.51)$$

where the upper sign is for $\theta_s < \pi - \theta_i$ and the lower sign is for $\theta_s > \pi - \theta_i$. Now the incident wave ψ_i is continuous across the transition angle, but ψ_r is discontinuous. For a soft surface, we have

$$\psi_i = \begin{cases} -A_0 \dfrac{e^{-jk(\rho+s)}}{\sqrt{\rho+s}} & \text{if } \theta_s < \pi - \theta_i, \\[2mm] 0 & \text{if } \theta_s > \pi - \theta_i. \end{cases} \qquad (13.52)$$

Adding (13.51) and (13.52), we have the continuous total field. See Fig. 13.6 for a pictorial explanation. Summarizing this section, the total field is given by (13.45)–(13.47), and (13.49). We now have two choices. The ordinary GTD discussed in Sections 13.1–13.3 is simple and useful outside the transition regions. The UTD discussed in this section makes use of Fresnel integrals and therefore requires more numerical work than the GTD. However, it can be used in all regions and gives continuous and useful results even in the transition regions.

13.5 EDGE DIFFRACTION FOR A POINT SOURCE

Up to this point, we have considered only two-dimensional problems of a knife edge excited by a plane wave or a line source. If a knife edge is illuminated by a point source, we need to take into account the angle of incidence (Fig. 13.7). According to Fermat's principle, the diffracted rays should be along the optical length, which

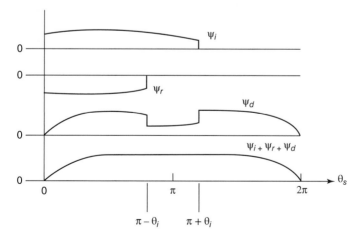

FIGURE 13.6   Pictorial explanation of UTD and cancellation of discontinuities of ψ_i, ψ_r, and ψ_d at the reflection ($\pi - \theta_i$) and the shadow ($\pi + \theta_i$) boundaries.

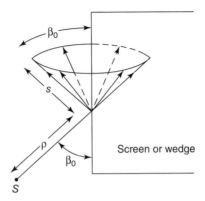

FIGURE 13.7   Cone of diffracted rays from an edge illuminated by a point source S.

is stationary among all paths from the source to the observation point. From this, we conclude that the incident ray and the diffracted rays make equal angles with the edge. This angle is denoted by β_0 in Fig. 13.7.

Now we write the GTD solution for the diffracted field due to a point source. The scalar diffracted field ψ_0 at the observation point is given by

$$\psi_d = A_i \, D(\theta_s, \theta_i) \left[\frac{\rho}{s(s+\rho)} \right]^{1/2} e^{-jks}, \tag{13.53}$$

where $A_i = A_0 e^{-jk\rho}/\rho$ is the incident wave at the edge.

The *divergence factor* $[\rho/s(s+\rho)]^{1/2}$ is obtained by observing the conservation of power in a tube of geometric optical rays. In Fig. 13.8, the total power through the small cross-sectional area dA is conserved.

$$\left| \psi_d^2 \right| dA = |\psi_0|^2 \, dA_0. \tag{13.54}$$

Note the ratio of the cross-sectional area is

$$\frac{dA}{dA_0} = \frac{(\rho+s)(\rho'+s)}{\rho\rho'}. \tag{13.55}$$

Therefore, we get

$$|\psi_d|^2 = \frac{\rho\rho'}{(\rho+s)(\rho'+s)} |\psi_0|^2. \tag{13.56}$$

As $\rho' \to 0$, $|\psi_0|$ becomes infinite, but $\rho'|\psi_0|^2$ is finite. Therefore, we conclude that ψ_d is proportional to $[\rho/s(\rho+s)]^{1/2}$ as $\rho' \to 0$.

The diffraction coefficient $D(\theta_s, \theta_i)$ for the region not close to the transition region can be obtained by examining the exact solution for the field when a plane wave is obliquely incident on the edge. Let the incident field be (Fig. 13.9)

$$\psi_i = A_i \exp[jkr \, \sin \beta_0 \cos(\theta - \theta_i) + jkz \cos \beta_0]. \tag{13.57}$$

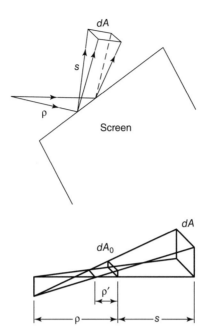

FIGURE 13.8   Conservation of power.

Then the exact total field at (r, θ, z) is given by

$$\psi = \frac{A_i\ e^{jkz\ \cos\beta_0 + j(\pi/4)}}{\sqrt{\pi}} \left[e^{jkr\ \sin\beta_0\ \cos(\theta-\theta_i)} F(a_1) \mp e^{jkr\ \sin\beta_0\ \cos(\theta+\theta_i)} F(a_2) \right], \qquad (13.58)$$

where

$$a_1 = -(2kr\ \sin\beta_0)^{1/2} \cos\frac{\theta - \theta_i}{2},$$

$$a_2 = -(2kr\ \sin\beta_0) \cos\frac{\theta + \theta_i}{2},$$

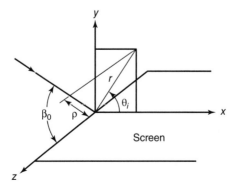

FIGURE 13.9   Plane wave incident on an edge from the direction (β_0, θ_i).

and the minus sign in front of the second term is for Dirichlet's problem and the plus sign is for Neumann's problem. If we consider the far-field approximation of (13.58) as we did in Section 13.1, and note that $s = r \sin \beta_0 - z \cos \beta_0$, we get the diffraction coefficient

$$D(\theta_s, \theta_i) = -\frac{e^{-j(\pi/4)}}{2(2\pi k)^{1/2} \sin \beta_0} \left\{ \frac{1}{\cos[(\theta_s - \theta_i)/2]} \mp \frac{1}{\cos[(\theta_s + \theta_i)/2]} \right\}. \quad (13.59)$$

The UTD solution applicable to all angles, including the transition regions, is therefore given by

$$\begin{aligned}
\psi &= \psi_i + \psi_r + \psi_d && \text{in region I} \\
&= \psi_i + \psi_d && \text{in region II} \\
&= \psi_d && \text{in region III,} \\
\psi_d &= A_i \, D \, (\theta_s, \theta_i) \left[\frac{\rho}{s(s+\rho)} \right]^{1/2} e^{-jks}, && (13.60) \\
A_i &= A_0 \frac{e^{-jk\rho}}{\rho}, \\
D \, (\theta_s, \theta_i) &= D \, (\theta_s - \theta_i) \mp D \, (\theta_s + \theta_i),
\end{aligned}$$

where the upper sign is for Dirichlet's surface and the lower sign is for Neumann's surface.

$$\begin{aligned}
D \, (\beta) &= \frac{e^{j(\pi/4)}}{\sin \beta_0} \left(\frac{L}{\pi} \right)^{1/2} e^{jX} \, F(\sqrt{X}) \, \mathrm{sgn}(\beta - \pi), \\
X &= 2kL \, \cos^2 \frac{\beta}{2}, \\
\sqrt{X} &= (2kL)^{1/2} \left| \cos \frac{\beta}{2} \right|, && (13.61) \\
F(a) &= \int_a^\infty e^{-j\tau^2} |d\tau, \\
L &= \frac{s\rho \sin^2 \beta_0}{\rho + s}.
\end{aligned}$$

Equation (13.59) gives the GTD solution applicable to the regions outside the transition region and (13.60) is the UTD solution applicable to all angles. Both apply to the region not too close to the edge, $kL > 1.0$.

The scalar solutions above can easily be extended to the electromagnetic problem. We note that the components of the electric field parallel and perpendicular to the plane of incidence $(E_{\beta 0}^i, E_\phi^i)$ are proportional to E_z^i and H_z^i, respectively, and therefore

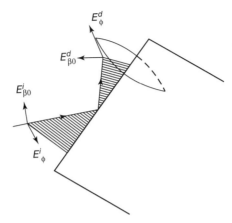

FIGURE 13.10 Electromagnetic diffraction by an edge.

we get the diffracted field $(E^d_{\beta 0}, E^d_\phi)$ using the diffraction coefficients for soft D_s and hard D_h surfaces (Fig. 13.10).

$$
\begin{bmatrix} E^d_{\beta 0} \\ E^d_\phi \end{bmatrix} = \begin{bmatrix} -D_s & 0 \\ 0 & -D_h \end{bmatrix} \begin{bmatrix} E^i_{\beta 0} \\ E^i_\phi \end{bmatrix} \left[\frac{\rho}{s(\rho + s)} \right]^{1/2} e^{-jks}. \tag{13.62}
$$

13.6 WEDGE DIFFRACTION FOR A POINT SOURCE

Up to this point, we have discussed GTD and UTD for a knife edge. We now consider GTD and UTD for a wedge with the angle $(2 - n)\pi$ (Fig. 13.11). Exact solutions for a wedge illuminated by a point source, a line source, and a plane wave are available in the form of infinite series and Fourier integrals. However, they are not in a convenient closed form. For the field point far from the wedge, the exact solution can be evaluated to give an asymptotic closed-form expression. In this section, we summarize the results for GTD and UTD.

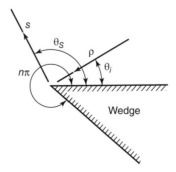

FIGURE 13.11 Wedge diffraction.

The GTD solution for the diffracted field for a wedge is applicable to the region far from the edge and excluding the transition region. For a two-dimensional problem, the diffracted field is given by

$$\psi_d = A_i D\,(\theta_s,\theta_i)\frac{e^{-jks}}{\sqrt{s}}, \tag{13.63}$$

where A_i is the incident wave at the edge, and for a line source it is given by

$$A_i = A_0 \frac{e^{-jk\rho}}{\sqrt{\rho}}.$$

For a wedge illuminated by a point source, the diffracted field is given by (Fig. 13.7)

$$\psi_d = A_i D\,(\theta_s,\theta_i)\left[\frac{\rho}{s\,(s+\rho)}\right]^{1/2} e^{-jks}, \tag{13.64}$$

where $A_i = A_0 e^{-jk\rho}/\rho$. The GTD diffraction coefficient is given by

$$D\,(\theta_s,\theta_i) = D\,(\theta_s - \theta_i) \mp D\,(\theta_s + \theta_i), \tag{13.65}$$

where the upper (lower) sign is for a soft (hard) surface wedge, and

$$D\,(\beta) = \frac{e^{-j(\pi/4)}\sin(\pi/n)}{n(2\pi k)^{1/2}\sin\beta}\,\frac{1}{\cos(\pi/n) - \cos(\beta/n)}. \tag{13.66}$$

Note that for a knife edge, $n = 2$ and $D(\beta)$ reduces to the knife-edge form. Note also that $D(\theta_s,\theta_i)$ is singular at the reflection and shadow boundaries.

The UTD diffraction coefficient is valid for all regions and is given by Kouyoumjian and Pathak (1974, p. 1453).

$$D(\theta_s,\theta_i) = D_+(\theta_s - \theta_i) + D_-(\theta_s - \theta_i) \mp [D_+(\theta_s + \theta_i) + D_-(\theta_s + \theta_i)], \tag{13.67}$$

where the upper (lower) sign is for a soft (hard) surface,

$$D_\pm(\beta) = \frac{e^{-j(\pi/4)}}{2n(2\pi k)^{1/2}\sin\beta_0}\cos\frac{\pi \pm \beta}{2n}F(X_\pm).$$

$$X_\pm = 2kL\,\cos^2\frac{2n\,\pi N^\pm - \beta}{2},$$

$N^\pm$ are integers that most nearly satisfy $2\pi n N^\pm - \beta = \pm\pi$,

$$F(X) = 2j\sqrt{|X|}\,e^{jX}\int_{\sqrt{|X|}}^{\infty} e^{-j\tau^2}\,d\tau.$$

13.7 SLOPE DIFFRACTION AND GRAZING INCIDENCE

We have discussed in Section 13.3 that if the incident wave is zero at the edge, we need to include the derivative of the incident wave. In general, if the incident wave is not slowly varying, the derivative or higher-order terms need to be included. We therefore write for a soft surface

$$\psi_d = \left(\psi_i D_s + \frac{\partial \psi_i}{\partial n} d_s \right) \left[\frac{\rho}{s(\rho + s)} \right]^{1/2} e^{-jks}, \tag{13.68}$$

where

$$d_s = \frac{1}{jk \sin \beta_0} \frac{\partial}{\partial \theta_i} D_s(\theta_s, \theta_i),$$

and $D_s(\theta_s, \theta_i)$ is given in (13.67). For example, if the wave is incident on a soft wedge at the grazing angle, $\theta_i = 0$, the incident wave consists of the direct wave and the reflected wave. They are canceled on the surface and therefore the incident wave is zero. In this case, the slope term is dominant and is given by

$$\psi_d = \frac{1}{2} \frac{\partial \psi_i}{\partial n} \frac{1}{jk \sin \beta_0} \frac{\partial}{\partial \theta_i} D_s \left[\frac{\rho}{s(\rho + s)} \right]^{1/2} e^{-jks}. \tag{13.69}$$

The factor $\frac{1}{2}$ is needed because only half of the incident wave is the direct wave incident on the edge.

If the wedge is hard and the wave is incident at the grazing angle, we have

$$\psi_d = \frac{1}{2} \psi_i D_h \left[\frac{\rho}{s(s + \rho)} \right]^{1/2} e^{-jks}. \tag{13.70}$$

Here the factor of $\frac{1}{2}$ is also needed.

13.8 CURVED WEDGE

If the wedge has a curved edge with a radius of curvature a, the divergence factor $[\rho/s(s + \rho)]^{1/2}$ must be modified to include the effects of the radius of curvature. The diffracted ray should appear as originating at a distance ρ' rather than ρ. To derive ρ', first consider the case when the incident and diffracted rays are in the plane formed by the edge and the normal $\hat{n}$ (or the radius of curvature) (Fig. 13.12). We can then easily see that

$$\Delta\theta = \Delta\theta_1 + 2\Delta\theta_2. \tag{13.71}$$

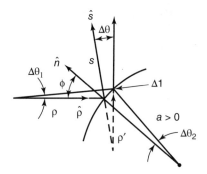

FIGURE 13.12   Divergence factor for a curved wedge.

Since $\Delta\theta = \Delta l \cos \phi/\rho'$, $\Delta\theta_1 = \Delta l \cos \phi/\rho$, and $\Delta\theta_2 = \Delta l/a$, we get

$$\frac{1}{\rho'} = \frac{1}{\rho} + \frac{2}{a \cos \phi}. \tag{13.72}$$

If we consider the incident and diffracted rays that are not in this plane, we need to consider the projection of these rays on this plane. Thus we get

$$\frac{1}{\rho'} = \frac{1}{\rho} - \frac{\hat{n} \cdot (\hat{\rho} - \hat{s})}{a \sin^2 \beta_0}. \tag{13.73}$$

The divergence factor should then be given by

$$\left[\frac{\rho'}{s(s + \rho')}\right]^{1/2}. \tag{13.74}$$

ρ' is therefore the distance between the edge and the caustic (focal point) of the diffracted ray.

For example, if a plane wave is normally incident on a circular aperture of radius a, $\rho \to \infty$, $\beta_0 = \pi/2$, $\hat{n} \cdot \hat{\rho} = 0$, and therefore (Fig. 13.13)

$$\frac{1}{\rho'} = \frac{\hat{n} \cdot \hat{s}}{a} = -\frac{\sin \theta_1}{a}. \tag{13.75}$$

The divergence factor is therefore

$$\left[\frac{\rho'}{s(s + \rho')}\right]^{1/2} = \frac{1}{\{s[1 - (s \sin \theta_1)/a]\}^{1/2}}. \tag{13.76}$$

The diffraction coefficients for the curved wedge are the same as those for the straight wedge since it is not affected by the radius of curvature.

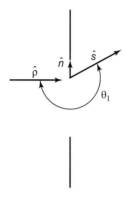

FIGURE 13.13   Wave incident on a circular aperture.

13.9 OTHER HIGH-FREQUENCY TECHNIQUES

We have discussed the GTD developed by Keller and the UTD developed by Kouy-oumjian and Pathak. In addition to UTD, the UAT has been developed by Lewis, Boersma, Ahluwalia, Lee, and Deschamps (Lee, 1977, 1978). In this section, we outline some basic ideas of UAT, but we will omit the detailed description.

We start with Keller's GTD and write it as

$$\psi(\text{Keller}) = \psi_g + \psi_d, \tag{13.77}$$

where ψ_g is the geometric optical solution and ψ_d is Keller's diffracted field. As we discussed in Sections 13.1 and 13.4, ψ_d is singular at the shadow and reflection boundaries. The UTD can be written as

$$\psi(\text{UTD}) = \psi_g + \psi(K - P), \tag{13.78}$$

where $\psi(K - P)$ is the diffracted field developed by Kouyoumjian and Pathak and is continuous in all regions. The UAT can be written as

$$\psi(\text{UAT}) = \psi(A) + \psi_d, \tag{13.79}$$

where ψ_d is Keller's GTD diffracted field and $\psi(A)$ is the new asymptotic term. $\psi(\text{UAT})$ is then continuous in all regions, and it may also be applicable in the near field of the edge, although its mathematical expressions may be somewhat complex.

Ufimtsev (1975) developed a technique called the physical theory of diffraction (PTD) in 1962. This is an extension of physical optics to include the field produced by the fringe current. For a perfectly conducting body, the PTD can be written as

$$\psi(\text{PTD}) = \psi(\text{PO}) + \psi(\text{fringe}), \tag{13.80}$$

where ψ(PO) is the physical optics field produced by the current $2\hat{n} \times \bar{H}_i$ ($\bar{H}_i$ is the incident magnetic field and $\hat{n}$ is the unit vector normal to the surface) and ψ(fringe) is the fringe field produced by the fringe current. Far from the edge, ψ(PO) approaches ψ_g and ψ(fringe) approaches ψ_d (Knott and Senior, 1974).

13.10   VERTEX AND SURFACE DIFFRACTION

We have shown that the GTD diffracted field from an edge or wedge is given by

$$\psi_d = \psi_i D \left[\frac{\rho}{s(s+\rho)} \right]^{1/2} e^{-jks}, \tag{13.81}$$

where ψ_i is the incident field at the edge, D the diffraction coefficient, $[\rho/s(s+\rho)]^{1/2}$ the divergence factor, s the distance from the edge to the observation point, and ρ the distance between the edge and the caustic of the diffracted ray [see (13.53)].

For the diffracted wave from a vertex, we should have a spherical wave,

$$\psi_d = \psi_i D \frac{e^{-jks}}{s}. \tag{13.82}$$

The diffraction coefficient D can be determined from the study of the appropriate canonical problems, such as the corner of an edge or the corner of a cube. Some studies have been done on this difficult problem (Bowman et al., 1969).

The GTD diffraction field for a smooth convex surface is obtained in the following manner. Consider the wave originating at P and observed at Q (Fig. 13.14). We assume that the diffracted ray obeys Fermat's principle that the total optical path length between P and Q is minimum. Thus the diffracted ray follows a geodesic path from P to A and A' and Q. The wave originating at P is tangentially incident on the surface at A and is a creeping wave from A to A' behaving at each point as if it were a creeping wave on a circular cylinder with the radius equal to the radius of curvature of the actual surface. The wave then leaves the surface at A' and propagates to Q. The diffracted field at Q is therefore given by

$$\psi_d = \psi_i(A)D \left[\frac{\rho_3}{(\rho_3 + s)s} \right]^{1/2} e^{-jk(\tau+s)}, \tag{13.83}$$

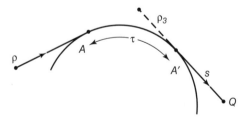

FIGURE 13.14   Diffracted ray for a smooth convex surface.

where $\psi_i(A)$ is the incident field at A; the diffraction coefficient includes the attenuation constant $\alpha(\tau')$ of the creeping wave from A to $A'[\exp(-\int_0^\tau \alpha(\tau')\,d\tau')]$. The radius of curvature ρ_3 in (13.83) represents the curvature in the plane perpendicular to the plane containing the ray P–A–A'–Q. For a detailed discussion on this, see James (1976, Chapter 6).

13.11 LOW-FREQUENCY SCATTERING

Up to this point in this chapter, we have considered the GTD and some of its variations. These are an improvement over geometric optics. In general, the high-frequency approximations of the field (ψ, $\bar{E}$, and $\bar{H}$) may be obtained by expanding the field in inverse powers of $k = \omega/c$.

$$\bar{E} = \sum_{n=0}^{\infty} \frac{\bar{E}_n}{(-jk)^n}. \tag{13.84}$$

The first term ($n = 0$) represents geometric optics. In contrast, the low-frequency approximations are obtained by expanding the field in powers of $k = \omega/c$.

Let us consider the scattered field from a dielectric object whose size is small compared with a wavelength. We write Maxwell's equations as where $k = \omega/c$ is the free-space wave number, $\eta_0 = \mu_0/\varepsilon_0$ is the free-space characteristic impedance, and ε_r is the relative dielectric constant. We now expand the field in powers of k.

$$\nabla \times \bar{E} = (-jk)\eta_0\bar{H},$$

$$\nabla \times \bar{H} = (-jk)\left(-\frac{\varepsilon_r}{\eta_0}\right)\bar{E},$$

$$\nabla \cdot (\varepsilon_r \bar{E}) = 0, \tag{13.85}$$

$$\nabla \cdot \bar{H} = 0,$$

$$\bar{E} = \sum_{n=0}^{\infty} (-jk)^n \bar{E}_n,$$

$$\bar{H} = \sum_{n=0}^{\infty} (-jk)^n \bar{H}_n. \tag{13.86}$$

Substituting (13.86) into (13.85) and equating the like powers of $(-jk)$, we get the zeroth-order equations

$$\nabla \times \bar{E}_0 = 0,$$

$$\nabla \times \bar{H}_0 = 0,$$

$$\nabla \cdot (\varepsilon_r \bar{E}_0) = 0, \tag{13.87}$$

$$\nabla \cdot \bar{H}_0 = 0.$$

Note that the electric and magnetic fields are not coupled, and they are identical to the equations for the static case.

The first-order equations are

$$\nabla \times \bar{E}_1 = \eta_0 \bar{H}_0,$$

$$\nabla \times \bar{H}_1 = -\frac{\varepsilon_r}{\eta_0} \bar{E}_0,$$

$$\nabla \cdot (\varepsilon_r \bar{E}_1) = 0,$$

$$\nabla \cdot \bar{H}_1 = 0.$$

(13.88)

Here $\bar{E}_1$ and $\bar{H}_1$ are generated by $\bar{H}_0$ and $\bar{E}_0$, respectively.

As an example, consider a plane wave incident on a dielectric ellipsoid. The incident wave $(\bar{E}_i, \bar{H}_i)$ is expanded in powers of k.

$$\bar{E}_i = \sum_{n=0}^{\infty} (-jk)^n \bar{E}_n,$$

$$\bar{H}_i = \sum_{n=0}^{\infty} (-jk)^n \bar{H}_n.$$

(13.89)

The zeroth-order solution for $\bar{E}_0$ is identical to the electrostatic case (Stratton, 1941, p. 211) and is given in Section 10.6. The zeroth-order solution for $\bar{H}_0$ is obtained by (13.87). The incident field is $\bar{H}_i = E_0 / \eta_0$, and these equations are the same as those in free space and therefore $\bar{H}_0 = \bar{H}_i = \bar{E}_0 / \eta_0$. Higher-order solutions are given in van Bladel (1964, p. 279).

Note that for the zeroth-order electric field $\bar{E}_0$, we used the dielectric constant ε_r, which is in general complex. For a lossy medium with conductivity σ, we have $\varepsilon_r = \varepsilon_r' - j\varepsilon_r'' = \varepsilon_r' - j\sigma / \omega \varepsilon_0$. If the frequency is low but not zero, we should use the complex ε_r rather than separating ε_r into ε_r' and σ. If dc fields are involved, we should rewrite Maxwell's equations as

$$\nabla \times \bar{E} = (-jk)\eta_0 \bar{H},$$

$$\nabla \times \bar{H} = (-jk)\left(-\frac{\varepsilon_r'}{\eta_0}\right)\bar{E} + \sigma\bar{E},$$

$$\nabla \cdot \bar{H} = 0,$$

$$\nabla \cdot \left[\sigma\bar{E} - (-jk)\frac{\varepsilon_r'}{\eta_0}\bar{E}\right] = 0.$$

(13.90)

We can then expand $\bar{E}$ and $\bar{H}$ in powers of k and obtain the zeroth-order equations.

$$\nabla \times \bar{E}_0 = 0,$$

$$\nabla \times \bar{H}_0 = \sigma \bar{E}_0,$$

$$\nabla \cdot (\sigma \bar{E}_0) = 0, \qquad (13.91)$$

$$\nabla \cdot \bar{H}_0 = 0.$$

This is the equation for magnetostatics. Solutions for an ellipsoidal conductor are given by Stratton (1941, p. 207).

Let us next consider the dc current distribution in a conducting body. Since $\nabla \times \bar{E}_0 = 0$ from (13.91), we get

$$\bar{E}_0 = -\nabla V. \qquad (13.92)$$

Substituting this into the divergence equation in (13.91), we get

$$\nabla \cdot (\sigma \nabla V) = 0. \qquad (13.93)$$

The current density $\bar{J}$ is given by

$$\bar{J} = \sigma \bar{E}_0 = -\sigma \nabla V. \qquad (13.94)$$

As an example, consider the conducting body shown in Fig. 13.15. At the surface S_1, the voltage V_0 is applied and at S_2, the voltage is zero. The voltage V then satisfies

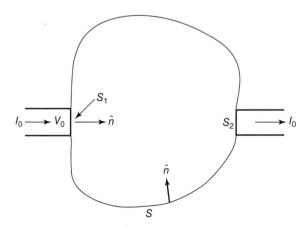

FIGURE 13.15 DC current distribution in a conducting body.

(13.93). The boundary condition at the surface S is that the current normal to the surface is zero. From (13.94), this means that

$$\frac{\partial V}{\partial n} = 0 \qquad \text{on } S.$$

The total current I_0 is then given by

$$I_0 = \int_{S_1} \bar{J} \cdot \hat{n} \, ds = - \int_{S_1} \sigma \frac{\partial V}{\partial n} \, ds. \qquad (13.95)$$

Numerical techniques such as the finite-element method are often employed to solve (13.93) with appropriate boundary conditions.

PROBLEMS

13.1   A conducting knife edge is illuminated by an electric line source I_z as shown in Fig. P13.1. Find and plot the field $E_z(x)$ at $y = 0$, $h = d = 5\lambda$ as a function of x. $\lambda = 300$ m.

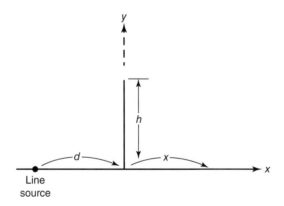

FIGURE P13.1   Diffraction by a knife edge.

13.2   Show that the transmission cross section of a slit for Dirichlet screen is given by (13.34).

13.3   Show that the transmission cross section for a slit on Neumann screen is given by (13.44).

13.4   There are two mountains between a radio transmitter and a receiver as shown in Fig. P13.4. Assuming that two mountain ridges can be approximated by two knife edges, calculate the received field normalized to the direct field in free space as a function of the distance x. Assume that the edges are soft.

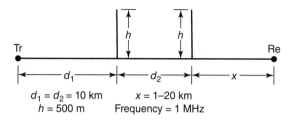

$d_1 = d_2 = 10$ km $x = 1–20$ km
$h = 500$ m   Frequency = 1 MHz

FIGURE P13.4   Double ridge diffraction.

13.5 A point source is located near a 90° wedge as shown in Fig. P13.5. Find the diffracted field $\psi(x).h = d = 5\lambda$.

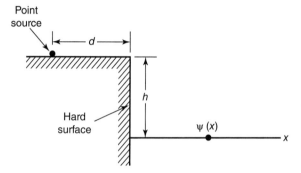

FIGURE P13.5   90° hard wedge.

13.6 Find the total resistance of the square plate shown in Fig. P13.6. Assume that the two terminal wires A and B are perfect conductors.

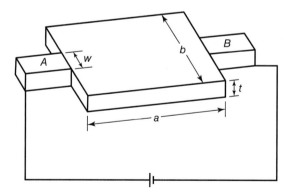

FIGURE P13.6   Square conducting plate with conductivity σ.

CHAPTER 14

PLANAR LAYERS, STRIP LINES, PATCHES, AND APERTURES

Planar dielectric layers and slabs are found in many applications. Examples are a dielectric coating on conducting surface, propagation in the atmosphere and the ocean, strip lines, periodic patches, and apertures embedded in dielectric layers. In this chapter, we first discuss excitation of waves in planar layers. We then consider strip lines, patches, and apertures in dielectric layers which are useful in microwave and millimeter wave applications. See Itoh (1987) for excellent collections of papers on planar structures, and Unger (1977) for planar waveguides. The spectral domain method is discussed in Yamashita and Mittra (1968), Itoh (1980), and Scott (1989).

14.1 EXCITATION OF WAVES IN A DIELECTRIC SLAB

Let us consider the excitation of TM waves on a dielectric slab placed on a conducting plane. This is a two-dimensional problem, and the field components are E_x, E_z, and H_y. Let us assume that the wave is excited by a small two-dimensional slot shown in Fig. 14.1. At the slot ($x = 0$), $E_z = E_0$ is given. Now according to the equivalence theorem (Section 6.10), the field excited by the aperture field $\bar{E}_a$ on an aperture in a conducting surface is the same as the field excited by the following magnetic surface current density $\bar{K}_s$ placed on the conducting surface

$$\bar{K}_s = \bar{E}_a \times \hat{n}, \tag{14.1}$$

Electromagnetic Wave Propagation, Radiation, and Scattering: From Fundamentals to Applications, Second Edition. Akira Ishimaru.

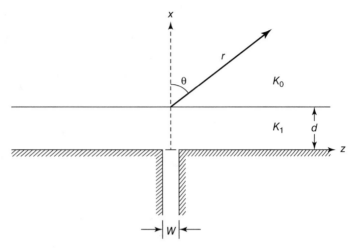

FIGURE 14.1   Dielectric slab on a conducting plane excited by a slot.

where $\hat{n}$ is the unit vector directed into the observation point (Fig. 14.2). Therefore, the field is the same as that excited by a magnetic surface current density $\bar{K}_s = E_0 \hat{y}$.

We assume that the slot width W is small compared with a wavelength, and therefore we can approximate the magnetic source current density $\bar{J}_{ms}$ by

$$\bar{J}_{ms} = \bar{K}_s \delta(x) = E_0 \, W \delta(x) \delta(z) \hat{y}. \tag{14.2}$$

Note that the unit for $\bar{J}_{ms}$ is volt/m^2, and the unit for $\bar{K}_s$ is volt/m.

Now from Maxwell's equations,

$$\nabla \times \bar{E} = -j\omega\mu\bar{H} - \bar{J}_{ms},$$
$$\nabla \times \bar{H} = j\omega\varepsilon\bar{E}, \tag{14.3}$$

and letting $\partial/\partial y = 0$, we get

$$\left(\frac{\partial^2}{\partial x^2} + \frac{\partial^2}{\partial z^2} + k^2 \right) H_y = j\omega\varepsilon E_0 \, W \delta(x) \delta(z), \tag{14.4}$$

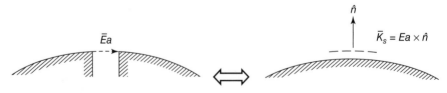

FIGURE 14.2   Equivalence of the aperture field and magnetic current.

where $k = k_0$ in air and $k = k_1$ in dielectric. The boundary conditions are $E_z = 0$ at $x = 0$ and the continuity of E_z and H_y at $x = d$. In terms of H_y, the boundary conditions are

$$\frac{\partial}{\partial x} H_y = 0 \quad \text{at } x = 0,$$

$$\frac{1}{\varepsilon} \frac{\partial}{\partial x} H_y \text{ and } H_y \text{ are continous at } x = d.$$

(14.5)

Let us start with the delta functions in (14.4) located at an arbitrary point (x', z') in the slab and then let x' and z' be zero later. We then have $H_y = -j\omega\varepsilon_1 E_0 WG$ and

$$\left(\frac{\partial^2}{\partial x^2} + \frac{\partial^2}{\partial z^2} + k_1^2\right) G = -\delta(x - x')\delta(z - z') \quad \text{in slab,}$$

$$\left(\frac{\partial^2}{\partial x^2} + \frac{\partial^2}{\partial z^2} + k_0^2\right) G = 0 \qquad\qquad\qquad \text{outside.}$$

(14.6)

We now take the Fourier transform of (14.6) in the z direction.

$$g(x, \beta) = \int_{-\infty}^{\infty} G(x, z) e^{j\beta z} \, dz,$$

$$G(x, z) = \frac{1}{2\pi} \int g(x, \beta) e^{-j\beta z} \, d\beta.$$

(14.7)

Inside the slab, we get

$$\left(\frac{d^2}{dx^2} + k_1^2 - \beta^2\right) g(x, \beta) = -\delta(x - x') e^{j\beta z'}.$$

(14.8)

The solution is then given by a sum of the primary field g_p and the secondary field g_s. The primary field is the field that is excited by the source in the infinite homogeneous space in the absence of any boundaries, and the secondary field represents all the effects of the boundaries. The primary field may also be called the incident field and is the particular solution of the inhomogeneous differential equation (14.8). The secondary field can be called the scattered field and is the complementary solution of (14.8).

The primary field is easily obtained following Section 5.5.

$$g_p(x, \beta) = \frac{\exp[-jq_1|x - x'|]}{2jq_1},$$

(14.9)

where

$$
q_1 = \begin{cases} \left(k_1^2 - \beta^2\right)^{1/2} & \text{for } |\beta| < |k_1|, \\ -j\left(\beta^2 - k_1^2\right)^{1/2} & \text{for } |\beta| > |k_1|. \end{cases}
$$

The secondary field, g_s, is given by two waves traveling in the $+x$ and $-x$ directions.

$$
g_s(x, \beta) = B \exp(-jq_1 x) + C \exp(+jq_1 x). \tag{14.10}
$$

In air $x > d$, we have

$$
\left(\frac{d^2}{dx^2} + k_0^2 - \beta^2\right) g(x, \beta) = 0. \tag{14.11}
$$

Therefore, the solution is

$$
g(x, \beta) = A \, \exp[-jq_0(x - d)], \tag{14.12}
$$

where

$$
q_0 = \begin{cases} \left(k_0^2 - \beta^2\right)^{1/2} & \text{for } |\beta| < |k|, \\ -j\left(\beta^2 - k_0^2\right)^{1/2} & \text{for } |\beta| > |k|. \end{cases}
$$

Note that, in general, (14.12) should consist of both the outgoing wave satisfying the radiation condition and the incoming wave $\exp(\,+jq_0 x)$, but since there is no incoming wave, this term is zero.

We now satisfy the boundary conditions. At $x = 0$, we have $(\partial/\partial x)G = 0$. Thus we get

$$
\frac{\exp(-jq_1 x')}{2jq_1} - B + C = 0. \tag{14.13}
$$

At $x = d$, we have G and $(1/\varepsilon)(\partial/\partial x)G$ being continuous. Thus we get

$$
\frac{\exp[-jq_1(d - x')]}{2jq_1} + B \, \exp(-jq_1 d) + C \exp(+jq_1 d) = A,
$$

$$
\frac{q_1}{\varepsilon_1} \left\{ \frac{\exp[-jq_1(d - x')]}{2jq_1} + B \, \exp(-jq_1 d) - C \, \exp(+jq_1 d) \right\} = \frac{q_0}{\varepsilon_0} A. \tag{14.14}
$$

Equations (14.13) and (14.14) are easily solved for three unknowns A, B, and C. Here we write the field H_y in air $(x > d)$ when the slot is at $x = 0$ $(x' = 0)$

$$G = \frac{1}{2\pi} \int_c A(\beta) \exp[-jq_0(x - d) - j\beta z] \, d\beta.$$

$$A(\beta) = \frac{e^{-jq_1 d}}{jq_1} \frac{T}{1 - R \, \exp(-j2q_1 d)}, \tag{14.15}$$

$$R = \frac{(q_1/\varepsilon_1) - (q_0/\varepsilon_0)}{(q_1/\varepsilon_1) + (q_0/\varepsilon_0)}, \quad T = 1 + R.$$

The physical interpretation of this solution is that $\exp(-jq_1 d)/jq_1$ is twice the direct primary wave $\exp(-jq_1 d)/(2jq_1)$ and represents a sum of the waves from the magnetic current source and from its image. The images for electric and magnetic current sources can easily be seen noting that the electric field tangential to the conducting surface is zero (Fig. 14.3). T is the transmission coefficient from dielectric to air, and R is the reflection coefficient. If we expand $[1 - R \exp(-j2q_1 d)]^{-1}$ in a series, we get

$$A(\beta) = \frac{e^{-jq_1 d}}{jq_1} \sum_{n=0}^{\infty} TR^n \, \exp(-j2q_1 nd). \tag{14.16}$$

Each term can be identified as the multiple reflected wave (Fig. 14.4).

Let us consider the integral in (14.15). We note first that there are poles located at the roots $\beta = \beta_s$ of the denominator

$$1 - R \, \exp(-j2q_1 d) = 0. \tag{14.17}$$

In addition, there are branch points at $\beta = \pm k_0$ due to $q_0 = (k_0^2 - \beta^2)^{1/2}$. There is no branch point, however, at $\beta = \pm k_1$ even if $q_1 = (k_1^2 - \beta^2)^{1/2}$ appears in the integral. This is because q_1 represents the propagation constant in the x direction within the slab, and since the slab is bounded at $x = 0$ and d, changing $+q_1$ to $-q_2$ simply

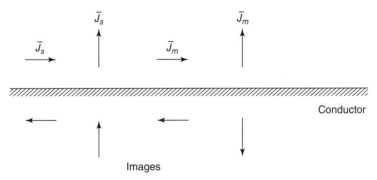

FIGURE 14.3   Images for a conducting plane.

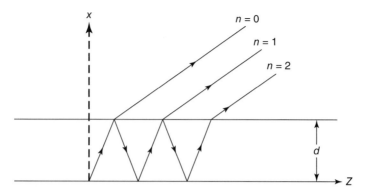

FIGURE 14.4   Multiple reflected waves.

interchanges the positive-going and negative-going waves inside the slab and the results are unchanged. Of course, this can be verified mathematically by changing q_1 to $-q_1$ in $A(\beta)$.

Let us next examine the poles given in (14.17). This equation can easily be converted to

$$\frac{q_1}{\varepsilon_1} \tan q_1 d = \frac{jq_0}{\varepsilon_0}. \tag{14.18}$$

This is identical to the eigenvalue equation (3.84) for the trapped surface wave.

The integrand in (14.15) therefore contains a finite number of trapped surface wave poles β_p and these are located on the real axis (Fig. 14.5). It is often convenient to use the transformation from β to α.

$$\beta = k_0 \sin \alpha. \tag{14.19}$$

We also let

$$x - d = r \cos \theta,$$
$$z = r \sin \theta. \tag{14.20}$$

Then (14.15) becomes

$$G = \int_c F(\alpha) \exp[k_0 \, rf(\alpha)] \, d\alpha, \tag{14.21}$$

where

$$F(\alpha) = \frac{A \, (\alpha) k_0 \cos \alpha}{2\pi},$$
$$f(\alpha) = -j \cos(\alpha - \theta).$$

$F(\alpha)$ has poles at $\alpha_p (\beta_p = k_0 \sin \alpha_p)$.

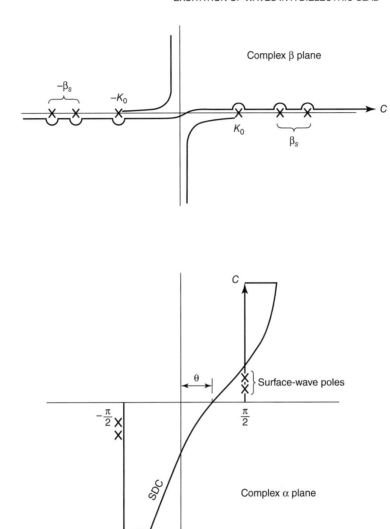

FIGURE 14.5   Original contour C, SDC, and surface-wave poles.

Let us first consider the radiation field far from the source and the surface. The radiation field in the form of (14.21) has been evaluated in Section 11.5. According to (11.48), which is obtained by the saddle-point technique, the radiation field is given by

$$G_r = \frac{A\,(\theta)k_0\cos\theta}{(2\pi k_0 r)^{1/2}}\, e^{-jk_0 r + j(\pi/4)}. \tag{14.22}$$

This is valid when $k_0 r \gg 1$ and $\theta \neq \pi/2$. The saddle point α_s is located at $\alpha_s = \theta$ in the complex α plane (Fig. 14.5).

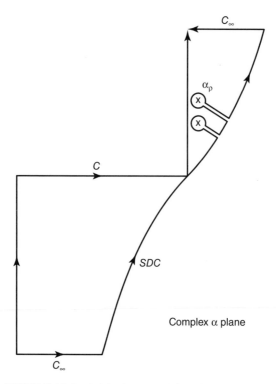

FIGURE 14.6   Original contour C is deformed to SDC.

Let us next consider the field on the surface where $\theta = \pi/2$ (Fig. 14.1). In the complex α plane, the steepest descent contour (SDC) now goes through the saddle point $\alpha_s = \pi/2$. Then the original contour c is deformed as shown in Fig. 14.6. Therefore, we have

$$\int_C g(\alpha)\, d\alpha = \int_{SDC} g(\alpha)\, d\alpha - 2\pi j \sum \text{residue at } \alpha_p \qquad (14.23)$$

where $g(\alpha)$ is the integrand in (14.21). The first term is (14.22) which vanishes on the surface $\theta = \pi/2$. The residue terms can be evaluated from (14.15). The surface wave poles are given by (14.18). Let β_s be the propagation constant of the surface wave and $q_0 = -j\alpha_t$. Then the residue term in (14.23) is given by

$$- 2\pi j \sum \text{residue} = -j \sum \frac{N(\beta_s)}{D'(\beta_s)}\, e^{-\alpha_t(x-d)-j\beta_s z}, \qquad (14.24)$$

where $A(\beta) = N(\beta)/D(\beta)$ (see Appendix 11.D).

In summary, the radiation field for $\theta \neq \pi/2$ is given by (14.22) and the surface wave at $\theta = \pi/2$ is given by (14.24). In Fig. 14.7 we also show a leaky wave pole at $\alpha = \alpha_l$, which can be found in some structures. If leaky wave poles exist, then as the observation angle θ increases beyond θ_l, the leaky wave contribution appears. It is

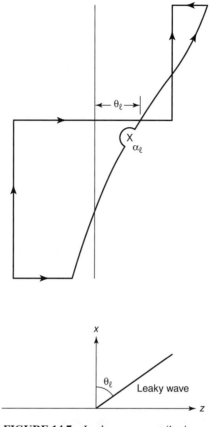

FIGURE 14.7   Leaky wave contributions.

seen in Section 3.10 that the leaky wave is an improper wave and cannot exist by itself. However, as seen in the above, leaky waves can exist in a part of the space $\theta > \theta_l$.

In the above, the leaky wave appears discontinuously at $\theta = \theta_l$. This is, of course, not physical, and is caused by our separate evaluation of the saddle-point and the surface-wave contributions. These two contributions should be evaluated together to obtain a continuous total field. This is the modified saddle-point technique to be discussed in Chapter 15.

14.2 EXCITATION OF WAVES IN A VERTICALLY INHOMOGENEOUS MEDIUM

Let us consider the excitation of waves by a point source located in an inhomogeneous medium whose refractive index $n(z)$ varies as a function of z only (Fig. 14.8). An example is acoustic excitation in the ocean. We formulate this problem using the WKB approximation.

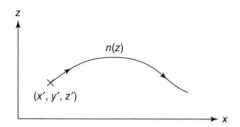

FIGURE 14.8   Wave excitation in a vertically inhomogeneous medium.

Let us consider the scalar Green's function $G(\bar{r}, \bar{r}')$, $\bar{r} = (x, y, z)$ and $\bar{r}' = (x, y', z')$ in an inhomogeneous medium whose refractive index n is a function of z only.

$$\left[\nabla^2 + k_0^2 n^2(z)\right] G(\bar{r}, \bar{r}') = -\delta(\bar{r} - \bar{r}'). \tag{14.25}$$

First, we take the Fourier transform in the x and y directions.

$$g(q_1, q_2, z) = \int\int G(\bar{r}, \bar{r}') \, \exp(jq_1 x + jq_2 y) \, dx \, dy. \tag{14.26}$$

Equation (14.25) is then transformed to

$$\left[\frac{d^2}{dz^2} + q^2(z)\right] g(q_1, q_2, z) = -\delta(z - z')e^{jq_1 x' + jq_2 y'}, \tag{14.27}$$

where $q^2(z) = k_0^2 n^2(z) - q_1^2 - q_2^2$.

The WKB solution to (14.27) is given by (Sections 3.14 and 5.5)

$$g(q_1, q_2, z) = \begin{cases} \dfrac{y_1(z)y_2(z')e^{jq_1 x' + jq_2 y'}}{\Delta} & \text{if } z > z' \\[4mm] \dfrac{y_1(z')y_2(z')e^{jq_1 x' + jq_2 y'}}{\Delta} & \text{if } z < z' \end{cases}, \tag{14.28}$$

where

$$\Delta = y_1 y_2' - y_1' y_2 = 2j,$$

$$y_1(z) = q^{-1/2} \exp\left(-j \int_{z'}^z q \, dz\right),$$

$$y_2(z) = q^{-1/2} \exp\left(+j \int_{z'}^z q \, dz\right).$$

Note that $y_1(z)$ and $y_2(z)$ represent the outgoing waves in the $+z$ and $-z$ directions satisfying the radiation condition as $z \to +\infty$ and $z \to -\infty$, respectively, and Δ is the

Wronskian of y_1 and y_2. Green's function $G(\bar{r}, \bar{r}')$ is then given by the inverse Fourier transform of $g(q_1, q_2, z)$

$$G(\bar{r}, \bar{r}') = \frac{1}{(2\pi)^2} \int \int \frac{\exp[-jf(z, q_1, q_2)]}{2jq(z)^{1/2}q(z')^{1/2}} \, dq_1 \, dq_2 \quad \text{for } z > z', \quad (14.29)$$

where $f = \int_{z'}^{z} q \, dz + q_1(x - x') + q_2(y - y')$.

Green's function for $z < z'$ is given by the same expression except that f is replaced by

$$f = -\int_{z'}^{z} q \, dz + q_1(x - x') + q_2(y - y'). \quad (14.30)$$

Equation (14.29) represents the WKB solution to Green's function. It is applicable to the case away from the turning point.

Let us evaluate (14.29) approximately using the saddle-point method (stationary phase). The saddle points $q_1 = q_{1s}$ and $q_2 = q_{2s}$ are given by

$$\frac{\partial f}{\partial q_1} = 0 \quad \text{and} \quad \frac{\partial f}{\partial q_2} = 0. \quad (14.31)$$

Green's function is then given by (14A.9) in Appendix 14.A,

$$G = \frac{1}{(2\pi)^2} \frac{e^{-jf_s}}{2jq(z)^{1/2}q(z')^{1/2}} \frac{(2\pi)e^{j(\pi/2)}}{\left(f_{11}f_{22} - f_{12}^2\right)^{1/2}}, \quad (14.32)$$

where

$$f_s = f(z, q_{1s}, q_{2s}), \quad f_{11} = \frac{\partial^2}{\partial q_1^2} f, \quad f_{22} = \frac{\partial^2}{\partial q_2^2} f, \quad f_{12} = \frac{\partial^2}{\partial q_1 \, \partial q_2} f,$$

and f_{11}, f_{22}, and f_{12} are evaluated at q_{1s} and q_{2s}. Equation (14.32) gives Green's function at (x, y, z) when the source is located at (x', y', z') in the medium with the refractive index $n(z)$. This requires that the saddle points q_{1s} and q_{2s} be found from (14.31). It is easier, however, to assume q_{1s} and q_{2s} first and then to determine (x, y, z) for these saddle points, as will be shown shortly. The saddle points give the direction of the ray at the source point, and the ray equations below can be used to determine the ray path $x = x(z)$ and $y = y(z)$.

Using the definition of f in (14.29), the saddle points q_{1s} and q_{2s} in (14.31) are given by

$$x - x' = \int_{z'}^{z} \frac{q_{1s}dz}{\left[k_0^2 n^2(z) - q_{1s}^2 - q_{2s}^2\right]^{1/2}},$$

$$y - y' = \int_{z'}^{z} \frac{q_{2s}dz}{\left[k_0^2 n^2(z) - q_{1s}^2 - q_{2s}^2\right]^{1/2}}. \tag{14.33}$$

These two equations will be shown to be identical to the ray equation of geometric optics (Chapter 15). We can also determine the physical meaning of the saddle points q_{1s} and q_{2s} as follows: Consider the ray for a given q_{1s} and q_{2s}. From (14.33), a small distance ds along this ray is given by

$$ds^2 = dx^2 + dy^2 + dz^2, \tag{14.34}$$

where

$$dx = \frac{q_{1s}dz}{q_s}, \quad q_s = \left[k_0^2 n^2(z) - q_{1s}^2 - q_{2s}^2\right]^{1/2},$$

$$dy = \frac{q_{2s}dz}{q_s}.$$

Therefore, we have

$$ds^2 = \frac{k_0^2 n^2 dz^2}{q_s^2} = \frac{k_0^2 n^2 dx^2}{q_{1s}^2} = \frac{k_0^2 n^2 dy^2}{q_{2s}^2}.$$

If we let

$$\frac{dx}{ds} = \sin \theta \cos \phi, \quad \frac{dy}{ds} = \sin \theta \sin \phi, \tag{14.35}$$

then we get

$$q_{1s} = k_0 n(z) \sin \theta \cos \phi,$$

$$q_{2s} = k_0 n(z) \sin \theta \sin \phi. \tag{14.36}$$

Equation (14.36) shows that for given saddle points q_{1s} and q_{2s}, $n(z) \sin \theta \cos \phi$ and $n(z) \sin \theta \sin \phi$ are constant. Since $q_{2s}/q_{1s} = \tan \phi = $ constant, ϕ is constant along the ray, and we have $n(z) \sin \theta(z) = $ constant along the ray. (14.37).

This is identical to Snell's law. The stationary phase solution is therefore equivalent to the geometric optics solution.

The procedure for obtaining Green's function is as follows. Suppose that the ray is started at (x', y', z') in the direction (θ_0, ϕ_0). Then q_{1s} and q_{2s} are given by

$$q_{1s} = k_0 n(z') \sin \theta_0 \cos \phi_0,$$
$$q_{2s} = k_0 n(z') \sin \theta_0 \sin \phi_0. \qquad (14.38)$$

These q_{1s} and q_{2s} are used in the ray equation (14.33) to determine (x, y) at z. These (x, y, z) and q_{1s} and q_{2s} are then used in (14.32) to obtain Green's function.

As an example, consider the parabolic refractive index profile

$$n^2(z) = n_0^2 \left[1 - \left(\frac{z}{z_0} \right)^2 \right]. \qquad (14.39)$$

The source is located at $x' = 0$ and $z' = 0$, and the ray is launched in the direction θ_0 and $\phi_0 = 0$. Then the ray equation (14.33) can be integrated to give

$$z = z_0 \cos \theta_0 \sin \left[\frac{x}{z_0 \sin \theta_0} \right]. \qquad (14.40)$$

Green's function is then given by

$$G = \frac{e^{-jfs}}{4\pi} \frac{q_{1s} q_0^{1/2}}{[x^2 q_s q_0^2 + xz q_{1s}^3]^{1/2}}, \qquad (14.41)$$

where $q_0 = q(z') = q(0)$ and q_{1s} and q_s are given in (14.38) and (14.34), respectively. If $n(z) = 1$, (14.41) reduces to $\exp(-jkr)/(4\pi r)$, as it should.

14.3 STRIP LINES

We now turn to a different topic. Strip lines have many attractive features. They are compact, inexpensive, and can easily be produced as printed circuits. The most common types have a strip of conductor on a grounded dielectric slab (Fig. 14.9). Even though this is a two-conductor line, this structure cannot support TEM modes because the dielectric occupies only a part of the cross section (see Section 4.9). Thus, in general, both E_z and H_z are nonzero, and we need to consider hybrid modes consisting of both TE and TM modes.

Even though the hybrid mode analysis is required for exact analysis of strip lines, simpler techniques can be used for approximate analysis of strip lines. One useful method is the quasi-static approximation, in which E_z and H_z are neglected and the field is assumed to be TEM. The error introduced by this approximation is usually a fraction of a percent. One drawback of the quasi-static approximation is that since this is a TEM solution, the propagation constant is independent of frequency (no

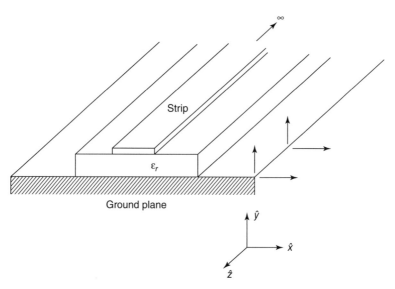

FIGURE 14.9   Strip line.

modal dispersion) except for the frequency dependence of the dielectric material itself. In this section, we start with the TEM solution and then discuss the quasi-TEM approximation and the exact hybrid solution. Important papers on this topic are compiled by Itoh (1987).

14.3.1 TEM Solution

We have discussed in Section 4.9 that if we have a transmission line consisting of two conductors and a homogeneous dielectric medium, this line can support a TEM mode. An example is that of boxed strip lines (Fig. 14.10). As shown in Section 4.9, TEM modes are given by the following:

$$\bar{E}_t = -(\nabla_t V)e^{-jkz},$$

$$\bar{H}_t = \frac{1}{Z}\hat{z} \times \bar{E}_t,$$

$$(14.42)$$

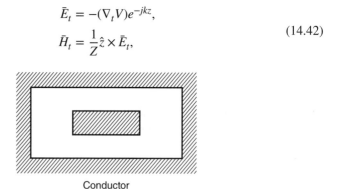

Conductor

FIGURE 14.10   Boxed strip line.

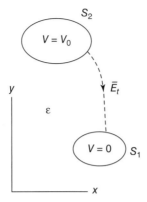

FIGURE 14.11   Cross-sectional view of a two-conductor line with a homogeneous medium.

where $\bar{E}_t$ and $\bar{H}_t$ are the electric and magnetic fields transverse to the direction of wave propagation $\hat{z}$ and are given by $k = \omega/v$, $v = (\mu\varepsilon)^{-1/2}$, and $Z = (\mu/\varepsilon)^{1/2}$, which are the wave number, the velocity of the electromagnetic wave, and the characteristic impedance of the medium, respectively. The electrostatic potential $V(x, y)$ in (14.12) satisfies the two-dimensional Laplace equation

$$\left(\frac{\partial^2}{\partial x^2} + \frac{\partial^2}{\partial y^2}\right) V(x, y) = 0. \tag{14.43}$$

We next define the voltage V_0, the current I_0, and the characteristic impedance Z_0 of the transmission line. The voltage V_0 is the electrostatic potential at one conductor with respect to the other and is given by the line integral of the electric field (Fig. 14.11).

$$V_0 = -\int_{s_1}^{s_2} \bar{E}_t \cdot d\bar{l} = \int_{s_1}^{s_2} \nabla_t V \cdot d\bar{l}. \tag{14.44}$$

The current is given by the integral of the surface current density J_s around one conductor,

$$I_0 = \oint |\bar{J}_s|\, dl. \tag{14.45}$$

The current density $\bar{J}_s$ is related to $\bar{H}_t$ by $\bar{J}_s = \hat{n} \times \bar{H}_t$, where $\hat{n}$ is the unit vector normal to the surface of the conductor, and $\bar{H}_t$ is related to $\bar{E}_t$. Thus we have

$$\bar{J}_s = \hat{n} \times \bar{H}_t = \frac{1}{Z}\,\hat{n} \times (\hat{z} \times \bar{E}_t)$$

$$= \frac{\hat{z}}{Z}(\hat{n} \cdot \bar{E}) = \frac{\hat{z}}{Z}\frac{\rho_s}{\varepsilon} = \hat{z}v\,\rho_s, \tag{14.46}$$

where ρ_s is the surface charge density. The current I_0 is therefore related to the total charge Q,

$$I_0 = \oint v \rho_s dl = vQ. \qquad (14.47)$$

The characteristic impedance Z_c of the transmission line is then given by

$$Z_c = \frac{V_0}{I_0} = \frac{V_0}{vQ} = \frac{1}{vC_e}, \qquad (14.48)$$

where C_e is the electrostatic capacitance per unit length of the line. It is, therefore, clear that once the electrostatic capacitance is found, the characteristic impedance is given by (14.48) and that the solution to the TEM line problem is reduced to finding the capacitance C_e per unit length of the line.

The capacitance C_e can be expressed using the time-averaged electric stored energy W_e

$$W_e = \frac{1}{4} C_e V_0^2 = \frac{1}{4} \frac{Q^2}{C_e}. \qquad (14.49)$$

Now W_e can be expressed either in terms of the energy density in the medium or the charge on the conductor

$$W_e = \frac{1}{4} \int_a \varepsilon |\bar{E}|^2 \, dx \, dy = \frac{1}{4} \int_a \varepsilon |\nabla_t V|^2 \, dx \, dy, \qquad (14.50)$$

$$W_e = \frac{1}{4} \oint_{s_2} \rho_s V \, dl, \qquad (14.51)$$

where a is the area between two conductors and (14.51) is the integral around the conductor s_2.

Using (14.49) and (14.50), we write

$$C_e = \frac{\displaystyle\int_a \varepsilon |\nabla_t V|^2 \, dx \, dy}{V_0^2}$$

$$= \frac{\displaystyle\int_a \varepsilon |\nabla_t V|^2 \, dx \, dy}{\left| \displaystyle\int_{s_1}^{s_2} \nabla_t V \cdot d\bar{l} \right|^2}. \qquad (14.52)$$

If the electrostatic potential $V(x, y)$ is found, the formula above can be used to calculate the capacitance C_e. We can also use (14.49) and (14.51) to obtain

$$
\frac{1}{C_e} = \frac{\oint \rho V \, dl}{Q^2}
$$

$$
= \frac{\oint dl \oint dl' \rho_s(x, y) G(x, y; \; x', y') \rho_s(x', y')}{\left| \oint \rho_s(x, y) \, dl \right|^2},
\tag{14.53}
$$

where G is Green's function satisfying the equation

$$
\nabla^2 G(\bar{r}, \bar{r}') = -\frac{\delta(\bar{r}, \bar{r}')}{\varepsilon}.
\tag{14.54}
$$

It has been shown (Collin, 1966, Chapter 4) that (14.52) and (14.53) are the variational expressions for the capacitance C_e and that (14.52) gives the upper bound on C_e while (14.53) gives the lower bound on C_e. Once the solution for $V(x, y)$ or for $\rho_s(x, y)$ is obtained, C_e is obtained by either (14.52) or (14.53).

Let us also note that the time-averaged magnetic stored energy W_m is equal to

$$
W_m = \frac{L_e I_0^2}{4} = W_e.
\tag{14.55}
$$

Therefore, the characteristic impedance Z_c, the propagation constant k, and the total transmitted power P can also be given by

$$
Z_c = \left(\frac{L_e}{C_e} \right)^{1/2},
$$

$$
k = \frac{\omega}{v} = \omega(\mu\varepsilon)^{1/2} = \omega(L_e C_e)^{1/2},
\tag{14.56}
$$

$$
P = \frac{V_0 I_0}{2} = \frac{Z_0 I_0^2}{2}.
$$

There are several techniques to obtain the solutions for $V(x, y)$ or $\rho_s(x, y)$ or the capacitance C_e. Analytical techniques include conformal mapping and the Fourier transform. The variational technique is also used with numerical techniques and Fourier transforms.

14.3.2 Quasi-Static Approximation

The TEM solutions discussed above are applicable only to two-conductor lines with a homogeneous medium. If the medium is inhomogeneous and consists of different

dielectric slabs as shown in Fig. 14.9, the TEM wave cannot exist and both TE
and TM modes are needed. However, the z components of electric and magnetic
fields are often negligibly small, and the characteristics of the transmission line
with an inhomogeneous dielectric medium can be approximated by the quasi-static
approximation to be described below. The error introduced may be less than a fraction
of a percent for most applications.

In the quasi-static approximation, we assume that the transverse electric and
magnetic field distributions are approximately equal to that of the static electric and
magnetic field distributions. The time-averaged electric and magnetic stored energies
are assumed to be

$$W_e = W_m = \tfrac{1}{4}C_e V_0^2 = \tfrac{1}{4}L_e I_0^2. \tag{14.57}$$

The characteristic impedance Z_c is given approximately by $(L_e/C_e)^{1/2}$. However, since
the inductance L_e does not depend on the dielectric material, it is the same as that of
free space. Therefore, we write

$$Z_c = Z_0 \left(\frac{C_0}{C_e}\right)^{1/2}, \tag{14.58}$$

where C_e is the actual capacitance per unit length of the line, $Z_0 = (L_e/C_0)^{1/2}$ is the
characteristic impedance of the transmission line, and C_0 is the capacitance per unit
length of the line if the medium is free space. We also approximate the propagation
constant k by

$$k = \omega(L_e C_e)^{1/2} = k_0 \left(\frac{C_e}{C_0}\right)^{1/2}, \tag{14.59}$$

where $k_0 = 2\pi/\lambda_0$ is the free-space wave number.

Under the quasi-static approximation, we only need to calculate C_e and C_0 in
(14.58) and (14.59). To find C_e in an inhomogeneous medium, we first note that $\nabla_t \times
\bar{E} = 0$, where $\nabla_t = \hat{x}(\partial/\partial x) + \hat{y}(\partial/\partial y)$, and $\bar{E}$ is given by the electrostatic potential V.

$$\bar{E} = -\nabla_t V. \tag{14.60}$$

Substituting this into $\nabla_t \cdot \bar{D} = \nabla_t \cdot (\varepsilon \bar{E}) = 0$, we get $\nabla_t \cdot (\varepsilon \nabla_t V) = 0$, which we write
as

$$\left[\frac{\partial}{\partial x}\left(\varepsilon_r \frac{\partial}{\partial x}\right) + \frac{\partial}{\partial y}\left(\varepsilon_r \frac{\partial}{\partial y}\right)\right] V(x, y) = 0, \tag{14.61}$$

where $\varepsilon_r(x, y)$ is the relative dielectric constant of the medium.

Once (14.61) is solved for $V(x, y)$ satisfying the boundary condition that $V = V_1 = $ constant on one conductor and $V = V_2 = $ constant on the other conductor, the capacitance C_e is obtained by (14.52). We can also use the other variational form (14.53) if we use

$$\nabla_t \cdot (\varepsilon \nabla_t) G(\bar{r}, \bar{r}') = -\delta(\bar{r} - \bar{r}'), \tag{14.62}$$

in place of (14.54). This is of course the same as solving (14.61) with the delta function as the source term.

14.3.3 Exact Hybrid Solution

Both the TEM solution and the quasi-static solution depend only on the electrostatic capacitance and therefore are nondispersive as long as the frequency dependence of the dielectric constant is negligible. However, these TEM solutions are approximate, and the exact solutions are not TEM but a combination of TE and TM modes. Only this exact hybrid solution exhibits the complete dispersion characteristics of a strip line.

Let us start with the exact formulation of the problem. We need both TM and TE modes as shown in (4.11) and (4.16). Thus we write the fields in air ($i = 1$) and in dielectric ($i = 2$)

$$
\begin{aligned}
E_{zi} &= k_{ci}\phi_i(x, y)e^{-j\beta z}, \\
H_{zi} &= k_{ci}\psi_i(x, y)e^{-j\beta z}, \\
\bar{E}_{ti} &= [-j\beta\nabla_t\phi_i(x, y) + j\omega\mu_i\hat{z} \times \nabla_i\psi_i(x, y)]e^{-j\beta z}, \\
\bar{H}_{ti} &= [-j\omega\varepsilon_i\hat{z} \times \nabla_t\phi_i(x, y) - j\beta\nabla_t\psi_i(x, y)]e^{-j\beta z},
\end{aligned}
\tag{14.63}
$$

where

$$
\begin{aligned}
\left(\nabla_t^2 + k_{ci}^2\right)\phi_i &= 0, \\
\left(\nabla_t^2 + k_{ci}^2\right)\psi_i &= 0, \\
k_{ci}^2 &= k_i^2 - \beta^2.
\end{aligned}
$$

The boundary conditions are that the tangential components of the electric and magnetic fields are continuous across the air–dielectric boundary, the tangential electric field is zero on the conductor, and the field satisfies the radiation condition at infinity. Several methods have been proposed to solve this exact hybrid problem. The spectral domain approach has been used successfully to solve this problem. We outline this approach in a later section. A numerical technique such as the finite-element method has also been used to solve this problem (*IEEE Trans. MTT*, October 1985).

14.4 WAVES EXCITED BY ELECTRIC AND MAGNETIC CURRENTS PERPENDICULAR TO DIELECTRIC LAYERS

In Section 14.1, we considered the two-dimensional problem of wave excitation in dielectric layers. In this section, we present formulations for a general three-dimensional problem. Consider the dielectric layers shown in Fig. 14.12. It can be shown that the vertical electric dipole in the presence of dielectric layers excites only the TM waves, which are generated by the electric Hertz vector $\Pi'\hat{z}$. Also, the vertical magnetic dipole excites only the TE waves, which are generated by the magnetic Hertz vector $\Pi''\hat{z}$.

Let us first consider the waves excited by the electric current density $J_z(x', y', z')$ pointed in the z direction. In layer 1, the electric Hertz vector $\Pi'\hat{z}$ satisfies the scalar wave equation

$$\left(\nabla^2 + k_1^2\right) \Pi' = -\frac{J_z}{j\omega\varepsilon_1}. \tag{14.64}$$

Let us take Fourier transform in the x and y directions.

$$\begin{aligned}
\tilde{\Pi}'(\alpha, \beta, z) &= \int \Pi'(x, y, z)e^{j\alpha x + j\beta y} \, dx \, dy, \\
\tilde{J}_z(\alpha, \beta, z) &= \int J_z(x, y, z)e^{j\alpha x + j\beta y} \, dx \, dy.
\end{aligned} \tag{14.65}$$

Equation (14.64) can then be transformed to

$$\left(\frac{d^2}{dz^2} + \gamma_1^2\right) \tilde{\Pi}' = -\frac{\tilde{J}_z}{j\omega\varepsilon_1}, \tag{14.66}$$

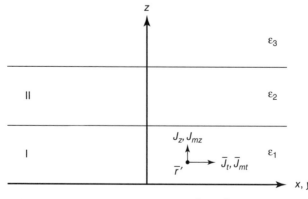

FIGURE 14.12 Dielectric layers excited by $\bar{J}$ and $\bar{J}_m$ located at $\bar{r}'(x', y', z')$

where

$$\gamma_1 = \begin{cases} \left(k_1^2 - \alpha^2 - \beta^2\right)^{1/2} & \text{if } k_1^2 > \alpha^2 + \beta^2 \\ -j\left(\alpha^2 + \beta^2 - k_1^2\right)^{1/2} & \text{if } k_1^2 < \alpha^2 + \beta^2. \end{cases}$$

Now the tangential electric and magnetic fields are given by

$$\bar{E}_t = \nabla_t \frac{\partial}{\partial z}\, \Pi', \tag{14.67}$$

$$\bar{H}_t = j\omega\varepsilon(\nabla_t \Pi' \times \hat{z}).$$

Fourier transforms of $\bar{E}_t$ and $\bar{H}_t$ are obtained from (14.67) by letting $\nabla_t = -j\alpha\hat{x} - j\beta\hat{y} = -j\bar{\alpha}$. We also note that while $\bar{H}_t$ is proportional to Π', $\bar{E}_t$ is proportional to $\partial\Pi'/\partial z$. This means that when a delta function source is located at z', $\bar{E}_t$ changes its sign depending on whether $z > z'$ or $z < z'$. Thus it is more convenient to express $\bar{E}_t$ by the current, which changes its sign in the same manner. We introduce the magnetic voltage V'_m, which is proportional to $\bar{H}_t$, and the magnetic current I'_m, which is proportional to $\bar{E}_t$, and then write the Fourier transforms of $\bar{E}_t$ and $\bar{H}_t$ as follows:

$$\bar{e} = \int \bar{E}_t e^{j\alpha x + j\beta y}\, dx\, dy,$$

$$\bar{h} = \int \bar{H}_t e^{j\alpha x + j\beta y}\, dx\, dy,$$

$$\bar{e} = -j\bar{\alpha}\left(-I'_m\right)\frac{\tilde{J}_z}{j\omega\varepsilon_1}, \tag{14.68}$$

$$\bar{h} = -j\hat{z} \times \bar{\alpha}\left(-V'_m\right)\frac{\tilde{J}_z}{j\omega\varepsilon_1},$$

where

$$\bar{\alpha} = \alpha\hat{x} + \beta\hat{y},$$

$$V'_m = j\omega\varepsilon_i\, G,$$

$$Z'_m = \frac{\omega\varepsilon_i}{\gamma_i}.$$

G is Green's function given by

$$\left(\frac{d^2}{dz^2} + \gamma_1^2\right) G = -\delta(z - z'). \tag{14.69}$$

The boundary conditions that the tangential electric and magnetic fields are continuous are equivalent to the continuity of the voltage V'_m and the current I'_m at the

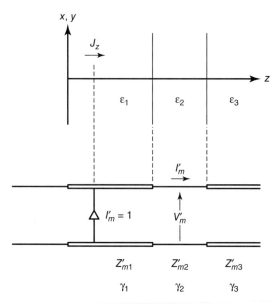

FIGURE 14.13 Equivalent transmission line for J_z. Perfect conducting plane is "open terminal."

junctions. Thus V'_m and I'_m can be formed by using the equivalent transmission line shown in Fig. 14.13. Note that at perfectly conducting plane, $\bar{e} = 0$ and thus $I'_m = 0$. Therefore, the perfectly conducting plane is represented by an open terminal in the transmission line. Note also that the choice of V_m and I_m in (14.68) are made so that the current injected at the source point is $I'_m = 1$ when $\tilde{J}_z = 1$, as shown in Fig. 14.13.

The procedure of finding $\bar{E}_t$ and $\bar{H}_t$ is as follows: First we solve the transmission-line problem in Fig. 14.13 with the unit current $I_m = 1$ at the source point z'. Each layer has the characteristic impedance $Z'_{mi} = \omega \varepsilon_i / \gamma_i$ $(i = 1, 2, ...)$, and the propagation constant γ_i. Once we find the voltage and current distribution V'_m and I'_m, $\bar{e}$ and $\bar{h}$ are obtained in (14.68). We then take the inverse Fourier transform to obtain $\bar{E}_t$

$$\bar{E}_t = \frac{1}{(2\pi)^2} \int \bar{e}\, e^{-j\alpha x - j\beta y}\, d\alpha\, d\beta.$$

Similarly, we get $\bar{H}_t$ from $\bar{h}$. The Hertz vector $\Pi' \hat{z}$ is given by

$$\Pi' = \frac{1}{(2\pi)^2} \int \frac{G\tilde{J}_z}{j\omega\varepsilon_1} e^{-j\alpha x - j\beta y}\, d\alpha\, d\beta. \tag{14.70}$$

As an example, consider an electric dipole $J_z = I_0 L_0 \delta(\bar{r} - \bar{r}')$ located in air $\varepsilon = \varepsilon_0$ at $\bar{r}' = (0, 0, h)$ above the ground with dielectric constant ε_2. From Fig. 14.13, the voltage incident on the boundary $z = 0$ is

$$V_{in} = \frac{Z_1'}{2} e^{-j\gamma_1 h}.$$

Then the reflected wave is

$$V_{ref} = \left| \frac{Z_2' - Z_1'}{Z_2' + Z_1'} \right| \frac{Z_1'}{2} e^{-j\gamma_1 h - j\gamma_1 z}.$$

The reflected Hertz vector Π_r' in air is then given by (14.70)

$$\Pi_r' = \frac{1}{(2\pi)^2} \int \frac{V_{ref}}{j\omega\varepsilon_1} \left[\frac{I_0 L_0}{j\omega\varepsilon_1} \right] e^{-j\alpha x - j\beta y} \, d\alpha \, d\beta.$$

When converted to a cylindrical system, this is identical to the expression (15.23) for the Sommerfeld dipole problem.

Next consider the magnetic current source J_{mz}. We use the duality principle and find the following:

$$\bar{E}_t = j\omega\mu(\hat{z} \times \nabla_t \Pi''),$$

$$\bar{H}_t = \nabla_t \frac{\partial}{\partial z} \Pi''. \tag{14.71}$$

The Fourier transforms $\bar{e}$ and $\bar{h}$ are therefore given by

$$\bar{e} = -j\bar{\alpha} \times \hat{z}(-V'') \frac{\tilde{J}_{mz}}{j\omega\mu},$$

$$\bar{h} = -j\bar{\alpha}(-I'') \frac{\tilde{J}_{mz}}{j\omega\mu},$$

$$\bar{\alpha} = \alpha\hat{z} + \beta\hat{y}, \tag{14.72}$$

$$V'' = j\omega\mu G,$$

$$Z_i'' = \frac{\omega\mu}{\gamma_i}.$$

$\tilde{J}_{mz}$ is the Fourier transform of J_{mz}. The voltage V'' and current I'' are obtained by solving the transmission-line problem (Fig. 14.14). The magnetic Hertz vector $\Pi''\hat{z}$ is given by

$$\Pi'' = \frac{1}{(2\pi)^2} \int \frac{G\tilde{J}_{mz}}{j\omega\mu} e^{-j\alpha x - j\beta y} \, d\alpha \, d\beta. \tag{14.73}$$

14.5 WAVES EXCITED BY TRANSVERSE ELECTRIC AND MAGNETIC CURRENTS IN DIELECTRIC LAYERS

Microstrip lines carry electric currents in the x–y plane embedded in layers. Frequency-selective surfaces are composed of periodic patches and apertures carrying electric and magnetic currents in the x–y plane in dielectric layers. It is therefore necessary to formulate the problem for waves excited by the currents $\bar{J}_t$ and $\bar{J}_{mt}$ which are transverse to the z direction.

First we note that Π_x and Π_y are simply related to J_x and J_y. However, Π_x and Π_y are coupled at the dielectric interface and the formulation will become quite involved. A simpler technique is to use $\Pi'\hat{z}$ and $\Pi''\hat{z}$, because there is no coupling between Π' and Π'' at the dielectric interface. This formulation was proposed by Itoh (1980) and by Felsen and Marcuvitz (1973). Both give the same results. We will follow Felsen–Marcuvitz formulation. First we note that $\bar{E}$ and $\bar{H}$ are given by

$$\bar{E} = \nabla \times \nabla \times (\Pi'\hat{z}) - j\omega\mu\nabla \times (\Pi''\hat{z}),$$

$$\bar{H} = j\omega\varepsilon\nabla \times (\Pi'\hat{z}) + \nabla \times \nabla \times (\Pi''\hat{z}). \tag{14.74}$$

The Hertz potentials Π' and Π'' are produced by $\bar{J}$ and $\bar{J}_m$.

$$\Pi' = g_0' \frac{J_z}{j\omega\varepsilon} + \left(\frac{1}{j\omega\varepsilon} \bar{J}_t \frac{\partial}{\partial z'} + \bar{J}_{mt} \times \hat{z} \right) \cdot \nabla_t' g',$$

$$\Pi'' = g_0'' \frac{J_{mz}}{j\omega\mu} + \left(\hat{z} \times \bar{J}_t + \frac{1}{j\omega\mu} \bar{J}_{mt} \frac{\partial}{\partial z'} \right) \cdot \nabla_t' g'',$$

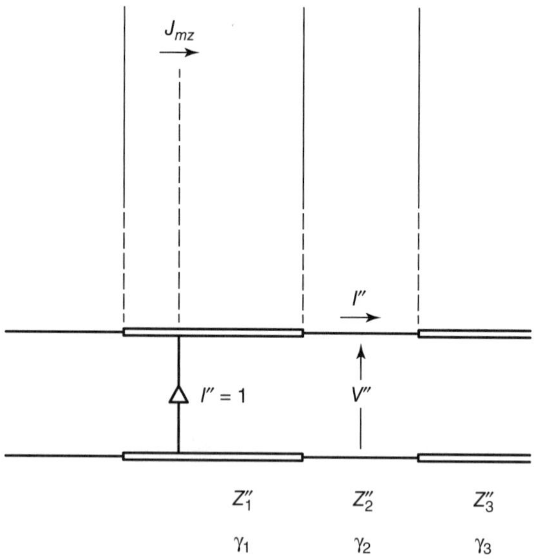

FIGURE 14.14 Transmission line for J_{mz}. Perfect conducting plane is a short circuit.

$$(\nabla^2 + k^2)g'_0 = -\delta(\bar{r} - \bar{r}'),$$

$$(\nabla^2 + k^2)g''_0 = -\delta(\bar{r} - \bar{r}'),$$

$$-\nabla_t^2 g' = g'_0,$$

$$-\nabla_t^2 g'' = g''_0,$$

$$\bar{J} = J_z \hat{z} + \bar{J}_t,$$

$$\bar{J}_m = J_{mz} \hat{z} + \bar{J}_{mt}.$$

(14.75)

The first terms with J_z and J_{mz} were discussed in Section 14.4. In the above, ∇'_t is the transverse gradient operator on the source $\bar{r}'$ coordinate and $\partial/\partial z'$ is the derivative with respect to z'. The derivation of (14.75) is given by Felsen and Marcuvitz (1973, p. 445). Now we take Fourier transform of the transverse electric and magnetic field

$$\bar{e}_t = \int \bar{E}_t e^{-j\alpha x - j\beta y} \, dx \, dy$$

$$= \bar{\bar{e}} \cdot \bar{J},$$

$$\bar{h}_t = \int \bar{H}_t e^{j\alpha x + j\beta y} \, dx \, dy$$

$$= \bar{\bar{h}} \cdot \bar{J},$$

(14.76)

where $\bar{J}$ is the Fourier transform of $\bar{J}_t$

$$\bar{J} = \int \bar{J}_t e^{j\alpha x + j\beta y} \, dx \, dy$$

and $\bar{\bar{e}}$ and $\bar{\bar{h}}$ are Green's dyadic. Once we find $\bar{e}_t$ and $\bar{h}_t$, we get $\bar{E}_t$ and $\bar{H}_t$ by taking the inverse Fourier transform

$$\bar{E}_t = \frac{1}{(2\pi)^2} \int \bar{e}_t e^{-j\alpha x - j\beta y} \, dx \, d\beta,$$

$$\bar{H}_t = \frac{1}{(2\pi)^2} \int \bar{h}_t e^{-j\alpha x - j\beta y} \, d\alpha \, d\beta.$$

(14.77)

Green's dyadic $[\bar{\bar{e}}]$ and $[\bar{\bar{h}}]$ can be obtained based on (14.75). We give the final results for $\bar{\bar{e}}$ and $\bar{\bar{h}}$ using the equivalent transmission-line voltages and currents

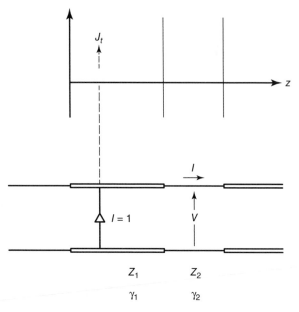

FIGURE 14.15   Equivalent transmission line excited by transverse current source J_t. Use γ_i and Z_i' to find V' and I', and Z_i to find V'' and I''.

shown in Fig. 14.15.

$$[\bar{\bar{e}}] = \begin{bmatrix} \alpha^2 & \alpha\beta \\ \alpha\beta & \beta^2 \end{bmatrix} \frac{-V'}{\alpha^2 + \beta^2} + \begin{bmatrix} \beta^2 & -\alpha\beta \\ -\alpha\beta & \alpha^2 \end{bmatrix} \frac{-V''}{\alpha^2 + \beta^2},$$

$$[\bar{\bar{h}}] = \begin{bmatrix} \alpha\beta & \alpha^2 \\ -\alpha^2 & -\alpha\beta \end{bmatrix} \frac{I'}{\alpha^2 + \beta^2} + \begin{bmatrix} \alpha\beta & -\alpha^2 \\ \beta^2 & -\alpha\beta \end{bmatrix} \frac{-I''}{\alpha^2 + \beta^2}, \tag{14.78}$$

where

$$V' = -\frac{\partial}{\partial z}\left(\frac{1}{j\omega\varepsilon_1}\frac{\partial}{\partial z'}g'\right),$$

$$I' = \frac{\varepsilon}{\varepsilon_1}\frac{\partial}{\partial z'}g',$$

$$V'' = j\omega\mu g'',$$

$$I'' = -\frac{\partial}{\partial z}g''.$$

These definitions of V', V'', I', and I'' are used to derive $[\bar{\bar{e}}]$ and $[\bar{\bar{h}}]$. They are defined so that the boundary conditions at dielectric interfaces, that the tangential electric and magnetic fields are continuous, are automatically satisfied.

In actual calculation, we solve the transmission line in Fig. 14.15 with the unit current source as shown. We get V' and I' if we use the characteristic impedance Z', and we get V'' and I'' if we use the characteristic impedance Z''. These are then substituted in (14.78) to obtain Green's dyadic $[\bar{\bar{e}}]$ and $[\bar{\bar{h}}]$.

The characteristic impedance and the propagation constant for the ith region are given by

$$
\begin{aligned}
Z'_i &= \frac{\gamma_i}{\omega \varepsilon_i}, \\[6pt]
Z''_i &= \frac{\omega \mu}{\gamma_i}, \\[6pt]
\gamma_i &= \left(k_i^2 - \alpha^2 - \beta^2\right)^{1/2} \quad \text{if } k_i^2 > \alpha^2 + \beta^2 \\[4pt]
&= -j \left(\alpha^2 + \beta^2 - k_i^2\right)^{1/2} \quad \text{if } k_i^2 < \alpha^2 + \beta.
\end{aligned}
\tag{14.79}
$$

At conducting surfaces, V' and V'' are zero (short-circuited); the Fourier transform e_z of E_z is obtained by using $\nabla \cdot \bar{E} = 0$.

$$
e_z = -\frac{\alpha e_x + \beta e_y}{\gamma},
\tag{14.80}
$$

where $\bar{e}_t = \hat{x} e_x + \hat{y} e_y$.

Let us next consider the excitation by apertures embedded in dielectric layers (Fig. 14.16). The Fourier transforms of the transverse electric and magnetic fields are given by

$$
\begin{aligned}
\bar{e}_t &= \bar{\bar{e}}_m \cdot \bar{e}_a, \\[4pt]
\bar{h}_t &= \bar{\bar{h}}_m \cdot \bar{e}_a,
\end{aligned}
\tag{14.81}
$$

where $\bar{e}_a$ is the Fourier transform of the aperture field $\bar{E}_a$. See Fig. 14.16.

Using the equivalent transmission line in Fig. 14.16, we get V'_m and I'_m using Z'_{mi} and get V''_m and I''_m using Z''_{mi}. Green's dyadic $\bar{\bar{e}}_m$ and $\bar{\bar{h}}_m$ in matrix form are given by

$$
\begin{aligned}
[\bar{\bar{e}}_m] &= \begin{bmatrix} \alpha^2 & \alpha\beta \\ \alpha\beta & \beta^2 \end{bmatrix} \frac{I'_m}{\alpha^2 + \beta^2} + \begin{bmatrix} \beta^2 & -\alpha\beta \\ -\alpha\beta & \alpha^2 \end{bmatrix} \frac{I''_m}{\alpha^2 + \beta^2}, \\[10pt]
[\bar{\bar{h}}_m] &= \begin{bmatrix} \alpha\beta & \alpha^2 \\ -\alpha^2 & -\alpha\beta \end{bmatrix} \frac{-V'_m}{\alpha^2 + \beta^2} + \begin{bmatrix} \alpha\beta & -\alpha^2 \\ \beta^2 & -\alpha\beta \end{bmatrix} \frac{V''_m}{\alpha^2 + \beta^2}, \\[10pt]
Z'_m &= \frac{\omega\varepsilon}{\gamma_i}, \quad Z''_m = \frac{\gamma_i}{\omega\mu}.
\end{aligned}
\tag{14.82}
$$

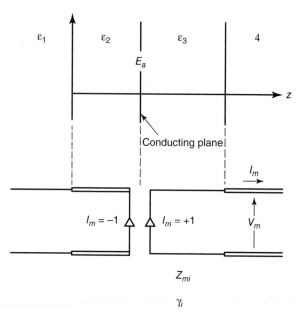

FIGURE 14.16   Equivalent transmission line for J_{mt} (the aperture field $\bar{E}_a$). Use γ_i and Z'_{mi} to find V'_m and I'_m, and γ_i and Z''_{mi} to find V''_m and I''_m.

At conducting surfaces, I'_m and I''_m are zero (open-circuited). The voltages and currents are defined by

$$V'_m = j\omega\varepsilon g',$$

$$I'_m = -\frac{\partial}{\partial z} g',$$

$$V''_m = -\frac{1}{j\omega\mu} \frac{\partial}{\partial z} \frac{\partial}{\partial z'} g'',$$

$$I''_m = \frac{\partial}{\partial z'} g''.$$

(14.83)

These definitions are needed to derive (14.82), but they are not needed to find Green's dyadic.

14.6 STRIP LINES EMBEDDED IN DIELECTRIC LAYERS

Making use of the formulation given in Section 14.5, we now discuss the method to calculate the propagation constant of a strip line (Fig. 14.17). At $z = 0$, the tangential electric field is zero on the strip. We also note that the unknown in the problem is

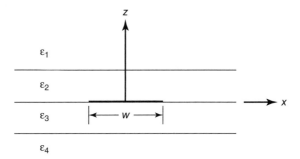

FIGURE 14.17   Strip lines. Wave propagates in the y direction.

β, the propagation constant in the y direction. Thus we only have the integral with respect to α. We then have

$$\bar{E}_t = \frac{1}{2\pi} \int \bar{\bar{e}} \cdot \bar{J} e^{-j\alpha x} \, d\alpha = 0 \tag{14.84}$$

at $z = 0$ and $|x| < w/2$.

Now we use the moment method. We express $\bar{J}$, the Fourier transform of $\bar{J}_t$ in a series of N basis functions $\bar{J}_n$ and N unknown coefficients C_n.

$$\bar{J} = \sum_{n=1}^{N} \bar{J}_n \, C_n, \tag{14.85}$$

where

$$\bar{J}_n = \begin{bmatrix} J_{nx} \\ J_{ny} \end{bmatrix},$$

$$\bar{C}_n = \begin{bmatrix} C_1 \\ C_2 \\ \vdots \\ C_N \end{bmatrix},$$

and $\bar{J}_n$ is the Fourier transform of the basis function $\bar{J}_{tn}$

$$\bar{J}_n = \int \bar{J}_{tn} e^{+j\alpha x} \, dx. \tag{14.86}$$

We then multiply (14.84) from the left by the transpose of $\bar{J}_{tm}$ and integrate over the strip with respect to x. This gives the transpose of the conjugate of $\bar{J}_m$, $m = 1, 2, \ldots, N$. Thus we get the following $N \times N$ matrix equation

$$[K_{mn}][C_n] = 0, \tag{14.87}$$

where

$$K_{mn} = \int \left[J_{mx}^* \, J_{my}^* \right] [\bar{e}] \begin{bmatrix} J_{nx} \\ J_{ny} \end{bmatrix} d\alpha.$$

In order to have nonzero solution for $[C_n]$, the determinant of $[K]$ must be zero.

$$|K_{mn}| = 0. \tag{14.88}$$

The solution of (14.88) gives the propagation constant β. The foregoing technique of using the basis function as the weighting function is called *Galerkin's method*.

The basis functions satisfying the edge condition have been proposed (Itoh, 1980). For example, we can use for $n = 1, 2, \dots, N$,

$$J_{tnx} = \frac{\sin(2n\,\pi x/w)}{[1 - (2x/w)^2]^{1/2}},$$
$$J_{tny} = 0, \tag{14.89}$$

and for $n = N + 1, N + 2, \dots, 2N$,

$$J_{tnx} = 0,$$
$$J_{tny} = \frac{\cos[2(n-1)\pi x/w]}{[1 - (2x/w)^2]^{1/2}}. \tag{14.90}$$

The corresponding Fourier transforms are

$$J_{nx} = \frac{\pi w}{4j} \left[J_0 \left(\left| \frac{w\alpha}{2} + n\pi \right| \right) - J_0 \left(\left| \frac{w\alpha}{2} - n\pi \right| \right) \right],$$
$$J_{ny} = \frac{\pi w}{4j} \cdot \left[J_0 \left(\left| \frac{w\alpha}{2} + (n-1)\pi \right| \right) + J_0 \left(\left| \frac{w\alpha}{2} - (n-1)\pi \right| \right) \right]. \tag{14.91}$$

14.7   PERIODIC PATCHES AND APERTURES EMBEDDED IN DIELECTRIC LAYERS

Let us consider periodically placed patches and apertures embedded in dielectric layers. These structures are often used as frequency-selective surfaces. For periodic structures, we can use all the formulations developed in the preceding sections if we replace Fourier integrals with space harmonic Floquet representations (see Chapter 7). Specifically, we make the following changes:

$$\frac{1}{(2\pi)^2} \int \int f(\alpha, \beta) e^{-j\alpha x - j\beta y} \, d\alpha \, d\beta \rightarrow \sum_m \sum_n \frac{f(\alpha_m, \beta_n)}{l_x \, l_y} e^{-j\alpha_m x - j\beta_n y}, \tag{14.92}$$

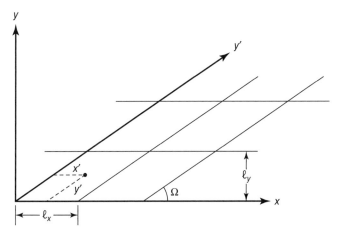

FIGURE 14.18   Skewed periodic structure.

where

$$\alpha_m = \alpha_0 + \frac{2m\pi}{l_x},$$

$$\beta_n = \beta_0 + \frac{2n\pi}{l_y},$$

$$\alpha_0 = k \sin \theta_0 \cos \phi_0,$$

$$\beta_0 = k \sin \theta_0 \sin \phi_0.$$

The periods in the x and the y directions are l_x and l_y, respectively, and (θ_0, ϕ_0) is the direction of the incident wave.

For skewed periodic structure (Fig. 14.18), noting that it is periodic in x' and y', we have

$$y' \sin \Omega = y,$$

$$x' + y' \cos \Omega = x. \tag{14.93}$$

From this, we get $y' = y/\sin \Omega$ and $x' = x - y \cot \Omega$. The Floquet modes, which are periodic in x' and y', are given by

$$\exp\left[-j\alpha_0 x - j\frac{2\pi x}{l_x} x' - j\beta_0 y - j\frac{2\pi n \sin \Omega}{l_y} y' \right]$$

$$= \exp\left[-j \left(\alpha_0 + \frac{2\pi m}{l_x} \right) x - j \left(\beta_0 + \frac{2\pi n}{l_y} - \frac{2m\pi}{l_x} \cot \Omega \right) y \right].$$

Therefore, we use for skew periodic structure,

$$\alpha_m = \alpha_0 + \frac{2m\pi}{l_x},$$

$$\beta_{mn} = \beta_0 + \frac{2n\pi}{l_y} - \frac{2m\pi}{l_x}\cot\Omega. \tag{14.94}$$

Let us consider the current distribution on the patch in the periodic structure. The total tangential electric field is the sum of the incident and the scattered waves, and this must be zero on the patch. Therefore, we write

$$\bar{E}_{\text{inc}} + \bar{E}_t = 0 \quad \text{on patch.} \tag{14.95}$$

In Floquet mode representations, we have

$$\bar{E}_{\text{inc}} + \frac{1}{l_x l_y}\sum_m\sum_n \bar{\bar{e}}\bar{J}e^{-j\alpha_m x - j\beta_n y} = 0. \tag{14.96}$$

We use the moment method and express the current $\bar{J}_t$ in a series of the basis functions $\bar{J}_{tj}$ and its Fourier transform $\bar{J}_j$.

$$\bar{J}_t = \sum_j \bar{J}_{tj}\, C_j,$$

$$\bar{J} = \sum_j \bar{J}_j\, C_j. \tag{14.97}$$

We substitute (14.97) into (14.96), multiply (14.96) by the transpose of $\bar{J}_{ti}$, and integrate over the patch as we did in Section 14.6. We then get

$$[K_{ij}][C_j] + [E_i] = 0, \tag{14.98}$$

where

$$K_{ij} = \frac{1}{l_x l_y}\sum_m\sum_n [J_{ix}^* J_{iy}^*]\,[\bar{\bar{e}}]\begin{bmatrix} J_{jx} \\ J_{jy} \end{bmatrix},$$

$$[E_i] = [J_{i0}^*]^t[\bar{E}_0],$$

where $[J_{i0}^*]^t$ is the transpose of the conjugate of J_i evaluated at $\alpha = \alpha_0$ and $\beta = \beta_0$, and the incident wave $\bar{E}_{\text{inc}}$ is given by $\bar{E}_{\text{inc}} = \bar{E}_0\, e^{-j\alpha_0 x - j\beta_0 y}$. The solution of (14.98) gives the coefficient C_j and the current distribution is given by (14.97).

Next we consider the field of aperture embedded in dielectric layers. On one side of the aperture, the tangential magnetic field is given by $\bar{H}_{\text{inc}} + \bar{H}_1$, where $\bar{H}_1$ is the contribution from the aperture field $\bar{E}_a$. On the other side we have the magnetic field $\bar{H}_2$ contributed by the same aperture field $\bar{E}_a$. These two magnetic fields must be equal at the aperture and therefore we write

$$\bar{H}_{\text{inc}} + \bar{H}_1 = \bar{H}_2 \quad \text{on aperture.} \tag{14.99}$$

Making use of the formulations given in (14.81) we express $\bar{H}_1$ and $\bar{H}_2$ in terms of the aperture field $\bar{E}_a$.

$$\bar{H}_1 = \frac{1}{l_x \, l_y} \sum_m \sum_n \bar{\bar{h}}_m \cdot \bar{e}_a \, e^{-j\alpha_m x - j\beta_n y}. \tag{14.100}$$

$\bar{H}_2$ is given in the same form except that $\bar{\bar{h}}_m$ is chosen for the other side of the aperture.

Next, the aperture field $\bar{E}_a$ and its Fourier transform $\bar{e}_a$ are expressed in series of basis functions $\bar{E}_j$ and its Fourier transform $\bar{e}_j$ with unknown coefficients C_j.

$$\begin{aligned} \bar{E}_a &= \sum_j \bar{E}_j \, C_j, \\ \bar{e}_a &= \sum_j \bar{e}_j \, C_j. \end{aligned} \tag{14.101}$$

We then take the vector product of the aperture basis functions $\bar{E}_i$ and (14.99), and integrate over the aperture

$$\int dx \, dy \bar{E}_i \times (\bar{H}_{\text{inc}} + \bar{H}_1) = \int dx \, dy \bar{E}_i \times \bar{H}_2. \tag{14.102}$$

The above is the equivalent of multiplying (14.99) by $\hat{z} \times \bar{E}_i$ and integrating over the aperture. Thus we obtain the following matrix equation:

$$[H_i] + [L_{ij}][C_j] = [M_{ij}][C_j], \tag{14.103}$$

where

$$L_{ij} = \frac{1}{l_x \, l_y} \sum_m \sum_n \left[-e_{iy}^* \; e_{ix}^* \right] \, [\bar{\bar{h}}] \begin{bmatrix} e_{jx} \\ e_{jy} \end{bmatrix}.$$

M_{ij} is the same as L_{ij} except that $[\bar{\bar{h}}]$ is on the other side of the aperture.

$$\begin{aligned} [H_i] &= \left[-e_{iy0}^* \; e_{ix0}^* \right] [\bar{H}_0], \\ \bar{H}_{\text{inc}} &= \bar{H}_0 \, e^{-j\alpha_0 x - j\beta_0 y}. \end{aligned} \tag{14.104}$$

The matrix equation (14.103) is now solved for the unknown coefficients $[C_j]$ and the final expression for the aperture field is given by (14.101).

PROBLEMS

14.1   Find the field E_y where an electric line source I_0 is located as shown in Fig. P14.1. Also find the radiation field.

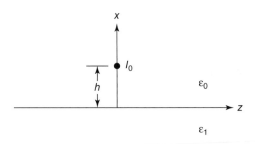

FIGURE P14.1   Line source above a dielectric half-space.

14.2   A magnetic line source I_m is located above the impedance plane where $E_z/H_y = Z$ is given (Fig. P14.2). Find the field for $x > 0$ and the radiation field, where $Z = R_s + jX_s = [j\omega(\mu_0/\sigma)]^{1/2}$. Do this when $Z = jX$.

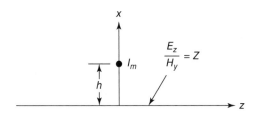

FIGURE P14.2   Impedance surface.

14.3   Consider Green's function for the parabolic profile shown in (14.39). If $\theta_0 = 45°$, $\phi_0 = 0$, $z_0 = 1$, and $n_0 = 2$, trace the ray and calculate the magnitude of Green's function.

14.4   Find the characteristic impedance of the strip line shown in Fig. P14.4. First find Green's function and use (14.53). The charge density ρ_s satisfies the edge condition and is given by

$$\rho_s = \frac{\rho_0 \delta(y)}{[1 - (2x/W_0)^2]^{1/2}}.$$

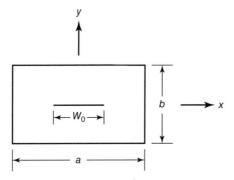

FIGURE P14.4   Strip line.

14.5   Use the quasi-static approximation to find the normalized phase velocity and the characteristic impedance of the coaxial line filled with two concentric dielectric materials as shown in Fig. P14.5, where a is the radius of the outer conductor and b is the radius of the inner conductor. $a = 0.5$ cm, $b = 0.2$ cm, $c = 0.3$ cm, $\varepsilon_1/\varepsilon_0 = 1.5$, and $\varepsilon_2/\varepsilon_0 = 2$.

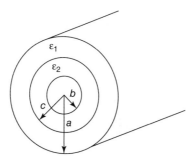

FIGURE P14.5   Coaxial line.

14.6   A vertical electric dipole is located in dielectric layers as shown in Fig. P14.6. Find the radiation field in air.

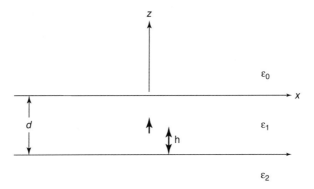

FIGURE P14.6   Electric dipole in a dielectric layer.

14.7 The vertical dipole in Fig. P14.6 is replaced by a horizontal dipole pointed in the x direction. Find the radiation field in air.

14.8 A plane wave is normally incident from the left on the layer shown in Fig. 14.16. Find an integral equation for the aperture field $\bar{E}_a$.

14.9 For the strip line shown in Fig. 14.17, find an equation to determine the propagation constant.

14.10 Find an equation to determine the current on the periodic patches shown in Fig. P14.10.

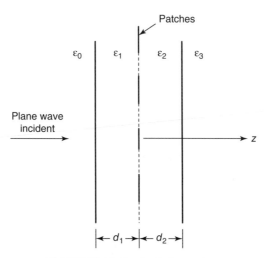

FIGURE P14.10   Periodic patches.

14.11 Find an equation to determine the aperture field on the periodic aperture shown in Fig. P14.11.

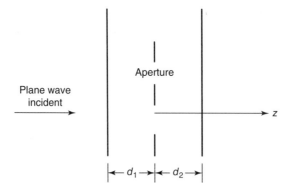

FIGURE P14.11   Periodic apertures.

CHAPTER 15

RADIATION FROM A DIPOLE ON THE CONDUCTING EARTH

15.1 SOMMERFELD DIPOLE PROBLEM

One of the important practical problems in radio is the determination of the characteristics of radio-wave propagation over the earth. The theoretical investigation of this problem, however, involved some subtle mathematical analyses that attracted attention from a great number of mathematicians and scientists over the past several decades. Historically, the problem dates back to the work by Zenneck in 1907, in which he investigated the characteristics of the wave propagating over the earth's surface, now called the *Zenneck wave*. Sommerfeld in 1909 investigated the excitation of the Zenneck wave by a dipole source. One portion of this solution had all the characteristics of the Zenneck wave on the surface and thus was called the *surface wave*.

The work of Sommerfeld, however, initiated a widely known controversy which was not completely resolved until the works of Ott, Van der Waerden, and Banos in the late 1940s. The controversy ranged over the existence of the Zenneck surface wave, the definition of the surface wave, the choice of the branch cuts, and the poles in right or wrong Riemann surfaces. In addition, the 1909 paper by Sommerfeld contained an error in a sign, adding considerable confusion, even though the 1926 paper by Sommerfeld was correct (see Banos, 1966; Brekhovskikh, 1960; Sommerfeld, 1949; Wait, 1962, 1981; see also related topics in Bremmer, 1949, and Wait, 1982). In this chapter, we present a systematic study of this problem, and whenever appropriate, we indicate the source of the historical controversies.

Electromagnetic Wave Propagation, Radiation, and Scattering: From Fundamentals to Applications,
Second Edition. Akira Ishimaru.
© 2017 by The Institute of Electrical and Electronic Engineers, Inc. Published 2017 by John Wiley & Sons, Inc.

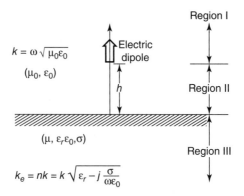

FIGURE 15.1 Vertical dipole above the ground.

15.2 VERTICAL ELECTRIC DIPOLE LOCATED ABOVE THE EARTH

Let us consider the radiation from a vertical dipole above the earth (Fig. 15.1). The earth is characterized by the relative dielectric constant ε_r and the conductivity σ or, equivalently, the complex relative dielectric constant $\varepsilon/\varepsilon_0$ or complex index of refraction n.

$$n^2 = \frac{\varepsilon}{\varepsilon_0} = \varepsilon_r - j\frac{\sigma}{\omega\varepsilon_0}. \tag{15.1}$$

The dipole is located at $z = h$ in air, where the wave number is k; the wave number within the earth is k_e. In this example k is real and k_e is complex, but the formulation is equally applicable to cases where k is complex (such as in the ionosphere) and k_e is real (air), or both k and k_e are complex (ice and water or ground).

Let us first consider the field in air ($z > 0$). We recognize that the field can be described by the Hertz vector $\bar{\pi}$, whose rectangular components satisfy a scalar wave equation,

$$(\nabla^2 + k^2)\pi_z = -\frac{J_z}{j\omega\varepsilon_0}, \tag{15.2}$$

and the electric dipole with the equivalent length L carrying a current I located at r' is given by

$$J_z = IL\delta(r - r'). \tag{15.3}$$

For convenience, we let

$$\frac{IL}{j\omega\varepsilon_0} = 1, \tag{15.4}$$

and thus in air we have

$$(\nabla^2 + k^2)\pi_z = -\delta(r - r'). \tag{15.5}$$

Within the ground ($z < 0$) we have

$$\left(\nabla^2 + k_e^2\right)\pi_z + 0. \tag{15.6}$$

The boundary condition at $z = 0$ is that both tangential electric and magnetic fields are continuous across the boundary. Equations (15.5) and (15.6) together with the boundary conditions and the radiation condition give a complete mathematical description of the problem. We write the solution of (15.5) as a sum of the primary wave and the secondary wave. The primary wave is the wave radiated from the dipole in an infinite space in the absence of the boundary and has the correct singularity at the location of the antenna. The secondary wave represents the effects of the boundary but has no singularity at the antenna location.

Let us consider the primary wave π_p. This is obviously given by

$$\pi_p = \frac{e^{-jk|r-r'|}}{4\pi |r - r'|}. \tag{15.7}$$

To satisfy the boundary conditions, we need to express (15.7) in terms of cylindrical waves that have the same radial wave number in air and within the ground. This is done by the Fourier–Bessel transform shown below.

Let us write (15.5) in a cylindrical coordinate system

$$\left\{\frac{1}{\rho}\frac{\partial}{\partial\rho}\left(\rho\frac{\partial}{\partial\rho}\right) + \frac{1}{\rho^2}\frac{\partial^2}{\partial\phi^2} + \frac{\partial^2}{\partial z^2} + k^2\right\}\pi_p = \frac{\delta(\rho - \rho')\delta(\phi - \phi')\delta(z - z')}{\rho} \tag{15.8}$$

First, we expand π_p in a Fourier series in ϕ. Noting that π_p is a function of $\phi - \phi'$, we write

$$\pi_p = \sum_{m=-\infty}^{\infty}\pi_m(\rho, z)e^{-jm(\phi-\phi')}, \tag{15.9}$$

where

$$\pi_m(\rho, z) = \frac{1}{2\pi}\int_0^{2\pi}\pi_p e^{jm(\phi-\phi')}d\phi.$$

We obtain

$$\left[\frac{1}{\rho}\frac{\partial}{\partial\rho}\left(\rho\frac{\partial}{\partial\rho}\right) - \frac{m^2}{\rho^2} + \frac{\partial^2}{\partial z^2} + k^2\right]\pi_m = -\frac{\delta(\rho - \rho')\delta(z - z')}{2\pi\rho} \tag{15.10}$$

We now express π_m in a Fourier–Bessel transform

$$\pi_m(\rho, z) = \int_0^{\infty}g_m(\lambda, z)J_m(\lambda\rho)\lambda\, d\lambda, \tag{15.11}$$

where

$$g_m(\lambda, z) = \int_0^\infty \pi_m(\rho, z) J_m(\lambda\rho)\rho \, d\rho.$$

Equation (15.11) is the Fourier–Bessel transform, which is the cylindrical equivalent of the usual Fourier transform. Note that λ is a complex variable commonly used in Fourier–Bessel transforms and is not a wavelength here.

Applying the Fourier–Bessel transform to both sides of (15.10), we obtain

$$\left(\frac{d^2}{dz^2} + k^2 - \lambda^2 \right) g_m(\lambda, z) = -\frac{\delta(z - z')J_m(\lambda\rho')}{2\pi}. \tag{15.12}$$

To obtain (15.12) from (15.10), we use integration by parts and note the behaviors of π_m as $\rho \to 0$ and $\rho \to \infty$. Equation (15.12) is easily solved to give

$$g_m(\lambda, z) = \frac{e^{-jq|z-z'|}}{2jq} \frac{J_m(\lambda\rho')}{2\pi}, \tag{15.12a}$$

where $\lambda^2 + q^2 = k^2$.

Substituting (15.12a) into (15.11) and then into (15.9), we obtain

$$\pi_p = \frac{1}{4\pi} \sum_{m=-\infty}^{\infty} e^{-jm(\phi-\phi')} \int_0^\infty J_m(\lambda\rho)J_m(\lambda\rho')e^{-jq|z-z'|}\frac{\lambda \, d\lambda}{jq}. \tag{15.13}$$

Equation (15.13) is, of course, equal to (15.7), but it is written in terms of the cylindrical wave with the propagation constant λ (i.e., $J_m(\lambda\rho)$). By expanding the field above and below the boundary in terms of the same wave number λ, we can satisfy the boundary conditions at any ρ. In contrast, when the boundary is in parallel to the z axis, such as in a dielectric cylinder, we use the Fourier transform in the z direction and express the field inside and outside the cylinder in terms of the same wave number in the z direction, e^{-jhz}.

In particular, when the antenna is located at $\rho' = 0$ and $z' = h$, (15.13) becomes

$$\pi_p(\rho, z) = \frac{1}{4\pi} \int_0^\infty J_0(\lambda\rho)e^{jq|z-h|}\frac{\lambda \, d\lambda}{jq}, \tag{15.14}$$

where $\lambda^2 + q^2 = k^2$.

Let us now examine the problem pictured in Fig. 15.1. In air (regions I and II), π_z satisfies the differential equation

$$(\nabla^2 + k^2)\pi_z = -\delta(r - r'), \tag{15.15}$$

where r' is at $\rho = 0$ and $z = h$. We write π_z as a sum of the primary wave π_p and the scattered wave π_s

$$\pi_z = \pi_p + \pi_s. \tag{15.16}$$

We write the primary wave π_p in region I as

$$\pi_p = \frac{1}{4\pi} \int_0^\infty J_0(\lambda\rho)e^{-jq(z-h)} \frac{\lambda \, d\lambda}{jq}, \tag{15.17a}$$

and in region II as

$$\pi_p = \frac{1}{4\pi} \int_0^\infty J_0(\lambda\rho)e^{-jq(h-z)} \frac{\lambda \, d\lambda}{jq}, \tag{15.17b}$$

where $\lambda^2 + q^2 = k^2$.

The difference in exponents $(z - h)$ and $(h - z)$ in (15.17a) and (15.17b) represents the singularity at $z = h$. The scattered wave π_s has no singularity at $z = h$ and it satisfies the homogeneous wave equation. Therefore, for both regions I and II, we write

$$\pi_s = \frac{1}{4\pi} \int_0^\infty R(\lambda)J_0(\lambda\rho)e^{-jq(z+h)} \frac{\lambda \, d\lambda}{jq}. \tag{15.18}$$

In region III there is no primary wave, and thus the scattered wave π_s satisfying the wave equation

$$\left(\nabla^2 + k_e^2\right)\pi_s = 0, \tag{15.19}$$

can be written as

$$\pi_s = \frac{1}{4\pi} \int_0^\infty T(\lambda)J_0(\lambda\rho)e^{+jq_e z - jqh} \frac{\lambda \, d\lambda}{jq}, \tag{15.20}$$

where $\lambda^2 + q_e^2 = k_e^2$. The choice of $(-jq)$ instead of $(+jq)$ for (15.17) and (15.18) is made so that this represents the outgoing wave in the $+z$ direction satisfying the "radiation condition" when q is in the fourth quadrant. Similarly, $(+jq_e z)$ is chosen for (15.20) to represent the outgoing wave in the $-z$ direction when q_e is in the fourth quadrant. Other commonly used notations are μ or γ in place of jq. We choose jq because q represents the wave number in the z direction in the same sense that λ is the wave number in the ρ direction.

Equations (15.16)–(15.18), and (15.20) represent the complete expressions of the fields, which are expressed in terms of two unknown functions, $R(\lambda)$ and $T(\lambda)$. These two functions are now determined by applying the boundary conditions at $z = 0$. The conditions are that the tangential electric and tangential magnetic fields are continuous across the boundary. Because of the symmetry of the problem, the only tangential electric field is E_ρ, and the only magnetic field is H_ϕ. Thus, noting that

$$E_\rho = \frac{\partial^2}{\partial\rho \, \partial z}\pi_z \quad \text{and} \quad H_\phi = -j\omega\varepsilon\frac{\partial}{\partial\rho}\pi_z,$$

the boundary conditions are given by

$$\frac{\partial}{\partial z}\pi_z^{(2)} = \frac{\partial}{\partial z}\pi_z^{(3)}$$
at $z = 0$,
$$\pi_z^{(2)} = n^2\pi_z^{(3)}$$

(15.21)

where $\pi_z^{(2)}$ and $\pi_z^{(3)}$ are π_z in regions II and III, respectively.

Applying (15.21) to (15.17), (15.18), and (15.20), we get

$$R(\lambda) = \frac{n^2\,q - q_e}{n^2\,q + q_e},$$

$$T(\lambda) = \frac{2q}{n^2\,q + q_e}.$$

(15.22)

Therefore, the solution to the original problem pictured in Fig. 15.1 is

$$+ \frac{1}{4\pi}\int_0^\infty \frac{n^2\,q - q_e}{n^2\,q + q_e}J_0(\lambda\rho)e^{-jq(z+h)}\frac{\lambda\,d\lambda}{jq}\Bigg] \quad \text{for } z > 0$$

(15.23)

and

$$\pi_z = \left(\frac{IL}{j\omega\varepsilon_0}\right)\frac{1}{4\pi}\int_0^\infty \frac{2}{n^2\,q + q_e}J_0(\lambda\rho)e^{jq_e z - jqh}\frac{\lambda\,d\lambda}{j} \quad \text{for } z < 0$$

(15.24)

where $\lambda^2 + q^2 = k^2$ and $\lambda^2\,q_e^2 = k_e^2$ and q and q_e are chosen to be in the fourth quadrant, and $IL/j\omega\varepsilon_0$ is restored in this expression. Equations (15.23) and (15.24) are the general expressions of the Hertz potential, and the other field components are obtained by differentiation.

$$E_z = \left(\frac{\partial^2}{\partial z^2} + k^2\right)\pi_z,$$

$$E_\rho = \frac{\partial^2}{\partial\rho\,\partial z}\pi_z,$$

(15.25)

$$H_\phi = -j\omega\varepsilon\frac{\partial}{\partial\rho}\pi_z.$$

15.3 REFLECTED WAVES IN AIR

Let us consider (15.23) without $IL/j\omega\varepsilon_0$. We write

$$\pi_z = \pi_p + \pi_s$$

(15.26)

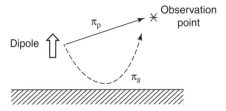

FIGURE 15.2   Direct and reflected waves.

and

$$\pi_p = \frac{ee^{-jk|r-r'|}}{4\pi\,|r-r'|},$$

$$\pi_s = \frac{1}{4\pi}\int_0^\infty \frac{n^2 q - q_e}{n^2 q + q_e} J_0(\lambda_\rho) e^{-jq(z-h)}\frac{\lambda\,d\lambda}{jq}.$$

π_p represents the direct wave from the antenna to the observation point, and π_s represents the reflected wave as pictured in Fig. 15.2.

To evaluate the integral, it is convenient to convert the range of the integral from $(0 \to +\infty)$ to $(-\infty \to +\infty)$. To this end, we use the identity

$$J_0(\lambda\rho) = \tfrac{1}{2}\left[H_0^{(1)}(\lambda\rho) + H_0^{(2)}(\lambda\rho)\right] \tag{15.27}$$

and write the integral as

$$\int_{W_1} f(\lambda)J_0(\lambda\rho)\lambda\,d\lambda = \tfrac{1}{2}\int_{W_1} f(\lambda)H_0^{(1)}(\lambda\rho)\lambda\,d\lambda + \tfrac{1}{2}\int_{W_1} f(\lambda)H_0^{(2)}(\lambda\rho)\lambda\,d\lambda, \tag{15.28}$$

where the contour W_1 is shown in Fig. 15.3. We now convert the first integral into the contour W_2, which is symmetric to W_1 about the origin. Thus writing

$$\lambda = \lambda',$$

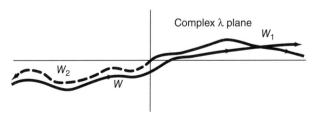

FIGURE 15.3   The integral from 0 to ∞ is converted to the integral from − ∞ to + ∞.

the first integral becomes

$$\frac{1}{2}\int_{W_1} f(\lambda)H_0^{(1)}(\lambda\rho)\lambda\,d\lambda = \frac{1}{2}\int_{W_1} f(-\lambda')H_0^{(1)}(-\lambda')H_0^{(1)}(-\lambda'\rho)\lambda'\,d\lambda'. \quad (15.29)$$

Noting the identity

$$H_\nu^{(1)}(e^{\pi j}Z) = -e^{-\nu\pi j}H_\nu^{(2)}(Z),$$

(15.29) becomes

$$-\frac{1}{2}\int_{W_2} f(-\lambda')H_0^{(2)}(\lambda'\rho)\lambda'\,d\lambda'.$$

We reverse the path W_2 to $-W_2$, and noting that in our problem $f(\lambda)$ is an even function of λ, we finally obtain

$$\int_{W_1} f(\lambda)J_0(\lambda\rho)\lambda\,d\lambda = \frac{1}{2}\int_{-W_1} f(-\lambda)H_0^{(2)}(\lambda\rho)\lambda\,d\lambda + \frac{1}{2}\int_{W_1} f(\lambda)H_0^{(2)}(\lambda\rho)\lambda\,d\lambda$$

$$= \frac{1}{2}\int_{W_1} f(\lambda)H_0^{(2)}(\lambda\rho)\lambda\,d\lambda. \quad (15.30)$$

Using (15.30), we write the reflected wave equation (15.26) as

$$\pi_s = \frac{1}{8\pi}\int_W \frac{n^2q - q_e}{n^2q - q_e}H_0^{(2)}(\lambda\rho)e^{-jq(z+h)}\frac{\lambda\,d\lambda}{jq}. \quad (15.31)$$

The contour W is shown in Fig. 15.4.

It is not possible to evaluate the integral (15.31) in terms of elementary functions in a closed form. Thus it is necessary to evaluate the integral by approximate techniques. The approximate evaluation of the integral (15.31) may be classified into the following three cases:

1. *Radiation field using the saddle-point technique.* We may evaluate (15.31) for a large distance from the image point using the saddle-point technique. This yields a simple expression for the radiation pattern of the reflected wave.

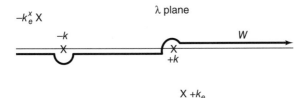

FIGURE 15.4   Contour W for (15.31).

2. *The field along the surface.* The evaluation of the field on the earth surface is the central point of this Sommerfeld problem. This is accomplished by the saddle-point technique modified by the presence of the pole (Sommerfeld pole).

3. *Lateral wave.* If the index of refraction n is smaller than unity, the integration along the branch cut must be considered. This results in the lateral wave.

Mathematically, the three cases above result from the *saddle-point* evaluation, the effect of the *pole*, and the *branch point*. The consideration of these three points is most important when evaluating a complex integral of this type. In addition to the three analytical methods above, we can also carry out the integration *numerically.* A significant improvement in efficiency results when the integration path is deformed into the steepest descent path. This is discussed in Section 15.7.

15.4 RADIATION FIELD: SADDLE-POINT TECHNIQUE

When the observation point is far from the antenna, it is possible to evaluate the integral in (15.31) and obtain a simple expression for the radiation field. This far-field solution is not valid for the region close to the surface, where it is necessary to employ the modified saddle-point technique described in the next section.

Let us evaluate (15.31) by means of the saddle-point technique. We first approximate the Hankel function by its asymptotic form

$$H_0^{(2)}(\lambda\rho) = \left(\frac{2}{\pi\lambda\rho}\right)^{1/2} e^{-j\lambda\rho + j(\pi/)}. \tag{15.32}$$

This is valid only for $|\lambda\rho| \gg 1$. But λ ranges from $-\infty$ to $+\infty$. Note, however, that the main contribution to the integral comes from the neighborhood of $\lambda = k \sin\theta$ and thus $|\lambda\rho| \sim kR_2 \sin^2\theta$. Therefore, (15.32) can be used as long as θ is not too small (the radiation close to the vertical axis) (see Banos, 1966, p. 76). If θ is small, the double-saddle-point method should be used. Then the first term of the rigorous asymptotic solution is identical to that obtained by (15.32). Using (15.32), we get

$$\pi_s = \frac{1}{8\pi} \int_W \frac{n^2 q - q_e}{n^2 q - q_e} \left(\frac{2}{\pi\lambda\rho}\right)^{1/2} \frac{e^{j(\pi/4)}}{jq} e^{-j\lambda\rho - jq(z+h)} \lambda \, d\lambda. \tag{15.33}$$

Using the saddle-point evaluation of the integral

$$\int_W f(\lambda) e^{-j\lambda\rho - jq(z+h)} \, d\lambda \approx f(k \sin\theta) \sqrt{\frac{2\pi}{kR_2}} \, k \, \cos\theta \, e^{-jkR_2 + j(\pi/4)}, \tag{15.34}$$

where

$$R_2 = \sqrt{\rho^2 + (z+h)^2}, \quad z + h = R_2 \cos\theta, \quad \rho = R_2 \sin\theta,$$

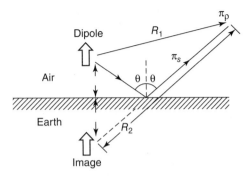

FIGURE 15.5   Radiation field.

we get

$$\pi_2 = \left(\frac{n^2 \cos \theta - \sqrt{n^2 - \sin^2 \theta}}{n^2 \cos \theta - \sqrt{n^2 - \sin^2 \theta}} \right) \frac{e^{-hjR_2}}{4\pi R_2}, \tag{15.35}$$

where $\lambda = k \sin \theta$ is substituted in $q = \sqrt{k^2 - \lambda^2}$ and $q_e = \sqrt{k^2 n^2 - \lambda^2}$, and R_2 is the distance from the image point (Fig. 15.5).

Equation (15.35) is the spherical wave originating at the image point, but the magnitude is multiplied by the reflection coefficient $R(\theta)$.

$$R(\theta) = \frac{n^2 \cos \theta - \sqrt{n^2 - \sin^2 \theta}}{n^2 \cos \theta + \sqrt{n^2 - \sin^2 \theta}}. \tag{15.36}$$

This reflection coefficient is exactly the same as the plane-wave reflection coefficient when the incident angle is θ. This is to be expected because the saddle-point technique is valid for the field far from the source and thus the wave is expected to behave locally as a plane wave.

The total field is then given by

$$\begin{aligned} \pi &= \pi_p + \pi_s \\ &= \frac{e^{-jkR_1}}{4\pi R_1} + R(\theta) \frac{e^{-jkR_2}}{4\pi R_2}. \end{aligned} \tag{15.37}$$

We note, however, that on the surface of the earth, $R_1 = R_2 = R$ and at a large distance, θ approaches $\pi/2$, and thus $R(\theta)$ approaches -1. Therefore the expression (15.37) reduces to zero on the surface far from the source. This is to be expected because far from the source the wave should behave essentially as a spherical wave, but on the surface the air and the earth have two different wave numbers, k and k_e. Thus it is impossible to have two different spherical waves in air and within the earth having the same phase relationship to satisfy boundary conditions.

In general, a spherical wave of the type e^{-jkR}/R should vanish along the interface between two different media. This of course does not imply that other wave types, such as $1/R^2$, vanish, and in fact, the Sommerfeld problem in Section 15.5 is essentially that of finding the field on the surface.

15.5 FIELD ALONG THE SURFACE AND THE SINGULARITIES OF THE INTEGRAND

The evaluation of the field on the surface due to a dipole source is central to the Sommerfeld problem. This is an important practical problem of radio-wave propagation. It also presents a rather involved mathematical subtlety; therefore, a thorough understanding of the basic techniques is very important.

Let us first write the total field as a sum of the field when the earth is perfectly conducting and the term representing the finiteness of the conductivity of the earth.

$$\pi = \frac{e^{-jkR_1}}{4\pi R_1} + \frac{e^{-jkR_2}}{4\pi R_2} - 2P. \tag{15.38}$$

The first term is the incident wave, the second term is the reflected wave when the earth is perfectly conducting, and P is given by

$$P = \frac{1}{8\pi} \int_W \frac{q_e}{n^2 q + q_e} H_0^{(2)}(\lambda\rho) e^{-jq(z+h)} \frac{\lambda \, d\lambda}{jq}, \tag{15.39a}$$

$$\approx \frac{1}{8\pi} \int_W \frac{q_e}{n^2 q + q_e} \left(\frac{2}{\pi\lambda\rho}\right)^{1/2} \frac{e^{j(\pi/4)}}{jq} e^{-j\lambda\rho - jq(z+h)} \lambda \, d\lambda, \tag{15.39b}$$

where the approximation (15.32) has been used in (15.39b).

We can work with this form (15.39b), but as will be shown shortly, this form contains branch points at $\lambda = \pm k$ and $\lambda = \pm k_e$. It is possible to eliminate the branch points at $\lambda = \pm k$ by the following transformation of the variable of integration from λ to α:

$$\lambda = k \sin \alpha. \tag{15.40}$$

Furthermore, using R_2 and θ instead of ρ and z through the transformation

$$\rho = R_2 \sin\theta,$$
$$z + h = R_2 \cos\theta, \tag{15.41}$$

Eq. (15.39b) becomes

$$P = \int_c F(\alpha) e^{-jkR_2 \cos(\alpha-\theta)} \, d\alpha, \tag{15.42}$$

where

$$F(\alpha) = \frac{ke^{-j(\pi/4)}}{4\pi} \left(\frac{1}{2\pi k R_2} \frac{\sin \alpha}{\sin \theta} \right)^{1/2} \frac{\sqrt{n^2 - \sin^2 \alpha}}{n^2 \cos \alpha + \sqrt{n^2 - \sin^2 \alpha}}.$$

Equations (15.39b) and (15.42) are the basic integrals that must be evaluated. The actual evaluation in this section is done in the α plane using (15.42) because the procedure is simpler and clearer in the α plane than in the λ plane. However, the corresponding discussion in the λ plane is given in Appendix 15.B, to clarify the situation, particularly with respect to the historical controversies of this Sommerfeld problem.

In evaluating complex integrals, it is essential first to examine all the singularities of the integrand. In general, there are three kinds of singularities: poles, essential singularities, and branch points. The singularities of a function $f(\lambda)$ are the points where $f(\lambda)$ is not analytic.

1. *Pole (isolated singularity)*. In the neighborhood of the pole, $f(\lambda)$ can be expanded in a Laurent series with *finite* negative powers.
2. *Essential singularities.* $f(\lambda)$ is expressed in an infinite series of negative powers. For example, $e^{1/\lambda} = 1 + 1/\lambda + 1/2!\lambda^2 + \cdots + 1/n!\lambda^n + \cdots$.
3. *Branch points*. The function $f(\lambda)$ has more than one value at a given λ, representing more than one branch. The points where these branches meet are the branch points. For example, $f(\lambda) = \lambda^{1/2}$, $f(\lambda) = \ln \lambda$.

In our problem there are poles where the denominator of the integrand vanishes,

$$n^2 q + q_e = 0 \qquad \text{in the } \lambda \text{ plane,} \qquad (15.43a)$$

$$n^2 \cos a + \sqrt{n^2 - \sin^2 \alpha} = 0 \qquad \text{in the } \lambda \text{ plane,} \qquad (15.43b)$$

and branch points

$$\text{at } \lambda = \pm k \quad \text{and} \quad \lambda = \pm k_e \quad \text{in the } \lambda \text{ plane,} \qquad (15.44a)$$

and

$$\text{at } \sin \alpha = \pm n \text{ in the } \alpha \text{ plane.} \qquad (15.44b)$$

In addition, there is a branch point at $\lambda = 0$ due to the Hankel function $H_0^{(2)}(\lambda \rho)$. However, since λ represents the propagation constant in the radial direction, the wave corresponding to this branch point $\lambda = 0$ does not propagate on the surface, and thus this branch point has practically no effect on the field.

The integration in (15.42) will be carried out using the saddle-point technique. However, the pole in the integrand is located close to the saddle point and thus the modified saddle-point technique must be used. It is therefore important to determine the location of the pole, particularly with respect to the different branches of the complex plane resulting from the branch points.

Let us consider the branch points in (15.42). Because of the square root in $\sqrt{n^2 - \sin^2 \alpha}$, there are two Riemann surfaces. We draw the branch cut in the α plane along

$$\text{Im}\sqrt{n^2 \sin^2 \alpha} = 0. \tag{15.44c}$$

We note that

$$q_e = k\sqrt{n^2 \sin^2 \alpha}$$

is the wave number in the $-z$ direction inside the earth and thus Im q_e must be negative because

$$\left|e^{jq_e z}\right| = e^{-(\text{Im } q_e)z} \quad \text{for } z < 0$$

vanishes as $z \to -\infty$. Therefore,

$$\text{Im}\sqrt{n^2 - \sin^2 \alpha} < 0 \tag{15.45a}$$

corresponds to the wave attenuating in the $-z$ direction satisfying the radiation condition, and thus this may be called the *proper Riemann surface*. On the other hand,

$$\text{Im}\sqrt{n^2 - \sin^2 \alpha} > 0 \tag{15.45b}$$

corresponds to the wave exponentially increasing toward $z \to -\infty$, and thus, this may be called the *improper Riemann surface*. Obviously, the original contour c in (15.42) is located in the proper Riemann surface. This is shown in Fig. 15.6 (see Appendix 15.B).

15.6 SOMMERFELD POLE AND ZENNECK WAVE

Let us consider the pole given by (15.43a).

$$n^2 q + q_e = 0. \tag{15.46}$$

This is identical to the equation used to determine the propagation constant for the Zenneck wave. To show this, consider the Zenneck wave propagating in the x direction along the surface at $z = 0$. Then for $z > 0$, we have

$$\pi_1 = Ae^{-j\lambda x - jqz}, \tag{15.47}$$

and for $z < 0$, we write

$$\pi_2 = Be^{-j\lambda x + jq_e z}. \tag{15.48}$$

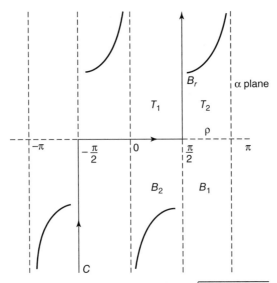

FIGURE 15.6   Riemann surface corresponding to IM $\sqrt{n^2 - \sin^2\alpha} < 0$. The branch points are at $\sin\alpha \pm n$ and the branch cuts are along IM $\sqrt{n^2 - \sin^2\alpha} = 0$.

Now satisfying the boundary condition,

$$n^2\,\pi = \text{continuous} \quad \text{and} \quad \frac{\partial\pi}{\partial z} = \text{continuous}$$

we get

$$A = n^2 B \quad \text{and} \quad -jqA = jq_e B.$$

From this, we get (15.46).

Let us find the solution λ_p from (15.46). Substituting $q^2 = k^2 - \lambda^2$ and $q_e^2 = k^2 n^2 - \lambda^2$ into (15.46) and squaring both sides of $n^2 q = -q_e$, we get $n^4(k^2 - \lambda_p^2) = k^2 n^2 - \lambda_p^2$. From this we get

$$\frac{1}{\lambda_p^2} = \frac{1}{k^2} + \frac{1}{k^2 n^2}. \tag{15.49}$$

We note here that by squaring $q_e = -n^2 q$, we introduced an additional solution corresponding to $q_e = +n^2 q$. This is the solution with $\exp(+jqz)$ for $z > 0$, and therefore it is in the improper Riemann surface. To see this more clearly, we obtain q using (15.49),

$$q = \pm\frac{k}{(n^2 + 1)^{1/2}}. \tag{15.50}$$

Now we need to choose the sign for q such that it satisfies (15.46). Noting that $-n^2q = q_e$ and that $\mathrm{Im}\,q_e$ must be negative, we require that $\mathrm{Im}(-n^2q)$ be negative.

However, considering that $\mathrm{Im}(n^2)$ is negative, we see that $\mathrm{Im}[n^2/(n^2 + 1)^{1/2}]$ is also negative. Thus, to make $\mathrm{Im}(-n^2q)$ negative, we must choose the minus sign in (15.50).

$$q = -\frac{k}{(n^2 + 1)^{1/2}},$$

$$\lambda_p = \frac{nk}{(n^2 + 1)^{1/2}}, \tag{15.51}$$

$$q_e = \frac{n^2 k}{(n^2 + 1)^{1/2}}.$$

For detailed behavior of the locations of the Sommerfeld poles, see Appendix 15.B.

For conducting earth, $|n|$ is normally much greater than unity, and therefore we can get the following approximate solution from (15.51):

$$\lambda_p \approx k\left(1 - \frac{1}{2n^2}\right). \tag{15.52}$$

Next we examine the Sommerfeld pole in the complex α plane:

$$\lambda_p = k \sin \alpha_p. \tag{15.53}$$

Following the procedure indicated above for q, we get

$$\cos \alpha_p = \frac{1}{(n^2 + 1)^{1/2}},$$

$$\sin \alpha_p = \frac{1}{(n^2 + 1)^{1/2}}. \tag{15.54}$$

Here $\cos \alpha_p$ is in the third quadrant.

$$\mathrm{Re}(\cos \alpha_p) < 0 \quad \text{and} \quad \mathrm{Im}(\cos \alpha_p) < 0.$$

Therefore, α_p must be located in T_2 of Fig. 15.6. Considering that $|n| \gg 1$, we see that α_p is located very close to $\pi/2$ and is slightly below the 45° line, as can be seen below and in Fig. 15.7.

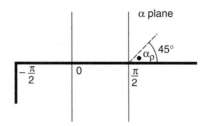

FIGURE 15.7 Location of Sommerfeld pole α_p.

$$\cos \alpha_p = \cos \left(\frac{\pi}{2} + \Delta \right) = -\sin \Delta = -\frac{1}{(n^2 + 1)^{1/2}},$$

$$\Delta = \sin^{-1} \frac{1}{(n^2 + 1)^{1/2}} = \left(\frac{\omega \varepsilon_0}{\sigma} \right)^{1/2} e^{j(\pi/4)},$$

$$n^2 = \varepsilon_r - j \frac{\sigma}{\omega \varepsilon_0} \approx -j \frac{\sigma}{\omega \varepsilon_0},$$

$$\alpha_p = \frac{\pi}{2} + \sin^{-1} \frac{1}{(n^2 + 1)^{1/2}} \approx \frac{\pi}{2} + \frac{1}{n}. \tag{15.55}$$

15.7 SOLUTION TO THE SOMMERFELD PROBLEM

We now evaluate the integral given in (15.42).

$$P = \int_c F(\alpha) e^{-jkR_2 \cos(\alpha - \theta)} d\alpha,$$

where $F(\alpha)$ has the pole at $\alpha = \alpha_p$ given in Section 15.6. The pole is located below the original contour c, and therefore, we can make use of the modified saddle-point technique discussed in Appendix 15.C.

$$I_0 = \int_c F(\alpha) e^{zf(\alpha)} d\alpha \tag{15.56}$$
$$= I_1 + I_2,$$

where

$$I_1 = e^{zf(\alpha_s)} \int_{-\infty}^{\infty} \left[F(s) \frac{d\alpha}{ds} - \frac{R_1(s_p)}{s - s_p} \right] \exp \left(-z \frac{s^2}{2} \right) ds,$$

$$I_2 = -j\pi R(s_p) \, \mathrm{erfc}(j\sqrt{z/2} s_p),$$
$$R(s_p) = R_1(s_p) \, \exp[sf(\alpha_p)]$$
$$= \text{residue of } F(\alpha) \exp[zf(\alpha)] \text{ at the pole } \alpha = \alpha_p,$$

$$f(\alpha) - f(\alpha_s) = -\frac{s^2}{2}, \quad \mathrm{erfc}(z) = 1 - \mathrm{erf}(z), \quad \mathrm{erf}(z) = \frac{2}{\sqrt{\pi}} \int_0^z e^{-t^2} \, dt.$$

The saddle point is located at $s = s_s$, and the pole is located at $s = s_p$.

For our problem (15.42), we have

$$f(\alpha) = -j \cos(\alpha - \theta),$$
$$z = kR_2.$$
$$(15.57)$$

Therefore, the saddle point is at $\alpha_s = \theta$, and

$$f(\alpha) - f(\alpha_s) = j[\cos(\alpha - \theta) - 1] = -\frac{s^2}{2}.$$

(15.58)

From this we get

$$s = 2e^{-j(\pi/4)} \sin \frac{\alpha - \theta}{2},$$

$$\frac{d\alpha}{ds} = \left(1 - j\frac{s^2}{4}\right)^{-1/2} e^{j(\pi/4)},$$

(15.59)

$$\alpha = \theta + 2 \sin^{-1}\left(\frac{s}{2}e^{j(\pi/4)}\right).$$

The path of integration for I_1 is on the steepest descent contour (SDC) and is along the real axis of s. The original contour of c and SDC in the α plane and the s plane are pictured in Fig. 15.8. Because of (15.59), two branch points (B_{r1} and B_{r2}) appear at $s = \pm 2e^{-j(\pi/4)}$ in the s plane, but these branch points are far from saddle point and have little effect on the integral. Note that the integrand for I_1 has no pole, and therefore I_1 can be evaluated numerically by expressing the integrand in (15.56) as a function

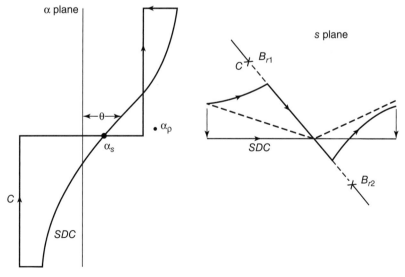

FIGURE 15.8   Original contour C and the steepest descent contour (SDC) in the α plane and s plane.

of s using (15.59). We can use the following integration formula (Abramowitz and Stegun, 1964, p. 890):

$$\int_{-\infty}^{\infty} e^{-x^2} f(x)\, dx = \sum_{i=1}^{n} W_i f(x_i) + R_n, \tag{15.60}$$

where x_i is the ith zero of the Hermite polynomial $H_n(x)$, W_i the weighting coefficient, and R_n the remainder. The table for x_i and W_i is available in Abramowitz and Stegun (1964, p. 924).

Alternatively, we can use the series expansion (15C.31) in Appendix 15.C. The series expansion is given by

$$I_1 = e^{-jkR_2} \sum_{n=0}^{\infty} B_{2n} \left(\frac{2}{kR_2} \right)^{n+1/2} \Gamma \left(n + \frac{1}{2} \right), \tag{15.61}$$

where

$$B_0 = \left[F(s) \frac{d\alpha}{ds} \right]_{s=0} + \frac{R_1(s_p)}{s_p}$$

$$= F(\alpha_s) e^{j(\pi/4)} + \frac{R_1(s_p)}{s_p}.$$

However, this term B_0 can be shown to be vanishingly small as $|n|$ becomes large, and $\theta \to \pi/2$. Therefore, the series (15.61) starts with the term $n = 1$, and thus for large R_2, I_1 is proportional to $1/R_2^2$. This term decays much faster than I_p and I_1 is negligible compared with I_p. Therefore, P is given approximately by I_2.

$$p \approx -j\pi R(s_p) \operatorname{erfc}\left(j \sqrt{\frac{z}{2}} s_p \right). \tag{15.62a}$$

For a large refractive index $|n|$ and $\theta \to \pi/2$, this can be simplified to

$$P = \frac{e^{-jkR_2}}{4\pi R_2} \left[\sqrt{\pi} j \sqrt{P_1} e^{-P_1} \operatorname{erfc}\left(j \sqrt{P_1} \right) \right], \tag{15.62b}$$

where

$$P_1 = kR_2 \frac{s_p^2}{2}.$$

Substituting this into (15.38), we finally obtain for the field on the surface,

$$\pi = 2 \frac{e^{-jkR}}{4\pi R} \left[1 - \sqrt{\pi} j \sqrt{P_1} \, e^{-P_1} \operatorname{erfc}(j \sqrt{P_1}) \right]. \tag{15.63}$$

The quantity

$$F = 1 - \sqrt{\pi} j \sqrt{P_1} \, e^{-P_1} \operatorname{erfc}(j \sqrt{P_1}) \tag{15.64}$$

represents the attenuation of the wave from the value for a perfectly conducting surface and is called the *attenuation function*.

Let us examine the attenuation function F. F depends only upon p_1. p_1 is given in (15.62b), which we write as

$$p_1 = jkR\cos(\alpha_p - \theta) - jkR. \tag{15.65}$$

The first term, $jkR\cos(\alpha_p - \theta)$, is the total complex phase from the origin to the observation point for the Zenneck wave, because

$$e^{-jkR\cos(\alpha_p - \theta)} = [e^{-j\lambda\rho - jq(z+h)}]_{\lambda=\lambda_p},$$

where λ_p is the radial propagation constant for the Zenneck wave given in (15.51). Therefore, p_1 represents the difference between the propagation constant for the Zenneck wave and the free-space wave.

It is clear, then, that the characteristics of the total wave do not depend on the distance (kR) alone, but depend on how large or small p_1 is. The magnitude of p_1 is called the *numerical distance* by Sommerfeld.

For small $|p_1|$ we have

$$F = 1 - j\sqrt{\pi}\sqrt{p_1}\,e^{-p_1} + \cdots, \tag{15.66}$$

which approaches 1 as $p_1 \to 0$.

For large $|p_1|$ ($|p_1| > 10$), we use the asymptotic expansion

$$\mathrm{erfc}(Z) = \frac{e^{-z^2}}{\sqrt{\pi}Z}\left[1 - \frac{1}{2Z^2} + \frac{1\cdot 3}{(2Z^2)^2} - \cdots\right], \tag{15.67}$$

valid for Re $Z > 0$, and we get

$$F = -\frac{1}{2p_1} + \cdots. \tag{15.68}$$

Note also that for large n,

$$j\sqrt{p_1} = j\sqrt{\frac{kR}{2}}\frac{e^{-j(\pi/4)}}{n}, \tag{15.69}$$

and thus the angle of $j\sqrt{p_1}$ is close to but slightly less than $+90°$.

Equation (15.63) may be written in the following form:

$$\pi = 2\frac{e^{-jkR}}{4\pi R}\left(1 - j\sqrt{p_1\pi}\,e^{-p_1} - 2\sqrt{p_1}\,e^{-p_1}\int_0^{\sqrt{p_1}}e^{\alpha^2}\,d\alpha\right). \tag{15.70}$$

This is the form given in Sommerfeld's 1926 paper. (See Appendix 15.A, for the controversies over the sign error.) Equation (15.70) is valid for large kR and moderate numerical distance. It is not valid for small kR, where the quasi-static approach must be used. It is also not valid for extremely large kR and large numerical distances in which case the effect of the pole is negligible and the conventional saddle-point technique is applicable. Here the field on the surface behaves as

$$\frac{e^{-jKR}}{R^2} \tag{15.71}$$

and there is no term exhibiting the Zenneck wave characteristics.

15.8 LATERAL WAVES: BRANCH CUT INTEGRATION

As noted in Section 15.3, we need to consider three points in the complex integration: the saddle point, poles, and branch points. As may be noted from Section 15.7 for the Sommerfeld problem of wave propagation over the earth, the pole is very close to the saddle point, but the branch points are far from the saddle point. Thus, only the effect of the pole needs to be considered for the Sommerfeld problem for dipoles on earth.

However, for other problems, where the index of refraction n is smaller than unity, the pole is far from the saddle point, but the branch point is very close to the saddle point. Thus the effect of the pole can be neglected, but the effect of the branch points needs to be taken into account. We first examine the mathematical technique of handling the branch point and later discuss the physical significance.

Let us examine the reflected wave given in (15.26).

$$\pi_z = \pi_p + \pi_s,$$
$$\pi_p = \frac{e^{-jk|r-r'|}}{4\pi|r-r'|}, \tag{15.72}$$
$$\pi_s = \frac{1}{8\pi}\int_W \frac{n^2 q - q_e}{n^2 q - q_e} H_0^{(2)}(\lambda\rho) e^{-jq(z+h)} \frac{\lambda\, d\lambda}{jq}.$$

We use the approximation (15.32) and the transformation (15.40) from λ to α and write

$$\pi_s = \frac{1}{4\pi}\left(\frac{k}{2\pi\rho}\right)^{1/2} e^{-j(\pi/4)} \int_C R(\alpha) e^{-jkR_2 \cos(\alpha-\theta)} \sqrt{\sin\alpha}\, d\alpha, \tag{15.73}$$

where

$$R(\alpha) = \frac{n^2 \cos\alpha - \sqrt{n^2 - \sin^2\alpha}}{n^2 \cos\alpha + \sqrt{n^2 - \sin^2\alpha}},$$

is the reflection coefficient.

As is evident from (15.73), the branch point is located at

$$\sin\alpha_b = n. \tag{15.74}$$

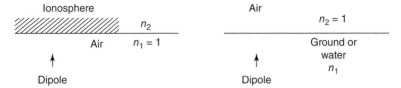

FIGURE 15.9   Dipole is located in the medium n_1 and $|n| = |n_2/n_1| < 1$.

If $|n| > 1$ as in the case of the earth, α_b is located far from the location of the saddle point $\alpha_s = \theta(0 \le \theta \le \pi/2)$, and this branch point α_b has little effect on the wave propagation over the earth.

However, if n is real and $0 < n < 1$, then α_b is real and $0 \le \alpha_b \le \pi/2$, and therefore the branch cut integration for α_b must be taken into account when the saddle-point integration is performed. For example, consider an antenna located in air below the ionosphere or an antenna buried underground. In these cases, the index of refraction n_2 of the medium on the other side of the boundary from the antenna is less than that of the medium n_1 where the antenna is located, and thus $|n| = |n_2/n_1| < 1$ (Fig. 15.9).

We evaluate (15.73) by means of the saddle-point technique. The saddle-point contour is shown in Fig. 15.10. We note, first, that if the observation angle θ is smaller than α_b,

$$\theta < \alpha_b, \alpha_b = \sin^{-1}n,$$

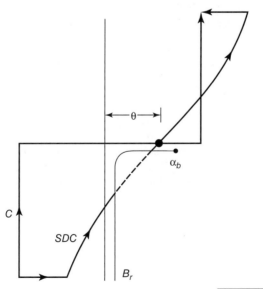

FIGURE 15.10   Branch cut B_r is drawn along IM $\sqrt{n^2 - \sin^2\alpha} = 0$.

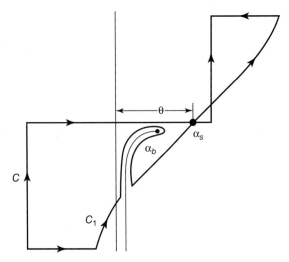

FIGURE 15.11   Path of integration C_1.

then the branch point α_b has practically no effect and thus

$$\pi_s = \int_C = \int_{SDC}$$
$$= R(\theta)\frac{e^{-jkR_2}}{4\pi R_2},$$

(15.75)

which is the reflected wave previously obtained.

On the other hand, if the observation angle θ is greater than α_b,

$$\theta > \alpha_b,$$

the integral along C becomes (Fig. 15.11)

$$\int_C = \int_{C_1}.$$

However, the contour C_1 can be deformed to the branch cut integration $(A–B–C)$ and SDC $(D–E)$ shown in Fig. 15.12. Note that a portion of the SDC is in the bottom Riemann surface, as shown by the dotted line

$$\int_C = \int_{C_1} = \int_{Br} + \int_{SDC}.$$

Therefore, for $\theta > \alpha_b$,

$$\pi_s = R(\theta)\frac{e - jkR_2}{4\pi R_2} + \pi_l,$$

(15.76)

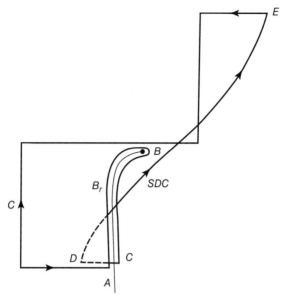

FIGURE 15.12   The integral along C_1 is equal to the branch cut integration along B_r and the steepest descent contour (SDC).

where π_l is the branch cut integration

$$\pi_l = \frac{1}{4\pi} \left(\frac{k}{2\pi\rho} \right)^{1/2} e^{-j(\pi/4)} \int_{Br} R(\alpha) e^{-jkR_2 \cos(\alpha-\theta)} \sqrt{\sin \alpha}\, d\alpha. \quad (15.77)$$

The physical significance of the angle α_b is shown in Fig. 15.13. When the observation point is within region $A(\theta < \alpha_b)$, the wave consists of the primary wave and the reflected wave. Note that the transmitted wave makes an angle θ_t with the vertical axis as given by Snell's law.

$$\sin \theta = n \sin \theta_t. \quad (15.78)$$

When θ exceeds the angle that makes $\theta_t = \pi/2$, total reflection occurs. This critical angle is α_b,

$$\sin \alpha_b = n. \quad (15.79)$$

Therefore, regions A and B correspond to the cases where the angle of incidence θ is less than or greater than the critical angle for the total reflection.

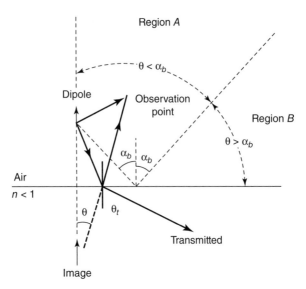

FIGURE 15.13   The critical angle α_b for the total reflection.

Let us evaluate (15.77) for large kR_2,

$$\pi_l = \frac{1}{4\pi} \left(\frac{k}{2\pi\rho} \right)^{1/2} e^{-j(\pi/4)} I,$$

$$I = \int_{Br} F(\alpha) e^{-f(\alpha)} d\alpha, \tag{15.80}$$

where

$$f(\alpha) = jkR_2 \cos(\alpha - \theta),$$

$$F(\alpha) = R(\alpha) \sqrt{\sin \alpha}.$$

The integral is along the paths A–B and B–C shown in Fig. 15.12. We then note that the value of $F(\alpha)$ along A–B is different from that along B–C because $F(\alpha)$ contains $\sqrt{n^2 - \sin^2 \alpha}$. Note that along A–B,

$$F_1(\alpha) = \frac{n^2 \cos \alpha - \sqrt{n^2 - \sin^2 \alpha}}{n^2 \cos \alpha + \sqrt{n^2 - \sin^2 \alpha}} \sqrt{\sin \alpha}. \tag{15.81}$$

where $\mathrm{Re} \sqrt{n^2 - \sin^2 \alpha} > 0$, but along B–C, the sign of $\sqrt{n^2 - \sin^2 \alpha}$ must be changed as it is on the other side of the branch cut. Thus we should write

$$F_2(\alpha) = \frac{n^2 \cos \alpha + \sqrt{n^2 - \sin^2 \alpha}}{n^2 \cos \alpha - \sqrt{n^2 - \sin^2 \alpha}} \sqrt{\sin \alpha}. \tag{15.82}$$

Using (15.81) and (15.82), we get for (15.80),

$$I = \int_A^B [F_1(\alpha) - F_2(\alpha)]e^{-f(\alpha)}d\alpha. \tag{15.83}$$

To evaluate this integral, let us first expand $f(\alpha)$ about $\alpha = \alpha_b$ and keep the first term. As will be shown shortly, this is justified because the major contribution to the integral comes from the immediate neighborhood of α_b.

$$f(\alpha) = f(\alpha_b) + (\alpha - \alpha_b)\left(\frac{\partial f}{\partial \alpha}\right)_{\alpha b} \tag{15.84}$$
$$= jkR_2 \cos(\theta - \alpha_b) + [jkR_2 \sin(\theta - \alpha_b)](\alpha - \alpha_b).$$

Next we deform the contour A–B such that the exponent in (15.84) decays exponentially in a steepest descent path from the branch point α_b. To this end, we write the exponent as

$$e^{-jkR_2 \sin(\theta - \alpha_b)(\alpha - \alpha_b)} = e^{-[kR_2 \sin(\theta - \alpha_b)]s},$$

where s is real, representing the distance from the branch point and obviously

$$\alpha - \alpha_b = -js. \tag{15.85}$$

We then get

$$I = e^{-jkR_2 \cos(\theta - \alpha_b)} \int_0^\infty [-F_1(\alpha) + F_2(\alpha)]e^{-[kR_2 \sin(\theta - \alpha_b)]s}(-j)\,ds, \tag{15.86}$$

$$[-F_1(\alpha) + F_2(\alpha)] = \frac{4n^2 \cos\alpha \sqrt{n^2 - \sin^2\alpha}}{n^2 \cos\alpha - (n^2 - \sin^2\alpha)} \sqrt{\sin\alpha}. \tag{15.87}$$

Because the integrand decays exponentially in (15.86), most of the contribution comes from the neighborhood of $s = 0$. It is therefore necessary to examine the behavior of (15.87) in the neighborhood of $s = 0$.

Near $s = 0$, $(\alpha = \alpha_b)$ we expand $(n^2 - \sin^2\alpha)$ in a Taylor's series about $\alpha = \alpha_b$ and keep the first term.

$$n^2 - \sin^2\alpha \simeq -2 \sin\alpha_b \cos\alpha_b(\alpha - \alpha_\alpha)$$
$$= [2jn \cos\alpha_b]s. \tag{15.88}$$
$$\sqrt{n^2 - \sin^2\alpha} = (2n \cos\alpha_b)e^{-j(\pi/4)}\sqrt{s}.$$

Thus, near $s = 0$, (15.87) may be approximated by

$$[-F_1(\alpha) + F_2(\alpha)] \simeq \frac{4\sqrt{2}}{n(\cos\alpha_b)^{1/2}}e^{-j(\pi/4)}\sqrt{s}. \tag{15.89}$$

The integral (15.86) then becomes

$$I = e^{-jkR_2 \cos(\theta-\alpha_b)} \frac{4\sqrt{2}e^{-j(\pi/4)}(-j)}{n(\cos \alpha_b)^{1/2}} \int_0^\infty \sqrt{s}\, e^{-[kR_2 \sin(\theta-\alpha_b)]s}\, ds. \qquad (15.90)$$

Using the integral

$$\int_0^\infty \sqrt{s}\, e^{-as}\, ds = \frac{\sqrt{2\pi}}{(2a)^{3/2}},$$

we finally get

$$I = e^{-jkR_2 \cos(\theta-\alpha_b)} \frac{2\sqrt{2\pi}e^{j(\pi/4)}(-j)}{n(\cos \alpha_b)^{1/2}} \frac{1}{[kR_2 \sin(\theta-\alpha_b)]^{3/2}}. \qquad (15.91)$$

Substituting this in (15.80), we get the expression for the branch cut contribution

$$\pi_l = \frac{(-j2)}{4\pi kn (\rho \cos \alpha_b)^{1/2}} \frac{e^{-jkR_2 \cos(\theta-\alpha_b)}}{[R_2 \sin(\theta-\alpha_b)]^{3/2}}. \qquad (15.92)$$

We now explain the physical meaning of π_l by rewriting this in terms of the distances pictured in Fig. 15.14. We note that

$$kR_2 \cos(\theta - \alpha_b) = k(L_0 + L_2) + knL_1,$$

and

$$R_2 \sin(\theta - \alpha_b) = L_1 \cos \alpha_b,$$

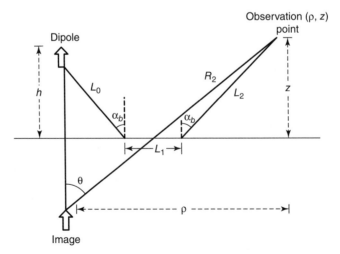

FIGURE 15.14   Lateral wave contribution.

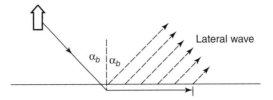

FIGURE 15.15   Lateral wave.

and therefore

$$\pi_1 = \frac{(-j2)}{4\pi kn(1-n^2)} \frac{1}{\rho^{1/2}L_1^{3/2}} e^{-jk(L_0+L_2)-jknL_1}. \tag{15.93}$$

Let us examine the physical meaning of (15.93). First, we note that the phase front is given by

$$kL_0 + knL_1 + kL_2 = \text{constant}.$$

This shows that this wave first propagates over the distance L_0 with free-space propagation constant k and is incident on the surface at the critical angle α_b. Then the wave propagates just below the surface with the propagation constant of the lower medium kn, and at the same time, radiating into the upper medium in the direction of the critical angle. This is pictured in Fig. 15.15. This particular wave is called the *lateral wave, headwave (kopfwelle)*, or *flank wave (flankenwelle)*. The lateral wave is related to the following seismic phenomenon. Let the seismic impulse (the earthquake) originate at a point on the surface at $t = 0$. The wave propagates faster in the ground than in the air. ($v(\text{air}) < v(\text{ground})$ and thus $n = v(\text{air})/v(\text{ground}) < 1$.) Thus the wavefronts at t in the air and in the ground are different on the surface (Fig. 15.16). But there should be no discontinuity in the wavefront, and therefore there is a wave whose wavefront is tangential from the wavefront in the ground to the wavefront in the air. This particular wavefront propagates in exactly the same direction as the lateral wave. Moreover, on the ground this wave arrives first and thus is called the head wave.

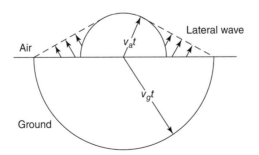

FIGURE 15.16   Lateral wave and seismic wave.

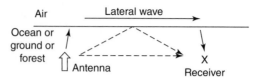

FIGURE 15.17 Communication by lateral wave.

In addition to this particular phase characteristic, the lateral wave has a distinct amplitude characteristic given by

$$\frac{1}{\rho^{1/2}} \frac{1}{L_1^{3/2}}.$$

Because of this dependence, π_l becomes infinite as $L_1 \to 0$. This is because of the approximation used to obtain (15.93). In fact, as $L_1 \to 0$, the saddle point approaches the branch point ($\theta \to \alpha_b$), and thus the saddle point and the branch cut integration cannot be performed separately. The field in the neighborhood of $L_1 \to 0$ (or $\theta \to \alpha_b$) is examined in Brekhovskikh (1960) showing the "caustic" behavior.

The lateral wave described above occurs when $0 < n < 1$. An example is the ionosphere, in which $n = \sqrt{1 - (\omega_p / \omega)^2}$. Another important case where the lateral wave plays a dominant role is the problem of communication between two points in the absorbing medium such as the ocean or the earth (Fig. 15.17). In this case, k is complex and its magnitude is large, but kn is the free-space wave number, and thus $|n| < 1$. Due to absorption, the direct wave and the reflected wave are almost completely attenuated in the medium. But the lateral wave propagates from the antenna to the surface suffering some attenuation, then propagates in the air without attenuation over a long distance and then arrives at the receiver. Thus the communication is only through the lateral wave in this case. The communication between two points in the forest is often done by the lateral wave.

15.9 REFRACTED WAVE

We now return to (15.24) and examine the field in the lower medium ($z < 0$).

$$\pi_z = \frac{1}{4\pi} \int_0^\infty \frac{2}{n^2 q + q_e} J_0(\lambda\rho) e^{jq_e z - jqh} \frac{\lambda \, d\lambda}{j}. \tag{15.94}$$

We use (15.30) to obtain the integral along $W(-\infty$ to $+\infty)$ and use the asymptotic form (15.32) for $H_0^{(2)}(\lambda\rho)$. We thus have

$$\pi_z = \frac{1}{8\pi} \int_W \frac{2}{n^2 q + q_e} \left(\frac{2}{\pi\lambda\rho}\right)^{1/2} \frac{e^{-j(\pi/4)\lambda}}{j} e^{-j\lambda\rho - jqh + jq_e z} \, d\lambda. \tag{15.95}$$

We evaluate this by means of the saddle-point technique. We use the first term of the asymptotic series given by

$$\int_W F(\lambda) e^{-f(\lambda)}\, d\lambda \simeq F(\lambda_s) e^{-f(\lambda_s)} \sqrt{\frac{2\pi}{f''(\lambda_s)}}, \tag{15.96}$$

where λ_s is the saddle point given by

$$f'(\lambda_s) = \left.\frac{\partial f}{\partial \lambda}\right|_{\lambda=\lambda_s} = 0.$$

The saddle point for (15.95) is given by

$$\begin{aligned}
\frac{\partial f}{\partial \lambda} &= \frac{\lambda}{\partial \lambda}(j\lambda\rho + jqh - jq_e z)s \\
&= j\left(\rho + \frac{\partial q}{\partial \lambda}h - \frac{\partial q_e}{\partial \lambda}z\right) \\
&= 0.
\end{aligned}$$

But

$$\frac{\partial q}{\partial \lambda} = -\frac{\lambda}{q} \quad \text{and} \quad \frac{\partial q_e}{\partial \lambda} = -\frac{\lambda}{q_e}.$$

Thus the saddle point λ_s is given by

$$\rho - \frac{\lambda_s}{q_s}h + \frac{\lambda_s}{q_{es}}z = 0, \tag{15.97}$$

where q_s and q_{es} are q and q_e evaluated at λ_s.

The physical meaning of (15.97) is clear when we use the transformation

$$\lambda = k \sin \alpha.$$

Then (15.97) becomes (see Fig. 15.18)

$$\rho - (\tan \alpha_1)h + (\tan \alpha_2)z = 0, \tag{15.98}$$

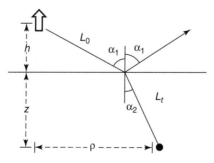

FIGURE 15.18   Refracted wave.

where α_1 and α_2 satisfy Snell's law,

$$\sin \alpha_1 = n \sin \alpha_2.$$

Thus the wave from the dipole arrives at the surface and refracts according to Snell's law and propagates to the observation point. The total phase is then

$$\begin{aligned}
f(\lambda_s) &= j(\lambda\rho + qh - q_e z)_{\lambda_s} \\
&= j(kL_0 + knL_t).
\end{aligned} \tag{15.99}$$

Therefore, the complete refracted field is given by using (15.96).

$$\pi_z = \frac{1}{4\pi} \frac{T(\alpha_1)(\sin \alpha_1)^{1/2} e^{-j(kL_0 + knL_t)}}{\rho^{1/2} \cos \alpha_1 [h/\cos^3 \alpha_1 + (-z)/n \cos^3 \alpha_2]^{1/2}}, \tag{15.100}$$

where $T(\alpha_1)$ is the transmission coefficient for the incident angle α_1,

$$T(\alpha_1) = \frac{2 \cos \alpha_1}{n^2 \cos \alpha_1 + \sqrt{n^2 - \sin^2\alpha_1}}. \tag{15.101}$$

Equation (15.100) is based on the saddle-point technique, and therefore it is valid when the dipole and the observation point are sufficiently far from the surface. It can be shown that expression (15.100) is identical to that obtained by the application of geometric optical techniques. In general, use of the saddle-point technique leads to the geometric optical solution.

15.10   RADIATION FROM A HORIZONTAL DIPOLE

In contrast with the azimuthally uniform radiation from a vertical dipole, the radiation from a horizontal dipole is directional. Furthermore, on the surface of the ground, most radiation is in the direction of the dipole axis. This is in contrast with the radiation in free space, where the radiation is broadside (in the direction perpendicular to the axis). One important practical case where a horizontal dipole is essential is the radiation from a buried antenna (antenna buried underground, in ice, or submerged underwater). In this case it can be shown that the horizontal dipole is most effective and practical and that a vertical dipole is an ineffective radiator.

Let us now consider a horizontal dipole located at $(0, 0, h)$ and oriented in the x direction (Fig. 15.19). The primary field can be easily obtained from the x component of the Hertz potential π_x

$$(\nabla^2 + k^2)\pi_x = \frac{I_x L}{j\omega\varepsilon_0} \delta(r - r'). \tag{15.102}$$

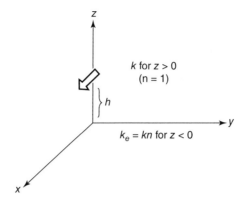

FIGURE 15.19   Horizontal dipole.

The primary field π_{xp} is then given by

$$
\begin{aligned}
\pi_{xp} &= \frac{e^{-jk|r-r'|}}{4\pi\,|r-r'|} \\
&= \frac{1}{4\pi} \int_0^\infty J_0(\lambda\rho)e^{-jq|z-h|}\frac{\lambda\,d\lambda}{jq} \\
&= \frac{1}{8\pi} \int_W H_0^{(2)}(\lambda\rho)e^{-jq|z-h|}\frac{\lambda\,d\lambda}{jq}.
\end{aligned} \tag{15.103}
$$

Here we omit $I_xL/j\omega\varepsilon_0$ for convenience. The final results must be multiplied by $I_xL/j\omega\varepsilon_0$ to obtain the true field.

Now we write the secondary fields in the air and the ground

$$
\begin{aligned}
\pi_{xs1} &= \frac{1}{4\pi} \int_0^\infty R(\lambda)J_0(\lambda\rho)e^{-jq|z+h|}\frac{\lambda\,d\lambda}{jq} \quad \text{for } z > 0, \\
\pi_{xs2} &= \frac{1}{4\pi} \int_0^\infty T(\lambda)J_0(\lambda\rho)e^{+q_ez-jqh}\frac{\lambda\,d\lambda}{jq} \quad \text{for } z < 0.
\end{aligned} \tag{15.104}
$$

At this point it may appear that these two functions $R(\lambda)$ and $T(\lambda)$ may be determined by applying the boundary conditions as was done for a vertical dipole. However, this is not possible. Note that, in general, a complete description of the electromagnetic field requires two scalar functions (such as π_z and π_z^* for TM and TE modes). Therefore, we need another scalar function in addition to π_{xs1} and π_{xs2}. Convenient choice, which was first used by Sommerfeld, is the z component of π, π_{zs1}, and π_{zs2}.

Let us now examine the boundary conditions at $z = 0$. At $z = 0$, the tangential electric fields E_x and E_y and the tangential magnetic field, H_x and H_y must be continuous. We write these conditions in terms of π_x and π_z, noting that

$\bar{E} = \nabla(\nabla \cdot \bar{\pi}) + k^2\bar{\pi}$ and $\bar{H} = j\omega\varepsilon\nabla \times \bar{\pi}$.

Continuity of E_x : $\quad \pi_{x1} = n^2\,\pi_{x2},$ (15.105a)

Continuity of E_y : $\quad \dfrac{\partial}{\partial x}\pi_{x1} + \dfrac{\partial}{\partial z}\pi_{z1} = \dfrac{\partial}{\partial x}\pi_{x2} + \dfrac{\partial}{\partial z}\pi_{z2},$ (15.105b)

Continuity of H_x : $\quad \pi_{z1} = n^2\,\pi_{z2},$ (15.105c)

Continuity of H_y : $\quad \dfrac{\partial}{\partial z}\pi_{x1} = n^2\dfrac{\partial}{\partial z}\pi_{x2}.$ (15.105d)

We use (15.105a) and (15.105d) to determine $R(\lambda)$ and $T(\lambda)$ for π_x in (15.104).

$$R(\lambda) = \frac{q - q_e}{q + q_e},$$

$$T(\lambda) = \frac{1}{n^2}\frac{2q}{q + q_e}.$$

(15.106)

Therefore, we write

$$\pi_{x1} = \frac{1}{4\pi}\int_0^\infty J_0(\lambda\rho)e^{-jq|h-h|}\frac{\lambda\,d\lambda}{jq}$$
$$+\frac{1}{4\pi}\int_0^\infty \frac{q-q_e}{q+q_e}J_0(\lambda\rho)e^{-jq|z+h|}\frac{\lambda\,d\lambda}{jq},\quad z>0,$$
$$\pi_{x2} = \frac{1}{4\pi}\int_0^\infty \frac{2q}{n^2(q+q_e)}J_0(\lambda\rho)e^{jq_ez-jqh}\frac{\lambda\,d\lambda}{jq},\quad z<0.$$

(15.107)

We note that if the second medium is perfectly conducting, Equation (15.107) reduces to a correct image representation. Now we use (15.105b) and (15.105c) to obtain π_{z1} and π_{z2}. First we note from (15.105b) that

$$\frac{\partial}{\partial z}(\pi_{z1} - \pi_{z2}) = \frac{\partial}{\partial x}(\pi_{x2} - \pi_{x1}).$$

(15.108)

Note that

$$\frac{\partial}{\partial x} = \frac{\partial\rho}{\partial x}\frac{\partial}{\partial\rho} + \frac{\partial\phi}{\partial x}\frac{\partial}{\partial\phi} \quad\text{and}\quad \frac{\partial\rho}{\partial x} = \cos\phi \quad\text{and}\quad \frac{\partial}{\partial\phi} = 0$$

for our problem.

Therefore, we write the right side of (15.108), using (15.107) and $J_0'(\lambda\rho) = -J_1(\lambda\rho)$,

$$\frac{\partial}{\partial x}(\pi_{x2} - \pi_{x1}) = \frac{\cos\phi}{4\pi}\int_0^\infty [T(\lambda) - 1 - R(\lambda)][-J_1(\lambda\rho)]e^{-jph}\frac{\lambda^2\,d\lambda}{jq}.$$

Since this must be equal to the left side of (15.108), π_z on the left side must have the same form as (15.109).

$$\cos \phi \int_0^\infty J_1(\lambda\rho) \cdots .$$

Thus we write

$$\pi_{z1} = \cos \phi \int_0^\infty A(\lambda)J_1(\lambda\rho)e^{-jq|z+h|} \lambda^2 \, d\lambda \quad \text{for } z > 0,$$

$$\pi_{z2} = \cos \phi \int_0^\infty B(\lambda)J_1(\lambda\rho)e^{q_e z - jqh}\lambda^2 \, d\lambda \quad \text{for } z < 0.$$

(15.109)

Now satisfying (15.108) and (15.105c), we determine $A(\lambda)$ and $B(\lambda)$

$$A(\lambda) = -\frac{2}{k^2} \frac{q - q_e}{n^2 \, q + q_e},$$

$$B(\lambda) = -\frac{2}{n^2 k^2} \frac{q - q_e}{n^2 \, q + q_e}.$$

(15.110)

Equations (15.107) and (15.109) constitute the complete expressions for the field of a horizontal dipole.

We note that the radiation from (15.107) is mostly in the direction perpendicular to the axis, as this is similar to the radiation from a horizontal dipole and its image. On the surface, however, this field is extremely small because the radiation from the image tends to cancel the direct radiation.

On the other hand, the field due to π_z as given in (15.109) is directional because of the $\cos \phi$ factor. Also, the coefficients $A(\lambda)$ and $B(\lambda)$ have the same Sommerfeld denominator $(n^2 q + q_e)$ as in the vertical dipole. Thus π_x produced by a horizontal dipole does not contribute much to the field on the surface, but it gives rise to π_z, which propagates mostly in the direction of the dipole axis and behaves in a manner similar to that of the vertical dipole.

15.11 RADIATION IN LAYERED MEDIA

In many practical problems we need to consider more than one interface. For example, wave propagation over the earth is greatly affected by the presence of the ionosphere, particularly for VLF and lower frequencies. Also, the thermal ionosphere, which will be discussed later, may be represented by layers. Another example would be the radiation from slot or dipole antennas on the surface of a spacecraft or high-speed vehicle, which may be covered by some protective material (Fig. 15.20). It is therefore important to study the effects of layers on the radiation field.

Let us consider a vertical dipole located within two interfaces, as shown in Fig. 15.21. As was shown in Section 15.3, the field between two boundaries consists of the primary wave π_{p0} and the secondary wave π_{s0}.

$$\pi_{z0} = \pi_{p0} + \pi_{s0}, \quad 0 < z < h,$$

(15.111)

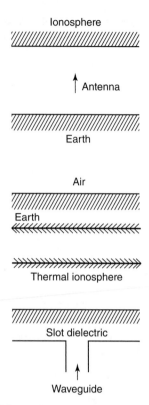

FIGURE 15.20   Radiation in layered media.

where

$$\pi_{p0} = \frac{1}{4\pi} \int_0^\infty e^{-jq|z-z_0|} J_0(\lambda\rho) \frac{\lambda\, d\lambda}{jq}, \tag{15.112}$$

$$\pi_{s0} = \frac{1}{4\pi} \int_0^\infty [a_0(\lambda)e^{jqz} + b_0(\lambda)e^{+jqz}] J_0(\lambda\rho) \frac{\lambda\, d\lambda}{jq}, \tag{15.113}$$

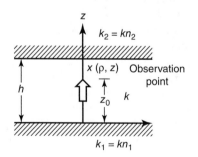

FIGURE 15.21   Vertical dipole in a layer.

Here we again omit $IL/j\omega\varepsilon_0$ for convenience. Note also that a_0 and b_0 in (15.113) represent the upward and downward waves.

The field below the lower boundary is then given by

$$\pi_{s1} = \frac{1}{4\pi} \int_0^\infty [b_1(\lambda)e^{+jq_1 z}]J_0(\lambda\rho)\frac{\lambda\,d\lambda}{jq}, \quad z < 0, \tag{15.114}$$

and the field above the upper boundary is given by

$$\pi_{s2} = \frac{1}{4\pi} \int_0^\infty [a_2(\lambda)e^{+jq_2 z}]J_0(\lambda\rho)\frac{\lambda\,d\lambda}{jq}, \quad z > h, \tag{15.115}$$

where

$$\lambda^2 + q^2 = k^2,$$
$$\lambda^2 + q_1^2 = k_1^2, \tag{15.116}$$
$$\lambda^2 + q_2^2 = k_2^2.$$

Here b_1 and a_2 represent the outgoing waves.

The boundary conditions at $z = 0$ and $z = h$ are

$$n^2\pi = \text{continuous},$$
$$\frac{\partial\pi}{\partial z} = \text{continuous}. \tag{15.117}$$

Applying these four conditions, we can determine four constants a_0, b_0, b_1, and a_1. Instead of simply applying (15.117), however, we wish to express these constants in terms of the reflection and transmission coefficients for the upper and lower media.

$$R_2 = \frac{n_2^2\,q - q_2}{n_2^2\,q + q_2}, \quad T_2 = \frac{2q}{n_2^2\,q + q_2},$$
$$R_1 = \frac{n_1^2\,q - q_1}{n_1^2\,q + q_1}, \quad T_1 = \frac{2q}{n_1^2\,q + q_1}. \tag{15.118}$$

To do this, we note that for the upper medium, the incident wave is given by the Fourier–Bessel transform of

$$[e^{-jq(z-z_0)} + a_0 e^{-jqz}],$$

and the reflected wave is given by

$$\left[b_0\, e^{jqz}\right].$$

The ratio of these two at $z = h$ must be the reflection coefficient R_2. Thus

$$R_2[e^{-jq(h-z_0)} + a_0\, e^{-jqh}] = [b_0\, e^{jqh}]. \tag{15.119}$$

Similarly, at $z = 0$, we have

$$R_1[e^{-jqz_0} + b_0] = a_0. \tag{15.120}$$

Solving these two equations for a_0 and b_0, we obtain

$$a_0 = \frac{R_1 e^{-jqz_0} + R_1 R_2 e^{-jq(2h-z_0)}}{1 - R_1 R_2 e^{-j2qh}}, \tag{15.121}$$

$$b_0 = \frac{R_2 e^{-jq(h-z_0)} + R_1 R_2 e^{-jq(h-z_0)}}{1 - R_1 R_2 e^{-j2qh}}. \tag{15.122}$$

Similarly, we can obtain $b_1(\lambda)$ and $a_2(\lambda)$ by using T_1 and T_2.

$$b_1 = \frac{T_1 e^{-jqz_0} + T_1 R_2 e^{-jq(2h-z_0)}}{1 - R_1 R_2 e^{-j2qh}}, \tag{15.123}$$

$$a_2 = \frac{T_2 e^{-jqz_0} + T_2 R_1 e^{-jq(2h+z_0)}}{1 - R_1 R_2 e^{-j2qh}} e^{jq2h}. \tag{15.124}$$

Substituting (15.121) to (15.124) into (15.112) to (15.115), we get a complete expression of the field.

Let us now examine these expressions. We first look for singularities in the integrand. It is clear that there are poles located at the roots of

$$1 - R_1 R_2 e^{-j2qh} = 0. \tag{15.125}$$

In general, there are infinite number of roots for (15.125) and thus there is a series of poles in the λ plane. This will be discussed further later.

It is also noted that the integrand contains q, q_1, and q_2, given by

$$q = \sqrt{k^2 - \lambda^2},$$

$$q_1 = \sqrt{k_1^2 - \lambda^2},$$

$$q_2 = \sqrt{k_2^2 - \lambda^2},$$

and thus it *appears* that there are three branch points in the λ plane: at $\lambda = \pm k$, $\lambda = \pm k_1$, and $\lambda = \pm k_2$. However, closer examination reveals that there is no branch point at $\lambda = \pm k$. This can be proved by showing that the integrand is unchanged when q is changed to $-q$. Another point of view is to recognize that in describing the field in (15.113), there is no need to differentiate $+q$ and $-q$. For example, in (15.113), e^{-jqz} represents the upward wave, but by changing it to e^{+jqz}, we simply interchange the role of a_0 (and b_0) from upward (downward) to downward (upward). However, for q_1 and q_2, we do need to differentiate $+q_1$ and $-q_1$, $+q_2$ and $-q_2$ because one represents the outgoing wave satisfying the radiation condition, and

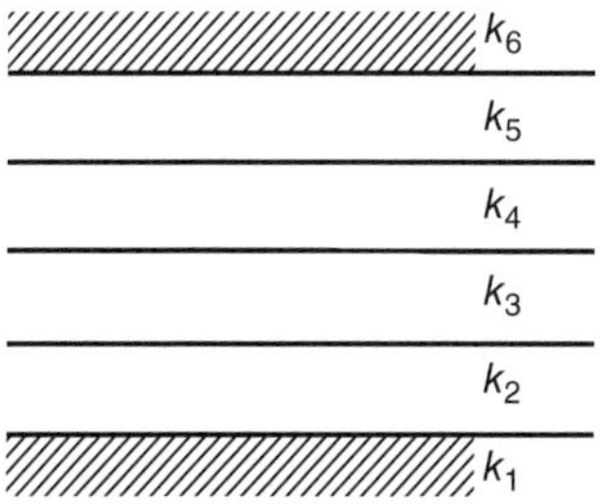

FIGURE 15.22 There is no branch point at $\lambda = \pm k_2$, $\pm k_3$, $\pm k_4$, and $\pm k_5$. The only branch points are at $\lambda = \pm k_1$ and $\pm k_6$.

the other represents the incoming wave. In general, if there are many layers as shown in Fig. 15.22, even though the integrand contains

$$q_i = \sqrt{k_i^2 - \lambda^2}, \quad i = 1, 2, \dots, 6,$$

the only branch points are at

$$\lambda = \pm k_1 \quad \text{and} \quad \lambda = \pm k_6,$$

and the integrands are even functions of q_2, q_3, q_4, and q_5 and have no branch points at k_2, k_3, k_4, and k_5.

15.12 GEOMETRIC OPTICAL REPRESENTATION

Let us now evaluate the integrals in the expressions (15.111)–(15.115). We may proceed along the following two lines:

1. *Geometric optical representation.* This is based on the saddle-point technique and is applicable to the case where the height h is many wavelengths high.
2. *Mode and lateral wave representation.* This is based on the residue series and the branch cut integration and is useful when the height h is small. We now show the details of these two approaches.

As an example, we take the field in the upper medium $z > h$, which is given by (15.115) and (15.124). Noting the two exponential terms in (15.124), we write π_{s2} as follows:

$$\pi_{s2} = \pi'_{s2} + \pi''_{s2}, \tag{15.126}$$

where

$$\pi'_{s2} = \frac{1}{4\pi} \int_0^\infty \left[\frac{T_2 e^{-jq(h-z_0)-jq_2(z-h)}}{1 - R_1 R_2 e^{-j2qh}} \right] J_0(\lambda\rho) \frac{\lambda \, d\lambda}{jq}, \tag{15.127}$$

$$\pi_{s2}'' = \frac{1}{4\pi} \int_0^\infty \left[\frac{T_2 R_1 e^{-jq(h+z_0)-jq_2(z-h)}}{1 - R_1 R_2 e^{-j2qh}} \right] J_0(\lambda\rho) \frac{\lambda \, d\lambda}{jq}. \qquad (15.128)$$

Let us first consider π'_{s2} and its z-dependent term in the brackets. First we expand

$$\frac{1}{1 - R_1 \, R_2 e^{-j2qh}} = \sum_{n=0}^\infty (R_1 \, R_2)^n e^{-j2qnh}, \qquad (15.129)$$

and write the quantity in brackets in (15.127) in a series form

$$[\cdot] = \sum_{n=0}^\infty u_n, \qquad (15.130)$$

where

$$u_n = T_2 e^{-jq(h-z_0)-jq_2(z-h)} (R_1 \, R_2)^n e^{-j2qnh}.$$

We now show that each of u_n represents the wave successively refracted and reflected at the boundaries.

We note that the first term,

$$u_0 = T_2 \, e^{-jq(h-z_0)-jq(z-h)},$$

represents the wave originating from the dipole, traveling over the distance $(h - z_0)$ with the propagation constant q, reaching the upper surface, traveling over the distance $(z - h)$ with the propagation constant q_2 and arriving at the observation point (Fig. 15.23).

The second term,

$$u_1 = T_2 \, e^{-jq(h-z_0)-jq_2(z-h)} \, R_1 \, R_2 e^{-j2qh},$$

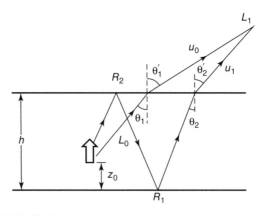

FIGURE 15.23   Geometric optical representation for u_0 and u_1.

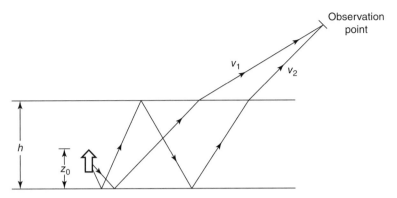

FIGURE 15.24   Geometric optical representation for v_1 and v_2.

represents the wave propagating over an additional distance of $2h$ with the propagation constant q and reflected by both sides once $(R_1 R_2)$. Similarly, the rest of the terms represent the multiple reflections of the wave between two boundaries.

Similarly, for π''_{s2} we write

$$[\cdot] = \sum_{n=0}^{\infty} v_n, \tag{15.131}$$

where

$$v_n = T_2\, R_1\, e^{-jq(h-z_0)-jq_2(z-h)}\, (R_1\, R_2)^n e^{-j2qnh}.$$

Each of v_n represents the wave shown in Fig. 15.24.

Each of the waves above can be evaluated by means of the saddle-point technique shown below.

$$\frac{1}{4\pi}\int_0^\infty A(\lambda)e^{-jqH-jq_2Z}J_0(\lambda\rho)\frac{\lambda\,d\lambda}{jq} = \frac{1}{8\pi}\int_W A(\lambda)e^{-jqH-jq_2Z}H_0^{(2)}(\lambda\rho)\frac{\lambda\,d\lambda}{jq}$$

$$\simeq \frac{1}{8\pi}\int_W \left[A(\lambda)\left(\frac{2}{\pi\lambda\rho}\right)^{1/2}\frac{\lambda e^{j(\pi/4)}}{jq}\right]e^{-if(\lambda)}\,d\lambda \tag{15.132}$$

where

$$f = qH + q_2Z + \lambda\rho.$$

Using the saddle-point technique, we get

$$\frac{1}{8\pi}\left\{A(\lambda)\left(\frac{2}{\pi\lambda\rho}\right)^{1/2}\frac{\lambda e^{j(\pi/4)}}{jq}\right\}e^{j(\pi/4)}\sqrt{\frac{2\pi}{-f''(\lambda)}},$$

evaluated at the saddle point given by

$$\frac{\partial f}{\partial \lambda} = 0.$$

Thus we obtain

$$\pi'_{s2} = \sum_{n=0}^{\infty} U_n,$$

$$\pi''_{s2} = \sum_{n=0}^{\infty} V_n, \qquad (15.133)$$

where

$$U_n = \frac{1}{4\pi} \int_0^{\infty} u_n J_0(\lambda\rho) \frac{\lambda \, d\lambda}{jq},$$

$$V_n = \frac{1}{4\pi} \int_0^{\infty} v_n J_0(\lambda\rho) \frac{\lambda \, d\lambda}{jq},$$

and the saddle-point evaluations of U_0 and U_1 are

$$U_0 \simeq 4\pi \frac{T_2(\theta_1(\sin\theta_1)^{1/2}}{\rho^{1/2}\cos\theta_1} \frac{e^{-j(kL_0+k_2L_1)}}{\sqrt{L_0/\cos^2\theta_1 + L_1/\cos^2\theta'_1}},$$

$$U_0 \simeq 4\pi \frac{T_2(\theta_1(\sin\theta_1)^{1/2}}{\rho^{1/2}\cos\theta_1} \frac{e^{-j(kL_0+k_2L_1)}}{\sqrt{L_0/\cos^2\theta_1 + L_1/\cos^2\theta'_1}}, \qquad (15.134)$$

$$U_1 \simeq 4\pi \frac{T_2(\theta_2)R_1(\theta_2)R_2(\theta_2)(\sin\theta_2)^{1/2}}{\rho^{1/2}\cos\theta_2} \frac{e^{-j(kL_2+k_2L_3)}}{\sqrt{L_2/\cos\theta_2 + L_3/\cos\theta'_2}},$$

where θ_1, θ'_1 and θ_2, θ'_2 are the angles corresponding to the saddle point. L_0 and L_2 are the total path lengths from the dipole to the point where the ray leaves the surface, and L_1 and L_2 are the path lengths from the surfaces to the observation point. All U_n and V_n can be expressed in a similar manner.

Since the saddle-point technique is used to obtain (15.134), it is valid only when the distances L_0, L_1, L_2, and L_3 are large. This occurs when the height h is large and the observation point is far from the surface.

On the other hand, if h is small but the observation point is far from the surface, we can apply the saddle-point technique to (15.126) directly. We write (15.127) and (15.128) as follows:

$$\pi'_{s2} = \frac{1}{4\pi} \int_0^{\infty} (A_1) e^{-jq2(z-h)} J_0(\lambda\rho) \frac{\lambda \, d\lambda}{jq},$$

$$\pi''_{s2} = \frac{1}{4\pi} \int_0^{\infty} (A_2) e^{-jq2(z-h)} J_0(\lambda\rho) \frac{\lambda \, d\lambda}{jq}, \qquad (15.135)$$

assume A_1 and A_2 to be slowly varying functions of λ and obtain the far field in terms of the distance $R = \sqrt{(z-h)^2 + \rho^2}$. This is the usual technique of obtaining the radiation pattern from an antenna covered with a dielectric layer.

15.13 MODE AND LATERAL WAVE REPRESENTATION

Let us consider the field within the layer $0 < z < h$ given in (15.111). We note that π_{z0} can be written in the following manner:

$$\pi_{z0} = \frac{1}{8\pi} \int_{-\infty}^{\infty} \frac{A(\lambda)e^{-jqz} + B(\lambda)e^{+jqz}}{1 - R_1 R_2 e^{-j2qh}} H_0^{(2)}(\lambda\rho) \frac{\lambda\,d\lambda}{jq}. \tag{15.136}$$

The integrand has a series of poles at the roots of

$$1 - R_1 R_2 e^{-j2qh} = 0 \tag{15.137}$$

and two branch points at

$$\lambda = \pm k_1 \quad \text{and} \quad \pm k_2.$$

Therefore, the integral can be expressed as

$$\pi_{z0} = -2\pi j \sum_{n=1}^{\infty} \text{residue at poles } \lambda_n + \int_{B_{r1}} + \int_{B_{r2}}, \tag{15.138}$$

where B_{r1} and B_{r2} are the branch cuts for K_1 and K_2 (Fig. 15.25).

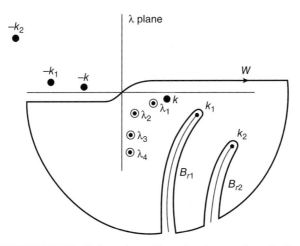

FIGURE 15.25   Poles at λ_n and branch points at $\pm k_1$ and $\pm k_2$.

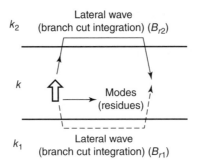

FIGURE 15.26   Lateral waves and waveguide modes.

In general, the residues have a radial dependence of the form

$$e^{-j\lambda_n \rho} \tag{15.139}$$

and represent waveguide modes. On the other hand, the branch cut integration has the characteristics of the lateral wave as shown in Fig. 15.26. Therefore, if k_1 and k_2 are lossy, the waveguide modes dominate as we would expect in the case of a lossy waveguide. On the other hand, if k_1 and k_2 are lossless, the lateral wave contributes most to the field.

It is obvious that if the height is small, there are small numbers of propagating modes and the other modes are almost cut off, implying that the residue series is highly convergent. On the other hand, if h is large, there may be a great number of propagating modes and the residue series is slowly convergent, and thus the geometric optical approach is more useful.

PROBLEMS

15.1 Find expressions for $\bar{E}$ and $\bar{H}$ for the problem shown in Fig. 15.1, and find the radiation field. The relative dielectric constant of the ground is 10 and the conductivity is $\sigma = 5 \times 10^{-3}$. The frequency is 1 MHz.

15.2 A vertical dipole is located on the ground operating at 1 MHz. The ground conductivity is $\sigma = 5 \times 10^{-3}$(S/m), and the relative dielectric constant (real part) is 10.

(a) Find the Sommerfeld pole α_p in the α plane.

(b) Find the numerical distance p at a distance of 10 km on the ground.

(c) Calculate and plot the attenuation factor as a function of $|p|(10^{-2} \le |p| \le 10^2)$.

(d) Calculate the ratio of $|E_z|$ on the ground to $|E_z|$ in free space at a distance of 10 km.

15.3 A short vertical dipole of length 1 m is carrying a current of 1 A at 50 MHz. The dipole is located on flat ground. The relative dielectric constant of the ground is 15 and the conductivity is 5×10^{-3}.

(a) Find the propagation constant of the Zenneck wave.

(b) Find the numerical distance at a distance of 3 km.

(c) Find the field strength as a function of distance.

15.4 Find the field in the ground when a vertical magnetic dipole is located on the ground.

15.5 Find the solution for the Sommerfeld problem when a magnetic horizontal dipole is located at height h above the ground.

15.6 A vertical dipole is located at height $h = 10$ km above a semi-infinite lossless plasma medium with $0 < n_2 < 1$ as shown in Fig. 15.9. The plasma frequency is 1 MHz and the operating frequency is 2 MHz. Find the field in air.

15.7 A vertical dipole is located above the interface between air $n_1 = 1$ and a lossless medium $n_2 = 2$ as shown in Fig. 15.18. Find the field transmitted in the second medium. $h = 5$ km, and the frequency is 1 MHz.

PART II

APPLICATIONS

CHAPTER 16

INVERSE SCATTERING

In the usual scattering problem, we specify the object and the incident wave, and then we attempt to find the scattered wave. This is the *direct problem*. In contrast, in the *inverse problem*, we measure the scattered wave for a given incident wave, and then we attempt to determine the properties of the object. Two considerations are important in the inverse problem. First, our measurements are normally limited and we can only measure certain quantities within some ranges. Second, we need an effective inverse method so that we can determine the object characteristics with limited measured data. It is clear, therefore, that the inverse solution may not be unique and that the existence of the solution may not be apparent. It is also common that the inverse solution is unstable, so that a slight error in the measurement may create a large error in the unknown. In this chapter, we outline several inversion techniques and their advantages and disadvantages. Also in this chapter, we use the convention $\exp(-i\omega t)$ commonly used in optics and acoustics, rather than $\exp(j\omega t)$ used in electrical engineering and by the IEEE, because there are many references in acoustics and physics on this topic.

16.1 RADON TRANSFORM AND TOMOGRAPHY

In a CT scanner (computed tomography scanner or X-ray tomography), an object is illuminated by an X-ray and the intensity of the transmitted X-ray is recorded for various angles of illumination. These recorded data are then used to reconstruct the

Electromagnetic Wave Propagation, Radiation, and Scattering: From Fundamentals to Applications,
Second Edition. Akira Ishimaru.
© 2017 by The Institute of Electrical and Electronic Engineers, Inc. Published 2017 by John Wiley & Sons, Inc.

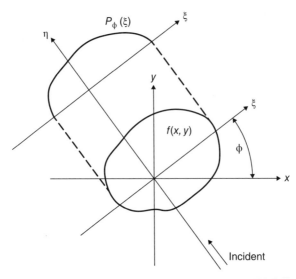

FIGURE 16.1   Radon transform or projection $P_\phi(\xi)$ of the object $f(x, y)$.

image of the object. Let us consider a two-dimensional cross section of an object whose attenuation coefficient is given by $f(x, y)$ (Fig. 16.1). The object is illuminated uniformly in the direction of $\hat{\eta}$ with the intensity I_0. The transmitted power $I_t(\xi)$ is reduced by the total attenuation through the object and is given by

$$I_t(\xi) = I_0 \exp\left[-\int f(x, y)d\eta\right]. \tag{16.1}$$

Letting $\ln(I_0/I_t) = P_\phi(\xi)$, we have

$$P_\phi(\xi) = \int f(x, y)d\eta. \tag{16.2}$$

The function $P_\phi(\xi)$ represents the total attenuation at ξ when the object is illuminated at the angle ϕ and is called the *projection*. The inverse problem is then to find $f(x, y)$ from the measured projection $P_\phi(\xi)$. We can also rewrite (16.2) as follows:

$$P_\phi(\xi) = \int f(x, y)\delta(\xi - \bar{r} \cdot \hat{\xi}) \, dx \, dy, \tag{16.3}$$

where $\bar{r} = x\hat{x} + y\hat{y}$ and $\bar{r} \cdot \hat{\xi} = x \cos\phi + y \sin\phi$ and $0 \le \phi < \pi$, $-\infty \le \xi \le +\infty$. Equation 16.3 can be considered a type of transform from $f(x, y)$ to $P_\phi(\xi)$ and is called the *Radon transform*. Therefore, the inverse problem is that of finding the *inverse Radon transform*. This was first studied by Radon in 1917 (Devaney, 1982; Kak, 1979; Herman, 1979).

Let us first consider a one-dimensional Fourier transform of the projection.

$$\bar{P}_\phi(K) = \int_{-\infty}^{\infty} P_\phi(\xi)e^{-jK\xi}\,d\xi. \tag{16.4}$$

Substituting (16.3) and integrating with respect to ξ, we get

$$\bar{P}_\phi(K) = \int f(x,y)e^{-iK\hat{\xi}\cdot\vec{r}}\,dx\,dy. \tag{16.5}$$

This is the two-dimensional Fourier transform of $f(x,y)$ evaluated at $\bar{K} = K\hat{\xi}$

$$\bar{P}_\phi(K) = F(K\hat{\xi}), \tag{16.6}$$

where

$$F(\bar{K}) = \int f(x,y)e^{-i\bar{K}\cdot\vec{r}}\,dx\,dy,$$

and $0 \le \phi < \pi, -\infty \le K \le \infty$.

We may think of $\bar{P}_\phi(K)$ as the Fourier transform of $f(x,y)$ along a slice at angle ϕ (Fig. 16.2). Therefore, we state that the one-dimensional Fourier transform of the projection of an object $f(x,y)$ is a slice at angle ϕ of the Fourier transform $F(\bar{K})$ of $f(x,y)$. This is called the *projection slice theorem.*

Now the complete Fourier transform of $f(x,y)$ is obtained by summing the slices over all the K space and $f(x,y)$ is obtained by the inverse Fourier transform. To do this, consider the inverse Fourier transform

$$f(x,y) = \frac{1}{(2\pi)^2} \int F(\bar{K})e^{i\bar{K}\cdot\vec{r}}\,d\bar{K}, \tag{16.7}$$

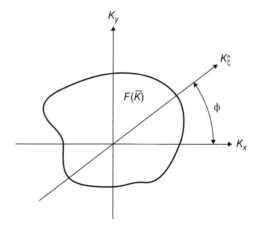

FIGURE 16.2   Projection slice theorem.

where $F(\bar{K}) = \bar{P}_\phi(K)$ and $d\bar{K} = K \, dK \, d\phi$. In order to use the range $0 \le \phi < \pi$, we write the integral as

$$\int_0^\infty K \, dK \int_0^\pi d\phi + \int_0^\infty K \, dK \int_\pi^{2\pi} d\phi.$$

Then using $\phi' = \phi - \pi$ and $K' = -K$, the second integral becomes

$$-\int_{-\infty}^0 K' \, dK' \int_0^\pi d\phi' = \int_{-\infty}^0 |K'| \, dK' \int_0^\pi d\phi'.$$

We therefore get

$$f(x, y) = \frac{1}{(2\pi)^2} \int_{-\infty}^\infty |K| \, dK \int_0^\pi d\phi \bar{P}_\phi(K) e^{iK\hat{\xi}\cdot\bar{r}}. \tag{16.8}$$

We rewrite this in the following form:

$$f(x, y) = \frac{1}{2\pi} \int_0^\pi d\phi Q_\phi(t).$$

$$Q_\phi(t) = \frac{1}{2\pi} \int_{-W}^W |K| \, dK \bar{P}_\phi(K\hat{\xi}) e^{iKt}, \tag{16.9}$$

$$t = \hat{\xi} \cdot \bar{r} = x \cos \phi + y \sin \phi.$$

Note that $Q_\phi(t)$ is the Fourier transform of the product of $\bar{P}_\phi$ and $|K|$ and therefore $|K|$ acts as a filter function $\bar{h}(K) = |K|$. We also used the highest spatial frequency W, because the projection can be measured only over a finite bandwidth $|K| \le W$. $Q_\phi(t)$ is called the *filtered projection*. It is a function of $t = \xi$ and is independent of η.

Since $Q_\phi(t)$ is given by the Fourier transform of the product of two Fourier transforms, we can express $Q_\phi(t)$ as the following convolution integral:

$$Q_\phi(t) = \int_{-\infty}^\infty P_\phi(\xi') h(t - \xi') \, d\xi', \tag{16.10}$$

where

$$h(t) = \frac{1}{2\pi} \int_{-W}^W |K| e^{iKt} \, dK$$

$$= \frac{W^2}{\pi} \frac{\sin Wt}{Wt} - \frac{W^2}{2\pi} \left[\frac{\sin(Wt/2)}{(Wt/2)} \right]^2, \tag{16.11}$$

where W is the highest spatial frequency.

Summarizing this section, we reconstruct the object $f(x, y)$ from the projection $P_\phi(\xi)$ as follows:

$$f(x, y) = \frac{1}{2\pi} \int_0^\pi d\phi Q_\phi(t), \quad t = x \cos \phi + y \sin \phi. \tag{16.12}$$

The filtered projection $Q_\phi(t)$ is given in the following two alternative forms:

$$Q_\phi(t) = \frac{1}{2\pi} \int_{-W}^{W} \bar{h}(K) \bar{P}_\phi(K) e^{iKt} \, dK$$

$$= \int_{-\infty}^{\infty} h(t - \xi') P_\phi(\xi') \, d\xi', \tag{16.13}$$

where

$$h(t) = \frac{1}{2\pi} \int_{-W}^{W} |K| \, e^{iKt} \, dK,$$

$$\bar{h}(K) = |K| \,.$$

Note that the filtered projection $Q_\phi(t)$ is obtained from the projection $P_\phi(\xi)$ either by the inverse Fourier transform of $\bar{P}_\phi(K)$ filtered with $\bar{h}(K)$ or by the convolution integral. Once $Q_\phi(t)$ is obtained for a given ϕ, it is summed over all ϕ from 0 to π as shown in (16.12) and the object is reconstructed. This process is called *back projection*. Therefore, we speak of the entire process as the *back projection of the filtered projection*. Equation (16.12) can be regarded as an inverse Radon transform to obtain $f(x, y)$ from the Radon transform $P_\phi(\xi)$. Note that if the object is a delta function $f(x, y) = \delta(x)\delta(y)$, the reconstructed image is $[W/(2\pi r)]J_1(Wr)$, where $r = (x^2 + y^2)^{1/2}$.

In practice, the calculations of (16.13) are made by digital processing (Kak, 1979). For example, the sampling interval of $P_\phi(\xi)$ must be $\tau = 1/(2W)$ and

$$Q_\phi(n\tau) = \tau \sum_m P_\phi(m\tau)h[(n - m)\tau]. \tag{16.14}$$

This can also be done by using FFT in the frequency domain in (16.13).

16.2 ALTERNATIVE INVERSE RADON TRANSFORM IN TERMS OF THE HILBERT TRANSFORM

In Section 16.1, we gave an inverse Radon transform in the form of the back projection of the filtered projection. It is also possible to express the inverse Radon transform using the Hilbert transform as was done by Radon in 1917. This inversion formula, however, contains a derivative and is more sensitive to noise.

Let us first consider the Fourier transform of the projection

$$\bar{P}_\phi(K) = \int P_\phi(\xi) e^{-iK\xi} \, d\xi .$$ (16.15)

Note that we are using $i = \sqrt{-1}$ rather than $j = \sqrt{-1}$ in this chapter.

We integrate this by parts and noting $P_\phi(\xi)$ vanishes as $\xi \to \pm \infty$, we obtain

$$\bar{P}_\phi(K) = \int \frac{1}{iK} \frac{\partial}{\partial \xi} P_\phi(\xi) e^{-iK\xi} \, d\xi.$$ (16.16)

Substituting this in (16.13), we get

$$Q_\phi(t) = \frac{1}{2\pi} \int \frac{|K|}{iK} \, dK \int \frac{\partial}{\partial \xi} P_\phi(\xi) e^{iK(t-\xi)} \, d\xi.$$ (16.17)

Now the integration with respect to K can be performed using the following (see Appendix 16.A):

$$\frac{1}{2\pi} \int i\,\mathrm{sgn}(K) e^{iKt} \, dK = -\frac{1}{\pi t},$$

$$\mathrm{sgn}(K) = \frac{|K|}{K} = \begin{cases} 1 & \text{if } K > 0 \\ -1 & \text{if } K < 0 \end{cases}.$$ (16.18)

The results can be expressed using the following Hilbert transform (see Appendix 16.A):

$$F_h(t) = \frac{1}{\pi} \int_{-\infty}^{\infty} \frac{f(t')}{t' - t} \, dt',$$ (16.19)

where the integral is the Cauchy principal value. Thus we get

$$f(x, y) = \frac{1}{2\pi} \int_0^\pi d\phi\, Q_\phi(t)$$

$$Q_\phi(t) = -\text{Hilbert transform of } \frac{\partial}{\partial \xi} P_\phi(\xi)$$ (16.20)

$$= \int_{-\infty}^{\infty} \frac{1}{\pi(t - \xi)} \frac{\partial}{\partial \xi} P_\phi(\xi) \, d\xi.$$

This is an alternative form of the inverse Radon transform.

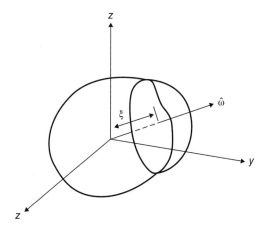

FIGURE 16.3 Three-dimensional radon transform.

It is also possible to consider a three-dimensional Radon transform (Fig. 16.3)

$$R(\xi, \hat{\omega}) = \int f(\bar{r})\delta(\xi - \bar{r} \cdot \hat{\omega})\, dx\, dy\, dz. \tag{16.21}$$

The inverse Radon transform is then given by

$$f(\bar{r}) = -\frac{1}{8\pi^2} \int \frac{\partial^2 R(\xi, \hat{\omega})}{\partial \xi^2}\, d\omega, \tag{16.22}$$

where $d\omega$ is the differential element in the solid angle.

16.3 DIFFRACTION TOMOGRAPHY

The X-ray tomography described in Section 16.2 deals with the reconstruction of the image of an object from the projection obtained by illuminating the object with uniform X-rays. All rays are assumed to travel in straight lines, and this is true at the limit of zero wavelengths. If acoustic or electromagnetic waves are used to reconstruct the object, diffraction effects due to finite wavelengths cannot be ignored. In general, the formulations of inverse problems, including diffraction, are extremely complicated and general solutions are not yet available. However, if the scattering is weak, it is possible to formulate the inverse problem in a manner similar to X-ray tomography and general solutions can be obtained. This will be presented in this section (see Devaney, 1982).

For a weakly scattering medium, the field U is represented by the Born approximation or the Rytov approximation. Let us assume that the field U satisfies the wave equation

$$[\nabla^2 + k^2 n^2(x, y)]U(x, y) = 0, \tag{16.23}$$

where k is the wave number of the background medium and $n(x, y)$ is the refractive index of the object. For example, the background medium is water in the case of ultrasound imaging of an object in water, and n represents the deviation of the refractive index of the object from water. We rewrite (16.23) as

$$(\nabla^2 + k^2)U = -f(x, y)U, \tag{16.24}$$

where

$$f(x, y) = k^2(n^2 - 1).$$

This can be converted to the following integral equation for U:

$$U(\bar{r}) = U_i(\bar{r}) + \int G(\bar{r}, \bar{r}')f(\bar{r}')U(\bar{r}')\, d\bar{r}', \tag{16.25}$$

where U_i is the incident wave and $G(\bar{r}, \bar{r}')$ is Green's function satisfying the equation

$$(\nabla^2 + k^2)G = -\delta(\bar{r} - \bar{r}').$$

The first Born approximation is obtained by approximating U in the integrand by the incident wave

$$U(\bar{r}) = U_i(\bar{r}) + \int G(\bar{r}, \bar{r}')f(\bar{r}')U_i(\bar{r}')\, d\bar{r}'. \tag{16.26}$$

This is valid when

$$|k(n - 1)D| \ll 1, \tag{16.27}$$

where D is a typical size of the object.

The Rytov approximation is obtained by considering the total complex phase Ψ of U. We let

$$U(x, y) = U_i(\bar{r})e^{\Psi(x,y)}. \tag{16.28}$$

The first Rytov approximation is then given by (see Appendix 16.B)

$$\Psi(x, y) = \frac{1}{U_i(\bar{r})} \int G(\bar{r}, \bar{r}')f(\bar{r}')U_i(\bar{r}')\, d\bar{r}'. \tag{16.29}$$

Note that if we expand the exponent in (16.28) and keep the first term, we obtain the first Born approximation, and therefore the Rytov approximation contains more scattering terms and is a better approximation than the first Born approximation. In either the Born or Rytov solution, we can use the scattered field U_s

$$U_s(\bar{r}) = \int G(\bar{r}, \bar{r}')f(\bar{r}')U_i(\bar{r}')\, d\bar{r}'. \tag{16.30}$$

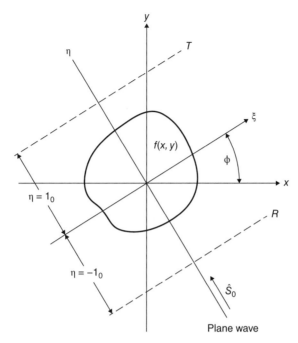

FIGURE 16.4   Diffraction tomography. Plane wave is incident on the object $f(x, y)$. T is the receiver plane for transmission tomography and R is the receiver plane for reflection tomography.

In the Rytov solution, for a given incident field U_i we measure the total field U and obtain the field U_s by calculating $U_i \ln(U/U_i)$. The function $f(x, y)$ represents the object, and the inverse problem is then reduced to finding the object $f(x, y)$ from U_s.

Let us consider a two-dimensional cross section of the object $f(x, y)$ illuminated by a plane wave propagating in the direction $\hat{s}_0$ (Fig. 16.4).

$$U_i(\bar{r}) = e^{ik\hat{s}_0 \cdot \bar{r}} = e^{ik\eta'}. \tag{16.31}$$

Note that we are using the convention $\exp(-i\omega t)$ in this chapter.

The two-dimensional Green's function is given by

$$
\begin{aligned}
G(\bar{r}, \bar{r}') &= \frac{i}{4} H_0^{(1)}(K|\bar{r} - \bar{r}'|) \\
&= \frac{1}{2\pi} \int \frac{i}{2K_2} e^{iK_1(\xi - \xi') + iK_2(\eta - \eta')} \, dK_1,
\end{aligned} \tag{16.32}
$$

where

$$
K_2 = \begin{cases}
(k^2 - K_1^2)^{1/2} & \text{if } |K_1| < k \\
i(K_1^2 - k^2)^{1/2} & \text{if } |K_1| > k
\end{cases}
$$

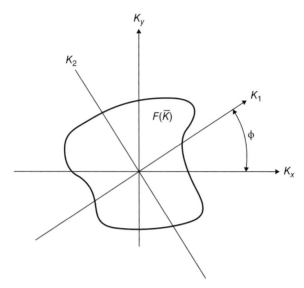

FIGURE 16.5   Spatial frequency space K.

The spatial frequency K_1 and K_2 are chosen parallel to $\hat{\xi}$ and $\hat{\eta}$ axes, respectively (Fig. 16.5).

Let us first consider U_s for the transmission tomography. Substituting (16.31) and (16.32) into (16.30), and noting that $|\eta - \eta'| = l_0 - \eta'$, we get

$$U_s(\xi, l_0) = \frac{1}{2\pi} \int \frac{i}{2K_2} e^{iK_1\xi + iK_2l_0} F(K_1, K_2 - k) \, dK_1, \tag{16.33}$$

where F is the Fourier transform of $f(x, y)$.

$$F(K_1, K_2 - k) = \int f(x', y') e^{-iK_1\xi' - i(K_2-k)\eta'} \, d\xi' \, d\eta'.$$

Note that (16.33) can be seen as a one-dimensional Fourier transform, and therefore, if we take the one-dimensional Fourier transform of the observed data $U_s(\xi, l_0)$ at $\eta = l_0$, we get from (16.33),

$$\bar{U}_s(K_1, l_0) = \int U_s(\xi, l_0) e^{-iK_1\xi} \, d\xi$$

$$= \frac{ie^{iK_2l_0}}{2K_2} F(K_1, K_2 - k). \tag{16.34a}$$

Note that K_1 and K_2 are related through

$$K_1^2 + K_2^2 = k^2, \tag{16.34b}$$

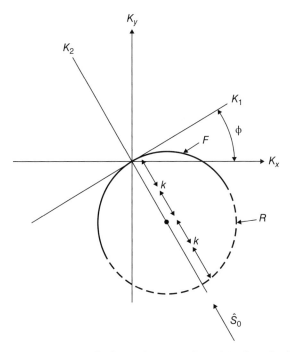

FIGURE 16.6   Semicircle arc (F) for forward tomography and (R) for reflection tomography.

as given in (16.32), and therefore, if we limit ourselves to the real K_1 and K_2, this represents a circle with radius k in the $\bar{K}$ space. The real K_1 and K_2 means that only the propagating wave is considered and the evanescent wave is neglected. We may rewrite (16.34) using

$$K_1\hat{\xi} + K_2\hat{\eta} = k\hat{s}, \quad K\hat{\eta} = k\hat{s}_0, \quad \bar{U}_s(K_1, l_0) = \frac{ie^{iK_2 l_0}}{2K_2}F[k(\hat{s} - \hat{s}_0)]. \quad (16.35)$$

This means that the Fourier transform of the observed data $U_s(\xi, l_0)$ at $\eta = l_0$ is proportional to the Fourier transform $F(\bar{K})$ of the object $f(x, y)$ evaluated at $\bar{K} = k(\hat{s} - \hat{s}_0)$. This is a semicircle in $\bar{K}$ space where $k > K_2 > 0$, shown in Fig. 16.6. This is a generalization of the projection slice theorem for conventional tomography discussed in Section 16.1. Here, instead of a slice at an angle θ, we have a semicircle centered at $\bar{K} = -k\hat{s}_0$. Note that in the high-frequency limit, $k \to \infty$ and the semicircle is stretched to a slice for conventional tomography.

For reflection tomography, we note that $|\eta - \eta'| = \eta' + l_0$ and the Fourier transform of the observed data $U_s(\xi, -l_0)$ at $\eta = -l_0$ is given by

$$\bar{U}_s(K_1, -l_0) = \frac{ie^{iK_2 l_0}}{2K_2}F(K_1, -K_2 - k). \quad (16.36)$$

This is the lower half of the semicircle as indicated in Fig. 16.6.

From the discussion above, it is clear that the Fourier transform of the observed data for a given incident direction $\hat{s}_0$ is proportional to the Fourier transform of the object evaluated on a semicircle. If we do this process for all angles $0 \le \phi < \pi$, we can cover the K space within the circle of radius $\sqrt{2}k$ for transmission tomography and within the band between the radii from $\sqrt{2}k$ to $2k$ for reflection tomography. We can then invert the Fourier transform to reconstruct the object. This reconstruction process is similar to the back projection for conventional tomography, but since it includes the propagation effect, it is called back propagation and is explained below. We also note that we can cover a larger K space by varying the frequency $k = \omega/c$.

Let us consider the forward tomography. $U_s(\xi, l_0)$ is measured and its one-dimensional Fourier transform $\bar{U}_s(K_1, l_0)$ is calculated. We then calculate the function

$$F(K_1, K_2 - k) = -i2K_2 e^{iK_2 l_0} \bar{U}_s(K_1, l_0). \tag{16.37}$$

From this, we get the object by the inverse Fourier transform:

$$f(x, y) = \frac{1}{(2\pi)^2} \int F(\bar{K}) e^{i\bar{K}\cdot\bar{r}} dK_x dK_y, \tag{16.38}$$

where $\bar{K} = k(\hat{s} - \hat{s}_0)$.

Now we note that

$$\bar{K} = K_1 \hat{\xi} + (K_2 - k)\hat{\eta}$$
$$\hat{\xi} = \cos\phi\hat{x} + \sin\phi\hat{y} \tag{16.39}$$
$$\hat{\eta} = -\sin\phi\hat{x} + \cos\phi\hat{y}.$$

Therefore, we get

$$\bar{K} = K_x\hat{x} + K_y\hat{y}$$
$$K_x = K_1 \cos\phi - (K_2 - k)\sin\phi \tag{16.40}$$
$$K_y = K_1 \sin\phi + (K_2 - k)\cos\theta.$$

Changing the variables from (K_x, K_y) to (K_1, ϕ), and noting $K_1^2 + K_2^2 = k^2$, we get

$$dK_x \, dK_y = \begin{vmatrix} \dfrac{\partial K_x}{\partial K_1} & \dfrac{\partial K_x}{\partial \phi} \\ \dfrac{\partial K_y}{\partial K_1} & \dfrac{\partial K_y}{\partial \phi} \end{vmatrix} dK_1 \, d\phi = \frac{kK_1}{K_2} dK_1 \, d\phi \tag{16.41}$$

Following the procedure in Section 16.1 and using (16.37), we get

$$f(x,y) = \frac{1}{2\pi} \int_0^\pi d\phi Q_\phi(\xi,\eta),$$

$$Q_\phi(\xi,\eta) = \frac{1}{2\pi} \int_{-k}^{+k} h(K_1,\eta)\bar{U}_s(K_1,l_0)e^{iK_1'\xi}\, dK_1,$$

(16.42)

where $h(K_1,\eta)$ is the filter function given by

$$h(K_1,\eta) = -i2k|K_1|e^{-ikl_0+i(K_2-k)(\eta-l_0)},$$

where

$$\xi = x\cos\phi + y\sin\phi$$

$$\eta = -x\sin\phi + y\cos\phi,$$

$$K_2 = \left(k^2 - K_1^2\right)^{1/2}.$$

We can conclude here that Q_ϕ is a generalization of the filtered projection and the reconstruction process is called filtered back propagation. It is also clear that Q_ϕ can also be written as a convolution integral as shown in Section 16.1.

We also note that if the object is a delta function $f(x,y) = \delta(x)\delta(y)$, the reconstructed image is an Airy disk given by

$$f(x,y) = \frac{k}{\sqrt{2\pi r}} J_1(\sqrt{2}kr).$$

(16.43)

16.4 PHYSICAL OPTICS INVERSE SCATTERING

Based on the physical optics approximation, it is possible to derive an inversion formula that gives the size and shape of a conducting object from the knowledge of the monostatic scattering for all frequencies and all aspect angles. This inverse scattering formula was first obtained by Bojarski in 1967 and is now known as *Bojarski's identity* (see the survey paper by Bojarski, 1982b and Lewis, 1969).

Let us consider the scattered far field $\mathbf{E}_s$ from a conducting object.

$$\mathbf{E}_s = -jk\eta_0 \frac{e^{-jkR}}{4\pi R} \int_s [-\hat{o} \times (\hat{o} \times \mathbf{J}_s)]e^{jk\hat{o}\cdot\vec{r}'}\, ds',$$

(16.44)

where $\eta_0 = (\mu_0/\varepsilon_0)^{1/2}$, R is the distance from the reference point on the object, $\mathbf{J}_s$ is the surface current density, and $\hat{o}$ is the direction of observation.

For monostatic scattering, $\hat{o} = -\hat{i}$ where $\hat{i}$ is the direction of the incident wave. For physical optics approximation, we have $\mathbf{J}_s = 2(\hat{n} \times \mathbf{H}_i)$ where $\hat{n}$ is the unit vector normal to the surface. We note that

$$\hat{i} \times (\hat{n} \times \mathbf{H}_i) = -\mathbf{H}_i(\hat{i} \cdot \hat{n}),$$
$$-\hat{o} \times (\hat{o} \times \mathbf{J}_s) = \hat{i} \times \mathbf{H}_i(\hat{i} \cdot \hat{n}).$$

(16.45)

The scattered wave $\mathbf{E}_s$ can then be expressed as

$$\mathbf{E}_s = jk\eta_0\hat{e}_i \frac{e^{-jkR}}{4\pi R} \int_s 2(\hat{i} \cdot \hat{n})e^{j2k\hat{o}\cdot\mathbf{r}'} ds'$$

(16.46)

where $\hat{e}_i$ is the unit vector in the direction of the polarization of the incident wave $\mathbf{E}_i = E_i \hat{e}_i$. This formula shows that the monostatic scattered wave is polarized in the direction $\hat{e}_i$, and therefore there is no cross polarization in the monostatic scattered wave based on physical optics.

Let us now define the normalized complex far-field scattering amplitude $\rho(\mathbf{K})$ when illuminated in the direction $\hat{i}$. We also use $\exp(-i\omega t)$ dependence rather than $\exp(j\omega t)$ in this section, as this is commonly used in this work.

$$\rho(\mathbf{K}) = \frac{i}{\sqrt{4\pi}} \int_{\mathbf{K}\cdot\hat{n}>0} e^{-i\mathbf{K}\cdot\mathbf{r}'} \mathbf{K} \cdot d\mathbf{s}',$$

(16.47)

where $\mathbf{K} = 2k\hat{o} = -2k\hat{i}$, $d\mathbf{s}' = \hat{n}\, ds'$, and $\mathbf{K} \cdot \hat{n} > 0$ means the illuminated surface (Fig. 16.7). The normalized complex scattering amplitude $\rho(\mathbf{K})$ is defined so that the back-scattering cross section σ_b is given by [see (10.101)]

$$\sigma_b = \rho\rho^*.$$

(16.48)

Now we consider the scattered wave when illuminated from the opposite direction $(\mathbf{K} \to -\mathbf{K})$. We also take the complex conjugate and obtain

$$\rho^*(-\mathbf{K}) = \frac{i}{\sqrt{4\pi}} \int_{\mathbf{K}\cdot\hat{n}<0} e^{-i\mathbf{K}\cdot\mathbf{r}'} \mathbf{K} \cdot d\mathbf{s}'.$$

(16.49)

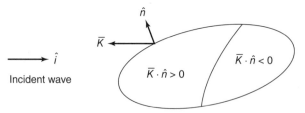

FIGURE 16.7   Illuminated surface $\bar{\mathbf{K}} \cdot \hat{n} > 0$.

Adding $\rho(\mathbf{K})$ and $\rho^*(-\mathbf{K})$, we get

$$\rho(\mathbf{K}) + \rho^*(-\mathbf{K}) = \frac{i}{\sqrt{4\pi}} \int_s e^{-i\mathbf{K}\cdot\mathbf{r}'} \mathbf{K} \cdot d\mathbf{s}', \qquad (16.50)$$

where s is the complete surface of the object.

By using the divergence theorem, the right side of (16.50) becomes

$$\frac{i}{\sqrt{4\pi}} \int_v \nabla \cdot (e^{-i\mathbf{K}\cdot\mathbf{r}'} \mathbf{K}) \, dv' = \frac{2k^2}{\sqrt{\pi}} \int_v e^{-i\mathbf{K}\cdot\mathbf{r}'} \, dv', \qquad (16.51)$$

where v is the volume of the object.

We can now define the complex scattering amplitude $\Gamma(\mathbf{K})$ and the characteristic function $\gamma(\mathbf{r}')$ of the scatterer.

$$\Gamma(\mathbf{K}) = \frac{\sqrt{\pi}}{2k^2} [\rho(\mathbf{K}) + \rho^*(-\mathbf{K})],$$

$$\gamma(\mathbf{r}') = \begin{cases} 1 & \text{if } \mathbf{r}' \text{ is inside the scatterer} \\ 0 & \text{if } \mathbf{r}' \text{ is outside.} \end{cases} \qquad (16.52)$$

We then get the following three-dimensional Fourier transform relationship:

$$\Gamma(\mathbf{K}) = \int_v \gamma(\mathbf{r}') e^{-i\mathbf{K}\cdot\mathbf{r}'} \, dv' \qquad (16.53)$$

For scatterers of finite volume, we can invert this Fourier transform and obtain

$$\gamma(\mathbf{r}) = \frac{1}{(2\pi)^3} \int \Gamma(\mathbf{K}) e^{i\mathbf{K}\cdot\mathbf{r}} \, d\mathbf{K}. \qquad (16.54)$$

This is Bojarski's identity and shows that the object shape can be determined by measuring the back-scattered far field over all $\mathbf{K}$. This requires that the back scattering be known for all frequencies and all angles, which presents a practical measurement difficulty. Also, the identity is based on the physical optics approximation, which is valid only for high frequencies and should not be used for lower frequencies or resonance regions. This is another source of theoretical difficulty. Therefore, much of the work is directed to obtaining solutions even if the scattering information is incomplete (see Bojarski, 1982b; Lewis, 1969).

If the measurement of the scattered wave can be made only in some portion D of the $\mathbf{K}$ space, we write

$$A(\mathbf{K}) = \begin{cases} 1 & \text{if } \mathbf{K} \text{ is in } D, \\ 0 & \text{if } \mathbf{K} \text{ is outside.} \end{cases} \qquad (16.55)$$

Then we can measure $A(\mathbf{K})\Gamma(\mathbf{K})$. The inverse Fourier transform is then given by

$$f(\mathbf{r}) = \frac{1}{(2\pi)^3} \int A(\mathbf{K})\Gamma(\mathbf{K})e^{i\mathbf{K}\cdot\mathbf{r}}\,d\mathbf{K}. \qquad (16.56)$$

This is given by

$$\begin{aligned} f(\mathbf{r}) &= \int A(\mathbf{r}-\mathbf{r}')\gamma(\mathbf{r}')\,dv' \\ &= \int A(\mathbf{r}')\gamma(\mathbf{r}-\mathbf{r}')\,dv', \end{aligned} \qquad (16.57)$$

where

$$A(\mathbf{r}) = \frac{1}{(2\pi)^3} \int A(\mathbf{K})e^{i\mathbf{K}\cdot\mathbf{r}}\,d\mathbf{K}.$$

In (16.57), $f(\mathbf{r})$ is calculated from the measurement and $A(\mathbf{r})$ is also known, and therefore this constitutes an integral equation for $\gamma(\mathbf{r})$.

As an example, suppose that the measurement is conducted only from one direction $\hat{z}$. Then D is the line $K_x = K_y = 0$ and therefore

$$A(K) = \delta(K_x)\delta(K_y). \qquad (16.58)$$

We then get

$$A(\mathbf{r}) = \frac{1}{(2\pi)^3} \int e^{iK_z z}\,dK_z = \frac{1}{(2\pi)^2}\delta(z).$$

Substituting this in (16.57), we get

$$f(\mathbf{r}) = \frac{1}{(2\pi)^2} \int\int \gamma(x, y, z)\,dx\,dy. \qquad (16.60)$$

This means that $f(\mathbf{r})$ calculated from the measured data gives the cross-sectional area of the target as a function of z.

16.5 HOLOGRAPHIC INVERSE SOURCE PROBLEM

If the field Ψ and its normal derivative $\partial\Psi/\partial n$ produced by a source $\rho(\mathbf{r})$ are measured over a surface S, it should be possible to reconstruct the source distribution $\rho(\mathbf{r})$ from the knowledge of Ψ and $\partial\Psi/\partial n$ on S. This inversion technique has been developed independently by Porter and Bojarski. This is based on the same principle as conventional holography, in which the scattered field is recorded on a photographic film which, when illuminated, reproduces the source distribution (see Bojarski, 1982a, for a review; see also Porter and Devaney, 1982; and Tsang et al., 1987).

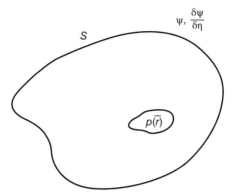

FIGURE 16.8   Holographic inversion.

Let us first write Green's theorem for the volume enclosed by the surface S (Fig. 16.8).

$$\int_V (u\,\nabla^2 v - v\,\nabla^2 u)\,dV = \int_S \left(u\frac{\partial v}{\partial n} - v\frac{\partial u}{\partial n}\right) ds, \qquad (16.61)$$

where $\partial/\partial n$ is the outward normal derivative. We now let the scalar function u be the field Ψ. The field Ψ is produced by the source ρ and satisfies the following inhomogeneous wave equation:

$$(\nabla^2 + k^2)\psi = -\rho. \qquad (16.62)$$

We then back propagate the field toward the source. To do this, we use $v = G^*$, where G is Green's function satisfying

$$(\nabla^2 + k^2)G = -\delta(\mathbf{r} - \mathbf{r}'), \quad G = \frac{\exp(ik\,|\mathbf{r} - \mathbf{r}'|)}{4\pi\,|\mathbf{r} - \mathbf{r}'|} \qquad (16.63)$$

Substituting (16.62) and (16.63) into (16.61), we get

$$\left(\Psi\frac{\partial G^*}{\partial n} - G^*\frac{\partial \Psi}{\partial n}\right) ds = -\Psi + \int G^*\rho\,dV. \qquad (16.64)$$

Also, if·we use $v = G$, we get

$$\int_S \left(\Psi\frac{\partial G}{\partial n} - G\frac{\partial \Psi}{\partial n}\right) ds = -\Psi + \int G\rho\,dV = 0. \qquad (16.65)$$

Note that the surface integral vanishes, as it produces no scattered field inside S. Subtracting (16.64) from (16.65), we get the following integral equation for ρ:

$$\Gamma(\mathbf{r}) = \int K(\mathbf{r}, \mathbf{r}')\rho(\mathbf{r}')dV', \tag{16.66}$$

where

$$\Gamma(\mathbf{r}) = \int_S \left\{ \left[\Psi(\mathbf{r}_s) \frac{\partial g(\mathbf{r},\mathbf{r}_s)}{\partial n} \right] - \left[g(\mathbf{r},\mathbf{r}_s) \frac{\partial \Psi(\mathbf{r}_s)}{\partial n} \right] \right\} ds,$$

$$K(\mathbf{r},\mathbf{r}') = \frac{\sin k|\mathbf{r} - \mathbf{r}'|}{k|\mathbf{r} - \mathbf{r}'|} = G - G^*,$$

$$g(\mathbf{r},\mathbf{r}_s) = \frac{\sin k|\mathbf{r} - \mathbf{r}_s|}{k|\mathbf{r} - \mathbf{r}_s|} = G - G^*.$$

From knowledge of the surface field $\Psi(\mathbf{r}_s)$ and $\partial\Psi(\mathbf{r}_s)/\partial n$, $\Gamma(\mathbf{r})$ can be calculated, and by solving the integral equation, the source distribution $\rho(\mathbf{r}')$ can be obtained.

An alternative form of the integral equation is possible. We may use the conjugate field Ψ^* and obtain

$$\nabla^*(\mathbf{r}) = \int K(\mathbf{r}, \mathbf{r}')r^*(r')dV'. \tag{16.67}$$

The solutions to the integral equations (16.66) and (16.67) are, however, not unique. The source ρ consists of the nonradiating sources ρ_n and the radiating sources ρ_r. The nonradiating source generates a field that is identically zero outside the source region and does not contribute to the field on S. It has been shown that if we minimize the source energy E,

$$E = \int |\rho|^2 \, dV, \tag{16.68}$$

it yields a unique solution. For a detailed discussion on the uniqueness and nonradiating sources, see Porter and Devaney (1982) and Cohen and Bleistein (1979). Also see Devaney and Porter (1985) and Tsang et al. (1987) for inversion of inhomogeneous and attenuating media.

16.6 INVERSE PROBLEMS AND ABEL'S INTEGRAL EQUATION APPLIED TO PROBING OF THE IONOSPHERE

Consider the ionosphere whose electron density profile $N_e(z)$ is known as a function of height z. The refractive index is given by

$$N(\omega) = \left(1 - \frac{\omega_p^2}{\omega^2} \right)^{1/2}, \tag{16.69}$$

$$\omega_p^2 = \frac{e^2 N_e}{m\,\varepsilon_0},$$

where $f_p = \omega_p/2\pi$ is plasma frequency, e and m are the charge and mass of an electron, ε_0 is the free-space permittivity, and the loss is neglected.

If we send up a radio wave of angular frequency ω, it reaches the height h where ω_p is equal to ω and returns to the ground. The time $T(\omega)$ for the wave to travel to $z = h$ and return to $z = 0$ is given by

$$T(\omega) = 2 \int_0^h \frac{dz}{v_g}, \tag{16.70}$$

where v_g is the group velocity given by

$$\frac{1}{v_g} = \frac{\partial k}{\partial \omega} = \frac{\omega}{c} \frac{1}{\left(\omega^2 - \omega_p^2\right)^{1/2}}$$

$$k = \frac{\omega}{c} n(\omega)$$

and the height h at the turning point at a frequency ω is given by

$$\omega = \omega_p(h).$$

For a given profile of $N_e(z)$, and therefore $\omega_p(z)$, we can calculate $T(\omega)$ from (16.70). This is called the *direct problem*.

Now consider the inverse problem. We send up radio waves at various frequencies and measure $T(\omega)$ as a function of ω. From the measured data $T(\omega)$, we attempt to determine the plasma frequency profile $\omega_p(z)$ and the electron density profile $N_e(z)$. This is the *inverse problem* (Fig. 16.9).

Let us rewrite (16.70) as

$$g(\omega) = \frac{T(\omega)}{2\omega/c} = \int_0^h \frac{dz}{\left[\omega^2 - \omega_p^2(z)\right]^{1/2}}. \tag{16.71}$$

Here $g(\omega)$ is the measured data and $\omega_p(z)$ is the unknown. This is a nonlinear equation for $\omega_p(z)$. To simplify (16.71), we let $\omega^2 = E$ and $\omega_p^2(z) = V(z)$, and write (16.71) in the following form:

$$\begin{aligned} g(E) &= \int_0^h \frac{dz}{[E - V(z)]^{1/2}} \\ &= \int_0^E \frac{1}{(E - V)^{1/2}} \frac{dz}{dV}\, dV, \end{aligned} \tag{16.72}$$

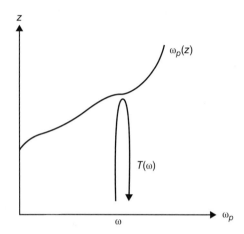

FIGURE 16.9   Probing of the electron density profile.

where the upper limit of integration is determined by $E = V(h)$. We assume that $V(z)$ is a monotonic function of z. Here we used E and V, as this problem is identical to the problem of sliding a particle up a frictionless hill with initial kinetic energy E and measuring the time $T(E)$ required for the particle to return. The time $T(E)$ is measured for different E and the shape of the hill represented by the potential energy $V(h) = mgh$ is to be determined. This is the problem solved by Abel in 1826.

Equation (16.72) is Abel's integral equation,

$$g(E) = \int_0^E \frac{f(V)}{(E - V)^{1/2}} \, dV, \tag{16.73}$$

where $g(E)$ is known and $f(v)$ is the unknown function. This is a Volterra integral equation of the first kind with the kernel $(E - V)^{-1/2}$. Its solution is given by (see Appendix 16.C)

$$z = \int_0^V f(V) \, dV = \frac{1}{\pi} \int_0^V \frac{g(E)}{(V - E)^{1/2}} \, dE. \tag{16.74}$$

Converting $g(E)$, E, and V to $T(\omega)$, ω, and ω_p, we get

$$z = z(\omega_p) = \frac{c}{\pi} \int_0^{\omega_p} \frac{T(\omega) \, d\omega}{\left(\omega_p^2 - \omega^2\right)^{1/2}}. \tag{16.75}$$

By measuring $T(\omega)$, we can determine $\omega_p(z)$ from (16.75). For example, if $T(\omega) = (T_0/\omega_0)\omega$, then $z = z(\omega_p) = (cT_0/\pi)(\omega_p/\omega_0)$ and therefore $\omega_p(z) = (\omega_{0\pi}/cT_0)z$.

16.7 RADAR POLARIMETRY AND RADAR EQUATION

In Section 10.2 we discussed the conventional radar equation in the following form:

$$\frac{P_r}{P_t} = \frac{\lambda^2}{(4\pi)^3} \frac{G_t G_r}{R_1^2 R_2^2} \sigma_{bi} m, \tag{16.76}$$

where P_r is the received power; P_t the transmitted power; G_t and G_r the gain of the transmitter and receiver; R_1 and R_2 are distance from transmitter to object and distance from object to receiver, respectively; σ_{bi} the bistatic cross section of the object; and m the mismatch factor. If both impedance and polarizations are matched, $m = 1$, but $0 < m < 1$ otherwise. This conventional radar equation deals with the total power received, but it gives no specific information about the relationships among the polarization characteristics of the transmitter, the object, and the receiver. However, recent advances in measurement techniques have made possible acquisition of more detailed polarization information and thus have stimulated intensive research on radar polarimetry, which is the utilization of the complete polarization characteristics in radar. The polarimetric techniques are also applicable to the remote sensing of the terrain and the discrimination of signals from clutter, interference, and jamming (Boerner, 1985; Huynen, 1978).

Let us now reexamine the radar equation. Following Section 10.2, we first consider a field incident on the object. We assume that the object is in the far field of both the transmitting and receiving antennas. Since the incident flux density S_i at the object is $S_i = |\bar{E}_t|^2 / 2\eta$, where $\bar{E}_t$ is the electric field, we write

$$\bar{E}_t = (2\eta S_i)^{1/2} \bar{E}_{tn}, \quad S_i = \frac{G_t P_t}{4\pi R_1^2}, \tag{16.77}$$

where E_{tn} is the normalized transmitted field with $|\bar{E}_{tn}| = 1$. We now express $\bar{E}_t$ in an orthogonal coordinate system

$$\bar{E}_t = E_{t1}\hat{x}_1 + E_{t2}\hat{x}_2. \tag{16.78}$$

For example, in a spherical system, we have

$$\bar{E}_t = E_{t\theta}\hat{\theta} + E_{t\phi}\hat{\phi}. \tag{16.79}$$

This wave, $\bar{E}_t$, is incident on the target. The scattered wave, $\bar{E}_s$, at the receiver is then given by

$$[E_s] = \frac{e^{ikR_2}}{R_2}[F][E_t], \tag{16.80}$$

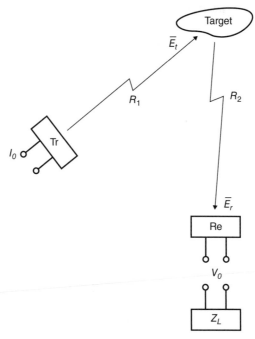

FIGURE 16.10   Transmitter, target and receiver.

where we used matrix notation

$$[E_s] = \begin{bmatrix} E_{s1} \\ E_{s2} \end{bmatrix}, \quad [F] = \begin{bmatrix} f_{11} & f_{12} \\ f_{21} & f_{22} \end{bmatrix}, \quad [E_t] = \begin{bmatrix} E_{t1} \\ E_{t2} \end{bmatrix}. \qquad (16.81)$$

Now consider the received power P_r when $\bar{E}_s$ is incident on the receiving antenna. This problem has been investigated in detail. (See Collin and Zucker, 1969, Chapter 4; Lo and Lee, 1988, Chapter 6). First, it has been shown that when the wave $\bar{E}_s$ is incident on the receiving antenna, the open-circuit voltage V_0 is given by

$$V_0 = \bar{h} \cdot \bar{E}_s, \qquad (16.82)$$

where $\bar{h}$ is called the *complex effective height* of the antenna (Fig. 16.10). It is given by the radiation field $\bar{E}_r$ when the receiver is used as a transmitter. If the receiver is fed by the current $\bar{I}_r$, the far field is given by (9.26) of Section 9.2.

$$\bar{E}_r = -\frac{j\omega\mu_0}{4\pi R} e^{-jkR} \bar{N}. \qquad (16.83)$$

The complex effective height $\bar{h}$ is defined by

$$\bar{h} = \frac{\bar{N}}{\bar{I}_r}. \qquad (16.84)$$

If the input impedance of the antenna is Z_i and the load impedance is $Z_l = R_l + iX_l$, the received power P_r is given by

$$P_r = \frac{1}{2}\frac{|V_0|^2 R_l}{|Z_i + Z_l|^2}.\tag{16.85}$$

Considering that the maximum received power is obtained when $Z_l = Z_i^*$, we write

$$\frac{R_l}{|Z_i + Z_l|^2} = \frac{q}{4R_i}, \quad q = 1 - \left|\frac{Z_l - Z_i^*}{Z_l + Z_i}\right|^2\tag{16.86}$$

We then write

$$P_r = A_r S_r, \quad S_r = \frac{|E_s|^2}{2\eta},\tag{16.87}$$

where S_r is the power flux density and the receiving cross section A_r is given by

$$A_r = \frac{\eta}{4R_i}\frac{|\bar{h}\cdot\bar{E}_s|^2}{|\bar{E}_s|^2}q.\tag{16.88}$$

If the impedance is matched, $q = 1$, and if the polarization is matched, $|\bar{h}\cdot\bar{E}_s|^2$ takes its maximum value $|\bar{h}|^2|\bar{E}_s|^2$.

If both the impedance and polarization are matched, the receiving cross section is equal to $(\lambda^2/4\pi)G_r$, where G_r is the gain of the receiving antenna. Therefore, we write

$$A_r = \frac{\lambda^2}{4\pi}G_r pq,\tag{16.89}$$

where

$$p = \frac{|\bar{h}\cdot\bar{E}_s|^2}{|\bar{h}|^2|\bar{E}_s|^2} = |\bar{h}_n\cdot\bar{E}_{sn}|^2;$$

$\bar{h}_n$ and E_{sn} are normalized such that $|\bar{h}_n| = 1$ and $|\bar{E}_{sn}| = 1$. Combining (16.77), (16.80), and (16.89), we get

$$\frac{P_r}{P_t} = \frac{\lambda^2 G_r qp}{4\pi}\frac{|\bar{E}_s|^2}{2\eta},\tag{16.90}$$

where

$$\bar{E}_s = \frac{1}{R^2}\bar{\bar{F}}\cdot\bar{E}_t.$$

This can be rewritten in the following form of radar equation:

$$\frac{P_r}{P_t} = \frac{\lambda^2 G_r G_t q}{(4\pi)^2 R_1^2 R_2^2} |\bar{h}_n \cdot \bar{E}_{sn}|^2 |\bar{\bar{F}} \cdot \bar{E}_{tn}|^2, \tag{16.91}$$

where $\bar{h}_n$, $\bar{E}_{sn}$, and $\bar{E}_{tn}$ are all normalized so that $|\bar{h}_n|^2 = 1$, $|\bar{E}_{sn}|^2 = 1$, and $|\bar{E}_{tn}|^2 = 1$. In matrix notation, we write

$$\begin{aligned} V_{0n} &= \bar{h}_n \cdot \bar{E}_{sn} = [h_n]^t [E_{sn}] \\ V &= \bar{\bar{F}} \cdot \bar{E}_{tn} = [F][E_{tn}]. \end{aligned} \tag{16.92}$$

The sense of polarization used in V_{0n} can be confusing, and therefore it is important to test them for known physical problems. First, if LHC is transmitted to a specular reflector such as a conducting plate, the scattered wave will be RHC propagating toward the receiver. Then the received voltage is null. For this case

$$\bar{h}_n = \frac{1}{\sqrt{2}}(\hat{x} - i\hat{y}),$$

$$\bar{E}_{sn} = \frac{1}{\sqrt{2}}(\hat{x} - i\hat{y}).$$

Therefore, $V_{0n} = 0$.

Another example is two identical helical antennas facing each other. Then if LHC is transmitted, $\bar{E}_{sn}$ at the receiver is

$$\bar{E}_{sn} = \frac{1}{\sqrt{2}}(\hat{x} + i\hat{y}).$$

The complex effective height for the identical antennas is then

$$\bar{h}_n = \frac{1}{\sqrt{2}}(\hat{x} - i\hat{y}).$$

Thus $V_{0n} = 1$ and they are polarization matched.

16.8 OPTIMIZATION OF POLARIZATION

Let us now consider the optimization problem. We attempt to find the polarization of the transmitter and the polarization of the receiver such that the received power is maximum. This will be done in three stages (Kostinski and Boerner, 1986). First, we attempt to find the polarization of the transmitter to maximize $|V|^2$ in (16.92). We have, using matrix notation,

$$|V|^2 = V^*V = [E_{tn}]^+[F]^+[F][E_{tn}]$$
$$= [E_{tn}]^+[G][E_{tn}], \qquad (16.93)$$

where + means "adjoint" (complex conjugate of transpose). $[G] = [F]^+ [F]$ is called *Graves power matrix* and is Hermitian (see Appendix 8.A),

$$[G]^+ = [G]. \qquad (16.94)$$

To maximize $|V|^2$, consider the eigenvalue equation

$$[G][X] = \lambda[X], \qquad (16.95)$$

where λ is the eigenvalue and $[X]$ is normalized so that $[X]^+ [X] = 1$. Now we multiply (16.95) from the left by $[X]^+$. We then get

$$[X]^+[G][X] = \lambda[X]^+[X] = \lambda. \qquad (16.96)$$

Therefore, the maximum value of $[X]^+ [G][X]$ is given by the maximum eigenvalue λ. Furthermore,

$$\lambda^* = \{[X]^+[G][X]\}^+ = [X]^+[G]^+[X] = [X]^+[G][X] = \lambda. \qquad (16.97)$$

Therefore, the eigenvalue λ is real. The eigenvalue λ is easily found from (16.95)

$$\begin{vmatrix} g_{11} - \lambda & g_{12} \\ g_{21} & g_{22} - \lambda \end{vmatrix} = 0. \qquad (16.98)$$

The polarization of the transmitted wave is therefore given by (16.95)

$$[E_{tn}] = \frac{1}{[1 + |a|^2]^{1/2}} \begin{bmatrix} 1 \\ a \end{bmatrix}$$
$$a = \frac{\lambda_1 - g_{11}}{g_{12}}, \quad \lambda_1 > \lambda_2. \qquad (16.99)$$

The second stage consists of calculating $[E_s] = [F][E_t]$ using the optimum polarization for $[E_t]$ obtained in (16.99), and in the third stage we adjust the receiver polarization state $\bar{h}_n$ to maximize $|V_{0n}|^2 = |\bar{h}_n \cdot \bar{E}_{sn}|^2$. This optimum polarization state is given by (Collin and Zucker, 1969, p. 108)

$$\bar{h}_n = \bar{E}_{sn}^*. \qquad (16.100)$$

16.9 STOKES VECTOR RADAR EQUATION AND POLARIZATION SIGNATURE

Let us reformulate the radar equation developed in Section 16.7 using Stokes' vector formulation. First we write transmitted flux density $[S_i]$ at the object using the normalized Stokes vector I_{tn}

$$[S_i] = \frac{G_t P_t}{4\pi R_1^2}[I_{tn}], \tag{16.101}$$

where

$$[I_{tn}] = \begin{bmatrix} I_{tn1} \\ I_{tn2} \\ U_{tn} \\ V_{tn} \end{bmatrix}$$

and $[I_t]$ is normalized so that $I_{tn1} + I_{tn2} = 1$. This is incident on the object and the scattered Stokes vector $[I_s]$ is given by

$$[I_s] = [M][S_i], \tag{16.102}$$

where $[M]$ is a 4×4 Mueller matrix (Section 10.10). Now consider the received power P_r. Since this is proportional to $|\bar{h} \cdot \bar{E}_s|^2$, we first express $|\bar{h} \cdot \bar{E}_s|^2$ using the Stokes vector. $\bar{h} = h_1\hat{e}_1 + h_2\hat{e}_2$ is the transmitted wave and $\bar{E}_s = E_{s1}\hat{e}_1 + E_{s2}(-\hat{e}_2)$ is the incoming wave (Fig. 16.11). Note that $\bar{h}$ is directed to $\hat{e}_1 \times \hat{e}_2$, but $\bar{E}_s$ is directed to $\hat{e}_1 \times (-\hat{e}_2)$. Therefore, we get

$$
\begin{aligned}
|\bar{h} \cdot \bar{E}_s|^2 &= |h_1 E_{s1} - h_2 E_{s2}|^2 \\
&= I_{h1}I_{s1} + I_{h2}I_{s2} - \tfrac{1}{2}U_h U_s + \tfrac{1}{2}V_h V_s,
\end{aligned}
\tag{16.103}
$$

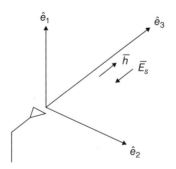

FIGURE 16.11 $\bar{h}$ is directed to $\hat{e}_1 \times \hat{e}_2$ while $\bar{E}_s$ is directed to $\hat{e}_1 \times (-\hat{e}_2)$.

where $[I_h]$ and $[I_s]$ are the Stokes vectors for $\bar{h}$ and $\bar{E}_s$, respectively.

$$[I_h] = \begin{bmatrix} I_{h1} \\ I_{h2} \\ U_h \\ V_h \end{bmatrix}, \quad [I_s] = \begin{bmatrix} I_{s1} \\ I_{s2} \\ U_s \\ V_s \end{bmatrix}.$$

We therefore get the Stokes vector radar equation

$$\frac{P_r}{P_t} = \frac{\lambda^2 G_r G_t q}{(4\pi)^2 R_1^2 R_2^2} [\widetilde{h}_n][I_s], \tag{16.104}$$

where $[\widetilde{h}_n]$ is the normalized effective height Stokes vector defined by

$$[\widetilde{h}_n] = \frac{1}{I_{h1} + I_{h2}} \left(I_{h1}, I_{h2}, -\frac{U_h}{2}, \frac{V_h}{2} \right),$$

$$[I_s] = [M][I_{tn}].$$

If a single antenna is used as both transmitter and receiver, we can express the transmitting Stokes vector using the orientation angle ψ and the ellipticity angle χ (see Section 10.8).

$$[I_{tn}] = \begin{bmatrix} \frac{1}{2}(1 + \cos 2\chi \cos 2\psi) \\ \frac{1}{2}(1 - \cos 2\chi \cos 2\psi) \\ \cos 2\chi \sin 2\psi \\ \sin 2\chi \end{bmatrix},$$

$$[h_n] = \begin{bmatrix} \frac{1}{2}(1 + \cos 2\chi \cos 2\psi) \\ \frac{1}{2}(1 - \cos 2\chi \cos 2\psi) \\ -\frac{1}{2}\cos 2\chi \sin 2\psi \\ \frac{1}{2}\sin 2\chi \end{bmatrix}. \tag{16.105}$$

The quantity $P_s = [\widetilde{h}_n][M][I_{tn}]$ is called the *polarization signature* and is displayed as a function of the ellipticity angle χ and the orientation angle ψ. It is used for identification of scattering mechanisms of the objects, terrain, vegetation, and so on, in imaging radar (Fig. 16.12). As a simple example, consider RHC wave transmitted

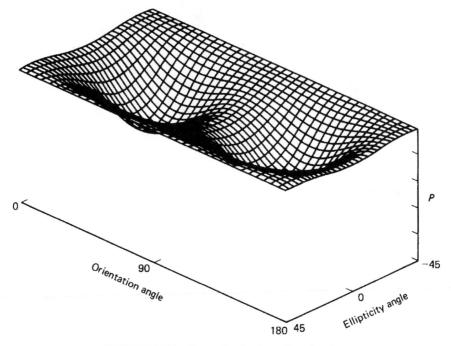

FIGURE 16.12 Example of polarization signature.

toward a specular reflector; then

$$
[h_n] = \begin{bmatrix} \frac{1}{2} \\ \frac{1}{2} \\ 0 \\ -\frac{1}{2} \end{bmatrix}, \quad [I_{tn}] = \begin{bmatrix} \frac{1}{2} \\ \frac{1}{2} \\ 0 \\ 1 \end{bmatrix},
$$

Therefore, we get $P_s = 0$.

16.10 MEASUREMENT OF STOKES PARAMETER

Stokes parameter (I_1, I_2, U, V) can be obtained by measuring the amplitude and phase of E_1 and E_2 and computing $I_1 = \langle |E_1|^2 \rangle$, $I_2 = \langle |E_2|^2 \rangle$, $U = 2\mathrm{Re}\langle E_1 E_2^* \rangle$, and $V = 2\mathrm{IM}\langle E_1 E_2^* \rangle$. This is called *coherent* measurement, as it involves the measurement of the phase. When E_1 and E_2 are randomly varying in time, the coherent measurement must be done within a fraction of the coherent time and therefore requires fast and accurate measurement.

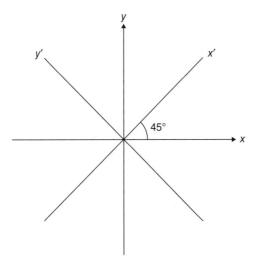

FIGURE 16.13   Measurement of Stokes parameters.

Incoherent measurement uses the power measurement and therefore need not be fast; however, it may be more susceptible to noise. I_1 and I_2 can be obtained directly by measuring the power in the x and y components (Fig. 16.13). Next measure the power for the component at 45°. Then we get

$$P_{45} = \langle |E_{x'}|^2 \rangle$$
$$= \frac{1}{2} \left[\langle |E_x|^2 \rangle + \langle |E_y|^2 \rangle + 2\mathrm{Re}\langle E_x E_y^* \rangle \right] \qquad (16.106)$$
$$= \frac{1}{2}[I_1 + I_2 + U]$$

where $E_{x'} = 1/\sqrt{2}[E_x + E_y]$. Similarly for 135°, we get

$$P_{135} = \frac{1}{2}[I_1 + I_2 - U]. \qquad (16.107)$$

From these two, we get

$$\frac{U}{I} = \frac{P_{45} - P_{135}}{P_{45} + P_{135}}, \quad I = I_1 + I_2. \qquad (16.108)$$

The denominator is to ensure the normalization.

Next we use the receiver that accepts the RHC wave. This has $\bar{h}_n = 1/\sqrt{2}(\hat{x} + i\hat{y})$ and therefore the power measured with this antenna is

$$
\begin{aligned}
P_R &= \left\langle |\bar{h}_n \cdot \bar{E}_s|^2 \right\rangle \\
&= \left\langle \left| \tfrac{1}{2} |E_x + iE_y|^2 \right\rangle \right. \\
&= \tfrac{1}{2}[I_1 + I_2 + V].
\end{aligned}
\tag{16.109}
$$

Similarly for the receiver that accepts the LHC wave, the power measured is

$$
P_L = \tfrac{1}{2}[I_1 + I_2 - V].
\tag{16.110}
$$

Thus we get

$$
\frac{V}{I} = \frac{P_R - P_L}{P_R + P_L}.
\tag{16.111}
$$

PROBLEMS

16.1 Show that if the object is a delta function $f(x, y) = \delta(x)\delta(y)$, the reconstructed image is $[W/(2\pi r)]J_1(W\gamma)$.

16.2 Prove the three-dimensional inverse Radon transform shown in (16.21) and (16.22).

16.3 Show that if the object is a delta function, the reconstructed image using the diffraction tomography is given by (16.43).

16.4 If the target is a sphere of radius a, find the complex scattering amplitude $\Gamma(\bar{K})$ and the characteristic function $\gamma(\bar{r})$ discussed in Section 16.4.

16.5 Derive the holographic inverse source solution for the one-dimensional source distribution $\rho(x)$.

16.6 Assume that the plasma frequency is given by $\omega_p^2 = A(z - z_0)^2$ for $z > z_0$ and $\omega_p = 0$ for $z < z_0$. Find the transit time $T(\omega)$ in (16.70) and the height h where $\omega = \omega_p(h)$. Show that $T(\omega)$ and the height satisfy (16.25).

16.7 A left-handed helical antenna is transmitting an LHC wave, which is normally incident on a conducting plate. The scattered wave is then received by the same helical antenna. Find $\bar{h}_n$, $\bar{E}_{tn}$, $\bar{E}_{sn}$, V_{0n}, and V given in Section 16.7. If the scattered wave is received by a right-handed helical antenna, find the power received.

16.8 Assume that the scattering matrix $[F]$ is given by

$$[F] = \begin{bmatrix} 2j & \frac{1}{2} \\ \frac{1}{2} & j \end{bmatrix}.$$

Find the eigenvalue λ and the optimum polarization $[E_{tn}]$ of the transmitting wave. Next find the complex effective height of the receiver to maximize the power received.

16.9 Assume that a target is a corner reflector that has the scattering matrix

$$[F] = \begin{bmatrix} 1 & 0 \\ 0 & -1 \end{bmatrix}.$$

Find the Mueller matrix and the polarization signature.

CHAPTER 17

RADIOMETRY, NOISE TEMPERATURE, AND INTERFEROMETRY

In this chapter we first discuss radiometry, which is the passive detection of natural radiation from various media, targets, and objects. Included in this discussion are brightness, antenna temperature, radiative transfer, and emissivity. The effects of the receiving system on the system noise temperature and the minimum detectable temperature are discussed, and the chapter concludes with the use of interferometry for mapping the brightness distribution (see Brookner, 1977; King, 1970; Kraus, 1966; Skolnik, 1970, 1980; Tsang et al., 1985; and Ulaby et al., 1981).

17.1 RADIOMETRY

All natural and man-made objects, terrain, and atmospheric media emit electromagnetic energy. Thermal emission is generally dominant. They also scatter the radiation incident on them. A radiometer is a very sensitive, low-noise receiver that detects natural incoherent radiation from these objects. Typical radiometers monitor broadband continuous radiation and the received power is proportional to the bandwidth of the receiver. Radiometers are used on earth-orbiting satellites and on the ground to probe atmospheric conditions and terrain and to detect targets at microwave, millimeter wave, and infrared frequencies. They are also used for medical applications to probe radiation from biological media.

Electromagnetic Wave Propagation, Radiation, and Scattering: From Fundamentals to Applications,
Second Edition. Akira Ishimaru.
© 2017 by The Institute of Electrical and Electronic Engineers, Inc. Published 2017 by John Wiley & Sons, Inc.

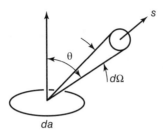

FIGURE 17.1   Brightness $B\,(\hat{s})$ and the received power dP.

17.2 BRIGHTNESS AND FLUX DENSITY

The fundamental quantity in radiometry is the brightness B. Consider a small area da and the power flux density incident on this area from the direction $\hat{s}$ within a unit frequency band centered at frequency ν within a unit solid angle (Fig. 17.1). This quantity $B\,(\hat{s})$ is called the *brightness* and is measured in W m^{-2}Hz^{-1}sr^{-1} (sr = steradian = unit solid angle). The brightness B is also called the *specific intensity* in radiative transfer theory and *radiance* in infrared radiometry. The amount of power dP flowing through the area da within the solid angle $d\Omega$ in a frequency interval $(\nu, \nu + d\nu)$ is therefore given by

$$dP = B \cos\theta \, da \, d\Omega \, d\nu. \tag{17.1}$$

At a given location, the brightness B is a function of the direction $\hat{s}$, and therefore we may call $B\,(\hat{s})$ the *brightness distribution*. The variation of the brightness B with the frequency is called the *brightness spectrum*.

The brightness B of the solar disk at $\lambda = 0.5$ μm is 1.33×10^{-12} and the brightness of rough ground at $\lambda = 3.9$ cm is 5.4×10^{-24}. Coherent sources such as lasers and radars present a completely different situation. For example, the brightness of an argon ion laser at $\lambda = 0.5145$ μm is 7.1×10^{3}, and the brightness of a Haystack X-band radar at $\lambda = 3.9$ cm is 4.8×10^{3} (Skolnik, 1970).

Let us consider a receiving antenna whose receiving cross section is given by $A_r(\theta, \phi)$. The received power P_r in W/Hz is then given by (Fig. 17.2)

$$P_r = \frac{1}{2} \int_{4\pi} B(\theta, \phi) A_r(\theta, \phi) \, d\Omega, \tag{17.2}$$

where the integral is over all solid angles, and the factor $\frac{1}{2}$ is introduced because normally the brightness radiation is incoherent and unpolarized while any antenna receives only one polarization. In general, however, if the brightness is partially polarized, the factor can be between 0 and 1.

If we normalize the receiving cross section A_r to the maximum value A_{rm}, we write

$$A_r(\theta, \phi) = A_{rm} P_n(\theta, \phi). \tag{17.3}$$

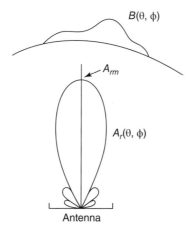

FIGURE 17.2   Receiving cross section and brightness.

We then get

$$P_r = \tfrac{1}{2}A_{rm}S, \tag{17.4}$$

where $S(\text{W m}^{-2}\text{Hz}^{-1})$ is called the *observed flux density*.

$$S = \int_{4\pi} B(\theta, \phi)P_n(\theta, \phi)\, d\Omega. \tag{17.5}$$

Note also that the maximum receiving cross section A_{rm} and the maximum gain G_m are related by

$$A_{rm} = \frac{\lambda^2}{4\pi}G_m. \tag{17.6}$$

The total flux density from the source is given by

$$S_s = \int_{4\pi} B(\theta, \phi)\, d\Omega, \tag{17.7}$$

and is called the *source flux density*. The unit for flux density is $\text{Wm}^{-2}/\text{Hz}^{-1}$; in radio astronomy, this is called 1 jansky ($= 1 \ \text{Wm}^{-2}/\text{Hz}^{-1}$) after the pioneer radio astronomer K. G. Jansky. The flux density of most radio sources in radio astronomy is on the order of 10^{-26} jansky.

17.3   BLACKBODY RADIATION AND ANTENNA TEMPERATURE

All objects emit electromagnetic energy. They may also absorb and scatter the energy incident on them. According to Kirchhoff, a good absorber of electromagnetic energy is also a good emitter. A perfect absorber that absorbs electromagnetic energy at all wavelengths is called a *blackbody* and is also a perfect emitter.

The brightness of the electromagnetic radiation from a blackbody depends only on its temperature and frequency and is given by Planck's radiation law:

$$B \text{ (blackbody)} = \frac{2hv^3}{c^2} \frac{1}{\exp(hv/kT) - 1}, \tag{17.8}$$

where h is Planck's constant (6.63×10^{-34} J·s), v the frequency (Hz), c the velocity of light (3×10^8 m s^{-1}), K is Boltzmann's constant (1.38×10^{-23} J/K), and T is the temperature (K).

For microwave and millimeter waves, hv is much smaller than KT; only in infrared and shorter wavelengths does hv become comparable to or greater than KT, exhibiting quantum effects. For microwave and millimeter waves, therefore, we can approximate $\exp(hv/KT)$ by $1 + hv/KT$ and obtain the *Rayleigh–Jeans law*

$$B \text{ (blackbody)} = \frac{2K}{\lambda^2} T. \tag{17.9}$$

Note that brightness is proportional to temperature (Fig. 17.3).

For an actual object that is not a blackbody, the brightness is not proportional to its actual temperature as given in (17.9). However, we can define the equivalent black-body temperature T_s which gives the brightness identical to the actual brightness B.

$$B = \frac{2K}{\lambda^2} T_s. \tag{17.10}$$

This equivalent temperature T_s is called the *source temperature*. Since B is a function of the direction $B(\theta, \phi)$, the source temperature is also a function of the direction $T_s(\theta, \phi)$.

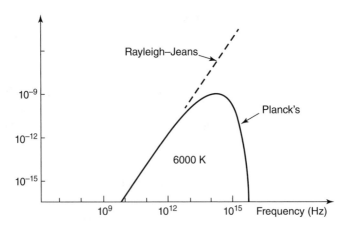

FIGURE 17.3   Planck's law and Rayleigh–Jeans law.

The received power P_r in (17.2) is then given by

$$P_r = \frac{K}{\lambda^2} \int T_s(\theta, \phi) \, A_r(\theta, \phi) \, d\Omega. \tag{17.11}$$

Nothing that

$$\int A_r(\theta, \phi) \, d\Omega = \lambda^2, \tag{17.12}$$

as shown in Section 9.1, we get

$$P_r = KT_A, \tag{17.13}$$

where K is Boltzmann's constant and

$$T_A = \frac{\displaystyle\int T_s(\theta, \phi) \, P_n(\theta, \phi) \, d\Omega}{\displaystyle\int P_n(\theta, \phi) \, d\Omega}$$

This quantity T_A is called the *antenna temperature*. As an example, if the source is uniform over a small solid angle Ω_s and the antenna receiving pattern is confined within a solid angle Ω_A, we have

$$T_A = \begin{cases} \dfrac{T_s \Omega_s}{\Omega_A} & \text{if } \Omega_s < \Omega_A, \\[2mm] T_s & \text{if } \Omega_s > \Omega_A. \end{cases} \tag{17.14}$$

This ratio Ω_s/Ω_A is called the *fill factor*.

The antenna temperature T_A is also equal to the temperature of a resistor that produces the same noise power as the actual power P_r (Fig. 17.4). According to Nyquist's formula, the open-circuit rms noise voltage (Johnson noise or thermal noise) across a resistor R at temperature T in the frequency band dv is given by

$$V = (4RKT \, dv)^{1/2}. \tag{17.15}$$

The available power W per unit frequency band produced by this resistor when it is transmitted to a matched load is $\frac{1}{2} (V^2/2R)$, and therefore

$$W = KT. \tag{17.16}$$

To show that this is equal to the antenna temperature, note that the antenna temperature is also equal to the temperature of a blackbody enclosure in which the antenna is embedded (Fig. 17.4). In this case T_s is constant, and therefore, from (17.13)

$$P_r = KT. \tag{17.17}$$

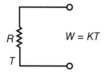

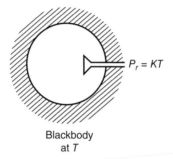

Blackbody
at T

FIGURE 17.4   Antenna temperature.

17.4 EQUATION OF RADIATIVE TRANSFER

In Section 17.3, we discussed the source temperature and the antenna temperature. The source temperature may originate outside the atmosphere, such as the sun. Also, the source temperature may come from the atmosphere itself or the ground. This section and Section 17.5 deal with the emission, propagation, absorption, and scattering of the temperature in the atmosphere and the ground.

Let us consider the propagation of the brightness in a medium emitting electromagnetic radiation. An example is the brightness in the atmosphere. Since the brightness B is proportional to the equivalent temperature T, we may use the temperature in place of the brightness. As the brightness B (or T) propagates in the direction $\hat{s}$ through the medium, it is partly absorbed and partly scattered (Fig. 17.5). We therefore write

$$\frac{dT}{ds} = -\gamma T, \qquad (17.18)$$

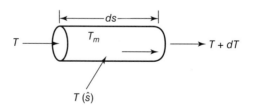

FIGURE 17.5   Equation of radiative transfer.

where

$$\gamma = \text{Extinction coefficient}$$
$$= \gamma_a + \gamma_s,$$
$$\gamma_a = \text{Absorption coefficient,}$$
$$\gamma_s = \text{Scattering coefficient.}$$

The brightness is also increased by the scattering into the direction $\hat{s}$ from the brightness coming from all other directions

$$\frac{1}{4\pi} \int_{4\pi} p(\hat{s}, \hat{s}') \, T(\hat{s}') \, d\Omega',$$

where $p(\hat{s}, \hat{s}')$ is called the *phase function* (Ishimaru, 1978). The brightness is also increased by the emission from the medium with temperature T_m. The emission is equal to the absorption according to Kirchhoff's law; therefore, we have $\gamma_a T_m$. Collecting all these terms, we get the equation of radiative transfer (Fig. 17.5)

$$\frac{dT}{ds} = -\gamma T + \frac{1}{4\pi} \int p(\hat{s}, \hat{s}') \, T(\hat{s}) \, d\Omega' + \gamma_a T_m. \qquad (17.19)$$

In most microwave radiometry, the scattering effect is negligible compared with the absorption, and therefore we write

$$\frac{dT}{ds} = -\gamma T + \gamma_a T_m, \qquad (17.20)$$

where $\gamma \approx \gamma_a$.

Let us consider a ground-based radiometer pointed upward (Fig. 17.6). The absorption coefficient of the atmosphere, clouds, or rain at height z is given by $\gamma_a(z)$ and their temperature by $T_m(z)$. The temperature of the external source such as the sun is T_e. The radiometer measures a temperature given by the solution of (17.20):

$$T = \int_0^\infty \gamma_a(z) \, T_m(z) \exp\left[-\int_0^z \gamma(z') \sec\theta \, dz'\right] \sec\theta \, dz + T_e \exp\left[-\int_0^\infty \gamma(z') \sec\theta \, dz'\right]. \qquad (17.21)$$

Note also that if $T_e = 0$ and T_m, γ_a, and γ are uniform over the height H, we get

$$T = T_m \frac{\gamma_a}{\gamma} [1 - \exp(-\gamma H \sec\theta)]. \qquad (17.22)$$

For example, rain may be approximated by $T_m = 273°$, $H \approx 3$ km, and $\gamma_a = $ absorption coefficient of rain.

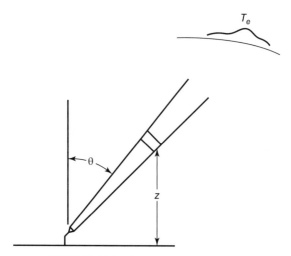

FIGURE 17.6   Ground-based radiometer.

The total attenuation τ over a distance l due to the scattering and absorption is called *optical depth*

$$\tau = \int_0^l \gamma \, ds. \tag{17.23}$$

The path loss L is defined as

$$L = \exp(\tau). \tag{17.24}$$

17.5 SCATTERING CROSS SECTIONS AND ABSORPTIVITY AND EMISSIVITY OF A SURFACE

All surfaces absorb, scatter, and emit electromagnetic radiation. Let us consider the plane wave $\bar{E}^i$ incident on a surface in the direction $\hat{i}$ and the wave $\bar{E}^s$ scattered in the direction $\hat{s}$ at a large distance R from the surface (Fig. 17.7). The scattering cross section per unit area of the surface is given by

$$\sigma_{\beta\alpha}^0(\hat{s}, \hat{i}) = \lim_{R \to \infty} \frac{4\pi \left| \bar{E}_\beta^s \right|^2 R^2}{\left| \bar{E}_\alpha^i \right|^2 A}, \tag{17.25}$$

where α and β represent the polarization states of the incident and the scattered wave, respectively, and α and β can be v or h for vertical and horizontal polarizations. Thus we have σ_{vv}^0, σ_{hh}^0, σ_{vh}^0, and σ_{hv}^0.

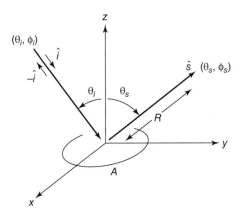

FIGURE 17.7   Scattering from surface A.

The scattering coefficient $\gamma_{\beta\alpha}$ is defined in terms of the projected area of the incident wave

$$\gamma_{\beta\alpha}(\hat{s}, \hat{i}) = \lim_{R \to \infty} \frac{4\pi |\bar{E}_\beta^s|^2 R^2}{\cos \theta_i |\bar{E}_\alpha^i|^2 A}. \tag{17.26}$$

Thus we have

$$\sigma_{\beta\alpha}^0(\hat{s}, \hat{i}) = \cos \theta_i \gamma_{\beta\alpha}(\hat{s}, \hat{i}). \tag{17.27}$$

From the reciprocity theorem, we get

$$\sigma_{\beta\alpha}^0(\hat{s}, \hat{i}) = \sigma_{\beta\alpha}^0(\hat{i}, \hat{s}). \tag{17.28}$$

When a wave with the polarization state α is incident on a surface in the direction $\hat{i}$, the total scattered power when a unit power flux density is incident in the projected area $A \cos \theta_i$ is given by (Fig. 17.7)

$$\begin{aligned}
\Gamma_\alpha(\hat{i}) &= \frac{1}{4\pi} \int_{2\pi} [\gamma_{\beta\alpha}(\hat{s}, \hat{i}) + \gamma_{\alpha\alpha}(\hat{s}, \hat{i})] \, d\Omega_s \\
&= \frac{1}{4\pi \cos \theta_i} \int_{2\pi} \left[\sigma_{\beta\alpha}^0(\hat{s}, \hat{i}) + \sigma_{\alpha\alpha}^0(\hat{s}, \hat{i}) \right] d\Omega_s,
\end{aligned} \tag{17.29}$$

where the scattered power in both α and β polarizations are included. The integration is over the upper hemisphere and 4π comes from the definition (17.26). This fraction

$\Gamma_\alpha(\hat{i})$, called the *albedo*, is the ratio of the total scattered power to the incident power in the projected area $A \cos \theta_i$. The fractional power absorbed by the surface is therefore

$$a_\alpha(\hat{i}) = 1 - \Gamma_\alpha(\hat{i}). \qquad (17.30)$$

This quantity $a_\alpha(\hat{i})$ is called *absorptivity*.

Let us next consider the brightness or the temperature $T_\alpha(\hat{i})$ emitted in the direction $(-\hat{i})$ in the polarization state α from a surface kept at the temperature T (Fig. 17.7). If the surface is a blackbody, the source temperature should be T. The ratio $\varepsilon_\alpha(\hat{i})$ of the actual emission $T_\alpha(\hat{i})$ to that of a blackbody at temperature T is called *emissivity*.

$$T_\alpha(\hat{i}) = \varepsilon_\alpha(\hat{i})T. \qquad (17.31)$$

According to Kirchhoff's law, if the surface is in thermal equilibrium, the absorption must be equal to the emission, and therefore emissivity is equal to absorptivity,

$$\varepsilon_\alpha(\hat{i}) = a_\alpha(\hat{i}). \qquad (17.32)$$

A rigorous proof of Kirchhoff's law (17.32) can be made by considering thermodynamic equilibrium when the surface is surrounded by a blackbody (Tsang et al. 1985).

Equation (17.31) gives the emission from the surface at temperature T, and (17.29), (17.30), and (17.32) relate the emissivity to the scattering characteristics of the surface. For example, if the surface is perfectly rough and scatters the radiation in all directions, it is called a *Lambertian surface*. In this case, the scattered power $|\bar{E}^s|^2$ is proportional to the projected surface area $A \cos \theta_s$ for a given incident flux in the projected area $A \cos \theta_i$ and $\gamma_{\beta\alpha}(\hat{s}, \hat{i})$ is independent of the polarization. Thus we have

$$\sigma^0_{\beta\alpha}(\hat{s}, \hat{i}) + \sigma^0_{\alpha\alpha}(\hat{s}, \hat{i}) = \sigma^0_0 \cos \theta_s \cos \theta_i. \qquad (17.33)$$

Substituting this into (17.29), we get

$$\Gamma = \frac{\sigma^0_0}{4}$$

$$\varepsilon = a = 1 - \frac{\sigma^0_0}{4}. \qquad (17.34)$$

If the surface is smooth, the scattered wave is in the specular direction and there is no cross polarization. The albedo $\Gamma_\alpha(\hat{i})$ is then given by the Fresnel reflection coefficient $R_\alpha(\hat{i})$

$$\Gamma_\alpha(\hat{i}) = |R_\alpha(\hat{i})|^2, \qquad (17.35)$$

where

$$R_v = \frac{n_1 \cos\theta_2 - n_2 \cos\theta_1}{n_1 \cos\theta_2 + n_2 \cos\theta_1} \quad \text{for vertical polarization,}$$

$$R_h = \frac{n_1 \cos\theta_1 - n_2 \cos\theta_2}{n_1 \cos\theta_1 + n_2 \cos\theta_2} \quad \text{for horizontal polarization,}$$

and the wave is incident from the medium with refractive index n_1 to the medium with n_2 at the angle of incidence θ_1 and $\cos\theta_2 = [1 - (n_1/n_2)^2 \sin^2\theta_1]^{1/2}$. If the incident wave is completely unpolarized, we have

$$\Gamma(\hat{i}) = \tfrac{1}{2}\left[|R_v|^2 + |R_h|^2\right]. \tag{17.36}$$

As an example, consider a radiometer looking down on the earth (Fig. 17.8). The temperature T at the radiometer consists of the atmospheric emission T_1, the contribution from the ground T_2, and the scattering by the ground T_3:

$$T = T_1 + T_2 + T_3. \tag{17.37}$$

The atmospheric emission T_1 is given by the emission from the medium with the temperature $T_m(z)$ attenuated through the atmosphere.

$$T_1 = \int_0^h \gamma_a(z) T_m(z) \exp\left[-\int_z^h \gamma(z') \sec\theta \, dz'\right] \sec\theta \, dz. \tag{17.38}$$

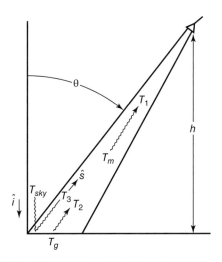

FIGURE 17.8   Radiometer looking down on the earth.

The contribution T_2 from the ground emission attenuated through the atmosphere is given by

$$T_2 = \varepsilon T_g \exp\left[-\int_0^h \gamma(z') \sec\theta \, dz'\right],\qquad(17.39)$$

where ε is the emissivity of the earth and T_g is the actual temperature of the ground. The emissivity depends on the angle. The approximate values of the emissivity normal to the surface are close to unity for grass, 0.93 for dry soil, 0.68 for moist soil, 0.85 for concrete, 0.45 for water, 0.9 for asphalt, and close to zero for metal.

The scattered component T_3 is the temperature scattered by the ground when the sky temperature T_{sky} is incident on the ground and then attenuated through the atmosphere. Thus we have

$$T_3 = T_{3s} \exp\left[-\int_0^h \gamma(z') \sec\theta \, dz'\right],\qquad(17.40)$$

where T_{3s} is the temperature scattered by the ground when the sky temperature T_{sky} is incident. If the ground is a Lambertian surface, we have (Fig. 17.7)

$$
\begin{aligned}
T_{3s}(\hat{s}) &= \frac{1}{4\pi\cos\theta_s} \int_{2\pi} T_{sky}(\hat{i})\sigma_0^0 \cos\theta_i \cos\theta_s \, d\Omega_i \\
&= \frac{\sigma_0^0}{4\pi} \int_{2\pi} T_{sky}(\hat{i}) \cos\theta_i \, d\Omega_i,
\end{aligned}
\qquad(17.41a)
$$

where the integration is over the hemisphere. If the sky temperature is uniform, we get

$$T_{3s} = \frac{\sigma_0^0}{4} T_{sky},\qquad(17.41b)$$

where $\sigma_0^0/4$ is the albedo of the surface. More generally, we have

$$T_\alpha(\hat{s}) = \frac{1}{4\pi\cos\theta_s} \int_{2\pi} \left[\sigma_{\alpha\beta}^0(\hat{s}, \hat{i}) \, T_\beta(\hat{i}) + \sigma_{\alpha\alpha}^0(\hat{s}, \hat{i}) \, T_\alpha(\hat{i})\right] d\Omega_i,\qquad(17.42)$$

where $T_\alpha(\hat{i})$ and $T_\beta(\hat{i})$ are the sky temperatures incident on the surface with the polarization state α and β.

17.6 SYSTEM TEMPERATURE

In Sections 17.1 through 17.5 we have discussed how the noise temperature is emitted, propagated, absorbed, and scattered, and then reaches the antenna. In this section, we

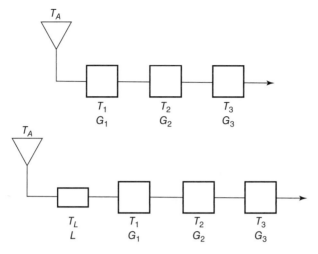

FIGURE 17.9 System temperature.

discuss what happens in the receiver. In addition to the antenna temperature given in (17.13), the receiver itself contributes to the total noise of the receiving system. Thus we have the system noise temperature T_{sys}.

$$T_{sys} = T_A + T_r, \qquad (17.43)$$

where T_r is the receiver noise temperature. If the receiver consists of several cascaded stages with temperature and gain as shown in Fig. 17.9, the gain of the previous stages effectively reduces the contribution of each stage to the overall system temperature. Therefore, the system temperature is given by

$$T_{sys} = T_A + T_r,$$
$$T_r = T_1 + \frac{T_2}{G_1} + \frac{T_3}{G_1 G_2}. \qquad (17.44)$$

Note that the total available noise power at the output is given by $T_{out} = G_1 G_2 T_{sys}$. If the first stage is a transmission line with loss $L(\text{gain} = \varepsilon = L^{-1} < 1)$ kept at a physical temperature of T_L, the effective noise temperature at the end of the transmission line is $T_L(1 - \varepsilon)$. This is seen by noting that the noise generated in dx is equal to the absorbed power $\alpha T_L dx$, where α is the attenuation constant. At the end of the line of length l, we have

$$\int_0^l \alpha T_L e^{-\alpha(l-x)} \, dx = T_L(1 - \varepsilon),$$

where $\varepsilon = e^{-\alpha l}$. We therefore have the noise temperature $T_L(1 - \varepsilon) + T_r$ for the attenuator with loss $L(\text{gain} = \varepsilon)$ followed by the receiver. The total system temperature is then given by

$$
\begin{aligned}
T_{\text{sys}} &= T_A + \frac{1}{\varepsilon}[T_L(1 - \varepsilon) + T_r] \\
&= T_A + T_L(L - 1) + LT_r,
\end{aligned}
\tag{17.45}
$$

where T_r is given in (17.44).

17.7 MINIMUM DETECTABLE TEMPERATURE

A receiver that measures the total noise power is called the "*total power receiver*" (Fig. 17.10). The detector is normally a square-law device and the output voltage is proportional to the output noise power. The incident power consists of the broadband system noise temperature T_{sys} and the signal noise temperature ΔT. If the bandwidth of the receiver is B, the output power for the system noise temperature is proportional to $(KT_{\text{sys}}B)^2$ and the output power for the signal is proportional to $(K\Delta TB)^2$. The system noise, however, is B effectively independent noise pulses per second. These pulses are averaged over the integration time τ, and therefore there are $B\tau$ independent pulses. These independent contributions are largely canceled by each other over this period, and thus the system noise contribution is reduced by $B\tau$.

$$
W_{\text{sys}} \sim \frac{(KT_{\text{sys}}B)^2}{B\tau}.
\tag{17.46}
$$

This means that the system temperature is effectively reduced to $T_{\text{sys}}/(B\tau)^{1/2}$. The signal output W_s is

$$
W_s \sim (K\Delta TB)^2.
\tag{17.47}
$$

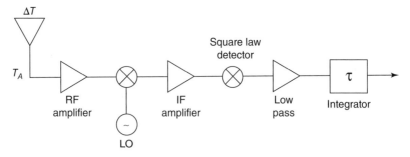

FIGURE 17.10   Total power receiver.

The *sensitivity* or the *minimum detectable signal* $\Delta T_{\min}$ is defined as ΔT, which produces the output W_s equal to W_{sys}. Thus we have

$$\Delta T_{\min} = \frac{K_s T_{\text{sys}}}{\sqrt{B\tau}},\qquad(17.48)$$

where K_s is the sensitivity constant of the order of unity. For example, $K_s = 1$ for the total power receiver and $K_s = \pi/\sqrt{2}$ for a Dick receiver.

The total power receiver is affected by the variation of the system gain $\Delta G/G$. The *Dick receiver* switches at a constant rate between the receiving antenna and a reference load, thus canceling out gain instability (Kraus, 1966, Chapter 7; Skolnik, 1970).

17.8 RADAR RANGE EQUATION

When the noise temperature is included in a radar equation, there is a maximum range in which a radar can detect a target. Radar equations give the received power P_r for a given transmitted power P_t (see Section 10.2)

$$\frac{P_r}{P_t} = \frac{\lambda^2}{(4\pi)^3}\frac{G^2}{R^4}\sigma.\qquad(17.49)$$

The system noise power P_n for the system with temperature T_{sys} and bandwidth B is given by

$$P_n = K T_{\text{sys}} B.\qquad(17.50)$$

The signal-to-noise ratio is therefore

$$\frac{S}{N} = \frac{P_r}{P_n}.\qquad(17.51)$$

We therefore obtain the minimum detectable value of P_r

$$P_r = K T_{\text{sys}} B \frac{S}{N}.\qquad(17.52)$$

Substituting this into (17.49), we get the maximum range in which the radar can detect a target in terms of S/N, T_{sys}, and B. If the output can be integrated over the time τ, the system temperature is effectively reduced to $T_{\text{sys}}/(B\tau)^{1/2}$.

17.9 APERTURE ILLUMINATION AND BRIGHTNESS DISTRIBUTIONS

In visible wavelengths, the atmosphere is transparent, and in wavelengths between 1 cm and 100 m the atmosphere is again transparent. For wavelengths longer than about 10 m, radio waves may not penetrate the ionosphere. Between 1 cm and around 10 µm, there are a considerable number of molecular absorptions. Also for wavelengths shorter than 0.1 µm, there is considerable molecular absorption. Therefore, we have *optical windows* and *radio windows* through the atmosphere. Radio astronomy makes use of this radio window, whereas optical astronomy makes use of the optical window (see Kraus, 1966).

Let us consider a radio telescope pointed to the sky (Fig. 17.11). The aperture field distribution is given by $A(x, y)$ and the radiation pattern $g(\theta, \phi)$ is related to the aperture distribution through a Fourier transform (see Section 9.5)

$$g(\theta, \phi) = \int A(x, y)e^{ik_x x + jk_y y} \, dx \, dy, \qquad (17.53)$$

where

$$k_x = k \sin \theta \cos \phi,$$
$$k_y = k \sin \theta \sin \phi.$$

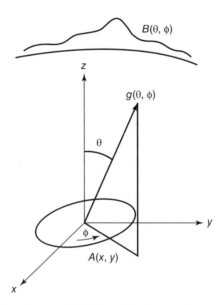

FIGURE 17.11 Aperture distribution $A(x,y)$ and radiation pattern $g(\theta, \phi)$.

We now consider the antenna beam scanning the sky in the x–z plane ($y = 0$, $\phi = 0$). Letting $g(\theta, 0) = g(\beta)$ and $\int A(x, y)\, dy = A(x)$, we rewrite (17.53):

$$g(\beta) = \int A(x)e^{j\beta x}dx, \tag{17.54}$$

where $\beta = k \sin \theta$.

Next we consider the power pattern $p(\beta)$, which is given by

$$p(\beta) = g(\beta)g^*(\beta)$$
$$= \iint A(x)A^*(x')e^{j\beta(x-x')}\, dx\, dx'. \tag{17.55}$$

The Fourier transform of the power pattern $P(x'')$ is then given by

$$P(x'') = \frac{1}{2\pi}\int p(\beta)e^{-j\beta x''}\, d\beta$$
$$= \int A(x)A^*(x - x'')\, dx. \tag{17.56}$$

This is a convolution integral of the aperture distribution. These relationships are expressed conveniently in Fig. 17.12.

Next we let the antenna beam be pointed to the direction θ_0 and let the beam scan the sky with brightness distribution $B(\theta) = B(\beta)$. This scanning is often accomplished by the rotation of the earth. The output $S(\beta_0)$ then depends on the scanning angle θ_0 and thus (Fig. 17.13)

$$S(\beta_0) = \int B(\beta)p(\beta_0 - \beta)d\beta, \quad \beta_0 = k \sin \theta_0. \tag{17.57}$$

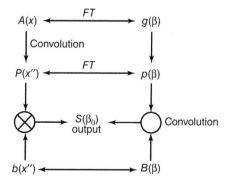

FIGURE 17.12 Relationships among aperture distribution $A(x)$, radiation pattern $p(\beta)$, brightness distribution $B(\beta)$, and output $S(\beta_0)$.

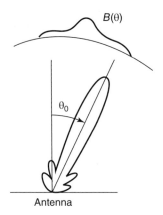

FIGURE 17.13   Antenna beam scanning the sky.

The Fourier transform of the output $S(\beta_0)$ is then given by the following product:

$$S(x) = \frac{1}{2\pi} \int S(\beta_0)e^{-j\beta_0 x}\, d\beta_0$$
$$= b\,(x)P(x),$$

(17.58)

where

$$b(x) = \frac{1}{2\pi} \int B\,(\beta)e^{-j\beta x} d\beta,$$

$$P(x) = \frac{1}{2\pi} \int p\,(\beta)e^{-j\beta x} d\beta.$$

The foregoing relationships among the brightness distribution $B(\beta)$, power pattern $p(\beta)$, and output $S(\beta_0)$ are shown in Fig. 17.12. Equation (17.58) shows that the spectrum of the brightness distribution is filtered by the antenna pattern and produces the output.

17.10 TWO-ANTENNA INTERFEROMETER

As an example of the use of the relationships shown in Section 17.9, we consider a two-antenna interferometer. Two point receiving antennas are separated by a distance a (Fig. 17.14). The aperture distribution $A(x)$ is given by

$$A\,(x) = \frac{1}{2}\delta\left(x - \frac{a}{2}\right) + \frac{1}{2}\delta\left(x + \frac{a}{2}\right).$$

(17.59)

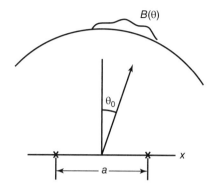

FIGURE 17.14   Two-antenna interferometer.

The field pattern, power pattern, and Fourier transform are given by

$$g(\beta) = \cos \frac{\beta a}{2}$$

$$p(\beta) = \frac{1}{2}(1 + \cos \beta a). \qquad (17.60)$$

The observed output $S(\beta_0)$ as a function of the scan angle $\theta_0 (\beta_0 = k \sin \theta_0)$ is

$$S(\beta_0) = \frac{S_0}{2}[1 + V(\beta_0, a)], \qquad (17.61)$$

where

$$S_0 = \int B(\beta) \, d\beta,$$

$$V(\beta_0, a) = \frac{1}{S_0} \int B(\beta) \cos[(\beta_0 - \beta)a] \, d\beta.$$

Now we consider the Fourier transform of the brightness distribution $B(\beta)$

$$V_c(a) = \frac{1}{S_0} \int B(\beta) e^{-j\beta a} \, d\beta. \qquad (17.62)$$

This function $V_c(a)$ is called the *complex visibility function* and is the normalized correlation function of the wave at the receiver as a function of the separation a. When the separation a is zero, V_c is unity, and as the separation a increases, the correlation decreases. The function $V_c(a)$ is the normalized *mutual coherence function* or *degree of coherence*. The Fourier transform relationship (17.62) between the source brightness distribution of the incoherent source and the degree of coherence at the observation point is a special case of the van Cittert–Zernike theorem when the source and the observation point are far from each other (Born and Wolf, 1970, Chapter 10).

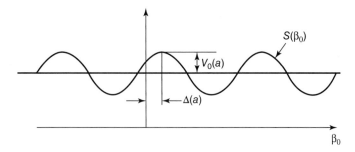

FIGURE 17.15   Observed output $S\left(\beta_0\right)$ and the complex visibility function.

The complex visibility function $V_c(a)$ can be expressed as

$$V_c(a) = V_0(a)\, e^{-j\Delta(a)}.$$ (17.63)

Then the observed output $S(\beta_0)$ is given by

$$S(\beta_0) = \frac{S_0}{2}\,\{1 + V_0(a)\cos[\beta_0 a - \Delta(a)\,]\},$$ (17.64)

where

$$V(\beta_0, a) = \text{Re}[V_c(a)e^{j\beta_0 a}].$$

The observed output $S(\beta_0)$ as a function of the scan angle β_0 is sketched in Fig. 17.15. In radio astronomy, the scanning is accomplished by the rotation of the earth.

It is now possible to determine the complex visibility function from measurement of the observed output $S(\beta_0)$. The maximum and minimum values of $S(\beta_0)$ are given by

$$S_{\text{max}} = \frac{S_0}{2}[1 + V_0(a)],$$
$$S_{\text{min}} = \frac{S_0}{2}[1 - V_0(a)].$$ (17.65)

From this we get

$$V_0(a) = \frac{S_{\text{max}} - S_{\text{min}}}{S_{\text{max}} + S_{\text{min}}}.$$ (17.66)

This magnitude $V_0(a)$ is called the *fringe visibility* or *visibility*. Also at $\beta_0 = 0$, we get

$$V(0, a) = V_0(a)\cos \Delta(a).$$ (17.67)

From (17.66) and (17.67) we get $V_0(a)$ and $\Delta(a)$, and thus $V_c(a)$. Once we get $V_c(a)$ for all separations a, it should be possible to obtain the brightness distribution $B(\beta)$ by the inverse Fourier transform of (17.62).

The Fourier transform relationship (17.62) can be generalized to two dimensions as indicated in (17.53). The technique of measuring the complex visibility function and then obtaining the map of the brightness distribution of the radio sources in the sky is called the *aperture synthesis*. To obtain the complex visibility function, the measurement must be made at a sufficient number of separations. The Very Large Array (VLA) constructed in New Mexico can have a separation up to 35 km for frequencies ranging from 1.4 to 24 GHz.

PROBLEMS

17.1 Assume that the sun has an angular size of $1°$ and is equivalent to a blackbody with 6000 K. An antenna at $f = 10$ GHz has a beamwidth of $0.5°$ and the receiver bandwidth is 1 MHz. This antenna is directed toward the sun. Calculate

(a) The gain of the antenna (in dB)

(b) The source flux density

(c) The observed flux density

(d) The source temperature

(e) The antenna temperature

17.2 Do Problem 17.1 with a beamwidth of $2°$.

17.3 Assume that raindrops are uniformly distributed from the ground to a height of 3 km and that the rain attenuation is 0.3 dB/km at 10 GHz when the precipitation rate is 12.5 mm/hr. The rain temperature is assumed to be 273 K. If an antenna is always pointed toward the sun, find the antenna temperature as a function of angle θ. Assume that the scattering effect is negligible.

17.4 Consider the radiometer shown in Fig. 17.8, located at $h = 3$ km and $\theta = 30°$. The antenna has a beamwidth of $1°$ at 10 GHz. Assume a rain medium as described in Problem 17.3. Find the rain emission T_1. Find the ground emission contribution T_2 assuming a ground temperature of 283 K for moist soil. Find the scattered component T_3 assuming a uniform sky temperature of 273 K and a Lambertian surface for the ground. Calculate the total temperature $T_1 + T_2 + T_3$. If the ground is covered by a smooth metallic surface, what is the temperature?

17.5 Consider the system shown in Fig. 17.9. Assume that $T_A = 40$ K, $T_L = 290$ K, $L = 0.5$ dB, $T_1 = T_2 = T_3 = 290$ K, and $G_1 = G_2 = G_3 = 20$ dB. Find the system temperature.

17.6 A microwave antenna operating at 10 GHz has a beamwidth of $1°$ and a bandwidth of 1 MHz. Assuming that the system noise temperature of the antenna is 290 K, the radar cross section of the object is 10 m², the required signal-to-noise ratio is at least 10 dB, and the maximum range at which this object should be detected is 100 km, find the transmitting power required.

17.7 Assume that an extraterrestrial civilization (ETC) exists at a distance of R light-years. An antenna 26 m in diameter is used to detect the signal from the ETC. Assume that the ETC is sending a signal using an antenna with a transmitting power of 1 MW and a diameter of 100 m. Assume that the receiver has a system noise of 20 K, and that $\lambda = 12$ cm, $\Delta f = 1$ Hz and $\tau = 1$ s. Calculate the maximum distance R (light-year) at which the signal from the ETC can be detected.

17.8 Consider an antenna with aperture distribution $A(x)$ pointed to the sun (Fig. P17.8). Assume that the sun has an equivalent angular size of 1° at an equivalent blackbody temperature of 6000 K. The wavelength is 3 cm and the receiver bandwidth is 1 MHz.
   Calculate
   (a) The source flux density
   (b) The observed flux density
   (c) The source temperature
   (d) The antenna temperature
   Use appropriate approximations, if necessary. Also find
   (e) $g(\lambda)$, $\lambda = k \sin \theta$
   (f) $P(\lambda)$
   (g) $\bar{P}(x'')$
   (h) $B(\lambda)$
   (i) $\bar{B}(x'')$
   (j) $S(\lambda_0)$

$A(x)$

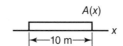

FIGURE P17.8   Aperture antenna.

17.9 If a star has the brightness distribution shown in Fig. P17.9, calculate and plot $S(\lambda_0)$ for two-antenna interferometry. The frequency is 1 GHz and $a = 500$ m.

17.10 Assume that a certain nebula has an angular size of $1'$ at 1 GHz and a source flux density of 1000×10^{-26} jansky. The receiving antenna has a diameter of 10 m and a bandwidth of 5 MHz. Find the antenna temperature. If the integration time is 1 s, what is the maximum receiver noise temperature allowed to detect this nebula?

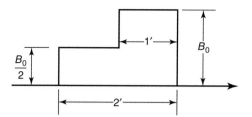

FIGURE P17.9   Brightness distribution.

17.11 For Problem 7.10, two identical receiving antennas are used to determine the angular size of the nebula. Calculate and plot the visibility as a function of the distance between the two antennas.

CHAPTER 18

STOCHASTIC WAVE THEORIES

Many of the problems we deal with can be classified as either "deterministic" or "random." For most of the problems in this book, we deal with media and objects that are well defined in shape, positions, and material characteristics such as dielectric constant, and we determine the field quantities such as electric and magnetic fields and the Poynting vectors for various applications. These are called "deterministic" problems. In contrast, there are many cases where the media characteristics, positions, and shape may vary randomly in space and time. Examples include most geophysical and biological media. They are called "random media," and the waves propagating in the random media vary randomly in space and time. It must be noted, however, that even though the medium and the waves vary randomly, there are well-defined theories underlining the random phenomena. Statistical wave theories are therefore aimed at discovering and making use of these well-defined theories governing random phenomena.

We discussed many examples and applications of statistical wave theories in several chapters and therefore we do not duplicate those discussions here. Rather, in this chapter, we present a summary of the historical development of these theories and new ideas, and key questions that may be outstanding or may need further attention.

Electromagnetic Wave Propagation, Radiation, and Scattering: From Fundamentals to Applications,
Second Edition. Akira Ishimaru.
© 2017 by The Institute of Electrical and Electronic Engineers, Inc. Published 2017 by John Wiley & Sons, Inc.

18.1 STOCHASTIC WAVE EQUATIONS AND STATISTICAL WAVE THEORIES

Random media include characteristics that are continuous random functions of space and time, such as the dielectric constant, and can therefore be considered as a random continuum. An example is turbulence in the atmosphere, ionosphere, or troposphere, ocean turbulence, interplanetary and interstellar turbulence, and heterogeneous earth (Keller, 1964). Random media also include the random boundary surfaces that are a continuous random function of space and time, such as ocean surfaces. Random media can also be considered as a random distribution of discrete scatterers such as rain, fog, snow, and ice particles. This also includes a diffuse medium such as tissues and blood cells (Twersky, 1964).

Waves in random media are governed by stochastic wave equations. Solutions are usually sought for statistical averages and statistical moments of the fields. Therefore, "stochastic wave theories" can also be called "statistical wave theories."

18.2 SCATTERING IN TROPOSPHERE, IONOSPHERE, AND ATMOSPHERIC OPTICS

Early studies of radio wave scattering in the troposphere were conducted by Booker and Gordon (1950), who made use of the exponential correlation function for refractive index fluctuations giving relationships among the angular dependence, correlation distance, and wavelength. The results are related to the Bragg diffraction and are well known (Ishimaru, 1997). Other spectra such as the Gaussian model and modified Kolmogorov model have been studied and applied to scattering in the troposphere, in the planetary atmosphere, and the solar wind. Scattering by ionospheric turbulence has been studied extensively, including the dispersion, the scintillation index, the double-passage effects, and synthetic aperture radar (SAR) imaging through the ionosphere (Jin, 2004; Rino, 2011; Tatarskii et al., 1992; Yeh and Lin, 1992). Atmospheric optics has been investigated extensively (Tatarskii, 1961, 1971). In particular, studies have included backscattering enhancement, the fourth-order moments, speckles, localization, tissue optics, imaging through turbulence, and adaptive optics (Ishimaru, 1997). Also noted are "thin screen" and "extended Huygen–Fresnel" techniques. The "path-integral" approach has been applied by Tatarskii and others (1993). Also noted are Useinski's fundamental solutions and two-scale solutions (Tatarskii, 1993).

One of the important problems is the pulse propagation through random media. This requires the study of a two-frequency mutual coherence function (MCF) for random media, and several techniques are proposed (Ishimaru, 1997).

18.3 TURBID MEDIUM, RADIATIVE TRANSFER, AND RECIPROCITY

We have already discussed (and will further discuss) radiative transfer theory and their applications in Chapters 17, 19, 20, and 24. Early work on radiative transfer was done by Schuster (1905) to study radiation through a foggy atmosphere. The definitive work was reported by Chandrasekhar (1950) to study the transport and scattering of radiation in planetary and stellar atmospheres. The equation of transfer is equivalent to Boltzmann's equation used in the diffusion of neutrons. The radiative transfer theory is also used in scattering in geophysical media and in biological media (Chapters 19 and 20).

The fundamental equations for waves in multiple scattering diffuse media are the Dyson equations for the first moment and Bethe–Salpeter equations for the second moment. This is discussed in detail in Section 24.4. Here we discuss the reciprocity relations for the radiative transfer. Green's function $G(r_s r_i)$ is the field at r_s when the point source is at r_i. Because of the reciprocity, we have

$$G(r_s r_i) = G(r_i r_s). \tag{18.1}$$

The intensity Green's function L is given by

$$L\left(r_s r_i r_s' r_i'\right) = \left\langle G(r_s r_i) G^* \left(r_s' r_i'\right)\right\rangle. \tag{18.2}$$

Now consider the following four cases:

$$\begin{aligned}
L_1 &= \left\langle G(r_s r_i) G^* \left(r_s' r_i'\right)\right\rangle, \\
L_2 &= \left\langle G(r_i r_s) G^* \left(r_i' r_s'\right)\right\rangle, \\
L_3 &= \left\langle G(r_s r_i) G^* \left(r_i' r_s'\right)\right\rangle, \\
L_4 &= \left\langle G(r_i r_s) G^* \left(r_s' r_i'\right)\right\rangle.
\end{aligned} \tag{18.3}$$

Without the statistical average, these four are the same. However, we need to consider the interference. L_1 and L_2 correspond to interchanging the source and observation points and they are equal, and this shows the reciprocity of radiative transfer (Fig. 18.1). However, L_3 is exactly the same as L_1 only if this is exact backscattering and the forward and the time-reversed counter propagating path are the same. Therefore at the exact backscattering, the intensity becomes $L_1 + L_3 = 2L$. Away from the exact backscattering, the interference destroys the magnitude of L_3, leaving a sharp peak with the angular width of λ/l_{tr}, where $l_{tr} =$ transport mean free path (Fig. 18.2).

18.4 STOCHASTIC SOMMERFELD PROBLEM, SEISMIC CODA, AND SUBSURFACE IMAGING

In Chapter 15, we discussed radiation from a dipole on the conducting earth. These are related problems that deal with wave excitation and propagation over earth's surface. Examples are rough surface scattering in Chapter 23, and seismic acoustic pulse scattering and propagation over the heterogeneous earth and the wave train called

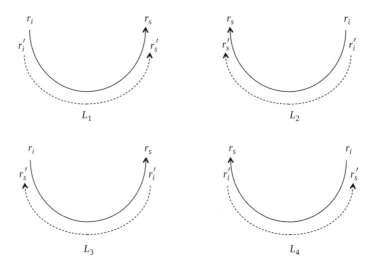

FIGURE 18.1 L_1 and L_2 constitute the reciprocity of radiative transfer. At exact back-scattering, the interference in L_3 is constructive and exactly the same as L_1.

the "coda." They are different problems, but can be examined from more unified and broad viewpoints.

The Sommerfeld problem deals with radio waves over a flat earth. However, if we consider imaging of objects near the ocean surface or terrain, it may be necessary to study the effects of roughness of the surface. This is a study of "stochastic Green's function" for rough surfaces. This problem has been studied using Dyson and Bethe–Salpeter equations for coherent and incoherent fields using the smoothed diagram method (Ishimaru et al., 2000) similar to Watson–Keller studies (1984). Sommerfeld and Zenneck waves, and the effects of rough surfaces on the Sommerfeld pole can be found in Ishimaru et al. (2000). These studies, however, are not complete as they deal with one-dimensional rough surfaces, and small root mean square heights. Seismic pulse excitation for a flat earth surface has been studied extensively (de

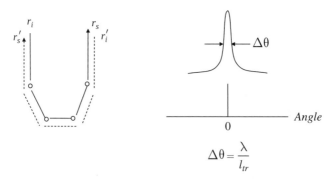

FIGURE 18.2 Exact backscattering.

Hoop, 1985). However, recent studies are directed to the study of the effects of random heterogeneous earth, which gives the wave trains called the "coda." The characteristics of the coda have been studied, but most of the studies assume the medium extending to all directions in three dimensions, and the rough surface effects of the boundary between air and the earth are not included (Sato et al., 2012, Chapter 26). The earth–air boundary has been included in the pulse over a flat earth surface. However, we are interested in the heterogeneous earth, which gives the seismic coda. It may, therefore, be useful to consider the excitation and propagation of the seismic pulse in the heterogeneous earth with a flat air–earth boundary. The boundary conditions at the air–earth boundary are given in Appendix 26 (Eq. 26.A.9) where normal stress and tangential stress are zero. Using the space–time Fourier transform, the formal solutions can be obtained. It is necessary to calculate both coherent and incoherent waves. The Fourier transform includes the pole due to the Rayleigh surface wave and the reflection at the air–earth boundary with P-wave and S-wave incident waves. This study has not been fully explored and reported. There have been recent studies indicating that deep subsurface roughness on plasmonic thin film can create evanescent waves which make it possible to enhance subwavelength imaging (Tsang et al., 2015).

18.5 STOCHASTIC GREEN'S FUNCTION AND STOCHASTIC BOUNDARY PROBLEMS

Green's function is well known among students of electrical engineering. George Green was born a son of a baker in Nottingham. He had limited education. His first published work in 1828 contained Green's function, Green's theorem, and other mathematical tools that became the essential link between classical field theory and quantum electrodynamics of Schwinger, Feynman, Tomonaga, the Feynman diagram, Dyson and Bethe–Salpeter equations, and the physics of condensed matter (Dyson, 1993). Green's function $G(r, t, r', t')$ describes the response of the field at a given point r at a later time t due to an impulse excitation (delta function) of the field at another given point r' at an earlier time t'. Here we briefly summarize basic formulations.

For a time-harmonic deterministic problem in free space, the scalar Green's function is given by

$$G(r, r') = \frac{\exp(ik_0 r)}{4\pi r}, \tag{18.4}$$

where k_0 is the free space wave number.

If the medium is random with relative dielectric constant $\varepsilon = \langle \varepsilon \rangle (1 + \varepsilon')$, where $\langle \varepsilon \rangle$ is the average dielectric constant and ε' is its fluctuation, Green's function is a random function of space and time. Therefore we need to consider the average and the fluctuating Green's function. We have the first moment and the second moment.

$$\langle G \rangle = \text{first moment,}$$
$$\langle G(r_1)G^*(r_2) \rangle = \text{second moment.}$$
(18.5)

The first moment is the average or the coherent Green's function. The fluctuating Green's function (also called the incoherent Green's function) can be written as

$$G_f = G - \langle G \rangle.$$
(18.6)

The second moment is called the MCF.

$$\Gamma(r_1, r_2) = \langle G(r_1)G^*(r_2) \rangle = |\langle G \rangle|^2 + \langle G_f(r_1)G_f^*(r_2) \rangle.$$
(18.7)

Our task is therefore finding the first and second moments, including the boundary conditions and the object characteristics for imaging, communications, and other applications. As an example, consider a point target (Fig. 18.3). $G_1 G_2$ are forward and backward waves and $G_3 G_4$ are their conjugates. The received field is $G_1 G_2$ and the received power is $(G_1 G_2)(G_3^* G_4^*)$. Each Green's function consists of the average $\langle G_i \rangle$ and the fluctuation v_i, $i = 1, 2, 3$, and 4. We now look at the received power P_r.

$$P_r = \langle (G_1 G_2)(G_3^* G_4^*) \rangle$$
$$= \langle (G_1 + v_1)(G_2 + v_2)(G_3 + v_3)^*(G_4 + v_4)^* \rangle.$$
(18.8)

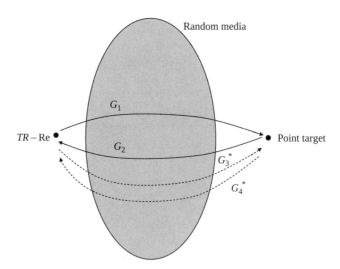

FIGURE 18.3 Green's functions for a point target. $G_1 G_2$ are forward and backward Greens' functions and $[G_3 G_4]^*$ are its conjugates.

This is the fourth-order moment. We now make an assumption that the random Green's functions are circular complex Gaussian random variables (Appendix 23.A). From this, we obtain

$$
\begin{aligned}
P_r = &\langle G_1 \rangle \langle G_3^* \rangle \langle G_2 \rangle \langle G_4^* \rangle \\
&+ \langle G_1 \rangle \langle G_3^* \rangle \langle v_2 v_4^* \rangle \\
&+ \langle G_2 \rangle \langle G_4^* \rangle \langle v_1 v_3^* \rangle \\
&+ \langle v_1 v_3^* \rangle \langle v_2 v_4^* \rangle \\
&+ \langle G_1 \rangle \langle G_4^* \rangle \langle v_2 v_3^* \rangle \\
&+ \langle G_2 \rangle \langle G_3^* \rangle \langle v_1 v_4^* \rangle \\
&+ \langle v_1 v_4^* \rangle \langle v_2 v_3^* \rangle .
\end{aligned}
\tag{18.9}
$$

From this, noting $G_1 = G_2$ and $G_3 = G_4$, $v_1 = v_2$ and $v_3 = v_4$, we get

$$
P_r = \left[\langle G_1 \rangle \langle G_3^* \rangle \right]^2 + 4 \langle G_1 \rangle \langle G_3^* \rangle \langle v_1 v_3^* \rangle + 2 \langle v_1 v_3^* \rangle^2
\tag{18.10}
$$

Note that this includes the correlation between the forward and backward waves. If we ignore this correlation, we get

$$
P_r = \langle G_1 \rangle \langle G_3^* \rangle^2 + 2 \langle G_1 \rangle \langle G_3^* \rangle \langle v_1 v_3^* \rangle + \langle v_1 v_3^* \rangle^2
\tag{18.11}
$$

The difference between (18.10) and (18.11) clearly shows the effects of the correlation between the forward and backward waves. This effect equals the increase of the backscattered power shown in (18.10) because of the correlation, over that in (18.11) if we ignore the correlation. This is an indication of the double-passage effect.

Let us next look at an example of stochastic integral equations. Consider a target in a random medium (Fig. 18.4). We take a simple case that the boundary condition

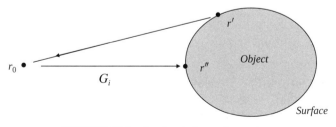

FIGURE 18.4 Stochastic integral equation.

of the surface of the target is the Dirichlet type. This is given in Section 12.1. We write the surface integral equation

$$G_i(r, r_0) = \int_S G(\bar{r}, \bar{r}')J(\bar{r}')ds', \tag{18.12}$$

where $J(\bar{r}') = \frac{\partial \psi(r')}{\partial n}$ is equivalent to the surface current, G is Green's function, and G_i is the incident Green's function originating at r_0 and observed at r. Here we are assuming that the incident wave G_i is excited by a delta function at $r = r_0$. The scattered field $\psi_s(r')$ is then given by

$$\psi_s(r') = -\int_S G(\bar{r}, \bar{r}')J(\bar{r}')ds' \tag{18.13}$$

Note that when r is on the surface at $r = r_s$, we have

$$\psi_i(r_s) + \psi_s(r_s) = 0, \tag{18.14}$$

satisfying the Dirichlet boundary conditions.

The target is in a random medium and therefore G_i, ψ_s, G, and J are all random functions. Let us first consider the first moment.

$$\langle G_i \rangle = \int \langle G(rr')J(r^*) \rangle ds'. \tag{18.15}$$

The incident wave $\langle G_1 \rangle$ can be calculated if we know the random medium characteristics.

Let us write J, unknown current, as

$$J(r') = \int_S S(r', r'')\psi_i(r'')ds''. \tag{18.16}$$

The function S is called the "transition operator" or the "scattering operator" (Frisch, 1968). It is also called the "current generator" (Meng-Tateiba, 1996). The integral equation (18.12) is then written as

$$G_i(r, r_0) = \int_S ds' \int_S ds'' G(r, r')S(r, r')G_i(r'', r_0). \tag{18.17}$$

The scattered field (18.13) is given by

$$\psi_s = -\int_S ds' G(rr') \int_S ds'' S(r'r'')\psi_i(r''). \tag{18.18}$$

The second moment of the backscattered wave at $r = r_0$ is then given by

$$\langle |G_s|^2 \rangle = ds_1'' ds_2' ds_2'' \langle [S(r_1', r_1'')S^*(r_2', r_2'')] G(r_0, r_1')G_i(r_1'' r_0)G^*(r_0, r_2')G_i^*(r_2'' r_0) \rangle.$$
(18.19)

Note that S is a random function. However, this is the current at r' produced by the field at r'' and therefore it may be possible to assume that to calculate S, the medium surrounding the surface has the average dielectric constant. Under this assumption, we can approximate S by a deterministic average value $\langle S \rangle$.

We then get

$$\langle |G_s|^2 \rangle = ds_1'' ds_2' ds_2'' \langle S \rangle \langle S^* \rangle \langle GG_i G^* G_i^* \rangle.$$
(18.20)

The fourth-order moment can be expressed by the second moment if we assume they are complex circular Gaussian variables (Appendix 23.A).

We then write

$$\langle GG_i G^* G_i^* \rangle = \langle GG^* \rangle \langle G_i G_i^* \rangle + \langle GG_i^* \rangle \langle G_i G^* \rangle.$$
(18.21)

These second-order moments have been calculated and therefore the radar cross section can be calculated from $\langle |G_s|^2 \rangle$.

18.6 CHANNEL CAPACITY OF COMMUNICATION SYSTEMS WITH RANDOM MEDIA MUTUAL COHERENCE FUNCTION

Channel capacities of communications through various environments have been extensively studied. The channel transfer matrix H is often assumed to have the elements that are zero mean circularly complex Gaussian random variable (Paulraj et al., 2003). On the other hand, the propagation through random media has been extensively studied and MCF has been calculated for various random media environments. This section shows the relationship between the channel capacity and the MCF.

Consider a multiple-input, multiple-output (MIMO) propagation channel with M transmitters and N receivers. The output y ($N \times 1$ matrix), the input x ($M \times 1$ matrix), and the noise N ($N \times N$) are written in the conventional model (Chizhik et al., 2002; Ishimaru et al., 2010; Paulraj et al., 2003).

$$y = Hx + N.$$
(18.22)

$H(N \times M)$ is the channel transfer matrix normalized as

$$T_r \langle HH' \rangle = MN,$$
(18.23)

where H' is the conjugate transpose of H.

The channel capacity is given by

$$C = \sum_i \log_2 \left(1 + \frac{(P_{SR})\lambda_i}{M} \right). \tag{18.24}$$

P_{SR} is the signal-to-noise ratio and λ_i are the eigenvalues (or the singular values) of HH'.

Let us now look at the key idea of this section: how to express H using MCF. H is the transfer matrix from n receivers to m transmitters. Therefore its element H_{nm} is proportional to the stochastic Green's function from the transmitter at r_m to the receiver at r_n.

$$H_{nm} \propto G_{nm}. \tag{18.25}$$

From this we obtain

$$[HH']_{nn'} \propto \sum_m G_{nm} G^*_{mn'} = \sum_m \Gamma_{nn'}. \tag{18.26}$$

Note that $\Gamma_{nn'}$ is the MCF at r_n and $r_{n'}$ when a delta function is at r_m (Fig. 18.5). Note that the figure shows possible diffraction effects including keyhole (Chizhik et al., 2002).

Normalization of H is done by choosing the constant in front of (18.26) such that

$$T_r \langle HH' \rangle = \sum_i \lambda_i = MN. \tag{18.27}$$

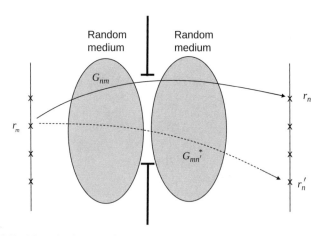

FIGURE 18.5   Mutual coherence function $T_{nn'}$, $T_{nn'} = G_{nm} G^*_{mn'}$. This also shows diffraction due to keyhole.

With the above formulation, it is now possible to calculate the channel capacity (Eq. 18.24) once we calculate the MCF in (18.26). Note that this applies to any random medium and diffracting walls or keyholes since the formulation is based on the statistical Green's function and MCF, which may be obtained in many situations.

The general formulation given in this section has been used to calculate the channel capacity of MIMO model in a random medium including 60 GHz 500 m link in rain (Ishimaru, 2010).

18.7 INTEGRATION OF STATISTICAL WAVES WITH OTHER DISCIPLINES

Stochastic wave theories have been developed over many years and are now established as an important discipline. However, it is important to keep in mind that while statistical wave theories can be viewed as separate from many other deterministic problems, integration of statistical wave theories with other physical disciplines has been recognized as an important area of research and applications. In this section, we give and highlight some examples of this integration.

(1) **Statistical Signal Processing**

   In Section 18.6, we discussed one of the examples of the inclusion of statistical wave theory in a communication problem. A key in the problem is how to express the channel transfer function H for communication through random media. As stated in Section 18.6, the transfer function H can be given by a stochastic Green's function in random media. In Section 18.6, we expressed H in terms of the stochastic Green's function in homogenous random media. However, this can be generalized to include the effects of keyhole or any diffraction effects as indicated there. This problem of communication in media that includes random media and deterministic objects including statistical signal processing has not been fully investigated.

(2) **Heat Transfer and Multi-Physics**

   If a wave propagates through a random medium and is incident on an object, the wave energy may be absorbed by the object and transformed to heat energy, which will diffuse in the medium creating time and space variations of the temperature distribution. We discuss this in Section 20.2. This is also one of the examples of "multi-physics" problems, which deals with simultaneous phenomena of more than one physical model. In this case, we are dealing with electromagnetics and temperature diffusion equations.

(3) **Seismic Fluctuations and Coda, Acoustic-Electromagnetic Interaction**

   In chapter 26, we discuss the problem of seismic coda, which is the fluctuation of acoustic waves in random heterogeneous solid earth, including Rayleigh surface waves. Here, it is important to understand the basic theories and formulations of acoustic waves in solids. This is also applicable to electromagnetic excitation of acoustic waves in a solid (Sarabandi, 2003).

(4) **Porous Medium and Mixtures**

In Section 8.6, we discussed the use of the "mixing formula" to determine the effective dielectric constant of a mixture of two or more materials with different dielectric constants. A mixture can also be treated as a porous medium, where the pore and porosity, and the relation to the fractional volume, are the concepts of interest. It is also related to "percolation theory." It may be noted that for a porous medium such as shale, the production of oil depends on whether the porosity is greater than or smaller than the critical porosity. Further discussion of the porous medium is given in Chapter 26.

(5) **Quantum Electromagnetics, Casimir Force, and Graphene**

This book is based on classical electromagnetics; quantum electrodynamics is outside the scope of this book. However, recently numerical classical electromagnetics have been used to calculate the Casimir force, which may be potentially useful for microelectromagnetic systems (MEMS) and nanoelectromagnetic systems (NEMS). The Casimir force was predicted by Casimir in 1948. It exists among charge-neutral bodies due to the quantum fluctuations of electromagnetic fields in vacuum. The force is extremely small, but it becomes significant between charge-neutral objects when the separation is below microns. (Xiong, Tong, Atkins, and Chew 2010) (Atkins, Chew et al., 2014). Computational electromagnetic techniques have been used to calculate Casimir forces.

Graphene is a monolayer of carbon atoms packed into a dense honeycomb crystal structure. It was discovered by Geim and Novoselov in 2004, and they were awarded the Nobel Prize in Physics in 2010. Graphene has attracted much attention because at present it is considered the strongest component, the best conductor of heat, the best conductor of electricity, and has strong broadband absorption characteristics. Its many potential applications include solar cells, broadband THz absorbers, plasmic antennas, cloaking, and many others (Chen and Alu, 2011, 2013; Novoselov and Geim et al., 2005; Sailing He, 2013; Yao et al., 2013).

18.8 SOME ACCOUNTS OF HISTORICAL DEVELOPMENT OF STATISTICAL WAVE THEORIES

The history of statistical wave theories is rooted over a hundred years. In this section, we outline some key developments.

(a) **Random Continuum**

Early work on microwave scattering by tropospheric turbulence includes the Booker–Gordon formula (1950), scattering of ionospheric turbulence (Yeh-Liu, 1972), atmospheric optics (Tatarski, 1961, 1971), and astrophysics wave scintillations in solar wind, planetary atmosphere, and pulsar scintillations. Fundamental to these studies are the "Dyson equation" for coherent waves and the "Bethe–Salpeter" equation for incoherent wave correlations. These equations can be expressed using a Feynman diagram. In fact, the field of statistical wave theory has benefited from the studies of stochastic Green's functions and MCFs in astrophysics.

(b) **Turbid Medium**

Closely related to the work on waves in a random continuum is the early work on the propagation of radiation in a foggy atmosphere (Schuster, 1905). Radiative transfer theory (Chandrasekhar, 1950) has been extensively used to describe waves in fog, rain, snow, and terrain. Radiative transfer has been applied extensively to geophysical remote sensing and scattering. It has been used in biomedical tissue optics, imaging, and ultrasound imaging of tissues. Basic formulations of radiative transfer are closely related to neutron transport and Boltzmann's transport equation.

(c) **Partially Coherent Waves**

There were considerable studies made on partially coherent waves, which are not strictly monochromatic in space and time and polarizations. Born and Wolf in *Principle of Optics* (1964) discussed in detail coherence time, degree of coherence, the MCF, and interference. In fact, when we talk about the coherent and the incoherent field, we need to keep in mind that the incoherent field is actually a partially coherent field and much of the study of partially coherent waves applies to the waves in random media.

(d) **Rough Surface and Rough Interface Scattering**

Research on scattering by a rough surface dates back to Rayleigh (1898), who conducted a study of scattering by a sinusoidal corrugated surface based on what is now called the Rayleigh hypothesis, which assumes that the scattered wave within the corrugation is expressed by the outgoing wave only. This hypothesis has been proved correct if the maximum slope is less than 0.448. Further studies have been reported by Rice (1951) and others, as shown in Chapter 23.

(e) **Enhanced Backscattering and Memory Effects**

In Chapter 24, we discuss enhanced backscattering and memory effects. Both represent the coherence effects in multiple scattering.

CHAPTER 19

GEOPHYSICAL REMOTE SENSING AND IMAGING

Geophysical remote sensing and imaging is a key engineering discipline dealing with the remote sensing of the earth, oceans, atmosphere, and space. It covers vast areas as represented by the Geoscience and Remote Sensing Society (GRSS) of IEEE (Ulaby et al., 1981; Tsang and Kong, 2001). Remote sensing of the earth's environment is one of the important societal needs. Satellite or aircraft observations of geophysical media, the atmosphere, and the ocean give useful information on our environment. In this chapter, we discuss several techniques and applications of remote sensing and imaging. Active radar technologies make use of the transmitted radar signals and the waves scattered by the geophysical media. Passive sensors detect the thermal radiation emitted by the geophysical media. We can also make use of the active and passive sensors to detect and image objects hidden in the geophysical environment. In both active and passive sensors, we can measure the amplitude or the intensity of the received signal. More recently, however, the polarization characteristics and the space–time correlations have been measured and processed to give more complete information on the environment and objects. In this chapter, we emphasize selected topics. First, we review polarimetric radars as applied to the decomposition theorem, which helps to obtain medium characteristics from radar backscattering. Nonspherical particles and differential reflectivity and general eigenvalue formulations of nonspherical scattering are discussed next. More fundamental formulations of space–time vector radiative transfer and Wigner distributions are presented. Passive radars are discussed in relation to polarization and ocean wind directions. We discuss the inclusion of antenna temperature in the van Cittert–Zernike theorem. The

Electromagnetic Wave Propagation, Radiation, and Scattering: From Fundamentals to Applications,
Second Edition. Akira Ishimaru.
© 2017 by The Institute of Electrical and Electronic Engineers, Inc. Published 2017 by John Wiley & Sons, Inc.

final section covers ionospheric dispersion and Faraday rotation effects on synthetic aperture radar (SAR) images.

19.1 POLARIMETRIC RADAR

In Section 16.7, we discussed radar polarimetry for a single target including the complex effective height of the antennas and the matched impedance and polarization. In Section 16.8, we discussed the optimum polarization and its relation to the eigenvalue problem. We then presented the Stokes vector radar equation and polarization signature in Section 16.9. In this section, we discuss the fundamentals of polarimetric radar, including the decomposition of target scattering matrices to extract physical information about the target.

It was noted in early 1950s that it was necessary to include the polarization characteristics in the transmission and reception (Sinclair, 1950). This was followed by extensive research conducted by Kennaugh and Huynen (1978) and Boerner (1983) pointing out that new information on targets can be obtained through the use of the vector nature of electromagnetic waves, the scattering matrix, the Mueller matrix, and the coherency matrix. Cloude (1985), Kostinski and Boerner (1986), Cloude and Pottier (1996), and Cloude (2009) performed pioneering and comprehensive work on target decomposition theorems, which show the decomposition of the scattering or the Mueller matrix into a sum of a set of matrices representing the physical basis for each component. Expanding the 2×2 scattering matrix in terms of Pauli matrices is one example of the decomposition. The polarization information can be displayed in the form of polarization signature, showing different signatures identified for different physical objects (van Zyl et al., 1987; Zebker et al., 1991). Comprehensive reviews were given by Zebker et al. (1991). Freeman and Durden (1998) developed a scattering model which gives a physically based three-component scattering mechanism of canopy layer, double bounce, and rough surface scattering. A tutorial introduction of radar polarimetry was given by Luneburg (1995), and Boerner (2003) described recent advances in radar polarimetry and radar interferometry in synthetic aperture remote sensing. The three-component model by Freeman and Durden was extended to the four-component model by Yamaguchi et al. (2005) with the addition of helix scattering. Lopez-Martinez, Pottier, and Cloude (2005) discussed the target decomposition theorems based on the eigenvector decomposition with the entropy H, the anisotropy A, and the α and β angles. A review of radar polarimetry and polarimetric SAR was given by Boerner (2003). Arii, van Zyl, and Kim (2010) presented a generalization of the scattering from a vegetation canopy and adaptive model-based decomposition techniques. In this section, we give an introduction to the fundamental theory of radar polarimetry.

19.1.1 Coordinate System

In discussing the scattering matrix of geophysical media, it is necessary to define two coordinate systems commonly used in polarimetry. Consider bistatic scattering

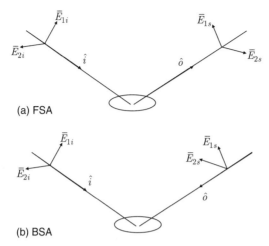

(a) FSA

(b) BSA

FIGURE 19.1   Bistatic scattering. (a) FSA (forward scattering alignment). (b) BSA (backward scattering alignment).
Note: $\bar{E}_{1i} \times \bar{E}_{2i}$ is pointed in the direction $\hat{i}$ and $\bar{E}_{1s} \times \bar{E}_{2s}$ is pointed in the direction $\hat{o}$. If $\bar{E}_{1s}$ is the same for both FSA and BSA, $\bar{E}_{2s}(BSA) = -\bar{E}_{2s}(FSA)$.

shown in Fig. 19.1. For forward scattering alignment (FSA), the wave is incident on a scatterer or a surface in a direction $\hat{i}$ and the scattered wave in a direction $\hat{o}$ is received by a receiving antenna. For backward scattering alignment (BSA), the scattered wave $(\bar{E}_{1s}, \bar{E}_{2s})$ is expressed such that $(\bar{E}_{1s} \times \bar{E}_{2s})$ is pointed toward the scatterer. The advantage of BSA is that, for backscattering, the same basis can be used for both incident $\hat{i}$ and scattered $\hat{o}$ waves. For most geophysical imaging, BSA is commonly used because of this advantage. However, FSA is commonly used in most other applications. In particular, FSA is used for vector radiative transfer, as is shown in Section 19.6. In this section, we use the BSA formulation.

19.1.2 Scattering Matrix [S] and Mueller Matrix [M]

As we discussed in Section 16.7, the scattered wave $\bar{E}_s$ can be expressed by (16.80).

$$[\bar{E}_s] = \frac{e^{ikR_2}}{R_2}[F][\bar{E}_t].$$
(19.1)

In geophysical remote sensing, it is common to use the scattering matrix [S] instead of the scattering amplitude matrix [F].

$$[\bar{E}_s] = \frac{e^{ikR_2}}{kR_2}[S][\bar{E}_i],$$
(19.2)

where

$$[E_s] = \begin{bmatrix} E_{s1} \\ E_{s2} \end{bmatrix}, \quad [E_i] = \begin{bmatrix} E_{i1} \\ E_{i2} \end{bmatrix},$$

$$[S] = \begin{bmatrix} S_{11} & S_{12} \\ S_{21} & S_{22} \end{bmatrix}.$$

In geophysical applications, we often use the horizontal and vertical polarizations. Then we have

$$[E_s] = \begin{bmatrix} E_{h'} \\ E_{v'} \end{bmatrix}, \quad [E_i] = \begin{bmatrix} E_h \\ E_v \end{bmatrix},$$

$$[S] = \begin{bmatrix} S_{h'h} & S_{h'v} \\ S_{v'h} & S_{v'v} \end{bmatrix}. \tag{19.3}$$

h' and v' represent the horizontal and vertical components of a scattered wave and h and v represent the horizontal and vertical components of the incident wave. Note that here we used BSA, and as it is commonly used in many remote sensing applications, we used the time dependence $\exp(-i\omega t)$ rather than IEEE convention of $\exp(j\omega t)$.

It has already been noted in Section 10.10 that the Stokes parameters (I_{1s}, I_{2s}, U_S, and V_s) of the scattered waves are related to the Stokes parameters (I_{1i}, I_{2i}, U_i, and V_i) of the incident wave by the 4×4 Mueller matrix $[M]$ (Ishimaru, 1997).

$$[I_s] = \frac{1}{(kR)^2} [M][I_i], \tag{19.4}$$

where

$$I_s = \begin{bmatrix} I_{1s} \\ I_{2s} \\ U_s \\ V_s \end{bmatrix}, \quad I_i = \begin{bmatrix} I_{1i} \\ I_{2i} \\ U_i \\ V_i \end{bmatrix},$$

$$[M] = \begin{bmatrix} |S_{11}|^2 & |S_{12}|^2 & \mathrm{Re}(S_{11}S_{12}^*) & -\mathrm{Im}(S_{11}S_{12}^*) \\ |S_{21}|^2 & |S_{22}|^2 & \mathrm{Re}(S_{21}S_{22}^*) & -\mathrm{Im}(S_{21}S_{22}^*) \\ 2\mathrm{Re}(S_{11}S_{21}^*) & 2\mathrm{Re}(S_{12}S_{22}^*) & \mathrm{Re}(S_{11}S_{22}^* + S_{12}S_{21}^*) & -\mathrm{Im}(S_{11}S_{22}^* - S_{12}S_{21}^*) \\ 2\mathrm{Im}(S_{11}S_{21}^*) & 2\mathrm{Im}(S_{12}S_{22}^*) & \mathrm{Im}(S_{11}S_{22}^* + S_{12}S_{21}^*) & \mathrm{Re}(S_{11}S_{22}^* - S_{12}S_{21}^*) \end{bmatrix}. \tag{19.5}$$

For bistatic scattering, there are only seven independent parameters in $[S]$, with four amplitude and three phases, excluding the absolute phase. In the monostatic case, $S_{12} = S_{21}$ because of the reciprocity and therefore there are only five independent parameters. There are 16 elements in the Mueller matrix. However since only seven independent parameters are in the scattering matrix and the Mueller matrix is equivalent to the scattering matrix, there are 9 (=16 − 7) relations among 16 Mueller elements. In the monostatic case, [M] has 10 parameters. Since there are only 5 independent parameters, there are 5 (=10 − 5) relations among Mueller elements. The above are for deterministic scatterers.

In most geophysical media, all the elements in the scattering and Mueller matrices are random and therefore it is necessary to take statistical averages. It is then noted that the average of the element of [S] represents the coherent field, which may become negligible for many applications, while the average of the element of [M] represents the intensity and the correlations. This clearly shows that the Mueller matrix is essential in geophysical remote sensing.

19.1.3 Decomposition of Scattering Matrix and Pauli Matrices

The 2×2 target scattering matrix [S] contains information about the target. There are several ways to extract physical information from [S]. One is the use of the Pauli matrices. It is known that any 2×2 matrix can be expressed as a linear sum of four basic matrices consisting of the unit matrix $[\sigma_0]$ and three Pauli spin matrices $[\sigma_1]$, $[\sigma_2]$, and $[\sigma_3]$ (Cloude and Pottier, 1996; Goldstein, 1981).

$$[S] = e_o[\sigma_0] + e_1[\sigma_1] + e_2[\sigma_2] + e_3[\sigma_3], \tag{19.6}$$

where e_0, e_1, e_2, and e_3 are complex constants and

$$[\sigma_0] = \begin{pmatrix} 1 & 0 \\ 0 & 1 \end{pmatrix}, \quad [\sigma_1] = \begin{pmatrix} 0 & 1 \\ 1 & 0 \end{pmatrix},$$

$$[\sigma_2] = \begin{pmatrix} 0 & -i \\ i & 0 \end{pmatrix}, \quad [\sigma_3] = \begin{pmatrix} 1 & 0 \\ 0 & -1 \end{pmatrix}. \tag{19.7}$$

These matrices σ_i ($i = 0, 1, 2, 3$) are independent and σ_1, σ_2, and σ_3 have zero traces (the sum of the diagonal elements). Note that

$$Tr(e_i) = 0, \ i = 1, 2, 3,$$
$$Tr(\sigma_i \sigma_j) = 0, \ i \neq j, \tag{19.8}$$
$$\sigma_i \sigma_i = \text{unit matrix} = \sigma_0, \ i = 1, 2, 3.$$

Using these, we can obtain all the coefficients e_i in (19.6) by the following.

$$e_i = \frac{1}{2} Tr([S][\sigma_i]). \tag{19.9}$$

Note that the decomposition (19.6) is not unique and there are many other ways of decomposition. The process of decomposing the scattering matrix [S] as shown in

(19.6) with (19.7) is one of several examples of "target decomposition theorem." This theorem is developed showing that the scattering matrix is a sum of independent elements to associate a physical mechanism and meaning with each element, as discussed by Cloude and Pottier (1996).

Let us discuss the physical interpretation of the basis matrices $[\sigma_i]$ and their physical mechanisms (Boerner et al., 1981). $[\sigma_0]$ represents simple scattering from a plane surface or a sphere. $[\sigma_3]$ represents the scattering by a 90 corner reflector (trough) and $[\sigma_1]$ is the same as $[\sigma_2]$ except that the axis is rotated by 45. $[\sigma_2]$ represents scattering by helix. For example,

$$[\sigma_+] = [\sigma_3] + [\sigma_2][\sigma_3] = \begin{pmatrix} 1 & i \\ -i & 1 \end{pmatrix}, \tag{19.10}$$

transforms a linearly polarized wave (E_1, E_2) to a right-handed circular polarized wave (E_x, iE_x).

$$\begin{pmatrix} 1 & i \\ -i & 1 \end{pmatrix} \begin{pmatrix} E_1 \\ E_2 \end{pmatrix} = \begin{pmatrix} E_1 + iE_2 \\ i(E_1 + iE_2) \end{pmatrix} = \begin{pmatrix} E_x \\ iE_x \end{pmatrix}. \tag{19.11}$$

19.2 SCATTERING MODELS FOR GEOPHYSICAL MEDIUM AND DECOMPOSITION THEOREM

Observations of polarimetric scattering by a geophysical medium reveal scattering mechanisms and physical characteristics of the medium. The backscattering from a forest, trees, ground roughness, dielectric constant, and terrain can be obtained from the medium characteristics and this is called the "forward problem." The "inverse" of this process to find a medium characteristic from radar backscattering is called the "decomposition" of the observed scattering matrix. These methods are known as the "target decomposition theorem" and comprehensive reviews have been given (Cloude and Pottier, 1996). In this section, we outline the three-component model by Freeman and Durden (1998) and the eigenvector-based composition.

19.2.1 Three-Component Scattering Model

In 1998, Freeman and Durden developed a method based on a combination of three simple scattering mechanisms. They are (1) the canopy layer model, (2) the double-bounce model, and (3) the rough surface model, illustrated in Fig. 19.2. The scattering matrix $[S]$ is given by

$$[S] = \begin{bmatrix} S_{vv} & S_{vh} \\ S_{hv} & S_{hh} \end{bmatrix}. \tag{19.12}$$

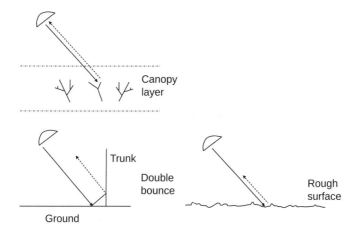

FIGURE 19.2   Three scattering mechanisms. Top: canopy scatter, middle: double-bounce scatter, bottom: rough surface scatter (Freeman and Darden, 1998).

Since scattering is due to multiple scattering, it is necessary to consider the second-order moments, which can be written as

$$\left\langle |S_{hh}|^2 \right\rangle = f_s\,|\beta|^2 + f_d\,|\alpha|^2 + f_v,$$
$$\left\langle |S_{vv}|^2 \right\rangle = f_s + f_d + f_v,$$
$$\left\langle S_{hh}S_{vv}^* \right\rangle = f_s\beta + f_d\alpha + f_v(^1\!/_3), \qquad (19.13)$$
$$\left\langle |S_{hv}|^2 \right\rangle = f_v(^1\!/_3),$$
$$\left\langle S_{hh}S_{hv}^* \right\rangle = \left\langle S_{hv}S_{vv}^* \right\rangle = 0,$$

where f_s, f_d, and f_v are the surface, double-bounce, and volume (canopy) contributions.

The canopy contribution f_v needs to include the random orientation of the scatterers (branches) from the vertical, resulting in (1/3). The double-bounce contribution includes the difference between the vertical and the horizontal polarization represented by α. The surface scattering includes β representing the difference between the horizontal and the vertical polarization.

Extensive studies have been conducted to show the usefulness of this three-component model (Freeman and Durden, 1998) and this has been extended to a four-component model including helix scattering contributions (Yamaguchi et al., 2005).

19.2.2   Target Decomposition Based on Eigenvectors

Cloude and Pottier (1996) gave a comprehensive review of three types of target decomposition theorems.

The first type is based on the Mueller matrix and Stokes vector including statistical fluctuations, and coherent backscattering in multiple scattering. The second is the decomposition proposed by Huynen (1970), and the third is eigenvector-based decomposition.

For backscattering, the scattering matrix $[S]$ consists of three elements, S_{11}, S_{12}, and S_{22}, which forms the target vector $[k_{3L}] = [S_{11}, \sqrt{2}S_{12}, S_{22}]^T$ where T denotes the transpose. $\sqrt{2}$ is required to keep the total scattered power $T_r(SS^+) = |S_{11}|^2 + |S_{22}|^2 + 2|S_{12}|^2$ constant. The coherency matrix $[T]$ is formed by

$$[T] = [k_{3L}][k_{3L}]^{*T}. \tag{19.14}$$

Cloude was the first to consider the decomposition based on the eigenvalues λ_1, λ_2, and λ_3 of $[T]$. He defined the target entropy H, which is a measure of randomness of the scatterer, the anisotropy A, and the α angle. Different values of H, A, and α, can be used to identify shallow water, rough surfaces, vegetated areas, forested areas, and other geophysical characteristics. Extensive work with examples is shown by Cloude and Pottier (1996).

19.3 POLARIMETRIC WEATHER RADAR

Weather radar is extensively used to determine rainfall rate, drop size distribution, and other parameters (Bringi and Chandrasekar, 2001; Doviak and Zrnic, 1984). This section presents an outline of fundamental ideas. Let us first consider the radar radiating a short pulse into a region of precipitation (rainfall) (see Fig. 19.3). The radar has a narrow beam with angular half-power beam width θ_b and the pulse duration of T_o. This includes interferometric radar, planar and cylindrical phased arrays, and disdrometer observations (Zhang, et al., 2006, 2007, and 2011). First, we assume that the antenna is linearly polarized. The wave propagates through rain. In general, raindrops may not be spherical, and the propagation constant depends on the antenna

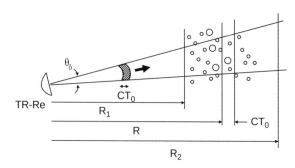

FIGURE 19.3   Narrow beam radar with half-power beam width θ_b sends out a short pulse of duration T_0, illuminating rain region.

polarization in relation to the geometric shapes of the raindrop. This propagation effect will be discussed in Sections 19.4 and 19.5. In this section, we assume that the raindrops are spherical and there is one propagation constant and no depolarization.

The radar equation for the problem shown in Fig. 19.3 is well known. The received signal P_r is given by (Ishimaru, 1997, Eq. (5.35))

$$P_r(t) = \frac{\lambda^2 G^2}{(4\pi)^3} \left(\frac{\pi \theta_b^2}{8 \ln 2} \right) \int_{R_1}^{R_2} \frac{\eta_b}{R^2} e^{-2\gamma} \left| U_i \left(t - \frac{2R}{C} \right) \right|^2 dR, \qquad (19.15)$$

where G is the gain of the antenna, θ_b is the half-power beam width, and $\eta_b =$ backscattering coefficient of raindrops per unit volume.

$$\eta_b = \int_0^\infty \sigma_b(D) n(D) dD, \quad \text{m}^2/\text{m}^3, \qquad (19.16)$$

where $\sigma_b(D)$ = backscattering cross section of a single raindrop (m^2) of diameter D, and $n(D)$, size distribution, is the number of particles within the diameter range D to $D + dD$. γ is the attenuation coefficient from R_1 to R.

$$\gamma = \int_{R_1}^{R} \eta_t dR,$$
$$\eta_t = \int_0^\infty \sigma_t(D) n(D) dD, \qquad (19.17)$$

where η_t is the total extinction cross-section coefficient per unit volume of rain. The input for a square pulse with duration T_0 is given by

$$|u_i(t)|^2 = \begin{cases} P_t & \text{for } 0 < t < T_0, \\ 0 & \text{for } t < 0 \text{ and } t > T_0. \end{cases} \qquad (19.18)$$

For the square pulse, (19.15) can be evaluated, noting that

$$R_2 = (ct)/2 \quad and \quad R_1 = c(t - T_0)/2. \qquad (19.19)$$

We then get

$$P_r(t) = \frac{\lambda^2 G^2}{(4\pi)^3} \left(\frac{\pi \theta_b^2}{8 \ln 2} \right) \frac{\eta_b}{2\eta_t} e^{-2\eta_t \left[\frac{c}{2}(t - T_0) - R_1 \right]} [1 - e^{-\eta_t c T_0}]. \qquad (19.20)$$

This is valid within the time $\frac{2R_1}{c} < t < \frac{2R_2}{c}$. Here we assume that $(1/R^2)$ in the integrand within R_1 to R_2 can be approximated by $(1/R_2^2)$. Equation (19.20) is sketched in Fig. 19.4. Note that if cT_0 is much smaller than the mean free path $= \frac{1}{\eta_t}$, then

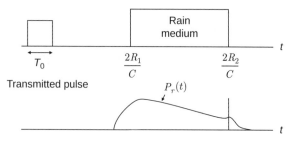

FIGURE 19.4   A square pulse of duration T_0.

(19.20) can be well represented by

$$P_r(t) = \frac{\lambda^2 G^2}{(4\pi)^3} \left(\frac{\pi \theta_b^2}{8 \ln 2} \right) \frac{\eta_b}{2} c T_0 e^{-2\eta_t \left[\frac{c}{2}(t-T_0)-R_1 \right]}.$$ (19.21)

For the special case of Rayleigh scattering of a spherical particle when the particle diameter is less than 0.1λ, we have

$$\sigma_b = \frac{k^4}{(4\pi)} \left| \frac{3(\varepsilon_r - 1)}{\varepsilon_r + 1} \right|^2 V^2, \quad V = \frac{4\pi a^3}{3}.$$ (19.22)

We can then rewrite (19.16) as

$$\begin{aligned}
\zeta_b &= \frac{\pi^5}{\lambda^4} K^2 \int_0^\infty D^6 n(D)dD \\
&= \frac{\pi^5}{\lambda^4} K^2 Z, \quad K = \frac{\varepsilon_r - 1}{\varepsilon_r + 1},
\end{aligned}$$ (19.23)

where Z is called the reflectivity factor. In our case, however, we are interested in nonspherical raindrops, as will be shown in Section 19.4.

19.4   NONSPHERICAL RAINDROPS AND DIFFERENTIAL REFLECTIVITY

In Section 19.3, the weather radar equation is formulated assuming spherical rain droplets. The output can be used to determine the reflectivity factor Z and the backscattering coefficient η_b using (19.21) and (19.23). However, the rain rate R cannot be directly determined from Z. The differential reflectivity Z_{dr} has been successfully used to estimate the rainfall rate R (Seliga and Bringi, 1976). This is done by making simultaneous measurements of horizontal and vertical reflectivities (Z_H and Z_V) and deriving the differential reflectivity Z_{dr}.

$$Z_{dr} = 10 \log(Z_H/Z_V)$$
$$= 10 \log(\eta_H/\eta_V), \tag{19.24}$$

where

$$\eta_H = \int_0^\infty \sigma_{bH}(D)n(D)dD,$$

$$\eta_V = \int_0^\infty \sigma_{bV}(D)n(D)dD.$$

σ_{bH} and σ_{bV} are the backscattering cross sections of a single raindrop when the incident wave is horizontally and vertically polarized. The rain rate R is given by

$$R = \int_0^\infty v(D)V(D)n(D)dD, \tag{19.25}$$

where $v(D)$ is the terminal velocity (m/s), $V(D)$ is the volume of a single droplet, and $n(D)$ is the number of particles per unit volume having diameter between D and $D + dD$. Here D is the diameter of a sphere whose volume is the same as the actual particle. The rain rate R is normally given in mm/hr and ranges from 0.25 mm/hr (drizzle), 10 mm/hr (moderate), 10–50 mm/hr (heavy rain), to >50 mm/hr (extremely heavy rain). The backscattering cross sections σ_{bH} and σ_{bV} can be calculated using the Rayleigh scattering (Section 10.6). For spheroidal raindrops with major and minor axes oriented in the horizontal and vertical directions, we have (Section 10.6)

$$\sigma_{bH} = \frac{k^4}{4\pi} \left[\frac{\varepsilon_r - 1}{1 + (\varepsilon_r - 1)L_H} \right]^2 V^2,$$

$$\sigma_{bV} = \frac{k^4}{4\pi} \left[\frac{\varepsilon_r - 1}{1 + (\varepsilon_r - 1)L_V} \right]^2 V^2,$$

$$L_V = \frac{1 + f^2}{f^2} \left(1 - \frac{1}{f} \arctan f \right), \tag{19.26}$$

$$L_H = \frac{1}{2}(1 - L_V),$$

$$f^2 = \left(\frac{a}{c}\right)^2 - 1, \qquad V = \frac{4\pi}{3}b^2 c.$$

For drops smaller than 1 mm in diameter, the shapes are spherical. For larger particles, the shape is approximated by oblate spheroids, which have been shown by Pruppacher and Pitter (1971).

The size distribution $n(D)$ of rain is well represented by the Gamma distribution

$$n(D) = N_0 D^m \exp(-\Lambda D). \tag{19.27}$$

There has been extensive work on the estimation of rain rate using the differential reflectivity.

19.5 PROPAGATION CONSTANT IN RANDOMLY DISTRIBUTED NONSPHERICAL PARTICLES

In Section 19.4, we discussed scattering by nonspherical raindrops whose major and minor axes are oriented in the horizontal and vertical directions (Fig. 19.5). In general, however raindrops are canted with a mean canting angle of approximately 10 (Oguchi, 1983). The wave propagating in such randomly distributed nonspherical particles consists of the coherent and the incoherent components. The coherent components $\langle E_1 \rangle$ and $\langle E_2 \rangle$ satisfy the following differential equation (Cheung and Ishimaru, 1982; Ishimaru and Cheung, 1980; Ishimaru and Yeh, 1984):

$$\frac{d}{ds} \begin{bmatrix} \langle E_1 \rangle \\ \langle E_2 \rangle \end{bmatrix} = [M] \begin{bmatrix} \langle E_1 \rangle \\ \langle E_2 \rangle \end{bmatrix}, \tag{19.28}$$

where

$$[M] = [M_0] + [M']$$

$$[M_0] = ik \begin{bmatrix} 1 & 0 \\ 0 & 1 \end{bmatrix}$$

$$[M'] = i\frac{2\pi}{k} \begin{bmatrix} \rho f_{11} & \rho f_{12} \\ \rho f_{21} & \rho f_{22} \end{bmatrix} = \begin{bmatrix} M_{11} & M_{12} \\ M_{21} & M_{22} \end{bmatrix}.$$

f_{ij} are the scattering amplitudes for $\langle E_i \rangle$ in the forward direction when the incident wave is $\langle E_j \rangle$ and ρf is the integration over the size distribution.

$$\rho f_{ij} = \int_0^\infty f_{ij} n(D) dD. \tag{19.29}$$

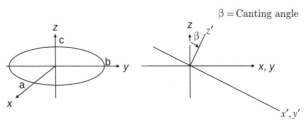

FIGURE 19.5 Oblate spheroidal particle with surface. $\dfrac{x^2}{a^2} + \dfrac{y^2}{b^2} + \dfrac{z^2}{c^2} = 1, \quad a = b > c.$

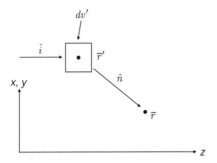

FIGURE 19.6   Coherent field $\langle E \rangle$ is obtained by summing the contribution from $-\infty < x < \infty, -\infty < y < \infty$.

Equation (19.28) is obtained by summing contributions from a slab of particles (Fig. 19.6). The integral equation for the coherent field $\langle E \rangle$ is therefore given by

$$\langle E \rangle = E_{in} + \int dv' \left\langle E(\bar{r}') \right\rangle f(\hat{n} \cdot \hat{i}) \frac{e^{ik|\bar{r}-\bar{r}'|}}{|\bar{r} - r'|}. \qquad (19.30)$$

We apply $(\nabla^2 + k^2)$ to both sides of (19.30) and obtain

$$(\nabla^2 + k^2) \langle E \rangle = 0$$
$$K^2 = k^2 + 4\pi f(\hat{i} \cdot \hat{i})\rho, \qquad (19.31)$$

where K is the propagation constant for the coherent field $\langle E \rangle$. Note that

$$(\nabla^2 + k^2)E_m = 0,$$
$$(\nabla^2 + k^2)\frac{e^{ik|\bar{r}-\bar{r}'|}}{|\bar{r} - \bar{r}'|} = -4\pi\delta(\bar{r} - r'). \qquad (19.32)$$

When the particle density ρ is low such that

$$k^2 \gg 4\pi f(\hat{i} \cdot \hat{i})\rho, \qquad (19.33)$$

then we get approximately

$$K = k + \frac{2\pi f(\hat{i} \cdot \hat{i})}{k}\rho. \qquad (19.34)$$

Equation (19.28) is obtained under this assumption. We also note that under the approximation (19.34), we get the coherent intensity

$$|\langle E \rangle|^2 = e^{-I_m(\frac{4\pi f \rho}{k})z} = e^{-\rho\sigma_t z} \qquad (19.35)$$

using the forward scattering theorem

$$I_m\left(\frac{4\pi f \rho}{k}\right) = \sigma_s + \sigma_a. \tag{19.36}$$

The extinction matrix for the coherent stokes vector $[I_c]$ is then given by

$$\frac{d}{ds}[I_c] = -[T][I_c],$$

$$-[T] = \begin{bmatrix} 2\mathrm{Re}M_{11} & 0 & \mathrm{Re}M_{12} & \mathrm{Im}M_{12} \\ 0 & 2\mathrm{Re}M_{22} & \mathrm{Re}M_{21} & \mathrm{Im}M_{21} \\ 2\mathrm{Re}M_{21} & 2\mathrm{Re}M_{12} & \mathrm{Re}(M_{11}+M_{22}) & -\mathrm{Im}(M_{11}-M_{22}) \\ -2\mathrm{Im}M_{21} & 2\mathrm{Im}M_{12} & \mathrm{Im}(M_{11}-M_{22}) & \mathrm{Re}(M_{11}+M_{22}) \end{bmatrix}.$$
$$\tag{19.37}$$

Let us consider the coherent field given by (19.28) for nonspherical particles. There are two propagation constants γ_1 and γ_2 which are the eigenvalues of the matrix $[M]$. We let $\langle E \rangle$ be $\exp(\gamma s)$ and obtain the following:

$$\begin{vmatrix} ik + M_{11} - \gamma & M_{12} \\ M_{21} & ik + M_{22} - \gamma \end{vmatrix} = 0. \tag{19.38}$$

The coherent field $\langle E_c \rangle$ is then given by

$$\langle E_c \rangle = c_1 [A_1] e^{\gamma_1 s} + c_2 [A_2] e^{\gamma_2 s}, \tag{19.39}$$

where A_1 and A_2 are the eigenvectors for γ_1 and γ_2.

$$[M][A_i] = \gamma_i [A_i], \quad i = 1, 2. \tag{19.40}$$

These two eigenvalues represent two fundamental characteristic polarizations for the coherent field.

The matrix $[M]$ in (19.28) can be calculated by the scattering amplitude f_{ij}. From small ellipsoidal objects, Rayleigh scattering is obtained by the static solution as given in Section 10.6. Section 10.7 gives Rayleigh–Debye scattering including randomly oriented objects.

19.6 VECTOR RADIATIVE TRANSFER THEORY

In Section 19.5, we showed the propagation constant in nonspherical particles. This is the coherent field. This can be combined with the incoherent field to form the complete vector radiative transfer theory. Let us start with the scalar radiative transfer equation (17.19). The brightness B, which is also called the specific intensity I, satisfies the equation

$$\frac{dI}{ds} = -\gamma I + \frac{1}{4\pi} \int P(\hat{s},\hat{s}')I(\hat{s}')d\Omega', \tag{19.41}$$

where γ is the extinction coefficient, and P is the phase function.

To include all polarization characteristics, we use the 4×1 Stokes specific intensity vector $[I]$.

The complete vector radiative transfer equation is then given by (Cheung, 1982; Ishimaru, 1980)

$$\frac{d}{ds}[I] = -[T][I] + \int [S(\hat{s},\hat{s}')][I(\hat{s}')]d\Omega', \tag{19.42}$$

where $[T]$ is the extinction matrix given in (19.37) and $[S]$ is the 4×4 phase function scattering matrix given by

$$[S] = \begin{bmatrix} \rho|f_{11}|^2 & \rho|f_{12}|^2 & \rho\mathrm{Re}(f_{11}f_{12}) & -\rho\mathrm{Im}(f_{11}f_{12}) \\ \rho|f_{21}|^2 & \rho|f_{22}|^2 & \rho\mathrm{Re}(f_{21}f_{22}) & -\rho\mathrm{Im}(f_{21}f_{22}) \\ \rho2\mathrm{Re}(f_{11}f_{21}) & \rho2\mathrm{Re}(f_{12}f_{22}) & \rho\mathrm{Re}(f_{11}f_{22}+f_{12}f_{21}) & -\rho\mathrm{Im}(f_{11}f_{22}-f_{12}f_{21}) \\ \rho2\mathrm{Im}(f_{11}f_{21}) & \rho2\mathrm{Im}(f_{12}f_{22}) & \rho\mathrm{Im}(f_{11}f_{22}+f_{12}f_{21}) & \rho\mathrm{Re}(f_{11}f_{22}+f_{12}f_{21}) \end{bmatrix}. \tag{19.43}$$

The Stokes vector $[I]$ is given by

$$I = \begin{bmatrix} I_1 \\ I_2 \\ U \\ V \end{bmatrix} = \begin{bmatrix} \langle E_1 E_1^* \rangle \\ \langle E_2 E_2^* \rangle \\ 2\mathrm{Re}\langle E_1 E_2^* \rangle \\ 2\mathrm{Im}\langle E_1 E_2^* \rangle \end{bmatrix}. \tag{19.44}$$

Solutions for the vector radiative transfer (19.42) have been obtained numerically (Ishimaru, 1997) for a plane wave incident on a slab of random medium of spherical particles (Ishimaru, Cheung, 1980). The circularly polarized incident waves, the first-order solutions, the coherent and incoherent components, and the cross-polarization discrimination are calculated for rain at 30 GHz.

19.7 SPACE–TIME RADIATIVE TRANSFER

In Section 19.6, we discussed CW incident waves, which are also applicable to the narrow band case. However, if the incident wave is a broadband short pulse, the CW theory may not be appropriate and we need to consider the complete space–time radiative transfer. Let us start with the scalar radiative transfer equation (19.41). For a narrow band short pulse, the specific intensity is a function of time and we write the equation of transfer including the time dependence.

$$\frac{d}{ds}I = -\gamma I - \frac{\partial}{c\partial t}I + \int S(\hat{s}, \hat{s}')I(\hat{s}', t)d\Omega', \qquad (19.45)$$

where $I(t, \hat{s}')$ is the time-dependent specific intensity, and γ and S are the extinction coefficient and the phase function at the carrier frequency. This time-dependent radiative transfer equation (19.45) is commonly used in pulse propagation problems. Extensive work has been conducted based on this equation (Ishimaru et al., 2001). It should be noted, however, that (19.45) is a narrow band approximation and a more general formulation should start with the two-frequency specific intensity $I(\omega_1, \omega_2)$.

$$I(t_1, t_2) = I(t_c, t_d)$$

$$= \frac{1}{(2\pi)^2} \iint I(\omega_1, \omega_2)e^{-i\omega_1 t_1 + i\omega_2 t_2}d\omega_1 d\omega_2 \qquad (19.46)$$

$$= \frac{1}{(2\pi)^2} \iint I(\omega_c, \omega_d)e^{-i(\omega_c t_d + \omega_d t_c)}d\omega_c d\omega_d,$$

where $I(\omega_1, \omega_2)$ is given by

$$I(\omega_1, \omega_2) = \langle E(\omega_1)E^*(\omega_2)\rangle,$$
$$I(t_1, t_2) = \langle E(t_1)E^*(t_2)\rangle,$$
$$\omega_c = \frac{1}{2}(\omega_1 + \omega_2) = \text{center frequency},$$
$$\omega_d = \omega_1 - \omega_2 = \text{difference frequency},$$
$$t_c = \frac{1}{2}(t_1 + t_2) = \text{center time},$$
$$t_d = t_1 - t_2 = \text{difference time}.$$

The specific intensity $I(t_1, t_2)$ represents the correlation of the field at t_1 and t_2.

The radiative transfer equation for $I(\omega_1, \omega_2)$ is given by

$$\frac{d}{ds}I = i[K(\omega_1) - K^*(\omega_2)]I + \int S(\omega_1, \omega_2)I'd\Omega'. \qquad (19.47)$$

Noting that $K(\omega) = K_r(\omega) + iK_i(\omega)$, we get

$$K(\omega_1) - K^*(\omega_2) = [K_r(\omega_1) - K_r(\omega_2)] + i[K_i(\omega_1) + K_i(\omega_2)]. \qquad (19.48)$$

For a sparse distribution, we can use (19.34), and obtain approximately

$$K(\omega_1) - K^*(\omega_2) = \omega_d \left.\frac{\partial K}{\partial \omega}\right|_{\omega_c} + i\rho\sigma_t. \qquad (19.49)$$

The radiative transfer (19.47) is then reduced to

$$\frac{d}{ds}I = \left(i\left(\omega_d/v_g\right) - \rho\sigma_t\right)I + \int SI' d\Omega', \qquad (19.50)$$

where v_g is the group velocity.

Noting that $i\omega_d \rightarrow -\frac{\partial}{\partial t}$, we get the time-dependent radiative transfer (19.45).

$$\frac{d}{ds}I = -\rho\sigma_t I - \frac{1}{c}\frac{\partial}{\partial t}I + \int SI' d\Omega'.$$

We note however that (19.45) is an approximation and a more exact formulation needs to start with the two-frequency radiative transfer (19.47). We also note that at a given time t_c with $t_1 = t_2$, $I(t_c, t_d = 0)$ gives the specific intensity for a pulse intensity at t_c.

$$I(t_c) = \frac{1}{(2\pi)}\int I(\omega_d)e^{-i\omega_d t_c} d\omega_d. \qquad (19.51)$$

Thus $I(t_c)$ satisfies (19.45).

19.8 WIGNER DISTRIBUTION FUNCTION AND SPECIFIC INTENSITY

In Section 19.7, we discussed the two-frequency specific intensity and its relation with the two-frequency radiative transfer theory. It should, however, be noted that the formulations given in the last section can be viewed from a more fundamental formulation of the Wigner distribution.

Let us now start with the CW mutual coherence function (MCF) given by

$$\Gamma(\bar{r}_1, \bar{r}_2) = \langle U(\bar{r}_1)U^*(\bar{r}_2)\rangle$$
$$= \Gamma(\bar{r}, \bar{r}_d), \qquad (19.52)$$

where $\bar{r} = \frac{1}{2}(\bar{r}_1 + \bar{r}_2)$ and $\bar{r}_d = \bar{r}_1 - \bar{r}_2$.

The Wigner distribution function $W(\bar{r}, \bar{K})$ is the Fourier transform of Γ with respect to $\bar{r}_d$.

$$W(\bar{r}, \bar{K}) = \int \Gamma(\bar{r}, \bar{r}_d)e^{-i\bar{K}\bar{r}_d} dV_d$$
$$= \int \left\langle U\left(\bar{r} + \frac{\bar{r}_d}{2}\right)U^*\left(\bar{r} - \frac{\bar{r}_d}{2}\right)\right\rangle e^{-i\bar{K}\bar{r}_d} dV_d. \qquad (19.53)$$

The Wigner function W has certain properties as shown here. First, the integral of W with respect to the wave vector $\bar{K}$ is the "energy density."

$$\frac{1}{(2\pi)^3} \int W(\bar{r}, \bar{K})d\bar{K} = \Gamma(\bar{r}, \bar{r}_d = 0) \tag{19.54}$$

$$= \langle |U(\bar{r})|^2 \rangle.$$

Note that

$$\int e^{ti\bar{K}\bar{r}_d}d\bar{K} = (2\pi)^3 \delta(\bar{r}_d). \tag{19.55}$$

Also note that W is real.

$$W^*(\bar{r}, \bar{K}) = \int \left\langle U^* \left(\bar{r} + \frac{\bar{r}_d}{2} \right) U \left(\bar{r} - \frac{\bar{r}_d}{2} \right) \right\rangle e^{i\bar{K}\bar{r}_d}dV_d. \tag{19.56}$$

We let $\bar{r}_d = -\bar{r}'_d$ and get

$$W^*(\bar{r}, \bar{K}) = \int \left\langle U^* \left(\bar{r} - \frac{\bar{r}'_d}{2} \right) U \left(\bar{r} + \frac{\bar{r}'_d}{2} \right) \right\rangle e^{-i\bar{K}\bar{r}_d}dV_d, \tag{19.57}$$

$$= W(\bar{r}, \bar{K})$$

showing that $W(\bar{r}, \bar{K})$ is real.

The Wigner distribution function W is real and is closely related to the specific intensity $I(\bar{r}, \hat{s})$. The difference is that the specific intensity I is real and positive as it represents the real power flow in the direction of $\hat{s}$. $W(\bar{r}, \bar{K})$ is identical to $I(\bar{r}, \hat{s})$ if $\bar{K}$ is limited to $\bar{K} = k\hat{s}$ where $\hat{s}$ is the unit vector with $|\hat{s}| = 1$. While I is real and positive, W is real, but need not be positive.

Next consider the energy flux $\bar{F}$. For the specific intensity, we have

$$\bar{F} = \int I(\bar{r}, \hat{s})\hat{s}d\Omega. \tag{19.58}$$

The corresponding flux using Wigner function is given by

$$\frac{1}{(2\pi)^3} \int W(\bar{r}, \bar{K})\bar{K}d\bar{K}. \tag{19.59}$$

To show that (19.59) gives the Poynting vector, we start with the inverse Fourier transform (19.53).

$$\frac{1}{(2\pi)^3} \int W(\bar{r}, \bar{K})e^{i\bar{K}\bar{r}_d}d\bar{K}$$

$$= \left\langle U \left(\bar{r} + \frac{\bar{r}_d}{2} \right) U^* \left(\bar{r} - \frac{\bar{r}_d}{2} \right) \right\rangle \tag{19.60}$$

$$= \langle U_1 U_2^* \rangle.$$

We take the divergence of both sides and let $r_d = 0$. We get

$$\frac{1}{(2\pi)^3} \int W(\bar{r}, \bar{K}) \left(i\bar{K}\right) d\bar{K}$$

$$= \frac{1}{2}(\nabla U)U^* - \frac{1}{2}U(\nabla U^*). \tag{19.61}$$

Note that $\nabla_{rd}U_1 = \nabla U$ and $\nabla_{rd}U_2 = -\nabla U$.

The right side of (19.61) is the Poynting vector. The Poynting vector for a scalar acoustic pressure field p is given by (Section 2.12)

$$\bar{S} = \text{Re} \left[\frac{1}{2}p\bar{V}^*\right],$$

$$\bar{V} = \text{particle velocity} = \frac{1}{i\omega\rho_0}\nabla p. \tag{19.62}$$

For a scalar field U which is proportional to the pressure field p, we have, except for a constant,

$$\bar{S} = \frac{1}{(-i)}[U\nabla U^* - U^*\nabla U]. \tag{19.63}$$

From (19.61), we finally get

$$\frac{1}{(2\pi)^3} \int W(\bar{r}, \bar{K})\bar{K}d\bar{K} = \bar{S}. \tag{19.64}$$

The temporal Wigner distribution function can be obtained similarly. We have

$$W(t, \omega) = \int_{-\infty}^{\infty} \langle U(t_1)U^*(t_2)\rangle e^{i\omega t_d} dt_d, \tag{19.65}$$

where

$$t_1 = t + \frac{t_d}{2}, \quad t_2 = t - \frac{t_d}{2}.$$

We can express this using the Fourier transform.

$$U(t) = \frac{1}{2\pi} \int \bar{U}(\omega)e^{-i\omega t} d\omega,$$

$$W(t, \omega) = \frac{1}{(2\pi)^2} \iint \langle \bar{U}(\omega_1)\bar{U}^*(\omega_2)\rangle e^{-i\omega_1 t_1 + i\omega_2 t_2 + i\omega t_d} d\omega_1 d\omega_2 dt_d, \tag{19.66}$$

where $\langle \bar{U}(\omega_1)\bar{U}^*(\omega_2)\rangle$ is the two-frequency MCF.

Noting that $\omega_1 t_1 - \omega_2 t_2 = \omega_c t_d + \omega_d t_c$, we get

$$W(t, \omega) = \frac{1}{(2\pi)} \int \langle \bar{U}\left(\omega + \frac{\omega_d}{2}\right)\bar{U}^*\left(\omega - \frac{\omega_d}{2}\right)\rangle e^{-i\omega_d t} d\omega_d. \tag{19.67}$$

The Wigner distribution function $W(t, \omega)$ is related to the "ambiguity function" $\chi(\omega_d, t_d)$.

$$\chi(\omega_d, t_d) = \frac{1}{2\pi} \int U(t_1)U^*(t_2)e^{i\omega_d t}dt. \tag{19.68}$$

From this, we can show that

$$W(t, \omega) = \int \chi(\omega_d, t_d)e^{-i\omega_d t + i\omega t_d}d\omega_d dt_d. \tag{19.69}$$

19.9 STOKES VECTOR EMISSIVITY FROM PASSIVE SURFACE AND OCEAN WIND DIRECTIONS

In Section 17.5, the brightness temperature $T_\alpha(\hat{i})$ emitted in the direction $(-\hat{i})$ in the polarization state α from a surface kept at temperature T is shown to be

$$T_\alpha(\hat{i}) = \varepsilon_\alpha(\hat{i})T, \tag{19.70}$$

where $\varepsilon_\alpha(\hat{i})$ is called the "emissivity." It is also shown that if the surface is in thermal equilibrium, the absorption must be equal to the emission and therefore, the emissivity is equal to the absorptivity.

$$\varepsilon_\alpha(\hat{i}) = a_\alpha(\hat{i}). \tag{19.71}$$

On the other hand, if a power is incident upon a surface, a part of the power is scattered and the other part is absorbed. The ratio of the total scattered power to the incident power in the projected area $A \cos \theta_i$ is called the "albedo" denoted by $\Gamma_\alpha(i)$. The absorptivity a_α is then the fractional power absorbed by the surface and is given by

$$a_\alpha(\hat{i}) = 1 - \Gamma_\alpha(\hat{i}) = \varepsilon_\alpha(\hat{i}). \tag{19.72}$$

This shows that the emissivity of a passive surface is related to the scattering problem when a wave is incident on a surface. The albedo $\Gamma_\alpha(\hat{i})$ is given in (17.29) of Section 17.5.

This relationship (19.72) can be generalized to the Stokes vector formulation. It is shown, (Tsang et al., 2000) that T_α and ε_α (19.70) are generalized to the Stokes vector. We have

$$\begin{bmatrix} T_1(\hat{i}) \\ T_2(\hat{i}) \\ U(\hat{i}) \\ V(\hat{i}) \end{bmatrix} = T \begin{bmatrix} 1 - \Gamma_1 \\ 1 - \Gamma_2 \\ 1 - \Gamma_U \\ 1 - \Gamma_V \end{bmatrix}, \tag{19.73}$$

where 1 and 2 denote the vertical and the horizontal polarization.

$$\Gamma_1 = \frac{1}{4\pi}\int_{2\pi}\left[|f_{11}(\hat{s},\hat{i})|^2 + |f_{21}(\hat{s},\hat{i})|^2\right]d\Omega_s,$$

$$\Gamma_2 = \frac{1}{4\pi}\int_{2\pi}\left[|f_{12}(\hat{s},\hat{i})|^2 + |f_{22}(\hat{s},\hat{i})|^2\right]d\Omega_s,$$

$$\Gamma_U = \frac{1}{4\pi}\int_{2\pi}2\mathrm{Re}\left[f_{11}(\hat{s},\hat{i})f_{12}^*(\hat{s},\hat{i}) + f_{21}(\hat{s},\hat{i})f_{22}^*(\hat{s},\hat{i})\right]d\Omega_s,$$

$$\Gamma_V = \frac{1}{4\pi}\int_{2\pi}2\mathrm{Im}\left[f_{11}(\hat{s},\hat{i})f_{12}^*(\hat{s},\hat{i}) + f_{21}(\hat{s},\hat{i})f_{22}^*(\hat{s},\hat{i})\right]d\Omega_s.$$

If we can measure T_1, T_2, U, and V, then the scattering characteristics of the surface and the ocean wind directions may be determined as shown in (19.73). The measurement of the Stokes vector can be made by using the technique described in Section 16.10. T_1 and T_2 can be obtained by the intensities in vertical and horizontal polarizations. U can be obtained by measuring the power in 45° and 135°. V can be obtained by the power for the right-handed and left-handed circular polarizations.

It has been noted that for a sea surface, T_v and T_h are even functions of the azimuth angle ϕ and U and V are the odd functions of ϕ, where $\phi = \phi_w - \phi_r$, ϕ_w = wind direction and ϕ_r is the radiometer look direction (Fig. 19.7). Using up to the second

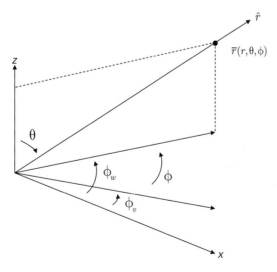

FIGURE 19.7   Thermal emission in the direction (r, θ, ϕ). ϕ_W = wind direction, ϕ_r = radiometer look direction, $\phi = \phi_W - \phi_r$.

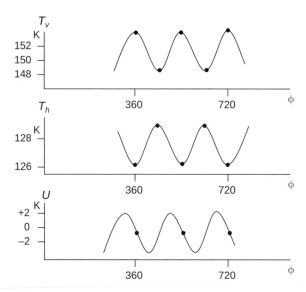

FIGURE 19.8 T_v and T_h are even functions of ϕ and U is an odd function of ϕ.

harmonic of ϕ we have

$$
\begin{aligned}
T_v &\approx T_{vo} + T_{v1} \cos \phi + T_{v2} \cos 2\phi, \\
T_h &\approx T_{ho} + T_{h1} \cos \phi + T_{h2} \cos 2\phi, \\
U &\approx U_1 \sin \phi + U_2 \sin 2\phi, \\
V &\approx V_1 \sin \phi + V_2 \sin 2\phi.
\end{aligned}
\tag{19.74}
$$

It has been experimentally and numerically observed that T_v, T_h, and U behave as shown in Fig. 19.8 which are consistent with theoretical predictions (Yueh, 1999). It is therefore possible to determine the ocean wind direction by measuring the Stokes vectors T_1, T_2, U, and V.

19.10   VAN CITTERT–ZERNIKE THEOREM APPLIED TO APERTURE SYNTHESIS RADIOMETERS INCLUDING ANTENNA TEMPERATURE

In Sections 17.9 and 17.10, we discussed a special case of the Fourier transform relationship between the source brightness distribution of the incoherent source and the degree of coherence. This is a special case of the van Cittert–Zernike theorem. In this section, we revisit this problem as applied to passive microwave remote sensing. In particular, we include the effects of the antenna temperature (Corbella et al., 2004; Camps et al., 2008; Wedge and Rutledge, 1991).

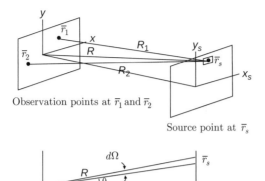

Observation points at $\bar{r}_1$ and $\bar{r}_2$

Source point at $\bar{r}_s$

FIGURE 19.9 van Cittert–Zernike theorem relating the source brightness distribution $T(\bar{r}_s, \hat{s})$ to the visibility function $V(\bar{r}_1, \bar{r}_2)$.

Let us start with the van Cittert–Zernike theorem. It relates the MCF $V(\bar{r}_i, \bar{r}_j)$ of electromagnetic waves at two points $\bar{r}_i$ and $\bar{r}_j$ and the source brightness distribution $T(\bar{r}_s, \hat{s})$ of the thermal source (Fig. 19.9).

$$V(\bar{r}_1, \bar{r}_2) = \int G(\hat{s})T(\bar{r}_s, \hat{s})\frac{e^{ik(R_1-R_2)}}{R_1 R_2}d\Omega, \qquad (19.75)$$

where $\hat{s}$ is the unit vector in the direction to $\bar{r}_s$, and $G(\hat{s})$ is the directivity pattern of the antenna.

$V(\bar{r}_1, \bar{r}_2)$ is called the "visibility function." $d\Omega$ is the elementary solid angle $d\Omega = \sin\theta\, d\theta\, d\phi$.

In the parabolic approximation, we have

$$R_1 = R + \frac{|\bar{r}_1 - \bar{r}_s|^2}{2R},$$
$$R_2 = R + \frac{|\bar{r}_2 - \bar{r}_s|^2}{2R}. \qquad (19.76)$$

We then have

$$V(\bar{r}_1, \bar{r}_2) = \int G(\hat{s})T(\bar{r}_s, \hat{s})\exp\left[ik\frac{r_1^2 - r_2^2}{2R} - ik\frac{(r_1 - r_2) \cdot \bar{r}_s}{R}\right]d\Omega, \qquad (19.77)$$

where we let

$$\frac{1}{R_1 R_2} \approx \frac{1}{R^2},$$

which can be absorbed in $G(\hat{s})$.

We also note that

$$d\zeta d\eta = \cos\theta\, d\Omega = \sqrt{1 - \xi^2 - \eta^2}\, d\Omega,$$

where $\xi = \sin\theta\cos\phi$ and $\eta = \sin\theta\sin\phi$ are the direction cosines of $\hat{s}$. The extra phase $\left[ik\frac{(r_1^2 - r_2^2)}{2R}\right]$ is the difference between $\bar{r}_s = 0$ to $\bar{r}_1$ and $\bar{r}_2$ in wavelength and is independent of $\hat{s}$ and can be absorbed in G. We finally get the following Fourier transform representation of the van Cittert–Zernike theorem commonly used in remote sensing.

$$V(u, v) = \int_{\xi^2 + \eta^2 \leq 1} G(\xi, \eta) T(\xi, \eta) \frac{\exp[-ik(u\xi + v\eta)]}{\sqrt{1 - \xi^2 - \eta^2}} d\xi d\eta, \qquad (19.78)$$

where $u = x_1 - x_2$, $v = y_1 - y_2$.

This Fourier transform relationship means that if we make measurements of the visibility function, which is the spatial correlation of the waves at different pairs of two points $\bar{r}_1$ and $\bar{r}_2$, then in principle, we can invert the measured visibility at different spacings by inverse Fourier transform and obtain the temperature distribution $T(\bar{r}_s, \hat{s})$.

19.10.1 The Effects of Antenna Temperature and Bosma Theorem

The van Cittert–Zernike theorem (19.78) does not include the effects of *antenna temperature*. The Bosma theorem relates the correlation of the noise temperature to the antenna temperature and the scattering matrix. Consider two receiving antennas with antenna temperature T_a with reflectionless terminations (Fig. 19.10). The incident wave (a_1, a_2), the output (b_1, b_2), and the noise wave (b_{s1}, b_{s2}) are related by the scattering matrix S.

$$[b] = [S][a] + [b_s], \qquad (19.79)$$

FIGURE 19.10 Bosma theorem $\begin{aligned} b &= Sa + b_s \\ \langle b_s b_s^+ \rangle &= kT_a(1 - SS^+) \end{aligned}$.

where

$$[b] = \begin{bmatrix} b_1 \\ b_2 \end{bmatrix}, \quad [S] = \begin{bmatrix} S_{11} & S_{12} \\ S_{21} & S_{22} \end{bmatrix},$$

$$[a] = \begin{bmatrix} a_1 \\ a_2 \end{bmatrix}, \quad [b_s] = \begin{bmatrix} b_{s1} \\ b_{s2} \end{bmatrix}.$$

The incident wave $[a]$ emanates from the termination and is uncorrelated with the noise $[b_s]$. The incident power is due to the thermal noise from terminations at temperature T_a and therefore we have, per unit bandwidth (1 Hz),

$$\begin{aligned} \langle |a_1|^2 \rangle &= \langle |a_2|^2 \rangle = KT_a, \\ \langle a_1 a_2^* \rangle &= 0, \end{aligned} \tag{19.80}$$

where K is Boltzmann's constant.

From (19.79), we get

$$\begin{aligned} \langle bb^+ \rangle &= \langle Saa^+ S^+ \rangle + \langle b_s b_s^+ \rangle \\ &= KT_a(SS^+) + \langle b_s b_s^+ \rangle, \end{aligned} \tag{19.81}$$

where b^+ is the conjugate transpose of b.

We may write the components of (19.81).

$$\begin{aligned} \langle b_1 b_1^+ \rangle &= KT_a \left[|S_{11}|^2 + |S_{12}|^2 \right] + \langle |b_{s1}|^2 \rangle, \\ \langle b_1 b_2^+ \rangle &= KT_a \left[S_{11} S_{21}^* + S_{12} S_{22}^* \right] + \langle b_{s1} b_{s2}^* \rangle. \end{aligned} \tag{19.82}$$

This shows the correlation of the output waves b_1 and b_2.

The correlation of the wave b_{s1} and b_{s2} from the thermal noise is the visibility function given in (19.78).

$$\langle b_{s1} b_{s2}^* \rangle = V(u, v). \tag{19.83}$$

Equation (19.83) shows that there is a difference between the correlation of the output wave $\langle b_1 b_2^* \rangle$ and the visibility function $V(u, v)$.

In order to find the difference $KT_a[S_{11}S_{21}^* + S_{12}S_{22}^*]$, we use the Bosma theorem, which relates the noise wave and the scattering matrix under thermodynamic equilibrium, in which case, the powers $|a|^2$ and $|b|^2$ need to be balanced, and we have

$$\begin{aligned} |a_1|^2 &= |a_2|^2 = |b_1|^2 = |b_2|^2 = KT_a, \\ \langle a_1 a_2^* \rangle &= 0, \\ \langle b_1 b_2^* \rangle &= 0. \end{aligned} \tag{19.84}$$

Under this thermodynamic equilibrium, we can write (19.82) as

$$\langle b_s b_s^+ \rangle = KT_a(1 - SS^+) \tag{19.85}$$

This is the Bosma theorem.

If we surround these two antennas with a microwave absorber with constant temperature T_a, then we get the thermodynamic equilibrium and the Bosma theorem applies. We can write (19.82) as

$$0 = KT_a \left[S_{11}S_{21}^* + S_{12}S_{22}^* \right] + V(\text{at } T_a), \tag{19.86}$$

where $V(\text{at } T_a)$ is the visibility function (19.78) with $T = T_a$. Thus we get

$$KT_a \left[S_{11}S_{21}^* + S_{12}S_{22}^* \right] = -V(\text{at } T_a). \tag{19.87}$$

Substituting this into (19.82) we finally get

$$\langle b_1 b_2^* \rangle = \int G(\zeta, \eta)(T(\zeta, \eta) - T_a) \frac{\exp \left[-ik(u\zeta + v\eta) \right]}{\sqrt{1 - \zeta^2 - \eta^2}} d\zeta d\eta. \tag{19.88}$$

$\langle b_1 b_2^* \rangle$ is the visibility function with the antennas at $\bar{r}_1$ and $\bar{r}_2$. Therefore, this is the generalized van Cittert–Zernike theorem including the temperature T_a. This was derived by Corbella (2004).

19.11 IONOSPHERIC EFFECTS ON SAR IMAGE

Space-borne SAR has been used for measuring forest biomass. However, microwave frequencies such as c-band (4–8 GHz, wavelength 7.5–3.8 cm) used in conventional SAR cannot penetrate into foliage and therefore lower frequencies, typically P-band (250–500 MHz) are employed for biomass studies. At these low frequencies, how-ever, ionospheric effects become significant. These effects include reduced azimuthal resolution due to the reduced coherent length, dispersion causing group delay, a shift of image in the range, pulse broadening, and Faraday rotation. In this section, we present analytical studies on these effects (Ishimaru et al., 1999).

The coherent received signal $v(\bar{r}_0)$ of SAR focused on $\bar{r}_0$ is given by

$$v(\bar{r}_0) = \int S(\bar{r})\chi(\bar{r}, \bar{r}_0)ds, \tag{19.89}$$

where $S(\bar{r})$ is the surface reflectivity at $\bar{r}$ and χ is the generalized ambiguity function (system point target response) given by (Fig. 19.11)

$$\chi(\bar{r}, \bar{r}_0) = \sum_n \int g_n(t, \bar{r}_n)f_n^*(t, \bar{r}_{0n})dt. \tag{19.90}$$

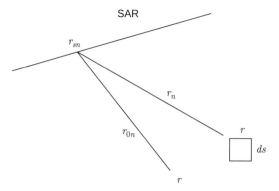

FIGURE 19.11 Synthetic aperture radar (SAR) focused on r_0. Here r_{sn} is the receiver position, and r_{0n} and r_0 are the distance from the receiver to r_0 and r, respectively.

This is the coherent sum of the signal $g_n(t, \bar{r}_n)$ received at $\bar{r}_{sn}$ due to a point target at $\bar{r}$ illuminated by an incident pulse $u_i(t)$. $f_n(t, \bar{r}_{0n})$ is a matched filter function focused at $\bar{r}_0$. It may be noted that in the imaging studies, $f_n(t, \bar{r}_{0n})$ is also called the "steering vector" or "focusing function" focused on the "search point" $\bar{r}_0$ (Ishimaru et al., 2012).

We take the Fourier transform of (19.90) and obtain in the frequency domain

$$\chi(\bar{r}, \bar{r}_0) = \sum_n \frac{1}{2\pi} \int \bar{g}_n(\omega, r_n) \bar{f}_n^*(\omega, r_{0n}) d\omega$$
$$= \sum_n \chi_n(r_n, r_{0n}),$$

$\qquad\qquad(19.91)$

where

$$\bar{g}_n(\omega, r_n) = \int g_n(t, r_n) e^{i\omega t} dt,$$

$$\bar{f}_n^*(\omega, r_{0n}) = \int f_n(t, r_{0n}) e^{i\omega t} dt.$$

The signal $\bar{g}_n$ consists of the incident pulse spectrum $u_i(\omega)$ and the two-way Green's function G^2.

$$\bar{g}_n(\omega, r_n) = \bar{u}_{iI}(\omega) \bar{G}_0(\omega, r_n),$$

$\qquad\qquad(19.92)$

where

$$\bar{G}_0 = G^2 = \frac{\exp\left(i2\int \beta ds + \psi_d + \psi_u\right)}{(4\pi r_n)^2}.$$

The focusing matched filter is also given by

$$\bar{f}_n(\omega, r_{0n}) = \bar{U}_i(\omega) \left[\frac{\exp(ikr_{0n})}{4\pi r_{0n}} \right]^2$$

$$= \bar{U}_i(\omega) G_s(\omega, \bar{r}_{0n})^2.$$

(19.93)

Note that β in (19.92) gives the dispersion and $\psi_d + \psi_u$ gives the effects of the ionospheric turbulence.

The input pulse is normally a chirp expressed by

$$U_i(t) = \exp\left(-i\omega_0 t - i\frac{B}{4T_0}t^2 \right), \quad |t| < T_0.$$

(19.94)

The chirp can also be approximated by the following for mathematical convenience.

$$U_i(t) = \exp(-i\omega_0 t - \alpha t^2), \quad |t| < \infty,$$

$$\alpha = \frac{\pi}{4T_0^2} + i\frac{B}{4T_0} = \alpha_1 + i\alpha_2,$$

$$\bar{U}_i(\omega) = \sqrt{\frac{\pi}{\alpha}} \exp\left[-\frac{(\omega - \omega_0)^2}{4\alpha} \right],$$

(19.95)

$$B = \text{bandwidth}.$$

Let us examine χ_n in (19.91).

$$\chi(\bar{r}, \bar{r}_{0n}) = \frac{1}{2\pi} \int \bar{U}_i(\omega) \bar{U}_i^*(\omega) G^2(\omega, \bar{r}_n) G_s^2(\omega, r_{0n}).$$

(19.96)

Note that β in G^2 given in (19.92) gives the ionospheric dispersion. Thus we write

$$\beta(\omega) = \beta(\omega_0) + (\omega - \omega_0)\beta'(\omega_0) + \frac{(\omega - \omega_0)^2}{2}\beta''(\omega_0) + \cdots$$

(19.97)

$\beta'(\omega_0)$ gives the group delay and $\beta''(\omega_0)$ gives the pulse broadening. $\psi_d + \psi_u$ in (19.92) is caused by the fluctuations of the ionospheric turbulence and gives the broadening of the SAR image.

In Fig. 19.12, we show these three effects, azimuthal resolution $\Delta y'$, the pulse broadening $\Delta x'$, and the group delay ΔU.

We can now write χ_n in (19.96) in the following form:

$$\chi(\bar{r}_n, \bar{r}_{0n}) = \frac{1}{2\pi} \int d\omega \frac{\exp\left[i\Phi_0 + i(\omega - \omega_0)\Phi_1 - (\omega - \omega_0)^2\Phi_2 \right]}{(4\pi r_n)^2 (4\pi r_{0n})^2},$$

(19.98)

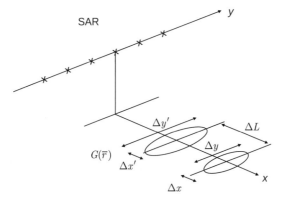

FIGURE 19.12   Ionospheric effects on SAR. Free space image is given by $\Delta x - \Delta y$. $\Delta x' =$ pulse broadening, $\Delta y' =$ azimuthal resolution, $\Delta L =$ group delay.

where

$$\Phi_0 = 2\frac{\omega_0}{c}(r_n - r_{0n} - d_i) + 2\beta(\omega_0)d_i + \psi_d + \psi_u,$$

$$\Phi_1 = \frac{2}{c}(r_n - r_{0n}) + 2\left[\beta'(\omega_0) - \frac{1}{c}\right]d_i,$$

$$\Phi_2 = \frac{\alpha_2}{2|\alpha|^2} - i\beta''(\omega_0)d_i.$$

d_i is the ionospheric thickness, Φ_0 gives the azimuthal resolution $\Delta y'$, Φ_1 represents the group delay U, and Φ_2 gives the pulse broadening $\Delta x'$ (Fig. 19.12). $\psi_d + \psi_u$ in Φ_1 represents the ionospheric fluctuations.

The detailed derivation of these effects is given by Ishimaru et al. (1999).

Let us next consider the fluctuation $\psi_d + \psi_u$. Because of the reciprocity, $\psi = \psi_u = \psi$. Letting the one-way field $u = \exp(\psi)$ and assuming that u is the complex Gaussian field, we get

$$\langle u^2 \rangle \approx \langle u \rangle^2,$$
$$\langle u_1^2 u_2^{*2} \rangle \approx 2\langle u_1 u_2^* \rangle^2 - \langle u_1 \rangle^2 \langle u_2^* \rangle^2. \tag{19.99}$$

We then get the coherent component

$$\langle \exp(2\psi) \rangle = \langle u \rangle^2 = \exp(-2\alpha_i d_i), \tag{19.100}$$

where

$$\alpha_i = 2\pi^2 k^2 \int_0^\infty \Phi_n(k, \kappa)d\kappa.$$

Also we get

$$\langle u_1^2 u_2^{*2} \rangle = 2e^{-D_s} - e^{-4\alpha_i d_i}, \tag{19.101}$$

where

$$D_s = 8\pi^2 k^2 d_i \int_{0k}^{\infty} [1 - J_0(\kappa\rho)] \Phi_n(z, \kappa) \kappa d\kappa,$$

$$\rho = y_s - y_s' + \left[y - y' - (y_s - y_s') \frac{z}{r_0} \right].$$

The details of the derivation are given in Ishimaru et al. (1999).

19.11.1 Faraday Rotation

When a linearly polarized radio wave propagates through the ionosphere, the medium becomes anisotropic as discussed in Chapter 8, and the wave will be split into two circular polarized waves with two different propagation constants. After propagation, these two waves recombine to give a linearly polarized wave with the plane of polarization rotated in proportion to the geomagnetic field and the distance. Figure 19.9 shows a wave propagating in the z direction with the geomagnetic field H_{dc} pointed in the direction of the masking angle θ. At $z = 0$, the wave is given by $E_x(0)$ and $E_y(0)$, and at $z \neq 0$, the wave is given by $E_x(z)$ and $E_y(z)$. They are related through $T(z)$.

$$\begin{bmatrix} E_x(z) \\ E_y(z) \end{bmatrix} = [T(z)] \begin{bmatrix} E_x(0) \\ E_y(0) \end{bmatrix}, \tag{19.102}$$

where $[T(z)]$ is given by

$$[T] = \frac{1}{R_1 - R_2} \begin{bmatrix} R_1 & -1 \\ 1 & -R_2 \end{bmatrix} \exp(ikn_1 z) + \frac{1}{R_1 - R_2} \begin{bmatrix} -R_2 & 1 \\ -1 & R_1 \end{bmatrix} \exp(ikn_2 z),$$

$$R_{1,2} = -\frac{i}{Y_z} \left\{ \frac{Y_y^2}{2(X - U)} \mp \left[\frac{Y_y^4}{4(X - U)^2} - Y_z^2 \right]^{1/2} \right\},$$

$$X = \omega_p^2 / \omega^2, \quad Y = e\mu_0 H_{dc} / m\omega,$$

$$Y_z = Y \cos \theta, \quad Y_y = Y \sin \theta,$$

$$U = 1 - j\frac{\nu}{\omega},$$

$\nu =$ collision frequency,

$\omega =$ plasma frequency.

For P-band (500 MHz) at 400–600 km altitude, the azimuthal resolution of SAR can be seriously affected in an average ionosphere (30 TECU, variance 10%). TEC is the total electron content, the integration of the electron density over the vertical path from the ground to the upper ionosphere. TECU is the unit $= 10^{16} \mathrm{m}^{-2}$ electrons. For the range resolution at P-band, the image shift can be significant. Faraday rotation is significant at P-band.

CHAPTER 20

BIOMEDICAL EM, OPTICS, AND ULTRASOUND

Imaging and detection of malignant tissue are important in health care. Applications of EM energy for medical purposes are important but require careful attention to unintended deleterious biological effects. EM fields from static to terahertz are used in these applications, but they require careful study of safety and health precautions. In this chapter, we discuss some key formulations in biomedical electromagnetics. Extensive work has been reported on static through terahertz coupling to bodies near fields, cell phones, specific absorption rate (SAR), narrow and UWB pulse in biological systems (Lin, 2012). In this chapter, we consider SAR and heat diffusion in tissues. For bio-optics, we discuss optical scattering and imaging in tissues (Tuchin and Thompson, 1994) and optical diffusion and photon density waves. Optical coherence tomography (OCT) and low coherence interferometry are extensively used in medical imaging (Fujimoto, 2001). Fundamentals of ultrasound scattering and imaging of tissues and blood are discussed in this chapter (Shung and Thieme, 1993). An overview of acoustical and optical scattering and imaging of tissues is given by Ishimaru (2001).

As we examine these formulations, it is important to clearly understand the differences among EM, optics, and ultrasound characteristics in biological media. For EM, whose frequency may be of the order of 1 GHz, the wavelength is comparable to geometric sizes of bodies. Therefore the electric field distribution in a biological medium depends very much on the size and shape of the medium, which requires

Electromagnetic Wave Propagation, Radiation, and Scattering: From Fundamentals to Applications,
Second Edition. Akira Ishimaru.
© 2017 by The Institute of Electrical and Electronic Engineers, Inc. Published 2017 by John Wiley & Sons, Inc.

numerical study of boundary value problems. It is also necessary to use a higher frequency if we wish to reduce the wavelength and to improve the resolution. However, higher frequency (shorter wavelength) increases absorption and reduces skin depth, resulting in smaller illuminating power and reduced signal-to-noise ratio.

For optics, the wavelength in tissues is of the order of 1 μm. However, the tissues have small loss and large scattering, resulting in multiple scattering and diffusion. The resolution is poor, requiring the use of techniques such as photon density waves. For short distances, other techniques such as OCT can be used, which will be discussed later. For ultrasound of 1 MHz to 10 MHz and to 50 MHz, the wavelength (in water) is 1.5 mm, 0.15 mm, and 30 μm. Absorption is significant, but scattering is small. Therefore, the single scattering process is dominant, helping resolution and mathematical analysis. Multiple scattering by tissues are negligibly small for ultrasound. It may be noted that high-frequency ultrasound imaging up to 50 MHz is considered a new frontier in tissue imaging though the penetration depth may be 8–9 mm (Shung et al., 2009).

20.1 BIOELECTROMAGNETICS

Electromagnetic fields in the frequency range of 1 MHz to 100 GHz can be transmitted through a biological medium with varying degrees of absorption, reflection, and scattering. Reflection occurs at tissue boundaries, and scattering is caused by inhomogeneities of the order of a wavelength in tissues (Ishimaru, 1997; Lin, 2012; Johnson and Guy, 1972).

Microwave frequencies of 27.12, 915, and 2450 MHz are used for diathermy in the United States, and 433 MHz is used in Europe. Other medical applications of microwaves include warming of refrigerated bank blood to body temperature prior to transfusion, and selective heating of a cancer or tumor area for administration of anticancer drugs. Microwave ovens are typically operated at 2450 MHz at a normal power of 2 kW with a maximum of 5.25 kW. The frequency bands used for cellular phones cover 800 MHz to 2.6 GHz and their biological effects have been extensively studied (Lin, 2012).

The effect of microwaves on biological systems may be thermal or nonthermal. The power absorbed by the tissues produces a temperature rise that depends on the cooling and heat diffusion mechanisms of the tissues. When the thermoregulatory capability of the system is exceeded, tissue damage and death can result. The maximum recommended safe power density for long-term human exposures is 10 mW/cm^2 in the United States. Soviet scientists reported that the central nervous system is sensitive to microwaves at intensities below thermal thresholds and have set the safety power level at 0.01 mW/cm^2.

The dielectric constants of tissues with high water content, such as muscle and skin, have been investigated thoroughly (Johnson and Guy, 1972) and are shown in Fig. 20.1. The dielectric constants of tissues with low water content, such as fat and bone, are also shown in Fig. 20.1. The values of ε and σ vary with temperature at the rate of +2 and $-0.5\%/^\circ C$, respectively.

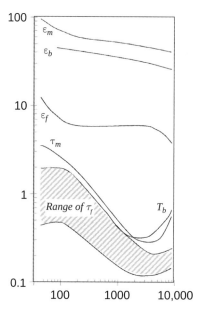

FIGURE 20.1   Dielectric constants ε_f, ε_m, and ε_b and loss tangent τ_f, τ_m, and τ_b for fat, muscle, and brain, respectively, as functions of frequency (Johnson and Guy, 1972).

A key quantity in bio-EM is "SAR," which is the power loss per unit mass of biological media when the incident power flux density is 1 mW/cm² (= 10 W/m²).

$$SAR = \frac{L}{\rho}(\text{watts/kg}),$$

where

$$L = \frac{\omega\sigma_0\varepsilon''}{2}|E|^2 = \frac{\sigma|E|^2}{2}. \tag{20.1}$$

L is the power dissipation per unit volume of the medium (W/ m³), E is the electric field (volt/m), σ is the conductivity (S/m) of the medium, and ρ is the density of the medium (kg/m³). The density ρ for SAR (20.1) is usually taken to be that of water ($\rho = 10^3$ kg/m³).

There has been extensive research on EM fields in biological media (Lin, 2012) covering extremely low-frequency waves (ELF) to radio frequency waves (RF) to microwaves.

20.2 BIO-EM AND HEAT DIFFUSION IN TISSUES

As tissues are illuminated by EM waves, some of the EM power is dissipated in the lossy medium and converted into heat energy. The heat energy is then diffused in the medium and the associated temperature varies. This section gives a brief introduction to the heat diffusion in a biological medium.

Let us consider a biological medium illuminated by an incident power flux of 1 mW/cm². The power loss L per unit volume can be calculated by finding the electric field E and knowing the conductivity σ (or ε'') of the biological medium. The power loss L is converted into heat, which raises the temperature. If the temperature varies from point to point in a medium, then the heat flow per unit area is proportional to the temperature gradient.

$$\bar{F} = -K\nabla T, \tag{20.2}$$

where K (W/m°C) is the thermal conductivity and $\bar{F}$ (W/m²) is the heat flux per unit area.

The internal energy per unit mass u (J/kg) in differential form is given by the specific heat C_s (J/kg°C) and the temperature ΔT.

$$du = C_s dT. \tag{20.3}$$

Note that energy is in joules J (Ws = Nm = m²kg/s²). The internal energy per unit volume ρu (J/m³) and the heat flow $\bar{F}$ satisfy the conservation of power; ρ is the density (kg/m³).

$$\rho C_s \frac{\partial T}{\partial t} + \nabla \cdot \bar{F} = q, \tag{20.4}$$

where q is the energy introduced or subtracted from the medium per unit volume (W/m³), which will be discussed later. Substituting (20.2) into (20.4), we finally get the diffusion equation for the temperature

$$\rho C_s \frac{\partial T}{\partial t} = \nabla \cdot (K\nabla T) + q. \tag{20.5}$$

This diffusion equation is often written as

$$\frac{\partial T}{\partial t} = a^2 \nabla^2 T + \frac{q}{\rho C_s}, \tag{20.6}$$

where a is called the "diffusion constant,"

$$a = \left[\frac{K}{\rho C_s} \right]^{1/2}. \tag{20.7}$$

In a biological medium, the energy is introduced by the EM power loss L, which is converted into heat, and the energy C, subtracted from the medium by blood flow. We can then write

$$q = L - C. \tag{20.8}$$

L is given in (20.1), and C is the cooling function and can be approximated by

$$C = h_b(T - T_b),\qquad(20.9)$$

where T is the tissue temperature, T_b is the average local arterial blood temperature, and h_b is the heat transfer coefficient.

The diffusion equation (20.6) with (20.8) is a simplified form of the bioheat equation developed by Pennes in (1948). More complete forms of the bioheat equation include the porosity of the tissue and blood velocity (Khaled and Vafai, 2003; Nakayama and Kuwahara, 2008). Van den Berg and others (1983) gave a computational model of EM heating of tissues including the cooling function. The cooling function (20.9) is a first-order approximation, which may be applicable when the temperature is below a critical temperature (Chan et al., 1973).

The solution of the diffusion equation (20.6) requires the boundary conditions. On the surface, the outward power flux must be proportional to the difference between the surface temperature and the environmental temperature T_e.

$$-K\frac{\partial T}{\partial n} = E(T - T_e).\qquad(20.10)$$

This is called the "Newton cooling law" (Pennes, 1948). This is consistent with the boundary conditions used by Ayappa et al. (1991) where the constant (E/K) is proportional to the Biot number (Chen and Peng, 2005). This is also consistent with that used by Taflove and Brodwin (1975). It should be noted, however, that the boundary condition for the diffusion equation is inherently inaccurate because it takes some distance from the boundary for any physical quantities such as waves or particles to diffuse.

Mathematically, we can have at the boundary

(1) Uniform temperature,

$$T = \text{constant.}\qquad(20.11)$$

(2) No heat transfer

$$\frac{\partial T}{\partial n} = 0.\qquad(20.12)$$

It may be added that light diffusion in a multiple scattering medium obeys the same diffusion equation. If the medium is an isotropic scatterer, the boundary condition is given by (Ishimaru, 1997)

$$U = 0.7104\, l_s \frac{\partial}{\partial n} U,\qquad(20.13)$$

where l_s is the scattering mean free path. U is the average intensity, which satisfies the diffusion equation (Ishimaru, 1997)

$$\frac{\partial U}{\partial t} = a^2 \nabla^2 U + Q.\qquad(20.14)$$

The average intensity U in a multiple scattering medium is therefore equivalent to the temperature.

The diffusion equation (20.6) needs to be solved with boundary conditions. It is, however, instructive to find the diffusion solution in an infinite medium. For an arbitrary excitation $(q/\rho c_s)$, we can use Green's function. For a one-dimensional diffusion equation, we have

$$\frac{\partial T}{\partial t} = a^2 \frac{\partial^2}{\partial x^2} T + \left(\frac{q}{\rho c_s}\right). \tag{20.15}$$

We let

$$\frac{q}{\rho c_s} - = f_1(t)f_2(x), \tag{20.16}$$

and write

$$T(t,x) = \int_0^t dt' \int_0^x dx' \, G(t - t', x - x') f_1(t') f_2(x'). \tag{20.17}$$

Green's function $G(t,x)$ satisfies

$$\frac{\partial}{\partial t} G = a^2 \frac{\partial^2}{\partial x^2} G + S(t)S(x). \tag{20.18}$$

We can take the Fourier transform in t and x, and obtain

$$\bar{G}(w, \lambda) = \frac{1}{iw - a^2\lambda^2} \tag{20.19}$$

Taking the inverse Fourier transform, we write

$$G(t,x) = \frac{1}{(2\pi)^2} \int e^{-iwt} dw \int e^{+i\lambda x} d\lambda \bar{G}(w, \lambda). \tag{20.20}$$

The integration with dw can be performed by noting the pole at $w_p = -ia^2\lambda^2$ and taking the residue. We then perform $\lambda-$ integration and obtain

$$G(t,x) = \frac{1}{\sqrt{4\pi a^2 t}} \exp\left(-\frac{x^2}{4a^2 t}\right). \tag{20.21}$$

This is the well-known diffusion solution.

For an actual problem, we need to apply the boundary condition (20.10) or approximations (20.11) or (20.12). Equation (20.11) is equivalent to the case of the surface temperature being equal to the surrounding temperature. If the medium is semi-infinite, and if we use this boundary condition, we get Green's function

$$G(t, x) = G(t, x - x_0)G(t, x + x_0). \tag{20.22}$$

Formulations in this section are concerned with the microwave heating of tissues, and the heat diffusion. This same technique can be used to study the heat diffusion and the temperature rise in objects illuminated by high-power optical beam. The CO_2 optical beam with varied pulse duration may be focused through atmosphere and incident upon copper, water, or composite targets, causing temperature rise (Stonebeck et al., 2013).

20.3 BIO-OPTICS, OPTICAL ABSORPTION AND SCATTERING IN BLOOD

Optical propagation in biological materials is dominated by scattering because the inhomogeneities of cellular structures and particle sizes are of the order of an optical wavelength. Cells are commonly several microns in diameter. Muscle cells may be a few millimeters long and nerve cells may be over a meter long. A cell consists of a thin membrane, $\approx 75\text{Å}$ thick, cytoplasm, and a nucleus. Epithelial tissues consist of cells in layered membranes that cover or line a surface and perform the functions of protection and regulation of secretion. Connective tissues support and connect cellular tissue to the skeleton. Muscle tissues consist of cells that are 1–40 mm long and up to 40 μm in diameter. Nervous tissues consist of nerve cells called neurons that transmit information between the central nervous system and muscle, organs, glands, etc. (Johnson and Guy, 1972, p. 709; Ishimaru, 1997).

Light has been used to determine the oxygen content in blood. Red blood cells (erythrocytes) have the shape of a biconcave disk with a broad diameter of about 7 μm and a thickness of about 1 μm in the center and about 2 μm near the edges. Erythrocytes are continuously formed in the bone marrow, fed into the bloodstream, and absorbed and reprocessed in the liver. Normally there are about 5×10^6 erythrocytes/mm^3. About 40% of the volume of whole blood consists of erythrocytes, and this percentage is called hematocrit H. Therefore, normal blood has $H = 0.4$. The remaining 60% is a nearly transparent solution of water and salt called plasma.

Red blood cells contain hemoglobin molecules Hb, which are easily oxygenated to oxyhemoglobin molecules HbO_2. Oxygen saturation OS is defined as the ratio of oxyhemoglobin $[HbO_2]$ to total hemoglobin $[HbO_2] + [Hb]$. In order to obtain the optical absorption characteristic of hemoglobin itself, the erythrocyte membrane is ruptured and the hemoglobin is released into solution. This solution is called hemolyzed blood and is a homogeneous absorbing medium. The decrease of light intensity dI in this medium is proportional to the intensity I and the elementary distance dz

$$dI = -\alpha I dz. \tag{20.23}$$

The intensity therefore decays exponentially

$$I(z) = I(0) \exp(-\alpha z). \tag{20.24}$$

This is called the Lambert–Beer law. The absorption constant α is dependent on the molecular concentration C (moles/cm^3) and the specific absorption coefficient κ(cm^2/mole)

$$\alpha = C\kappa. \tag{20.25}$$

The specific absorption coefficients κ_h and κ_o of hemoglobin Hb and oxyhemoglobin HbO$_2$ of hemolyzed blood are shown in Fig. 20.2. Oxyhemoglobin HbO$_2$ has low absorption in the red region of the spectrum and thus blood looks red when oxy-hemoglobin is predominant. The absorption of Hb and that of HbO$_2$ are equal at $\lambda = 0.548$, 0.568, 0.587, and 0.805 μm. These wavelengths are called isosbestic points.

Consider a light beam sent through a slab of thickness D. Transmittance T and reflectance R are defined as the ratios of transmitted to incident intensity and reflected to incident intensity, respectively. The optical density OD is defined as

$$OD = \log(T^{-1}). \tag{20.26}$$

The optical density of hemolyzed blood with oxygen saturation OS is given by

$$OD = 0.4343C\kappa d, \tag{20.27}$$

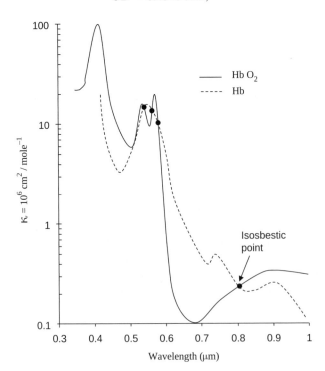

FIGURE 20.2   Optical absorption spectrum of hemoglobin (H_b) and oxyhemoglobin (H_bO_2).

TABLE 20.1   Optical Properties of Tissues In Vivo (Tuchin and Thompson, 1994)

	λ nm	μ_t cm^{-1}	μ_a cm^{-1}	μ_s cm^{-1}	$\mu_s(1-g)$ cm^{-1}	g
Aorta	632.8	316	0.52	316	41.0	0.87
Whole blood	665	1247	1.3	1246	6.11	0.995
Lung	635	332	8.1	324	81.0	0.75

where $C\kappa$ is an average of κ_o and κ_h in proportion to their concentrations and is given by $C\kappa = C_o\kappa_o + C_h\kappa_h$. C_o and C_h are the concentrations of HbO$_2$ and Hb, respectively, and κ_o and κ_h are the specific absorption coefficients of HbO$_2$ and Hb, respectively. Noting that $OS = C_o/(C_o + C_h)$, we can also write

$$C = C_o + C_h, \quad \kappa = \kappa_h + OS(\kappa_o - \kappa_h). \tag{20.28}$$

By making measurements of the optical density OD of hemolyzed blood at two wavelengths and knowing κ_o and κ_h at these wavelengths from Fig. 20.2, we can determine the oxygen saturation OS.

The optical density of whole blood is significantly different from that of hemolyzed blood because in whole blood, hemoglobin is packaged in erythrocytes and considerable light scattering takes place. The scattering and absorption characteristics of erythrocytes (red blood cells) are most conveniently represented by cross sections. Typical cross sections are shown in Table 20.1.

The hematocrit H is related to the number density ρ and the volume V_e of a single erythrocyte

$$\rho = H/V_e. \tag{20.29}$$

Therefore the absorption coefficient $\rho\sigma_a$ is given by

$$\mu_a = \rho\sigma_a = H\sigma_a/V_e. \tag{20.30}$$

The scattering coefficient $\rho\sigma_s$ is also given by

$$\mu_s = \rho\sigma_s = H\sigma_s/V_e, \tag{20.31}$$

if H is sufficiently small ($H < 0.2$). For $H > 0.5$, the particles are densely packed and the medium becomes almost homogeneous with absorbing hemoglobin material. In this case, the whole blood may be viewed as a homogeneous hemoglobin medium with the scattering particles made of the plasma between red blood cells. In the limit $H \to 1$, the "plasma particles" disappear and $\rho\sigma_s$ should approach zero. This consideration leads to the following approximate representation of $\rho\sigma_s$

$$\mu_s = \rho\sigma_s = H(1-H)\sigma_s/V_e, \tag{20.32}$$

where the multiplying factor (1-H) takes care of the disappearance of scattering as $H \to 1$. Often complete packing ($H = 1$) cannot be reached and the effect of packing cannot be expressed by a simple factor such as (1-H). For example, if the particles are rigid spheres, H cannot exceed 0.64. We should then write

$$\mu_s = \rho\sigma_s = (H\sigma_s/V_e)f(H),\qquad(20.33)$$

where $f(H)$ should monotonically decrease from 1 at $H = 0$ to 0 at a certain value of H. The function $f(H)$, called the "packing factor," for rigid spheres is given by the Percus–Yevick packing factor

$$f(H) = \frac{(1-H)^4}{(1+2H)^2}.\qquad(20.34)$$

20.4   OPTICAL DIFFUSION IN TISSUES

As noted in Section 20.3, the optical propagation in whole blood depends on not only the absorption and scattering characteristics of a single blood cell, but how much the particles are packed. This is given by the fractional volume, H, which is the ratio of the volume occupied by the particles to the total volume. If H is much smaller than one, the scattering is small and single scattering takes place. As H increases, double scattering and multiple scattering need to be considered. As H increases further, the wave scatters in all directions and its behaviors approach "random walk," and this is equivalent to the "diffusion" (Fig. 20.3).

In this section, we discuss diffusion approximation (Ishimaru, 1997, Section 7.3 and Chapter 9). The transport of optical energy in tissue and blood is given by the radiative transfer equation in Section 17.4. In terms of the specific intensity $I(\bar{r}, \hat{s})$,

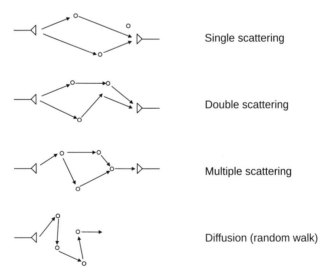

Single scattering

Double scattering

Multiple scattering

Diffusion (random walk)

FIGURE 20.3   Single, double, and multiple scattering and diffusion.

we write the equation of transfer.

$$\frac{dI}{ds} = -\mu_t I + \frac{\mu_t}{4\pi} \int p(\hat{s}, \hat{s}') I(\hat{s}') d\Lambda', \tag{20.35}$$

where $\mu_t = \rho \sigma_t$ is the extinction coefficient, $p(\hat{s}, \hat{s}')$ is the phase function, and ρ is the number density (#/vol) and σ_t is the extinction cross section of a single particle.

The specific intensity I is given by the coherent component called the "reduced incident intensity" I_{ri} and the diffuse component I_d. I_{ri} satisfies the equation

$$\frac{dI_{ri}}{ds} = -\mu_t I_{ri}. \tag{20.36}$$

The diffuse component I_d satisfies the equation

$$\frac{dI_d}{ds} = -\mu_t I_d + \frac{\mu_t}{4\pi} \int p(\hat{s}, \hat{s}') I'_d d\Omega' + \Sigma_{ri}(\hat{r}, \hat{s}), \tag{20.37}$$

where Σ_{ri} = source function is given by

$$\Sigma_{ri}(\bar{r}, \hat{s}) = \frac{\mu_t}{4\pi} \int p(\hat{s}, \hat{s}') I_{ri}(\bar{r}, \hat{s}) d\Omega'. \tag{20.38}$$

The phase function $p(\hat{s}, \hat{s}')$ is related to the scattering amplitude $f(\hat{s}, \hat{s})$ of a single particle as shown in Section (10.1).

$$|f(\hat{s}, \hat{s}')|^2 = \frac{\sigma_t}{4\pi} p(\hat{s}, \hat{s}'), \tag{20.39}$$

$\frac{1}{4\pi} \int p(\hat{s}, \hat{s}') d\Omega = W_o = \frac{\sigma_s}{\sigma_t} =$ albedo of a single particle.

We also note that

$$\mu_t = \rho \sigma_t = \frac{1}{l_t} = \text{extinction coefficient},$$

$$\mu_a = \rho \sigma_a = \frac{1}{l_a} = \text{absorption coefficient},$$

$$\mu_s = \rho \sigma_s = \frac{1}{l_s} = \text{scattering coefficient},$$

$$\mu_{tr} = \rho \sigma_{tr} = \frac{1}{l_{tr}} = \text{transport coefficient},$$

l_t, l_a, l_s, and l_{tr} are extinction, absorption, scattering, and transport mean free path, respectively. The transport coefficient will be explained further in (20.50) together with g (anisotropy factor or the mean cosine of the scattering angle).

As shown in Ishimaru (1997, Chapter 9), the diffusion approximation is based on the assumption that the diffuse intensity encounters many particles and is scattered

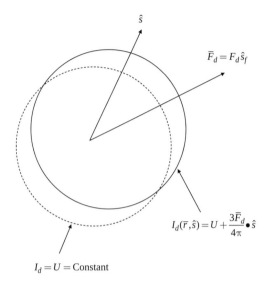

$$\bar{F}_d = F_d \hat{s}_f$$

$$I_d(\bar{r}, \hat{s}) = U + \frac{3\bar{F}_d}{4\pi} \bullet \hat{s}$$

$$I_d = U = \text{Constant}$$

FIGURE 20.4   Diffusion intensity $I_d(\bar{r}, \hat{s})$ for diffusion approximation.

almost uniformly in all directions. The diffuse intensity has a magnitude in the slightly more forward direction giving the net forward flux (Fig. 20.4). This is described by the following formula (Ishimaru, 1997):

$$I_d(\bar{r}, \hat{s}) = U_d(\bar{r}) + \left(\frac{3}{4\pi}\right) \bar{F}_d(\bar{r}) \bullet \hat{s}, \tag{20.40}$$

where $U_d = \frac{1}{4\pi} \int I_d(\bar{r}, \hat{s}) d\Omega$ is the average diffuse intensity, which is also called the "light fluence rate" (Watt/m^2).

First, we integrate (20.37) over 4π and obtain the following equation of power conservation.

$$\nabla \bullet \bar{F}_d = -4\pi\mu_a U_d(\bar{r}) + 4\pi\mu_s U_{ri} + E(\bar{r}),$$

where

$$E = \int_{4\pi} E(\bar{r}, \hat{s}) d\Omega. \tag{20.41}$$

$\bar{F}_d$ is the flux vector given by

$$\bar{F}_d(\bar{r}, \hat{s}_f) = \int_{4\pi} I_d(\bar{r}, \hat{s}) \hat{s} \bullet \hat{s}_f d\Omega,$$

and

$$U_{ri} = \frac{1}{4\pi} \int_{4\pi} I_{ri}(\bar{r}, \hat{s}) d\Omega. \tag{20.42}$$

Next, we substitute (20.40) into (20.37) and obtain

$$grad U_d = -\frac{3}{4\pi}\mu_{tr}\bar{F}_d + \frac{3}{4\pi}\int_{4\pi}E_{ri}(\bar{r},\hat{s})\hat{s}d\Omega \qquad (20.43)$$

We can now combine (20.41) and (20.43) and obtain

$$\left(\nabla^2 + K_d^2\right)U_d = -3\mu_s\mu_{tr}U_{ri} + \frac{3}{4\pi}\nabla\int E_{ri}\hat{s}d\Omega,$$

$$K_d^2 = -3\mu_a\mu_{tr}, \qquad (20.44)$$

where σ_{tr} is called the "transport cross section" and is given by

$$\sigma_{tr} = \sigma_t - g\sigma_s,$$

$$g = \frac{\displaystyle\int_{4\pi}p(\hat{s},\hat{s}')\cos\theta d\Omega'}{\displaystyle\int_{4\pi}p(\hat{s},\hat{s}')d\Omega'}. \qquad (20.45)$$

This process of first obtaining $\nabla \bullet \bar{F}$ and then expressing $\bar{F}$ as a gradient of U and combining these two to obtain the fundamental equation (20.44) is commonly used in many physical problems.

The transport cross section σ_{tr} shows that if scattering is not isotropic, the equivalent scattering cross section and the attenuation due to scattering are reduced by a factor g, which is called the "anisotropy factor" or "the mean cosine of the scattering angle."

Note that K_d^2 is negative and therefore K_d is purely imaginary giving a diffusion characteristic.

The exact boundary condition for the diffuse intensity I_d is that at the surface there should be no diffuse intensity entering the medium from outside. However, the diffusion approximation is based on the approximation (20.40) and therefore the exact boundary condition cannot be satisfied. One approximate boundary condition is that the total diffuse flux directed inward must be zero. This can be expressed as (Ishimaru, 1997, Section 9.2)

$$U_d(\bar{r}_s) - \frac{2}{3\rho\sigma_{tr}}\frac{\partial}{\partial n}U_d(\bar{r}_s) + \frac{2\hat{n}\bullet Q_1(\bar{r}_s)}{4\pi} = 0,$$

where $\qquad (20.46)$

$$Q_1(\bar{r}) = \frac{\sigma_t}{\sigma_{tr}}\int_{4\pi}d\Omega'\left[\frac{1}{4\pi}\int_{4\pi}p(\hat{s},\hat{s}')\hat{s}d\Omega\right]I_{ri}(\bar{r},\hat{s}').$$

The diffusion equation and the boundary conditions for a plane wave incident on a slab of scatterers, a pencil beam incident on a slab, and a point source inside the diffuse medium have been investigated and applied to optical diffusion in tissues and development of oximeter using optical fibers (Ishimaru, 1997).

20.5 PHOTON DENSITY WAVES

In Section 20.4, we discussed the diffusion approximation of the radiative transfer theory. In the diffusion approximation, the specific intensity has a broad angular spectrum and therefore the diffused wave cannot be focused into a small volume. There is a way to focus the diffused wave, but the focusing requires constructive interference near the focal point and destructive interference elsewhere. This is a wave-like behavior including phase. The diffusion is the diffusion of the intensity, not a field with amplitude and phase. There is a way to include the phase information in the diffusion approximation of the intensity. If the intensity consists of a constant background plus a modulated intensity, so that the total intensity is always positive, then the modulated intensity behaves as if it is a wave (Ishimaru, 1978; Jaruwatanadilok et al., 2002).

Let us write the total specific intensity as

$$I(\bar{r}, t) = I_c(\bar{r}) + I_m(\bar{r})e^{-i\omega_m t}, \tag{20.47}$$

where ω_m is the modulation frequency. Let us write the time-dependent radiative transfer equation for the component I_m.

$$\frac{dI_m}{ds} = -\mu_t I_m - \frac{1}{c_b}\frac{\partial}{\partial t}I_m + \frac{\mu_t}{4\pi}\int p(\hat{s}, \hat{s}')I_m(\hat{s}')d\Omega', \tag{20.48}$$

where c_b is the light velocity in the background medium. Writing the time dependence of $\exp(-i\omega_m t)$, we replace $\frac{\partial}{\partial t}$ in (20.48) by $(-i\omega_m)$. We then follow the procedure in Section 20.4 of obtaining $\nabla \bullet \bar{F}_d$ and $grad\ U_d$ and combining these two. We get the following diffusion equation for the modulation part of the intensity.

$$\left(\nabla^2 + K_d^2\right)U = 0,$$

where

$$K_d^2 = -3\mu_a\mu_{tr} + i\frac{\omega_m}{c_b}3\mu_{tr} + \frac{3}{c_b^2}\omega_m^2. \tag{20.49}$$

This is consistent with that given by Boas et al. (1994) except the term $\frac{3}{c_b^2}\omega_m^2$ representing the wave propagation. For most biological applications, this term is small and therefore we have

$$K_d^2 = -3\mu_a\mu_{tr} + i\frac{\omega_m}{c_b}3\mu_{tr}. \tag{20.50}$$

It may be noted that the diffusion equation (20.49) is similar to the equation for the thermal diffusion for temperature T.

$$\left(\nabla^2 - \frac{1}{D}\frac{\partial}{\partial t}\right)T = 0. \tag{20.51}$$

It is also similar to the equation for electromagnetic waves in a conducting medium.

$$\left[\nabla^2 - \mu\varepsilon\frac{\partial^2}{\partial t^2} - \mu\sigma\frac{\partial}{\partial t}\right]E = 0. \tag{20.52}$$

Thus, the diffusion of the photon density wave is similar to the waves in (20.51) and (20.52).

It may be instructive to examine the case of small absorption. If we assume that μ_a is negligibly small, we have

$$\left(\nabla^2 + i\frac{\omega_m}{c_b}3\mu_{tr}\right)U = 0. \tag{20.53}$$

This is identical to the wave in conductive medium with large conductivity. We then have

$$U = \exp(iK_d x),$$

where

$$K_d = \frac{1}{\sqrt{2}}(1+i)\left[\left(\frac{2\pi}{\lambda_m}\right)3\mu_{tr}\right]^{1/2}, \quad \frac{\omega_m}{c_b} = \frac{2\pi}{\lambda_b}. \tag{20.54}$$

We see that the attenuation constant α_d and the phase constant β_d are the same

$$U = \exp(-\alpha_d x + i\beta_d)x,$$

$$\alpha_d = \beta_d = \frac{1}{\sqrt{2}}\left[\left(\frac{2\pi}{\lambda_b}\right)3\mu_{tr}\right]^{1/2}. \tag{20.55}$$

Note that λ_b and c_b are the wavelength and the light velocity in the background medium. The modulated intensity I_m propagates with the velocity v_m.

$$v_m = \frac{\omega_m}{\beta_d} = \left[\frac{2\omega_m c_b}{3\mu_{tr}}\right]^{1/2} = \left[\frac{2\omega_m}{c_b 3\mu_{tr}}\right]^{1/2}c_b, \tag{20.56}$$

which can be much slower than the medium speed c_b.

And the equivalent wavelength of the modulated wave is

$$\lambda_m = \frac{2\pi}{\beta_d} = \left[\frac{4\pi}{\lambda_b 3\mu_{tr}} \right]^{1/2} \lambda_b, \qquad (20.57)$$

which can be considerably smaller than the background wavelength λ_b, giving a better resolution with the modulated intensity. For example, using 200 MHz for aorta, we have approximately $\mu_{tr} \approx 41$, and λ_b(in water) = 2.14×10^8 m/s. We then get $\lambda_m \approx 3.3$ cm. However, this is based on an approximation (20.53). More accurate estimates should be obtained by (20.49).

20.6 OPTICAL COHERENCE TOMOGRAPHY AND LOW COHERENCE INTERFEROMETRY

OCT was introduced to obtain noninvasive images of biological images of biological tissues with resolutions of 1–15 μm (Fujimoto, 2001; Gabriele, et al., 2011; Huang et al., 1991). The images are generated by measuring the echo time delay and the light intensity backscattered from tissues. The echo time delay is measured by using low coherence light. The backscattered light from the sample is interfered with low coherence light. The interference occurs only when the two path lengths match to within the coherence length. By varying the path length of the low coherence light, the echo time delay can be varied. This is equivalent to short pulses but it uses continuous-wave light without the need for ultrashort pulses. A sketch of the schematic of OCT is shown in Fig. 20.5.

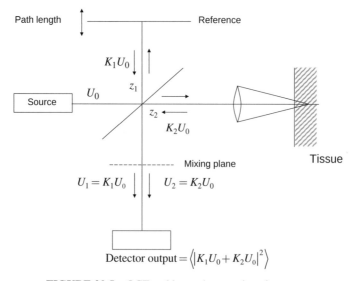

FIGURE 20.5 OCT and low coherence interferometry.

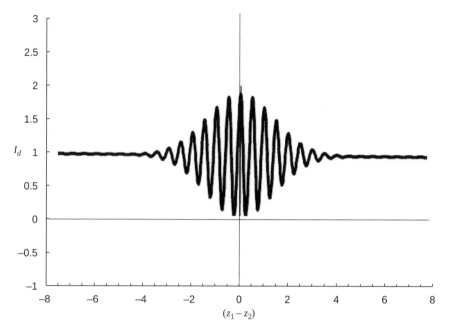

FIGURE 20.6   Intensity incident on the detector.

In this section, we discuss a theory of OCT following Schmitt and Knuttel (1997), Thrane et al. (2000), Ishimaru (1997, 2012), and Ralston et al. (2006). The detector produces the heterodyne photo current i_s (Fig. 20.6). The amplitude of the photo current is proportional to the correlation of the reference beam U_r and the sample beam U_s (Schmitt and Knittel, 1997), and is given by

$$i_s = (2\zeta q) \int_A U_r(\bar{\rho}, \omega) U_s^*(\bar{\rho}, \omega) d\bar{\rho}, \qquad (20.58)$$

where ζ is the quantum efficiency and q is the electron charge (Fig. 20.5). A is an aperture area at $z = 0$. Here we start with frequency domain. Time-domain expression is obtained later by Fourier transform.

We first discuss the axial resolution. There are many papers and books written on OCT including Izatt and Choma (2008) and Schmitt (1999). As shown in Fig. 20.5, the wave U_0 is incident on a Michelson interferometer. The incident wave is from a low coherent source and a polychromatic wave. We express U_0 in the analytic signal shown in Section 4.10. (Goodman, 1985).

$$U_0(t) = u_0(t) \exp(-i\omega_0 t), \qquad (20.59)$$

where u_0 is the complex envelope and ω_0 is the center frequency (carrier frequency). It is assumed that this is a low coherence wave and that the complex envelope u_0 is substantially constant within the coherence time τ_c. The mutual coherence function $\Gamma(\tau)$ of the incident wave is given by

$$\Gamma(\tau) = \langle u_0(t + \tau)u_0^*(t) \rangle. \tag{20.60}$$

The normalized spectrum $S(\omega)$ is then given by

$$S(\omega) = \frac{\displaystyle\int_{-\infty}^{\infty} \Gamma(\tau)e^{+i\omega\tau}d\tau}{\displaystyle\int_{-\infty}^{\infty} \Gamma(\tau)d\tau}. \tag{20.61}$$

If we assume that $S(\omega)$ is a Gaussian spectrum, we have

$$S(\omega) = \left(\frac{\sqrt{\pi 4}\sqrt{\ln 2}}{\Delta\omega} \right) \exp\left(\frac{-4\ln 2(\omega - \omega_0)^2}{\Delta\omega^2} \right), \tag{20.62}$$

where $S(\omega)$ is normalized so that

$$\frac{1}{2\pi} \int S(\omega)d\omega = 1. \tag{20.63}$$

Note that $\Delta\omega$ is the half-power bandwidth such that

$$S\left(\omega_0 \pm \frac{\omega}{2} \right) = \frac{1}{2}S(\omega_0). \tag{20.64}$$

The self-coherence function $\Gamma(\tau)$ is related to the spectrum.

$$\Gamma(\tau) = \frac{1}{2\pi} \int S(\omega)e^{-i\omega\tau}d\omega. \tag{20.65}$$

For the Gaussian spectrum (20.62), we have

$$\Gamma(\tau) = \exp\left(-\frac{\tau^2\Delta\omega^2}{4\cdot 4\ln 2} - i\omega_0\tau \right) = \exp\left(-\left(\frac{\pi}{2}\right)\frac{\tau^2}{\tau_c^2} - i\omega_0\tau \right),$$

where the coherence time τ_c is defined by (Goodman, 1985)

$$\tau_c = \int_{-\infty}^{\infty} |\Gamma(\tau)|^2 \, d\tau. \tag{20.66}$$

For the Gaussian spectrum, we have

$$\tau_c = \sqrt{\frac{2 \ln 2}{\pi}} \frac{1}{\Delta\nu} = \frac{0.664}{\Delta\nu}, \quad \Delta\nu = \frac{1}{2\pi} \Delta\omega_0. \tag{20.67}$$

The coherence length l_c is then given by

$$l_c = c\tau_c, \tag{20.68}$$

which can be written as

$$\frac{l_c}{\lambda_0} = \sqrt{\frac{2 \ln 2}{\pi}} \left(\frac{\lambda_0}{\Delta\lambda_0} \right) = 0.664 \left(\frac{\lambda_0}{\Delta\lambda_0} \right). \tag{20.69}$$

A typical OCT coherence source is a light emitting diode (Schmitt, 1999) with center wavelength $\lambda_0 = 1300$ nm and bandwidth $\Delta\lambda_0 = 50\text{--}100$ nm which gives the coherence length $l_c \approx 22 \sim 11$ μm.

The incident field U_0 with coherence time (20.67) and the coherence length l_c (20.68) is now incident on the Michelson interferometer shown in Fig. 20.5. The incident wave $U_0(t)$ propagates and splits into U_1 and U_2 in Fig. 20.5. U_1 travels to the reference mirror and is reflected back to become $K_1 U_0$, and U_2 is reflected back by the sample and becomes $K_2 U_0$. K_1 and K_2 include the effects of the beam splitter and reflections. The total $U_1 + U_2$ is incident on the detector. The output from the detector I_d is then given by

$$\begin{aligned} I_d &= \left\langle |K_1 U_0 + K_2 U_0|^2 \right\rangle \\ &= |U_0|^2 \left(K_1^2 + K_2^2 + 2\text{Re}(K_1 K_2)\Gamma_{12} \right), \end{aligned} \tag{20.70}$$

where Γ_{12} is the mutual coherence function.

Let us examine Γ_{12}. Note that U_1 traveled a distance of $2z_1$ while U_2 traveled a distance of $2z_2$. Therefore we have, except for the common distance from the beam splitter to the detector,

$$\begin{aligned} U_1 &= U_0 e^{ik(2z_1)}, \\ U_2 &= U_0 e^{ik(2z_2)}. \end{aligned} \tag{20.71}$$

We then have

$$\Gamma_{12} = \left\langle |U_0|^2 e^{ik2(z_1 - z_2)} \right\rangle, \quad k = (\omega/c). \tag{20.72}$$

The detected output I_d is then given by

$$I_d = |U_0|^2 \left[K_1^2 + K_2^2 + 2K_1 K_2 \cos(k2(z_1 - z_2)) \right]. \tag{20.73}$$

This is sketched in Fig. 20.6. Note that in the neighborhood of $k = k_0$, the peak appears when $k2(z_1 - z_2) \approx k_0 2(z_1 - z_2)$ with $k \approx k_0$ as multiples of 2π and the first minimum is when the path length difference $|z_1 - z_2|$ is $\lambda_0/4$. Note that the term with $k_1^2 + k_2^2$ is the DC term and the second term represents the cross correlation between two beams U_1 and U_2 which is the desired component. The main peak occurs when $|z_1 - z_2|$ and I_d diminishes as $|z_1 - z_2|$ increases. In particular, when $|z_1 - z_2|$ exceeds the coherence length l_c, then I_d becomes negligibly small. Therefore the axial resolution is given by the path difference $|z_1 - z_2| \approx l_c$.

Let us next consider the transverse resolution. If the aperture with the diameter $2W_0$ transmits a beam wave focused at R_f, the transverse spot size and the focal plane $(z = R_f)$ is given by (Section 6.6)

$$W_s = \frac{\lambda R_f}{\pi W_0}. \tag{20.74}$$

The axial spot size (resolution) is given by the coherence length l_c (20.67)

$$\Delta z = l_c = 0.664 \left(\frac{\lambda_0^2}{\Delta \lambda_0} \right). \tag{20.75}$$

In addition, the depth of focus Δz_d is given by

$$\Delta z_d = \frac{\lambda R_f^2}{\pi W_0^2}. \tag{20.76}$$

This is obtained by noting that the intensity of a beam wave on axis is given by (6.66)

$$I = I_0 \frac{W_0^2}{W^2} \tag{20.77}$$

and Δz_d is obtained by finding the distance at which the beam intensity becomes $(1/2)$ of the intensity at the focal point.

$$I(z = R_f + \Delta z) = \frac{1}{2} I(Z = R_f). \tag{20.78}$$

If $R_f \gg W_0$, then Δz_d may become greater than $\Delta z = l_c$ and the coherence length determines the axial resolution (Fig. 20.7).

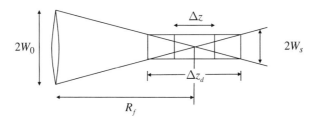

FIGURE 20.7   Focused beam spot size W_0, axial spot size Δz, depth of focus Δz_d.

20.7   ULTRASOUND SCATTERING AND IMAGING OF TISSUES

In Sections 20.1 through 20.6, we discussed propagation and scattering characteristics of electromagnetic and optical waves in biological media and their applications to imaging of the biological medium. Ultrasound imaging has advantages over other electromagnetic and optical imaging. It is noninvasive and relatively inexpensive to collect anatomical and blood flow information. However, there are ultrasound speckles or texture that are not well understood and therefore it is necessary to study the scattering phenomena for proper interpretation and development of ultrasound imaging. In this section, we start with the basic formulation of the ultrasound scattering theory (Ishimaru, 1997; Ishimaru, 2001; Shung and Thieme, 1993; Shung et al., 2009).

Acoustic scattering was discussed in Section 10.11 and the radar equation in Section 10.2 can be used to describe the scattered power from tissues illuminated by ultrasound.

For ultrasound, tissues can be considered a "random continuum," which means that the density ρ and the compressibility κ are continuous random functions of position. Under this assumption, we first obtain the scattering cross section per unit volume of the tissue.

Consider a volume δv of the tissues with density ρ_e and compressibility κ_e which are different from the surrounding average density ρ and compressibility κ. Under the assumption that the medium ρ_c and κ_c are only slightly different from ρ and κ, we can use the Born approximation to obtain the following well-known formula for the scattering amplitude:

$$f(\hat{o},\hat{i}) = \frac{k^2}{4\pi}\int_{\delta v}(\gamma_\kappa+\gamma_\rho\cos\theta)e^{i\bar{k}_s\cdot\bar{r}'}dv',$$

where

$$\gamma_\kappa = \frac{\kappa_e-\kappa}{\kappa} = \text{compressibility fluctuation},$$

$$\gamma_\rho = \frac{\rho_e-\rho}{\rho_e} = \text{density fluctuation}.$$

(20.79)

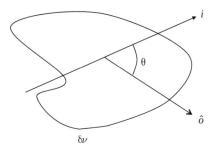

FIGURE 20.8   Incident wave is propagating in the direction of $\hat{i}$ and the scattered wave is observed in the direction of $\hat{o}$.

We then obtain the differential scattering coefficient σ_d or the differential cross section per unit volume of the tissue (Fig. 20.8).

$$\sigma_d(\hat{o}, \hat{i}) = \frac{\langle ff^* \rangle}{\delta v} = \left(\frac{k^2}{4\pi}\right)^2 \frac{1}{\delta v} \int \int \langle \gamma(\bar{r}_1)\gamma(\bar{r}_2) \rangle e^{i\bar{k}_s \cdot (\bar{r}_1 - \bar{r}_2)} dv_1 dv_2,$$

where

$$\gamma(\bar{r}) = \gamma_\kappa(\bar{r}) + \gamma_\rho(\bar{r})\cos\theta,$$ (20.80)

and

$$\bar{k}_s = k(\hat{i} - \hat{o}).$$

If the medium is assumed to be statistically homogenous and isotropic, then the covariance $\langle \gamma(\bar{r}_1)\gamma(\bar{r}_2) \rangle$ is a function of the magnitude of the difference $|\bar{r}_1 - \bar{r}_2|$. Then, the double integral in (20.80) can be expressed as a Fourier transform of the covariance function $\langle \gamma(\bar{r}_1)\gamma(\bar{r}_2) \rangle$ which is called the spectral density as noted by the Wiener–Khinchin theorem.

We can express (20.80) using the spectral densities

$$S_\gamma(\bar{k}_s) = \frac{1}{(2\pi)^3} \int B_\gamma(\bar{r}_d) e^{i\bar{k}_s \cdot \bar{r}_d} dv_d,$$ (20.81)

and $B_\gamma(\bar{r}_d)$ is the correlation function given by

$$B_\gamma(\bar{r}_d) = \langle \gamma(\bar{r}_1)\gamma(\bar{r}_2) \rangle = B_\kappa(\bar{r}_d) + B_\rho(\bar{r}_d)\cos^2\theta + 2B_{\kappa\rho}(\bar{r}_d)\cos\theta.$$ (20.82)

We therefore have the expression for σ_d

$$\sigma_d(\hat{o}, \hat{i}) = \left(\frac{\pi}{2}\right) k^4 \left[S_\kappa(k_s) + S_\rho(k_s)\cos^2\theta + 2S_{\kappa\rho}\cos\theta\right].$$ (20.83)

The unit commonly used for σ_d (differential cross section per unit volume) of the tissue is $cm^2/(cm^3 sr) = cm^{-1} sr^{-1}$ where $sr =$ steradian (unit solid angle).

Tissues such as myocardium are often anisotropic. For example, they may be elongated in one direction. This can be expressed using a Gaussian correlation function as

$$B_K(\bar{r}_d) = \sigma_\kappa^2 \exp\left(-\frac{x_d^2}{l_1^2} - \frac{y_d^2}{l_2^2} - \frac{z_d^2}{l_3^2}\right). \tag{20.84}$$

We can also assume that

$$B_\rho(\bar{r}_d) \approx \frac{1}{2} B_\kappa(\bar{r}_d),$$
$$B_{\kappa\rho}(\bar{r}_d) \approx 0. \tag{20.85}$$

And typically, $\sigma_\kappa^2 \approx 10^{-4}$, $l_1 \approx l_2 \approx 30\,\mu m$, and $l_3 \approx 200\,\mu m$.

We then get

$$S_\kappa(\bar{k}_s) = \frac{\sigma_\kappa^2 l_1 l_2 l_3}{8\pi\sqrt{\pi}} \exp\left[-\frac{1}{4}\left(k_{s1}^2 l_1^2 + k_{s2}^2 l_2^2 + k_{s3}^2 l_3^2\right)\right],$$

where

$$k_{s1} = k(\sin\theta_i \cos\phi_i - \sin\theta_0 \cos\phi_0),$$
$$k_{s2} = k(\sin\theta_i \sin\phi_i - \sin\theta_0 \sin\phi_0),$$
$$k_{s3} = k(\cos\theta_i - \cos\theta_0).$$

$$\tag{20.86}$$

It is known that anisotropic tissues such as those shown above exhibit the double peaks in the scattering pattern.

The Gaussian spectrum (20.66) is often used since it is mathematically simple and includes the essential parameters σ_κ, l_1, l_2, and l_3. However, other spectra which may be more representative of the actual tissues have been proposed including fluid spheres, exponentials, and modified exponentials. Here, we add the following power-law spectrum.

$$S_\kappa(\bar{k}_s) = S_\kappa(0)\left[1 + (k_{s1}l_1)^2 + (k_{s2}l_2)^2 + (k_{s3}l_3)^2\right]^{-n/2}, \tag{20.87}$$

where k_{s1}, k_{s2}, and k_{s3} are given in (20.86), and n is called "spectral index."

If the spectral index n is 3, (20.87) reduces to the "Henyey–Greenstein" formula and if $n = 4$, it reduces to the spectrum for the exponential correlation function. In general, for the isotropic case, we write

$$B(r_d) = B(0)\frac{1}{2^{\nu-1}\Gamma(\nu)}\left(\frac{r_d}{l}\right)^{\nu}K_{\nu}\left(\frac{r}{l}\right),$$

and

$$S(K_s) = B(0)\frac{\Gamma\left(\nu+\frac{3}{2}\right)}{\pi\sqrt{\pi}\Gamma(\nu)}\frac{l^3}{\left(1+k_s^2 l^2\right)^{\nu+\frac{3}{2}}},$$

(20.88)

valid when $\nu > -3/2$.

Using the differential cross section of the elementary volume shown above, we express the radar equation giving the received power P_r due to the transmitted power P_t illuminating the tissue.

$$P_r = A_r\frac{\sigma_d}{R_2^2}S_i.$$

(20.89)

The acoustic attenuation coefficient per unit length α for tissues generally varies as f^n, where f is the frequency and n ranges from 1 to 2. The attenuation coefficients α of muscle, liver, kidney, brain, and fat are approximately proportional to the frequency f and are in the range $(0.5 \times 10^{-6}$ to $2 \times 10^{-6})f$ dB/cm. Typical values of the attenuation coefficient α, sound velocity c, and density ρ_0 are shown in Ishimaru (1997).

20.8 ULTRASOUND IN BLOOD

Extensive studies have been made on ultrasonic properties of blood. For the normal frequency range of a few hundred kilohertz to 10 MHz used in biological media, the wavelength is much greater than the size of red blood cells. Therefore the Rayleigh formula for a sphere with the same volume as that of a red blood cell should give a good approximation of the absorption and scattering characteristics. The scattering amplitude $f(\hat{0}, \hat{i})$ is given by

$$f(\hat{0}, \hat{i}) = \frac{k^2 a^3}{3}\left(\frac{\kappa_e - \kappa}{\kappa} + \frac{3\rho_e - 3\rho}{2\rho_e + \rho}\cos\theta\right),$$

(20.90)

where $k = 2\pi/\lambda$, λ is the wavelength in the surrounding medium (plasma), a the radius of the equivalent sphere, κ_e and ρ_e the adiabatic compressibility and density of the red blood cell, respectively, κ and ρ are those of the plasma, and θ is the angle between $\hat{0}$ and $\hat{i}$. For a normal blood cell, the volume is 87 μm^3, and the equivalent radius a = 2.75 μm. The compressibility and density of a red blood cell are $\kappa_e = 34.1 \times 10^{-12}$ cm^2/dyne and $\rho_e = 1.092$ g/cm^3 and those of plasma are $\kappa = 40.9 \times 10^{-12}$ cm^2/dyne and $\rho_e = 1.021$ g/cm^3.

The scattering cross section σ_s is given by

$$\frac{\sigma_s}{\pi a^2} = \frac{4(ka)^4}{9} \left| \left| \frac{\kappa_e - \kappa}{\kappa} \right|^2 + \frac{1}{3} \left| \frac{3\rho_e - 3\rho}{2\rho_e + \rho} \right|^2 \right|. \tag{20.91}$$

Using the values of κ_e, ρ_e, κ, ρ, and a, we get

$$\sigma_s = 0.47 \times 10^{-16} f^4 \text{cm}^2, \tag{20.92}$$

where f is the frequency measured in megahertz. The backscattering cross section is approximately given by $\sigma_b \cong 1.86\sigma_s$.

The absorption cross section σ_a is proportional to the frequency and is given by

$$\frac{\sigma_a}{\pi a^2} = \frac{4ka}{3} \text{Im} \left(\frac{\kappa_e - \kappa}{\kappa} + \frac{3\rho_e - 3\rho}{2\rho_e + \rho} \right), \tag{20.93}$$

where Im designates the "imaginary part of." The imaginary parts of κ_e, ρ_e, κ, and ρ are not known. However, it is known that the absorption cross section σ_a is much greater than the scattering cross section σ_s in the frequency range 0.1–10 MHz, and therefore the attenuation through blood is mainly due to the absorption, and not the scattering. It is also known that the attenuation of a plane wave in random scatterers is given by $\rho\sigma_a$ nepers per unit distance (Np/cm). The attenuation constant is proportional to the hematocrit H and the frequency f (in megahertz) and is approximately given by

$$\alpha = (5 - 7) \times 10^{-2} \ Hf \ \text{Np/cm} = 0.3 \ Hf \ \text{dB/cm}. \tag{20.94}$$

Since $\alpha = \rho\sigma_a = (H/V_e)\sigma_a$, we obtain the absorption cross section of a single erythrocyte

$$\sigma_a = 6 \times 10^{-12} f \ \text{cm}^2, \tag{20.95}$$

where f is measured in megahertz. The effect of viscosity of the plasma and the erythrocyte on the scattering and absorption characteristics may be significant (Ahuja, 1970; Ishimaru, 1997).

The differential scattering cross section per unit volume of the blood is therefore

$$\sigma(\hat{o}, \hat{i}) = \frac{Hf_P(H)}{V_e} |f(\hat{o}, \hat{i})|^2, \tag{20.96}$$

where H is the hematocrit (0.4 for human), V_e is the volume of the single cell ($4\pi a^3/3$), and $f_P(H)$ is the packing factor. The Percus–Yevick packing factor for hard spheres is often used as an approximation

$$f_p(H) = \frac{(1 - H)^4}{(1 + 2H)^2}. \tag{20.97}$$

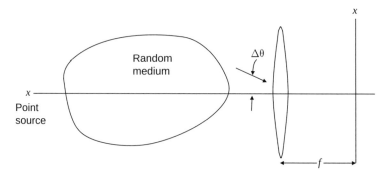

FIGURE 20.9 Wave propagated through random medium is incident on a focusing lens.

As a wave propagates through a random medium, the wave at any point is a mixture of coherent and incoherent waves. If we observe this wave with a lens or an array of detectors, we no longer obtain the Airy disk (Figs. 20.9 and 20.10). The coherent intensity P_c is the Airy disk with its magnitude diminished by the optical depth $\exp(-\tau_0)$. The incoherent intensity P_i is spread out due to the angular spread $\Delta\theta$, which is related to the correlation distance (coherence length) ρ_0 of the wave (Fig. 20.10).

$$\Delta\theta \sim \frac{\lambda}{\rho_0} \sim \frac{1}{k\rho_0}. \tag{20.98}$$

The coherence length ρ_0 is an important quantity not only giving the angular spread, but also the pulse spreads Δt

$$\Delta t \sim \frac{L}{C}\frac{\Delta\theta^2}{2} \sim \left(\frac{L}{C}\right)\frac{1}{2k^2\rho_0^2}, \tag{20.99}$$

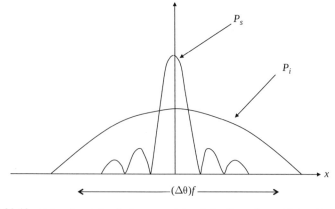

FIGURE 20.10 Intensity at focal plane consists of P_c, the coherent intensity, and P_i, the incoherent intensity.

where L is the propagation distance. Note that the coherence length ρ_0 is approximately given by Ishimaru (1997).

$$\rho_0 \sim \frac{l}{\sqrt{\tau_0}},$$

$l = $ correlation distance,

$\tau_0 = $ optical depth.

(20.100)

An overview of scattering and imaging of tissues is discussed by Ishimaru (2001) including beam propagation and scattering, pulse propagation, interface effects, Doppler shift due to blood cell motion, image resolution, and Wigner distribution.

CHAPTER 21

WAVES IN METAMATERIALS AND PLASMON

There has been an increasing interest in the development of new materials whose characteristics may not be found in nature. Metamaterials are an example of these materials. Meta means "beyond" indicating the material is beyond those found in nature. An example of metamaterials is artificial dielectrics, which have been developed and used since the 1940s. Artificial magnetic materials have magnetic properties not found in nature even though the materials are nonmagnetic. Split-ring resonators (SRRs) are a typical example. A man-made chiral material consisting of a collection of metal helices is another example. Another example that we will discuss later is bianisotropic media. These metamaterials have a broad range of applications including lenses, absorbers, antenna structures, composite materials, and frequency selective surfaces.

In 1968, Veselago published a paper describing the electrodynamics of "substances with simultaneously negative values of μ and ε" that showed some peculiar characteristics of waves, even though no physical material or devices were found having negative ε and μ until 1999, when Pendry and others proposed periodically stacked SRRs at microwave frequencies which exhibit simultaneous negative ε and μ.

Since that time, many investigations have been reported on related topics of perfect lenses, and potential applications in lenses, absorbers, antennas, optical and microwave components, and sensors. In this chapter, we discuss some of the fundamentals of metamaterials including transformation EM and cloaking. In this chapter, we use the convention $\exp(-i\omega t)$ commonly used in optics and acoustics, rather than $\exp(j\omega t)$ used in electrical engineering and by the IEEE, as there are many references

Electromagnetic Wave Propagation, Radiation, and Scattering: From Fundamentals to Applications,
Second Edition. Akira Ishimaru.
© 2017 by The Institute of Electrical and Electronic Engineers, Inc. Published 2017 by John Wiley & Sons, Inc.

in optics, acoustics, and physics on this topic. Extensive literature have been published including Solymar and Shamonina (2011), Eleftheriades and Balmain (2005), Sihvola (2007), Engheta and Ziolkowski (2006), and Caloz and Itoh (2006).

21.1   REFRACTIVE INDEX n AND μ–ε DIAGRAM

Most materials we encounter in practice have relative permittivity greater than one, except plasma whose relative permittivity is less than one, and relative permeability equal to one, except magnetic materials. Defining the relative permittivity and relative permeability, we write

$$
\begin{aligned}
\varepsilon &= \frac{\varepsilon}{\varepsilon_0} = \varepsilon_r' + i\varepsilon_r'', \\
\mu &= \frac{\mu}{\mu_0} = \mu_r' + i\mu_r''.
\end{aligned}
\tag{21.1}
$$

For a linear passive isotropic medium, we have

$$
\begin{aligned}
\varepsilon_r'' &= \operatorname{Im}\varepsilon_r > 0, \\
\mu_r'' &= \operatorname{Im}\mu_r > 0.
\end{aligned}
\tag{21.2}
$$

The ordinary dielectric material has

$$
1 < \varepsilon_r' < \infty, \quad \mu_r' = 1
\tag{21.3}
$$

and plasma has $\varepsilon_r' < 1, \mu_r' = 1$.

Note also that magnetic materials can be categorized by permeability as

$$
\begin{aligned}
\mu_r' &\le 1 \quad \text{diamagnetic,} \\
\mu_r' &\ge 1 \quad \text{paramagnetic,} \\
\mu_r' &\gg 1 \quad \text{ferromagnetic.}
\end{aligned}
\tag{21.4}
$$

The ordinary materials shown above are indicated in Fig. 21.1.

If we extend the regions to outside the ordinary materials, we can in principle cover all the regions of the μ_r'–ε_r' diagram. In particular, the region for simultaneously $\mu_r' < 0$ and $\varepsilon_r' < 0$, which Veselago noted is called negative index material (NIM), is also called left-handed medium (LHM), negative index of refraction medium (NIR) and double negative medium (DNG).

It should also be noted that the refractive index n is given by

$$
n = \sqrt{\varepsilon_r \mu_r}
\tag{21.5}
$$

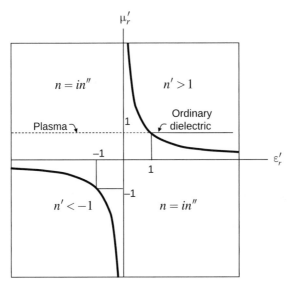

FIGURE 21.1 $\mu'_r - \varepsilon'_r$ diagram.

Therefore, we need to choose the correct sign for the square root. The choice should be made on the basis that with $\exp(-i\omega t)$ dependence (Fig. 21.2),

$$\text{Im}\, n = n'' > 0, \qquad (21.6)$$

and the characteristic impedance Z should be chosen such that

$$Z = \sqrt{\mu_r/\varepsilon_r} \quad \text{and} \quad \text{Re}\, Z > 0. \qquad (21.7)$$

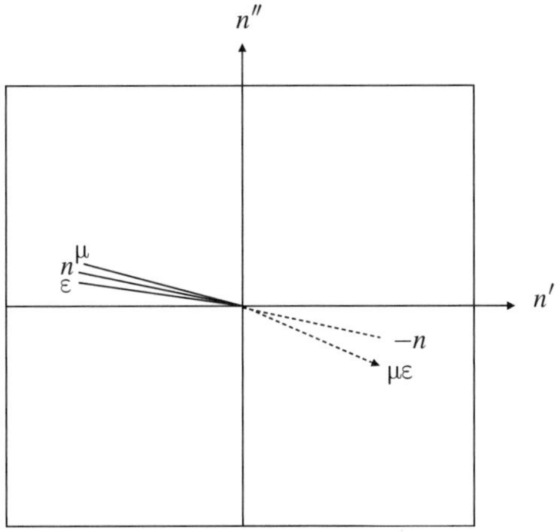

FIGURE 21.2   Refractive index $n = n' + in''$ for NIM showing n and $-n$. $\text{Im}\, n = n''$ needs to be positive.

Let us also note that as stated in Section 2.8, Maxwell's equations are unchanged if we interchange $\bar{E}$, $\bar{H}$, $\bar{J}$, $\bar{J}_m$, ρ, ρ_m, and ε and μ to the new primed fields as shown here.

$$\bar{E} \to \bar{H}' \quad \bar{J} \to \bar{J}'_m \quad \rho \to \rho'_m \quad \mu \to \varepsilon',$$
$$\bar{H} \to \bar{E}' \quad \bar{J}_m \to -\bar{J}' \quad \rho_m \to -\rho' \quad \varepsilon \to \mu'. \tag{21.8}$$

This means that we can change μ and ε to μ' and ε' if we also change $\bar{E}$, $\bar{H}$, $\bar{J}$, and ρ to the primed fields as shown in (21.8). Specifically, if we have TM solutions in the unprimed system with μ and ε, then we also have TE solutions with μ' and ε' with the above changes of the fields. This will be useful and important.

21.2 PLANE WAVES, ENERGY RELATIONS, AND GROUP VELOCITY

Let us consider the characteristics of a plane wave propagating in NIM. The wave is propagating in the z direction and we assume that $\bar{E} = E_x \hat{x}$ and $\bar{H} = H\hat{y}$. We then have

$$E_x = E_0 e^{ik_0 nz}$$
$$= E_0 e^{ik_0 n'z - k_0 n''z}, \tag{21.9}$$
$$Hy = \frac{E_x}{Z}, \quad Z = Z_0 \sqrt{\mu/\varepsilon},$$

where $n' < 0$ and $k_0 n' < 0$ for NIM.

The Poynting vector $\bar{S}$ is given by

$$\bar{S} = \mathrm{Re} \left(\frac{1}{2} E x H^* \right)$$
$$= \mathrm{Re} \left(\frac{1}{2} \frac{|E_x|^2}{Z^*} \right) \hat{z}. \tag{21.10}$$

Noting that $\mathrm{Re}\, Z > 0$, we have

$$\mathrm{Re} \left(\frac{1}{Z^*} \right) = \mathrm{Re} \left(\frac{Z}{ZZ^*} \right) > 0 \tag{21.11}$$

and therefore $\bar{S}$, is pointed in the $\hat{z}$ direction.

However, $\bar{K}_r = k_0 n' \hat{z}$ is pointed in the $(-\hat{z})$ direction. Therefore, the phase velocity is in the direction $(-\hat{z})$ opposite to the Poynting vector $(\hat{z})$.

NIM is known to be lossy and dispersive. Let us examine the time-averaged stored energy. For a dispersive medium such as NIM, the stored energy is given by (Section 2.5)

$$W = \frac{\varepsilon_0}{4} \frac{\partial}{\partial \omega} (\omega \varepsilon_r) |E|^2 + \frac{\mu_0}{4} \frac{\partial}{\partial \omega} (\omega \mu_r) |H|^2. \tag{21.12}$$

Note that if

$$\frac{\partial}{\partial \omega} (\omega \varepsilon_r) > 0 \quad \text{and} \quad \frac{\partial}{\partial \omega} (\omega \mu_r) > 0, \tag{21.13}$$

then W becomes positive.

If NIM were nondispersive, we would have

$$W = \frac{\varepsilon_0}{4} \varepsilon_r |E|^2 + \frac{\mu_0}{4} \mu_r |H|^2. \tag{21.14}$$

This becomes negative if $\varepsilon_r < 0$ and $\mu_r < 0$ and is not physical.

Group index of refraction for lossless NIM with $\varepsilon_r < 0$ and $\mu_r < 0$ is given by

$$
\begin{aligned}
n_g &= \frac{\partial}{\partial \omega} (\omega n) = \frac{c}{v_g} \\
&= \frac{n}{2} \left[\frac{1}{\varepsilon_r} \frac{\partial}{\partial \omega} (\omega \varepsilon_r) + \frac{1}{\mu_r} \frac{\partial}{\partial \omega} (\omega \mu_r) \right],
\end{aligned} \tag{21.15}
$$

v_g = group velocity.

For NIM, $n < 0$, $\varepsilon_r < 0$, $\mu_r < 0$, $\frac{\partial}{\partial \omega} (\omega \varepsilon_r) > 0$, and $\frac{\partial}{\partial \omega} (\omega \mu_r) > 0$, therefore

$$n_g > 0 \quad \text{and} \quad v_g > 0. \tag{21.16}$$

21.3 SPLIT-RING RESONATORS

As we already discussed, Veselago's investigation of materials with simultaneously negative values of ε and μ in 1968 did not attract much attention because no material existed with these characteristics at that time. This changed when Pendry showed in 1999 that it is possible to construct such materials. Smith and others (2000, 2004) conducted extensive studies on metamaterials, and a comprehensive overview was given by Sihvola (2007).

The SRR proposed by Pendry and others is a periodic structure with a unit cell of characteristic dimension (spacing) which is much smaller than a wavelength and therefore it can be regarded as a homogenous medium. This requires some averaging process, which is called "homogenization." It has effective permittivity and permeability that are both negative. SRR has been studied (Hardy, 1981) for use in magnetic

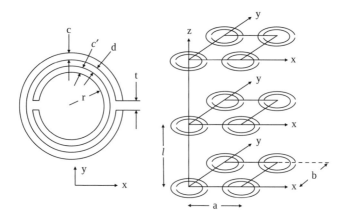

FIGURE 21.3   Metamaterial consists of three-dimensional array of nonmagnetic inclusions such as split-ring resonators (SRRs) shown here. Spacing in the x, y, and z directions are a, b, and l respectively. x-y is the plane of the inclusions and the z axis is perpendicular to the x-y plane.

resonance. Its size can be small and can be used as a homogeneous medium. The resonant frequency is proportional to $(\text{gap spacing/width})^{1/2} = (t/w)^{1/2}$.

SRRs for metamaterials consist of a three-dimensional periodic structure of flat disks of split ring (Fig. 21.3). Its permittivity and permeability have resonance characteristics sketched in Fig. 21.4, which was calculated numerically (Ishimaru et al., 2003). The figure shows the resonant frequency slightly under 5 GHz. Note that both μ_{zz} and ε_{xx} are negative in the frequency range slightly above the resonant frequency, and the refractive index $n = (\varepsilon_{yy}/\mu_{zz})^{1/2}$ is also negative. This also shows that the medium is anisotropic, as will be discussed in Section 21.4.

The key quantity is the resonance frequency. Pendry gave an approximate formula (1999).

$$\omega_0^2 = \frac{3lc_0^2}{\pi \left(\ln \frac{2c}{d} \right) r^3} \tag{21.17}$$

This is near the numerically calculated value of Fig. 21.4. Shamonin and others (2004) studied equivalent circuits of a single split resonator and the SRR consisting of the series inductances for outer and inner rings, and the mutual inductance between rings, the gap capacitances, the capacitances between rings, and emf induced in the inner and outer rings. The resonant frequency is that of an LC circuit and the capacitance includes the two gap capacitances. These circuit representations are useful to understand the physical interpretation of the SRRs.

It is possible to express the dispersive characteristics of the refractive index $n(\omega)$ taking into account the group refractive index. Note that the behavior of $n(\omega)$ near ω_0 is given by

$$n(\omega) = n(\omega_0) + \frac{\partial n}{\partial \omega}\bigg|_{\omega_0} (\omega - \omega_0), \tag{21.18}$$

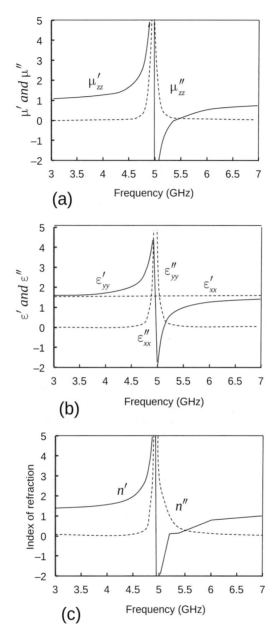

FIGURE 21.4   Numerical calculations of μ and ε of metamaterial of a split-ring resonator of Fig. 21.3. $r = 1.5$ mm, $c = 0.8$ mm, $d = 0.2$ mm, $a = 8$ mm, and $l = 3.9$ mm. The conductivity of the ring is 5.8×10^7 [s/m]. Thickness of the ring is much greater than the skin depth.

but noting

$$n_g(\omega) = \frac{\partial}{\partial \omega}(\omega n)$$

$$= n(\omega_0) + \omega_0 \left.\frac{\partial n}{\partial \omega}\right|_{\omega_0} \tag{21.19}$$

we get

$$n(\omega) = n(\omega_0) + \frac{[n_g(\omega_0) - n(\omega_0)]}{\omega_0}(\omega - \omega_0). \tag{21.20}$$

This becomes important when we discuss the space–time behavior of wave packet, as will be shown later.

21.4   GENERALIZED CONSTITUTIVE RELATIONS FOR METAMATERIALS

SRRs are usually studied using the plane of the rings as the x–y plane and the z-axis is perpendicular to the plane of the ring. However, it is clear that depending on the polarization of the incident wave, SRR behaves differently indicating that it is anisotropic. Therefore, it should be clear that the permittivity and the permeability must be 3×3 matrices. More generally, as was discussed in Section 8.22, there are couplings between electric and magnetic field excitations and responses, as exhibited in chiral materials. These couplings are also expressed in 3×3 matrices. Thus we have a 6×6 matrix.

$$\begin{bmatrix} \bar{D} \\ \bar{B} \end{bmatrix} = \begin{bmatrix} \bar{\bar{\varepsilon}} & \bar{\bar{\xi}} \\ \bar{\bar{\zeta}} & \bar{\bar{\mu}} \end{bmatrix} \begin{bmatrix} \bar{E} \\ \bar{H} \end{bmatrix}, \tag{21.21}$$

where $\bar{D}$, $\bar{B}$, $\bar{E}$, and $\bar{H}$ are 3×1 matrices and $\bar{\bar{\varepsilon}}$, $\bar{\bar{\xi}}$, $\bar{\bar{\zeta}}$, and $\bar{\bar{\mu}}$ are 3×3 matrices. This is the general constitutive relation of bianisotropic medium.

There are some fundamental relations among these matrices (Weiglhofer, 1995, 2002). The bianisotropic nongyrotropic mediums satisfy the following reciprocity relations (Kong 1986, 1972):

$$\begin{bmatrix} \bar{\bar{\varepsilon}} \end{bmatrix}^t = \begin{bmatrix} \bar{\bar{\varepsilon}} \end{bmatrix},$$

$$\begin{bmatrix} \bar{\bar{\mu}} \end{bmatrix}^t = \begin{bmatrix} \bar{\bar{\mu}} \end{bmatrix}, \tag{21.22}$$

$$\begin{bmatrix} \bar{\bar{\zeta}} \end{bmatrix}^t = -\begin{bmatrix} \bar{\bar{\xi}} \end{bmatrix},$$

where t denotes transpose.

It is also noted that to be compatible with Maxwell's equations, the following consistency constraint must hold (Kong, 1986):

$$\text{Trace}\left(\left[\bar{\bar{\xi}}\right]\left[\bar{\bar{\mu}}\right] + \left[\bar{\bar{\mu}}\right]^{-1}\left[\bar{\bar{\zeta}}\right]\right) = 0. \tag{21.23}$$

The constitutive relations given in (21.21) are called the E–H (or Tellegen) representations.

However, physically $\bar{E}$ and $\bar{B}$ are the fundamental fields and $\bar{D}$ and $\bar{H}$ are the responses. Thus, the following E–B (or Boys–Post) representation is more physically appropriate.

$$\begin{bmatrix} \bar{D} \\ \bar{H} \end{bmatrix} = \begin{bmatrix} \bar{\bar{\varepsilon}}_p & \bar{\bar{\alpha}}_p \\ \bar{\bar{\beta}}_p & \bar{\bar{\mu}}_p^{-1} \end{bmatrix} \begin{bmatrix} \bar{E} \\ \bar{B} \end{bmatrix} \tag{21.24}$$

These two representations (21.21) and (21.24) are equivalent for a linear medium, and related through the following:

$$\begin{aligned}
\bar{\bar{\varepsilon}} &= \bar{\bar{\varepsilon}}_p - \bar{\bar{\alpha}}_p \, \bar{\bar{\mu}}_p \, \bar{\bar{\beta}}_p, \\
\bar{\bar{\mu}} &= \bar{\bar{\mu}}_p, \\
\bar{\bar{\xi}} &= \bar{\bar{\alpha}}_p \, \bar{\bar{\mu}}_p, \\
\bar{\bar{\zeta}} &= -\bar{\bar{\mu}}_p \, \bar{\bar{\beta}}_p.
\end{aligned} \tag{21.25}$$

The E–H representation (21.21) is commonly used in most of the literature. It is convenient because Maxwell's equations appear symmetric and boundary conditions are generally given in terms of $\bar{E}$ and $\bar{H}$. Note, however, that even though the E–H representation is conventional, the E–B representation (21.24) is physically correct.

Let us now derive the constitutive relations for metamaterials consisting of a three-dimensional array of nonmagnetic inclusions (Fig. 21.3). As was discussed in Section 2.3, in isotropic and anisotropic media, we have

$$\begin{aligned}
\bar{D} &= \varepsilon_0 \bar{E} + \bar{P}, & \bar{P} &= \chi_e \varepsilon_0 \bar{E}, \\
\bar{B} &= \mu_0 (\bar{H} + \bar{M}), & \bar{M} &= \chi_m \bar{H},
\end{aligned} \tag{21.26}$$

where $\bar{P}$ and $\bar{M}$ are the electric and magnetic polarizations respectively, and χ_e and χ_m are the electric and magnetic susceptibilities. In this section, we discuss how (21.26) is generalized for metamaterials.

The inclusions shown in Fig. 21.3 are small in size in wavelength. The spacings are also much smaller than a wavelength, generally no more than 0.1λ and the inclusions are much smaller than the spacing. Under these conditions, the medium can be considered homogeneous and the Lorentz theory applies (Collin, 1991; Ishimaru et al., 2003). Consistent with the Lorentz theory, we assume that under the influence of an electromagnetic field, the inclusions produce electric and magnetic dipoles.

Using the E–H representation, we have in general

$$\bar{D} = \varepsilon_0 \bar{E} + \bar{P}(\bar{E}, \bar{H}),$$
$$\bar{B} = \mu_0 \bar{H} + \mu_0 \bar{M}(\bar{E}, \bar{H}),$$

(21.27)

where electric and magnetic polarizations $\bar{P}$ and $\bar{M}$ are produced by $\bar{E}$ and $\bar{H}$. Each inclusion is acted upon by the effective field $(\bar{E}_e, \bar{H}_e)$ to produce the electric $\bar{p}$ and magnetic $\bar{m}$ dipoles.

$\bar{P}$ and $\bar{M}$ are then given by

$$\bar{P} = N\bar{p},$$
$$\bar{M} = N\bar{m},$$

(21.28)

where $N = (abc)^{-1}$ is the number of inclusions per unit volume.

We can express $\bar{p}$ and $\bar{m}$ using the 6×6 generalized polarizability matrix $[\bar{\bar{\alpha}}]$.

$$\begin{bmatrix} \bar{P} \\ \bar{M} \end{bmatrix} = N \begin{bmatrix} \bar{p} \\ \bar{m} \end{bmatrix} = N \, [\bar{\bar{\alpha}}] \begin{bmatrix} \bar{E}_e \\ \bar{H}_e \end{bmatrix},$$

where

$$[\bar{\bar{\alpha}}] = \begin{bmatrix} \bar{\bar{\alpha}}_{ee} & \bar{\bar{\alpha}}_{em} \\ \bar{\bar{\alpha}}_{me} & \bar{\bar{\alpha}}_{mm} \end{bmatrix}.$$

(21.29)

Note that in Section 8.1, we used $\bar{E}' = \bar{E} + \bar{E}_p$, where $\bar{E}'$ is the local field, which is the same as $\bar{E}_e$, $\bar{E}$ is the external applied field, and $\bar{E}_p$ is the field produced by all the dipoles surrounding the dipole under consideration. $\bar{E}_p$ is the same as $\bar{E}_i$ (interaction field) in Section 21.5.

Here in this section we generalize $\bar{E}' = \bar{E} + \bar{E}_p$ and write the local field $(\bar{E}_l, \bar{H}_l)$.

$$\begin{aligned} \begin{bmatrix} \bar{E}_l \\ \bar{H}_l \end{bmatrix} &= \begin{bmatrix} \bar{E} \\ \bar{H} \end{bmatrix} + \begin{bmatrix} \bar{E}_i \\ \bar{H}_i \end{bmatrix} \\ &= \begin{bmatrix} \bar{E} \\ \bar{H} \end{bmatrix} + N \, [\bar{\bar{C}}] \begin{bmatrix} \bar{p} \\ \bar{m} \end{bmatrix} \\ &= \begin{bmatrix} \bar{E} \\ \bar{H} \end{bmatrix} + N \, [\bar{\bar{C}}] \, [\bar{\bar{\alpha}}] \begin{bmatrix} \bar{E}_l \\ \bar{H}_l \end{bmatrix}, \end{aligned}$$

(21.30)

where $[\bar{\bar{C}}]$ is the 6×6 matrix and is called the interaction constant matrix.

From (21.30), we get

$$\begin{bmatrix} \bar{E}_l \\ \bar{H}_l \end{bmatrix} = \left[\begin{bmatrix} \bar{\bar{U}} \end{bmatrix} - N \begin{bmatrix} \bar{\bar{C}} \end{bmatrix} [\bar{\alpha}] \right]^{-1} \begin{bmatrix} \bar{E} \\ \bar{H} \end{bmatrix}, \tag{21.31}$$

where $\left[\bar{\bar{U}} \right]$ is the 6×6 unit matrix.

We substitute this into (21.29) and noting

$$\begin{bmatrix} \bar{D} \\ \bar{B} \end{bmatrix} = \begin{bmatrix} \varepsilon_0 \bar{E} \\ \mu_0 \bar{H} \end{bmatrix} + \begin{bmatrix} \bar{P} \\ \mu_0 M \end{bmatrix}, \tag{21.32}$$

we get

$$\begin{aligned} \begin{bmatrix} \bar{D} \\ \bar{B} \end{bmatrix} &= \begin{bmatrix} \bar{\bar{\varepsilon}} & \bar{\bar{\xi}} \\ \bar{\bar{\zeta}} & \bar{\bar{\mu}} \end{bmatrix} \begin{bmatrix} \bar{E} \\ \bar{H} \end{bmatrix} \\ &= \left[\begin{bmatrix} \varepsilon_0 \bar{U} & 0 \\ 0 & \mu_0 \bar{U} \end{bmatrix} + N \begin{bmatrix} \bar{U} & 0 \\ 0 & -\bar{U} \end{bmatrix} [\bar{\alpha}] \left[\begin{bmatrix} \bar{U} \end{bmatrix} - N \begin{bmatrix} \bar{\bar{C}} \end{bmatrix} [\bar{\alpha}] \right]^{-1} \right] \begin{bmatrix} \bar{E} \\ \bar{H} \end{bmatrix}, \end{aligned} \tag{21.33}$$

where $\bar{U}$ is the 3×3 unit matrix.

That is the generalization of the Lorentz–Lorenz formula or Clausius–Mossotti formula (see Section 8.1).

The 6×6 interaction matrix [C] has been obtained (Collin, 1991; Ishimaru et al., 2003) and depends on the spacings in the $x, y,$ and z directions. For a three-dimensional array of inclusions of electric and magnetic dipoles, the interaction matrix is given by

$$\begin{bmatrix} \bar{\bar{C}} \end{bmatrix} = \begin{bmatrix} \dfrac{1}{\varepsilon_0} \bar{C} & \bar{O} \\ \bar{O} & \dfrac{1}{\mu_0} \bar{C} \end{bmatrix}, \tag{21.34}$$

where $\bar{O}$ is a 3×3 null matrix and $\bar{C}$ is a 3×3 diagonal matrix.

$$\bar{C} = \begin{bmatrix} C_x & 0 & 0 \\ 0 & C_y & 0 \\ 0 & 0 & C_z \end{bmatrix}.$$

For a cubic lattice of $a = b = c$, the matrix $\bar{C}$ is reduced to

$$\bar{C} = \frac{1}{3} \begin{bmatrix} 1 & 0 & 0 \\ 0 & 1 & 0 \\ 0 & 0 & 1 \end{bmatrix}. \tag{21.35}$$

The polarizability matrix $[\bar{\bar{\alpha}}]$ in (21.29) can be expressed in terms of the currents induced in the inclusions. The electric and magnetic dipoles $\bar{p}$ and $\bar{m}$ can be expressed by the electric currents $\bar{J}_e$ and $\bar{J}_m$ when the local electric and magnetic fields are $\bar{E}_l$ and $\bar{H}_l$, respectively (Jackson, 1962).

$$\begin{bmatrix} \bar{p} \\ \bar{m} \end{bmatrix} = [\bar{\bar{\alpha}}] \begin{bmatrix} \bar{E}_l \\ \bar{H}_l \end{bmatrix}. \tag{21.36}$$

where

$$[\bar{\bar{\alpha}}] = \begin{bmatrix} \bar{\bar{\alpha}}_{ee} & \bar{\bar{\alpha}}_{em} \\ \bar{\bar{\alpha}}_{me} & \bar{\bar{\alpha}}_{mm} \end{bmatrix} = 6 \times 6 \text{ matrix,}$$

$$\bar{\bar{\alpha}}_{ee} = \int dv \frac{1}{(-i\omega)} \bar{J}_e,$$

$$\bar{\bar{\alpha}}_{em} = \int dv \frac{1}{(-i\omega)} \bar{J}_m,$$

$$\bar{\bar{\alpha}}_{me} = \int dv \frac{\bar{r}}{2} \times \bar{J}_e,$$

$$\bar{\bar{\alpha}}_{mm} = \int dv \frac{\bar{r}}{2} \times \bar{J}_m.$$

$\bar{\bar{\alpha}}_{ee}$ is a 3×3 matrix, and $\bar{J}_e$ is also a 3×3 matrix. The first column is the current density in the x, y, and z directions for the x component of the effective field $\bar{E}_l$ and therefore the unit is $(A/m^2)/(V/m)$. Similarly $\bar{J}_m$ has the unit of $(A/m^2)/(A/m)$. Other components of $\bar{\bar{\alpha}}_{em}$, $\bar{\bar{\alpha}}_{me}$, and $\alpha_{mm}^{=}$ can be similarly defined.

As a simple example, consider an array of spherical inclusions of radius a_0 in a cubic lattice; we get (Section 8.1)

$$\bar{\bar{\alpha}}_{ee} = \frac{3(\varepsilon_r - 1)V}{\varepsilon_r + 2}, \quad V = \frac{1}{N} = \frac{4\pi a_0^3}{3}, \tag{21.37}$$

where N is the number of inclusions per unit volume, and $f = NV$ is the fractional volume.

Substituting this into (21.33), we get

$$\frac{\varepsilon}{\varepsilon_0} = 1 + N\alpha_{ee} \left[1 - \frac{N}{3} \alpha_{ee} \right]^{-1}$$

$$= \frac{1 + 2f \left(\dfrac{\varepsilon_r - 1}{\varepsilon_r + 2} \right)}{1 - f \left(\dfrac{\varepsilon_r - 1}{\varepsilon_r + 2} \right)} \quad f = NV. \tag{21.38}$$

This is the Maxwell–Garnett mixing formula (Section 8.6).

Similarly, for a spherical inclusion, the lowest order electric field inside the sphere in the uniform effective field $H_0\hat{z}$ is pointed in the ϕ direction and noting

$$(\pi\rho^2)(+i\omega)\mu_0 H_0 = 2\pi\rho E_\phi, \qquad (21.39)$$

we get

$$\bar{E}_\phi = (+i\omega\mu_0 H_0)\frac{\rho}{2}\hat{\phi},$$

The current $\bar{J}_m$ and α_{mm} are therefore

$$\bar{J}_m = -i\omega\varepsilon_0(\varepsilon_r - 1)E_\phi. \qquad (21.40)$$

We substitute this into (21.36) and get for $\bar{H} = H_0\hat{z}$

$$
\begin{aligned}
\alpha_{mm}H_0 &= \int dv \frac{\bar{r}}{2} \times \bar{J}_m \\
&= \int r^2 dr \sin\theta d\theta d\phi \bar{J}_m \qquad (21.41) \\
&= (k_0 a_0)^2 \frac{(\varepsilon_r - 1)}{10} V.
\end{aligned}
$$

Substituting this into (21.35), we get the permeability equivalent to the Maxwell–Garnett formula for a medium with spherical inclusions.

$$
\begin{aligned}
\frac{\mu}{\mu_0} &= 1 + N\alpha_{mm}\left(1 - \frac{N}{3}\alpha_{mm}\right)^{-1} \\
&= \frac{1 + 2f\dfrac{(k_0 a_0)^2(\varepsilon_r - 1)}{10}}{1 - f\dfrac{(k_0 a_0)^2(\varepsilon_r - 1)}{10}}. \qquad (21.42)
\end{aligned}
$$

The above is approximate, and a more complete analysis is given in Braunisch (2001).

21.5 SPACE–TIME WAVE PACKET INCIDENT ON DISPERSIVE METAMATERIAL AND NEGATIVE REFRACTION

We have already noted that NIM is generally dispersive and lossy. If a wave is incident upon a dispersive lossy medium, the wave is reflected and refracted. If the medium is NIM, then the angle of refraction can be negative according to Snell's law.

$$n_1 \sin\theta_1 = n_2 \sin\theta_2, \qquad (21.43)$$

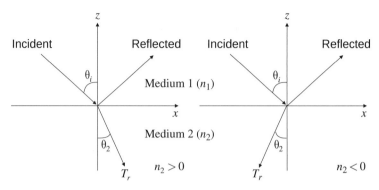

FIGURE 21.5   Wave is incident from medium 1 to medium 2, showing the negative refraction when $n_2 < 0$.

where n_1 and θ_1 are the refractive index and the angle of incidence in the first medium, and n_2 and θ_2 are those of the second medium. If n_2 is negative, then $\theta_2 < 0$ according to (21.43). This is called "negative refraction" (Fig. 21.5).

The negative refraction and Snell's law are normally used for a plane wave incident on a surface. Does this also apply to the incident wave such as a beam or a wave packet? How does the dispersive medium affect the negative refraction? To answer these questions, we need to study a space–time incident wave on a dispersive medium. Let us consider a Gaussian wave packet incident upon a half space of dispersive medium including NIM. We first consider the case where $-(\pi/2) < \theta_2 < (\pi/2)$. In later sections, we examine the cases where θ_1 is greater than the critical angle θ_c.

$$\sin \theta_c = \frac{n_2}{n_1}. \tag{21.44}$$

This gives rise to the Goos–Hanchen shift (Section 6.7) and the lateral wave (Section 15.8). For $n_2 < 0$, they become the "negative" Goos–Hanchen shift and the "backward" lateral waves which will be discussed later. For this section, we limit ourselves to the case $|\theta_2| < (\pi/2)$.

We consider the two-dimensional problem with TM polarization. The incident Gaussian wave packet using time dependence $\exp(-i\omega t)$ is given by

$$H_y(x, z, t) = \text{Re}\,\{\psi_i(x, z, t)\},$$

$$\psi_i = \exp\left[-i\omega_0 \left(t - \frac{z'}{c} \right) - \frac{\left(t - \dfrac{z'}{c} \right)^2}{T_0^2} - \frac{x'^2}{W_0^2} \right], \tag{21.45}$$

where ω_0 = carrier frequency, $T_0 = 2/\Delta\omega$, $\Delta\omega$ = frequency bandwidth, and W_0 is the full width of the packet (Fig. 21.6). The wave packet propagates along the z'

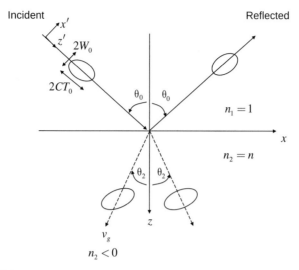

FIGURE 21.6   Gaussian wave packets incident from medium n_1 to medium n_2.

coordinate and x' in the transverse coordinate of the packet and (x', z') are related to (x, z) through rotation by the angle θ_0.

$$x' = x \cos \theta_0 - z \sin \theta_0,$$
$$z' = x \sin \theta_0 + z \cos \theta_0. \tag{21.46}$$

In the x–z plane, we can take a double Fourier transform with respect to x and time t. At $z = 0$, we get

$$\bar{\psi}_i(k_x, 0, \omega) = \int\limits_{-\infty}^{\infty} \int\limits_{-\infty}^{\infty} \psi_i(x, z, t) e^{+i\omega t - i k_x x} \, dt \, dx$$

$$= \frac{\pi W \omega_0 T_0}{\cos \theta_0} \exp\left[-\frac{(\omega - \omega_0)^2 T_0^2}{4} - \frac{\left(k_x - \dfrac{\omega}{c}\sin\theta_0\right)^2 W_0^2}{4\cos^2\theta_0} \right]. \tag{21.47}$$

The transmitted and the reflected waves are given by multiplying $\bar{\psi}_i$ by the transmission and reflected coefficients and the propagation factor. The transmitted wave is given by

$$\psi_t(x, z, t) = \frac{1}{(2\pi)^2} \int\limits_{-\infty}^{\infty} \int\limits_{-\infty}^{\infty} T(k_x, \omega) \bar{\psi}_i(k_x, 0, \omega) \exp(+i k_{z2} z + i k_x x - i\omega t) \, dk_z \, d\omega,$$

where

$$T = \frac{(2k_{z1}/\varepsilon_0)}{(k_{z1}/\varepsilon_1) + (k_{z2}/\varepsilon_2)},$$

$$k_{z1} = \sqrt{k^2 - k_x^2}, \quad k_{z2} = \sqrt{(kn)^2 - k_x^2}, \quad k = \frac{\omega}{c}. \tag{21.48}$$

The reflected wave is similarly given by

$$\psi_r(x, z, t) = \frac{1}{(2\pi)^2} \int\limits_{-\infty}^{\infty} \int\limits_{-\infty}^{\infty} R(k_x, \omega)\bar{\psi}_i(k_x, 0, \omega) \exp(-ik_{z1}z + ik_x x - i\omega t)dk_x d\omega,$$

where

$$R = \frac{(k_1/\varepsilon_1) - (k_{z2}/\varepsilon_2)}{(k_{z1}/\varepsilon_1) + (k_{z2}/\varepsilon_2)}. \tag{21.49}$$

The evaluation of (21.48) and (21.49) can be done numerically. However, it is more instructive to obtain the analytical solutions.

The analytical solutions can be obtained by using asymptotic solutions and the method of steepest descent. We first note that k_{z2} can be expressed using the first-order terms of the expansion in k_x and ω about the stationary positions k_{x0} and ω_0.

$$k_{z2} = k_{z0} + k_x'\left(\frac{\partial k_{z2}}{\partial k_x}\right)_0 + \omega'\left(\frac{\partial k_{z2}}{\partial \omega}\right)_0,$$

where

$$k_x' = k_x - k_{x0} = k_x - \frac{\omega}{c}\sin\theta_0,$$

$$\omega' = \omega - \omega_0. \tag{21.50}$$

$\left(\frac{\partial k_{z2}}{\partial k_x}\right)_0$ and $\left(\frac{\partial k_{z2}}{\partial \omega}\right)_0$ are evaluated at k_{x0} and ω_0. We also assume that T and R are slowly varying functions of k_0 and ω and therefore we approximate

$$T(k_x, \omega) \approx T(k_{x0}, \omega_0),$$

$$R(k_x, \omega) \approx R(k_{x0}, \omega_0). \tag{21.51}$$

Under these assumptions, we can perform the integration in (21.48) and (21.49) and obtain

$$\psi_t(x, z, t) = T(k_{x0}, \omega_0)e^{i\phi}F_b F_a,$$

$$\phi = +k_{z0}z + k_{x0}x - \omega_0 t,$$

$$F_b = \exp\left[-(x - (\tan\theta_z)z)^2 \frac{\cos^2\theta_0}{W_0^2}\right],$$

$$F_a = \exp\left[-\frac{(t - \bar{N} \cdot \bar{r})^2}{T_0^2}\right],$$

where

$$\bar{N} = \frac{\sin\theta_0}{c}\hat{x} + \frac{1}{c}\frac{(nn_g - \sin^2\theta_c)}{\sqrt{n^2 - \sin^2\theta_0}}\hat{z},$$

$$n_g = \text{Group refractive index} = \frac{\partial}{\partial\omega}(\omega n), \tag{21.52}$$

$$n(\omega) = n(\omega_0) + \frac{n_g(\omega_0) - n(\omega_0))\omega'}{\omega_0}.$$

The factors in (21.52) give detailed physical significance and wave behavior. The phase factor $\exp(i\phi)$ indicates that the phase progresses in the direction of θ_2 with the phase velocity of $v_p = (c/n)$. Note that in the direction of θ_2, $z = r\cos\theta_2$ and $x = r\sin\theta_2$, and ϕ becomes

$$\phi = knr - \omega_0 t.$$

Clearly this is the case for NIM. The amplitude variation is given by F_b. The peak of the Gaussian wave packet moves in the direction of θ_2, which means that in NIM, the wave packet moves in the direction of the negative refraction given by θ_2. The velocity of the wave packet is given by F_a. In the direction of the negative refraction, $\bar{r} = r\left[(\sin\theta_2)\hat{x} + (\cos\theta_2)\hat{z}\right]$ and if we substitute this in F_a, we get the velocity of the packet equal to the group velocity v_g. Thus the phase velocity v_p and v_g are pointed in opposite directions. Numerical calculations showing the phase velocity and the group velocity are presented in Ishimaru (2005).

21.6 BACKWARD LATERAL WAVES AND BACKWARD SURFACE WAVES

Let us first consider a two-dimensional problem (x–z plane) with a TM wave excited by a magnetic line source I_m. $H_y(x, z)$ satisfies the wave equation (Fig 21.7)

$$\left(\frac{\partial^2}{\partial x^2} + \frac{\partial^2}{\partial z^2} + k_i^2\right)H_y = -\delta(x)\delta(z - h) \tag{21.53}$$

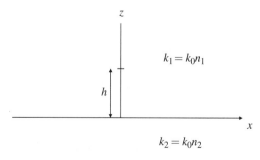

FIGURE 21.7 Magnetic line source at $x = 0$ and $z = h$ exciting TM wave in metamaterial ($z < 0$).

with the boundary condition that H_y and $(1/\varepsilon)(\partial/\partial z)H_y$ are continuous at $z = 0$. The magnetic line current I_m is normalized with $-j\omega\varepsilon_i I_m = 1$.

The well-known Fourier representation of the incident H_i, the reflected H_r, and the transmitted H_t waves are given by

$$
\begin{aligned}
H_{yi} &= \frac{1}{2\pi} \int \frac{\exp(-jk_{z1})\,|z-h| - jk_x x)}{2jk_{z1}}\,dk_x, \\
H_{yr} &= \frac{1}{2\pi} \int R(k_x)\frac{\exp(-jk_{z1}(z+h) - jk_x x)}{2jk_{z1}}\,dk_x, \\
H_{yt} &= \frac{1}{2\pi} \int T(k_x)\frac{\exp(-jk_{z1}h + jk_{z2}z - jk_x x)}{2jk_{z1}}\,dk_x,
\end{aligned}
\tag{21.54}
$$

where

$$
k_{zi} = \sqrt{k_i^2 - k_x^2}, \quad R(k_x) = (Z_1 - Z_2)/(Z_1 + Z_2),
$$

$$
T(k_x) = (2Z_1)/(Z_1 + Z_2), \quad Z_i = (k_{zi}/\omega\varepsilon_i), \ i = 1 \text{ and } 2.
$$

Let us now consider the surface waves. The pole for the surface wave is obtained by setting the denominator of the reflection coefficient to zero. We have

$$
Z_1 + Z_2 = 0 \tag{21.55}
$$

From this we get

$$
\frac{k_{z1}}{\varepsilon_1} = -\frac{k_{z2}}{\varepsilon_2}. \tag{21.56}
$$

Solving this for k_x, we get

$$
k_{xp} = k_1 S,
$$

$$
S^2 = \frac{\varepsilon^2 - n^2}{\varepsilon^2 - 1}, \quad \varepsilon = \frac{\varepsilon_2}{\varepsilon_1}, \quad n = \frac{n_2}{n_1}. \tag{21.57}
$$

Note that (21.56) gives the pole applicable to the surface wave pole as well as the Zenneck wave pole. The location of the pole given in (21.57) can be examined in the μ'–ε' diagram where μ' and ε' are the real part of μ and ε (Fig. 21.1). Assuming that the imaginary parts of μ and ε are negligibly small, and therefore S is real, we can examine S in the μ'–ε' plane.

If $S^2 > 1$, forward and backward surface waves can exist. If $0 < S^2 < 1$, there may be a Zenneck wave, and if $S^2 < 0$, the wave decays exponentially along the surface. When $S^2 > 1$, for a surface wave to exist, the pole must be in the proper Riemann surface. Depending on the imaginary part of k_{z1} and k_{z2}, we can have four cases, where the poles are located in the k_x complex plane. They are called the Riemann surfaces (Appendix 15.A).

$$\text{Riemann surface I:} \quad I_m(k_{z1}) < 0, I_m(k_{z2}) < 0,$$
$$\text{Riemann surface II:} \quad I_m(k_{z1}) > 0, I_m(k_{z2}) < 0,$$
$$\text{Riemann surface III:} \quad I_m(k_{z1}) < 0, I_m(k_{z2}) > 0,$$
$$\text{Riemann surface IV:} \quad I_m(k_{z1}) > 0, I_m(k_{z2}) > 0.$$

In Riemann surface I, the wave attenuates as $|z| \to \infty$ because both $I_m(k_{z1})$ and $I_m(k_{z2})$ are negative. Therefore this is called the "proper Riemann" surface. The surface wave exists if the pole is on Riemann surface I. To verify that the pole is in the proper Riemann surface with $I_m(k_{z1}) < 0$, we need to make sure to satisfy (21.56). To verify this, we first obtain $k_{z2} = \sqrt{k_2^2 - k_{xp}^2}$ and take the square root such that $I_m(k_{z2}) < 0$. This ensures that the pole is in Riemann surface I or II. Then we calculate k_{z1} from (21.56) and determine whether $I_m(k_{z1}) < 0$. If so, the pole is in Riemann surface I. Otherwise it is in Riemann surface II. As we see in Fig. 21.8, the forward surface wave and the backward surface wave exist in the regions denoted by SW^+ and SW^-.

As we see in Fig. 21.8, there is a small region where the backward surface wave SW^- exists. The phase velocity is pointed in the negative x direction as shown in Fig. 21.9.

The Poynting vector in air is pointed in the same direction as the phase velocity, but is pointed in the opposite direction in medium 2 (NIM).

The Poynting vector is given by

$$P_x = \frac{k_1 S}{2\omega\varepsilon_0\varepsilon_i}|H_y|^2 \qquad \begin{aligned} &i = 1 \text{ for medium 1,} \\ &i = 2 \text{ for medium 2,} \end{aligned} \tag{21.58}$$

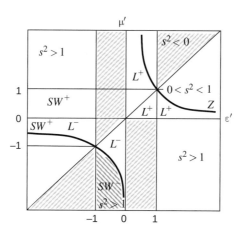

FIGURE 21.8 $\mu' - \varepsilon'$ diagram showing regions for forward surface wave S^+, backward surface wave S^-, forward lateral wave L^+, backward lateral wave L^-, and Zenneck wave Z.

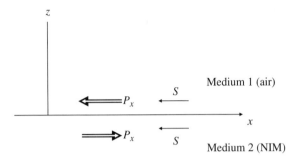

FIGURE 21.9 Backward surface waves. The phase velocity is pointed in the negative x direction, but the Poynting vector is pointed in the positive x direction in medium 2 and in the negative x direction in medium 1. The total power is pointed in the positive x direction.

and therefore the Poynting vector is pointed in the negative x direction in medium 1, but is pointed in the positive x direction in medium 2 since $S < 0$, $\varepsilon_2 < 0$. The total power in the x direction is given by

$$P_{\text{total}} = \int_0^\infty P_x dx + \int_{-\infty}^0 P_x dx$$
$$= \frac{1}{4} \frac{S}{\omega \varepsilon_0 \varepsilon_1} \frac{1}{\sqrt{S^2 - 1}} \left[1 - \frac{1}{\varepsilon^2} \right] \tag{21.59}$$

which becomes positive as shown in Fig. 21.9.

In the region where $0 < n < 1$, the conventional lateral wave can exist as shown in Section 15.8. If $-1 < n < 0$, then the backward lateral wave can exist. In Fig. 21.10, we show the conventional and the backward lateral waves. Note that for a backward lateral wave, the phase velocity v_p is negative and the branch point at $k_{x2} = k_0 n$ is located in the third quadrant of the k_x plane.

The branch cut integration gives rise to the lateral wave pointed in the negative direction as shown in Fig. 21.10a. If a wave packet is incident as shown in Fig. 21.10b, the wave packet in the medium 2 propagates with group velocity v_g in the x direction radiating the pulse in the negative direction as shown. This is obtained from Eqs. (21.48) and (21.49) in Section 21.5 and numerically calculating the integral. Figure 21.10c shows the branch cut for the complex k_x plane.

21.7 NEGATIVE GOOS–HANCHEN SHIFT

We have already discussed the "positive" Goos–Hanchen shift in Section 6.7. This shift occurs when the beam is incident from medium 1 with ε_1, μ_1, and n_1 to medium 2 with ε_2, μ_2, and n_2 with the incident angle that causes the total reflection. This requires that

$$\left(\frac{n_2}{n_1} \right)^2 > \sin^2 \theta_i. \tag{21.60}$$

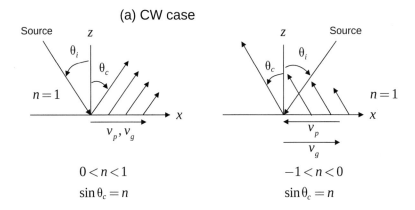

(a) CW case

$$0 < n < 1$$
$$\sin\theta_c = n$$

$$-1 < n < 0$$
$$\sin\theta_c = n$$

(b) Wave packet incidence

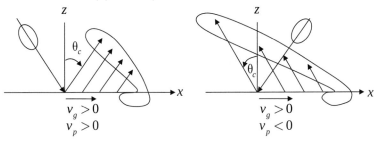

$$v_g > 0$$
$$v_p > 0$$

$$v_g > 0$$
$$v_p < 0$$

(c) Branch cut in complex k_x plane

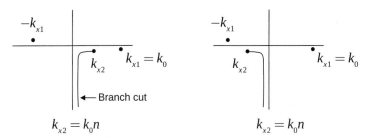

$$k_{x2} = k_0 n$$

$$k_{x2} = k_0 n$$

FIGURE 21.10 Left: conventional forward lateral wave. Right: backward lateral wave. Note that the angle of incidence θ_i is chosen to be negative for backward lateral wave to be consistent with branch point at $k_{x2} = k_0 n_2$.

In this case, the reflection coefficient R has the magnitude of one and the phase ϕ.

$$R = \exp(j\phi). \tag{21.61}$$

As shown in Section 6.7, the reflected beam $U_r(x, z)$ is given by

$$U_r(x, z) = R(\beta_a)e^{-j\beta_0\phi'(\beta_0)}U_{r0}(x - \phi'(\beta_0), z) \tag{21.62}$$

where U_{r0} is identical to the beam wave originated from the image source (see Section 6.7). The shift is given by $\phi'(\beta_0)$.

Now we examine the shift $\phi'(\beta_0)$ for all cases in the μ–ε diagram.

For s-polarization, we have

$$R_s = \frac{(q_1/\mu_1) - (q_2/\mu_2)}{(q_1/\mu_1) + (q_2/\mu_2)},$$

$$q_1 = \sqrt{k_1^2 - \beta^2}, \quad q_2 = \sqrt{k_2^2 - \beta^2},$$

$$\beta = k_1 \sin \theta_i - \beta_0.$$

(21.63)

For p-polarization, we have

$$R_p = \frac{(q_1/\varepsilon_1) - (q_2/\varepsilon_2)}{(q_1/\varepsilon_1) + (q_2/\varepsilon_2)}.$$

(21.64)

Under the total reflection, we have

$$q_2 = \sqrt{k_2^2 - \beta_0^2} = -j\sqrt{\beta_0^2 - k_2^2} = -j\alpha_2$$

(21.65)

We then get

$$R_s = \frac{(q_1/\mu_1) + j(\alpha_2/\mu_2)}{(q_1/\mu_1) - j(\alpha_2/\mu_2)} = \exp(j\phi_s),$$

$$R_p = \frac{(q_1/\varepsilon_1) + j(\alpha_2/\varepsilon_2)}{(q_1/\varepsilon_1) - j(\alpha_2/\varepsilon_2)} = \exp(j\phi_p),$$

(21.66)

where

$$\phi_s = 2\tan^{-1}\left(\frac{\mu_1 \alpha_2}{\mu_2 q_1}\right),$$

$$\phi_p = 2\tan^{-1}\left(\frac{\varepsilon_1 \alpha_2}{\varepsilon_2 q_1}\right).$$

The Goos–Hanchen shift is then given by (Fig. 21.11)

$$\phi'_s = \frac{\partial \phi_s}{\partial \beta} = \left(\frac{\mu_1}{\mu_2}\right) F_s \quad \text{for} \quad \text{s-pol},$$

$$\phi'_p = \frac{\partial \phi_p}{\partial \beta} = \left(\frac{\varepsilon_1}{\varepsilon_2}\right) F_p \quad \text{for} \quad \text{p-pol},$$

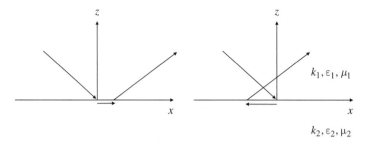

FIGURE 21.11 Positive and negative Goos–Hanchen shifts.

where

$$F_s = \frac{1}{\left[1 + \left(\dfrac{\mu_1}{\mu_2}\dfrac{\alpha_2}{q_1}\right)^2\right]} \frac{\partial}{\partial\beta}\left(\frac{\alpha_2}{q_1}\right),$$

$$F_p = \frac{1}{\left[1 + \left(\dfrac{\varepsilon_1}{\varepsilon_2}\dfrac{\alpha_2}{q_1}\right)^2\right]} \frac{\partial}{\partial\beta}\left(\frac{\alpha_2}{q_1}\right).$$

(21.67)

We can verify that F_s and F_p are positive.

From (21.67), we can obtain the following:

- Conventional positive Goos–Hanchen shift occurs in region a in Fig. 21.12, where $n^2 > \sin^2\theta_i$ and $n > 0$.
- Negative Goos–Hanchen shift occurs in region b, where $n^2 > \sin^2\theta_i$ and $n < 0$. Note that ϕ'_s and ϕ'_p are negative.
- In region c, negative Goos–Hanchen shift occurs for p-polarization. Note that $\varepsilon_2 < 0$.
- In region d, the negative shift occurs for p-polarization. Note that $\mu_2 < 0$.

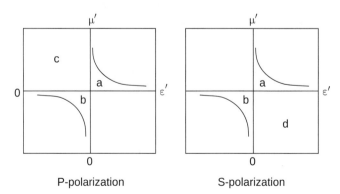

FIGURE 21.12 Negative Goos–Hanchen shift occurring in regions a, b, and c. S and p polarizations are symmetrically located in the μ–ε diagram (see Section 21.1).

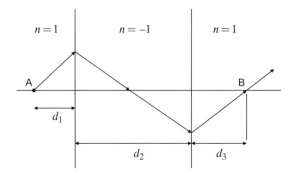

FIGURE 21.13   A point delta function source at A emits waves that are focused to a delta function image point at B. Normal lens can only focus to an image of the order of wavelength. Thus, this lens with $n = -1$ is called the "super resolution."

21.8 PERFECT LENS, SUBWAVELENGTH FOCUSING, AND EVANESCENT WAVES

As stated by Pendry in 2000, a slab of NIM with $n = -1$ can focus waves from a delta function point source to a delta function point image (Fig. 21.13). An ordinary lens cannot focus to a size less than the order of a wavelength. Therefore an $n = -1$ slab lens which can focus to a delta function image is called the "perfect lens" with "super resolution."

This is, of course, a mathematical idealization that does not exist. It is known that the mathematical formulations of any physical problem must be properly posed. This requires that the solution must satisfy three conditions: (1) uniqueness, (2) existence, and (3) it must depend continuously on the variation of the physical parameters. The slab with $n = -1$ is not properly posed and therefore the super resolution does not exist. We need to start with $\varepsilon = -1 + \delta_\varepsilon$ and $\mu = -1 + \delta_\mu$. We then find, instead of super resolution, subwavelength focusing.

As an example, consider a line magnetic source in front of the slab (Fig. 21.14). The magnetic field H_y for $z > d_2$ is given by

$$\left(\frac{\partial^2}{\partial x^2} + \frac{\partial^2}{\partial z^2} + k_i^2 \right) H_y = j\omega\varepsilon_0 I_m \delta(x)\delta(z + d_1) \tag{21.68}$$

where $k_1 = k_0(n = 1)$, and $k_2 = k_0 n$, and we normalize I_m so that $-j\omega\varepsilon_0 I_m = 1$. The solution is given in Fourier representation. For $z > d_2$, we have

$$H_y(x, z) = \frac{1}{2\pi} \int \frac{T(k_x)}{2jk_{z1}} \exp(-jk_{z2}(z - d_2 + d_1) - jk_x x) dk_x,$$

$$k_{z1} = \left(k_0^2 - k_x^2 \right)^{1/2}, \quad k_{z2} = \left((k_0 n)^2 - k_x^2 \right)^{1/2}. \tag{21.69}$$

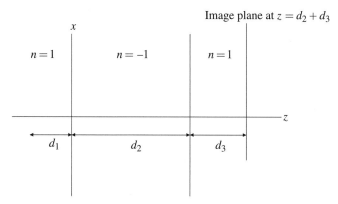

FIGURE 21.14 Magnetic line source is located at $z = -d_1$. The image is observed on the image plane $z = d_2 + d_3$.

The transmission coefficient $T(k_x)$ is obtained and given in the following form.

$$T(k_x) = \frac{2 \exp(-jk_{z2}dz)}{1 + M + \exp(-j2k_{z2}dz)(1 - M)},$$

where

$$M = \frac{1}{2}\left(\frac{Z_2}{Z_1} + \frac{Z_1}{Z_2}\right),$$

$$Z_1 = \frac{k_{z1}}{\omega\varepsilon_1}, \quad Z_2 = \frac{k_{z2}}{\omega\varepsilon_2}. \tag{21.70}$$

For s-polarization, E_g is given by the same expression as (12.69) with the normalization $-j\omega\mu I = 1$. The transmission coefficient (21.70) is the same except

$$Z_1 = \frac{\omega\mu_1}{k_{z1}}, \quad Z_2 = \frac{\omega\mu_2}{k_{z2}}. \tag{21.71}$$

Let us first examine the special cases. For $n = +1$ (free space), $Z_1 = Z_2$ and therefore $M = 1$. The transmission coefficient becomes $T = \exp(-jk_{z1}d_2)$ as expected. As we expect, for $n = 1$,

$$H_y = \frac{1}{2\pi}\int \frac{1}{2jk_{z1}}\exp(-jk_{z1}(d_1 + z) - jk_x x)dk_x. \tag{21.72}$$

That is the field at (x, z) due to a point source at $(x = 0, z = -d_1)$ and is a two-dimensional Green's function $-\frac{j}{4}H_0^{(2)}\left(k\sqrt{x^2 + (d_1 + z)^2}\right)$. If $n \to -1$, $\mu = -\mu_0$

and $\varepsilon = -\varepsilon_0$. Therefore, $Z_2 = -Z_1$, $M = -1$, and $k_{z2} = (k_2{}^2 n^2 - k_x{}^2)^{1/2} \to k_{z1}$. $T(k_x)$ becomes

$$T(k_X) = \exp(+jk_{z1}dz).$$

The field H_y then becomes

$$H_y(x, z) = \frac{1}{2\pi} \int \frac{1}{2jk_{z1}} \exp(-jk_{z1}(z + d_1 - 2d_2) - jk_x x)dk_x. \qquad (21.73)$$

This is an integral representation of the Hankel function and is equal to

$$H_y(x, z) = -j\frac{1}{4}H_0{}^{(2)}(kr),$$
$$r = \left[(z + d_1 - 2d_2)^2 + x^2\right], \qquad (21.74)$$

for $z > 2d_2 - d_1$. For $z < 2d_2 - d_1$, r becomes $-r$ and the Hankel function becomes the first kind.

$$H_y(x, z) = -j\frac{1}{4}H_0{}^{(1)}(k\,|r|). \qquad (21.75)$$

This means that near $x = 0$, $z = 2d_2 - d_1$, the field behaves as an incoming wave for $z < 2d_2 - d_1$ focused on a point and diverges for $z > 2d_2 - d_1$. All of these are simply a mathematical idealization that does not exist in physical problems. What is more important and useful is the question of what happens if the refractive index deviates from -1. Let us examine this. If n deviates from -1, and ε and μ deviate from $-\varepsilon_0$ and $-\mu_0$, M deviates from -1. We write this as

$$M \to -1 + \delta. \qquad (21.76)$$

The transmission coefficient T becomes

$$T = \frac{2\exp(-jk_{z2}d_2)}{\delta + \exp(-j2k_{z2}d_2)(2 - \delta)}. \qquad (21.77)$$

By noting $n^2 = 1 + \delta_n$, $\mu_2 = -1 + \delta_\mu$, and $\varepsilon_2 = -1 + \delta_\varepsilon$, we can examine $M = -1 + \delta$ carefully. We then get δ in terms of δ_n, δ_μ, and δ_ε. Detailed examination will show that the first order of δ_n, δ_μ, and δ_ε cancel out and δ is proportional to the second-order δ_μ^2 and δ_ε^2.

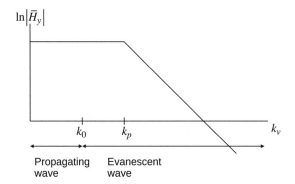

FIGURE 21.15   The spectrum $\bar{H}_y(k_x)$. For small k_x, it is flat up to a break point k_p and then decreases exponentially as k_x increases.

From (21.76), we note that the field H_y at the image plane $z = 2d_2 - d_1$ is

$$H_y(x) = \frac{1}{2\pi} \int \bar{H}_y e^{-jk_x x} dk_x,$$

$$\bar{H}_y = \frac{1}{2jk_{z1}} \frac{2 \exp(-jk_{z2}d_2 - jk_z d_2 0)}{\delta + \exp(-j2k_{z2}d_2)(2 - \delta)}. \tag{21.78}$$

The spectrum $\bar{H}_y$ has a general shape shown in Fig. 21.15. For $k_x \to 0$, $|\bar{H}_y|$ is nearly constant up to a break point k_p. Beyond k_p, it decreases exponentially.

Near k_0 and k_p, there may be some variations. The integral over k_x can be dominated by the portion of the spectrum for $k_x = 0$ to k_p and shows sinusoidal behavior proportional to $\sin(k_p x)/(k_p x)$, and thus the spot size k_p is approximately given by $k_p x_p \simeq \pi$. This gives the distance to the first minimum

$$x_p \approx \frac{\pi}{k_p}. \tag{21.79}$$

Here, k_p is approximately given by equating $\bar{H}_y$ for $k_x \to 0$ to $\bar{H}_y$ for large k_x. An approximate value for k_p is

$$k_p^2 = k^2 + \frac{1}{4d_2^2}\left[\ln\left|\frac{\delta}{2}\right|\right]^2. \tag{21.80}$$

The spot size x_p given by (21.79) shows that k_p is larger than a wavelength for small $|\delta|$. Note that as $|\delta|$ gets smaller, k_p increases and the spot size gets smaller.

Let us look at the spectrum shown in Fig. 21.15 for $k_x < k_0$, the propagation constant $k_{z1} = (k_0^2 - k_x^2)^{1/2}$ is real and the wave is propagating with a real wave number. If $k_x > k_0$, $k_{z1} = (k_0^2 - k_x^2)^{1/2}$ is purely imaginary and the wave is exponentially attenuating. This is called the "evanescent wave" and is the same as the cut-off mode (see Section 4.6). Also note that for $n = -1$, the evanescent wave does not attenuate, and the spectrum is constant for all k, and the spot size is infinitely small. This is an ideal, but an unphysical case. When $n = -1 + \delta$, the evanescent wave is attenuated exponentially for $k_x > k_p$, giving a finite spot size.

21.9 BREWSTER'S ANGLE IN NIM AND ACOUSTIC BREWSTER'S ANGLE

As stated in Section 3.6, Brewster's angle is defined as the angle of a p-polarized wave incident on the plane interface between two media, for which the reflection becomes zero. If the second medium is lossy, the reflection becomes a minimum at Brewster's angle. In Sections 3.9 and 15.1, the Zenneck wave was shown as the wave propagating over the earth's surface. Therefore, it appears that Brewster's angle and Zenneck wave are two different wave phenomena. However, mathematically these are closely related as will be shown below.

Let us consider an interface between two media. In p-polarization, the reflection coefficient for a plane wave incidence is given by

$$R_p(k_x) = \frac{(k_{z1}\varepsilon_1) - (k_{z2}/\varepsilon_2)}{(k_{z1}/\varepsilon_1) + (k_{z2}/\varepsilon_2)},$$

where

$$k_{z1} = \sqrt{k_1^2 - k_x^2}, \quad k_{z2} = \sqrt{k_2^2 - k_x^2}. \tag{21.81}$$

For a plane wave incidence at θ_i, $k_x = k_1 \sin\theta_i$. Therefore, Brewster's angle is obtained by the zero of the reflection coefficient.

$$\frac{k_{z1}}{\varepsilon_1} - \frac{k_{z2}}{\varepsilon_2} = 0. \tag{21.82}$$

On the other hand, Zenneck wave is obtained when the denominator is zero.

$$\frac{k_{z1}}{\varepsilon_1} + \frac{k_{z2}}{\varepsilon_2} = 0. \tag{21.83}$$

If medium 1 is free space, we can normalize ε and μ and write

$$\varepsilon_1 = \varepsilon_0, \quad \mu_1 = \mu_0,$$
$$\varepsilon = (\varepsilon_2/\varepsilon_0) \quad \mu = (\mu_2/\mu_0),$$
$$k_{z1} = k_0\sqrt{1 - S^2}, \quad S = \sin\theta_i, \tag{21.84}$$
$$k_{z2} = k_0\sqrt{n - S^2},$$
$$n^2 = \mu\varepsilon.$$

Brewster's angle θ_b is obtained from (21.82).

$$\sqrt{1 - S^2} = \frac{1}{\varepsilon}\sqrt{n^2 - S^2}, \quad S = \sin\theta_b.$$

By taking the square of both sides, we get

$$1 - S^2 = \frac{1}{\varepsilon^2}(n^2 - S^2), \quad n^2 = \mu\varepsilon. \tag{21.85}$$

From this, we get Brewster's angle θ_b for p-polarization.

$$S_b = \sin\theta_b = \left[\frac{\varepsilon^2 - \mu\varepsilon}{\varepsilon^2 - 1}\right]^{1/2}. \tag{21.86}$$

For ordinary material with $\mu = 1$, S_b is reduced to the well-known formula for Brewster's angle (Section 3.6).

$$S_b = \left[\frac{\varepsilon}{\varepsilon + 1}\right]^{1/2} = \left[\frac{n^2}{n^2 + 1}\right]^{1/2}. \tag{21.87}$$

Note that Brewster's angle (21.86) when $\mu \neq 1$ is different from that of free space (21.87).

Up to this point, we only considered p-polarization. For ordinary materials with $\mu = 1$, there is no Brewster's angle for s-polarization. However, there is Brewster's angle for p-polarization. The reflection coefficient (21.81) can be written for s-polarization

$$R_s = \frac{(k_{z1}\mu_1) - (k_{z2}/\mu_2)}{(k_{z1}/\mu_1) + (k_{z2}/\mu_2)}. \tag{21.88}$$

As we noted previously in (21.8), all wave characteristics in the p-polarization (TM) and the s-polarization (TE) are symmetric with the following change: ε and μ (TM) → μ and ε (TE).

Therefore Brewster's angle for s-polarization (TE) is given by (21.86) by exchanging ε and μ.

$$S_b = \left[\frac{\mu^2 - \mu\varepsilon}{\mu^2 - 1}\right]^{1/2}. \tag{21.89}$$

Brewster's angle when the medium is lossless is real and given by

$$0 < S_b^{\,2} < 1. \tag{21.90}$$

For p-polarization, this is shown in shaded areas in Fig. 21.16, and for s-polarization, it is shown in Fig. 21.17. Note that μ and ε in Fig. 21.16 are changed to ε and μ in Fig 21.17.

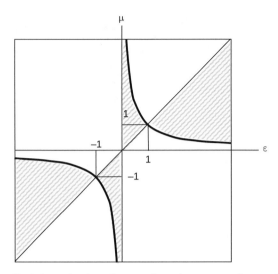

FIGURE 21.16   Shaded area is where Brewster's angle can occur for p-polarization. μ and ε are assumed real.

It may be instructive to compare the electromagnetic Brewster's angle with that of the acoustic Brewster's angle. We have discussed this in Section 3.8. We can compare the acoustic wave shown in (3.58).

$$(\nabla^2 + k^2)p = 0, \quad k^2 = \omega^2 \rho k,$$

$$\bar{v} = -\frac{1}{j\omega\rho}\nabla p. \tag{21.91}$$

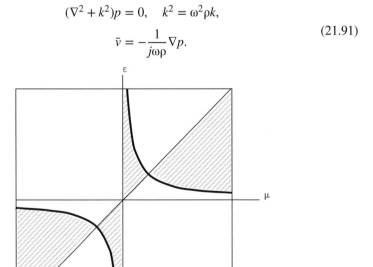

FIGURE 21.17   Shaded area is where Brewster's angle occurs for s-polarization. Note that there is no Brewster's angle for s-polarization when $\mu = 1$. Note $\mu - \varepsilon$ in Fig. 21.16 is switched to $\varepsilon - \mu$ in Fig. 21.17.

A plane wave is incident from the medium with ρ_0 and k_0 to the plane interface between two media from the medium (ρ_0, k_0) to the medium(ρ, k). To compare s-polarization with E_y, we note

$$(\nabla^2 + k^2)E_y = 0,$$

$$H_k = -\frac{1}{j\omega\mu}\frac{\partial}{\partial z}E_y, \quad k^2 = k_0^2\mu\varepsilon. \tag{21.92}$$

The plane wave is incident on the plane interface between the medium (μ_0, ε_0) and the medium (μ, ε). Noting the correspondence

$$\mu = (\rho/\rho_0),$$

$$\varepsilon = (k/k_0),$$

we see that the s-polarization corresponds to the acoustic wave. The difference is that while μ and ε can be $-\infty$ to $+\infty$, ρ and k can only be positive. Thus we have Fig. 21.18 and the range where acoustic Brewster's angle occurs. We then have

$$S_b = \left[\frac{(\rho/\rho_0)^2 - (c_0/c)^2}{(\rho/\rho_0)^2 - 1}\right]^{1/2}, \quad \frac{c_0^2}{c^2} = \frac{k\rho}{k_0\rho_0}, \tag{21.93}$$

which is already given in (3.62).

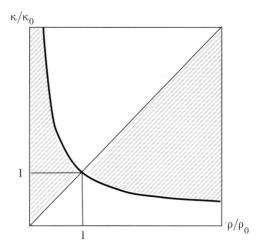

FIGURE 21.18   Acoustic Brewster's angle.

21.10 TRANSFORMATION ELECTROMAGNETICS AND INVISIBLE CLOAK

Maxwell's equations are form invariant under coordinate transformations. This means that the form of Maxwell's equations is valid in any arbitrary spatial coordinate systems. We normally deal with the simplest coordinate systems such as Cartesian rectangular or cylindrical or spherical systems. However, Maxwell's equations hold in other coordinate systems. For example, Maxwell's equations hold in a system (x, y, z) and in transformed system (x', y', z'). It is more convenient to write (x_1, x_2, x_3) for (x, y, z) and (x'_1, x'_2, x'_3) for (x', y', z') since we may be dealing with any other coordinate system. The 3×3 transformation matrix A is given by

$$
A = \begin{bmatrix} \dfrac{\partial x'_1}{\partial x_1} & \dfrac{\partial x'_1}{\partial x_2} & \dfrac{\partial x'_1}{\partial x_3} \\[2ex] \dfrac{\partial x'_2}{\partial x_1} & \dfrac{\partial x'_2}{\partial x_2} & \dfrac{\partial x'_2}{\partial x_3} \\[2ex] \dfrac{\partial x'_3}{\partial x_1} & \dfrac{\partial x'_3}{\partial x_2} & \dfrac{\partial x'_3}{\partial x_3} \end{bmatrix}.
\tag{21.94}
$$

Under this transformation, the material property of permittivity ε and permeability μ are changed to inhomogeneous and anisotropic. If the original space is free space, then the transformed medium has the relative permittivity matrix and the relative permeability matrix, $\bar{\bar{\varepsilon}}_r$ and $\bar{\bar{\mu}}_r$ given by

$$
\bar{\bar{\varepsilon}}_r = \bar{\bar{\mu}}_r = \frac{AA^T}{\det A},
\tag{21.95}
$$
$$
\det A = \text{determinant of } A.
$$

Thus, the transformed medium is inhomogeneous and anisotropic and $\bar{\bar{\varepsilon}}_r$ and $\bar{\bar{\mu}}_r$ are the same and symmetric. This also means that because $\bar{\bar{\varepsilon}}_r = \bar{\bar{\mu}}_r$, the impedance of free space is not changed in this transformed space. Thus, the transformed medium appears as anisotropic impedance matched media.

The derivation of (21.94) can be obtained by expressing Maxwell's equations in the transformed coordinate and identifying the constitutive equations (Leonhardt and Philbin, 2010; Pendry et al., 2006). Maxwell's equations in free space are

$$
\begin{aligned}
\nabla \times \bar{E} &= -j\omega\mu_0\bar{H}, \\
\nabla \times \bar{H} &= +j\omega\varepsilon_0\bar{E},
\end{aligned}
\tag{21.96}
$$

and the wave equations for $\bar{E}$ and $\bar{H}$ are

$$
\nabla \times \nabla \times \bar{E} = \omega^2\mu_0\varepsilon_0\bar{E}.
\tag{21.97}
$$

In the transformed space, Maxwell's equations are

$$
\begin{aligned}
\nabla' \times \bar{E}' &= -j\omega\bar{\bar{\mu}}\bar{H}', \\
\nabla' \times \bar{H}' &= +j\omega\bar{\bar{\varepsilon}}\bar{E}',
\end{aligned}
\tag{21.98}
$$

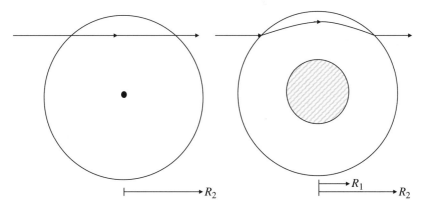

FIGURE 21.19 The original cylindrical system (ρ, ϕ, z) in free space is shown on the left, and the transformed system (ρ', ϕ', z') is shown on the right. The space $0 < \rho < R_2$ is transformed to the space $R_1 < \rho < R_2$. The origin at $\rho = 0$ is transformed to the space $0 < \rho' < R_1$. The space outside $R_2 < \rho$ and $R_2 < \rho'$ are identical and the wave behaves in the same manner. However, any object inside $\rho' < R_1$ is invisible to an outside observer, and the fields are excluded.

and the wave equations are

$$\nabla' \times (\bar{\bar{\mu}})^{-1} \nabla' \times \bar{E}' = \omega^2 \bar{\bar{\varepsilon}} \bar{E}', \tag{21.99}$$

where $\bar{\bar{\mu}} = \mu_0 \bar{\bar{\mu}}_r, \bar{\bar{\varepsilon}} = \varepsilon_0 \bar{\bar{\varepsilon}}_r$, and $\bar{\bar{\mu}}_r = \bar{\bar{\varepsilon}}_r . \nabla'$ is the operator in the transformed coordinate system.

Let us consider a simple cylindrical system shown in Fig. 21.19. The original system (ρ, ϕ, z) is transformed to (ρ', ϕ', z') given by

$$\rho' = \frac{R_2 - R_1}{R_2} \rho + R_1,$$

$$\phi' = \phi, \tag{21.100}$$

$$z' = z.$$

In a cylindrical system, we can write the transformation matrix A as follows.

$$A = \begin{bmatrix} \dfrac{\partial \rho'}{\partial \rho} & \dfrac{\partial \rho'}{\rho \partial \phi} & \dfrac{\partial \rho'}{\partial z} \\[2mm] \dfrac{\rho' \partial \phi'}{\partial \rho} & \dfrac{\rho' \partial \phi'}{\rho \partial \phi} & \dfrac{\rho' \partial \phi'}{\partial z} \\[2mm] \dfrac{\partial z'}{\partial \rho} & \dfrac{\partial z'}{\rho \partial \phi} & \dfrac{\partial z'}{\partial z} \end{bmatrix}. \tag{21.101}$$

Noting (21.99), we get

$$A = \begin{bmatrix} \dfrac{R_2 - R_1}{R_2} & 0 & 0 \\[2mm] 0 & \dfrac{\rho'}{\rho} & 0 \\[2mm] 0 & 0 & 1 \end{bmatrix},$$

$$\det A = \frac{R_2 - R_1}{R_2}\left(\frac{\rho'}{\rho}\right) = \text{determinant of A.}$$

(21.102)

Therefore $\bar{\bar{\varepsilon}}_r$ and $\bar{\bar{\mu}}_r$ in (21.94) are given by

$$\bar{\bar{\varepsilon}}_r = \bar{\bar{\mu}}_r = \begin{bmatrix} \dfrac{\rho' - R_1}{\rho'} & 0 & 0 \\[3mm] 0 & \dfrac{\rho'}{\rho' - R_1} & 0 \\[3mm] 0 & 0 & \left(\dfrac{R_2}{R_2 - R_1}\right)^2 \left(\dfrac{\rho' - R_1}{\rho'}\right) \end{bmatrix}$$

$$= [\varepsilon_{ij}].$$

(21.103)

Note that ε_{11}, ε_{22}, and ε_{33} in Fig. 21.20 show that the transformed medium is inhomogeneous and anisotropic. The above analysis is obtained based on the invariance of the distance ds given by

$$ds^2 = (d\rho)^2 + (\rho d\phi)^2 + (dz)^2.$$

(21.104)

This is specific to the cylindrical system. For a more complicate coordinate system, it requires the use of a "metric tensor."

As noted in Fig. 21.19, the waves outside the original and the transformed cylindrical space $\rho > R_2$, $\rho' > R_2$ are identical. The origin of $\rho = 0$ in the original system

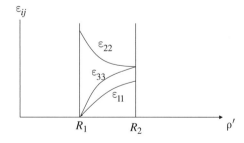

FIGURE 21.20 $\varepsilon_{ij} = \mu_{ij}$ for the cylindrical system shown in Fig. 21.19, showing the transformed medium is inhomogeneous and anisotropic.

is now transformed to the space $R_1 < \rho' < R_2$ and therefore any object in the transformed space $\rho' < R_1$ does not affect the waves outside the cylinder $\rho' > R_2$. Thus, the object is cloaked in this cylindrical space.

In spherical cloaking, if the spherical space (r, θ, ϕ) in $r \le R_2$ in free space is transformed to the space (r', θ', ϕ') with the following

$$r' = R_1 + \frac{R_2 - R_1}{R_2} r,$$
$$\theta' = \theta, \tag{21.105}$$
$$\phi' = \phi,$$

noting

$$(ds)^2 = (dr)^2 + (rd\theta)^2 + (r\sin\theta d\phi)^2,$$

we get

$$A = \begin{bmatrix} \dfrac{\partial r'}{\partial r} & \dfrac{\partial r'}{r\partial\theta} & \dfrac{\partial r'}{r\sin\theta\partial\phi} \\[2mm] \dfrac{r'\partial\theta'}{\partial r} & \dfrac{r'\partial\theta'}{r\partial\theta} & \dfrac{r'\partial\theta'}{r\sin\theta\partial\phi} \\[2mm] \dfrac{r'\sin\theta'\partial\phi'}{\partial r} & \dfrac{r'\sin\theta'\partial\phi'}{r\partial\theta} & \dfrac{r'\sin\theta'\partial\phi'}{r\sin\theta\partial\phi} \end{bmatrix}$$

$$= \begin{bmatrix} \dfrac{R_2 - R_1}{R_2} & 0 & 0 \\[2mm] 0 & \dfrac{r'}{r} & 0 \\[2mm] 0 & 0 & \dfrac{r'}{r} \end{bmatrix}, \tag{21.106}$$

$$\det A = \frac{R_2 - R_1}{R_2}\left(\frac{r'}{r}\right)^2.$$

From this, we get

$$\bar{\bar{\varepsilon}}_r = \bar{\bar{\mu}}_r = [\varepsilon_{rij}],$$
$$\varepsilon_{r11} = \frac{(r' - R_1)^2}{r'^2}\frac{R_2}{R_2 - R_1},$$
$$\varepsilon_{r22} = \frac{R_2}{R_2 - R_1}, \tag{21.107}$$
$$\varepsilon_{r33} = \frac{R_2}{R_2 - R_1},$$
and all others $\varepsilon_{rih} = 0.$

In the above, the possibility of making material to produce cloaking is noted. However, it requires that material with an inhomogeneous and anisotropic medium with specified $\bar{\bar{\varepsilon}} = \bar{\bar{\mu}}$ needs to be produced. Even though advances in metamaterials offer some possibilities, much more theoretical and experimental work is needed to realize possible cloaking materials and devices. Transformation electromagnetics may also be used for reshaping objects in scattering and miniaturizing waveguides (Ozgun and Kuzuoglu, 2010) and for flat focusing lenses, wave collimators, polarization splitters and rotators and beam benders (Kwon and Werner, 2010).

Another important point that needs to be considered is that mathematically, coordinate transformation is based on the "differential geometry," the theory of curves and surfaces, vector and tensor analysis, which are clearly beyond the scope of this book (Leonhardt and Philbin, 2010). It is, however, noted that in transforming from the original system (x_1, x_2, x_3) to the transformed space (x'_1, x'_2, x'_3), the following basic relations should be noted:

$$dx_1 = \frac{\partial x_1}{\partial x'_1} dx'_1 + \frac{\partial x_1}{\partial x'_2} dx'_2 + \frac{\partial x_1}{\partial x'_3} dx'_3,$$

$$\frac{\partial}{\partial x_1} = \frac{\partial}{\partial x'_1} \frac{\partial x'_1}{\partial x_1} + \frac{\partial}{\partial x'_2} \frac{\partial x'_2}{\partial x_1} + \frac{\partial}{\partial x'_3} \frac{\partial x'_3}{\partial x_1}. \tag{21.108}$$

This relation will be used in Section 21.11.

21.11 SURFACE FLATTENING COORDINATE TRANSFORM

In Section 21.10, we discussed "transformation electromagnetics." By transforming the coordinate system from (x, y, z) to the other coordinate system (x', y', z'), it is possible to transform the original space to the transformed space with the material property of ε and μ given in (21.95). If the original space is free space, the transformed medium is inhomogeneous and anisotropic, and it is possible to produce a cloaking. This interesting possibility created much interest. However, it should be noted that this requires the creation of new material such as metamaterial with specific inhomogeneous and anisotropic ε and μ. Though it is interesting, it has not been proven possible to physically produce such metamaterials.

In this section, we discuss a new and different application of the coordinate transformation, not for producing a new material, but for using the transformation to solve a wave scattering problem. We consider a wave propagating over a rough surface, such as ocean waves. The surface is rough and therefore the surface height is random. If we can transform the rough surface to a flat surface, then it may be possible to solve this scattering problem as a problem with a flat boundary. However, as we noted in Section 21.10, if the boundary becomes flat, then the medium becomes inhomogeneous and anisotropic, transforming a rough surface problem with free space above, to a flat surface problem with a complicated medium.

This approach called "surface flattening transform" has been proposed by Tappert (1979) and studied by Abarbanel (1998), Yeh (2001), Donohue (2000), and Wu (2005) using the path integral approach. In this section, we discuss key points of the flattening transform. The surface height is given by $\zeta(x, y)$ and the coordinate system (x, y, z) is now transformed to (x', y', z').

$$
\begin{aligned}
x' &= x, \\
y' &= y, \\
z' &= z - \zeta(x, y).
\end{aligned}
\tag{21.109}
$$

The wave equation is then transformed by using the coordinate transformation. The transformation matrix A in (21.93) is then given by

$$
A = \begin{bmatrix} \dfrac{\partial x'}{\partial x} & \dfrac{\partial x'}{\partial y} & \dfrac{\partial x'}{\partial z} \\[2mm] \dfrac{\partial y'}{\partial x} & \dfrac{\partial y'}{\partial y} & \dfrac{\partial y'}{\partial z} \\[2mm] \dfrac{\partial z'}{\partial x} & \dfrac{\partial z'}{\partial y} & \dfrac{\partial z'}{\partial z} \end{bmatrix} = \begin{bmatrix} 1 & 0 & 0 \\[2mm] 0 & 1 & 0 \\[2mm] -\dfrac{\partial \zeta}{\partial x} & -\dfrac{\partial \zeta}{\partial y} & 1 \end{bmatrix}.
\tag{21.110}
$$

The medium above $z' = 0$ is then transformed to an inhomogeneous and anisotropic medium with $\bar{\bar{\varepsilon}}$ and $\bar{\bar{\mu}}$ given below.

$$
\frac{\bar{\bar{\varepsilon}}}{\varepsilon_0} = \frac{\bar{\bar{\mu}}}{\mu_0} = \frac{AA^T}{\det A} = \begin{bmatrix} 1 & 0 & -\dfrac{\partial \zeta}{\partial x} \\[3mm] 0 & 1 & -\dfrac{\partial \zeta}{\partial y} \\[3mm] -\dfrac{\partial \zeta}{\partial x} & -\dfrac{\partial \zeta}{\partial y} & \left(\dfrac{\partial \zeta}{\partial x}\right)^2 + \left(\dfrac{\partial \zeta}{\partial y}\right)^2 + 1 \end{bmatrix}.
\tag{21.111}
$$

Note that A^T is the transpose of A and $\dfrac{\bar{\bar{\varepsilon}}}{\varepsilon_0}$ and $\dfrac{\bar{\bar{\mu}}}{\mu_0}$ are the same and symmetric. Maxwell's equations are form invariant under coordinate transformation. This means that we can write Maxwell's equations in (x, y, z) and (x', y', z') in the same form. We have

$$
\begin{array}{ll}
\nabla \times \bar{E} = i\omega\mu_0\bar{H} & \nabla' \times \bar{E}' = i\omega\bar{\bar{\mu}}\bar{H}' \\
\nabla \times \bar{H} = -i\omega\varepsilon_0\bar{E} & \Rightarrow \quad \nabla' \times \bar{H}' = -i\omega\bar{\bar{\varepsilon}}\bar{E}' \\
\nabla \times \nabla \times \bar{E} = \omega^2\mu_0\varepsilon_0\bar{E} & \nabla' \times (\bar{\bar{\mu}})^{-1}\nabla' \times \bar{E}' = \omega^2\bar{\bar{\varepsilon}}\bar{E}' \\
\text{Rough surface with free space} & \text{Flat surface with random media}
\end{array}
\tag{21.112}
$$

Note that in (x', y', z') coordinates, impedance is not changed giving "anisotropic impedance matched media." By solving Maxwell's equation on the right side of (21.111), we can solve the problem in free space with a rough surface.

Let us consider a scalar (acoustic) wave scattering by a rough surface. We use the coordinate transformation (21.108). Consider a scalar wave equation in free space above a rough surface with Dirichlet boundary condition.

$$(\nabla^2 + k^2)G = -\delta(x - x_0)\delta(y - y_0)\delta(z - z_0). \tag{21.113}$$

The transformation from the original space (x, y, z) to the transformed space (x', y', z') with a flat surface can be done by using the coordinate transformation.

$$\frac{\partial}{\partial x} = \frac{\partial}{\partial x'}\frac{\partial x'}{\partial x} + \frac{\partial}{\partial y'}\frac{\partial y'}{\partial x} + \frac{\partial}{\partial z'}\frac{\partial z'}{\partial x} \tag{21.114}$$

Noting (21.108), we get

$$\frac{\partial}{\partial x} = \frac{\partial}{\partial x'} - \frac{\partial \zeta}{\partial y'}\frac{\partial}{\partial z'}. \tag{21.115}$$

Continuing this process, we get

$$\frac{\partial^2}{\partial x^2} = \frac{\partial^2}{\partial x'^2} + 2\left(-\frac{\partial \zeta}{\partial x}\right)\frac{\partial^2}{\partial z'\partial x'} + \left(-\frac{\partial \zeta}{\partial x}\right)^2\frac{\partial^2}{\partial z'^2} + \left(-\frac{\partial^2 \zeta}{\partial x'^2}\right)\frac{\partial}{\partial z'},$$

$$\frac{\partial^2}{\partial y^2} = \frac{\partial^2}{\partial y'^2} + 2\left(-\frac{\partial \zeta}{\partial y}\right)\frac{\partial^2}{\partial z'\partial y'} + \left(-\frac{\partial \zeta}{\partial y}\right)^2\frac{\partial^2}{\partial z'^2} + \left(-\frac{\partial^2 \zeta}{\partial y'^2}\right)\frac{\partial}{\partial z'}, \quad (21.116)$$

$$\frac{\partial^2}{\partial z^2} = \frac{\partial^2}{\partial z'^2}.$$

We can then obtain the following expression for the wave equation in transformed space:

$$(\nabla^2 + k^2 + F)G = -\delta(x' - x_0)\delta(y' - y_0)\delta(z' - [z_0 - (x, y)]),$$

$$F = -2\nabla\zeta\nabla'\frac{\partial}{\partial z'} - (\nabla^2\zeta)\frac{\partial}{\partial z'} + |\nabla\zeta|^2\frac{\partial^2}{\partial z'^2}. \tag{21.117}$$

The boundary condition is

$$G = 0 \quad \text{on flat surface}$$

Note that F represents the inhomogeneous medium containing the slope $\nabla\zeta$, the radius of the curvature $\nabla^2\zeta$, and the magnitude of the slope $|\nabla\zeta|^2$.

This problem has been studied for rough surfaces, and imaging of objects near surface has been studied using the time-reversal technique (Ishimaru et al., 2013).

CHAPTER 22

TIME-REVERSAL IMAGING

The idea of time-reversal (TR) can be traced to the phase-conjugate mirror and retroreflectance. It is also called the "time-reversal mirror (TRM)" from a localized source incident on a TMR as reflected back to the original source. This is equivalent to producing a time-reversed field at the phase-conjugate mirror, and the reflected field propagates back to focus on the original source. In this back propagation process, the time-reversed field traces back through the original path even though the path may include complicated multiscattering media. The retrodirective array at microwave exhibits similar behavior.

The optical phase conjugation and microwave retrodirective array are mostly for narrow band operation where the time reversal is achieved by the phase conjugation. In acoustics, however, the waves are often broadband and it may be necessary to consider the time reversal for wide-band pulse operation. In this section, we give fundamental theories of the TRM, conjugate mirror, and space–time time-reversal focused imaging. We will discuss décomposition de l'opérateur de retournement temporel (DORT), singular value decomposition (SVD) and super resolution as applied to time-reversal imaging, TR-MUSIC (multiple signal classification), and related beam former and synthetic aperture radar (SAR) imaging. We will also include the use of SVD for communication in random media. Extensive discussions have been presented on time reversal. Basic theories have been developed by Fink and Prada (1994, 2003, 1997, 1995, 1996, 2000), Devaney (2003, 2005, 2005, 2012), Yavuz and Teixeira (2008, 2006), Borcea, Papanicolaou et al. (2002), and Ishimaru et al. (2012).

Electromagnetic Wave Propagation, Radiation, and Scattering: From Fundamentals to Applications,
Second Edition. Akira Ishimaru.
© 2017 by The Institute of Electrical and Electronic Engineers, Inc. Published 2017 by John Wiley & Sons, Inc.

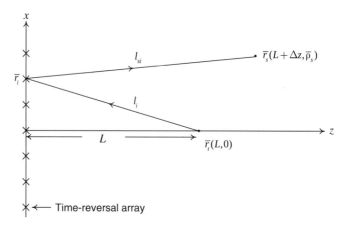

FIGURE 22.1   A point source at $\bar{r}_t$ emits a pulse which is received by a receiver located at $\bar{r}_i$, time reversed, and backpropagated into the same medium and observed at the point $\bar{r}_s$. The plane at $z = 0$ is called "time-reversal mirror" or "phase-conjugate mirror."

22.1 TIME-REVERSAL MIRROR IN FREE SPACE

Let us first consider the problem shown in Fig. 22.1. A pulse is emitted by a point source located at $\bar{r}_t$. This is received by the array element located at $\bar{r}_i$. This received pulse is time reversed and backpropagated into the same medium and observed at the point $\bar{r}_s$. The emitted pulse $f_t(t)$ and its spectrum $U(\omega)$ are given by

$$f_t(t) = \frac{1}{2\pi} \int U(\omega)\exp(-i\omega t)d\omega,$$

$$U(\omega) = \int f_t(t)\exp(+i\omega t)dt.$$

(22.1)

This pulse is propagated to $\bar{r}_i$ and is received at $\bar{r}_i$.

$$f(\bar{r}_i, t) = \frac{1}{2\pi} \int G_i(\bar{r}_i, \bar{r}_t, \omega)U(\omega)\exp(-i\omega t)d\omega,$$

(22.2)

where G_i is Green's function observed at $\bar{r}_i$ due to a point source at $\bar{r}_t$.

This pulse $f(\bar{r}_i, t)$ is now time reversed. The time-reversed pulse is given by

$$f_{TR}(\bar{r}_i, t) = f(\bar{r}_i, -t)$$

$$= \frac{1}{2\pi} \int G_i(\bar{r}_i, \bar{r}_t, \omega)U(\omega)e^{+i\omega t}d\omega.$$

(22.3)

The spectrum F_{TR} of the time-reversed pulse f_{TR} is then given by

$$
\begin{aligned}
F_{TR}(\bar{r}_i, \omega) &= \int f_{TR}(\bar{r}_i, t) e^{+i\omega t} dt \\
&= \frac{1}{2\pi} \int [G_i(\bar{r}_i, \bar{r}_t, \omega') U(\omega') e^{+i\omega' t} d\omega'] e^{+i\omega t} dt,
\end{aligned}
\tag{22.4}
$$

where we used ω' to differentiate ω in (22.3) and ω in (22.4).

Noting that

$$
\int e^{+i(\omega' + \omega)t} d\omega = 2\pi\delta(\omega' + \omega),
$$

we get

$$
\begin{aligned}
F_{TR}(\bar{r}_i, \omega) &= G_i(\bar{r}_i, \bar{r}_t, -\omega) U(-\omega) \\
&= G_i^*(\bar{r}_i, \bar{r}_t, \omega) U^*(\omega).
\end{aligned}
\tag{22.5}
$$

Note that the Fourier spectrum of $G_i(-\omega)$ and $U(-\omega)$ evaluated at $(-\omega)$ is equal to the complex conjugates $G_i^*(\omega)$ and $U^*(\omega)$.

The time-reversed pulse at $\bar{r}_i$ then propagates to $\bar{r}_s$, and the observed pulse at $\bar{r}_s$ is given by

$$
f_{TR}(\bar{r}_s, \bar{r}_i, t) = \frac{1}{2\pi} \int G_i(\bar{r}_s, \bar{r}_i, \omega) G_i^*(\bar{r}_i, \bar{r}_t, \omega) U^*(\omega) e^{-i\omega t} d\omega.
\tag{22.6}
$$

This is the final expression for the observed pulse at $\bar{r}_s$, due to the pulse emitted at $\bar{r}_t$, and time reversed at $\bar{r}_i$. Note that $G_i(\bar{r}_s, \bar{r}_i, \omega)$ is Green's function from $\bar{r}_i$ to $\bar{r}_s$.

The total pulse at $\bar{r}_s$ is then given by the sum of contributions from all elements at $\bar{r}_i$.

$$
f_{TR}(\bar{r}_s, t) = \sum_i f_{TR}(\bar{r}_s, \bar{r}_i, t).
\tag{22.7}
$$

This is the general formula for the time-reversal field at $\bar{r}_s$ when the TRM is illuminated by a source located at $\bar{r}_t$. This formula (22.7) can be compactly expressed using the following matrices.

We let the matrix g represent all the waves incident at $\bar{r}_i$ ($i = 1, 2, \ldots, N$) due to a source at $\bar{r}_t$.

$$
g = [G_1 \quad G_2 \quad \cdots \quad G_N]^T,
\tag{22.8}
$$

where $G_i = G_i(\bar{r}_i, \bar{r}_t, \omega)$ and $[\ ]^T$ is the transposed matrix.

We let

$$g_s = [G_{s1} \quad G_{s2} \quad \cdots \quad G_{sN}]^T, \tag{22.9}$$

where $G_{si} = G_{si}(\bar{r}_s, \bar{r}_i, \omega)$.

Note that G_i and G_{si} are Green's functions.

Using these matrix representations, we write (22.7) in the following compact matrix form:

$$f_{TR}(\bar{r}_s, t) = \sum_i \frac{1}{2\pi} \int g_s^T g_s^* U^* e^{-i\omega t} d\omega. \tag{22.10}$$

This matrix form will be used in the subsequent sections.

As an example, we take a Gaussian modulated pulse.

$$f_t(t) = A_o \exp\left(-\frac{t^2}{T^2} - i\omega_o t\right),$$

$$U(\omega) = A_o \frac{2\sqrt{\pi}}{\Delta\omega} \exp\left[-\frac{(\omega - \omega_o)^2}{\Delta\omega^2}\right], \tag{22.11}$$

where $\Delta\omega = 2/T_o$ is the bandwidth and ω_o is the carrier frequency. We also note that

$$G_i(\bar{r}_i, \bar{r}_t, \omega) = \frac{\exp(ik\ell_i)}{4\pi\ell_i},$$

$$G_{si}(\bar{r}_s, \bar{r}_i, \omega) = \frac{\exp(ik\ell_{si})}{4\pi\ell_{si}}, \tag{22.12}$$

$$k = \frac{\omega}{C}.$$

Substituting (22.11) and (22.12) in (22.6) and (22.7) and performing integration with respect to ω, we get the observed pulse at $\bar{r}_s$.

$$f_{TR}(\bar{r}_s, t) = \sum_i f_{TR}(\bar{r}_s, \bar{r}_i, t),$$

where

$$f_{TR}(\bar{r}_s, \bar{r}_i, t) = A_o \frac{\exp\left[ik_o(\ell_{si} - \ell_i - ct) - \frac{k_o^2}{4}(\ell_{si} - \ell_i - ct)^2\left(\frac{\Delta\omega}{\omega_o}\right)^2\right]}{(4\pi\ell_i)(4\pi\ell_{si})}, \tag{22.13}$$

$$k_o = \omega_o/c.$$

Note that the time of travel for the pulse from $\bar{r}_t$ to $\bar{r}_i$ is (ℓ_i/c) which is time reversed to become $-(\ell_i/c)$. And the time of travel from $\bar{r}_i$ to $\bar{r}_s$ is (ℓ_{si}/c). Therefore $t = 0$ corresponds to the time when the time-reversed pulse is observed when the pulse center is at the source point $\bar{r}_s = \bar{r}_t(\ell_{si} = \ell_i)$.

In order to examine the time-reversal pulse (22.13), let us consider a narrow band pulse. We can then ignore the second term of the exponent with $(\Delta\omega/\omega)^2$ and note that $(1/\ell_i)$ and $(1/\ell_{si})$ can be approximated by $1/L$.

In the plane of $z = L$, we can approximate ℓ_{si} and ℓ_i by

$$\ell_{si} = L + \frac{(x - x_i)^2}{2L},$$
$$\ell_i = L + \frac{x_i^2}{2L}, \tag{22.14}$$

with $\bar{\rho}_s = x\hat{x}$, and $\bar{r}_i = x_i\hat{x}$.

We then get

$$\ell_{si} - \ell_i \approx \frac{1}{2L}\left((x - x_i)^2 - x_i^2\right)$$
$$= \frac{1}{2L}\left(x^2 - 2xx_i\right). \tag{22.15}$$

With these approximations, we get the time-reversed pulse at $t = 0$.

$$f_{TR}(\bar{r}_s, t)1 \cong \sum_i \frac{A_o\exp\left[ik_o\dfrac{\left(x^2 - 2xx_i\right)}{2L}\right]}{(4\pi L)^2}. \tag{22.16}$$

Noting that $x_i = nd$, $n = -\frac{N}{2}$ to $\frac{N}{2}$, and d is the array element spacing, we get

$$f_{TR}(\bar{r}_s, t)1 \cong A_o \frac{e^{ik_o\frac{x^2}{2L}}}{(4\pi L)^2} \sum_n e^{ik_o\frac{x}{L}nd}$$
$$= A_o \frac{e^{ik_o\frac{x^2}{2L}}}{(4\pi L)^2} \frac{\sin\left(\dfrac{k_o xNd}{2L}\right)}{\sin\left(\dfrac{k_o xd}{2L}\right)}, \tag{22.17}$$

where x is the transverse distance. The spot size W of the time-reversed pulse is approximately given by

$$\frac{k_o WNd}{2L} \approx \pi.$$

This gives the transverse spot size of

$$W = \frac{L}{Nd} = \lambda_o(L/a). \tag{22.18}$$

As expected, λ_o is the wavelength at the carrier frequency, and $a = Nd$ is the total size of the aperture.

The axial spot size can be obtained approximately from (22.13). Note that on axis, using $\ell_i = L$, $\ell_{si} = L + \Delta z$, and $t = 0$, we get the axial spot size Δ.

$$\frac{k_o^2}{4}(\ell_{si} - \ell_i - ct)^2 \left(\frac{\Delta\omega}{\omega_o}\right)^2 \approx 1,$$

and the spot size Δz is approximately given by

$$\Delta z \approx \left| \frac{2c}{\Delta\omega} \right|. \tag{22.19}$$

The plane at $z = 0$ is called the "time-reversal mirror" or the "phase-conjugate mirror," because the wave incident on a point in the plane is time reversed and reemitted into the same medium. The spectrum of the time-reversed pulse is the complex conjugate of the incident pulse. The transverse spot size at the origin is the same as that of the wave from an aperture at the TRM focused on the pulse origin, and the axial spot size is given by the bandwidth of the pulse (Fig. 22.2).

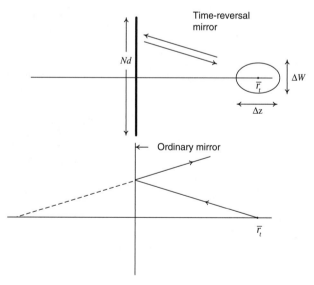

FIGURE 22.2   Time-reversal (conjugate) mirror and ordinary mirror. ΔW = transverse spot size, Δz = axial spot size.

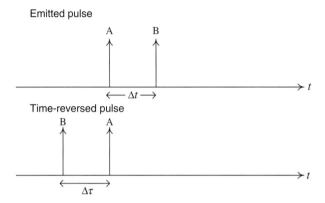

FIGURE 22.3   Two pulses A and B are emitted. The pulse B is emitted Δt time late. For the time-reversed pulse, the pulse B arrives first and the pulse A arrives $\Delta \tau$ time later.

We take one more example. Two delta function pulses are emitted at $t = 0$ and $t = \Delta t$ from the point source located at $\bar{r}_t$. Then we have

$$f_t(t) = A\delta(t) + B\delta(t - \Delta t),\tag{22.20}$$

where A and B are constant.
    The spectrum of $f_t(t)$ is

$$U(\omega) = A + Be^{i\omega\Delta t}.\tag{22.21}$$

Substituting this in (22.6) and (22.7), we get the time-reversed pulse at $\bar{r}_s$.

$$f_{TR}(\bar{r}_s, \bar{r}_i, t) = \frac{1}{(4\pi\ell_{si})(4\pi\ell_i)} \left[A\delta\left(t - \frac{\ell_{si} - \ell_i}{c}\right) + B\delta\left(t + \Delta t - \frac{\ell_{si} - \ell_i}{c}\right)\right].$$
$$\tag{22.22}$$

Note that the pulse A is emitted first, but the time-reversed pulse B arrives first (see Fig. 22.3). This represents the time-reversal process that the pulse emitted first at $\bar{r}_t$ is received last at $\bar{r}_s$ (first in, last out).

22.2 SUPER RESOLUTION OF TIME-REVERSED PULSE IN MULTIPLE SCATTERING MEDIUM

In Section 22.1, we discussed the time-reversed pulse in free space. It was shown in (22.18) that the transverse spot size is $\lambda_o(L/a)$ as expected. L is the distance from the array to the source point and a is the total size of the array.
    A question is what happens to the resolution if the medium is not free space, but is a multiple scattering medium.

Intuitively, we may expect that multiple scattering may make the resolution worse. However, numerical and experimental work show that the resolution improves under multiple scattering. In this section, we present a theoretical analysis of this improved resolution, called "super resolution."

Let us note from (22.10) that the time-reversed pulse at $(\bar{r}_s, t)$ is given by

$$f_{TR}(\bar{r}_s, t) = \sum_i \frac{1}{2\pi} \int g_s^T g^* U^* e^{-i\omega t} d\omega, \qquad (22.23)$$

where g is Green's function from $\bar{r}_t$ to $\bar{r}_i$ and g_s is Green's function from $\bar{r}_i$ to $\bar{r}_s$.

In a multiple scattering random medium, Green's functions g_s and g are random functions and therefore the time-reversed pulse f_{TR} is random.

Let us consider the average pulse f_{TR} which is given by Ishimaru et al. (2007).

$$\overline{f_{TR}(\bar{r}_s, t)} = \sum_i \frac{1}{2\pi} \int \overline{g_s^T g^*} U^* e^{-i\omega t} d\omega. \qquad (22.24)$$

The correlation function $\overline{g_s^T g^*}$ is called the mutual coherence function and represents the correlation of fields originated from $\bar{r}_i$ and observed at $\bar{r}_s$ and $\bar{r}_t$.

The mutual coherence function has been obtained for random distribution of particles and turbulence (Ishimaru, 1997).

Here we use the following approximate form of the mutual coherence function in the transverse plane at $z = L$.

$$\overline{g_s^T g^*} \approx \Gamma_o \exp\left(-\frac{\rho_s^2}{\rho_o^2}\right), \qquad (22.25)$$

where Γ_o is $\overline{g_s^T g^*}$ for free space given in (22.17), ρ_o is the coherence length, and ρ_s is the transverse distance between $\bar{r}_s$ and $\bar{r}_t$ at $z = L$.

The free space spot size $W_o = \lambda_o(L/a)$ is given in (22.18). Therefore approximately, the time-reversal average field $\overline{f_{TR}}$ is given by

$$\overline{f_{TR}} \approx \frac{1}{(4\pi L)^2} \exp\left[-\left(\frac{1}{W_o^2} + \frac{1}{\rho_o^2}\right)\rho_s^2\right]. \qquad (22.26)$$

From this, we conclude that the spot size W in random medium is given by

$$\frac{1}{W^2} = \frac{1}{W_o^2} + \frac{1}{\rho_o^2}. \qquad (22.27)$$

Noting that $W_o = \lambda_o(L/a)$, we can obtain the equivalent aperture size a_e resulting from the spot size W.

$$W_o = (\lambda_o L)/a,$$
$$W = (\lambda_o L)/a_e.$$

This results in the following equivalent effective aperture size a_e.

$$a_e^2 = a^2 + (\lambda_o L)^2/\rho_o^2. \tag{22.28}$$

Note, therefore, that if the coherence length is much greater than the free space spot size W_o, the equivalent aperture a_e is not much different from a and the random medium does not have much effect on the spot size W.

On the other hand, if ρ_o is much smaller than W_o, then the spot size W is smaller than free space spot size W_o. This is a mathematical explanation of super resolution.

Another physical interpretation is that when ρ_o is smaller than W_o, the angular spread $\Delta\theta = (\lambda/\rho_o)$ becomes greater than (λ/W_o) giving equivalent aperture size a_e greater than the actual array size a.

22.3 TIME-REVERSAL IMAGING OF SINGLE AND MULTIPLE TARGETS AND DORT (DECOMPOSITION OF TIME-REVERSAL OPERATOR)

As we discussed in Section 22.2, the TRM can refocus the incident wave to the original source. This characteristic can be used to form the image of targets. One such method is called the DORT.

Let us consider a TRM at $z = 0$ (Fig. 22.4).

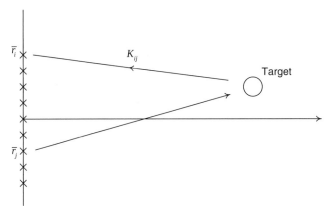

FIGURE 22.4   Multistatic data matrix $K = [K_{ij}]$, K is an $N \times N$ matrix, and K_{ij} is the field received at $\bar{r}_i$ when the wave is emitted at $\bar{r}_j$.

One of the matrices in time-reversal imaging is the "multistatic data matrix" $K = [K_{ij}]$. This is a symmetric $N \times N$ matrix and its element K_{ij} is the signal received at the ith antenna located at $\bar{r}_i$ when the signal wave is emitted from the jth antenna and reflected by targets. Consider the wave v ($N \times 1$ matrix) applied to the N elements. The wave propagates and is reflected by targets and received by the N receivers.

The received signal is then given by $N \times 1$ matrix Kv where K is the $N \times N$ multistatic data matrix. This received signal is then time reversed at each terminal giving the next transmitted signal $(Kv)^*$.

This process can continue and if this sequence converges, then except for a constant σ, we should get the transmitted signal $(Kv)^*$ equal to the original transmitted signal v. The constant σ represents the power lost by scattering and diffraction.

$$(Kv)^* = \sigma v. \tag{22.29}$$

Rewriting this, we have

$$Kv = \sigma^* v^*.$$

This can be converted to an eigenvalue equation by multiplying both sides by K^*.

$$K^* K v = \lambda v, \tag{22.30}$$

$$\lambda = |\sigma^2| = \text{eigenvalue}.$$

The $N \times N$ matrix $T = K^* K$ is called the time-reversal matrix and is a Hermitian matrix.

$$T' = T, \tag{22.31}$$

where T' is the transpose conjugate of T and K is symmetric, $K' = K^*$.

The eigenvalue equation can be written as

$$Tv = \lambda v. \tag{22.32}$$

Note that the eigenvalue λ for the Hermitian matrix is real and nonnegative. Since T is Hermitian, we have

$$v' T v = v' K^* K v = K v^2,$$
$$v' \lambda v = \lambda v^2, \tag{22.33}$$

and therefore, we have

$$\lambda = \frac{K v^2}{v^2} = \text{real and positive}, \tag{22.34}$$

where $v^2 = \text{vector norm} = \sum_n |v|^2$.

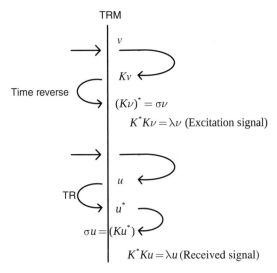

FIGURE 22.5 Excitation signal v is eigenvector of the time-reversal matrix $T = K^*K$. Received signal u is eigenvector of T^*. If the same array is used for excitation and receiver, $u = v^*$.

This shows that the eigenvalue λ is the ratio of the received power to the input power, and that the largest eigenvalue represents the largest target giving the largest reflected power.

By writing the eigenvalues in the decreasing order,

$$\lambda_1 > \lambda_2 > \lambda_3 > \cdots \lambda_M. \tag{22.35}$$

This shows that the order of the eigenvalues represents the decreasing reflecting power of targets.

Figure 22.5 sketches the process using v as the excitation signal. Alternatively, we can use u as the received signal and get an alternative eigenvalue equation.

$$T^*u = \lambda u. \tag{22.36}$$

The representations (22.33) and (22.36) are closely related to the SVD to be discussed in Section 22.5.

Once we obtain K (multistatic data matrix) and the eigenvectors v and eigenvalues λ, we can obtain the imaging function.

First note that the transmitted signal is given by $(Kv)^* = \sigma v$ (22.29). In order to obtain the image, we consider the imaging function ψ_{TR}.

$$\psi_{TR} = g_s^T \sigma v, \tag{22.37}$$

where $g_s^T = [G_{s1}(\bar{r}_1, \bar{r}_s) \cdots G_{sN}(\bar{r}_N, \bar{r}_s)]$ and v is the $N \times 1$ eigenvector obtained from (22.32).

The $N \times 1$ vector g_s is called the "steering vector."

It is equivalent to sending a wave from each receiver focused on the search point $\bar{r}_s$. In imaging, we may not know the location and the strength of the targets. We only know or measure the multistatic data matrix $K = [K_{ij}]$ and the eigenvectors.

In order to obtain the image, we use the signal σv and focus the wave to the search point $\bar{r}_s$ by using the steering vector g_s. If the steering vector is Green's function for the imaging problem, then the wave will be focused on the target. In general, however, we may not know Green's function and therefore we use an approximate g_s such as Green's function in homogeneous background. The steering function g_s is the same as the focusing matched filter function used in SAR.

As was discussed in Section 22.2, we can include the time dependence. We let $U(\omega)$ be the temporal spectrum as given in (22.1). We can use this pulse as the input signal to form K. Then, we obtain the space–time time-reversal imaging. Following the space–time formulation given in (22.6) and (22.7), we get the space–time imaging function

$$\Psi_{TR} = \frac{1}{2\pi} \int |U(\omega)|^2 \sigma g_s^T v e^{-i\omega t} d\omega, \qquad (22.38)$$

where the spectrum $U(\omega)$ is applied to the input as well as the matched filter function, resulting in $|U(\omega)|^2$. The time $t = 0$ corresponds to the time when the pulse for the steering vector g_s is observed at the target as shown in Section 22.1.

DORT is the French acronym for "décomposition de l'opérateur de retournement temporel" (decomposition of time-reversal operator), and was proposed by Prada and Fink (1994, 1995, 2001). It provides theoretical study of the iterative time-reversal process and shows that in a multitarget media, the brightest target is associated with the eigenvector of greatest eigenvalue. The discussion leading to (22.29) in this section can also be explained by DORT.

In Fig. 22.6, we show the iterative time-reversal process. Let us start with the first signal E_o ($N \times 1$ matrix) applied to the transmitting array. E_o is reflected by the targets and the reflected signal is then given by KE_o.

This received signal is time reversed at each terminal and becomes $E_1 = K^* E_o^*$.

This is emitted and received and given by KE_1. The process can continue and after n iterations, we get

$$E_{2n} = (K^* K)^n E_o = T^n E_o,$$
$$K_{2n+1} = (K^* K)^n K^* E_o^* = T^n (K^* E_o).$$

If we express the first signal E_o as a sum of eigenvector v_p, we have

$$E_o = v_1 + v_2 + \cdots + v_p.$$

Then after $2n$ iterations, we have

$$E_{2n} = T^n E_o$$
$$= \lambda_1^n v_1 + \lambda_2^n v_2 + \cdots + \lambda_p^n v_p.$$

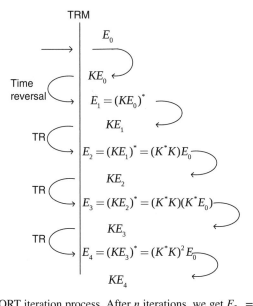

TRM

FIGURE 22.6   DORT iteration process. After n iterations, we get $E_{2n} = (K^*K)^n E_0$, $E_{2n+1} = (K^*K)^n K^* E_0$.

For a large number of n,

$$\lambda_1 > \lambda_2 > \lambda_3 \cdots > \lambda_p.$$

The first term becomes dominant, and we get

$$E_{2n} \approx \lambda_1^n v_1.$$

Therefore we conclude that after many iterations, the signal at each element of the array converges to the eigenvector v_1 of the largest eigenvector λ_1 shown in (22.30).

22.4 TIME-REVERSAL IMAGING OF TARGETS IN FREE SPACE

It may be instructive to examine the imaging of point targets in free space. Let us start with a single target in free space. The point target is located at $\bar{r}_t$. The multistatic data matrix K is easily obtained.

$$K = gg^T = [K_{ij}],$$
$$K_{ij} = G(\bar{r}_i, \bar{r}_t, \omega) G(\bar{r}_j, \bar{r}_t, \omega),$$

(22.39)

where g is $N \times 1$ matrix of Green's function.

$$g = [G(\bar{r}_1, \bar{r}_t, \omega) \quad G(\bar{r}_2, \bar{r}_t, \omega) \quad \cdots \quad G(\bar{r}_N, \bar{r}_t, \omega)]^T.$$

The time-reversal matrix T is then given by

$$T = K^*K = g^*g'gg^T, \tag{22.40}$$

where g' is the complex transpose of g.

Noting that $g'g$ is a constant, we write

$$T = g^2 g^* g^T, \tag{22.41}$$

where

$$g^2 = \text{vector norm} = g'g = \sum_i |G(\bar{r}_i, \bar{r}_t, \omega)|^2.$$

The eigenvector v and the eigenvalue λ can be formed from (22.41).

$$Tv = \lambda v, \tag{22.42}$$

where $v = g^*$ and $\lambda = g^4$. To obtain this, note $g^*g^Tg^* = g^2g^*$.

For a single target in free space, this is the only nonzero eigenvalue and all other eigenvalues are zero. However, this is true only for free space. In general, if a medium is random, the eigenvalues are not always zero.

Using this eigenvector and eigenvalue, we can construct the imaging function

$$\psi_{TR} = \frac{1}{2\pi} \int |U(\omega)|^2 \sqrt{\lambda} g_s^T v \exp(-i\omega t), \tag{22.43}$$

where $v = g^*$.

Note that the time-reversal imaging function (22.43) is identical to the refocused pulse (22.10) from the TRM, except an additional spectrum $U(\omega)$ as discussed after (22.38).

For steering vector g_s, we use (Fig. 22.7)

$$g_s = [G_s(\bar{r}_1, \bar{r}_s, \omega) \cdots G_s(\bar{r}_N, \bar{r}_s, \omega)]^T. \tag{22.44}$$

Noting that

$$G_s(\bar{r}_i, \bar{r}_t, \omega) = \frac{exp[ik\ell_i]}{4\pi\ell_i},$$
$$G_s(\bar{r}_i, \bar{r}_s, \omega) = \frac{exp[ik\ell_{si}]}{4\pi\ell_{si}}, \tag{22.45}$$

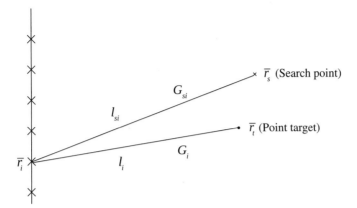

FIGURE 22.7   Imaging of a single target at $\bar{r}_t$. The search point is at $\bar{r}_s$.

and using the Gaussian spectrum (22.11), we get

$$
\Psi_{TR} = \sum_i \sqrt{\lambda} \frac{\exp\left[ik_o(\ell_{si} - \ell_i - ct) - \frac{k_o^2}{8}(\ell_{si} - \ell_i - ct)^2\left(\frac{\Delta\omega}{\omega_o}\right)^2\right]}{(4\pi\ell_i)(4\pi\ell_{si})}, \qquad (22.46)
$$

where

$$
\sqrt{\lambda} = g^2.
$$

As we note from (22.43) and (22.44), $v = g^*$ and $g_s = g(\bar{r} = \bar{r}_t)$. Therefore, $t = 0$ corresponds to the time when the observed pulse is centered at the target point. As we compare the TR imaging in free space (22.46) with that of the time-reversal pulse (22.13), they are substantially the same and therefore the resolution of the TR imaging in free space is the same as the TR-pulse from the TR mirror. The transverse spot size should be $\lambda_o(L/a)$ and the axial spot size is $|2c/\Delta\omega|$.

Let us next consider multiple targets located at $\bar{r}_m$, $m = 1, 2, \ldots, M$. The multi-static data matrix K is then given by

$$
K = \sum_m v_m g_m g_m^T, \qquad (22.47)
$$

where

$$
g_m = [G(\bar{r}_1, \bar{r}_m, \omega) \quad G(\bar{r}_2, \bar{r}_m, \omega) \quad \cdots \quad G(\bar{r}_N, \bar{r}_m, \omega)]^T,
$$

$$
v_m = \text{strength of the target at } \bar{r}_m.
$$

The time-reversal matrix T is then given by

$$T = K^*K$$

$$= \sum_m \sum_{m'} v_m v_{m'} g_m^* g_m' g_{m'} g_{m'}^T. \tag{22.48}$$

$H = g_m' g_{m'}^T$ is a constant and is given by

$$H = \sum_i G^*(\bar{r}_i, \bar{r}_m) G(\bar{r}_i, \bar{r}_{m'}). \tag{22.49}$$

This can be thought of as the correlation of the fields at $\bar{r}_m$ and $\bar{r}_{m'}$ where the point source is located at $\bar{r}_i$. This is called the "point-spread function."

Rewriting (22.48) as a sum of the terms with $m = m'$ and others, we get

$$T = \sum_m |v_m|^2 H(m = m') g_m^* g_m^T + \sum_m \sum_{\substack{m' \\ m \neq m'}} v_m v_m' H(m \neq m') g_m^* g_{m'}^T, \tag{22.50}$$

where

$$H(m, m') = \sum_i G^*(\bar{r}_i, \bar{r}_m) G(\bar{r}_i, \bar{r}_{m'}) = g^2.$$

Note that the first term of (22.50) represents the sum of the time-reversal matrix of each target at $\bar{r}_m$ and the second term represents the correlation between the target at $\bar{r}_m$ and $\bar{r}_{m'}$.

If the targets are well resolved and independent, $H(m \neq m') = 0$ and therefore we get only the first term of (22.50). The imaging function of the well-resolved multitargets is then given by the sum of all contributions from M targets given in (22.43).

The eigenvalues for $m = m'$ are given by $\lambda_{mm'} = v_m^2 H^2(m = m')$. The imaging function is therefore given by

$$\Psi_{TR} = \frac{1}{2\pi} \int |U|^2 g_s^T \sum_m v_m H(m = m') v_m e^{-i\omega t} d\omega, \tag{22.51}$$

where $v_m = g_m^*$.

For multiple targets, (22.51) gives the imaging function for M independent targets. The correlations between targets are included in the second term of (22.50). For the case $(m \neq m')$, we get the imaging function

$$\Psi_{TR} = \frac{1}{2\pi} \int |U|^2 g_s^T \sum_m \sum_{m'} \sqrt{v_m v_{m'}} H(m \neq m') v_m e^{-i\omega t} d\omega. \tag{22.52}$$

The complete imaging function is the sum of (22.51) and (22.52).

22.5   TIME-REVERSAL IMAGING AND SVD (SINGULAR VALUE DECOMPOSITION)

In Section 22.3, it was shown that the eigenvalue equation for the eigenvector v for the excitation signal is given by

$$K'Kv = \lambda v. \tag{22.53}$$

Similarly, the equation for the eigenvector u for the received signal is given by

$$K'Ku = \lambda u. \tag{22.54}$$

These two formulations are in fact identical to the SVD of the multistatic data matrix K.

According to SVD technique, any matrix $(m \times n)$ K can be expressed by the following.

$$K = uDv', \tag{22.55}$$

where u is an $(m \times k)$ matrix, D is a $(k \times k)$ diagonal matrix, and v is a $(k \times n)$ matrix. u and v are called the "left" and "right" singular vectors of K and orthonormal eigenvectors satisfying

$$\begin{aligned} U'U &= I_k, \\ V'V &= I_k. \end{aligned} \tag{22.56}$$

Note that the u $(m \times k)$ matrix is the eigenvector representing the m elements receiving array and the v $(k \times n)$ matrix represents the n elements transmitting array. If we use the same array for transmitter and receiver, then $m = n$.

D is the diagonal matrix with the elements $\sigma_1, \sigma_2, \ldots, \sigma_k$ and $\sigma_i^2 = \lambda_i$.

We also note the following:

$$Kv_p = \sigma_p u_p, \quad K'u_p = \sigma_p v_p, \tag{22.57}$$

and if $m = n$, $u_p = v_p^*$.

It is therefore noted that the eigenvector formulation in Section 22.3 can be considered as part of the SVD formulation.

22.6   TIME-REVERSAL IMAGING WITH MUSIC (MULTIPLE SIGNAL CLASSIFICATION)

In Section 22.4, we discussed time-reversal imaging of a single target and multiple targets in free space. For a point target, there is only one eigenvalue of the time-reversal matrix $T = K^*K$, and the eigenvector is given by Green's function matrix of (22.42), and the eigenvalue λ is given by $\|g^4\|$ (22.42). The eigenvalue for the other eigenvectors are all zeros. If there are M point targets, there are M nonzero eigenvalues and the other $(N - M)$ eigenvalues are zero, where N is the number of array elements (Devaney, 2005, 2015).

Those eigenvectors with nonzero eigenvalues are said to be in the signal space and the others with zero eigenvalues are in the noise space. It is also shown that the eigenvectors in the signal subspace and in the noise subspace are orthogonal and therefore the inner products of the eigenvectors in the signal subspace and the noise subspace are zero. Noting these, we can form the MUSIC pseudospectrum

$$\Psi_{TRMU} = \frac{1}{\sum_{p=M+2}^{N} \int d\omega |U|^2 g_s^T v_p}, \tag{22.58}$$

where $v_p, p = M, \ldots, N$, is the eigenvector in the noise subspace and if g_s is the eigenvector in the signal subspace, then the denominator becomes zero at the target points. However, we do not know the eigenvectors g in the signal subspace, and therefore we use an approximate g_s.

If we use a free space Green's function for g_s, the denominator is not zero, but it is small. The pseudospectrum (22.58) then gives a high image peak at the target point.

22.7   OPTIMUM POWER TRANSFER BY TIME-REVERSAL TECHNIQUE

The time-reversal technique described in Sections 22.3–22.5 can also be used to optimize power transfer between the transmitter and the receiver.

Consider a transmitter consisting of N antenna elements and a receiver consisting of M antenna elements (Fig. 22.8).

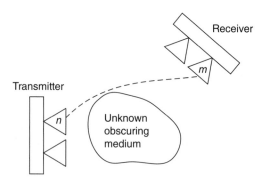

FIGURE 22.8 K_{mn} is the transfer function between the mth element of the receiver and the nth element of the transmitter. K_{mn} is measured and therefore known, even though the medium is unknown.

The following steps can be taken to obtain optimum power transfer:

1. Measure the transfer matrix K_{mn}.

    We measure the transfer function K_{mn} between the transmitter element n and the receiver m. These $M \times N$ elements K_{mn} are therefore assumed known, even though the environment is unknown.

2. Form the time-reversal transfer matrix.

    Using $M \times N$ matrix K, we form $N \times N$ transfer matrix T.

$$T = \bar{K}^* K, \tag{22.59}$$

where $\bar{K}^*$ denotes the conjugate of the transpose of K, called the adjoint matrix. Note that each element of K is reciprocal ($K_{mn} = K_{nm}$), but K is not symmetric unless $M = N$.

3. Calculate the eigenvectors and the eigenvalues.

    We calculate the eigenvectors V_i and the eigenvalue λ_i.

$$TV_i = \lambda_i V_i. \tag{22.60}$$

In general, there may be N eigenvalues. Since T is Hermitian ($\tilde{T}^* = T$), all λ_i are real and positive.

$$\lambda_i^* = \lambda_i. \tag{22.61}$$

The Hermitian matrix T is also called the Graves power matrix and is used to optimize the received power.

4. Determine the highest transmission efficiency. The largest eigenvalue λ_i is equal to the highest transmission efficiency. The receiver output V_r for the transmitter input V_i is given by

$$V_r = KV_i. \tag{22.62}$$

The total received power P_r and the total transmitted power P_i are

$$P_r = \tilde{V}_r^* V_r, \\ P_i = \tilde{V}_i^* V_i. \tag{22.63}$$

Therefore, the transmission efficiency η is given by

$$\eta = \frac{P_r}{P_i} = \frac{\tilde{V}_r^* V_r}{\tilde{V}_i^* V_i} = \frac{\tilde{V}_i^* \bar{K}^* K V_i}{\tilde{V}_i^* V_i} = \lambda. \tag{22.64}$$

The maximum efficiency is then given by the largest eigenvalue.

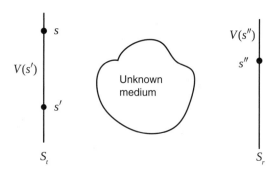

FIGURE 22.9 Aperture integral equation.

5. Use the eigenvector for the largest eigenvalue as the transmitter excitation V_i.

The transfer function K_{mn} is Green's function observed at the receiver n when a delta function source is located at the transmitter m. The medium need not be free space and there can be obstacles.

Similarly, the power transfer between transmitting and receiving aperture can be obtained.

The eigenvalue equation (22.60) can be rewritten for the continuous aperture distribution $V(s')$ on the transmitter aperture S_i and $V_r(s')$ on the receiver aperture S_r (Fig. 22.9).

$$\int_{S_i} T(s, s') V(s') ds' = \lambda V(s), \tag{22.65}$$

where $T(s, s') = \int_{S_i} K^*(s, s') K(s', s) ds', V_r(s'') = \int_{S_r} K(s'', s') V(s') ds'.$

Note that (22.65) is a homogeneous Fredholm integral equation of the second kind and its solution gives the eigenfunction $V_i(s)$ and the eigenvalue λ_i. The power consideration (22.64) can also be expressed as

$$\lambda = \frac{\int ds \int ds' T(s, s') V^*(s) V(s')}{\int ds V^*(s) V(s)}. \tag{22.66}$$

It may be interesting to note that this aperture integral formulation is similar to the aperture integral formulation for an open resonator where the largest eigenvalue corresponds to the least diffraction loss.

CHAPTER 23

SCATTERING BY TURBULENCE, PARTICLES, DIFFUSE MEDIUM, AND ROUGH SURFACES

In Chapter11, we have been concerned with scattering of waves by well-defined objects, such as radar cross sections and Mie scattering by spherical objects. There are, however, other important cases that require different treatments. Microwave and optical scattering by turbulence and particulate matter are examples that involve less well-defined conditions. Ultrasound scattering by tissues requires knowledge of tissue characteristics that can only be defined in statistical terms. Ocean surfaces and terrain are cases where the surface is constantly and randomly varying.

In all of the above examples, an important characteristic is that the medium can vary randomly in space and time, and we can describe the characteristics only in statistical terms. It is therefore important to understand statistical wave theory in contrast with deterministic wave theory.

In this chapter, we discuss basic aspects of statistical wave theory and its applications.

Most of the theories in this chapter are covered in Ishimaru (1997), where more details are presented. Note also that in this chapter, the time convention is $\exp(-i\omega t)$.

23.1 SCATTERING BY ATMOSPHERIC AND IONOSPHERIC TURBULENCE

The radar equation, which gives the ratio of the received power P_r to the transmitted power P_t in terms of the antenna gains, the distances and the bistatic cross section

Electromagnetic Wave Propagation, Radiation, and Scattering: From Fundamentals to Applications,
Second Edition. Akira Ishimaru.
© 2017 by The Institute of Electrical and Electronic Engineers, Inc. Published 2017 by John Wiley & Sons, Inc.

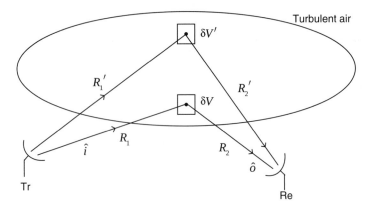

FIGURE 23.1   Transmitter Tr illuminates the turbulent air in the direction $\hat{i}$ and the scattered wave in the direction $\hat{o}$ is received by Receiver Re.

σ_{bi} of an object, was originally noted in (10.17) as

$$\frac{P_r}{P_t} = \frac{\lambda^2 G_t G_r \sigma_{bi}}{(4\pi)^3 R_1^2 R_2^2},$$ (23.1)

where all the quantities are explained in Section 10.2. If, instead of an object, turbulent air is illuminated by the transmitter, then what is the radar equation? (See Fig. 23.1.)

In turbulence, the bistatic cross section σ_{bi} for a discrete object (23.1) is replaced by $\sigma \delta V$ where $\sigma(\hat{o}, \hat{i})$ is the scattering cross section per unit volume of the turbulence. As noted in (10.5), σ is the differential cross section per unit volume and is related to the bistatic cross section by

$$\sigma(\hat{o}, \hat{i}) = \frac{1}{4\pi}\sigma_{bi}(\hat{o}, \hat{i}).$$ (23.2)

It is conventional, however, to use $\sigma(\hat{o}, \hat{i})$ for scattering from turbulence. Note also that the scattering amplitude $\bar{f}(\hat{o}, \hat{i})$ of the volume δV of turbulence is given by (10.32)

$$\bar{f}(\hat{o}, \hat{i}) = \frac{k^2}{4\pi} \int_{\delta V} [\bar{E} - \hat{o}(\hat{o} \cdot \bar{E})] \left[\epsilon_r(\bar{r}') - 1 \right] \exp(-ik\bar{r}' \cdot \hat{o}) dV,$$ (23.3)

and therefore $\sigma(\hat{o}, \hat{i})$ is given by

$$\sigma(\hat{o}, \hat{i}) = |f(\hat{o}, \hat{i})|^2 / \delta V.$$ (23.4)

It is, however, necessary to note the following. The electric field $\bar{E}$ in (23.3) is the total field at a point $\bar{r}'$ in turbulence. We assume that the relative dielectric constant ϵ_r is close to one, and therefore the average dielectric constant $\epsilon_r = 1$, and the fluctuation $\epsilon_1 = \epsilon_r - 1$ is small $|\epsilon_1| \ll 1$.

The electric field $\bar{E}$ is assumed to be equal to the incident wave $\bar{E}_i$. We also normalize it so that $|\bar{E}_i| = 1$ and $\bar{E}_i = \exp(ik\hat{i} \cdot \bar{r}')$.

We then write

$$f(\hat{o}, \hat{i}) = \hat{e}_s \sin \chi \left(\frac{k^2}{4\pi}\right) \int_{\delta V} \epsilon_1(\bar{r}')e^{i\bar{k}_s \cdot \bar{r}'} dV', \qquad (23.5)$$

where, as noted in Section 10.5,

$$\hat{e}_s \sin \chi = -\hat{o} \times \hat{o} \times \hat{e}_i$$
$$\bar{k}_s = k(\hat{i} - \hat{o}). \qquad (23.6)$$

We also note that the dielectric constant $\epsilon_r(\bar{r}')$ is a random function of $\bar{r}'$ and therefore $\sigma(\hat{o}, \hat{i})$ of (23.4) needs to be written as a statistical average.

$$\sigma(\hat{o}, \hat{i}) = \frac{\langle f(\hat{o}, \hat{i})f^*(\hat{o}, \hat{i})\rangle}{\delta V}. \qquad (23.7)$$

Substituting (23.5) into (23.7), we finally obtain the scattering cross section per unit volume of turbulence.

We get

$$\sigma(\hat{o}, \hat{i}) = \frac{k^4 \sin^2 \chi}{(4\pi)^2 \delta V} \int_{\delta V} dV_1' \int_{\delta V} dV_2' \left\langle \epsilon_1\left(\bar{r}_1'\right)\epsilon_1\left(\bar{r}_2'\right)\right\rangle \exp\left[i\bar{k}_s \cdot \left(\bar{r}_1' - \bar{r}_2'\right)\right], \qquad (23.8)$$

where $\bar{k}_s = k(\hat{i} - \hat{o})$.

If we assume that the turbulence is statistically homogeneous and isotropic, then the covariance $\langle \epsilon_1(\bar{r}_1')\epsilon_1(\bar{r}_2')\rangle$ is a function of the magnitude of the difference $r_d = |r_d| = |\bar{r}_1' - \bar{r}_2'|$.

We can then express the integral in (23.8) in the following:

$$\int_{\delta V} dV_1' \int_{\delta V} dV_2' B_\epsilon(r_d)e^{i\bar{k}_s \cdot \bar{r}_d} = \int dV_c \int dV_d B_\epsilon(r_d)e^{i\bar{k}_s \cdot \bar{r}_d}, \qquad (23.9)$$

where

$$B_\epsilon(\bar{r}_d) = \left\langle \epsilon_1\left(\bar{r}_1'\right)\epsilon_1\left(\bar{r}_2'\right)\right\rangle \qquad (23.10)$$

is the correlation function of fluctuation ϵ_1 of the dielectric constant.

We let the size of δV be much greater than the correlation distance, and since B_ϵ is negligibly small if r_d is much greater than the correlation distance, we get

$$\int dV_c \int dV_d \cong \int_{\delta V} dV_c \int_{\infty} dV_d.$$

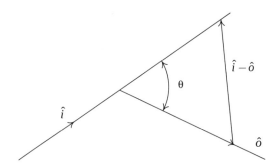

FIGURE 23.2   Scattering angle θ and $\bar{k}_s = k(\hat{i} - \hat{o})$. $|\hat{i} - \hat{o}| = 2\sin(\theta/2)$.

$\sigma(\hat{o}, \hat{i})$ of (23.8) can then be expressed as the Fourier transform of the correlation function $B_\epsilon(r_d)$.

$$\sigma(\hat{o}, \hat{i}) = \frac{\pi}{2}k^4 \sin^2 \chi \Phi_\epsilon(k_s), \tag{23.11}$$

$$\Phi_\epsilon(k_s) = \frac{1}{(2\pi)^3} \int B_\epsilon(r_d)e^{i\bar{k}_s \cdot \bar{r}_d} dV_d, \tag{23.12}$$

where $\bar{k}_s = k(\hat{i} - \hat{o})$ and $k_s = 2k\sin(\theta/2)$ and θ is the scattering angle (Fig. 23.2).

Equations (23.11) and (23.12) are the fundamental expressions of the scattering cross section per unit volume of the turbulence. Equation (23.12) shows the spectrum Φ_ϵ as the Fourier transform of the correlation function B_ϵ. This is called the "Wiener–Khinchin" theorem.

Using the above, we can finally obtain the Radar equation for the problem shown in Fig. 23.1

$$\frac{P_r}{P_t} = \int \frac{\lambda^2 G_t(\hat{i})G_r(\hat{o})}{(4\pi)^2 R_1^2 R_2^2} \sigma(\hat{o}, \hat{i}), \tag{23.13}$$

where $\sigma(\hat{o}, \hat{i})$ is given by (23.11) and (23.12).

Note that the fluctuation of the dielectric constant ϵ_1 is simply related to the refractive index fluctuation n_1.

$$1 + \epsilon_1 = (1 + n_1)^2 \approx 1 + 2n_1$$
$$B_\epsilon(r_d) = 4B_n(r_d), \Phi_\epsilon(k_s) = 4\Phi_n(k_s) \tag{23.14}$$

23.2 SCATTERING CROSS SECTION PER UNIT VOLUME OF TURBULENCE

As noted in Section 23.1, the scattering by turbulence is expressed by the scattering cross section per unit volume of turbulence, which is given by the correlation function or the spectrum of the fluctuation of dielectric constant or refractive index of the medium. Here are three commonly used spectra.

(a) Booker–Gorden Formula

Historically, scattering by atmospheric turbulence was studied by Booker and Gorden in 1950 using a simple exponential function for the refractive index fluctuation.

$$B_n(r_d) = \langle n_1^2 \rangle \exp(-r_d/\ell), \tag{23.15}$$

where ℓ is the correlation distance.

The spectrum is then given by

$$\begin{aligned}
\Phi_n(k_s) &= \frac{1}{(2\pi)^3} \int B_n(r_d) e^{i\vec{k}_s \cdot \vec{r}_d} dV_d \\
&= \frac{n_1^2 \ell^3}{\left[1 + (k_s \ell)^2\right]^2} \frac{1}{\pi^2}.
\end{aligned} \tag{23.16}$$

Using this we get the cross section σ.

$$\sigma(\theta) = \frac{2}{\pi} \frac{k^4 \ell^3 \sin^2 \chi n_1^2}{\left[1 + 4k^2 \ell^2 \sin^2 (\theta/2)\right]^2}. \tag{23.17}$$

This is the Booker–Gorden formula, which depends on two parameters of the turbulence, the variance n_1^2 and the correlation distance ℓ of the refractive index fluctuation.

For tropospheric turbulence, the variance is of the order of magnitude $n_1^2 \approx 10^{-12}$, and the correlation distance is approximately 50 m (20.130).

(b) Gaussian Spectrum

If we assume the correlation function is Gaussian, we get

$$B_n(r_d) = \langle n_1^2 \rangle \exp\left(-r_d^2/\ell^2\right),$$

$$\Phi_n(k_s) = \frac{\langle n_1^2 \rangle \ell^3}{8\pi \sqrt{\pi}} \exp\left[-\frac{k_s^2 \ell^2}{4}\right], \tag{23.18}$$

$$\sigma(\theta) = \frac{\langle n_1^2 \rangle k^4 \ell^3}{4\sqrt{\pi}} \sin^2 \chi \exp\left[-\frac{k_s^2 \ell^2}{4}\right].$$

(c) Kolmogorov Spectrum

The Booker–Gorden formula and the Gaussian formula are convenient to obtain a good estimate of the scattering by turbulence. However, they are not based on a consideration of the physical properties of turbulence.

A more realistic spectrum based on the actual turbulence characteristics has been proposed. According to Kolmogorov, the turbulent eddies may be characterized by two sizes: outer scale L_o and inner scale ℓ_o. The energy is introduced into the turbulence in the eddy size greater than L_o, and in the eddy sizes between L_o and ℓ_o the kinetic energy dominates over dissipation due to viscosity.

For an eddy size smaller than ℓ_o, the dissipation of energy due to viscosity dominates over the kinetic energy. The Kolmogorov spectrum is then approximately given by

$$\Phi_n(K) = 0.033 C_n^2 \left[K^2 + \left(\frac{2\pi}{L_o} \right)^2 \right]^{-11/6} \exp\left(-K^2/K_m^2\right), \quad (23.19)$$

where $K_m = 5.92/\ell_o$ and C_n is of the order of $10^{-7}(\text{m}^{-1/3})$ for strong turbulence and $10^{-9}(\text{m}^{-1/3})$ for weak turbulence.

23.3 SCATTERING FOR A NARROW BEAM CASE

If the transmitting and receiving patterns are confined within narrow solid angles, the gain function $G(\hat{i})$ can be approximated by the Gaussian function.

$$G(\hat{i}) = G(\hat{i}_o)\exp\left(-\ln 2 \left[(2\theta/\theta_1)^2 + (2\phi/\phi_1)^2\right]\right), \quad (23.20)$$

where θ_1 and ϕ_1 are the half-power beamwidths in two orthogonal directions, such that at $\theta = (\theta_1/2)$, the gain function becomes $(1/2)$. For bistatic case (Fig. 23.3), we

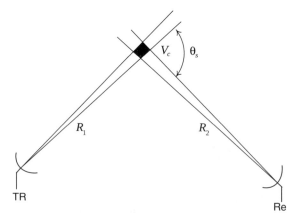

FIGURE 23.3   Bistatic radars and common volume V_c.

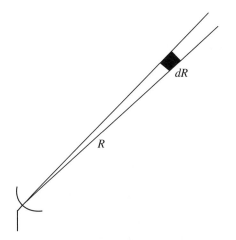

FIGURE 23.4 Monostatic radar.

perform integration with dV and obtain (Ishimaru, 1997)

$$\frac{P_r}{P_t} = \frac{\lambda^2 G_t(\hat{i}_o) G_r(\hat{o}_o)}{(4\pi)^2 R_1^2 R_2^2} \sigma(\hat{o}_o, \hat{i}_o) V_c exp(-\tau_1 - \tau_2), \qquad (23.21)$$

where

$$V_c = \frac{\pi \sqrt{\pi}}{8 (\ln 2)^{3/2}} \frac{R_1^2 R_2^2 \theta_1 \theta_2 \phi_1 \phi_2}{\left[R_1^2 \phi_1^2 + R_2^2 \phi_2^2 \right]^{1/2}} \frac{1}{\sin \theta_s}.$$

And the attenuation from τ_2 to δV to R_e is given by τ_1 and τ_2, which is also called the optical distance. For the monostatic case (Fig. 23.4), we get

$$\frac{P_r}{P_t} = A \int^R \frac{\sigma(\hat{i})}{R^2} e^{-2\tau} dR, \qquad (23.22)$$

where

$$A = \frac{\pi}{(4\pi)^2 (8 \ln 2)} \left(\lambda^2 G_t^2 \theta_1 \phi_1 \right).$$

If the radar transmits a short pulse with $E_i(t)$, then we get the received pulse

$$I_r(t) = A \int_{R_1}^{R_2} \frac{\sigma(\hat{i})}{R^2} e^{-2\tau} \left| E_i \left(t - \frac{2R}{c} \right) \right|^2 dR. \qquad (23.23)$$

The attenuation τ in this section comes from the medium attenuation τ_a, and the attenuation τ_s is due to the scattering. The attenuation τ_s is obtained by integrating $\sigma(\hat{o}, \hat{i})$ over the solid angle 4π. We have

$$\tau_s = \int \alpha_s dR,$$

$$\alpha_s = \int_{4\pi} \sigma(\hat{o}, \hat{i})d\Omega. \tag{23.24}$$

Noting that

$$d\Omega = \sin\theta \, d\theta \, d\phi = (k_s dk_s d\phi)/k^2,$$

with $k_s = 2k \sin(\theta/2)$, we get approximately

$$\alpha_s = 4\pi^2 k^2 \int_0^{2k} \Phi_n(k_s)k_s dk_s. \tag{23.25}$$

23.4   SCATTERING CROSS SECTION PER UNIT VOLUME OF RAIN AND FOG

In Section 23.1, we discussed the scattering cross section per unit volume of turbulence. It is described in (23.11) and (23.12) in terms of the correlation function of the fluctuations of the refractive index, or the dielectric constant.

For rain, the scattering cross section per unit volume is given by

$$\sigma(\hat{o}, \hat{i}) = \int_0^\infty n(p, a)\sigma_s(a)\,da, \tag{23.26}$$

where $n(p, a)$ is the size distribution, and $n(p, a)\,da$ is the number of droplets per unit volume having radius between a and $a + da$ at the precipitation rate p (in millimeter per hour), and $\sigma_s(a)$ is the scattering cross section of a single droplet with radius a.

The precipitation rate p in millimeter per hour is given by

$$p = (3600) \times 10^3 \int_0^\infty v(a)n(p, a)\left(\frac{4\pi a^3}{3}\right)da, \tag{23.27}$$

where $v(a)$ in m/s is the terminal velocity of a rain drop with radius a (m), nda is in m^{-3}.

There has been extensive work reported on the size distribution of rain drops. The most well known is the Laws–Parson distribution. The Marshall–Palmer distribution is given by

$$n(p, a) = n_o\exp(-\alpha a), \tag{23.28}$$

where $n_o = 8 \times 10^6$ m^{-4}, $\alpha = 8200p^{-0.21}$ m^{-1}, and p is in mm/hr. See Ishimaru (1997) for additional details. The terminal velocity $v(a)$ in m/s is approximately given by

$$v = 200.8a^{1/2} \ (a \text{ in m}). \tag{23.29}$$

In practice, exact size distribution and scattering characteristics are difficult to obtain, and therefore we make use of approximate formulas. The scattering characteristics of rain for different precipitation rates have been reported, and approximate formulas have been obtained. In Section 23.5, we discuss some of the useful formulas.

23.5 GAUSSIAN AND HENYEY–GREENSTEIN SCATTERING FORMULAS

The scattering function of particles with size distributions can often be approximated by a Gaussian function. The scattering cross section per unit volume of the medium is then given by

$$\begin{aligned} \sigma(\hat{o}, \hat{i}) &= \frac{\alpha_t}{4\pi} p(\theta) \\ &= \frac{\alpha_t}{4\pi} (4\alpha_p) e^{-\alpha_p s^2}, \quad s = 2\sin(\theta/2), \end{aligned} \tag{23.30}$$

where α_t = extinction coefficient (m^{-1}), and $p(\theta)$ is called the phase function.
   Note that the phase function $p(\theta)$ is normalized so that

$$\frac{1}{2} \int_0^\infty p(\theta)s \, ds = 1. \tag{23.31}$$

The Gaussian scattering function is mathematically convenient. It is, however, not directly based on the physical properties.
   The Henyey–Greenstein phase function is proposed to represent the optical diffuse scattering by small particles with size distribution. It has been shown to be useful for expressing optical scattering by fog and optical diffusion in tissues. The Henyey–Greenstein formula is given by

$$\begin{aligned} p(\theta) &= \frac{(1/2)(1-g^2)}{(1+g^2-2g\cos\theta)^{3/2}} \\ &= \frac{(1+g)(1/2)}{(1-g)^2 \left[1+\left(\dfrac{s}{S_0}\right)^2\right]^{3/2}}, \quad S_0 = \frac{(1-g)^2}{g} \end{aligned} \tag{23.32}$$

where $s = 2\sin(\theta/2)$ and g is called the "anisotropy factor."

The formula is normalized so that

$$\frac{1}{4\pi} \int p(\theta)d\Omega = 1. \tag{23.33}$$

Or

$$\frac{1}{2} \int_0^2 p(s)\, s \, ds = 1.$$

Note that $g = 0$ gives isotropic scattering, $p(\theta) = 1$. As g approaches 1, the scattering is increasingly peaked in the forward direction.

23.6 SCATTERING CROSS SECTION PER UNIT VOLUME OF TURBULENCE, PARTICLES, AND BIOLOGICAL MEDIA

In Section 23.1, we discussed $\sigma(\hat{o}, \hat{i})$, the scattering cross section per unit volume of turbulence. In Section 23.4, we discussed $\sigma(\hat{o}, \hat{i})$ for distributed particles such as rain. In Section 20.7, we discussed $\sigma(\hat{o}, \hat{i})$ for ultrasound scattering by tissues. These three discussions deal with three different physical problems. However, it should be noted that all three represent scattering cross section per unit volume of the medium. We can summarize these three cases as follows.

For turbulence, we have (23.11)

$$\sigma = \frac{\pi}{2}k^4 \Phi_\epsilon(k_s).$$

We neglect the polarization effect by letting $\sin^2 \chi = 1$.

For particles, we have (23.26)

$$\sigma = \int nda\sigma_s(a) = \int nda\,|f|^2.$$

For a biological medium, we have (20.82)

$$\sigma = \frac{\pi}{2}k^4 Sm,$$

$$S = S_k + S_\rho + 2S_{k\rho}.$$

Summarizing, the cross section per unit volume is

$$\sigma = \frac{\pi}{2}k^4 \Phi_\epsilon = 2\pi k^4 \Phi_n \qquad \text{(turbulence)}$$

$$= \int nda\,|f|^2 = \frac{\gamma_e}{4\pi}p(\theta) \qquad \text{(particles)}$$

$$= \frac{\pi}{2}k^4 S. \qquad \text{(biological medium)}$$

The unit for σ is $(\text{m}^2/\text{m}^3) = \text{m}^{-1}$.

23.7 LINE-OF-SIGHT PROPAGATION, BORN AND RYTOV APPROXIMATION

In the preceding sections, we considered the scattering of waves by turbulence and particles. Let us now consider the nature of the wave propagating through and scattered by turbulence. We consider the wave equation

$$\left[\nabla^2 + k^2(1 + \epsilon_1)\right] U = 0, \tag{23.34}$$

where ϵ_1 is the fluctuating part of the dielectric constant ϵ, and $k^2 = k_o^2\langle\epsilon\rangle$. $k_o =$ the free space wave number and $\langle\epsilon\rangle$ is the average of the dielectric constant.

$$\epsilon = \langle\epsilon\rangle(1 + \epsilon_1). \tag{23.35}$$

In terms of the refractive index n we have

$$\epsilon = \langle n\rangle^2(1 + n_1)^2 \approx \langle n\rangle^2(1 + 2n_1). \tag{23.36}$$

The solution of (23.34) can be expressed in the following two series:

$$U = U_0 + U_1 + U_2 + \cdots, \tag{23.37}$$
$$U = \exp(\psi_0 + \psi_1 + \psi_2 + \cdots). \tag{23.38}$$

The first series is the expansion of the field and is called the Neumann series, and each term is called the nth Born approximation.

Note that we can write (23.34) as

$$(\nabla^2 + k^2)U = -k^2\epsilon_1 U. \tag{23.39}$$

We convert this to the following integral equation for U:

$$U(\vec{r}) = U_0(\vec{r}) + \int G(\vec{r}, \vec{r}')V(\vec{r}')U(\vec{r}')dV', \tag{23.40}$$

where $V(\vec{r}') = k^2\epsilon_1(\vec{r}')$ is called the "potential function" and G is Green's function which satisfies

$$(\nabla^2 + k^2)G(\vec{r}, \vec{r}') = -\delta(\vec{r} - \vec{r}'). \tag{23.41}$$

Equation (23.40) is called the Lippmann–Schwinger integral equation. From (23.40), we can obtain the series expansion (23.37) assuming that the series converges,

$$U_n(\vec{r}) = U_0(\vec{r}) + \int G(\vec{r}, \vec{r}')V(\vec{r}')U_{n-1}(\vec{r}')dV'. \tag{23.42}$$

The first term is called the first Born approximation

$$U_1(\vec{r}) = U_0(\vec{r}) + \int G(\vec{r}, \vec{r}')V(\vec{r}')U_0(\vec{r}')dV'. \tag{23.43}$$

Note that in the absence of the fluctuation $\epsilon_1 = 0$ and in free space, we get

$$U_0(\bar{r}) = G(\bar{r}, \bar{r}') = \frac{\exp(ik|\bar{r} - \bar{r}'|)}{4\pi|\bar{r} - \bar{r}'|}. \tag{23.44}$$

In our discussion of scattering problems in preceding sections, we used the first Born approximation. We now consider the series expansion (23.38). First, let us note that $k^2\epsilon_1$ represents the fluctuation and therefore it is a small quantity.

Noting this we write

$$U = U_0 + \epsilon U_1 + \epsilon^2 U_2 + \cdots \tag{23.45}$$
$$= \exp\left(\psi_0 + \epsilon\psi_1 + \epsilon^2\psi_2 + \cdots\right), \tag{23.46}$$

where ϵ is a small parameter and the expansion (23.45) and (23.46) is the expansion in powers of ϵ.

We can write (23.46) in a series form

$$U = \exp(\psi_0)\left[1 + (\epsilon\psi_1 + \epsilon^2\psi_2 + \cdots) + \frac{1}{2}(\epsilon\psi_1 + \epsilon^2\psi_2 + \cdots)^2 + \cdots\right]. \tag{23.47}$$

Comparing that with (23.45) and equating the same order of the powers of ϵ up to ϵ^2, we get the following relations:

$$U_0 = \exp(\psi_0),$$
$$\frac{U_1}{U_0} = \psi_1,$$
$$\frac{U_2}{U_0} = \psi_2 + \frac{1}{2}\psi_1^2. \tag{23.48}$$

From this, we can see that the second-order ψ_2 is related to the first-order ψ_1^2 and (U_2/U_0).

It will be shown in Section 23.8 that the conservation of power imposes conditions that give expressions for the mutual coherence function (MCF) by making use of the first-order Rytov solution alone.

23.8 MODIFIED RYTOV SOLUTION WITH POWER CONSERVATION, AND MUTUAL COHERENCE FUNCTION

The first Rytov solution has been extensively used for weak fluctuation problems. For strong fluctuations, the moment equation method and the extended Fresnel–Huygens method have been used to derive MCF, and both methods give the same results for the second moment of the field. The first Rytov solution does not conserve power, but it is convenient for the log amplitude and phase fluctuations in the weak fluctuation case. In this section, we derive MCF using the first and second Rytov solutions in such a way that the final expression makes use of the first-order Rytov solution alone, but satisfies the conservation of power.

In general, the phase perturbation series as shown in (23.38) is consistent with the series expression of the WKB method (Section 3.14), which conserves the power by keeping the first- and second-order terms. This is also consistent with the geometric optics formulation where the first term satisfies the eikonal equation and the second term is needed to satisfy the conservation of power.

Let us consider the Rytov solution, keeping up to the second-order term,

$$U = U_0 \exp\left[\epsilon\psi_1 + \epsilon^2\psi_2\right], \tag{23.49}$$

where ϵ represents a small parameter for the perturbation series.

Let us then consider MCF.

$$\Gamma = \left\langle U_a U_b^* \right\rangle = \left(U_{a0}U_{b0}^*\right)\left\langle \exp\left[\epsilon\left(\psi_{a1} + \psi_{b1}^*\right) + \epsilon^2\left(\psi_{a2} + \psi_{b2}^*\right)\right]\right\rangle. \tag{23.50}$$

Now we assume that the exponent in (23.50) is normally distributed. Then the average of $\exp(\psi)$ is given by

$$\left\langle \exp(\psi)\right\rangle = \exp\left[\langle\psi\rangle + \frac{1}{2}\langle(\psi - \langle\psi\rangle)^2\rangle\right]. \tag{23.51}$$

Noting that

$$\langle\psi\rangle = \epsilon\left\langle\left(\psi_{a1} + \psi_{b1}^*\right)\right\rangle + \epsilon^2\left\langle\left(\psi_{a2} + \psi_{b2}^*\right)\right\rangle, \tag{23.52}$$

$$\langle\psi_{a1}\rangle = \left\langle\left(\frac{U_{a1}}{U_0}\right)\right\rangle = 0,$$

$$\left\langle\frac{1}{2}(\psi - \langle\psi\rangle)^2\right\rangle = \left\langle\frac{1}{2}\epsilon^2\left(\psi_{a1} - \psi_{b1}^*\right)^2\right\rangle.$$

We finally obtain

$$\Gamma = \left(U_{a0}U_{b0}^*\right)\exp\left[\epsilon^2\left(\langle\psi_{a2}\rangle + \langle\psi_{b2}^*\rangle + \frac{1}{2}\left\langle\left(\psi_{a1} + \psi_{b1}^*\right)^2\right\rangle\right)\right] \tag{23.53}$$

$$= \left(U_{a0}U_{b0}^*\right)\exp\left[\epsilon^2\left(\langle\psi_{a2}\rangle + \langle\psi_{b2}^*\rangle + \frac{1}{2}\langle\psi_{a1}^2\rangle + \frac{1}{2}\langle\psi_{b1}^{*2}\rangle + \langle\psi_{a1}\psi_{b1}^*\rangle\right)\right].$$

Now the conservation of power requires that

$$\Gamma(a = b) = |U_a|^2\exp\left[\epsilon^2\left(2Re\langle\psi_{a2}\rangle + Re\langle\psi_{a1}^2\rangle + \langle\psi_{a1}\psi_{b1}^*\rangle\right)\right] \tag{23.54}$$

$$= |U_a|^2.$$

Thus we get the condition for conservation of power

$$2Re\langle\psi_2\rangle + Re\langle\psi_1^2\rangle + \left\langle|\psi_1|^2\right\rangle = 0. \tag{23.55}$$

The sufficient condition for the conservation of power is that

$$2\langle \psi_2 \rangle + \langle \psi_1^2 \rangle + \langle |\psi_1|^2 \rangle = 0. \tag{23.56}$$

The imaging part of (23.56) does not affect the conservation of energy as shown in (23.54).

If we substitute (23.56) into (23.53), we finally get MCF.

$$\Gamma = \Gamma_o \exp\left(-\frac{1}{2}D\right), \tag{23.57}$$

$$\Gamma_o = \left(U_{a0}U_{b0}^*\right),$$

$$D = \langle |\psi_{a1}|^2 \rangle + \langle |\psi_{b1}|^2 \rangle - 2\langle \psi_{a1}\psi_{b1}^* \rangle.$$

This is the final expression for MCF using up to second-order perturbation (23.50) and power conservation (23.56).

ψ_{a1} and ψ_{b1} are the first Rytov solutions given by (23.48).

$$\psi_{a1} = \frac{U_1(\bar{r}_a)}{U_0(\bar{r}_a)}, \quad \psi_{b1} = \frac{U_1(\bar{r}_b)}{U_0(\bar{r}_b)}. \tag{23.58}$$

The modified Rytov solution is simple and useful. However, it is an approximate theory, and more complete theories have been presented including the extended Huygens–Fresnel formulation, the moment equation, the path integral approach, and the diagram method. These are given in Tatarskii, Ishimaru, and Zavorotny (1993).

23.9   MCF FOR LINE-OF-SIGHT WAVE PROPAGATION IN TURBULENCE

The final expression for MCF in turbulence is given in (23.57). We now consider MCF for a point source radiating in turbulence (Fig. 23.5).

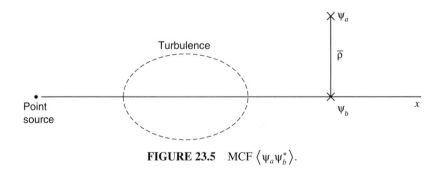

FIGURE 23.5   MCF $\langle \psi_a \psi_b^* \rangle$.

We note that for a line-of-sight propagation, the wave propagates in one direction and therefore the wave equation can be approximated by the following parabolic equation.

The wave equation for u

$$[\nabla^2 + k^2(1 + \epsilon_1)]u = 0 \qquad (23.59)$$

can be converted to the following by writing

$$u = U \exp(ikx),$$

$$2ik\frac{\partial U}{\partial x} + \nabla^2 U + k^2 \epsilon_1 U = 0. \qquad (23.60)$$

Since U is a slowly varying function of x, we note

$$|k\partial U/\partial x| \gg |\partial^2 U/\partial x^2|. \qquad (23.61)$$

And then we can obtain the following parabolic equation for U:

$$2ik\frac{\partial U}{\partial x} + \nabla_t^2 U + k^2 \epsilon_1 U = 0, \qquad (23.62)$$

where ∇_t^2 is the Laplacian in the transverse direction (y, z).

We now obtain the first Rytov solution based on the parabolic equation

$$\psi_1 = \frac{U_1}{U_0} \qquad (23.63)$$

where

$$U_0 = \frac{1}{4\pi x}\exp\left[ik\left(x + \frac{\rho^2}{2x}\right)\right],$$

$$U_1 = \int G(\bar{r}, \bar{r}')V(\bar{r}')U_0(\bar{r}')dV',$$

$$G = \frac{1}{4\pi |\bar{r} - \bar{r}'|}\exp\left\{ik\left[(x - x') + \frac{|\bar{\rho} - \bar{\rho}'|^2}{2(x - x')}\right]\right\},$$

$$V = k^2\epsilon_1 = 2k^2 n_1.$$

Let us now obtain the MCF given in (23.57).

To do this, we need to find

$$\Gamma_{ab} = \langle \psi_{a1}\psi_{b1}^* \rangle. \qquad (23.64)$$

Then we can get MCF Γ in (23.57)

$$\Gamma = \Gamma_0 \exp\left(-\frac{1}{2}D\right), \qquad (23.65)$$

where

$$\Gamma_0 = U_{a0}U_{b0}^*.$$

$$D = \Gamma_{aa} + \Gamma_{bb} - 2\Gamma_{ab}.$$

It is therefore necessary to obtain Γ_{ab} in (23.64) using (23.63). This process has been extensively discussed in Ishimaru (1997, Chapter 18) and therefore we can write the final expression. For spherical waves, we obtain the following expression for D:

$$D = 8\pi^2 k^2 \int_0^x dx' \int_0^\infty kdk \left[1 - J_0(k\rho)\right] \Phi_n(k), \qquad (23.66)$$

$$\rho = \left| \rho_d \frac{x'}{x} \right|.$$

Here we give examples that are commonly used (Ishimaru, 1997, 2004).

(a) Kolmogorov Spectrum

The Kolmogorov spectrum is given in (23.19). In many cases of turbulence encountered in optical propagation, the separation ρ is between the inner scale and the outer scale, and therefore we use (23.19) with $L_0 \to \infty$ and $l_0 \to 0$.

$$\Phi_n(k) = 0.033C_n^2 k^{-11/2}. \qquad (23.67)$$

We then get approximately

$$(D/2) \approx 0.547k^2 xC_n^2 \rho^{5/3}$$

$$= \left(\frac{\rho}{\rho_0} \right)^{5/3}, \qquad (23.68)$$

where $\rho_0 = \left[0.547k^2 xC_n^2\right]^{-3/5}$.

This distance ρ_0 is called the "correlation distance."

(b) Gaussian Spectrum

If the spectrum Φ_n can be approximated by a Gaussian function (23.18), we get

$$2\pi k^4 \Phi_n = \frac{\alpha_t}{4\pi} p(s). \qquad (23.69)$$

The phase function $p(s)$ is given by

$$p(s) = 4\alpha_p \exp\left(-\alpha_p s^2\right). \qquad (23.70)$$

Substituting this in (23.66), we get

$$\exp\left(-\frac{D}{2}\right) = \exp\left(-\frac{\rho^2}{\rho_0^2}\right),$$
(23.71)

where

$$\rho_0^2 = \left[\frac{k^2}{12\alpha_p}\tau_s\right]^{-1},$$

τ_s is the scattering optical depth.

(c) Henyey–Greenstein (HG) Phase Function

Using HG phase function given in (23.32), we get

$$\frac{D}{2} = \tau_s[1 - \exp(-ks_0\rho)].$$
(23.72)

The correlation distance ρ_0 is then approximately given by

$$\rho_0 = \left(\frac{\tau_s ks_0}{2}\right)^{-1}.$$
(23.73)

23.10 CORRELATION DISTANCE AND ANGULAR SPECTRUM

The MCF of a spherical wave is given by (23.65), and in turbulence that can be expressed as

$$\Gamma(\rho) = \Gamma_0\exp\left(-\frac{1}{2}D\right) = \Gamma_0\exp\left(-\left(\frac{\rho}{\rho_0}\right)^{5/3}\right).$$
(23.74)

The angular spectrum $\Gamma(\theta)$ is given by a Fourier transform of $\Gamma(\rho)$.

$$\Gamma(\theta) = \Gamma_0 \int \Gamma\left(\frac{\rho}{\rho_0}\right) e^{-ik\bar{\rho}\cdot\bar{\theta}} d\left(\frac{\bar{\rho}}{\rho_0}\right)$$

$$= \Gamma_0 \int \Gamma\left(\frac{\rho}{\rho_0}\right) e^{-i\left(\frac{\bar{\rho}}{\rho_0}\right)\cdot\frac{\bar{\theta}}{\theta_c}} d\left(\frac{\bar{\rho}}{\rho_0}\right),$$

where

$$\theta_c = \frac{1}{k\rho_0}.$$
(23.75)

It is thus seen that the correlation distance ρ_0 is related to the angular spread θ_c.

23.11 COHERENCE TIME AND SPECTRAL BROADENING

If the medium or transmitter or receiver is moving in time, the time correlation of the received signal is related to the motion. For example, if the medium is moving transverse to the line-of-sight propagation path, then the time correlation of the received signal at one point is related to the spatial correlation. As the medium moves with velocity v, the time correlation of the signal in time delay τ becomes equal to the spatial correlation in space $\rho = V\tau$. Therefore the temporal frequency spectrum $W(\omega)$ is a Fourier transform of MCF with $\rho = V\tau$.

We have

$$W(\omega) = \int \Gamma(\rho)e^{-i\omega\tau}\,d\tau, \tag{23.76}$$

with $\bar{\rho} = \bar{V}\tau$.

Note that

$$W(\omega) = \Gamma_0 \int \Gamma\left(\frac{V\tau}{\rho_0}\right) e^{-i\left(\frac{\omega}{\omega_c}\right)(\omega_c\tau)}\,d\tau. \tag{23.77}$$

The frequency spread ω_c, also called the spectral broadening, is given by (V/ρ_0).

Comparing (23.75) with (23.77), we note that the angular spread θ_c and the spectral broadening ω_c are closely related to the correlation distance ρ_0.

$$\theta_c = \frac{1}{k\rho_0}, \quad \omega_c = \frac{V}{\rho_0}. \tag{23.78}$$

The coherence time T_c is defined as the time lag for which the correlation reduces to a certain level, and it is inversely related to the frequency spread or spectral broadening ω_c.

$$T_c \sim \frac{1}{\omega_c}. \tag{23.79}$$

This is sketched in Fig. 23.6.

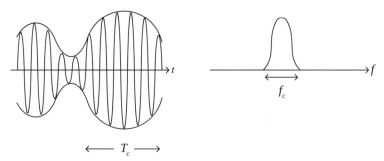

FIGURE 23.6 Coherence time T_c and spectral broadening f_c. $\left(T_c \sim \dfrac{1}{f_c}\right)$.

23.12   PULSE PROPAGATION, COHERENCE BANDWIDTH, AND PULSE BROADENING

In a multiple scattering environment with turbulence, a broadband pulse experiences pulse broadening. The pulse broadening can be studied using the two-frequency MCF. In general, MCF Γ_t in time can be expressed as

$$\Gamma_t = \langle U(\bar{r}_1, t_1)U^*(\bar{r}_2, t_2)\rangle$$
$$= \frac{1}{(2\pi)^2}\int d\omega_1 \int d\omega_2 \Gamma_\omega(\bar{r}_1, \bar{r}_2)e^{-i\omega_1 t_1 + i\omega_2 t_2}, \qquad (23.80)$$

where

$$\Gamma_\omega = \langle U(\bar{r}_1, \omega_1)U^*(\bar{r}_2, \omega_2)\rangle$$

is the two-frequency MCF.

The two-frequency MCF, Γ_ω, is the fundamental function to analyze pulse propagation and time delay and is the correlation function of the waves at two different frequencies ω_1 and ω_2.

Noting that the field $U(\bar{r}_1, t_1)$ is given by the Fourier transform

$$U(\bar{r}_1, t_1) = \frac{1}{2\pi}\int U(\bar{r}_1, \omega_1)e^{-i\omega_1 t_1}d\omega_1 \qquad (23.81)$$

we get the intensity $I(t)$

$$I(t) = |U(\bar{r}, t)|^2$$
$$= \frac{1}{(2\pi)^2}\int d\omega_1 d\omega_2 \Gamma_\omega(\omega_1, \omega_2)e^{-i(\omega_1 - \omega_2)t}. \qquad (23.82)$$

This shows that the pulse shape $I(t)$ is given by the Fourier transform of the two-frequency MCF with respect to $\omega_d = \omega_1 - \omega_2$.

Let us now consider the two-frequency MCF $\Gamma_\omega(\omega_1, \omega_2)$ based on the modified Rytov solution. Following Section 23.8, and noting (23.57) we have

$$\Gamma_\omega(\omega_1, \omega_2) = U_0(\omega_1)U_0^*(\omega_2)\exp\left(-\frac{D}{2}\right),$$
$$D = |\psi_1(\omega_1)|^2 + |\psi_1(\omega_2)|^2 - 2\psi_1(\omega_1)\psi_1^*(\omega_2). \qquad (23.83)$$

Let us then consider $\psi_1(\omega_1)\psi_1^*(\omega_2)$. Note that, for a plane wave,

$$\psi_1(\omega_1) = \int_0^L dx' \int_{-\infty}^\infty dy' \int_{-\infty}^\infty dz' h(L - x', y - y', z - z')n_1(x', y', z'). \qquad (23.84)$$

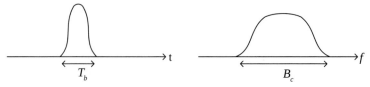

FIGURE 23.7   Pulse broadening T_b and coherence bandwidth B_c. $\left(T_b \sim \dfrac{1}{B_c}\right)$.

Using the Fourier transform of h (Ishimaru, 1997, Section 17.7) we get

$$H(L - x', k) = \int d\bar{\rho}e^{-i\bar{k}\cdot\bar{\rho}}h(L - x', \bar{\rho})$$

$$= ik \exp\left[-i\frac{(L - x')}{2k}k^2\right].$$

We get for a plane wave,

$$\Gamma_\omega(\omega_1, \omega_2) = U_0(\omega_1)U_0^*(\omega_2)\exp\left(-\frac{D}{2}\right),$$

$$U_0(\omega_1) = \exp(ik_1 x), \; U_0(\omega_2) = \exp(ik_2 x),$$

$$D = 8\pi^2 \int_0^L dx \left\{\int kdk \left\{\frac{k_1^2 + k_2^2}{2} - k_1 k_2 J_0\left(k\rho\right)\right.\right.$$

$$\left.\left. \times \exp\left[-\frac{i(L - x)}{2}\left(\frac{1}{k_1} - \frac{1}{k_2}\right)k^2\right]\right\}\Phi_n(k)\right\}.$$

(23.85)

For a spherical wave, ρ is replaced by $\rho_d\left(\frac{x'}{x}\right)$.

As a function of $\omega_d = \omega_1 - \omega_2$, the two-frequency MCF decreases as ω_d increases. The difference frequency where the MCF decreases to a certain level is called the "coherence bandwidth" B_c, and is related to the pulse broadening T_b (Fig. 23.7).

$$B_c \sim \frac{1}{T_b}.$$

23.13 WEAK AND STRONG FLUCTUATIONS AND SCINTILLATION INDEX

As a wave propagates through a random medium, the fluctuation of the wave increases with distance. Therefore, the fluctuation may be called "weak" or "strong." To classify

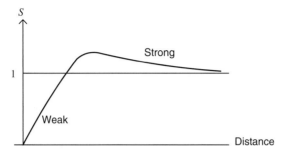

FIGURE 23.8   Weak and strong fluctuation. S = scintillation index.

the fluctuation in this way, let us first consider the average intensity $I = |U|^2$. The variance of the intensity fluctuation

$$\sigma_I^2 = \frac{\langle (I - \langle I \rangle)^2 \rangle}{\langle I \rangle^2} = \frac{\langle I^2 \rangle - \langle I \rangle^2}{\langle I \rangle^2} \tag{23.86}$$

is called the scintillation index. For a short distance, the fluctuation is weak and the scintillation index is small, and for a large distance, the fluctuation is strong and the scintillation index saturates to unity (Fig. 23.8).

The scintillation index involves the fourth-order moment $\langle I^2 \rangle$. In Section 23.8, we have been considering the MCF, which is the second-order moment. The fourth-order moment requires extensive studies of the wave propagation using several techniques including the moment equations, the extended Huygens–Fresnel principle, and others (Ishimaru, 1997). For this discussion, we use the method based on the assumption that the wave is a "circular complex Gaussian random variable" (Appendix 23).

Under this assumption, the fourth-order moments can be expressed by the second-order moments and the solution approaches the correct limits for both weak and strong fluctuations. To show this, we first note that under the circular complex Gaussian assumption,

$$\langle I^2 \rangle = 2\langle I \rangle^2 - \langle U \rangle^2 \langle U^* \rangle^2, \tag{23.87}$$

where $U = \langle U \rangle + v$, and $\langle U \rangle$ is the average field and v is the fluctuation.

Noting that

$$\langle I \rangle = \langle |U|^2 \rangle$$
$$= \langle U \rangle \langle U^* \rangle + \langle vv^* \rangle \tag{23.88}$$

we get

$$\sigma_I^2 = \frac{2\langle U \rangle \langle U^* \rangle \langle vv^* \rangle + \langle vv^* \rangle^2}{\langle U \rangle^2 \langle U^* \rangle^2 + 2\langle U \rangle \langle U^* \rangle \langle vv^* \rangle + \langle vv^* \rangle^2}. \tag{23.89}$$

For weak fluctuation at a short distance, we note

$$\langle U \rangle \langle U^* \rangle \gg \langle vv^* \rangle,$$

and therefore the scintillation index approaches

$$\sigma_I^2 \to \frac{2\langle vv^* \rangle}{\langle U \rangle \langle U^* \rangle}. \tag{23.90}$$

For strong fluctuation, we get

$$\langle U \rangle \langle U^* \rangle \ll \langle vv^* \rangle.$$

And therefore the scintillation index approaches

$$\sigma_I^2 \to 1. \tag{23.91}$$

Note that the circular complex Gaussian assumption leads to the correct behaviors of the scintillation index at weak and strong fluctuations, and gives approximate values for the region between the weak and the strong fluctuations.

It may be instructive to examine the behaviors of the wave $U = A\exp(i\phi)$ in weak and strong fluctuation regions. As shown in Fig. 23.9, for a short distance, the fluctuation is mainly in phase, as the phase is directly related to the distance. The amplitude fluctuation is small because the average amplitude is substantially unchanged since A^2 is substantially unchanged at short distance. The probability distribution of the amplitude fluctuation is close to the Nakagami–Rice distribution

$$W(A) = \frac{2A}{\sigma^2}\exp\left(-\frac{A^2 + \langle A \rangle^2}{\sigma^2}\right) I_0\left(\frac{2A\langle A \rangle}{\sigma^2}\right), \tag{23.92}$$

where σ^2 is the variance of $(A - \langle A \rangle)$.

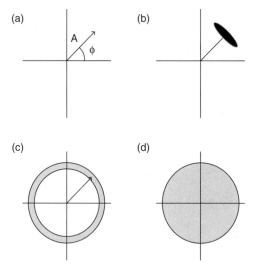

FIGURE 23.9   The behaviors of the wave $U = A\exp(i\phi)$ in turbulence at short and large distances. (a) Free space. (b) Short distance (Nakagami–Rice). (c) Intermediate distance. (d) Large distance (Rayleigh).

On the other hand, at a large distance, the probability distribution approaches the "Rayleigh distribution."

$$W(A) = \frac{2A}{\sigma^2} \exp\left(-\frac{A^2}{\sigma^2}\right).$$ (23.93)

23.14 ROUGH SURFACE SCATTERING, PERTURBATION SOLUTION, TRANSITION OPERATOR

Many natural surfaces, such as oceans, terrain, and biological surfaces and interfaces are rough to varying degrees, and this roughness affects the propagation and scattering characteristics of a wave. Propagation along a rough surface is different from that along a smooth surface. A wave incident on a rough surface is not only reflected in a specular direction but scattered in all directions. Imaging through a rough interface is different from the image through a smooth interface. It is therefore important to understand the effects of the roughness on the wave characteristics. In this section, and the following sections, we discuss the fundamental formulations of the different techniques dealing with rough surface scattering.

There are two standard methods of dealing with the scattering from rough surfaces. One is the "perturbation method" and the other is the "Kirchhoff approximation." The perturbation method is used for scattering from rough surfaces whose rms (root-mean-square) height $(k\zeta)^{21/2}$ is much smaller than $\lambda/8$ and the slopes $\left|\frac{\partial \zeta}{\partial x}\right|$ and $\left|\frac{\partial \zeta}{\partial y}\right|$ are much smaller than one, where $\zeta(x, y)$ is the rough surface height deviation from the average. In this chapter, we use $h(x, y)$ for the rough surface height. In the literature, ζ is also commonly used. The Kirchhoff approximation is for rough surfaces that are slowly varying so that the radius of curvature is much greater than a wavelength.

These two standard methods have been discussed in detail in Ishimaru (1997), Bass et al. (1979), De Santo and Brown (1986), Ogilvy (1991), Tsang et al. (1985), Voronovich (1994), and Fung (1994). They also include more advanced topics of rough surface scattering. In this section and the next, we focus on the perturbation method; the Kirchhoff method will be discussed in Section 23.16.

Let us first consider a scalar wave $\psi_i(\bar{r})$ incident on a rough surface with a Dirichlet boundary condition where height $h(x, y)$ is a random function of the surface coordinate (x, y) (Figs. 23.10 and 23.11).

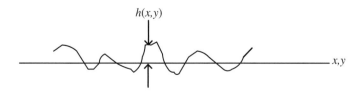

FIGURE 23.10 Rough surface with height $h(x,y)$.

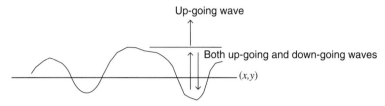

FIGURE 23.11 Rayleigh hypothesis. The wave is assumed to be up going only above the surface. For a sinusoidal surface, the Rayleigh hypothesis is valid if the maximum slope is less than 0.448.

The incident wave $\psi_i(\bar{r})$ can be expressed in Fourier transform $\bar{\psi}(\bar{r})$

$$\psi_i(\bar{r}) = \frac{1}{(2\pi)^2} \iint \bar{\psi}_i(\bar{k}) e^{i\bar{k}\cdot\bar{x} - ik_z z} d\bar{k}, \tag{23.94}$$

where $\bar{k} = k_x\hat{x} + k_y\hat{y}$, and $\bar{x} = x\hat{x} + y\hat{y}$.

Note that if ψ_i is a plane wave in the direction of (θ_i, ϕ_i), then

$$\psi_i(\bar{r}) = \exp(i\bar{k} \cdot \bar{x} - ik_{iz}z), \tag{23.95}$$

where $\bar{k}_i = k\sin\theta_i \cos\phi_i\hat{x} + k\sin\theta_i \sin\phi_i\hat{y}$, $k_{iz} = k\cos\theta_i$, and

$$\bar{\psi}_i(\bar{k}) = (2\pi)^2 \delta(\bar{k} - \bar{k}_i).$$

If ψ_i is originated by a point source at $\bar{r}_o$, we have

$$\psi_i(\bar{r}) = \frac{1}{4\pi R} \exp(ikR), R = |\bar{r} - \bar{r}_o|,$$
$$\bar{\psi}_i(\bar{k}) = \frac{i}{2k_z} \exp(-i\bar{k} \cdot \bar{x}_o + ik_z z_o) \text{ if } z_o > z. \tag{23.96}$$

The scattered field $\psi_s(\bar{r})$ can be written as

$$\psi_s(\bar{r}) = \frac{1}{(2\pi)^2} \int \bar{\psi}_s(\bar{k}) e^{i\bar{k}\cdot\bar{x} + ik_z z} dk. \tag{23.97}$$

This scattered field $\psi_s(\bar{r})$ is assumed to be an up-going wave only. This is called the "Rayleigh hypothesis."

In fact, this up-going wave is true only when z is greater than the highest point of the surface. Between the highest point and the lowest point of the surface, the wave should include both up-going and down-going waves (Fig. 23.12).

It has been proved that for a sinusoidal surface the Rayleigh hypothesis is correct if the maximum slope is less than 0.448. In most practical situations such as ocean surfaces, the maximum slope is less than 0.448 and the Rayleigh hypothesis can be used.

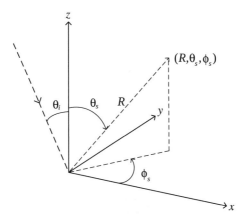

FIGURE 23.12   Scattering in the direction (θ_s, ϕ_s) from a rough surface $h(x,y)$ illuminated by a plane wave incident in the direction $(\theta_i, \phi_i = 0)$.

Let us express the spectrum $\bar{\psi}_s(\bar{k})$ of the scattered wave in terms of the transition operator T (also known as the scattering operator).

$$\bar{\psi}_s(\bar{k}) = \int T_s(\bar{k}, \bar{k}_i)\psi_i(\bar{k}_i)d\bar{k}_i. \tag{23.98}$$

We then apply the boundary condition

$$\psi_i(\bar{r}) + \psi_s(\bar{r}) = 0 \text{ at } z = h \qquad \text{(Dirichlet surface)}. \tag{23.99}$$

Therefore at $z = h$, we have

$$\frac{1}{(2\pi)^2}\int \bar{\psi}_i(\bar{k}_i)d\bar{k}_i e^{-ik_{zi}h + i\bar{k}_i \cdot \bar{x}} + \frac{1}{(2\pi)^2}\iint \bar{\psi}_i(k_i)d\bar{k}_i T_s(\bar{k}, \bar{k}_i)e^{ik_{zi}h + i\bar{k}\cdot\bar{x}}d\bar{k} = 0. \tag{23.100}$$

We now take the Fourier transform of this

$$\int e^{-i\bar{k}'\cdot\bar{x}}d\bar{x},$$

and obtain

$$\int e^{-i(\bar{k}'-\bar{k}_i)\cdot\bar{x}-ik_{zi}h}d\bar{x} + \int d\bar{k}T_s\left(\bar{k}, \bar{k}_i\right)\int e^{-i(\bar{k}'-\bar{k})\cdot\bar{x}+ik_z h}d\bar{x} = 0. \tag{23.101}$$

This is exact and our task now is to obtain the transition operator $T_s(\bar{k}, \bar{k}_i)$ for different rms height $[(kh)]^{1/2}$ and the correlation distance ℓ of the rough surface.

Equation (23.101) is an integral equation for T_s, and in general, an analytical solution has not been obtained. However, it is possible to obtain a solution using the perturbation method for small height h. Let us note that

$$\exp(ik_z h) = 1 + ik_z h + \cdots, \tag{23.102}$$

and we keep up to the first order in $k_z h$.

We then get

$$\int e^{-i(\bar{k}' - \bar{k}_i) \cdot \bar{x}} (1 - ik_z h + \cdots) d\bar{x}$$
$$+ \int d\bar{k} T_s(\bar{k}, \bar{k}_i) \int d\bar{x} e^{-i(\bar{k}' - \bar{k}) \cdot \bar{x}} (1 + ik_z h + \cdots) = 0. \tag{23.103}$$

If we let $\bar{H}$ be the Fourier transform of $h(x, y)$

$$\bar{H}(\bar{k}) = \int h e^{-i\bar{k} \cdot \bar{x}} d\bar{x}. \tag{23.104}$$

And substituting this in (23.103), we get

$$(2\pi)^2 \delta(\bar{k}' - \bar{k}_i) - ik_{zi} \bar{H}(\bar{k}' - \bar{k}_i) + (2\pi)^2 T_s(\bar{k}', \bar{k}_i)$$
$$+ \int d\bar{k} T_s(\bar{k}, \bar{k}_i) ik_z \bar{H}(\bar{k}' - \bar{k}) = 0. \tag{23.105}$$

Now noting that $\tilde{H}(\bar{k}) \sim \epsilon$ (a small number), we can obtain the expression

$$T_s = T_{s0} + \epsilon T_{s1} + \epsilon^2 T_{s2} + \cdots. \tag{23.106}$$

Substituting this in (23.105) we get the zero order

$$\delta(\bar{k}' - \bar{k}_i) + T_{s0}(\bar{k}', \bar{k}_i) = 0.$$

From this, we get

$$T_{s0}(\bar{k}', \bar{k}_i) = -\delta(\bar{k}' - \bar{k}_i). \tag{23.107}$$

Similarly, for the first order, we get

$$-ik_{zi} \bar{H}(\bar{k}' - k_i) + (2\pi)^2 T_{s1}(\bar{k}', k_i) - ik_{zi} \bar{H}(\bar{k}' - \bar{k}_i) = 0,$$

from which we get

$$T_{s1}(\bar{k}', k_i) = \frac{i2k_{zi} \bar{H}(\bar{k}' - \bar{k}_i)}{(2\pi)^2}. \tag{23.108}$$

We now note that the rough surface height $h(x, y)$ is a random function and therefore its Fourier transform H is also random. We can now consider the first and second moments of the scattered wave. Let us consider the coherent scattered wave. This is given by

$$\langle \psi_s(\bar{r}) \rangle = \frac{1}{(2\pi)^2} \int \langle \psi_s(\bar{k}) \rangle e^{i\bar{k}\cdot\bar{x}+ik_z z} d\bar{k}. \tag{23.109}$$

Noting that $\langle H \rangle = 0$, we get

$$\begin{aligned}\langle \psi_s(\bar{k}) \rangle &= (2\pi)^2 \langle T_s(\bar{k},\bar{k}_i) \rangle, \langle T_s \rangle = T_{s0}, \\ &= (2\pi)^2 T_{s0}(\bar{k},\bar{k}_i) \\ &= -(2\pi)^2 \delta(\bar{k}' - \bar{k}_i). \end{aligned} \tag{23.110}$$

The coherent scattered wave is then given by the inverse Fourier transform of (23.104)

$$\langle \psi_s(\bar{r}) \rangle = -e^{i\bar{k}_i\cdot\bar{x}+ik_{zi}z}, \tag{23.111}$$

which is a plane wave scattered in the direction of (θ_i, ϕ_i) as given in (23.95).

Next we consider the incoherent scattered wave, which is given by

$$\begin{aligned}\psi_f(\bar{r}) &= \psi_s(\bar{r}) - \langle \psi_s(\bar{r}) \rangle \\ &= \frac{1}{(2\pi)^2} \int \left[\bar{\psi}_s(\bar{k}) - \langle \bar{\psi}_s(\bar{k}) \rangle \right] e^{i\bar{k}\cdot\bar{x}+ik_z z} d\bar{k} \\ &= \frac{1}{(2\pi)^2} \int (2\pi)^2 T_{s1}(k) \, e^{i\bar{k}\cdot\bar{x}+ik_z z} dk, \end{aligned} \tag{23.112}$$

where

$$\begin{aligned}T_{s1} &= T_s - \langle T_s \rangle \\ &= \frac{i2k_{zi}\bar{H}(\bar{k}' - \bar{k}_i)}{(2\pi)^2}. \end{aligned}$$

Let us note that for a far field, we have

$$\begin{aligned}\psi_f(\bar{r}) &= \frac{1}{(2\pi)^2} \int \bar{\psi}_f(\bar{k}) e^{i\bar{k}\cdot\bar{x}+ik_z z} dk \\ &= \left[-i2k_z \bar{\psi}_f(\bar{k}_s) \right] \frac{e^{ikR}}{4\pi R}, \end{aligned} \tag{23.113}$$

where

$$\bar{k}_s = k\sin\theta\cos\phi\hat{x} + k\sin\theta\sin\phi\hat{y},$$

$$k_z = k\cos\theta,$$

$$\bar{\psi}_s(\bar{k}_s) = (2\pi)^2 T_{s1}(\bar{k}_s),$$

We therefore get the scattering cross section per unit area of the rough surface.

$$
\sigma^0 = \frac{4\pi R^2 \left\langle |\psi_f(\bar{r})|^2 \right\rangle}{|\psi_i|^2 \, d\bar{x}_c} = \frac{k_z^2 (2\pi)^4}{\pi} \frac{\left\langle |T_{s1}|^2 \right\rangle}{d\bar{x}_c}
$$

$$
= \frac{4k_z^2 k_{zi}^2}{\pi} \int \langle h(\bar{x}) h(\bar{x}') \rangle e^{-i(\bar{k}-\bar{k}_i)\cdot\bar{x}_d} \, dx_d \tag{23.114}
$$

$$
= \frac{4k_z^2 k_{zi}^2}{\pi} W(\bar{k} - \bar{k}_i),
$$

where

$$
|H(\bar{k} - \bar{k}_i)|^2 = W(\bar{k} - \bar{k}_i) d\bar{x}_c.
$$

Note that $W(\bar{k} - \bar{k}_i)$ is the Fourier transform of the correlation function of the surface height. If the correlation function is a Gaussian, we get

$$
\langle h(x, y) h(x', y') \rangle = \langle h^2 \rangle \exp\left[-\frac{|\bar{x} - \bar{x}'|^2}{\ell^2} \right], \tag{23.115}
$$

$$
W(\bar{k} - \bar{k}_i) = \langle h^2 \rangle (\pi l^2) \exp\left[-\frac{|\bar{k} - \bar{k}_i|^2 l^2}{4} \right]
$$

where

$[\langle h^2 \rangle]^{1/2}$ is the root mean square of the rough surface height, and

$$
\bar{k} = k \sin\theta \cos\phi \hat{x} + k \sin\theta \sin\phi \hat{y},
$$

$$
\bar{k}_i = k \sin\theta_i \cos\phi_i \hat{x} + k \sin\theta_i \sin\phi_i \hat{y},
$$

$$
\ell = \text{correlation distance.}
$$

Next we consider the rough surface with a Neumann boundary condition.

$$
\frac{\partial}{\partial n} \psi_i + \frac{\partial}{\partial n} \psi_s = 0 \text{ at } z = h. \tag{23.116}
$$

The procedure of finding the scattering cross section per unit area of the rough surface is similar to that discussed for a Dirichlet surface in this section. The difference is noted in the following:

$$
\frac{\partial}{\partial n} = \hat{N} \cdot \nabla, \tag{23.117}
$$

$$\hat{N} = \frac{-\dfrac{\partial h}{\partial x}\hat{x} - \dfrac{\partial h}{\partial y}\hat{y} + \hat{z}}{\left[1 + \left(\dfrac{\partial h}{\partial x}\right)^2 + \left(\dfrac{\partial h}{\partial y}\right)^2\right]^{\frac{1}{2}}}$$

= unit vector normal to the surface,

$$\frac{\partial}{\partial n} = -\frac{\partial h}{\partial x}\frac{\partial}{\partial x} - \frac{\partial h}{\partial y}\frac{\partial}{\partial y} + \frac{\partial}{\partial z} + \epsilon^2 + \cdots,$$

where ϵ is a small number representing a small height ($h \sim \epsilon$), and small slope ($\partial h/\partial x \sim \epsilon, \partial h/\partial y \sim \epsilon$).

We then get

$$\frac{\partial}{\partial n}\Psi_i = \int \bar{\Psi}_i (k) \left[-ik_z - i\bar{k} \cdot \nabla_t h\right] e^{i\bar{k}\cdot\bar{x} - ik_z h} d\bar{k},$$

$$\frac{\partial}{\partial n}\Psi_s = \int d\bar{k}_i T_s(\bar{k}, \bar{k}_i)\bar{\Psi}_i(k) \left[ik_z - i\bar{k} \cdot \nabla_t h\right] e^{i\bar{k}\cdot\bar{x} + ik_z h} d\bar{k}. \tag{23.118}$$

Following the perturbation procedure discussed in this section, we get

$$T_{s1}(\bar{k}, \bar{k}_i) = \frac{1}{ik_z(2\pi)^2} 2 \left[k^2 - \bar{k}_i \cdot \bar{k}\right] \bar{H}(\bar{k} - \bar{k}_i).$$

From this, we get the scattering cross section per unit area of the rough surface with Neumann boundary condition

$$\sigma^0 = \frac{k_z^2(2\pi)^4}{\pi} \frac{\langle T_{s1} T_{s1}^* \rangle}{d\bar{x}_c}$$
$$= \frac{4}{\pi} \left|k^2 - \bar{k}_i \cdot \bar{k}\right|^2 W(\bar{k} - \bar{k}_i). \tag{23.119}$$

23.15 SCATTERING BY ROUGH INTERFACES BETWEEN TWO MEDIA

The perturbation theory in Section 23.14 for Dirichlet and Neumann rough surfaces can be extended to a two-media problem. We describe incident, scattered, and transmitted waves using the Rayleigh hypothesis.

$$\Psi_i (\bar{r}) = \frac{1}{(2\pi)^2} \int d\bar{k}\bar{\Psi}_i (k) e^{-ik_{zi}z + i\bar{k}_i \cdot \bar{x}}, \tag{23.120}$$

$$\psi_s(\bar{r}) = \frac{1}{(2\pi)^2} \int d\bar{k}_i \int d\bar{k} T_s(\bar{k}, \bar{k}_i) \tilde{\psi}_i(\bar{k}_i) e^{ik_{z1}z + i\bar{k} \cdot \bar{x}},$$

$$\psi_t(\bar{r}) = \frac{1}{(2\pi)^2} \int d\bar{k}_i \int d\bar{k} T_t(\bar{k}, \bar{k}_i) \tilde{\psi}_i(\bar{k}_i) e^{-ik_{z2}z + i\bar{k} \cdot \bar{x}},$$

where T_s and T_t are the transition operators for scattering and transmitted waves, and

$$k_{zi} = \sqrt{k_1^2 - k_i^2}, \quad k_{z1} = \sqrt{k_1^2 - k^2}, \quad k_{z2} = \sqrt{k_2^2 - k^2}.$$

For acoustic waves, as discussed in Section 3.8,

$$k_1 = \frac{\omega}{c_1} \quad \text{and} \quad k_2 = \frac{\omega}{c_2}.$$

The boundary condition is the continuity of ψ and $(1/\rho_c)(\partial/\partial n)\psi$.
For electromagnetic waves, as shown in Sections 3.4 and 3.5,

$$k_1 = k_0 n_1 = k_0 (\mu_{r1} \epsilon_{r1})^{1/2},$$

$$k_2 = k_0 n_2 = k_0 (\mu_{r2} \epsilon_{r2})^{1/2}.$$

The boundary conditions for two-dimensional (x, z) problems are the continuity of ψ and $(\partial/\partial n)\psi$ for s-polarization and the continuity of ψ and $(1/\epsilon)(\partial/\partial n)\psi$ for p-polarization.

In this section, we consider the continuity of ψ and $(1/\epsilon)(\partial/\partial n)\psi$. We use the procedure similar to the perturbation solution in Section 23.14. We write the zero-order and the first-order transition operators

$$T_s = T_{s0} + T_{s1} + \cdots ,$$
$$T_t = T_{t0} + T_{t1} + \cdots . \tag{23.121}$$

We then get the zero-order solution.
Following the perturbation method in the Section 23.14, we get the equation for the zero-order transition operators T_{s0} and T_{t0} from the continuity of ψ.

$$\delta(k' - k_i) + T_{s0}(k', k_i) = T_{t0}(k', k_i),$$

$$k_{zi} \left[\delta(k' - k_i) - T_{s0}(k', k_i) \right] = \frac{k'_{z2}}{\epsilon_r} T_{t0}(k', k_i). \tag{23.122}$$

From this, we get

$$T_{s0}(\bar{k}, \bar{k}_i) = R(k_i, k_i)\delta(\bar{k} - \bar{k}_i),$$

$$T_{t0}(\bar{k}, \bar{k}_i) = T(k_i, k_i)\delta(\bar{k} - \bar{k}_i), \tag{23.123}$$

R and T are the reflection and transmission coefficients for a plane interface.

$$R(k_i, k_i) = \frac{(k_{z1}/\epsilon_1) - (k_{z2}/\epsilon_2)}{(k_{z1}/\epsilon_1) + (k_{z2}/\epsilon_2)},$$

$$T(k_i, k_i) = \frac{2(k_{z1}/\epsilon_1)}{(k_{z1}/\epsilon_1) + (k_{z2}/\epsilon_2)}.$$

(23.124)

Similarly, for the first-order solution, we have

$$2\pi T_{s1} - 2\pi T_{t1} = A(k, k_i)H(k - k_i),$$

$$ik'_{zi}2\pi T_{s1} - \frac{1}{\epsilon_r}(-ik'_{z2})2\pi T_{t1} = B(k, k_i)H(k - k_i).$$

(23.125)

From that, we get the first-order transition operator.

$$T_{s1}(k, k_i) = S_s(k, k_i)H(k - k_i),$$

$$T_{t1}(k, k_i) = S_t(k, k_i)H(k - k_i),$$

(23.126)

$$S_s = \frac{\left(\dfrac{ik_{z2}}{\epsilon_r}A + B\right)}{\left(\dfrac{ik_{z2}}{\epsilon_r} + ik_{zi}\right)},$$

$$S_t = \frac{B - \left(ik_{zi}A\right)}{\left(\dfrac{ik_{z2}}{\epsilon_r} + ik_{zi}\right)},$$

$$A = ik_{zi} - ik_{z1}R - ik_{z2}T,$$

$$B = -\left(k_i k - k_1^2\right)(1 + R) + \frac{1}{\epsilon_r}\left(k_i k - k_2^2\right)T,$$

$$H(k' - k_i) = \int he^{-i(k'-k_i)\bar{x}}d\bar{x}.$$

Noting the development in Section 23.14, we get the incoherent scattered field ψ_s in $z > 0$.

$$\psi_s = \int T_{s1}(\bar{k}, \bar{k}_i)e^{i\bar{k}\cdot\bar{x}+ik_{z}z}d\bar{k}.$$

(23.127)

The transmitted field ψ_t for $z < 0$ is given by

$$\psi_t = \int T_{t1}(\bar{k}, \bar{k}_i)e^{i\bar{k}\cdot\bar{x}-ik_{z2}z}d\bar{k}.$$

(23.128)

We therefore get the scattering cross section per unit area

$$\sigma_s^0 = \frac{k_z^2 (2\pi)^4}{\pi} \frac{\langle |T_{s1}|^2 \rangle}{d\bar{x}_c}. \tag{23.129}$$

The transmitted cross section per unit area is then given by

$$\sigma_t^0 = \frac{k_{z2}^2 (2\pi)^4}{\pi} \frac{\langle |T_{t1}|^2 \rangle}{d\bar{x}_c}. \tag{23.130}$$

23.16 KIRCHHOFF APPROXIMATION OF ROUGH SURFACE SCATTERING

As we stated at the beginning of Section 23.14, there are two conventional techniques for scattering by rough surfaces: the Perturbation technique and the Kirchhoff approximation. In this section, we outline the Kirchhoff approximation. This has been studied extensively by Ishimaru (1997) and therefore, we only give essential formulations in this section.

Consider a plane wave incident on a rough surface given by the surface height $h(x, y)$, which is measured from the average height, and therefore

$$\langle h(x, y) \rangle = 0. \tag{23.131}$$

The scattered field $\psi_s(\bar{r})$ is given by (6.17).

$$\psi_s(\bar{r}) = \int_s \left[\psi(\bar{r}') \frac{\partial G_0(\bar{r}, \bar{r}')}{\partial n'} - G_0(\bar{r}, \bar{r}') \frac{\partial \psi(\bar{r}')}{\partial n'} \right] ds'. \tag{23.132}$$

Under the Kirchhoff approximation given in Section 6.3, we have the surface field

$$\psi(\bar{r}') = (1 + R_f)\psi_i(\bar{r}'),$$
$$\frac{\partial \psi(\bar{r}')}{\partial n'} = (i\bar{k}_s \cdot \hat{N})(-R_f)\psi_i(\bar{r}'), \tag{23.133}$$

where R_f is the local reflection coefficient at $\bar{r}'$ and $\hat{N}$ is the unit vector normal to the surface, and

$$\bar{k}_s = (k \sin \theta_s \cos \phi_s)\hat{x} + (k \sin \theta_s \sin \phi_s)\hat{y} + (k \cos \theta_s)\hat{z},$$
$$\bar{k}_i = (k \sin \theta_i \cos \phi_i)\hat{x} + (k \sin \theta_i \sin \phi_i)\hat{y} - (k \cos \theta_i)\hat{z}. \tag{23.134}$$

Note that $k_{iz} = -k \cos \theta_i$, indicating that the incident wave is incoming.

We also note that

$$\hat{N} = N_x \hat{x} + N_y \hat{y} + N_z \hat{z}$$

$$= \frac{-\dfrac{\partial h}{\partial x} \hat{x} - \dfrac{\partial h}{\partial y} \hat{y} + \hat{z}}{\left[1 + \left(\dfrac{\partial h}{\partial x}\right)^2 + \left(\dfrac{\partial h}{\partial y}\right)^2\right]^{\frac{1}{2}}}, \tag{23.135}$$

$$ds' = \frac{dx'\,dy'}{N_z}.$$

Substituting (23.133) and (23.135) into (23.142), we get

$$\psi_s(\bar{r}) = \frac{i\exp(ikR)}{4\pi R} \int_S \left(\bar{V} R_f - \bar{W}\right) \cdot \hat{N} \exp\left(i\bar{V} \cdot \bar{V}'\right) ds', \tag{23.136}$$

where

$$\bar{V} = \bar{k}_i - \bar{k}_s,$$

$$\bar{W} = \bar{k}_i + \bar{k}_s,$$

$$\bar{V} \cdot \bar{V}' = v_x x' + v_y y' + v_z z',$$

We also note that if the observation point $\bar{r}(R_s, \theta_s, \phi_s)$ is in the far zone of the surface, we can approximate Green's function.

$$G_0 \cong \frac{\exp(ikR - i\bar{k}_s \cdot \bar{r}')}{4\pi R}, \qquad \frac{\partial G_0}{\partial n'} \cong -i\bar{k}_s \cdot \hat{N} G_0.$$

Note that

$$\bar{V} \cdot \hat{N} ds' = \left(-v_x \frac{\partial h}{\partial x} - v_y \frac{\partial h}{\partial y} + v_z\right) dx'\,dy',$$

$$\bar{W} \cdot \hat{N} ds' = \left(-w_x \frac{\partial h}{\partial x} - w_y \frac{\partial h}{\partial y} + w_z\right) dx'\,dy'.$$

The integral involving $\dfrac{\partial h}{\partial x}$ and $\dfrac{\partial h}{\partial y}$ can be handled by integration by parts.

$$\int \frac{\partial h}{\partial x} e^{iv_x x + iv_z h}\,dx' = \frac{e^{iv_x x + iv_z h}}{iv_z}\bigg|_{x_1}^{x_2} - \int dx' \frac{v_x}{v_z} e^{iv_x x + iv_z h}.$$

The first term represents the edge effect and is negligible compared with the second term for a large area. Therefore $\partial h/\partial x$ can be replaced by $(-v_x/v_z)$ in (23.136). Similarly we can replace $\partial h/\partial y$ by $(-v_y/v_z)$.

With this approximation we get

$$\psi_s(\bar{r}) = \frac{i\exp(ikR)}{4\pi R} FR_f \int_S \exp(i\bar{v} \cdot \bar{r}')dx'dy'. \qquad (23.137)$$

where F is given by

$$F = \frac{\bar{V} \cdot \bar{V}}{v_z} = \frac{2k[1 + \cos\theta_i \cos\theta_s - \sin\theta_i \sin\theta_s \cos(\phi_i - \phi_s)]}{\cos\theta_i + \cos\theta_s}.$$

Note that $\bar{V} \cdot \bar{W} = k_i^2 - k_s^2 = 0$.

The surface height h appears in the exponent $v_z h$ in (23.137). If the surface is flat, $h = 0$ and in the specular direction $v_x = v_y = 0$, $\theta_i = \theta_s$, and $\phi_i = \phi_s$ we get

$$\psi_s(\bar{r}) = \frac{i\exp(ikR)}{4\pi R}(2k\cos\theta_i)R_f \int_S dx'dy'. \qquad (23.138)$$

Normalizing ψ_s in (23.138) using the specularly reflected field ψ_o, we get

$$\psi_s(\bar{r}) = \psi_o(\bar{r})[f]\frac{1}{S} \int_S \exp(i\bar{V} \cdot \bar{r}')dx'dy', \qquad (23.139)$$

where

$$f = \frac{F}{2k\cos\theta_i} = \frac{1 + \cos\theta_i \cos\theta_s - \sin\theta_i \sin\theta_s \cos(\phi_i - \phi_s)}{\cos\theta_i [\cos\theta_i + \cos\theta_s]}.$$

This is the general expression of the scattered field in the Kirchhoff approximation. The rough surface height $h(x, y)$ is in the exponent of the integrant.

Using (23.137), we can get the coherent field in the specular direction

$$\langle \psi_s(\bar{r}) \rangle = \psi_o(\bar{r})[f]\langle e^{iv_z h} \rangle, \qquad (23.140)$$

where

$$\chi(v_z) = \langle e^{iv_z h} \rangle = \text{characteristic function of the random height } h.$$

If the height h is normally distributed with the variance σ_o^2, then the probability density function $W_o(h)$ is given by

$$W_o(h) = \frac{1}{(2\pi)^{1/2}\sigma_0} \exp\left(-\frac{h^2}{2\sigma_0^2}\right), \qquad (23.141)$$

and the characteristic function $\chi(v_z)$ is given by

$$\chi(v_z) = \int_{-\infty}^{\infty} W_o(h)\exp(v_z h)dh = \exp\left(-\frac{\sigma_0^2 v_z^2}{2}\right) = \exp\left(-2\sigma_0^2 k^2 \cos^2 \theta_i\right).$$

$$(23.142)$$

The coherent field is then given by

$$\psi_s(\bar{r}) = \psi_o(\bar{r})\chi(v_z) = \psi_o(\bar{r}) \exp\left(-2\sigma_0^2 k^2 \cos^2 \theta_i\right), \qquad (23.143)$$

where $[f] = 1$ in the specular direction $\theta_i = 0$.

$R_a = k\sigma_a \cos \theta_i$ is called the Rayleigh parameter. Rayleigh used this as a criterion that a surface may be considered rough or smooth depending on whether $2R_a$ is greater or smaller than $\pi/2$.

Let us next consider the incoherent scattered wave. We start with (23.139).

$$\psi_s(\bar{r}) = \psi_o(\bar{r}) [f] \frac{1}{S} \int_s \exp(i\bar{V} \cdot \bar{r}')dx'dy'.$$

The incoherent field $\psi_d(\bar{r})$ is given by

$$\psi_d(\bar{r}) = \psi_s(\bar{r}) - \langle \psi_s(\bar{r}) \rangle. \qquad (23.144)$$

Therefore the scattering cross section per unit area is given by

$$\sigma = \frac{4\pi R^2}{S} \langle |\psi_d|^2 \rangle.$$

Noting (23.152), we get

$$\sigma = \frac{k^2 \cos^2 \theta_i}{\pi} [f]^2 R_f^2 I,$$

where

$$I = \frac{1}{S} \int_s dx'dy' \int_s dx''dy'' \exp\left[iv_x(x'-x'') + iv_y(y'-y'')\right] \left[\chi_2(v_z,-v_z) - |\chi(v_z)|^2\right]$$

$$= \int_s dx_d dy_d \exp\left[iv_x x_d + iv_y y_d\right] \left[\chi_2(v_z,-v_z) - |\chi(v_z)|^2\right], \qquad (23.145)$$

and where $x_d = x' - x''$ and $y_d = y' - y''$. $\chi_2(v_z,-v_z)$ is a joint characteristic function of the height function h.

$$\chi_2(v_z,-v_z) = \langle \exp(iv_1 h_1 + iv_2 h_2) \rangle. \qquad (23.146)$$

Under the assumption that the surface is statistically homogeneous and isotropic, $\chi(v_1, v_2)$ is a function of $(x_d^2 + y_d^2)^{1/2}$ only.

If the surface is normally distributed, we have

$$\chi(v_z, -v_z) = \exp\left\{-v_z^2\sigma_0^2\left[1 - C(\rho)\right]\right\}, \tag{23.147}$$

where $C(\rho)$ is the correlation coefficient.

$$h(x', y')h(x'', y'') = \sigma_0^2 C(t),$$
$$\rho = [(x' - x'')^2 + (y' - y'')^2]^{1/2}. \tag{23.148}$$

Let us examine the scattering cross section (23.145) for the following two cases. First, if the rms height σ_0 is small, we can expand the following:

$$\chi_2(v_z, -v_z) - |\chi(v_z)|^2 = \exp\left[-v_z^2\sigma_0^2(1 - C(\rho))\right] - \exp\left(-v_z^2\sigma_0^2\right)$$

$$= \exp\left(-v_z^2\sigma_0^2\right)\sum_{m=0}^{\infty}\frac{\left[v_z^2\sigma_0^2 C(\rho)\right]^m}{m!} - \exp\left(-v_z^2\sigma_0^2\right)$$

$$= \exp\left(-v_z^2\sigma_0^2\right)\sum_{m=1}^{\infty}\frac{\left(v_z^2\sigma_0^2\right)^m}{m!}\exp\left(-m\frac{x_d^2}{\ell^2}\right). \tag{23.149}$$

Here we used the Gaussian correlation function.

$$C(\rho) = \exp\left(-x_d^2/\ell^2\right).$$

Substituting (23.149) into (23.145), we get

$$I = \exp\left(-v_z^2\sigma_0^2\right)\sum_{m=1}^{\infty}\frac{\left(v_z^2\sigma_0^2\right)^m}{m!}\frac{\pi\ell^2}{m}\exp\left(-\frac{v^2\ell^2}{4m}\right), \tag{23.150}$$

where

$$v_z^2 = k^2(\cos\theta_i + \cos\theta_s)^2,$$
$$v^2 = v_x^2 + v_y^2$$
$$= k^2\left[\left(\sin\theta_s\cos\phi_s - \sin\theta_i\right)^2 + \sin^2\theta_s\sin^2\phi_s\right]$$
$$= k^2\left[\sin\theta_s^2 + \sin^2\theta_i - 2\sin\theta_s\sin\theta_i\cos\phi_s\right].$$

The series in (23.150) is divergent if $v_z^2\sigma_0^2 \gg 1$. Then we use the following approximation for very rough surfaces.

$$\chi_2(v_z, -v_z) - |\chi(v_z)|^2 \approx \chi_2(v_z, -v_z), \tag{23.151}$$

$$\exp\left[-v_z^2\sigma_0^2(1 - C(\rho))\right] \approx \exp\left(-v_z^2\sigma_0^2\left(\rho^2/\ell^2\right)\right).$$

We then get

$$\sigma = \frac{k^2 \cos^2 \theta_i}{\pi} f^2 R_f^2 I,$$

$$I \approx \frac{\pi \ell^2}{v_z^2 \sigma_0^2} \exp\left(-\frac{v^2 \ell^2}{4 v_z^2 \sigma_0^2}\right),$$

$$(23.152)$$

where

$$v^2 = v_x^2 + v_y^2.$$

23.17 FREQUENCY AND ANGULAR CORRELATION OF SCATTERED WAVES FROM ROUGH SURFACES AND MEMORY EFFECTS

In Section 23.14, we gave a conventional perturbation solution of the scattered wave from rough surfaces. This is expressed as the scattering cross section per unit area of the surface given in (23.114) for Dirichlet surfaces and in (23.119) for Neumann surfaces. These represent the scattered power is the direction of $\bar{K}$ when the wave is incident from the direction of $\bar{K}_i$ where, as shown in (23.94) and (23.95),

$$\bar{k} = k \sin \theta \cos \phi \hat{x} + k \sin \theta \sin \phi \hat{y},$$

$$\bar{k}_i = k \sin \theta_i \cos \phi_i \hat{x} + k \sin \theta_i \sin \phi_i \hat{y}.$$

$$(23.153)$$

Now we seek a more general case where the two incident waves come in the directions $\bar{k}_i$ and $\bar{k}_i'$ and the scattered wave is observed in the directions $\bar{k}$ and $\bar{k}'$ (Fig. 23.13).

We wish to study the correlation of the scattered waves $\psi_s(\bar{k})$ and $\psi_s^*(\bar{k}')$. For a Dirichlet rough surface, the correlation can be expressed as the scattering correlation cross section of the illuminated area W_0^2 of the rough surface. It will be shown that

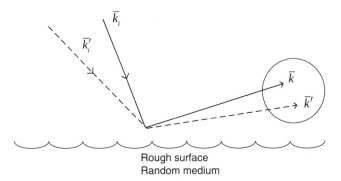

Rough surface
Random medium

FIGURE 23.13   Two waves in the direction $\bar{k}_i$ and $\bar{k}_i'$ are incident on the rough surface and produce the scattered waves in the direction of $\bar{k}$ and $\bar{k}'$, respectively. The correlation of the waves in $\bar{k}$ and $\bar{k}'$ is to be considered.

the resulting correlation cross section depends on the illumination area, rather than the cross section per unit area as shown in the previous section.

Following (23.114), we can obtain the following scattering correlation cross section of the illuminated area W_0^2 of the rough surface. For Dirichlet surfaces, we get

$$\sigma_c^0 \left(\bar{k}_i, \bar{k}_i', \bar{k}, \bar{k}' \right) = \frac{4\pi R^2 \psi_{f1} \psi_{f1}^*}{\psi_i \psi_i^*}, \tag{23.154}$$

where ψ_{f1} is the incoherent scattered wave in the direction of $\bar{k}$ when the incident wave ψ_i is coming from the direction of $\bar{k}_i$, and ψ_{f1}^* is the incoherent scattered wave in the direction of $\bar{k}'$ with the incident wave ψ_i^* coming from the direction of $\bar{k}_i$.

Following (23.114), we can write this as

$$\sigma_c^0 = \frac{4 k_z k_z' k_{zi} k_{zi}'}{\pi} W_c, \tag{23.155}$$

$$W_c = \iint h(\bar{x}) h(\bar{x}') \exp(-i\bar{v} \cdot \bar{x} + i\bar{v}' \cdot \bar{x}') d\bar{x}\, d\bar{x}',$$

where

$$\bar{v} = \bar{k} - \bar{k}_i, \quad \bar{v}' = \bar{k}' - \bar{k}_i'.$$

We can now use the center of gravity $\bar{x}_c$ and the difference $\bar{x}_d$, $\bar{v}_c$, and $\bar{v}_d$.

$$\bar{x}_c = \frac{1}{2}(\bar{x} + \bar{x}'), \bar{x}_d = \bar{x} - \bar{x}',$$
$$\bar{v}_c = \frac{1}{2}(\bar{v} + \bar{v}'), \bar{v}_d = \bar{v} - \bar{v}_d. \tag{23.156}$$

Also, we assume the height h is normally distributed, and get

$$\langle h(\bar{x}) h(\bar{x}') \rangle = \langle h^2 \rangle \exp \left[-\frac{x_d^2}{\ell^2} \right]. \tag{23.157}$$

We can then rewrite W_c.

$$W_c = \iint h^2 \exp \left(-\frac{x_d^2}{\ell^2} \right) \exp(-i\bar{v}_c \cdot \bar{x}_d + i\bar{v}_d \cdot \bar{x}_c) d\bar{x}_d d\bar{x}_c$$
$$= h^2 W_1 W_2, \tag{23.158}$$

$$W_1 = \int e^{-\frac{x_d^2}{\ell^2} - i\bar{v}_c \cdot \bar{x}_d} d\bar{x}_d$$

$$= \pi\ell^2 \exp\left(-\frac{v_c^2 \ell^2}{4}\right)$$

$$= \pi\ell^2 \exp\left[-\left|(\bar{k} - \bar{k}_i) + (\bar{k}' - \bar{k}'_i)\right|^2 (\ell^2/4)\right],$$

$$W_2 = \int e^{-i\bar{v}_d \cdot \bar{x}_c} d\bar{x}_c$$

$$= (2\pi)^2 \delta\left[(\bar{k} - \bar{k}_i)_x + (\bar{k}' - \bar{k}'_i)_x\right] \delta\left[(\bar{k} - \bar{k}_i)_y + (\bar{k}' - \bar{k}'_i)_y\right].$$

If we include the illumination area of W_0^2, we get

$$W_2 = \int e^{-\frac{x_c^2}{W_0^2} - i\bar{v}_d \cdot \bar{x}_c} d\bar{x}_c$$

(23.159)

$$= \pi W_0^2 \exp\left[-\left|(\bar{k} - \bar{k}_i) + (\bar{k}' - \bar{k}'_i)\right|^2 (W_0^2/4)\right].$$

Equation (23.155) with (23.158) and (23.159) give the correlation of the scattered field as shown in Fig. 23.13. In order to clarify the meaning of this correlation, let us consider a special case of $\phi_i = \phi = 0$ in (23.120).

Then, we have

$$\bar{k} = k \sin\theta\hat{x},$$

$$\bar{k}_i = k \sin\theta_i\hat{x}.$$

(23.160)

W_1 and W_2 in (23.158) can be written as

$$W_1 = \pi\ell^2 \exp\left[-\left|(\bar{k} - \bar{k}_i) + (\bar{k}' - \bar{k}'_i)\right|^2 (\ell^2/4)\right],$$

$$W_2 = \pi W_0^2 \exp\left[-\left|(\bar{k} - \bar{k}_i) + (\bar{k}' - \bar{k}'_i)\right|^2 (W_0^2/4)\right].$$

(23.161)

We examine W_2 with $k = k'$. W_2 has a peak value when

$$\sin\theta - \sin\theta_i = \sin\theta' - \sin\theta'_i.$$

(23.162)

This is shown in Figs. 23.14 and 23.15. Figure 23.16 shows the memory line when $k \neq k'$.

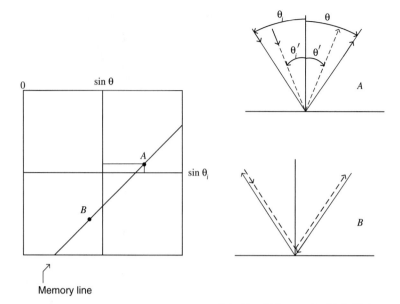

FIGURE 23.14 Memory line where the correlation peaks. The width of the line is approximately (λ/W_0), W_0 = illumination size.

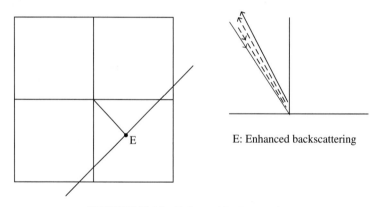

E: Enhanced backscattering

FIGURE 23.15 Enhanced backscattering.

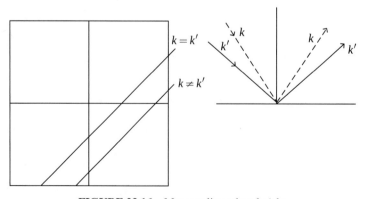

FIGURE 23.16 Memory line when $k \neq k\prime$.

This is called the "Memory effect," and shows that even under multiple scattering, the scattering wave "remembers" the direction of the incident wave as shown in these figures. These figures show the part of the memory effect under the perturbation theory of the rough surface scattering. The memory effect is more general than what is shown here. It was originally proposed by Feng (1985), and extensive research has been conducted.

CHAPTER 24

COHERENCE IN MULTIPLE SCATTERING AND DIAGRAM METHOD

As a wave propagates through random media such as turbulence, rain, fog, or biological media, the wave experiences multiple scattering, resulting in random variations in amplitude and phase. It is then normally expected that the scattered wave is mostly incoherent, and the wave scattered in the back direction is, in general, incoherent and scattered in all directions, more or less uniformly. This conventional view is consistent with and confirmed by the transport theory (radiative transfer theory).

It was therefore surprising that in an optical experiment with a beam incident on many micro latex spheres suspended in water, a sharp peak in the scattering was observed in the back direction. This optical experiment by Kuga and Ishimaru (1984) and its theoretical explanation by Tsang and Ishimaru (1984) are among several early works reporting on experimental and theoretical explanations of "coherent backscattering" or "enhanced backscattering." It is an optical equivalent of weak Anderson localization, which was discussed by Anderson (1958) to explain the absence of electron diffusion in multiple scattering in a disordered medium.

Enhanced backscattering occurs in many areas of engineering including scattering from particles, rough surfaces, and turbulence. It is related to "retro reflectance" or "opposition" effects, and observed enhanced backscattering by soils and vegetation. In this chapter, we present an introduction to "backscattering enhancement." Some historical and additional accounts are given by Ishimaru (1991).

Electromagnetic Wave Propagation, Radiation, and Scattering: From Fundamentals to Applications,
Second Edition. Akira Ishimaru.
© 2017 by The Institute of Electrical and Electronic Engineers, Inc. Published 2017 by John Wiley & Sons, Inc.

24.1 ENHANCED RADAR CROSS SECTION IN TURBULENCE

It may seem surprising that radar cross section (RCS) increases through turbulence, as we may have the conventional view that RCS may decrease in turbulence. However, this increase has been theoretically studied by noting that the backscattered intensity in turbulence is proportional to the fourth-order moment and approximately twice the multiple-scattered intensity. This has also been verified experimentally. In this section, we outline the basic idea of enhanced backscattering in turbulence.

Consider an object in turbulence (Fig. 24.1).

The radar equation can be written as

$$P_r = P_t \frac{\lambda^2 G^2}{(4\pi)^3 R^4} \sigma_{ap}. \tag{24.1}$$

The apparent target RCS σ_{ap} is different from the actual RCS σ_b and we can write

$$\sigma_{ap} = \sigma_b \langle |e_u e_d|^2 \rangle, \tag{24.2}$$

where e_u is the normalized random field on the up link and e_d is the normalized random field on the down link. Note that in the absence of turbulence, $\langle e_u \rangle = \langle e_d \rangle = 1$. Also, noting the reciprocity, we have $e_u = e_d = e$, and

$$\sigma_{ap} = \sigma_b \langle |e|^4 \rangle. \tag{24.3}$$

If the field through turbulence is Rayleigh distributed as is often the case, we have

$$\langle |e|^4 \rangle = \langle I^2 \rangle, \tag{24.4}$$

where $I = |e|^2$ is the normalized intensity.

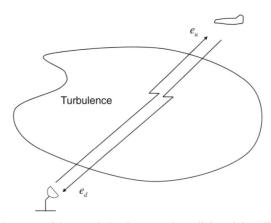

FIGURE 24.1   Because of the correlation between the uplink and downlink fields, the RCS is enhanced by a factor ranging from 1 to 2 (Rayleigh) to 3. The scintillation index ranges from 0 to 5 (Rayleigh) to 10.7.

We note that for the Rayleigh distribution, we have

$$\langle I^N \rangle = (N!) \langle I \rangle^N. \tag{24.5}$$

Therefore we get

$$\langle |e|^4 \rangle = 2 \langle |e|^2 \rangle. \tag{24.6}$$

The apparent RCS is then given by

$$\sigma_{ap} = 2\sigma_b. \tag{24.7}$$

This shows that the RCS is twice the actual cross section. The scintillation index S_4^2 is given by

$$S_4^2 = \frac{\langle (I^2 - \langle I^2 \rangle)^2 \rangle}{\langle I^2 \rangle^2} = \frac{\langle I^4 \rangle - \langle I^2 \rangle^2}{\langle I^2 \rangle^2}. \tag{24.8}$$

Now for the Rayleigh distribution, noting (24.5), we get

$$S_4^2 = \frac{4! - 2^2}{2^2} = 5. \tag{24.9}$$

Summarizing, the apparent cross section is twice the actual cross section and the scintillation index is 5 assuming the Rayleigh distribution. Experimental evidence for the enhanced RCS and the scintillation index has been given by Knepp and Houpis (1991).

The Rayleigh distribution is typical in waves in turbulence. However, more generally the distribution can be better represented by the Nakagami-m distribution (1960) where m ranges from 0.5 to 1 (Rayleigh)$\rightarrow \infty$. Then the apparent cross section ranges from 3 to 2 (Rayleigh) to 1, and the scintillation index ranges from (32/3) to 5 (Rayleigh) to 0. A more complete study is given by Fremouw and Ishimaru (1992).

Concerning the enhanced backscattering due to turbulence, a question may be asked about the relationship with power conservation. The conservation of power requires that the enhanced backscattered power needs to be balanced by the decrease of power. In our example, the decrease of power appears in the direction off the monostatic direction, and this decrease of power should be observed off the backscattering angle, where the correlation between the uplink and downlink waves can be negative.

24.2 ENHANCED BACKSCATTERING FROM ROUGH SURFACES

It is expected that the enhanced backscattering can occur, not only in turbulence, but for rough surfaces as well. For rough-surface scattering, there are two distinct enhancement phenomena, shown by E and SE in Fig. 24.2. E applies when the rms

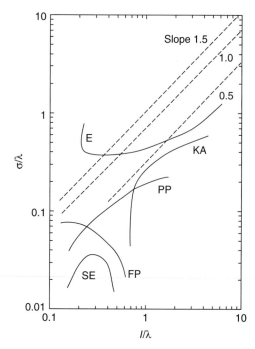

FIGURE 24.2   Ranges of validity for Kirchhoff (KA), phase perturbation (PP), and field perturbation (FP) theories. Enhanced backscattering occurs in the range labeled E. Also, the enhancement due to surface-wave modes occurs in the range labeled SE. σ is the rms height, and t is the correlation distance, of the rough surface.

height is close to a wavelength, and the slope is also close to unity. In that case, the two waves scattered off the sloped surface interfere constructively in the back direction, producing the enhanced peak. The angular width is broad, and approximately proportional to the slope (Fig. 24.3). An approximate theory has been proposed to explain this phenomenon by making use of the first- and second-order Kirchhoff approximations, with shadowing corrections (Ishimaru and Chen, 1991).

The second enhancement case, SE, occurs when the rms height is much smaller than a wavelength, but the second medium supports a surface wave. This occurs when an optical beam is scattered from a slightly rough metallic surface. If the incident wave is polarized (parallel to the plane of incidence), and if the dielectric constant of the second medium has a negative real part, then a surface wave is excited on the surface, and two surface waves traversing on the surface in opposite directions interfere constructively in the back direction, producing the enhancement. The angular width is very small and is proportional to (a wavelength)/(the decay distance of the surface wave) (Celli et al., 1985). The enhanced backscattering from rough surfaces is important in several applications, including the study of surface plasmon localization in rough-metal surfaces, phonon localization, and ocean acoustic applications.

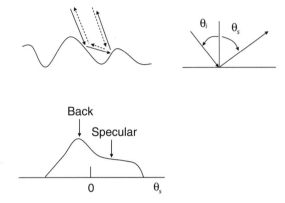

FIGURE 24.3   Backscattering enhancement from rough surfaces.

24.3 ENHANCED BACKSCATTERING FROM PARTICLES AND PHOTON LOCALIZATION

In 1984, an optical experiment was reported by Kuga and Ishimaru (1984) that showed that the scattering from latex microspheres is enhanced in the backward direction, with a sharp angular width of a fraction of a degree. It was explained theoretically, by Tsang and Ishimaru (1984), that the enhanced peak is caused by the constructive interference of two waves traversing through the same particles in opposite directions. Physicists have known that the transport of electrons in a strongly disordered material is governed by multiple scattering, and that multiple scattering leads to "Anderson localization" (1958, 1985) caused by "coherent backscattering" (John, 1990). It can then be shown that both electron localization in disordered material, and photon localization in disordered dielectrics, are governed by coherent backscattering, which is caused by the constructive interference of two waves traversing in opposite directions. The experimental work, in 1984, was followed by several independent optical experiments, showing that the backscattering enhancement is a weak-localization phenomenon. The enhanced peak value is close to 2, and the angular width is governed by the diffusion length in the medium.

Let us consider a plane wave, normally incident on a slab of randomly distributed particles. The first-order backscattering specific intensity, I_1, is approximately given by

$$
\begin{aligned}
I_1 &= \frac{\gamma_b I_0}{4\pi} \int_0^d e^{-2\gamma_t z} dz \\
&= \frac{\gamma_b I_0}{8\pi \gamma_t} (1 - e^{-2\gamma_t d}),
\end{aligned}
\tag{24.10}
$$

where I_0 is the incident intensity, and γ_b and γ_1 are the backscattering and extinction coefficients, respectively.

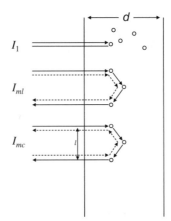

FIGURE 24.4   The first-order scattering I_1, the ladder term I_{ml}, and the cyclical scattering I_{mc}. t is the mean free path for a thin slab, and the transport mean free path for a thick slab.

The multiple scattering I_m consists of two terms, $I_{m\ell}$ and I_{mc}. One term, $I_{m\ell}$, corresponds to the wave multiply scattered through many particles, and is called the "ladder term." The other term, I_{mc}, corresponds to two waves traversing through the same particles in opposite directions. This is called the "cyclical" or the "maximally crossed" term, and has the same magnitude as $I_{m\ell}$ in the back direction, but diminishes away from the back direction (Fig. 24.4). The total intensity is therefore approximately as shown in Fig. 24.4. The enhancement factor is given by

$$\frac{I_1 + I_{m\ell} + I_{mc}}{I_1 + I_{m\ell}} = \frac{I_1 + 2I_{m\ell}}{I_1 + I_{m\ell}}. \tag{24.11}$$

The enhancement factor is therefore between 1 and 2.

The angular width, $\Delta\theta$, is (a wavelength)/(a mean free path), if the optical thickness of the slab is of the order of unity or less. If the optical thickness is much greater than unity and the particles are mostly scattering, then the wave is diffused, and the angular width, $\Delta\theta$, is (a wavelength)/(a transport mean free path). Note that the mean free path, ℓ, is normally much greater than a wavelength, and the transport mean free path, ℓ_{tr}, is normally many times the mean free path. Thus, this is a weak localization, in contrast with a strong localization, where the transport mean free path is of the order of a wavelength and where particle sizes are in the resonant region. It has also been observed, by Lagendijk and others, that the wave speed can be reduced to 10% of the light velocity in strong localization.

The coherent phenomena discussed above cannot be obtained from the transport theory (radiative transfer theory); these are new electromagnetic phenomena that have been overlooked in the past. They should be important in lidar applications and in geophysical remote sensing. It is also noted that the above optical experiment was on weak localization. Intensive work on strong localization is under way, which may have important applications in optical devices, optical components, spectroscopy, new lasers, and nonlinear optics.

24.4 MULTIPLE SCATTERING FORMULATIONS, THE DYSON AND BETHE–SALPETER EQUATIONS

In Sections 24.1–24.3, we presented simplified heuristic discussions of interesting coherent phenomena in multiple scattering. Exact mathematical formulations and solutions are very much involved and have been studied extensively. In this section, we present an elementary introduction to the important ideas and formulations.

Let us consider Green's function of a scalar wave equation for the medium with random fluctuation of relative dielectric constant $\varepsilon_r = 1 + \varepsilon_1$.

$$[\nabla^2 + k^2(1 + \varepsilon_1)]G = -\delta(r - r_o), \tag{24.12}$$

where ε_1 is the fluctuation with $\varepsilon_1 = 0$.

Rewriting this as

$$(\nabla^2 + k^2)G = -k^2\varepsilon_1 G - \delta(r - r_o), \tag{24.13}$$

we get the following integral equation for G.

$$G(r, r_o) = G_o(r, r_o) + \int G_o(r, r_1)V(r_1)G(r_1, r_o)dV_1, \tag{24.14}$$

where $V(r_1) = k^2\varepsilon_1(r_1)$ is called the random potential.

Rewriting this using the operator L_o and its inverse $-L_o^{-1} = M_o$, we get formally

$$L_o G = -VG - \delta, \quad L_o = \nabla^2 + k^2$$
$$G = -L_o^{-1}VG - L_o^{-1}\delta, \quad -L_o^{-1} = M_o$$
$$= G_o + M_o VG$$

$$M_o = \int dV' G_o.$$

Iterating this, we get the Neumann series

$$G = G_o + M_o VG_o + M_o VMVG$$
$$= G_o + \sum_{n=1}^{\infty}(M_o V)^n G_o. \tag{24.15}$$

We can express this in the following diagram.

$$G = \underline{\quad} + \underline{\quad\bullet\quad} + \underline{\quad\bullet\quad\bullet\quad} + \underline{\quad\bullet\quad\bullet\quad\bullet\quad} + \ldots,$$

$$\tag{24.16}$$

where ――――― is a free space Green's function G_o and ● is the random potential V. In more explicit form, we can write

$$G(r, r') = \underset{r \quad r'}{――} + \underset{r \quad r_1}{――●――} + \underset{r \quad r_2 \quad r_1 \quad r'}{――●――●――} + \; ... \tag{24.17}$$

The average Green's function $\langle G \rangle$ is therefore given by

$$\langle G \rangle = \langle \, ―― \, \rangle + \langle \, ――●―― \, \rangle + \langle \, ――●――●―― \, \rangle + \; ... \tag{24.18}$$

We can now assume that $V(r)$ is a Gaussian random function, and derive the Dyson equation for the average Green's function $\langle G \rangle$.

$$\langle G \rangle = G_o + G_o M \langle G \rangle, \tag{24.19}$$

where M is called the "mass operator."

In diagram form, we write

$$\blacksquare \; = \; ―― \; + \; ――●―― \tag{24.20}$$

where

$$\langle G \rangle = \; \blacksquare$$

$$● \; = \; \overset{\frown}{●――●} \; + \; \overset{\frown\!\!\frown}{●―●―●} \tag{24.21}$$

= mass operator.

Note that the dotted curve represents the correlation. For example,

$$\underset{r_2 \qquad r_1}{\overset{\frown}{●\quad●}} \; = \; G_o(r_2, r_1)\langle V(r_2)V(r_1)\rangle \, dV_2 dV_1. \tag{24.22}$$

Similarly, we write the Bethe–Salpeter equation.

$$\langle G(r)G^*(\rho)\rangle = \; \underset{\rho \quad \rho'}{\overset{r \quad r'}{\rule{1cm}{0pt}}} \; + \; \underset{\rho \quad \rho_2 \quad \rho'}{\overset{r \quad r_2 \quad r'}{\rule{1cm}{0pt}}} \; \langle G(r')G^*(\rho')\rangle \tag{24.23}$$

where

= Intensity operator.

The dotted lines represent the correlation.

The Dyson equation (24.19) for the average Green's function with the mass opera-tor (24.21), and the Bethe–Salpeter equation (24.23) for the covariance with the inten-sity operator, are two fundamental equations for the first and the second moments of Green's function in random media.

24.5 FIRST-ORDER SMOOTHING APPROXIMATION

It is clear that solutions for the exact formulation for the Dyson and the Bethe–Salpeter equations in Section 24.4 are difficult to find, and useful approximations are needed. The simplest approximation is called the "bilocal approximation," or the "Bourret approximation" or the "first-order smoothing approximation." Mass operator (24.21) is approximated by the first-order term. We have

$$(24.24)$$

For the scalar wave equation we are considering, the Dyson equation (24.20) becomes

$$\langle G(r, r')\rangle = G_o(r, r') + \int G_o(r, r_o) G_o(r_1, r_2)\langle V(r_1)V(r_2)\rangle\langle G(r_2, r')\rangle dV_1 dV_2.$$

$$(24.25)$$

This integral equation can be solved analytically by using a Fourier transform.

$$G(r, r') = \frac{\exp(iK|r - r'|)}{4\pi|r - r'|},$$

$$(24.26)$$

where K is the solution of the following equation.

$$K^2 = k^2 + \frac{1}{K} \int_0^\infty dr\langle V(r_1)V(r_2)\rangle e^{ikr} \sin Kr,$$

$$(24.27)$$

with $V(r_1)V(r_2) = B(|r_1 - r_2|)$.

The Bethe–Salpeter equation (24.23) becomes

$$\langle G(r)G^*(\rho)\rangle = \quad\quad + \quad\quad \langle G(r')G^*(\rho')\rangle \tag{24.28}$$

Analytically, we can write this as follows:

$$\langle G(r, r')G^*(\rho, \rho')\rangle = \langle G(r, r')\rangle\langle G^*(\rho, \rho')\rangle$$
$$+ \int \langle G(r, r_1)\rangle\langle G^*(\rho, \rho_1)\rangle\langle V(r_1)V(\rho_1)\rangle\langle G(r_1, r')G^*(\rho_1, \rho')\rangle d\bar{r}_1 d\bar{\rho}_1, \tag{24.29}$$

where $d\bar{r}_1 = dV$ at r_1 and $d\bar{\rho}_1 = dV$ at $\bar{\rho}_1$.

24.6 FIRST- AND SECOND-ORDER SCATTERING AND BACKSCATTERING ENHANCEMENT

The Dyson and the Bethe–Salpeter equations in Section 24.4 are exact integral equations for the average and the second-order moments. Though these equations are compact and reveal the physical scattering process clearly, actual analytical solutions are difficult to obtain. The first-order smoothing approximation in Section 24.5 can be solved analytically for a homogeneous random medium. Where it becomes difficult to solve for more general cases, it may be useful to use the first- and second-order scattering theory.

The first-order solution is obtained by considering only the first-order scattering. Therefore (24.24) becomes

$$\quad = \quad + \quad \tag{24.30}$$

Analytically, we write

$$\langle G(r, r')\rangle = G_o(r, r') + \int G_o(r, r_1)G_o(r_1, r_2)\langle V(r_1)V(r_2)\rangle G_o(r_2, r')dV_1 dV_2. \tag{24.31}$$

Note that $\langle G\rangle$ inside the integral in (24.20) is replaced by G_0.

The second-order solutions are given by the following:

$$\langle G(r)G^*(\rho)\rangle = \quad\quad + \quad\quad \langle G_o(r_1, r')G_o^*(\rho_1, \rho')\rangle. \tag{24.32}$$

Note also that $\langle GG^*\rangle$ inside the integral in (24.29) is replaced by $\langle G_0 G_0^*\rangle$.

We can follow (24.28) and (24.29) to obtain analytical expressions for the second moment.

It should be noted that the backscattering enhancement can be obtained by the second-order scattering theory. Considering a medium with random distributions of discrete scatterers, we can write

$$\langle \psi(r)\psi^*(\rho) \rangle = \quad \underline{\quad\quad} \quad + \quad \underline{\quad\quad} \quad \langle\langle \psi(r_1)\psi^*(\rho_1) \rangle\rangle$$

$$+ \quad \underline{\quad\quad} \quad \langle\langle \psi(r_1)\psi^*(\rho_1) \rangle\rangle \tag{24.33}$$

$$+ \quad \underline{\quad\quad} \quad \langle\langle \psi(r_1)\psi^*(\rho_1) \rangle\rangle.$$

The diagram represents the mean Green's functions, ⊗ is the transition operator and a solid line joining two ⊗ denotes that two ⊗ represent the same scatterer.

The first term of (24.33) is the intensity of the mean field, the second term is the single scattering. The third term is the second-order ladder term, which corresponds to the radiative transfer. The fourth term is called the "cyclical" term, which give a sharp peak in the back direction, but diminishes off the back direction. This has been discussed and shown in Fig. 24.5.

24.7 MEMORY EFFECTS

In Section 23.17, we made a brief remark about the "memory effects" using the perturbation solution of rough-surface scattering. As was noted, the memory effects are not limited to rough-surface scattering and apply to all multiple scattering from

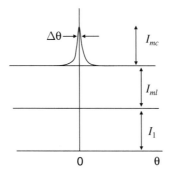

FIGURE 24.5   Backscattering enhancement (back direction is $\theta = 0$).

random media. More general memory effects proposed by Feng can be stated as follows:

A wave with the wave number vector $\bar{K}_i$ is incident on a random medium at frequency ω $\left(k = \dfrac{\omega}{c}\right)$ and the scattered wave in the direction of wave number vector $\bar{K}$ is observed at ω. Another wave with wave number vector $\bar{K}'_i$ at frequency ω' $\left(k' = \dfrac{\omega'}{c}\right)$ is incident upon the same random medium, and the scattered wave in the direction of the wave number $\bar{K}'$ is observed at ω'. Then there is a strong correlation between these two scattered waves at $\bar{K}$ and $\bar{K}'$ if the incident and scattered wave number vectors satisfy

$$(\bar{K} - \bar{K}_i)_t = (\bar{K}' - \bar{K}'_i)_t. \qquad (24.34)$$

The subscript (vector)$_t$ indicates the component of the vector along the surface.

We have already discussed the memory effects in Section 23.17. These interesting coherent interference effects with rough surfaces are also observed when the wave is incident on a random medium, as noted in Fig. 23.13.

CHAPTER 25

SOLITONS AND OPTICAL FIBERS

25.1 HISTORY

Solitons are a "singular and beautiful phenomenon" of a nonlinear wave; so noted John Scott Russell in 1844 in his "Report on Waves" to the British Association which described his observations of hydrodynamic solitary waves and extensive wave tank experiments in 1834 and 1835. He reported that when a boat in a narrow channel drawn by a pair of horses suddenly stopped, the mass of water rolled forward "assuming the form of a large solitary elevation, and a rounded, smooth and well-defined heap of water continued its course along the channel apparently without change of form or diminution of speed. (He) followed it on horseback and overtook it still rolling on at a rate of some eight or nine miles per hour, preserving its original figure some thirty feet long and a foot to a foot and a half in height... After a chase of one or two miles, (he) lost it in the windings of the channel..." That was the first observation of the solitary wave. In 1895, Korteweg and de Vries presented a study of solitary waves for shallow water waves that includes both nonlinear and dispersive effects. In 1965, Zabusky and Kruskal presented a study of the KdV equation for plasma waves and coined the term, "soliton."

The soliton is a solitary traveling wave solution of a wave equation that preserves its shape and velocity. If more than one soliton collides, then after collision, each

Electromagnetic Wave Propagation, Radiation, and Scattering: From Fundamentals to Applications,
Second Edition. Akira Ishimaru.
© 2017 by The Institute of Electrical and Electronic Engineers, Inc. Published 2017 by John Wiley & Sons, Inc.

soliton emerges without change of its shape except for a phase shift. Therefore we can write the solitons before the collision.

$$\phi(x, t) = \sum_i \phi_{st}(x - u_i t). \tag{25.1}$$

After collision, the solitons become

$$\phi(x, t) = \sum_i \phi_{st}(x - u_i t + \delta_i). \tag{25.2}$$

This is pictured in Fig. 25.1. It also shows that the larger the amplitude, the faster the speed, and the narrower the pulse width.

It is important to note that if the wave speed varies with the wavelength, it is called "dispersion.". If the signal velocity varies with the magnitude of the signal, this is a characteristic of a nonlinear wave. It is interesting to note that two solitons combine to balance the dispersion and the nonlinear effects, and produce the solitary wave given in (25.1) and (25.2) (Fig 25.2).

Today, solitons are observed in many fields of physical and geophysical environments. In 1973, Hasegawa and Tappert showed that an optical pulse in fiber forms an envelope soliton. This will be discussed later.

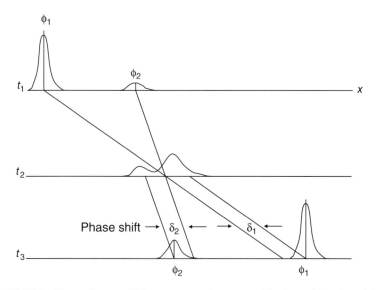

FIGURE 25.1 Two solitons collide at $t = t_2$ and emerge with phase shifts δ_1 and δ_2. The speed of each soliton is proportional to the amplitude. The higher the amplitude, the narrower the pulse and faster the speed.

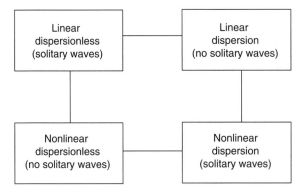

FIGURE 25.2   Linear and nonlinear waves, and dispersionless and dispersion (Scott et al., 1973).

25.2   KDV (KORTEWEG–DE VRIES) EQUATION FOR SHALLOW WATER

Surface gravity waves on water are dispersive. The velocity V depends on the wavelength and is given by (Lighthill 1978)

$$V^2 = \left(\frac{g}{k} + \frac{Tk}{\rho} \right) \tanh kd, \tag{25.3}$$

where g = gravitational constant = 9.8 m/s^2, T = surface tension = 0.074 N/m, ρ = density (water) = 1000 kg/m^3, d = depth, V = phase velocity = ω/k,   and group velocity = $\partial\omega/\partial k$. The first term in (25.3) is due to the gravity wave and the second term is due to the capillary wave. Therefore waves with wavelength exceeding 0.1 m are purely gravity waves. We then have

$$V^2 = \frac{g}{k} \tanh kd. \tag{25.4}$$

For deep water, $\tanh kd \rightarrow 1$ and we get

$$V = V_0 = \sqrt{(g/k)}. \tag{25.5}$$

For shallow water, we expand

$$\tanh kd \rightarrow kd - \left((kd)^3/3 \right) + \cdots$$

and we get

$$V = V_0 - \sigma k^2, \quad \sigma = \frac{1}{6} V_0 d^2 . \tag{25.6}$$

As seen in (25.3), the velocity V depends on the wavelength and this is a characteristic of dispersion.

As we already discussed in Section 25.1, the soliton is a combination of the dispersion and the nonlinearity. First we note that in the coordinate $\xi = x - vt$ moving with the wave, the wave height h does not change if there is no dispersion and nonlinearity. Then we can write

$$\frac{\partial h}{\partial \tau} = 0, \tag{25.7}$$

where $\tau = t$ and $\xi = x - vt$.

We can rewrite this in $x - t$ coordinates

$$\frac{\partial h}{\partial \tau} = \frac{\partial h}{\partial t}\frac{\partial t}{\partial \tau} + \frac{\partial h}{\partial x}\frac{\partial x}{\partial \tau} = \frac{\partial h}{\partial t} + v\frac{\partial h}{\partial x} = 0. \tag{25.8}$$

If we introduce small dispersion, we use V from (25.6). We note that since h is proportional to $\exp(ikx - i\omega t)$, $(\partial/\partial x) = ik$. Therefore we write (25.6) as

$$V = V_0 + \sigma\frac{\partial^2}{\partial x^2}. \tag{25.9}$$

Substituting this into (25.8), we get

$$\frac{\partial h}{\partial t} + \left(V_0 + \sigma\frac{\partial^2}{\partial x^2}\right)\frac{\partial h}{\partial x} = 0. \tag{25.10}$$

Next we include the nonlinear effect. V_0 is no longer $V_0 = \sqrt{(g/k)}$ and is slightly depending on height h. Writing this as $V_0 \rightarrow V_0 + \delta_1 h$, we finally get

$$\frac{\partial h}{\partial t} + (V_0 + \delta_1 h)\frac{\partial h}{\partial x} + \sigma\frac{\partial^3}{\partial x^3}h = 0, \tag{25.11}$$

where $\delta_1 = \dfrac{\partial V}{\partial h}$ at $h = 0$.

This is the KdV equation (Korteweg–de Vries). We normalize that equation by using time $t' = \sigma t$ and $h' = h + \frac{V_0}{\delta_1}$. We get (Hasegawa, 1989)

$$\frac{\partial h'}{\partial t'} + \left(\frac{\delta_1}{\sigma}\right)h'\frac{\partial h'}{\partial x} + \frac{\partial^3 h'}{\partial x^3} = 0. \tag{25.12}$$

It is usually expressed in the following normalized form.

$$\phi_t + \alpha\phi\phi_x + \phi_{xxx} = 0, \tag{25.13}$$

where the subscript t is $(\partial/\partial t)$, the subscript x is $(\partial/\partial x)$, and α is a constant.

The solution of (25.13) can be obtained by first noting that we look for a traveling wave. We consider the wave, which can be expressed as

$$\phi = \phi(\xi) = \phi(x - ut). \tag{25.14}$$

We then get

$$\phi_x = \phi_\xi,$$
$$\phi_t = -u\phi_\xi,$$

where the subscript ξ denotes $(\partial/\partial\xi)$.

The original equation (25.12) is then written as

$$(\alpha\phi - u)\phi_\xi + \phi_{\xi\xi\xi} = 0.$$

We integrate that once and get

$$\alpha\frac{\phi^2}{2} - u\phi + \phi_{\xi\xi} = \text{constant} = k_1.$$

We integrate again and get

$$\frac{\alpha}{6}\phi^3 - u\frac{\phi^2}{2} + \frac{1}{2}\phi_\xi^2 = k_1\phi + k_2 , \quad k_2 = \text{constant}.$$

In the above, we used

$$\left(\frac{\phi^2}{2}\right)_\xi = \phi\phi_\xi,$$

and

$$\left(\frac{\phi^3}{3}\right)_\xi = \phi^2\phi_\xi.$$

We can then write

$$\phi_\xi = \sqrt{2k_2 + 2k_1\phi + u\phi^2 - \frac{\alpha}{3}\phi^3}.$$

Now we use boundary conditions that $\phi_\xi = \phi_{\xi\xi} = 0$ as $\xi \to \pm\infty$. This gives $k_1 = k_2 = 0$. We then get

$$\int \frac{d\phi}{\phi\sqrt{u - \frac{\alpha}{3}\phi}} = \xi.$$

Performing integration, we get

$$\phi = \frac{3u}{\alpha} \text{sech}^2 \left[\frac{\sqrt{u}}{2} (x - ut) \right]. \tag{25.15}$$

This is the soliton solution to Eq. (25.13).

This shows two important characteristics of solitons. First, the amplitude $(3u/\alpha)$ is proportional to the velocity u and therefore the larger the amplitude, the faster the soliton travels. Second, the sech^2 term shows that the pulse width is given by $(x - ut) = \frac{2}{\sqrt{u}}$, showing that the faster the wave, the narrower the pulse width becomes. Third, the sign of the wave ϕ is the same as the sign of α, showing that if $\alpha > 0$ (or $\alpha < 0$), the pulse is positive (or negative).

There has been extensive literature on solitons. Historical reviews and many other works on solitons are given by Scott et al. (1973), Lonngren and Scott (1978), Dodd et al. (1982), and Uslenghi (1980).

25.3 OPTICAL SOLITONS IN FIBERS

In 1973, Hasegawa and Tappert showed that an optical pulse in a dielectric fiber forms an envelope soliton (Hasegawa, 1989; Yeh and Shimabukuro, 2008; Agrawal, 1989; Boyd, 2008).

Unlike the KdV soliton discussed in Section 25.2, the optical solitons are different in that they are the soliton waves of an envelope of light waves that are modulated pulse waves. Therefore this is sometimes called the "envelope soliton." As we discussed in Section 25.2, the soliton combines and balances the dispersion and the nonlinearity effects to form and propagate a stationary pulse. This is one of the most significant and useful characteristics of optical solitons in fibers.

Let us examine the nonlinear refractive index of fiber. Here we limit the discussion to linearly polarized waves, and therefore the displacement vector $\bar{D}$, the electric field $\bar{E}$, and the polarization vector $\bar{P}$ are all scalar and are given by

$$D = \varepsilon_0 E + P. \tag{25.16}$$

The polarization vector for nonlinear medium is given by

$$\begin{aligned} P &= P^{(1)} + P^{(2)} + P^{(3)} + \cdots \\ &= \varepsilon_0 [\chi^{(1)} E + \chi^{(2)} E^2 + \chi^{(3)} E^3 + \cdots], \end{aligned} \tag{25.17}$$

where $E = E(t)$ is scalar and a function of time.

Here we only deal with linear polarization, so $\chi^{(1)}$, $\chi^{(2)}$, $\chi^{(3)}$ are all scalar. In general, however, if we include all polarization, $\chi^{(1)}$ is the second-rank tensor, $\chi^{(2)}$ is the third-rank tensor, and $\chi^{(3)}$ is the fourth-rank tensor.

$P^{(1)}, P^{(2)}, P^{(3)}$ in (25.17) are the first-order, second-order, and third-order nonlinear polarization, respectively. $\chi^{(1)}$ is responsible for the dominant linear susceptibility, and $\chi^{(2)}$ is for the second harmonic generation, but for optical fibers, $\chi^{(2)}$ is generally zero. Therefore, the lowest-order nonlinear effects on optical fibers are due to $\chi^{(3)}$.

Next we examine $P^{(3)}$. For a monochromatic wave $E = E_0 \cos \omega t$, we have

$$P^{(3)} = \varepsilon_0 \chi^{(3)} E^3$$

$$= \varepsilon_0 \chi^{(3)} E_0^3 \cos^3 \omega t = \varepsilon_0 \chi^{(3)} E_0^3 \left(\frac{1}{4} \cos 3\omega t + \frac{3}{4} \cos \omega t \right). \qquad (25.18)$$

The first term is the third harmonic and the second term gives the nonlinear effect. Then the nonlinear term becomes

$$P^{(3)} \quad \text{(nonlinear)} \quad = \varepsilon_0 \frac{3}{4} \chi^{(3)} E_0^2 (E_0 \cos \omega t). \qquad (25.19)$$

This leads to the nonlinear term of the refractive index. The linear term n_0 and the nonlinear term n_{NL} of the refractive index are then given by

$$\varepsilon = \varepsilon_L + \varepsilon_{NL},$$

$$\varepsilon_L = 1 + \chi^{(1)} + i\alpha,$$

$$\varepsilon_{NL} = \left(\frac{3}{4} \right) \chi^{(3)} |E|^2,$$

$$n = (\varepsilon_L + \varepsilon_{NL})^{1/2} = n_0 + n_{NL} + i \frac{\alpha}{2n_0},$$

$$n_0 = (1 + \chi^{(1)})^{1/2},$$

$$n_{NL} = \frac{1}{2} \frac{\varepsilon_{NL}}{n_0} = n_2 |E|^2 = \frac{3}{8} \frac{\chi^{(3)}}{n_0} |E|^2 \quad \text{for} \quad \varepsilon_L \gg \varepsilon_{NL}. \qquad (25.20)$$

The nonlinear term $n_2 |E|^2$ represents the Kerr effect, and n_2 is called the "Kerr coefficient." n_2 is of the order of $10^{-20} (m^2 W^{-1})$ for silica optical fibers and therefore it requires intensities such as those of lasers to produce the nonlinear effects.

Using (25.20), we write the refractive index n.

$$n = n_0 + n_2 |E|^2 + i \frac{\alpha}{2n_0}, \qquad (25.21)$$

$$n_0 = 1 + \chi^{(1)},$$

$$n_2 = \frac{3}{8} \frac{\chi^{(3)}}{n_0}.$$

Using (25.21), we can now derive the nonlinear pulse propagation equation. This has been obtained by using a perturbation technique (Agrawal, 1989; Yeh and Shimabukuro, 2008; and Hasegawa, 1989).

Here we derive the equation using a parabolic equation, which is applicable to waves propagating mostly in the direction $\hat{x}$.

Let us start with the wave equation

$$(\nabla^2 + k^2)u = 0, \tag{25.22}$$

where k is the wave number given by

$$k^2 = k_f^2 n^2, \quad k_f = \text{free space wave number.}$$

Note that k includes the dispersion and the nonlinear term.

Let us write

$$u = U\exp(ik_0 x), \tag{25.23}$$

where $k_0 = k_f n_0$ is the wave number for the medium excluding dispersion, nonlinear effects, and attenuation.

Substituting this into (25.22) and noting

$$\nabla^2 u = \left[\nabla^2 U + 2\nabla U \cdot (ik_0 \hat{x}) + U\left(-k_0^2\right)\right]\exp(ik_0 x),$$

we get

$$\left[\frac{\partial^2}{\partial x^2} + \left(\frac{\partial^2}{\partial y^2} + \frac{\partial^2}{\partial z^2}\right) + i2k_0\frac{\partial}{\partial x} + \left(k^2 - k_0^2\right)\right]U = 0. \tag{25.24}$$

Now we note that

$$\left|\frac{\partial^2}{\partial x^2}U\right| \ll \left|k\frac{\partial U}{\partial x}\right|$$

as long as the slowly varying function U varies only over a distance much greater than the wavelength. We then get the following parabolic equation

$$i2k_0\frac{\partial U}{\partial x} + \nabla_t^2 U + \left(k^2 - k_0^2\right)U = 0, \tag{25.25}$$

where $\nabla_t^2 = \dfrac{\partial^2}{\partial y^2} + \dfrac{\partial^2}{\partial z^2}$ is the transverse Laplacian.

Let us consider a one-dimensional propagation ($\nabla_t^2 = 0$).

We have

$$i\frac{\partial U}{\partial x} + \frac{\left(k^2 - k_0^2\right)}{2k_0}U = 0. \tag{25.26}$$

Since k is only slightly different from k_0, we write

$$\frac{(k^2 - k_0^2)}{2k_0} = \frac{(k + k_0)(k - k_0)}{2k_0} \approx (k - k_0).$$

Thus, we get the following equation for U, where the dispersion, the normalization effects, and the attenuation are included in k.

$$i\frac{\partial U}{\partial x} + (k - k_0)U = 0. \tag{25.27}$$

Now we write

$$k - k_0 = k_0'(\Delta\omega) + \frac{k_0''}{2}(\Delta\omega)^2 + \Delta k, \tag{25.28}$$

$$\Delta k = k_0\left[n_2|E|^2 + i\frac{\alpha}{2n_0}\right]$$

$$= k_f\left[\frac{3}{8}\chi^{(3)}|E|^2 + \frac{i\alpha}{2}\right] , \quad |E|^2 = |U|^2,$$

where

$$k_0' = \frac{\partial k_0}{\partial\omega}, \quad k_0'' = \frac{\partial^2 k_0}{\partial\omega^2}.$$

Converting this to time domain using the following:

$$\int \frac{\partial}{\partial t}U(t)e^{i(\Delta\omega)t}dt = -i(\Delta\omega)U(\omega),$$

we get

$$\Delta\omega = i\frac{\partial}{\partial t} \text{ and } (\Delta\omega)^2 = \left(-\frac{\partial^2}{\partial t^2}\right).$$

Equation (25.27) is now transformed to the following:

$$i\frac{\partial U}{\partial x} + ik_0'\frac{\partial}{\partial t}U - \frac{k_0''}{2}\frac{\partial^2 U}{\partial t^2} + k_f\frac{3}{8}\chi^{(3)}|U|^2U + k_f\left(\frac{i\alpha}{2}\right)U = 0. \tag{25.29}$$

This is the final expression for the nonlinear pulse propagation equation. Note that U is the electric field E which includes dispersion and nonlinear effects as indicated in (25.23). If U is different from E, then the coefficient in the nonlinear term needs to be changed as was shown in Agrawal (1989).

This section gives a short sketch of the solitons in optical fiber, considering only a one-dimensional case where the transverse variation in the cross section of a fiber is not considered. There is extensive literature with a more complete discussion on nonlinear fiber optics (Agrawal, 1989) and (Yeh and Shimabukuro, 2008), If k'' in (25.29) is positive, the solitary wave appears as the absence of a light wave and is called a "dark" soliton (Hasegawa, 1989).

CHAPTER 26

POROUS MEDIA, PERMITTIVITY, FLUID PERMEABILITY OF SHALES AND SEISMIC CODA

In this chapter, we discuss the characteristics of porous media and seismic coda waves. A porous medium is a material containing pores or voids. The frame is the skeletal portion of the material, often called the "matrix." The pores are often filled with liquid (oil and gas), such as those found in shale formations during hydraulic fracturing (fracking) in petroleum engineering. In Section 8.6, we discussed a "mixing formula" to express the effective dielectric constant of a mixture of two or more materials with different dielectric constants. A porous medium is similar to this. In fact, for the mixing formula, we use the fractional volume of the ith species as f_i with

$$\sum_i f_i = 1. \tag{26.1}$$

For a porous medium, we use the porosity ϕ, which is related to the fractional volume of the matrix f_m.

$$\phi = 1 - f_m. \tag{26.2}$$

The porosity ϕ is therefore the fractional volume of the pore (or void).

One of the important examples of a porous medium is oil shale. We will discuss two subjects related to oil shale. One is the permittivity and conductivity of the porous medium as given by Archie's law. The other is the flow of liquid such as

Electromagnetic Wave Propagation, Radiation, and Scattering: From Fundamentals to Applications,
Second Edition. Akira Ishimaru.
© 2017 by The Institute of Electrical and Electronic Engineers, Inc. Published 2017 by John Wiley & Sons, Inc.

oil through the porous medium as given by Darcy's law and "fluid" permeability, which is different from the "magnetic" permeability concept we commonly use in electromagnetic studies.

26.1 POROUS MEDIUM AND SHALE, SUPERFRACKING

Even though our interest is not in petroleum engineering, for which extensive works have been reported already, we give a brief introduction to superfracking, a well-known technique for extracting oil or gas. Natural gas and oil production have increased in the United States in recent years, primarily due to hydraulic fracturing called "superfracking" in which large volumes of low-viscosity water are pumped into a low permeability shale formation (Turcotte et al., 2014). Note that permeability of the porous medium is not the "magnetic" permeability concept commonly used in electromagnetic studies. It is instead the rate of diffusion of a fluid under pressure through a porous medium (Revil and Cathles, 1999).

In Fig. 26.1, (a) traditional fracking and (b) superfracking are shown. In traditional fracking, a high-viscosity fluid creates a single hydraulic fracture through which oil or gas migrates to the production well. In superfracking, a horizontal production well is created to access a wide distribution of hydraulic fractures. In the United States, important shale producing locations include "Barnett shale" in Texas, "Bakken shale" in North Dakota, Montana, and Saskatchewan, and "Monterey shale" in California.

Shales are a rock equivalent of mud, just as sandstones are a rock equivalent of sand. Shales can extend horizontally for a thousand kilometers with a porosity of 2–20%. The shales that are a main source of hydrocarbons are known as "black shales" (Turcotte et al., 2014). Their pores are typically filled with 2–18% by weight of carbon in organic compounds. A representative grain in shale is less than 4 μm wide (Turcotte et al., 2014).

Fracking has several environmental concerns including reduction and contamination of available water, leakage of methane gas into the environment, and triggering of earthquakes.

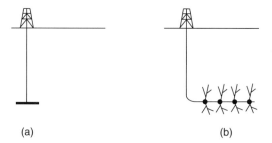

(a)   (b)

FIGURE 26.1 (a) Traditional fracking. (b) Superfracking with horizontal production well (Turcotte et al., 2014).

26.2 PERMITTIVITY AND CONDUCTIVITY OF POROUS MEDIA, ARCHIE'S LAW, AND PERCOLATION AND FRACTAL

The ground penetrating radar (GPR) technique is used for near surface geophysical subsurface imaging. Critical parameters for GPR include the dielectric constant, porosity, and water saturation, and frequencies are usually 25–1500 MHz. There have been extensive studies in modeling and experimental data (Martinez-Byrnes, 2001). It is important to study mixing models of geophysical materials, and the effective medium models such as Bruggeman are useful for studies of bulk dielectric constant, water saturation, and porosity. Carcione and Serian (2000) discuss detection of hydrocarbons in the subsoil. At radar frequencies (50 MHz to 1 GHz), hydrocarbons have a relative permittivity ranging from 2 to 30 while the permittivity of water is 80. The conductivity of hydrocarbons ranges from 0 to 10 mS/m while the conductivity of saltwater is 200 mS/m or more. (The unit for conductivity is (A/Vm) or (S/m), where S is siemens.) Therefore there is sufficient contrast between hydrocarbons and water for detection and mapping of hydrocarbons.

Determination of the dielectric constant of the porous medium in terms of the composition of the mixtures is of great importance for characterization, monitoring, and evaluation. In Fig. 26.2 a simplified sketch of a porous medium is shown, where the fractional volume of rock (matrix) is f_m. The pore or void is partly filled with hydrocarbon with a fractional volume of $f_h = (1 - S_w)\phi$, and partly filled with water with a volume fraction of $f_w = S_w\phi$, where S_w is the fraction of pore space saturated with water. This is a mixture of solid f_m and two liquids f_h and f_w. Additional complications include complex shapes, other minerals and materials. In this section,

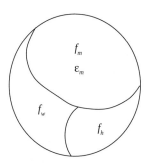

FIGURE 26.2   Simplified sketch of solid (matrix) and two fluids.
$f_m = \text{matrix} = 1 - \phi,$
$f_h = (1 - S_w)\phi = \text{hydrocarbon},$
$f_w = S_w\phi = \text{water},$
$S_w = \text{fraction of pore space saturated with water},$
$0 < S_w < 1,$
$f_h + f_w = \phi,$
$f_m + f_h + f_w = 1.$

we give a simplified theory of the mixture based on the Bruggeman model (Polder–van Santen) given in Section 8.6. We start with (8.31).

$$f_m \frac{\varepsilon_m - \varepsilon_l}{\varepsilon_m + 2\varepsilon_l} + f_h \frac{\varepsilon_h - \varepsilon_l}{\varepsilon_h + 2\varepsilon_l} + f_w \frac{\varepsilon_w - \varepsilon_l}{\varepsilon_w - 2\varepsilon_l} = 0, \tag{26.3}$$

where ε_m is the dielectric constant of the matrix, ε_h is the dielectric constant of hydrocarbon, and ε_w is the dielectric constant of water. ε_l is the effective dielectric constant of the medium.

Note that

$$\varepsilon_m = \varepsilon'_m + i\varepsilon''_m,$$

$$\varepsilon_h = \varepsilon'_h + i\varepsilon''_h,$$

$$\varepsilon_w = \varepsilon'_w + i\left(\varepsilon''_w + \frac{\sigma_w}{\omega\varepsilon_0}\right),$$

$$\varepsilon_l = \varepsilon'_l + i\left(\varepsilon''_l + \frac{\sigma_l}{\omega\varepsilon_0}\right) = \text{effective dielectric constant,} \tag{26.4}$$

$$\sigma_w = \text{conductivity of water,}$$

$$\sigma_l = \text{conductivity of effective medium.}$$

We can also convert (26.3) in the following alternative form. Noting for the ith medium,

$$f_i \frac{\varepsilon_i - \varepsilon_l}{\varepsilon_i + 2\varepsilon_l} = f_i \frac{\varepsilon_i + 2\varepsilon_l - 3\varepsilon_l}{\varepsilon_i + 2\varepsilon_l} = f_i - \frac{3\varepsilon_l f_i}{\varepsilon_i + 2\varepsilon_l}, \tag{26.5}$$

we get

$$\sum_i f_i - \sum_i \frac{3\varepsilon_l f_i}{\varepsilon_i + 2\varepsilon_l} = 0.$$

Therefore we get the following alternative Bruggeman form:

$$\frac{f_m}{\varepsilon_m + 2\varepsilon_l} + \frac{f_h}{\varepsilon_h + 2\varepsilon_l} + \frac{f_w}{\varepsilon_w + 2\varepsilon_l} = \frac{1}{3\varepsilon_l}. \tag{26.6}$$

It is then possible to calculate ε_l for a given ε_m, ε_h, and ε_w (or S_w and ϕ). As a special case, we consider a limiting case of dc when $\omega \to 0$. Then the conductivity of the effective dielectric constant ε_l is dominated by $(i\sigma_l/\omega\varepsilon_0)$ and the dialectic constant of water ε_w is dominated by $(i\sigma_w/\omega\varepsilon_0)$. As a limiting case $(\omega \to 0)$, we get

$$\sigma_e \approx \sigma_w \left(\frac{3}{2}\phi S_w - \frac{1}{2}\right). \tag{26.7}$$

This is valid only if $\phi S_w > (1/3)$.

The above simplified results do not include different processes of forming the porous medium such as adding oil and rock grains separately or simultaneously resulting in different results for (σ/σ_w) (Feng and Sen, 1985; Sen et al., 1981).

A more realistic model of effective conductivity σ_e was discovered by Archie (Gao, 2012). It is an empirical law relating porosity, electrical conductivity, and brine saturation of rocks.

$$\sigma = \frac{1}{a}\sigma_w\phi^m S_w^n, \tag{26.8}$$

where σ and σ_w are the conductivities of the medium and the formation saline water, m is the porosity/cementation exponent with common values of $1.8 < m < 2.0$, n is the saturation exponent, usually close to 2, and a is a tortuosity factor which corrects for variations in compaction, pore structure, and grain size, ranging from 0.5 to 1.5. (Moldrup et al., 2001) The bulk dielectric permittivity is affected by factors such as the air–solid–water interface (Chen and Or, 2006). A constitutive law for the electric response of saturated and unsaturated porous media is presented (Brovelli and Cassiani, 2010).

The study of porous media is closely related to the "percolation theory" (Stauffer and Aharony, 1991) and the "fractal" (Mandelbrot, 1977; Falconer, 1990). As an example, consider a random distribution of regions filled with hard rock, while the rest are pores filled with oil or gas. If the porosity ϕ is small, only a small amount of oil can be produced. If the porosity is well above the critical value, ϕ_c, it may be possible to produce a large amount of oil. However, the situation may be more complex if the porosity is close to ϕ_c. Oil fields and fractals are discussed by Stauffer and Aharony (1991).

Determination of porosity by dielectric permittivity measurements are discussed by Lina, Olivares et. al. (2000) including fractal geometry and porosity, instead of estimating the porosity with the common "mixture law."

26.3 FLUID PERMEABILITY AND DARCY'S LAW

The permeability of a porous medium such as shale is a measure of the resistance to the flow of a fluid through the medium (Revil and Cathles, 1999). Low permeability means it is difficult for fluid to pass through the medium. If fluid can pass through the porous medium easily, it has high permeability. The total discharge rate q per unit area obeys Darcy's law

$$q = -\frac{\kappa}{\mu}\nabla p, \tag{26.9}$$

where q is the Darcy flux (discharge per unit area per unit time $m^3/m^2 s$), κ is permeability expressed in Darcy (m^2), μ is viscosity ($P_a \bullet s$), and p is pressure (P_a). The fluid velocity v is then given by

$$v(m/s) = \frac{q}{\phi}, \quad \phi = \text{Porosity.} \tag{26.10}$$

This accounts for the fact that only a fraction of the total volume is available for flow. The permeability κ is usually expressed in millidarcy (10^{-3} darcy). For highly fractured rocks the permeability is $10^8 \sim 10^5$ millidarcy, for rocks it is $10^4 \sim 100$, and for sandstones it is $10 \sim 1$.

Shales have low permeability and therefore historically are a poor producer of hydrocarbons until horizontal drilling and advances in hydraulic fracking led to highly fractured rocks with high permeability.

For superfracking, large volumes of water with additives called "slick water" are injected with high pressure, which may create microseismic events, usually much too small to be felt at the surface. The flow of liquid in a porous medium whose structure is fractal is discussed by Adler (1996).

26.4 SEISMIC CODA, P-WAVE, S-WAVE, AND RAYLEIGH SURFACE WAVE

The upper 100 km of the earth consisting of the crust and the uppermost mantle is called the "lithosphere," which has been investigated using classical layered models. However, the crust is heterogeneous with scales of a few kilometers to tens of kilometers which creates the continuous wave trains in the tail portion of seismograms. These wave trains are called the "coda." The local earthquakes propagate through the lithosphere over distances of 100 km with frequencies from 1 to 30 Hz. The initial appearance of the seismic wave is followed by the wave train "coda," which is the incoherent wave scattered by the heterogeneous earth and is important to probe information on the earthquake source and media. (Sato et al., 2012).

The earth's crust can be considered a heterogeneous elastic solid, and seismic waves consist of a P-wave (pressure wave, longitudinal wave), S-wave (shear wave, transverse wave), and Rayleigh surface wave. The waves in an elastic solid (such as the earth) can be expressed in terms of the scalar potential ϕ and the vector potential $\hat{\psi}$ and they satisfy the following wave equation

$$\nabla^2 \phi = \frac{1}{c_p^2} \frac{\partial^2}{\partial t^2} \phi,$$

$$\nabla^2 \hat{\psi} = \frac{1}{c_s^2} \frac{\partial^2}{\partial t^2} \hat{\psi},$$

where   (26.11)

$$c_p = \sqrt{\frac{\lambda + 2\mu}{\rho}} = \text{velocity of P-wave,}$$

$$c_s = \sqrt{\frac{\mu}{\rho}} = \text{velocity of S-wave,}$$

λ and μ are Lame constants (N/m^2) and ρ is density (kg/m^3). The phase velocity c_r of a Rayleigh wave is given by

$$\left(2 - \frac{c_r^2}{c_s^2}\right)^2 - 4\left(1 - \frac{c_r^2}{c_p^2}\right)^{\frac{1}{2}}\left(1 - \frac{c_r^2}{c_s^2}\right)^{\frac{1}{2}} = 0. \qquad (26.12)$$

See Appendix 26.

The P-wave velocity is approximately $5 - 6$ km/s and the S-wave velocity is approximately 3 km/s, and the Rayleigh wave velocity is close to 90% of the S-wave. The Rayleigh wave is confined to the neighborhood of the earth–air boundary. This means that both P- and S-waves are in general spherical waves when excited by a point source. This is also called the "body wave." The Rayleigh wave is a cylindrical surface wave, and thus the magnitude of the Rayleigh wave can become greater than that of the P- and S-wave at large distances.

S-wave is the vector wave and therefore there are two polarized components. For a plane boundary, there are two types of transverse waves: SV-waves and SH-waves. The SV-waves have the displacement component in the plane of incidence, while the SH-waves have the component perpendicular to the incidence plane.

26.5 EARTHQUAKE MAGNITUDE SCALES

The earthquake source is called the "hypocenter" and the point on the earth's surface directly above the hypocenter is called the "epicenter" (Fig. 26.3). The earthquake magnitude is the most often cited measure of an earthquake size. There are several different types of earthquake magnitudes. The Richter scale M_L (local amplitudes) was developed in 1935 by C. F. Richter, based on the amplitude A in µm of the largest seismograin wave trace of the peak ground action at a period approximately 1 s. Thus

$$M_L \sim \log\left(\frac{A}{T}\right) + [correction\ factor],$$

where $A = 10$ (µm), $T \approx 1$ s. This gives $M_L \sim 1.0$. For amplitude of 100 µm $= 0.1$ mm, this gives $M_L \sim 2.0$. For amplitude of 1 mm, this gives $M_L \sim 3.0$. T is

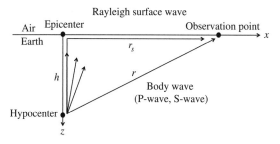

FIGURE 26.3 Body wave and Rayleigh surface wave.

the period of the measured signal in seconds, and is typically $T \sim 0.8$ s, which was chosen by Richter based on what we feel and what may damage buildings and other structures, using the standard short-period torsion seismometer. This is based on the peak ground motion due to the S-wave or surface wave. The Richter scale reflects the amount of seismic energy up to about $M_L = 6.5$, but it saturates beyond $M_L = 6.5$.

The surface wave magnitude, M_s, is based on ground motion due to a Rayleigh surface wave. There are many formulas that account for specific geographic regions. Typically, the standard surface wave formula is

$$M_s = \log_{10}\left(^A/_T\right) + [\text{correction factor due to distance}], \qquad (26.13)$$

where A (in microns) is the amplitude of ground motion, and T is the period (in seconds), typically 20 s.

The body-wave magnitude, m_b, (short period) is based on ground motion due to a P-wave over a period of 1 s, while M_b is for the body wave over a longer period. The body wave propagates through the earth, while the Rayleigh surface waves are confined near the surface.

Seismic moment M_O and moment magnitude M_W are related to the amount of energy released and are different from other magnitudes based on the ground motion at the observation point. The seismic moment M_O is defined as

$$M_O = DA\mu,$$

where D is the average displacement over the entire fault surface, A is the area of the fault surface and μ is the average shear rigidity of the faulted rocks (Kanamori, 1977; Pasyanos, 2010). The moment magnitude, M_W, is related to M_O. (Kanamori, 1977)

$$M_W = \left(^2/_3\right)M_O - 16.1, \qquad (26.14)$$

In addition to the above, the intensity in the surface shaking can be used. At the Japan Meteorological Agency (JMA), the accelerations measured by seismometers are used, with 10 levels of "Shindo" (Shindo is the degree of shaking).

26.6 WAVEFORM ENVELOPE BROADENING AND CODA

Figure 26.4 shows typical seismic waves observed at the receiver. Note that in general, a P-wave arrives first with the envelope, which includes the coda (wave train). Earthquake magnitude scales are measured in a period shown in the figure. In general, longer earthquakes require measuring over longer periods so that the measurements do not miss important features. However, this also means that the choice of periods needs to be made appropriately. Most magnitude scales depend on the observed waveform with the period, except the moment magnitudes, M_O and M_W, which are based on the energy released at the source.

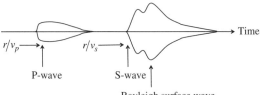

FIGURE 26.4   P-wave arrives first, S-wave second, Rayleigh surface wave arrives slightly later.

The total seismic energy radiated from the source propagates outward. The total propagating energy will experience attenuation due to energy absorption by intrinsic material characteristics expressed by the imaginary part of the dielectric constant. It also experiences scattering due to heterogeneous earth and the total energy is split into the "coherent" component and the partially coherent component. This partially coherent component is commonly called the "incoherent" and is the key component of the "wave train" called the "coda." It is therefore important and necessary to study the "coda" in order to understand the total energy radiation, rather than the data based on the amplitude in a period, which can be arbitrary. The study of the P-coda, S-coda, and Rayleigh-coda is therefore critical to the understanding of the total seismic radiation.

26.7   CODA IN HETEROGENEOUS EARTH EXCITED BY AN IMPULSE SOURCE

Let us first consider an idealized case of a point impulse radiator of a P-wave located in heterogeneous earth, radiating spherical wave with total energy W. In this section, we neglect the earth–air boundary. The earth is assumed to be homogenous and an isotropic random medium.

We first note that the scattering due to the heterogeneous earth is confined within small angles in the forward direction because the correlation length of the medium is much greater than a wavelength in most cases. Thus, the scattering effects can be approximated by plane wave scattering.

Let us study the pulse propagation in random heterogeneous earth. The intensity at r and t of a P-wave in homogeneous earth due to an impulse source $\delta(t)$ is given by

$$I_0(r, t) = \frac{W}{4\pi r^2} \delta[t - (r/c)], \tag{26.15}$$

where W is the total energy and c is the wave velocity in homogeneous earth.

In heterogeneous earth, the velocity c is changed to v_p, that of the coherent component of a P-wave, and the intensity is absorbed and scattered by the medium. We express this as

$$I_0(r, t) = \frac{W \exp(-\alpha_a r)}{4\pi r^2} [I_{coh} + I_{incoh}], \tag{26.16}$$

where α_a is the intrinsic attenuation loss due to the imaginary part of the medium dielectric constant. I_{coh} is the coherent component and I_{incoh} is the incoherent component. The coherent component diminishes due to the absorption and the scattering. The attenuation due to the absorption is included in $\exp(-\alpha_a r)$. The attenuation of the coherent component due to the scattering is represented by the total scattering in all directions. Thus we get

$$I_{coh} = \exp(-\alpha_s r)\delta\left[t - \left(^r/_v\right)\right], \tag{26.17}$$

where α_s is the "scattering cross section" per unit volume of the medium and is given by (Ishimaru, 1997, p. 335)

$$\alpha_s = \int_{4\pi} \sigma(\hat{o}, \hat{i})d\Omega$$

$$= 2\pi k^4 \int_0^{2k} \Phi_m(k_s)2\pi k_s dk_s, \tag{26.18}$$

$$d\Omega = \sin\theta d\theta d\phi = (k_s dk_s d\phi)/k^2,$$

where $\sigma(\hat{o}, \hat{i})$ is the differential cross section per unit volume of the medium given by

$$\sigma(\hat{o}, \hat{i}) = 2\pi k^4 \Phi_m(k_s), \quad k_s = k2\sin\left(^\theta/_2\right), \tag{26.19}$$

and θ is the scattering angle between the incident wave in the direction $\hat{i}$, unit vector, and the scattering wave in the direction $\hat{o}$, unit vector.

Note that the correlation function $B_n(r_d)$ of the refractive index $n(\bar{r})$ and its fluctuating component $n_0 n_1$ are shown in Section 23.1 and Ishimaru (1997).

$$n(\bar{r}) = n_0(1 + n_1), \quad |n_1| \ll 1,$$
$$B_n(r_d) = \langle n_1(r_1)n_1(r_2)\rangle, \quad r_d = |\bar{r}_1 - \bar{r}_2|. \tag{26.20}$$

It is assumed that the medium is a homogeneous and isotropic random medium and therefore the correlation function $B_n(r_d)$ is a function of the magnitude of the difference r_d. The spectrum is given by

$$\bar{\Phi}_n(k) = \frac{1}{(2\pi)^3}\int B_n(r_d)e^{i\bar{k}\cdot\bar{r}_d}d\bar{r}_d. \tag{26.21}$$

In this section, we use a Gaussian correlation function for B_n. Therefore we get

$$B_n(r_d) = \langle n_1^2\rangle \exp\left[-(r_d/l)^2\right],$$
$$\bar{\Phi}_n(k) = \frac{\langle n_1^2\rangle l^3}{8\pi\sqrt{\pi}} \exp\left[-\frac{(kl)^2}{4}\right], \tag{26.22}$$

where l is the correlation distance. Using (26.22) in (26.18), we get

$$\alpha_s = \sqrt{\pi} k^2 l \left\langle n_1^2 \right\rangle . \tag{26.23}$$

As the wave propagates over the distance L, the coherent component I_{coh} in (26.17) attenuates as

$$\exp(-\alpha_s L) = \exp([-(L/l_s)] = \exp(-\tau_s). \tag{26.24}$$

where τ_s is called the "optical scattering depth," and $l_s = (1/\alpha_s)$ is called the "scattering mean free path." The optical scattering depth and the scattering mean free path are the important key quantities to describe the amount of power scattered by the medium.

The coherent component diminishes due to the scattering. This scattered power is converted into the incoherent power. The total power is the sum of the coherent and the incoherent intensity and this will be preserved. In particular, we have the conservation of the energy.

$$\int_{r/v}^{\infty} I(r,t) 4\pi r^2 dt = W \exp(-\alpha_a r). \tag{26.25}$$

This shows that the total energy W, except for absorption, as a time integral of $I(r,t)$, is conserved. This includes the coda power.

Let us now examine the coherent I_{coh} and the incoherent intensities I_{incoh} in (26.16). If the optical scattering depth τ_s shown in (26.24) is smaller or not much greater than one, the coherent intensity decreases, but the increase of the incoherent intensity due to scattering is small. If the optical scattering depth τ_s is much greater than unity, the coherent intensity diminishes and is converted into the incoherent intensity (Fig. 26.5).

Let us consider the incoherent intensity in (26.16). The coda is the incoherent intensity as shown in Fig. 26.5 and is dominant as the optical scattering depth τ_s becomes much greater than unity. The shape of the coda envelope has been studied and for a medium with a Gaussian correlation function, the envelope shape takes a universal form. Mathematical expressions have been given by Ishimaru (1997, p. 316) and Sato and Fehler (1998, p. 250).

$$I_{incoh} = G(t)$$

$$= \left(\frac{\pi}{4 t_M} \right) \sum_{n=0}^{\infty} (2n+1)(-1)^n \exp\left\{ -\left[(2n+1)\frac{\pi}{4} \right]^2 T \right\}, \tag{26.26}$$

where

$$T = \frac{t - (r/v)}{t_M}.$$

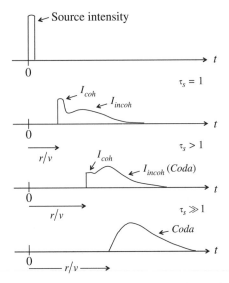

FIGURE 26.5   Propagation of pulse. Source intensity (coherent intensity) diminishes as $\exp(-a_s r)$. Incoherent intensity rises as coherent intensity diminishes and disappears, then Coda (I_{incoh}) dominates. t_s = Optical depth.

Noting that (Jolley, 1961)

$$\sum_{n=0}^{\infty} \frac{(-1)^n}{(2n+1)} = \frac{\pi}{4}.$$ (26.27)

we get the conservation of the power

$$\int_{r/v}^{\infty} G(t)dt = 1.$$ (26.28)

The characteristic time t_M gives the time scale of the pulse. Figure 26.6 shows the universal form of the pulse shape as a function of T. t_M is given by Ishimaru (1997) and Sato-Fehler (1998).

$$t_M = \frac{\tau_s}{2(kl)^2}\left(\frac{z}{v}\right) = \frac{\sqrt{\pi}}{2l}\langle n_1^2 \rangle z \left(\frac{z}{v}\right).$$ (26.29)

Note that as the optical scattering depth τ_s increases, the envelope stretches and the wave train (coda) becomes longer. However, the total power is conserved so that the integral (26.28) is unchanged, and the maximum value $G(t)_{max}$ decreases. The peak arrival time t_p is approximately given by

$$t_p = 0.67t_M.$$ (26.30)

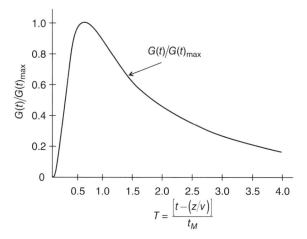

FIGURE 26.6   Universal form of the pulse shape of the impulse response.

26.8 S-WAVE CODA AND RAYLEIGH SURFACE WAVE

As shown in Fig. 26.3, P-waves and S-waves are the body waves. In Section 26.7, we described P-wave propagation. We did not include the boundary effects. The complete description of reflection and refraction of seismic pulse waves in homogeneous earth with the surface of a half-space, without heterogeneities, have been studied (Cagniard, 1962; de Hoop and van der Hijden, 1985). Complete solutions for a pulse in heterogeneous earth with a boundary have not been reported. Rayleigh surface waves have been extensively studied (Achenbach, 1980). Sato and Fehler (1998) discuss a detailed study of seismic coda.

As noted in Fig. 26.4, a P-wave arrives first followed by an S-wave. This time delay can be used to estimate the approximate distance to the hypocenter. If we have approximate velocities of P- and S-waves, the time difference Δt between first arrival time of the S-wave (r/v_s) and P-wave (r/v_p) is given by

$$r\left[\frac{1}{v_s} - \frac{1}{v_P}\right] = \Delta t. \tag{26.31}$$

This gives a circle centered at the observation point with radius r. If there are more than two observing points, we can get the location of the hypocenter from the intersection of the circles.

Appendix 26 gives a brief summary of the basic formulations of P- and S-waves and Rayleigh waves.

APPENDICES

APPENDIX TO CHAPTER 2

2.A MATHEMATICAL FORMULAS

2.A.1 Vector Formulas and Theorems

$$\nabla \cdot (\phi \bar{A}) = \phi \nabla \cdot \bar{A} + \bar{A} \cdot \nabla \phi,$$

$$\nabla \cdot (\bar{A} \times \bar{B}) = \bar{B} \cdot \nabla \times \bar{A} - \bar{A} \cdot \nabla \times \bar{B},$$

$$\nabla \times (\phi \bar{A}) = \nabla \phi \times \bar{A} + \phi \nabla \times \bar{A},$$

$$\nabla \times (\bar{A} \times \bar{B}) = \bar{A} \nabla \cdot \bar{B} - \bar{B} \nabla \cdot \bar{A} + (\bar{B} \cdot \nabla)\bar{A} - (\bar{A} \cdot \nabla)\bar{B},$$

$$\nabla \cdot \nabla \times \bar{A} = 0,$$

$$\nabla \times \nabla \phi = 0,$$

$$\bar{A} \cdot \bar{B} \times \bar{C} = \bar{B} \cdot \bar{C} \times \bar{A} = \bar{C} \cdot \bar{A} \times \bar{B},$$

$$\bar{A} \times (\bar{B} \times \bar{C}) = \bar{B}(\bar{A} \cdot \bar{C}) - \bar{C}(\bar{A} \cdot \bar{B}).$$

Divergence Theorem

$$\int_{v} \nabla \cdot \bar{A}\, dv = \int_{s} \bar{A} \cdot d\bar{a}.$$

Electromagnetic Wave Propagation, Radiation, and Scattering: From Fundamentals to Applications,
Second Edition. Akira Ishimaru.
© 2017 by The Institute of Electrical and Electronic Engineers, Inc. Published 2017 by John Wiley & Sons, Inc.

Stokes' Theorem

$$\int_a \nabla \times \bar{A} \cdot d\bar{a} = \oint_l \bar{A} \cdot d\bar{l}.$$

2.A.2 Gradient, Divergence, Curl, and Laplacian

Cartesian System

$$\nabla f = \left(\frac{\partial}{\partial x}\hat{x} + \frac{\partial}{\partial y}\hat{y} + \frac{\partial}{\partial z}\hat{z} \right) f,$$

$$\nabla \cdot A = \frac{\partial}{\partial x}A_x + \frac{\partial}{\partial y}A_y + \frac{\partial}{\partial z}A_z,$$

$$\nabla \times A = \begin{vmatrix} \hat{x} & \hat{y} & \hat{z} \\ \dfrac{\partial}{\partial x} & \dfrac{\partial}{\partial y} & \dfrac{\partial}{\partial z} \\ A_x & A_y & A_z \end{vmatrix},$$

$$\nabla^2 f = \left(\frac{\partial^2}{\partial x^2} + \frac{\partial^2}{\partial y^2} + \frac{\partial^2}{\partial z^2} \right) f.$$

Cylindrical System

$$\nabla f = \left(\frac{\partial}{\partial \rho}\hat{\rho} + \frac{1}{\rho}\frac{\partial}{\partial \phi}\hat{\phi} + \frac{\partial}{\partial z}\hat{z} \right) f,$$

$$\nabla \cdot \bar{A} = \frac{1}{\rho}\frac{\partial}{\partial \rho}(\rho A_\rho) + \frac{1}{\rho}\frac{\partial}{\partial \phi}A_\phi + \frac{\partial}{\partial z}A_z,$$

$$\nabla \times \bar{A} = \frac{1}{\rho}\begin{vmatrix} \hat{\rho} & \hat{\phi} & \hat{z} \\ \dfrac{\partial}{\partial \rho} & \dfrac{\partial}{\partial \phi} & \dfrac{\partial}{\partial z} \\ A_\rho & \rho A_\phi & A_z \end{vmatrix},$$

$$\nabla^2 f = \left[\frac{1}{\rho}\frac{\partial}{\partial \rho}\left(\rho\frac{\partial}{\partial \rho} \right) + \frac{1}{\rho^2}\frac{\partial^2}{\partial \phi^2} + \frac{\partial^2}{\partial z^2} \right] f.$$

Spherical System

$$\nabla f = \left(\hat{r}\frac{\partial}{\partial r} + \hat{\theta}\frac{\partial}{r\partial \theta} + \hat{\phi}\frac{1}{r\sin\theta}\frac{\partial}{\partial \phi} \right) f,$$

$$\nabla \cdot \bar{A} = \frac{1}{r^2}\frac{\partial}{\partial r}(r^2 Ar) + \frac{1}{r\sin\theta}\frac{\partial}{\partial \theta}(\sin\theta A_\theta) + \frac{1}{r\sin\theta}\frac{\partial}{\partial \phi}A_\phi,$$

$$\nabla \times \bar{A} = \frac{1}{r^2\sin\theta}\begin{vmatrix} \hat{r} & r\hat{\theta} & r\sin\theta\hat{\phi} \\ \dfrac{\partial}{\partial r} & \dfrac{\partial}{\partial \theta} & \dfrac{\partial}{\partial \phi} \\ A_r & rA_\theta & r\sin\theta A_\phi \end{vmatrix},$$

$$\nabla^2 f = \left[\frac{1}{r^2}\frac{\partial}{\partial r}\left(r^2\frac{\partial}{\partial r} \right) + \frac{1}{r^2\sin\theta}\frac{\partial}{\partial \theta}\left(\sin\theta\frac{\partial}{\partial \theta} \right) + \frac{1}{r^2\sin^2\theta}\frac{\partial^2}{\partial \phi^2} \right] f.$$

APPENDIX TO CHAPTER 3

3.A THE FIELD NEAR THE TURNING POINT

Consider the differential equation

$$\left[\frac{d^2}{dz^2} + q^2(z)\right] u(z) = 0, \tag{3.A.1}$$

with $q^2(z)$ varying as shown in Fig. 3.26. We will discuss the field in the neighborhood of the turning point z (region II). Here the WKB approximation fails, and we need a different approach. In this region, let us expand $q^2(z)$ about the turning point in a Taylor's series and retain its first term

$$q^2(z) = -a(z - z_0), \tag{3.A.2}$$

where a is the slope at z_0

$$a = -\left.\frac{d(q^2)}{dz}\right|_{z=z_0}.$$

Electromagnetic Wave Propagation, Radiation, and Scattering: From Fundamentals to Applications,
Second Edition. Akira Ishimaru.
© 2017 by The Institute of Electrical and Electronic Engineers, Inc. Published 2017 by John Wiley & Sons, Inc.

Then the differential equation becomes

$$\left[\frac{d^2}{dz^2} - a(z - z_0)\right] u(z) = 0. \tag{3.A.3}$$

We now convert this into the Stokes differential equation by using

$$t = a^{1/3}(z - z_0) = -a^{-2/3}q^2(z), \tag{3.A.4}$$

and obtain

$$\left(\frac{d^2}{dt^2} - t\right) u(t) = 0. \tag{3.A.5}$$

This is Stokes equation, and the solution satisfying the radiation condition as $t \to \infty$ is given by (Appendix 3.B)

$$u_a(t) = D_0 A_i(t), \tag{3.A.6}$$

where D_0 is constant and $A_i(t)$ is an Airy integral.

Let us examine the behavior of (3.A.6) toward region I. In this region, t is negative and large as seen from (3.A.4), and therefore, asymptotically,

$$\begin{aligned}
u_a(t) &= D_0 \frac{1}{\sqrt{\pi}(-t)^{1/4}} \sin\left[\frac{2}{3}(-t)^{3/2} + \frac{\pi}{4}\right] \\
&= \frac{D_0}{\sqrt{\pi}(-t)^{1/4}} \frac{e^{+j[2/3(-t)^{3/2}+\pi/4]} - e^{-j[2/3(-t)^{3/2}+\pi/4]}}{2j}.
\end{aligned} \tag{3.A.7}$$

Now we note from (3.A.2) that

$$-t = a^{-2/3}q^2(z) = -a^{1/3}(z - z_0),$$

and

$$\begin{aligned}
\frac{2}{3}(-t)^{3/2} &= \int_0^{(-t)} (-t)^{1/3} d(-t) \\
&= \int_{z_0}^{z} q(z) dz.
\end{aligned} \tag{3.A.8}$$

Therefore, we get

$$u_a(z) = \frac{D_0 e^{+j(\pi/4)}}{2j\sqrt{\pi}\, a^{-1/6} q^{1/2}} [e^{-j\int_{z_0}^{z} q(z)dz} - e^{+j\int_{z_0}^{z} q(z)dz - j(\pi/2)}]. \tag{3.A.9}$$

Now, we see that (3.A.9) can be smoothly connected to the WKB solution in region I

$$u_i(z) = \frac{A_0}{q^{1/2}} e^{-j\int_{z_0}^z q\,dz}. \tag{3.A.10}$$

The reflected WKB wave is

$$u_r(z) = \frac{B_0}{q^{1/2}} e^{+j\int_{z_0}^z q\,dz}. \tag{3.A.11}$$

Comparing this with the second term of (3.A.9), we get

$$B_0 = A_0\, e^{-j\int_{z_0}^z 2q\,dz + j(\pi/2)}. \tag{3.A.12}$$

The transmitted wave is obtained by examining $u_a(t)$ of (3.A.6) in region III where $t > 0$ and is large

$$u_a(t) = D_0 \frac{1}{2\sqrt{\pi}\, t^{1/4}} e^{-(2/3)t^{3/2}}, \tag{3.A.13}$$

which may be written as

$$u_a(t) = \frac{D_0 e^{-j(\pi/4)}}{2\sqrt{\pi}\, a^{-1/6}\, q^{1/2}} e^{-j\int_{z_0}^z q\,dz}. \tag{3.A.14}$$

Comparing this with the transmitted WKB solution,

$$u_t(z) = \frac{C_0}{q^{1/2}} e^{-j\int_{z_0}^z q\,dz}, \tag{3.A.15}$$

we get

$$C_0 = A_0\, e^{-j\int_{z_0}^z q\,dz}. \tag{3.A.16}$$

3.B STOKES DIFFERENTIAL EQUATION AND AIRY INTEGRAL

Consider the Stokes differential equation

$$\left(\frac{d^2}{dt^2} - t\right) u(t) = 0. \tag{3.B.1}$$

The two independent solutions can be written in the following form:

$$u(t) = w_1(t) = \frac{1}{\sqrt{\pi}} \int_{\Gamma_1} e^{t\lambda - \lambda^3/3} d\lambda, \tag{3.B.2}$$

$$u(t) = w_2(t) = \frac{1}{\sqrt{\pi}} \int_{\Gamma_2} e^{t\lambda - \lambda^3/3} d\lambda, \tag{3.B.3}$$

where Γ_1 and Γ_2 are two independent paths originating in one of the shaded regions and ending in another region in the Λ plane where the integrand vanishes (Fig. 3.B.1).

To show that $w_1(t)$ and $w_2(t)$ are solutions of the Stokes equation, we substitute them into (3.B.1). Then we get

$$\left(\frac{d^2}{dt^2} - t\right) w_1(t) = \frac{1}{\sqrt{\pi}} \int_{\Gamma_2} (\lambda^2 - t) e^{t\lambda - \lambda^3/3} d\lambda,$$

$$= \frac{-1}{\sqrt{\pi}} \int_{\Gamma_2} d(e^{t\lambda - \lambda^3/3}) \tag{3.B.4}$$

$$= \frac{-1}{\sqrt{\pi}} e^{t\lambda - \lambda^3/3} \Big|_A^B = 0.$$

Similarly, $w_2(t)$ also satisfies (3.B.1).

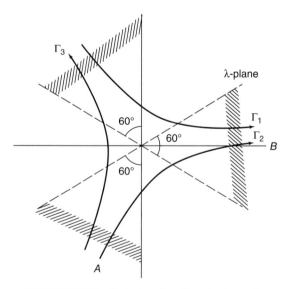

FIGURE 3.B.1   Complex plane for Airy integrals.

It is also possible to obtain two independent solutions by linear combinations of (3.B.2) and (3.B.3). Two commonly used functions are $A_i(t)$ and $B_i(t)$, defined by

$$A_i(t) = \frac{1}{2\sqrt{\pi}}[w_2(t) - w_1(t)]$$

$$= \frac{1}{\pi}\int_0^\infty \cos\left(\frac{x^3}{3} + tx\right) dx, \tag{3.B.5}$$

$$B_i(t) = \frac{1}{2\sqrt{\pi}}[w_2(t) + w_1(t)]$$

$$= \frac{1}{\pi}\int_0^\infty \left[e^{tx-x^3/3} + \sin\left(\frac{x^3}{3} + tx\right)\right] dx. \tag{3.B.6}$$

To show (3.B.5), we note that

$$w_2(t) - w_1(t) = \frac{1}{\sqrt{\pi}}\int_{\Gamma_3} e^{t\lambda - \lambda^3/3} d\lambda,$$

where Γ_3 is as shown in Fig. 3.B.1. We now choose the path along the imaginary axis by letting $\lambda = jx$ and

$$\int_{\Gamma_3} d\lambda = \int_{-\infty}^\infty j\,dx$$

and obtain (3.B.5). To obtain (3.B.6), we note that

$$\int_{\Gamma_2} d\lambda = j\int_{-\infty}^0 e^{j(tx+x^3/3)}dx + \int_0^\infty e^{tx-x^3/3}dx,$$

$$\int_{\Gamma_1} d\lambda = j\int_\infty^0 e^{j(tx+x^3/3)}dx + \int_0^\infty e^{tx-x^3/3}dx,$$

and combine these two.

Along the real axis of t, $A_i(t)$ and $B_i(t)$ behave as shown in Fig. 3.B.2. The asymptotic form of $A_i(t)$ and $B_i(t)$ are given below.

For $|t| \to \infty$ and $|\arg t| < \pi/3$,

$$A_i(t) \sim \frac{1}{2\sqrt{\pi}\,t^{1/4}}e^{-(2/3)t^{3/2}},$$

$$B_i(t) \sim \frac{1}{\sqrt{\pi}\,t^{1/4}}e^{(2/3)t^{3/2}}.$$

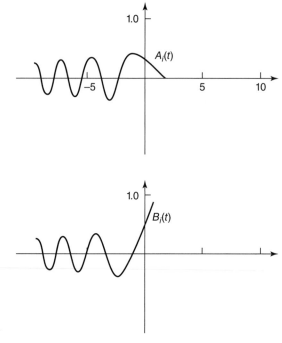

FIGURE 3.B.2 Airy functions.

For $|t| \to \infty$ and $|\arg(-t)| < 2\pi/3$,

$$A_i(t) \sim \frac{1}{\sqrt{\pi}(-t)^{1/4}} \sin\left[\frac{2}{3}(-t)^{3/2} + \frac{\pi}{4}\right],$$

$$B_i(t) \sim \frac{1}{\sqrt{\pi}(-t)^{1/4}} \cos\left[\frac{2}{3}(-t)^{3/2} + \frac{\pi}{4}\right].$$

APPENDIX TO CHAPTER 4

4.A GREEN'S IDENTITIES AND THEOREM

Consider the divergence theorem applied to vector $\bar{A}$

$$\int_V \nabla \cdot \bar{A} \, dV = \int_S \bar{A} \cdot d\bar{s}. \tag{4.A.1}$$

Letting $\bar{A} = u\nabla v$, where u and v are scalar fields, we get

$$\int_V \nabla \cdot (u\nabla v) \, dV = \int_S u\nabla v \cdot d\bar{s} = \int_S u\frac{\partial v}{\partial n} \, dS, \tag{4.A.2}$$

where $\partial/\partial n$ is the outward normal derivative. Using the vector identity on the left side of (4.A.2),

$$\nabla \cdot (u\nabla v) = \nabla u \cdot \nabla v + u\nabla^2 v, \tag{4.A.3}$$

we get *Green's first identity*,

$$\int_V (\nabla u \cdot \nabla v + u\nabla^2 v) \, dV = \int_S u\frac{\partial v}{\partial n} \, dS. \tag{4.A.4}$$

Electromagnetic Wave Propagation, Radiation, and Scattering: From Fundamentals to Applications,
Second Edition. Akira Ishimaru.
© 2017 by The Institute of Electrical and Electronic Engineers, Inc. Published 2017 by John Wiley & Sons, Inc.

Now we interchange u and v,

$$\int_V (\nabla u \cdot \nabla v + v\nabla^2 u)\, dV = \int_S v\frac{\partial u}{\partial n}\, dS. \tag{4.A.5}$$

Subtracting (4.A.5) from (4.A.4), we get *Green's second identity*, which is also called *Green's theorem*.

$$\int_V (u\nabla^2 v - v\nabla^2 u)\, dV = \int_S \left(u\frac{\partial v}{\partial n} - v\frac{\partial u}{\partial n} \right)\, dS. \tag{4.A.6}$$

The two-dimensional equivalents of Green's first and second identities are

$$\int_a \left(\nabla_t u \cdot \nabla_t v + v\nabla_t^2 u \right)\, da = \int_l v\frac{\partial u}{\partial n}\, dl, \tag{4.A.7}$$

$$\int_a \left(u\nabla_t^2 v - v\nabla_t^2 u \right)\, da = \int_l \left(u\frac{\partial v}{\partial n} - v\frac{\partial u}{\partial n} \right)\, dl, \tag{4.A.8}$$

where a is the area surrounded by a closed boundary curve l. The one-dimensional equivalents are

$$\int_{x_1}^{x_2} \left(\frac{\partial u}{\partial x}\frac{\partial v}{\partial x} + v\frac{\partial^2 u}{\partial x^2} \right)\, dx = v\frac{\partial u}{\partial x}\bigg|_{x_1}^{x_2}, \tag{4.A.9}$$

$$\int_{x_1}^{x_2} \left(u\frac{\partial^2 v}{\partial x^2} - v\frac{\partial^2 u}{\partial x^2} \right)\, dx = \left(u\frac{\partial v}{\partial x} - v\frac{\partial u}{\partial x} \right)\bigg|_{x_1}^{x_2}. \tag{4.A.10}$$

4.B   BESSEL FUNCTIONS $Z_V(x)$

Bessel functions $Z_v(z)$ are the solutions of the Bessel differential equation

$$\left[z^2\frac{d^2}{dz^2} + z\frac{d}{dz} + (z^2 - v^2) \right] Z_v(z) = 0.$$

J_v, N_v, $H_v^{(1)}$, and $H_v^{(2)}$ are called the Bessel, Neumann, and Hankel functions of the first kind, and the Hankel function of the second kind, respectively. They are related by the following:

$$\begin{aligned} H_v^{(1)}(z) &= J_v(z) + jN_v(z), \\ H_v^{(2)}(z) &= J_v(z) - jN_v(z). \end{aligned} \tag{4.B.1}$$

If v is a noninteger, J_v and J_{-v} are independent. When $v = n =$ integer, Z_n is proportional to Z_{-n}.

$$Z_{-n}(z) = (-1)^n Z_n(z). \tag{4.B.2}$$

For real $v \geq 0$, we have for $|z| \ll 1$:

$$J_v(z) \approx \frac{1}{\Gamma(v+1)} \left(\frac{z}{2}\right)^v,$$

$$N_v(z) \approx -\frac{\Gamma(v)}{\pi} \left(\frac{2}{z}\right)^v,$$

$$J_0(z) \approx 1 - \left(\frac{z}{2}\right)^2,$$

$$N_0(z) \approx -\frac{2}{\pi} \ln \frac{2}{\gamma z}, \tag{4.B.3}$$

where $\Gamma(v)$ is the gamma function, $\Gamma(n) = (n-1)!$, and $\lambda = 1.781072418$, Euler's constant. For $|z| \gg 1$, $|z| \gg v$,

$$J_v(z) \approx \left(\frac{2}{\pi z}\right)^{1/2} \cos\left(z - \frac{v\pi}{2} - \frac{\pi}{4}\right),$$

$$N_v(z) \approx \left(\frac{2}{\pi z}\right)^{1/2} \sin\left(z - \frac{v\pi}{2} - \frac{\pi}{4}\right). \tag{4.B.4}$$

APPENDIX TO CHAPTER 5

5.A DELTA FUNCTION

The delta function is defined by $\delta(\bar{r}, \bar{r}') = 0$ whenever $\bar{r} \neq \bar{r}'$ and $\int_V \delta(\bar{r} - \bar{r}')dV = 1$ when V includes $\bar{r}$.

Rectangular Coordinate

$$\delta(\bar{r} - \bar{r}') = \delta(x - x')\delta(y - y')\delta(z - z').$$

where $\delta(x - x') = 0$ whenever $x \neq x'$ and $\int \delta(x - x') = 1$, and thus $\delta(x - x')$ has a dimension of $(\text{length})^{-1}$.

Cylindrical Coordinate

$$\delta(\bar{r} - \bar{r}') = \frac{\delta(\rho - \rho')\delta(\phi - \phi')\delta(z - z')}{\rho},$$

where

$$\delta(\rho - \rho') = 0 \quad \text{whenever} \quad \rho \neq \rho',$$
$$\delta(\phi - \phi') = 0 \quad \text{whenever} \quad \phi \neq \phi',$$
$$\delta(z - z') = 0 \quad \text{whenever} \quad z \neq z',$$

Electromagnetic Wave Propagation, Radiation, and Scattering: From Fundamentals to Applications,
Second Edition. Akira Ishimaru.
© 2017 by The Institute of Electrical and Electronic Engineers, Inc. Published 2017 by John Wiley & Sons, Inc.

and

$$\int \delta(\rho - \rho')d\rho = 1,$$

$$\int \delta(\phi - \phi')d\phi = 1,$$

$$\int \delta(z - z')dz = 1,$$

$$dV = r^2 \sin\theta \, dr \, d\theta \, d\phi.$$

Note that $dV = \rho \, d\rho \, d\phi \, dz$.

Spherical Coordinate

$$\delta(\bar{r} - \bar{r}') = \frac{\delta(r - r')\delta(\theta - \theta')\delta(\phi - \phi')}{r^2 \sin\theta},$$

where

$$\delta(r - r') = 0 \quad \text{whenever} \quad r \neq r',$$
$$\delta(\theta - \theta') = 0 \quad \text{whenever} \quad \theta \neq \theta',$$
$$\delta(\phi - \phi') = 0 \quad \text{whenever} \quad \phi \neq \phi',$$

and

$$\int \delta(r - r')dr = 1,$$

$$\int \delta(\theta - \theta')d\theta = 1,$$

$$\int \delta(\phi - \phi')d\phi = 1,$$

$$dV = r^2 \sin\theta \, dr \, d\theta \, d\phi.$$

The delta function $\delta(r - r')$ has the following important characteristics:

$$\int_{V_0} f(\bar{r})\delta(\bar{r} - \bar{r}') \, dV = f(\bar{r}'),$$

for an arbitrary function $f(\bar{r})$, whenever V_0 includes $\bar{r}'$. The delta function may be thought of as a limiting case of a rectangular pulse $f(x)$ of height h and width W with a unit area as the width approaches zero.

$$\delta(x) = \lim_{W \to 0} f(x), \qquad \text{keeping} \quad hW = 1.$$

The exact shape of the function $f(x)$ is unimportant. More rigorously, the delta function must be interpreted in terms of the theory of distribution.

We note the following characteristics:

$$\int_{-\varepsilon}^{\varepsilon} \delta(x) \, dx = 1,$$

$$\int_{0}^{\varepsilon} \delta(x) \, dx = \frac{1}{2},$$

$$\delta(-x) = \delta(x),$$

$$\int_{-\varepsilon}^{\varepsilon} f(x)\delta'(x) \, dx = f(x)\delta(x)\big|_{-\varepsilon}^{\varepsilon} - \int_{-\varepsilon}^{\varepsilon} f'(x)\delta(x) \, dx = -f'(0).$$

Similarly,

$$\int f(x)\delta^{(n)}(x) \, dx = (-1)^n f^{(n)}(0),$$

$$\delta[g(x)] = \sum_{n=1}^{N} \frac{\delta(x - x_n)}{|g'(x_n)|},$$

where x_n are the zeros of $g(x)[g(x_n) = 0]$.

APPENDIX TO CHAPTER 6

6.A STRATTON–CHU FORMULA

To prove (6.110), (6.111), and (6.112), we start with the vector Green's theorem

$$\int_V (\bar{Q} \cdot \nabla \times \nabla \times \bar{P} - \bar{P} \cdot \nabla \times \nabla \times \bar{Q})\, dV = \int_S (\bar{P} \times \nabla \times \bar{Q} - \bar{Q} \times \nabla \times \bar{P}) \cdot d\bar{S}.$$

(6.A.1)

We let

$$\bar{P} = \hat{a}G(\bar{r}, \bar{r}'),$$
$$\bar{Q} = \bar{E}(\bar{r}),$$

(6.A.2)

where $\hat{a}$ is a constant unit vector and G is the scalar Green's function. We also have Maxwell's equations

$$\nabla \times \bar{E} = -j\omega\mu\bar{H} - \bar{J}_m,$$
$$\nabla \times \bar{H} = j\omega\varepsilon\bar{E} + \bar{J},$$
$$\nabla \cdot \bar{E} = \frac{\rho}{\varepsilon}, \quad \nabla \cdot \bar{H} = 0,$$

(6.A.3)

where we included the magnetic current density $\bar{J}_m$.

Electromagnetic Wave Propagation, Radiation, and Scattering: From Fundamentals to Applications,
Second Edition. Akira Ishimaru.
© 2017 by The Institute of Electrical and Electronic Engineers, Inc. Published 2017 by John Wiley & Sons, Inc.

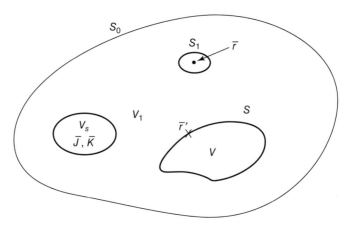

FIGURE 6.A.1 Proof of Stratton–Chu formula.

Now we apply Green's theorem to the volume V_1 surrounded by S_t consisting of S_1, S, and S_∞ (Fig. 6.A.1). The surface S_1 is the surface of a small sphere centered at $\bar{r}'$. (Later, $\bar{r}'$ will be identified as the observation point and $\bar{r}$ and $\bar{r}'$ will be interchanged.) We wish to transform (6.A.1) into the form

$$\hat{a} \cdot \int_{V_1} \cdots dV = \hat{a} \cdot \int_{S_t} \cdots dS,$$

and obtain

$$\int_{V_1} \cdots dV = \int_{S_t} \cdots dS.$$

For this purpose, we examine the left side of (6.A.1). We note that

$$\nabla \times \nabla \times \bar{P} = \nabla \times \nabla \times (\hat{a}G)$$

$$= \nabla(\nabla \cdot \hat{a}G) - \nabla^2(\hat{a}G)$$

$$= \nabla(\hat{a} \cdot \nabla G) + \hat{a}k^2 G,$$

$$\bar{E} \cdot \nabla(\hat{a} \cdot \nabla G) = \nabla \cdot (\bar{E}(\hat{a} \cdot \nabla G) - \hat{a} \cdot \nabla G(\nabla \cdot \bar{E}),$$

$$\int_{V_1} \nabla \cdot (\bar{E}\hat{a} \cdot \nabla G) \, dV = \int_{S_t} \hat{a} \cdot (\nabla G)\bar{E} \cdot \hat{n} \, dS.$$

Therefore, we get

$$\int_{V_1} \bar{Q} \cdot \nabla \times \nabla \times \bar{P} \, dV = \hat{a} \cdot \int_{S_t} (\nabla G)(\bar{E} \cdot \hat{n}) \, dS + \hat{a} \cdot \int_{V_1} \left(k^2 G\bar{E} - \frac{\rho}{\varepsilon} \nabla G \right) dV.$$

Similarly, we get

$$\bar{P} \cdot \nabla \times \nabla \times \bar{Q} = \hat{a} G \cdot \nabla \times \nabla \times \bar{E} = \hat{a} \cdot G(k^2 \bar{E} - j\omega\mu\bar{J} - \nabla \times \bar{J}_m).$$

We also note that

$$G\nabla \times \bar{J}_m = \nabla \times (G\bar{J}_m) + \bar{J}_m \times \nabla G$$

$$\int_{V_1} \nabla \times (G\bar{J}_m)\, dV = \int_{S_t} \hat{n} \times G\bar{J}_m dS.$$

Therefore, the left-hand side of (6.A.1) becomes

$$\text{L.H.} = \hat{a} \cdot \int_{V_1} \left(-\frac{\rho}{\varepsilon} \nabla G + j\omega\mu\bar{J} + \bar{J}_m \times \nabla G \right) dV$$

$$+ \hat{a} \cdot \int_{S_t} [\nabla G(\bar{E} \cdot \hat{n}) + \hat{n} \times (G\bar{J}_m)]\, dS. \qquad (6.A.4)$$

Now we examine the right-hand side of (6.A.1). We note that

$$\bar{P} \times \nabla \times \bar{Q} \cdot \hat{n} = \hat{a} G \times \nabla \times \bar{E} \cdot \hat{n}$$

$$= \hat{a} G \times (-j\omega\mu\bar{H} - \bar{J}_m) \cdot \hat{n}$$

$$= \hat{a} \cdot [\hat{n} \times (j\omega\mu G\bar{H} + G\bar{J}_m)],$$

$$\bar{Q} \times \nabla \times \bar{P} \cdot \hat{n} = \bar{E} \times \nabla \times (\hat{a} G) \cdot \hat{n}$$

$$= \hat{a} \cdot (\hat{n} \times \bar{E}) \times \nabla G.$$

Therefore, the right side of (6.A.1) becomes

$$\text{R.H.} = \hat{a} \cdot \int_{S_t} [-(\hat{n} \times \bar{E}) \times \nabla G + j\omega\mu G\hat{n} \times \bar{H} + \hat{n} \times G\bar{J}_m]\, dS. \qquad (6.A.5)$$

We now equate (6.A.4) to (6.A.5). We also interchange $\bar{r}$ and $\bar{r}'$ so that the source point is $\bar{r}'$ and the observation point is $\bar{r}$. We also use $\hat{n}' = -\hat{n}$. We then get

$$\int_{V_1} \bar{E}_v dV' + \int_{S_t} \bar{E}_s dS' = 0, \qquad (6.A.6)$$

where $\bar{E}_s$ is given in (6.113) and $\bar{E}_v$ is given by

$$\bar{E}_v = - \left(j\omega\mu G\bar{J} + \bar{J}_m \times \nabla' G - \frac{\rho}{\varepsilon} \nabla' G \right), \qquad (6.A.7)$$

where $\bar{J} = \bar{J}(\bar{r}')$ and $\bar{J}_m = \bar{J}_m(\bar{r}')$.

Similarly, for magnetic fields, we use the symmetry of Maxwell's equations (6.A.3) with the following interchanges of the field quantities:

$$\begin{aligned}
\bar{E} &\to \bar{H} & \rho &\to \rho_m, \\
\bar{H} &\to -\bar{E} & \rho_m &\to -\rho, \\
\bar{J} &\to \bar{J}_m & \varepsilon &\to \mu, \\
\bar{J}_m &\to -\bar{J} & \mu &\to \varepsilon.
\end{aligned} \tag{6.A.8}$$

We then get

$$\int_{V_1} \bar{H}_v \, dV' + \int_{S_t} \bar{H}_s \, dS' = 0, \tag{6.A.9}$$

where $\bar{H}_s$ is given in (6.113) and $\bar{H}_v$ is given by

$$\bar{H}_v = -\left(j\omega\varepsilon G \bar{J}_m - \bar{J} \times \Delta' G - \frac{\rho_m}{\mu} \nabla' G \right). \tag{6.A.10}$$

Note that the surface integral and the volume integral are in the same form if we define the following:

$$\begin{aligned}
\bar{J}_s &= \text{Surface current density} = \hat{n}' \times \bar{H}, \\
\bar{J}_{ms} &= \text{Surface magnetic current density} = -\hat{n}' \times \bar{E}, \\
\rho_s &= \text{Surface charge density} = \varepsilon(\hat{n}' \cdot \bar{E}), \\
\rho_{ms} &= \text{Surface magnetic charge density} = \mu(\hat{n}' \cdot \bar{H}).
\end{aligned} \tag{6.A.11}$$

Now we apply (6.A.6) to the problem shown in Fig. 12.10. First, we consider the case when the observation point $\bar{r}$ is outside S (Fig. 6.15). Consider the integral over S_1.

$$\int_{S_1} \bar{E}_s \, dS' = -\int_{S_1} [j\omega\mu G \hat{n}' \times \bar{H} - (\hat{n}' \times \bar{E}) \times \nabla' G - (\hat{n}' \cdot \bar{E}) \nabla' G] \, dS'. \tag{6.A.12}$$

S_1 is the surface of a small sphere of radius ε centered at $\bar{r}$. As $\varepsilon \to 0$, the first term contains only G, which varies as ε^{-1} and the surface is $4\pi\varepsilon^2$, and therefore, this term vanishes. The second and third terms, however, contain the gradient of G and thus $\nabla' G = -(4\pi\varepsilon^2)^{-1}\hat{n}'$, and

$$\lim_{\varepsilon \to 0} \int_{S_1} \bar{E}_s \, hdS' = \lim_{\varepsilon \to 0} [(\hat{n} \times \bar{E}) \times \hat{n}' + \hat{n}' \cdot \bar{E}\hat{n}] \left(-\frac{1}{4\pi\varepsilon^2} \right) 4\pi\varepsilon^2 \tag{6.A.13}$$

$$= -\bar{E}(\bar{r}).$$

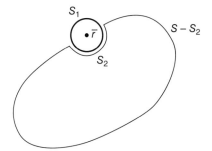

FIGURE 6.A.2 Observation point on the surface S.

The surface S_t consists of S_1, S_∞, and S. The integral over S_∞ vanishes as the fields diminish satisfying the radiation condition. Thus we get

$$\bar{E}(\bar{r}) = \int_{V_1} \bar{E}_v dv' + \int_S \bar{E}_s dS'. \tag{6.A.14}$$

Now the volume integral over the source can be identified as the incident wave $\bar{E}_i$ because if we remove the object, the integral over S disappears and $\bar{E}(\bar{r})$ should be equal to the incident wave $E_i(\bar{r})$. Therefore, we get (6.111),

$$\bar{E}_i(\bar{r}) + \int_S \bar{E}_s dS' = \bar{E}(\bar{r}). \tag{6.A.15}$$

Similarly, we get

$$\bar{H}_i(\bar{r}) + \int_S \bar{H}_s dS' = \bar{H}(\bar{r}).$$

If the observation point $\bar{r}$ is on the surface S, we consider S, S_2, and $S - S_2$ (Fig. 6.A.2). It is easy to see that

$$\int_{S_1} \bar{E}_s dS' = -\bar{E}(\bar{r}),$$

$$\int_{S_2} \bar{E}_s dS' = +\frac{1}{2}\bar{E}(\bar{r}). \tag{6.A.16}$$

Therefore, we get

$$\bar{E}_i(\bar{r}) + \oint_S \bar{E}_s dS' = \frac{1}{2}\bar{E}(\bar{r}). \tag{6.A.17}$$

Similarly,

$$\bar{H}_i(\bar{r}) + \fint_S \bar{H}_s dS' = \frac{1}{2}\bar{H}(\bar{r}),$$

where $\fint_S$ means the Cauchy principal value of the integral (over $S - S_2$ as $\varepsilon \to 0$).

If the observation point $\bar{r}$ is inside S, we can immediately obtain (6.112) by noting that the volume integral in (6.A.6) is $\bar{E}_i(\bar{r})$ and the surface integral is only on the surface S.

APPENDIX TO CHAPTER 7

7.A PERIODIC GREEN'S FUNCTION

Consider a periodic structure with period l_1 in the x direction and l_2 in the y direction, and consider one cell defined by $0 < x < l_1$, $0 < y < l_2$, and $-\infty \leq z \leq +\infty$. Within this cell, Green's function G satisfies the following:

$$(\nabla^2 + k^2)G = -\delta(x - x')\delta(y - y')\delta(z - z'). \tag{7.A.1}$$

Expanding G in Floquet series, we write

$$G = \sum_m \sum_n g_{mn}(z, z')e^{-j\alpha x - j\beta y}, \tag{7.A.2}$$

where $\alpha = \alpha_0 + 2m\pi/l_1$ and $\beta = \beta_0 + 2n\pi/l_2$. Substituting this into (7.A.1) and noting that

$$\delta(x - x') = \sum_m \frac{e^{-j\alpha(x-x')}}{l_1}, \tag{7.A.3}$$

we get

$$G = \sum_{m=-\infty}^{\infty} \sum_{n=-\infty}^{\infty} \frac{e^{-j\alpha(x-x')-j\beta(y-y')-j\gamma|z-z'|}}{2j\gamma l_1 l_2}, \tag{7.A.4}$$

Electromagnetic Wave Propagation, Radiation, and Scattering: From Fundamentals to Applications,
Second Edition. Akira Ishimaru.
© 2017 by The Institute of Electrical and Electronic Engineers, Inc. Published 2017 by John Wiley & Sons, Inc.

where $\alpha = \alpha_0 + 2m\pi/l_1$, $\beta = \beta_0 + 2n\pi/l_2$, and

$$
\gamma = \begin{cases} (k^2 - \alpha^2 - \beta^2)^{1/2} & \text{if } k^2 > \alpha^2 + \beta^2 \\ -j(\alpha^2 + \beta^2 - k^2)^{1/2} & \text{if } k^2 < \alpha^2 + \beta^2 \end{cases}.
$$

7.B   VARIATIONAL FORM

Note that in general we write

$$
\int_a^b dz \int_a^b dz' f^*(z) G(z, z', \beta) f(z') = 0, \tag{7.B.1}
$$

where G is Hermitian,

$$
G(z, z', \beta) = G^*(z', z, \beta). \tag{7.B.2}
$$

If the true f is not known, we use an approximate function $f_a = f + \delta f$, where δf is the variation. The corresponding propagation constant is $\beta_a = \beta + \delta\beta + \cdots$. If we show that $\delta\beta = 0$, the variation of β is of the second order and the error in β_a is much smaller than the error in f_a. Thus we get the value of β close to the true value from (7.B.1), even if we use a crude approximation for f.

Let us take the variation for (7.B.1). We get

$$
\int_a^b dz \int_a^b dz' \left[\delta f^*(z) G(z, z', \beta) f(z') + f^*(z) G(z, z', \beta) \delta f(z') + f^*(z) \frac{\partial G}{\partial \beta} \delta\beta f(z') \right] = 0,
$$

and thus interchanging z and z' in the second term and using (7.B.2), we get

$$
\int_a^b dz \int_a^b dz' f^*(z) \frac{\partial G}{\partial \beta} f(z) + \int_a^b dz \int_a^b dz' [\delta f^*(z) G(z, z', \beta) f(z')
$$
$$
+ \delta f(z) G^*(z, z', \beta) f^*(z')] = 0. \tag{7.B.3}
$$

But at the correct value of $f(z)$,

$$
\int_a^b dz' G(z, z', \beta) f(z') = 0,
$$

and thus the second and third terms of (7.B.3) vanish, proving that $\delta\beta = 0$.

7.C EDGE CONDITIONS

In the neighborhood of a conducting wedge, electric and magnetic field components behave in a particular manner dependent on the wedge angle. Let us consider a conducting wedge whose angle is $(2\pi - \phi_0)$ (Fig. 7.C.1). The field in the neighborhood of the tip can be represented by two different modes: TM with E_z, H_ϕ, and $H\rho$ and TE with H_z, E_ϕ, and $E\rho$. The TM modes may be given by

$$(\nabla^2 + k^2)E_z = 0, \tag{7.C.1}$$

$$H_\phi = \frac{1}{j\omega\mu}\frac{\partial}{\partial\rho}E_z, \quad \text{and} \quad H_\rho = -\frac{1}{j\omega\mu}\frac{1}{\rho}\frac{\partial}{\partial\phi}E_z. \tag{7.C.2}$$

A most general solution to (7.C.1) satisfying $E_z = 0$ at $\phi = 0$ and ϕ_0 is

$$E_z = \sum_{n=1}^{\infty} a_n Z_{v_n}(k\rho)\sin v_n\,\phi, \quad v_n = \frac{n\pi}{\phi_0}, \tag{7.C.3}$$

and $Z_{v_n}(\rho)$ is an appropriate Bessel function. Noting that $v_n \neq 0$ and noninteger, $Z_v(\rho)$ can be either $J_v(k\rho)$ or $J_{-v}(k\rho)$. For small $k\rho$,

$$J_v(k\rho) \sim \rho^v, \quad J_{-v}(k\rho) \sim \rho^{-v},$$

and thus taking the first term of (7.C.3) for small $k\rho$, we get

$$E_z \sim \rho^v \sin v\phi \quad \text{or} \quad \rho^{-v} \sin v\phi,$$

$$H_\phi \sim \rho^{v-1} \sin v\phi \quad \text{or} \quad \rho^{-v-1} \sin v\phi, \tag{7.C.4}$$

and

$$H\rho \sim \rho^{v-1} \cos v\phi \quad \text{or} \quad \rho^{-v-1} \cos v\phi$$

Now the choice between the two sets above must be resolved. This is done by noting that the total energy in the vicinity of the edge must be finite.

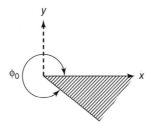

FIGURE 7.C.1 Edge conditions.

The total power entering the region within radius ρ_0 is given by

$$P = \int_0^{\phi_0} \tfrac{1}{2}\bar{E} \times \bar{H}^* \cdot (-\hat{r})\rho_0 d\phi$$

$$= \frac{1}{2}\int_0^{\phi_0} E_z H_\phi^* \rho_0 \, d\phi. \tag{7.C.5}$$

Using the first set of (7.C.4), we get

$$P \sim (\text{const.})\rho^{2\nu}. \tag{7.C.6}$$

In order that P be finite as $\rho \to 0$, we require that $\nu > 0$. On the other hand, if we use the second set of (7.C.4), we get

$$P \sim (\text{const.})\rho_0^{-2\nu}. \tag{7.C.7}$$

For P to be finite, ν must be negative, but this contradicts (7.C.3), and therefore this must be rejected. Thus the behavior of the field in the vicinity of the edge is

$$\begin{aligned} E_z &\sim \rho^\nu \sin \nu\phi, \\ H_\phi &\sim \rho^{\nu-1} \sin \nu\phi, \\ H_\rho &\sim \rho^{\nu-1} \cos \nu\phi \quad \text{with} \quad \nu = \frac{\pi}{\phi_0}, \end{aligned} \tag{7.C.8}$$

and the surface current density should behave as

$$J_z \sim \rho^{\nu-1}.$$

On the other hand, for TE modes, the same considerations lead to

$$\begin{aligned} E_z &\sim C_0 + C_1 \rho^\nu \cos \nu\phi, \\ H_\phi &\sim \rho^{\nu-1} \sin \nu\phi, \\ H_\rho &\sim \rho^{\nu-1} \cos \nu\phi \quad \text{with} \quad \nu = \frac{\pi}{\phi_0}, \end{aligned} \tag{7.C.9}$$

and the surface current behaves as

$$J\rho \sim C_0 + C_{1\rho}^\nu.$$

Equations (7.C.8) and (7.C.9) represent the behavior of the field components in the neighborhood of the edge.

As an example, take a knife edge ($\phi_0 = 2\pi$). Then

$$E_z \sim \rho^{1/2} \quad \text{and} \quad H_z \sim C_0 + C_{1\rho}^{1/2} \cos \nu\phi$$

but all the other field components become infinite at the edge. For example, $E\rho$ behaves as

$$E_\rho \sim \frac{1}{\rho^{1/2}} \sin \frac{\phi}{2},$$

and the current density on the surface becomes

$$J_z \sim H_\rho \sim \frac{1}{\rho^{1/2}}.$$

Even though the field and the current density themselves become infinite at the edge, the energy stays finite.

For a rectangular edge, $\phi_0 = 3\pi/2$, and thus

$$E_\rho \sim \frac{1}{\rho^{2/3}} \quad \text{and} \quad J_z \sim \frac{1}{\rho^{2/3}}.$$

APPENDIX TO CHAPTER 8

8.A MATRIX ALGEBRA

Consider a matrix given by

$$A = \begin{bmatrix} a_{11} & a_{12} & \cdots & a_{1n} \\ a_{21} & a_{22} & \cdots & a_{2n} \\ \vdots & & & \\ a_{m1} & a_{m2} & \cdots & a_{mn} \end{bmatrix}.$$

1. a_{ij} is called the element of the matrix A and to indicate the element a_{ij}, we write

$$A = (a_{ij}).$$

2. A has m rows and n columns and the *order* of the matrix is $m \times n$.
3. When $m = n$, A is called the *square matrix*.
4. Equality of matrices: $A = B$ means $a_{ij} = b_{ij}$. Addition of matrices: $A + B = C$ means $a_{ij} + b_{ij} = c_{ij}$. Matrix addition satisfies the commutative and associative laws:

$$A + B = B + A, (A + B) + C = A + (B + C).$$

Electromagnetic Wave Propagation, Radiation, and Scattering: From Fundamentals to Applications,
Second Edition. Akira Ishimaru.
© 2017 by The Institute of Electrical and Electronic Engineers, Inc. Published 2017 by John Wiley & Sons, Inc.

5. Multiplication by a scalar α:

$$\alpha A = B \text{ means } \alpha a_{ij} = b_{ij}.$$

6. Multiplication by matrix:

$$AB = C \text{ means } C_{ij} = \sum_{k=1}^{n} a_{ik} b_{kj}.$$

In order to have the product AB, the matrices must be conformable, which means that the number of columns of A must be equal to the number of rows of B. Thus, if the order of A is $m \times n$, the order of B must be $n \times p$ and the order of C becomes $m \times p$. The multiplication of matrices does not obey the commutative law, $AB \neq BA$. The associative law is valid, $(AB)C = A(BC)$.

7. A singular matrix is a square matrix with a zero determinant. A nonsingular matrix is a square matrix with a nonzero determinant.

8. The rank of the matrix is the highest order of a nonvanishing determinant in a matrix.

9. The reciprocal or inverse of A is given by

$$A^{-1} = \frac{(A_{ji})}{|A|},$$

where (A_{ji}) is the cofactor. The product of A and A^{-1} is the unit matrix

$$AA^{-1} = A^{-1}A = U.$$

The inverse of the product is given by

$$(AB)^{-1} = B^{-1}A^{-1}.$$

10. The transpose of A is given by

$$\tilde{A} = (a_{ji}).$$

11. A symmetric matrix is a matrix with $a_{ij} = a_{ji}$ and thus $\tilde{A} = A..$

12. An antisymmetric (or skew-symmetric) matrix is a matrix with $a_{ij} = -a_{ji}$, and thus $\tilde{A} = -A$ and all the diagonal elements are zero.

13. The transpose of the product is given by

$$\widetilde{(AB)} = \tilde{B}\tilde{A}.$$

14. Diagonal matrix: $a_{ij} = \delta_{ij}a_{ii}$, where

$$\delta_{ij} = \begin{cases} 1 & \text{if } i = j \\ 0 & \text{if } i \neq j \end{cases} \quad \text{(Kronecker's delta)}.$$

15. Unit matrix U: $a_{ij} = \delta_{ij}$. Zero matrix: $a_{ij} = 0$.

16. The adjoint matrix is the conjugate of the transpose:

$$A^{+} = \tilde{A}^{*}.$$

17. Orthogonal matrix A satisfies the following: $\tilde{A} = A^{-1}$ or $\tilde{A}A = U$.

18. Unitary matrix A satisfies: $A^{+} = A^{-1}$ or $A^{+}A = U$.

19. Hermitian matrix A satisfies: $A^{+} = A$. Skew-Hermitian matrix A satisfies: $A^{+} = -A$.

20. It is always possible to express A as a sum of symmetric and antisymmetric matrices:

$$A = \frac{A + \tilde{A}}{2} + \frac{A - \tilde{A}}{2},$$

or as a sum of Hermitian and skew-Hermitian matrices:

$$A = \frac{A + A^{+}}{2} + \frac{A - A^{+}}{2}.$$

APPENDIX TO CHAPTER 10

10.A FORWARD SCATTERING THEOREM (OPTICAL THEOREM)

Consider a linearly polarized wave $\bar{E}_i$ incident on an object and the incident Poynting vector $\bar{S}_i$.

$$\bar{E}_i = \hat{e}_i\, e^{-jk\bar{r}\cdot\hat{i}},$$

$$\bar{S}_i = \frac{1}{2}\bar{E}_i \times \bar{H}_i^* = \frac{|E_i|^2}{2\eta_0}\hat{i}, \quad \eta_0 = \left(\frac{\mu_0}{\varepsilon_0}\right)^{1/2}. \tag{10.A.1}$$

The total field $\bar{E}$ and $\bar{H}$ are given by

$$\bar{E} = \bar{E}_i + \bar{E}_s,$$

$$\bar{H} = \bar{H}_i + \bar{H}_s, \tag{10.A.2}$$

where $\bar{E}_s$ and $\bar{H}_s$ are the scattered waves.

First, we consider the total power P_a absorbed by the object. This is given by (Fig. 10.A.1)

$$P_a = S_i\sigma_a = -\int_s \operatorname{Re}\left[\tfrac{1}{2}\bar{E}\times\bar{H}^*\right]\cdot d\bar{s}. \tag{10.A.3}$$

Electromagnetic Wave Propagation, Radiation, and Scattering: From Fundamentals to Applications,
Second Edition. Akira Ishimaru.
© 2017 by The Institute of Electrical and Electronic Engineers, Inc. Published 2017 by John Wiley & Sons, Inc.

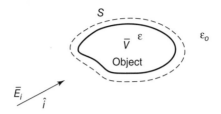

FIGURE 10.A.1 Forward scattering theorem.

Now, we note that

$$\bar{E} \times \bar{H}^* = \bar{E}_i \times \bar{H}_i^* + \bar{E}_s \times \bar{H}_s^* + \bar{E}_i \times \bar{H}_s^* + \bar{E}_s \times \bar{H}_i^*. \tag{10.A.4}$$

The scattered power P_s is given by

$$P_s = S_i \sigma_s = \int_s \mathrm{Re}\left[\tfrac{1}{2}\bar{E}_s \times \bar{H}_s^*\right] \cdot d\bar{s}. \tag{10.A.5}$$

We substitute (10.A.4) into (10.A.3) and note (10.A.5). We also use the following:

$$\int_s \bar{E}_i \times \bar{H}_i^* \cdot ds = 0,$$

$$\mathrm{Re}\left(\bar{E}_i \times \bar{H}_s^*\right) = \mathrm{Re}(\bar{E}_i^* \times \bar{H}_s),$$

$$\int_s \bar{E}_i \times \bar{H}_s^* \cdot d\bar{s} = \int_s \bar{E}_i \times \bar{H}^* \cdot d\bar{s}, \tag{10.A.6}$$

$$\int_s \bar{E}_s \times \bar{H}_i^* \cdot d\bar{s} = \int_s \bar{E} \times \bar{H}_i^* \cdot d\bar{s}.$$

We then get

$$S_i(\sigma_a + \sigma_s) = - \int_s \mathrm{Re}\tfrac{1}{2}\left[\bar{E}_i^* \times \bar{H} + \bar{E} \times \bar{H}_i^*\right] \cdot d\bar{s}. \tag{10.A.7}$$

Now noting

$$\int_s \bar{E}_i^* \times \bar{H} \cdot d\bar{s} = \int_v \nabla \cdot (\bar{E}_i^* \times \bar{H}) dV,$$

$$\nabla \cdot (\bar{E}_i^* \times \bar{H}) = \bar{H} \cdot \nabla \times \bar{E}_i^* - \bar{E}_i^* \cdot \nabla \times \bar{H}$$

$$= \bar{H} \cdot \left(j\omega\mu_0 \bar{H}_i^*\right) - \bar{E}_i^* \cdot (j\omega\varepsilon\bar{E}), \tag{10.A.8}$$

$$\nabla \cdot (\bar{E} \times \bar{H}_i^*) = \bar{H}_i^* \cdot \nabla \times \bar{E} - \bar{E} \cdot \nabla \times \bar{H}_i^*$$

$$= \bar{H}_i^* \cdot (-j\omega\mu_0 \bar{H}) - \bar{E} \cdot \left(-j\omega\varepsilon_0 \bar{E}_i^*\right).$$

We get

$$
\begin{aligned}
S_i(\sigma_a + \sigma_s) &= -\mathrm{Re} \int_v \frac{1}{2} [-j\omega(\varepsilon - \varepsilon_0)] \, \bar{E} \cdot \bar{E}_i^* \, dV \\
&= -\mathrm{Im} \int_v \frac{\omega(\varepsilon - \varepsilon_0)}{2} \, E \cdot \bar{E}_i^* \, dV.
\end{aligned}
\tag{10.A.9}
$$

However, the forward scattering amplitude $\bar{f}(\hat{i}, \hat{i})$ is given by

$$
\bar{f}(\hat{i}, \hat{i}) = \frac{k^2}{4\pi} \int_v \frac{\varepsilon - \varepsilon_0}{\varepsilon_0} \, \bar{E} e^{jk\bar{r}' \cdot \hat{i}} \, dV.
\tag{10.A.10}
$$

Noting $\bar{E}_i$ in (10.A.1), we get

$$
\bar{f}(\hat{i}, \hat{i}) \cdot \hat{e}_i = \frac{k^2}{4\pi} \int_v \frac{\varepsilon - \varepsilon_0}{\varepsilon_0} \, \bar{E} \cdot \bar{E}_i^* \, dV.
\tag{10.A.11}
$$

Combining (10.A.9) and (10.A.11), we finally get the forward scattering theorem (optical theorem)

$$
\sigma_a + \sigma_s = -\frac{4\pi}{k^2} \, \mathrm{Im} \, \bar{f}(\hat{i}, \hat{i}) \cdot \hat{e}_i.
\tag{10.A.12}
$$

APPENDIX TO CHAPTER 11

11.A BRANCH POINTS AND RIEMANN SURFACES

Let us examine the solution (11.34) obtained in Chapter 11. In particular, it is important to investigate the contour of the integration in more detail. To do this, we should first note that the integrand contains $\lambda = \sqrt{k^2 - h^2}$, and this raises the question of whether we should take $\lambda = +\sqrt{k^2 - h^2}$ or $-\sqrt{k^2 - h^2}$. To study this question in more detail, we consider a more general complex function of a complex variable h.

$$W = f(h). \tag{11.A.1}$$

If $f(h)$ is an analytic function within a region in the h plane, $f(h)$ is a single-valued function, and its value is uniquely determined by h (i.e., there is a one-to-one correspondence between W and h).

If $f(h)$ is a multivalued function of h, however, there are many different values of W for a given h. For example, letting $W = y$ (real) and $h = x$ (real), consider the case where

$$y^2 = x. \tag{11.A.2}$$

For a given value of x, there are two values of y : $y_1 = +\sqrt{x}$ and $y_2 = -\sqrt{x}$ (Fig. 11.A.1). We may call y_1 the first branch and y_2 the second branch. The point where these two branches meet is called the *branch point*.

Electromagnetic Wave Propagation, Radiation, and Scattering: From Fundamentals to Applications,
Second Edition. Akira Ishimaru.

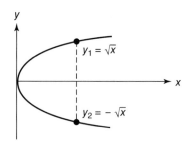

FIGURE 11.A.1 Two branches.

By using these two branches y_1 and y_2, we can keep the different branches apart so that we know exactly which branch we are dealing with. When W is a complex function of a complex variable h, each branch is represented by a different complex plane. We call these different planes of a multivalued function the *Riemann surfaces*. On each Riemann surface, the function is single valued and its value is uniquely determined.

Let us take an example.

$$W = \sqrt{h}. \tag{11.A.3}$$

Let us represent h, a complex variable, in the following polar form:

$$h = re^{j\theta}.$$

Then W is given by

$$W_1 = r^{1/2}e^{j(\theta/2)}.$$

But this is not the only value of W. In the h plane, we can go around the origin and thus add 2π to θ without changing the value of h.

$$h = re^{j\theta} = re^{j(\theta+2\pi)}.$$

But this second form of h gives a different value of W.

$$W_2 = r^{1/2}e^{j(\theta/2+\pi)} = -W_1.$$

If another 2π is added to θ, W will go back to W_1.

$$h = re^{j(\theta+4\pi)},$$
$$W = re^{j(\theta/2+2\pi)} = W_1.$$

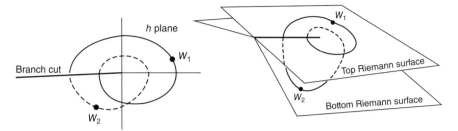

FIGURE 11.A.2   Riemann surfaces.

Therefore, for a given h, there are two W's: W_1 and W_2. We note that to transfer from W_1 to W_2, we must go around the origin once in the h plane. The origin in this example is the branch point.

To describe this situation more clearly and to tell which of these two values W_1 and W_2, we are dealing with, we introduce the idea of the *branch cut*. In the h plane, we imagine a cut extending from the branch point $h = 0$ to infinity. We say that as long as we do not cross the branch cut, we are on the first branch (the first Riemann surface), and the value of the function is W_1. To get W_2, we must cross the branch cut. The branch cut can be drawn in any direction from the origin, and it need not be a straight line. For convenience, we normally choose a cut along the negative real axis (Fig. 11.A.2). According to this choice of the branch cut, we define

$$\sqrt{h} = \begin{cases} r^{1/2}e^{j(\theta/2)} = W_1 & \text{for } -\pi < \theta < \pi, \\ -r^{1/2}e^{j(\theta/2)} = W_2 & \text{for } \pi < \theta < 3\pi, \\ W_1 & \text{for } 3\pi < \theta < 5\pi. \end{cases}$$

This situation is pictured in the two Riemann surfaces in Fig. 11.A.2. The top surface represents W_1 and the bottom surface W_2. These two surfaces are put together at the branch cut, indicating that when we cross the branch cut, W_2 must be used. As the branch cut is crossed the second time, we go back to the first Riemann surface and obtain W_1.

Another example of a multivalued function is

$$W = \ln h.$$

We note that for a given h,

$$h = re^{j\theta} = re^{j(\theta+2n\pi)}.$$

We get an infinite number of W's.

$$W_n = \ln r + j(\theta + 2n\pi), \quad n = 0, \pm 1, \pm 2, \dots, \pm\infty.$$

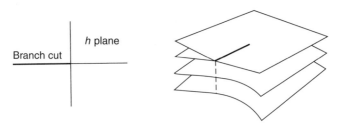

FIGURE 11.A.3 $W = \ln h$.

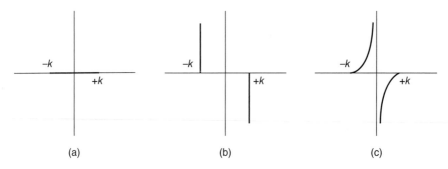

FIGURE 11.A.4   Three branch cuts in the h plane.

Therefore, there are an infinite number of Riemann surfaces. As we go around the branch point $h = 0$, we move to the next Riemann surface. This is pictured in Fig. 11.A.3.

Now consider $\lambda = \sqrt{k^2 - h^2}$. We note that there are two branch points $h = +k$ and $-k$. We may have the branch cuts as shown in Fig. 11.A.4. The cut in Fig. 11.A.4a is not acceptable on physical grounds. The cuts in both part (b) and part (c) are acceptable. Cut (b) is along the line Re $h = \pm$ Re k and cut (c) is along the curve Im $\lambda = 0$.

11.B   CHOICE OF THE CONTOUR OF INTEGRATION AND THE BRANCH CUT

Equation (11.34) has the general form

$$I = \int_c C_n(h) H_n^{(2)}(\lambda\rho) e^{-jhz} dh, \quad \lambda = \sqrt{k^2 - h^2}. \tag{11.B.1}$$

Because of $\lambda = \sqrt{k^2 - h^2}$, the integrand has two branch points at $h = \pm k$.

In this section we show how the contour c in (11.B.1) should be drawn and will discuss the reason for this choice. First, we note that due to the branch points, we can have two Riemann surfaces, and therefore we must choose the contour in such a manner that the inverse transform (11.B.1) exists. This choice must be made on the basis of physical argument.

First, we note that the integrand in (11.B.1) represents a wave with propagation constant h in the z direction. At a large distance from the origin, we approximate the Hankel function by

$$H_n^{(2)}(z) = \sqrt{\frac{2}{\pi z}} e^{-jz+j(2n+1)\pi/4}, \tag{11.B.2}$$

and we write

$$I \approx \int_c C_n(h) \sqrt{\frac{2}{\pi \lambda \rho}} e^{j(2n+1)\pi/4 - j\lambda\rho - jhz} dh. \tag{11.B.3}$$

If this equation is to represent a physical situation, the wave represented by the integrand must be an outgoing wave, and no wave should come back from infinity. This is the so-called *radiation condition*.

To satisfy this condition, the branch must be chosen so that Λ is in the fourth quadrant of the complex plane,

$$\text{Re}(\lambda) \geq 0 \quad \text{and} \quad \text{Im}(\lambda) \leq 0, \tag{11.B.4}$$

because the positive real part represents the outgoing phase progression, and the negative imaginary part represents the attenuation toward infinity.

$$e^{-j\lambda\rho} = e^{-j\lambda_r \rho + \lambda_i \rho}, \tag{11.B.5}$$

where $\lambda = \lambda_r + j\lambda_i$ and $\lambda_r \geq 0$ and $\lambda_i \leq 0$. Now the choice of the path of integration and branch cut is made such that λ is always in the fourth quadrant.

Let us consider a point Q in the h plane (Fig. 11.B.1). We note that

$$h - k = r_1 e^{j\theta_1},$$
$$h + k = r_2 e^{j\theta_2}. \tag{11.B.6}$$

Now consider λ.

$$\lambda_1 = \sqrt{k^2 - h^2} = \sqrt{(k-h)(k+h)}$$
$$= \sqrt{-r_1 r_2 e^{j(\theta_1 + \theta_2)}}. \tag{11.B.7}$$

Because of the square root, we can have two values for λ.

$$\lambda_1 = \sqrt{r_1 r_2} e^{j(\theta_1 + \theta_2)/2 - j(\pi/2)},$$
$$\lambda_2 = \sqrt{r_1 r_2} e^{j(\theta_1 + \theta_2)/2 + j(\pi/2)}. \tag{11.B.8}$$

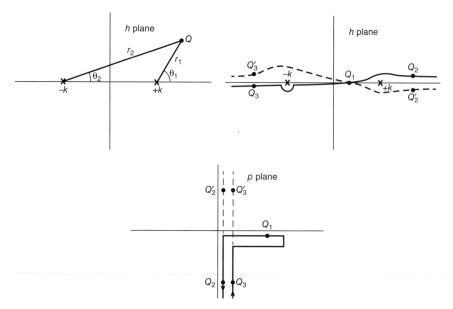

FIGURE 11.B.1   Choice of contour.

Let us first determine which of these two, λ_1 or λ_2, satisfies the radiation condition. To do this, let us consider a point Q_1 on the contour between $+k$ and $-k$. At Q_1, $\theta_1 = \pi$ and $\theta_2 = 0$ and thus

$$\lambda_1 = \sqrt{r_1 r_2} \quad \text{and} \quad \lambda_2 = \sqrt{r_1 r_2} e^{j\pi}.$$

λ_2 is not in the fourth quadrant, therefore λ_1 must be chosen. Next consider the path from Q_1 to Q_2 going around the branch point at $+k$ as shown. Then at Q_2, $\theta_1 = 0$ and $\theta_2 = 0$, and therefore

$$\lambda_1 = \sqrt{r_1 r_2} e^{-j(\pi/2)},$$

which is in the fourth quadrant and satisfies the radiation condition.

On the other hand, if the path going from Q_1 to Q'_2 is chosen, then at Q'_2, $\theta_1 = 2\pi$ and $\theta_2 = 0$. Thus at Q'_2,

$$\lambda_1 = \sqrt{r_1 r_2} e^{j(\pi/2)},$$

which is not in the fourth quadrant and does not satisfy the radiation condition.

In a similar manner, at Q_3, $\theta_1 = \pi$ and $\theta_2 = -\pi$ and therefore,

$$\lambda_1 = \sqrt{r_1 r_2} e^{-j(\pi/2)},$$

while at Q_3',

$$\lambda_1 = \sqrt{r_1 r_2} e^{j(\pi/2)}.$$

Thus to satisfy the radiation condition, the path $Q_3 - Q_1 - Q_2$ must be chosen (Fig. 11.B.1).

To ensure that the path of integration is as specified above, we draw branch cuts from the branch points and require that the path of integration not cross the cuts. We may have the cuts drawn vertically from the branch points (Fig. 11.A.4b) or any other choice of the branch cuts is satisfactory, as long as the original contour does not cross them. However, when we evaluate the integral, we must often deform the path of integration in some convenient manner. Then since the path may move off the real axis of the h plane, one choice of branch cuts is more convenient than the other.

One convenient choice of the cuts is to draw the cuts along the curve

$$\text{Im}(\lambda) = 0. \tag{11.B.9}$$

To clarify this situation, let us assume a small negative imaginary part of k.

$$k = k_r + jk_i. \tag{11.B.10}$$

Since λ and h must satisfy the relation

$$\lambda^2 + h^2 = k^2, \tag{11.B.11}$$

substituting (11.B.10) and $\lambda = \lambda_r + j\lambda_i$, and $h = h_r + jh_i$ in (11.B.11) and equating the real and imaginary parts of both sides, we obtain

$$\lambda_r^2 - \lambda_i^2 + h_r^2 - h_i^2 = k_r^2 - k_i^2,$$
$$\lambda_r \lambda_i + h_r h_i = k_r k_i. \tag{11.B.12}$$

Now let us consider the condition $\text{Im}(\lambda) = \lambda_i = 0$. In this case we must have

$$\lambda_r^2 + h_r^2 - h_i^2 = k_r^2 - k_i^2,$$
$$h_r h_i = k_r k_i. \tag{11.B.13}$$

Equation (11.B.13) is an equation for a hyperbola going through the point $h = k$ (Fig. 11.5). We note that for the curve from $+k$ to A, $|k_r| > |h_r|$ and $|h_i| > |k_i|$, and this satisfies (11.B.13) but for the curve from $+k$ to B, $|h_r| > |k_r|$ and $|h_i| < |k_i|$ and (11.B.13) cannot be satisfied. Thus the curve for $\text{Im}(\lambda) = 0$ should be drawn from $+k$ to A.

Similarly, the curve for the condition $\text{Re}(\lambda) = 0$ should be drawn from $+k$ to B. We need to keep in mind that this choice of the contour is based on the physical consideration that the wave must satisfy the radiation condition. If other physical situations are considered, the choice of the contour may be different.

For example, if h and $\pm k$ are replaced by ω and $\pm\omega_0$, this represents a transient wave in a waveguide or a magnetoplasma. In this case, the wave may be given by

$$u(x, t) = \int_C A(\omega)e^{j\omega t - jk(\omega)x}\,d\omega,$$

where

$$k(\omega) = \left[\left(\frac{\omega}{c}\right)^2 - \left(\frac{\pi}{a}\right)^2\right]^{1/2}.\tag{11.B.14}$$

The physical consideration in this case is that no wave should propagate faster than the light velocity, and thus $u(x, t) = 0$ for $t < x/c$. To ensure this condition, the contour C should be drawn to include all the singularities on one side (Fig. 11.B.2).

Note that for $|\omega|$ large, we have

$$\left|e^{j\omega t - j(w/c)\sqrt{1-(\omega_p^2/\omega^2)}x}\right| \Rightarrow e^{-\omega_i(t-(x/c))},$$

with $\omega = \omega_r + j\omega_i$, and this goes to zero along $C_-(\omega_i < 0)$ for $t < x/c$ and thus, since there are no singularities between C and C_-, the integral along C is equal to the integral along C_-, which is zero. This is, of course, the same as the Laplace transform ($j\omega = S$).

For spatial frequency h, there is no preferred direction, so the function should exist for $x < 0$ as well as for $x > 0$. This is the basis for the choice in Figs. 11.4 and 11.5.

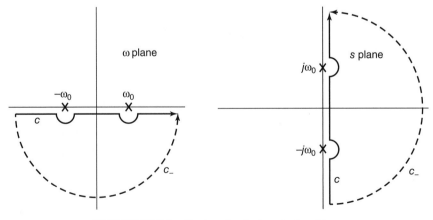

FIGURE 11.B.2 Contour for the transient problem.

11.C SADDLE-POINT TECHNIQUE AND METHOD OF STATIONARY PHASE

The complex integrals appearing in Section 11.B.1, such as

$$I_1 = \int_{-\infty}^{\infty} C_n(h) H_n^{(2)}(\sqrt{k^2 - h^2}\rho) e^{-jhz} dh, \qquad (11.C.1)$$

are in general extremely difficult to evaluate. However, this evaluation can be done quite easily for the field far from the source. This is usually referred to as the *radiation pattern* and is one of the most important characteristics of radiating systems. The technique used to obtain the radiation pattern is called the saddle-point technique. This technique is useful not only for particular cylindrical problems but for many other problems involving the evaluation of a contour integral. In this section we describe this technique.

Let us consider the following complex integral:

$$I = \int_c F(\alpha) \exp[Zf(\alpha)] d\alpha, \qquad (11.C.2)$$

where Z is a large positive real number and $f(\alpha)$ and $F(\alpha)$ are complex functions of the complex variable α. $F(\alpha)$ is a slowly varying function of α. Writing $f = f_r + jf_i$, we get

$$e^{Zf(\alpha)} = e^{Zf_r}(\cos Zf_i + j \sin Zf_i). \qquad (11.C.3)$$

Therefore, for a large Z, as f_i varies along the contour, the integrand oscillates very rapidly, and the evaluation of the integral would be extremely difficult.

However, there is a way to avoid this difficulty. If we deform the original contour in such a manner that the imaginary part f_i is constant along the new path, there is no rapid oscillation of the integrand. In this case, $\exp[Zf(\alpha)]$ varies slowly as $\exp(Zf_r)$ starts from zero, attains some value, and declines to zero again at the end of the contour.

The point in the complex plane where the real part becomes maximum should be the saddle point, as will be shown below. To study the behavior of $f(\alpha)$ near this point, let us first consider a point $\alpha = \alpha_s$ where

$$\frac{df(\alpha)}{d\alpha} = 0. \qquad (11.C.4)$$

We expand $f(\alpha)$ about this point,

$$f(\alpha) = f(\alpha_s) + (\alpha - \alpha_s)f'(\alpha_s) + \frac{(\alpha - \alpha_s)^2}{2!}f''(\alpha_s) + \cdots, \qquad (11.C.5)$$

and in the neighborhood of $\alpha = \alpha_s$, we can neglect the higher-order terms. Thus we get

$$f(\alpha) = f(\alpha_s) + \frac{(\alpha - \alpha_s)^2}{2!} f''(\alpha_s). \tag{11.C.6}$$

We note that as the angle of $\alpha - \alpha_s$ increases, the corresponding increase of the angle of $f(\alpha) - f(\alpha_s)$ is twice that for $\alpha - \alpha_s$, and thus the angle $\pi/2$ in the α plane around α_s corresponds to π in the f plane.

The valley region, where $f_r(\alpha) < f_r(\alpha_s)$ in the f plane, is represented by two regions in the α plane about the saddle point. The mountain region $f_r(\alpha) > f_r(\alpha_s)$ is also represented by two regions in the α plane, and these four regions meet at the saddle point, each occupying the angle of $\pi/2$.

It is then clear that the point $\alpha = \alpha_s$ is the saddle point as shown in Fig. 11.C.1. This is also clear considering that $f(\alpha)$ is an analytic function. An analytic function

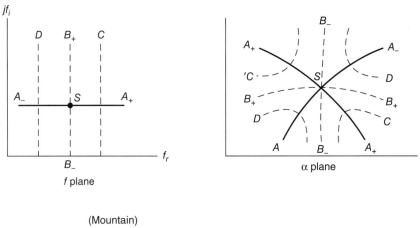

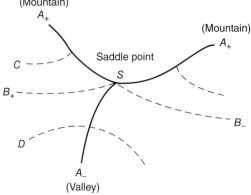

FIGURE 11.C.1 Saddle point S.

cannot have maxima or minima within the range of analyticity. This situation is mathematically identical to two-dimensional electrostatic problems. A two-dimensional electrostatic potential U and a flux function V can be represented by an analytic function $W = U + jV$ and U and V satisfy the Laplace equations, and it is clear that there is no maximum potential U in the region of analyticity. Only a singularity such as the electric charge can create maximum or minimum potentials. A point where $\partial f/\partial\alpha = 0$ is therefore a saddle point.

We now wish to deform the original contour into the path along which the imaginary part f_i is constant. As can be seen from Fig. 11.C.1, there are two perpendicular lines (SA_+ and SA_-) in the α plane along which f_i is constant. The path (SA_+) goes up the mountains on both sides of the saddle point, and the other (SA_-) goes down to the valleys on both sides. Obviously, the deformed path should be chosen to originate in one valley, cross the saddle point, and terminate in the other valley. This is the path $A_- - S - A_-$. As we note in Fig. 11.C.1, this path of constant f_i is in fact the path along which the real part f_r decreases fastest from the saddle point. Thus the path is called the *steepest descent path*, and the saddle-point technique is often called the *method of steepest descent*.

Let us evaluate the integral along this steepest descent contour (SDC):

$$I = \int_{SDC} F(\alpha)e^{Zf(\alpha)}d\alpha. \qquad (11.C.7)$$

We write

$$f(\alpha) = f(\alpha_s) + \frac{(\alpha - \alpha_s)^2}{2!}f''(\alpha_s),$$

where

$$\frac{df}{d\alpha} = 0 \quad \text{at } \alpha = \alpha_s \text{ saddle point.}$$

First, we assume that $F(\alpha)$ is a slowly varying function, and that its value near α_s is almost constant. Then, approximately, we get

$$I = F(\alpha_s)e^{Zf(\alpha_s)}\int_{SDC} \exp\left[Z\frac{(\alpha - \alpha_s)^2}{2!}f''(\alpha_s)\right]d\alpha. \qquad (11.C.8)$$

To study the exponent in the integral, let us note that along the contour of the constant phase, the imaginary part should be constant and equal to its value at α_s.

$$f_i(\alpha) = f_i(\alpha_s). \qquad (11.C.9)$$

Therefore, along the path, we have

$$\mathrm{Re}\, f(\alpha) = \mathrm{Re}f(\alpha_s) + \frac{(\alpha - \alpha_s)^2}{2!} f''(\alpha_s).$$

Now we note that $\mathrm{Re}f(\alpha)$ must decrease from the value at the saddle point $\mathrm{Re}f(\alpha_s)$ and therefore the path must be chosen such that

$$\frac{(\alpha - \alpha_s)^2}{2!} f''(\alpha_s)$$

must be not only real, but also negative. This choice can be made by properly orienting the path. We let

$$\alpha - \alpha_s = se^{j\Gamma},$$

where s is the distance measured from the saddle point and γ is the angle the path makes with the real axis. The exponent can be expressed by

$$\frac{Z(\alpha - \alpha_s)^2}{2} f''(\alpha_s) = Z\frac{s^2}{2} e^{j2\gamma} f''(\alpha_s)$$

$$= -\frac{Z[-e^{j2\gamma} f''(\alpha_s)]}{2} s^2.$$

We now choose γ such that

$$[-e^{j2}\Gamma f''(\alpha_s)] = P$$

is real and positive. With this choice, P is equal to $|f''(\alpha_s)|$. Then the integral becomes

$$I = F(\alpha_s)e^{Zf(\alpha_s)} \int_{SDC} \exp\left(-\frac{ZPs^2}{2} + j\gamma\right) ds. \qquad (11.\mathrm{C}.10)$$

This integral is in the form of the integral of a Gaussian curve, and for large Z, the integrand drops down very rapidly on both sides of $s = 0$. Therefore, the most contribution comes from the neighborhood of $s = 0$. Thus we approximate the integral by integrating over a small distance about $s = 0$.

$$I = F(\alpha_s)e^{Zf(\alpha_s)+j\gamma} \int_{-\varepsilon}^{\varepsilon} e^{-(ZP/2)s^2} ds.$$

After the change of variable from s to x,

$$\frac{ZP}{2} s^2 = x^2,$$

we get

$$I = F(\alpha_s)e^{Zf(\alpha_s)+j\gamma}\sqrt{\frac{2}{ZP}}\int_{\sqrt{ZP/2}\varepsilon}^{\sqrt{ZP/2}\varepsilon}e^{-x^2}dx.$$

For large Z, the limit of the integral can be extended to $-\infty$ and $+\infty$. Using

$$\int_{-\infty}^{\infty}e^{-x^2}dx = \sqrt{\pi},$$

we get finally

$$I = \int_c F(\alpha)e^{Zf(\alpha)}d\alpha$$

$$\cong F(\alpha_s)e^{Zf(\alpha_s)+j\gamma}\left[\frac{2\pi}{|f''(\alpha_s)|Z}\right]^{1/2}, \qquad (11.C.11)$$

where γ is chosen to make $[e^{j2\gamma}f''(\alpha_s)]$ real and negative, and $\partial f(\alpha_s)/\partial\alpha = 0$. Equation (11.C.11) is the final result of the approximate evaluation of the integral based on the method of steepest descent (the saddle-point technique) valid for large Z. There are two choices for γ. For example, if f'' has an angle of $\frac{1}{2}\pi$, then $2\gamma + \frac{1}{2}\pi = \pm\pi$, and we get $\gamma = \pi/4$ or $-3\pi/4$. The choice can be made by noting the path of integration (see Section 11.5).

We made several assumptions and approximations in deriving the result of the saddle-point integration. For example, we assumed that $F(\alpha)$ is a slowly varying function of α near $\alpha = \alpha_s$, and we approximated $F(\alpha)$ by $F(\alpha_s)$. We did not define what is meant by "slowly varying." We expanded $f(\alpha)$ in a Taylor's series and retained the first two terms. But what must be done if $f''(\alpha_s) = 0$? How do we justify taking the first two terms? Also, in one process we integrated from $-\varepsilon$ to $+\varepsilon$ under the assumption that the main contribution comes from this region. A more complete treatment must be done through the use of Watson's lemma.

Also, we did not account for the presence of singularities between the original contour and the deformed saddle-point contour. If the poles are present between these two contours, they may give rise to "surface waves" and "leaky waves." Furthermore, when the pole is close to the saddle point, we need a modified saddle-point technique. Wave propagation over the earth excited by a dipole, which is often referred to as a Sommerfeld problem, requires this technique. Also, the existence of a branch point may give rise to lateral waves. These are discussed in Chapter 15.

The method of stationary phase is equivalent to the method of steepest descent. As seen in Fig. 11.C.1, the steepest descent path is along the constant $f_i(SA_+, SA_-)$.

However, we may go through the saddle point along constant $f_r(SB_+, SB_-)$. Then writing

$$I = F(\alpha_s)e^{Zf(\alpha_s)} \int e^{Z[(\alpha-\alpha_s)^2/2]f''(\alpha_s)} d\alpha, \qquad (11.C.12)$$

we choose the path such that

$$f_r(\alpha) = f_r(\alpha_s) = \text{constant}. \qquad (11.C.13)$$

and then we choose $\alpha - \alpha_s = se^j\gamma$ such that

$$\frac{Z(\alpha - \alpha_s)^2}{2}f''(\alpha_s) = j\left[Z\frac{s^2}{2}e^{j2\gamma - j(\pi/2)}f''(\alpha_s)\right]$$

$$= j\frac{Z}{2}Qs^2,$$

is purely imaginary. Then

$$I = F(\alpha_s)e^{Zf(\alpha_s)+j\gamma} \int e^{j(Z/2)Qs^2} ds,$$

which is evaluated to give

$$I \sim F(\alpha_s)e^{Zf(\alpha_s)+j\gamma+j(\pi/4)} \left[\frac{2\pi}{|f''(\alpha_s)|Z}\right]^{1/2}, \qquad (11.C.14)$$

where γ is chosen so that

$$[e^{j2\gamma - j(\pi/2)}f''(\alpha_s)]$$

is real.

Obviously, $\gamma = \gamma' + \pi/4$, and the method of steepest descent and the method of stationary phase give the same result. The method of stationary phase is used, for example, to study reflections from curved surfaces where the main contribution comes from the portion of the reflector where the phase variation is stationary.

The method of stationary phase is usually used when the integral is expressed in the following form:

$$I = \int_c F(\alpha)e^{jZg(\alpha)} d\alpha. \qquad (11.C.15)$$

The stationary phase point $\alpha = \alpha_s$ is then given by

$$\frac{\partial}{\partial\alpha}g(\alpha) = 0. \qquad (11.C.16)$$

The integral is then evaluated to give

$$I \cong F(\alpha_s)e^{jZg(\alpha_s)+j\gamma'\pm j(\pi/4)}\left[\frac{2\pi}{|g''(\alpha_s)|Z}\right]^{1/2}, \qquad (11.C.17)$$

where γ' is chosen such that

$$[e^{j2\gamma'}g''(\alpha_s)]$$

is real and positive (negative) for the upper (lower) sign in the exponent. Thus if $g'' > 0$, $\gamma' = 0$ and $+j\frac{1}{4}\pi$ must be used, and if $g'' < 0$, $\gamma' = 0$ and $-j\frac{1}{4}\pi$ must be used.

11.D COMPLEX INTEGRALS AND RESIDUES

This appendix gives a short summary of important definitions and theorems for complex functions.

1. Analytic functions $w = f(z)$, $z = x + jy$. If $f(z)$ possess a derivation at $z = z_0$ and in its neighborhood, then $f(z)$ is called the *analytic function*.
2. If $w = f(z) = u(x, y) + jv(x, y)$ is analytic, then u and v satisfy the *Cauchy–Riemann equation*.

$$\begin{aligned}\frac{\partial u}{\partial x} &= \frac{\partial v}{\partial y}, \\ \frac{\partial u}{\partial y} &= -\frac{\partial v}{\partial x}.\end{aligned} \qquad (11.D.1)$$

Combining these two, we get

$$\nabla^2 u = 0 \quad \text{and} \quad \nabla^2 v = 0. \qquad (11.D.2)$$

The real part and the imaginary part of an analytic function satisfy the two-dimensional Laplace equation.

3. Cauchy's integral theorem. If $f(z)$ is analytic in a domain D, then for every closed path C in D (Fig. 11.D.1)

$$\oint f(z)dz = 0. \qquad (11.D.3)$$

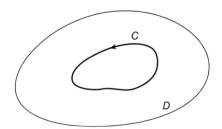

FIGURE 11.D.1   Cauchy's integral theorem.

4. Cauchy's integral formula. If $f(z)$ is analytic in D, then

$$f(z_0) = \frac{1}{2\pi j} \int_C \frac{f(z)}{z - z_0} dz,$$

$$f^{(n)}(z_0) = \frac{n!}{2\pi j} \int_C \frac{f(z)}{(z - z_0)^{n+1}} dz.$$

(11.D.4)

5. If $f(z)$ is analytic in D, $f(z)$ can be expanded in a Taylor's series.

$$f(z) = f(a) + f'(a)(z - a) + f''(a)\frac{(z - a)^2}{2!} + \cdots .$$

(11.D.5)

The radius of convergence of the series is the distance from a to the nearest singularity of $f(z)$.

6. Residue. If $f(z)$ has an isolated singularity at $z = a$, $f(z)$ can be expanded in a Laurent series,

$$f(z) = \sum_{n=0}^{\infty} c_n(z - a)^n + \frac{c_{-1}}{z - a} + \frac{c_{-2}}{(z - a)^2} + \cdots .$$

(11.D.6)

This is valid in the domain $0 < |z - a| < R$, where R is the distance from a to the nearest singularity of $f(z)$. The coefficient C_{-1} is called the *residue* of $f(z)$ at $z = a$.

7. If $f(z)$ has a simple pole at a, the residue is given by

$$c_{-1} = \lim_{z \to a}(z - a)f(z)$$

$$= \frac{N(a)}{D'(a)},$$

where $f(z) = N(z)/D(z)$.

8. If $f(t)$ has mth-order poles

$$f(z) = \sum_{n=0}^{\infty} c_n (z-a)^n + \frac{c_{-1}}{(z-a)} + \cdots + \frac{c_{-m}}{(z-a)^m},$$

then the residue is given by

$$c_{-1} = \frac{1}{(m-1)!} \lim_{z \to a} \frac{d^{m-1}}{dz^{m-1}} [(z-a)^m f(z)].$$

9. Residue theorem. If $f(z)$ is analytic inside a closed path C except for finite number of singular points $a_1, a_2, \ldots, a_m$, then

$$\int_C f(z)dz = 2\pi j \sum_{i=1}^{m} \text{residue at } a_i.$$

The integral is taken counterclockwise.

APPENDIX TO CHAPTER 12

12.A IMPROPER INTEGRALS

If the integrand is unbounded at several points in the interval of integration, the integral is called "improper." Consider the following improper integral:

$$I = \int_a^b \frac{dx}{x^\alpha} \quad \text{and } a < 0 \quad \text{and} \quad b > 0.$$

If $\alpha < 1$, the following limit exists and is finite.

$$\lim_{\varepsilon_1 \to 0} \int_a^{-\varepsilon_1} \frac{dx}{x^\alpha} + \lim_{\varepsilon_2 \to 0} \int_{\varepsilon_2}^b \frac{dx}{x^\alpha} = \frac{b^{1-\alpha} - a^{1-\alpha}}{1 - \alpha}.$$

This limit is taken as the value of the integral I. The integral is said to be convergent and *weakly singular*. If $\alpha = 1$, then the following limit exists:

$$\lim_{\varepsilon \to 0} \left(\int_a^{-\varepsilon} \frac{dx}{x} + \int_\varepsilon^b \frac{dx}{x} \right) = \lim_{\varepsilon \to 0} \left(\ln \frac{\varepsilon}{-a} + \ln \frac{b}{\varepsilon} \right) = \ln \frac{b}{|a|}.$$

Note that in the above, we used the same ε for both sides of $x = 0$. This is essential to obtain the finite limit. The integral is said to be *singular in a Cauchy sense*, and the

Electromagnetic Wave Propagation, Radiation, and Scattering: From Fundamentals to Applications,
Second Edition. Akira Ishimaru.
© 2017 by The Institute of Electrical and Electronic Engineers, Inc. Published 2017 by John Wiley & Sons, Inc.

value above is called the *Cauchy principal value*. If $\alpha > 1$, the integral is said to be *strongly singular*.

The integral $I(\bar{r})$,

$$I(\bar{r}) = \int_V f(\bar{r}, \bar{r}')\, dV', \tag{12.A.1}$$

is called the *improper integral* if $f(\bar{r}, \bar{r}')$ becomes unbounded at a point $\bar{r}' = \bar{r}$ within the volume V. The integral $I(\bar{r})$ is convergent if

$$\lim_{v \to 0} \int_{V-v} f(\bar{r}, \bar{r}')\, dV' \tag{12.A.2}$$

exists, where v is a small volume surrounding $\bar{r}' = \bar{r}$.

If we let $|\bar{r} - \bar{r}'| = R$,

$$\int_V \frac{dV'}{R^\beta} \quad \text{is convergent if } 0 < \beta < 3 \tag{12.A.3}$$

and

$$\int_S \frac{dS}{R^\alpha} \quad \text{is convergent if } 0 < \alpha < 2. \tag{12.A.4}$$

Next consider the normal derivative

$$I_+ = \frac{\partial}{\partial n} \int_S f(\bar{r}')G_0(\bar{r}, \bar{r}')\, dS' \tag{12.A.5}$$

as $\bar{r}$ approaches the surface S from the side $z > 0$. We let $S = (S - \sigma) + \sigma$, where σ is a small area shown in Fig. 12.A.1. Then $G_0 \to R^{-1} = (z^2 + x^2 + y^2)^{-1/2}$ and

$$\frac{\partial}{\partial n}\left(\frac{1}{R}\right) = \frac{\partial}{\partial z}\left(\frac{1}{R}\right) = -\frac{1}{R^2}\frac{z}{R},$$

$$dS' = R^2\, d\Omega(R/z),$$

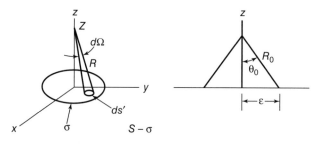

FIGURE 12.A.1 Improper integral.

where we used $dS' = 2\pi r dr = 2\pi R dR$, $R = z/\cos\theta$, and $d\Omega = 2\pi \sin\theta d\theta$. Therefore, we get

$$\frac{\partial}{\partial n}\int_\sigma fG_0\,dS' = -\int_\sigma f(\bar{r})\frac{d\Omega}{4\pi} = -\frac{f(\bar{r})}{2},$$

$$I_+ = -\frac{f(\bar{r})}{2} + \int_{S-\sigma} f(\bar{r}')\frac{\partial}{\partial n}G_0(\bar{r},\bar{r}')\,dS'.$$

(12.A.6)

The integral over $S - \sigma$ is called the Cauchy principal value and is expressed as

$$\fint f(\bar{r}')\frac{\partial}{\partial n}G_0(\bar{r},\bar{r}')\,dS'.$$

The integral is convergent because $(\partial/\partial n)G_0$ is zero near $\bar{r} = \bar{r}'$ and $z = 0$.

Note also that if $\bar{r}$ approaches the surface S from the side $z < 0$, we get

$$I_- = +\frac{f(\bar{r})}{2} + \int_{S-\sigma} f(\bar{r}')\frac{\partial}{\partial n}G_0(\bar{r},\bar{r}')\,dS'. \qquad (12.A.7)$$

Therefore, we get

$$I_+ - I_- = -f(\bar{r}). \qquad (12.A.8)$$

Consider the integral

$$I = \frac{\partial}{\partial n}\int_\sigma \frac{\partial}{\partial n'}G_0(\bar{r},\bar{r}')f(\bar{r}')\,dS', \qquad (12.A.9)$$

where σ is a small area with radius ε. Then using the results above and noting that $\partial/\partial n' = \partial/\partial z'$ rather than $\partial/\partial z$,

$$I = -\frac{\partial}{\partial n}\Omega f(\bar{r}), \qquad (12.A.10)$$

where Ω is the solid angle given by Fig. 12.A.1

$$\Omega = 2\pi(1 - \cos\theta_0) = 2\pi\left(1 - \frac{z}{R_0}\right).$$

Therefore, we get

$$I = -\frac{\partial}{\partial z}\Omega f = 2\pi\frac{\varepsilon^2}{R_0^3}f, \qquad (12.A.11)$$

where we used

$$\frac{1}{2\pi}\frac{\partial\Omega}{\partial z} = -\frac{1}{R_0} + \frac{z}{R_0^2}\frac{\partial R_0}{\partial z}.$$

As $z \to 0$, I approaches $(2\pi/\varepsilon)f(\bar{r})$, and thus I depends on the size of the area σ.

12.B INTEGRAL EQUATIONS

Integral equations are classified as follows:

1. Fredholm integral equation of the first kind.

$$\int_a^b K(x, x')f(x')\, dx' = g(x), \tag{12.B.1}$$

where $K(x, x')$ is a known function called the *kernel function*. $f(x')$ is the unknown and $g(x)$ is the given function. The solution depends greatly on the kernel function. Laplace transform is an example.

$$\int_0^\infty e^{-st}f(t)\, dt = \phi(s), \tag{12.B.2}$$

where $\exp(-st)$ is the kernel. The solution is given by

$$f(t) = \frac{1}{2\pi j}\int_C \phi(s)e^{st}\, ds. \tag{12.B.3}$$

For a more general kernel function, analytical solution is difficult to obtain.

2. Inhomogeneous Fredholm integral equation of the second kind.

$$\int_a^b k(x, x')f(x')\, dx' = f(x) + g(x). \tag{12.B.4}$$

3. Homogeneous Fredholm integral equation of the second kind.

$$\int_a^b K(x, x')f(x')dx' = \lambda f(x). \tag{12.B.5}$$

This constitutes an eigenvalue equation where Λ is the eigenvalue.

4. In the above, if the upper limit b is replaced by the variable x, this is called the *Volterra integral equation*. For example, the Volterra integral equation of the first kind is given by

$$\int_a^x K(x, x')f(x')\, dx' = g(x).$$

APPENDIX TO CHAPTER 14

14.A STATIONARY-PHASE EVALUATION OF A MULTIPLE INTEGRAL I

$$I = \int_{-\infty}^{\infty} dx_1 \int_{-\infty}^{\infty} dx_2 \cdots \int_{-\infty}^{\infty} dx_N A(x_1 \cdots x_N) e^{if(x_1 x_2 \cdots x_N)}. \qquad (14.A.1)$$

First we find a stationary-phase point $(x_{10}, x_{20}, x_{30}, \ldots, x_{N0})$ by satisfying N equations,

$$\frac{\partial f}{\partial x_1} = \frac{\partial f}{\partial x_2} = \frac{\partial f}{\partial x_3} = \cdots = \frac{\partial f}{\partial x_N} = 0. \qquad (14.A.2)$$

Then we expand f about this stationary-phase point.

$$f(x_1, x_2, \ldots, x_N) = f(x_{10}, x_{20}, \ldots, x_{N0})$$

$$+ \frac{1}{2!} \left[(x_1 - x_{10})\frac{\partial}{\partial x_1} + (x_2 - x_{20})\frac{\partial}{\partial x_2} + \cdots + (x_N - x_{N0})\frac{\partial}{\partial x_N} \right]^2 f \bigg|_{x_{10}, x_{20}, \ldots}$$

$$+ \text{higher-orderterms}.$$

Electromagnetic Wave Propagation, Radiation, and Scattering: From Fundamentals to Applications,
Second Edition. Akira Ishimaru.

These higher-order terms contribute little to the integral unless the second derivatives of f are small. Furthermore, we assume that the amplitude $A(x_1 \text{ L } x_N)$ is a slowly varying function of $x_1 \text{ L } x_N$, and thus we approximate

$$A(x_1 \ldots x_N) \approx A(x_{10}, x_{20}, \ldots, x_{N0}).$$

Then we write, letting

$$x_1 - x_{10} = x_1', \quad x_2 - x_{20} = x_2', \quad \text{etc.},$$

$$I = A(x_{10}, x_{20}, \ldots, x_{N0}) e^{if(x_{10}, x_{20}, \ldots, x_N)} \int_{-\infty}^{\infty} dx_1' \int_{-\infty}^{\infty} dx_2' \cdots \int_{-\infty}^{\infty} dx_N' e^{j(1/2)[T]},$$

$$(14.A.3)$$

where

$$[T] = \left(x_1' \frac{\partial}{\partial x_1'} + x_2' \frac{\partial}{\partial x_2'} + \cdots + x_N' \frac{\partial}{\partial x_N'} \right)^2 f.$$

we write $[T]$ in the following matrix form:

$$[T] = \tilde{x} F x$$

$$x = \begin{bmatrix} x_1' \\ \vdots \\ x_N' \end{bmatrix}, \quad F = \begin{bmatrix} f_{11} & f_{12} & \cdots & f_{1N} \\ f_{21} & & & \\ \vdots & & & \\ f_{N1} & \cdots & \cdots & f_{NN} \end{bmatrix}, \qquad (14.A.4)$$

where

$$f_{ij} = \left. \frac{\partial^2}{\partial x_i' \partial x_j'} f \right|_{x_1' = 0, x_2' = 0, \text{ etc.}}$$

Next we note that by the orthogonal transformation of X to Y,

$$X = PY \quad \text{and} \quad Y = \begin{bmatrix} y_1 \\ y_2 \\ \vdots \\ y_N \end{bmatrix}.$$

We can convert $[T]$ into the following diagonal form:

$$[T] = \tilde{X}FX$$
$$= \tilde{Y}\tilde{P}FPY \qquad (14.A.5)$$
$$= \tilde{Y}\alpha Y,$$

where

$$\alpha = \begin{bmatrix} \alpha_1^2 & 0 & 0 & 0 & 0 \\ 0 & \alpha_2^2 & 0 & 0 & 0 \\ 0 & 0 & \alpha_3^2 & 0 & 0 \\ 0 & 0 & 0 & \ddots & \alpha_N^2 \end{bmatrix}.$$

Thus we obtain

$$[T] = \alpha_1^2 y_1^2 + \alpha_2^2 y_2^2 + \cdots + \alpha_N^2 y_N^2. \qquad (14.A.6)$$

Also, note that the Jacobian of $x_1' \cdots x_N'$ with respect to $y_1 \, y_2 \, L \, y_N$ is 1.

$$dx_1' dx_2' \cdots dx_N' = \frac{\partial(x_1', \ldots, x_N')}{\partial(y_1, \ldots, y_N)} dy_1 \cdots dy_N,$$

$$\text{Jacobian} = \frac{\partial(x_1' \cdots x_N')}{\partial(y_1 \cdots y_N)}$$

$$= \begin{vmatrix} \dfrac{\partial x_1'}{\partial y_1} & \cdots & \dfrac{\partial x_N'}{\partial y_1} \\ \vdots & & \\ \dfrac{\partial x_1'}{\partial y_N} & \cdots & \dfrac{\partial x_N'}{\partial y_N} \end{vmatrix} = |P| = 1.$$

Therefore, we obtain

$$\int_{-\infty}^{\infty} dx_1' \int_{-\infty}^{\infty} dx_2' \cdots \int_{-\infty}^{\infty} dx_N' e^{j(1/2)[T]}$$

$$= \int_{-\infty}^{\infty} dy_1 \int_{-\infty}^{\infty} dy_2 \cdots \int_{-\infty}^{\infty} dy_N e^{j[\alpha_1^2 y_1^2 + \alpha_2^2 y_2^2 + \cdots + \alpha_N^2 y_N^2]} \qquad (14.A.7)$$

$$= \frac{(2\pi)^{N/2} e^{jN(\pi/4)}}{\sqrt{\alpha_1^2 \alpha_2^2 \cdots \alpha_N^2}}.$$

But since

$$\alpha_1^2 \alpha_2^2 \cdots \alpha_N^2 = |\alpha|$$
$$= |\tilde{P}FP|$$
$$= |\tilde{P}| |F| |P|$$
$$= |F|,$$

we obtain

$$I = A(x_{10}, x_{20}, \ldots, x_{N0}) e^{if(x_{10}, x_{20}, \ldots, x_{N0})} \frac{(2\pi)^{N/2} e^{jN(\pi/4)}}{\sqrt{\Delta}}, \qquad (14.\text{A}.8)$$

where Δ is the determinant of $F = |F|$, and is called *Hesse's determinant*. For $N = 2$, we have

$$I = \int_{-\infty}^{\infty} dx_1 \int_{-\infty}^{\infty} dx_2 A(x_1 \, x_2) e^{if(x_1 x_2)}$$
$$= A(x_{10} x_{20}) e^{if(x_{10} x_{20})} \frac{(2\pi) e^{j(\pi/2)}}{\sqrt{f_{11} f_{12} - f_{12}^2}}, \qquad (14.\text{A}.9)$$

where x_{10}, x_{20} are given by

$$\frac{\partial f}{\partial x_1} = 0 \quad \text{and} \quad \frac{\partial f}{\partial x_2} = 0.$$

APPENDIX TO CHAPTER 15

15.A SOMMERFELD'S SOLUTION

Sommerfeld's 1909 paper contains an error in sign for $\sqrt{p_1}$ (which is essentially $-s_p$ instead of $+s_p$, and this $-s_p$ corresponds to the pole on the bottom Riemann surface in the α plane). If we deliberately change $\sqrt{p_1}$ to $-\sqrt{p_1}$ in (15.63), we get

$$\pi_{er} = 2\frac{e^{-jkR}}{4\pi R}\left[1 + j\sqrt{p_1}\pi e^{-p_1}\ \mathrm{erfc}(-j\sqrt{p_1})\right]. \tag{15.A.1}$$

Using

$$\mathrm{erfc}(-z) = 2 - \mathrm{erfc}(z),$$

we write

$$\pi_{er} = 2\frac{e^{-jkR}}{4\pi R}\left[1 - j\sqrt{p_1}\pi e^{-p_1}\ \mathrm{erfc}\left(+j\sqrt{p_1}\right) + 2j\sqrt{p_1}\pi e^{-p_1}\right],$$

which differs from the correct π in (15.63) as follows:

$$\pi_{er} = \pi + 2\frac{e^{-jkR}}{4\pi R}\left(2j\sqrt{p_1}\pi e^{-p_1}\right)\Big]. \tag{15.A.2}$$

Electromagnetic Wave Propagation, Radiation, and Scattering: From Fundamentals to Applications,
Second Edition. Akira Ishimaru.
© 2017 by The Institute of Electrical and Electronic Engineers, Inc. Published 2017 by John Wiley & Sons, Inc.

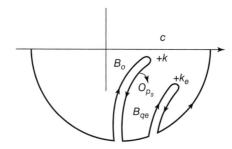

FIGURE 15.A.1 Branch cut integration and residue.

It is easy to show that the foregoing difference between π_{er} and π is precisely equal to $-2\pi j$ (the residue at the Sommerfeld pole, s_p). Thus we write

$$\pi = \pi_{er} - (-2\pi j \cdot \text{residue at } s_p). \qquad (15.\text{A}.3)$$

Now, in his 1909 paper, Sommerfeld evaluated the integral in the following form:

$$\pi_{er} = B_q + B_{qe} + (-2\pi j \cdot \text{residue at } s_p), \qquad (15.\text{A}.4)$$

where B_q and B_{qe} are the branch cut integration along $\text{Im} q = 0$ and $\text{Im} q_e = 0$, respectively (Fig. 15.A.1).

Now the term

$$(-2\pi j)(\text{residue at } s_p),$$

is precisely the Zenneck wave excited by the dipole source. But because of the sign error, the true field is π_{er} minus the Zenneck wave. Thus it *appeared* that the Zenneck wave contained in Sommerfeld's 1909 paper disappears if the correct sign is used. Therefore, it *appears* that the Zenneck wave does not exist.

As we pointed out earlier, this division (15.A.4) is arbitrary, and the fact that one term exhibits the Zenneck wave characteristic does not mean that the wave exists. In fact, the question of the existence of a wave based on one term of the total wave is meaningless, because what is important is the total wave, not a portion of the wave. However, this sign error created controversies over the existence of the Zenneck wave for many decades (see Banos, 1966, p. 154). This error in sign was not detected until 1935 by Norton, even though Sommerfeld has the correct sign in his 1926 paper. The entire question of the controversies was finally resolved around 1950 through work of van der Waeden, Ott, and Banos (see Banos, 1966).

15.B RIEMANN SURFACES FOR THE SOMMERFELD PROBLEM

In the book, the evaluation of the integral was performed in the α plane. It is instructive to see how the integration and the various Riemann surfaces may appear in the γ plane.

In the γ plane, the branch points appear at

$$\gamma = \pm k \quad \text{and} \quad \gamma = \pm k_e. \tag{15.B.1}$$

Let us now draw the branch cuts from these branch points along

$$\operatorname{Im} q = \operatorname{Im} \sqrt{k^2 - \lambda^2} = 0,$$

and

$$\operatorname{Im} q_e = \operatorname{Im} \sqrt{k_e^2 - \lambda^2} = 0. \tag{15.B.2}$$

As shown in Chapter 15, these branch cuts are hyperbolas passing through the branch points. It is obvious that the contour of the integration must be on the Riemann surface where

$$\operatorname{Im} q < 0 \quad \text{and} \quad \operatorname{Im} q_e < 0. \tag{15.B.3}$$

In this case, the wave attenuates as $z \to +\infty$ and $z \to -\infty$ and thus satisfies the *radiation condition*.

$$|e^{-jqz}| = e^{(\operatorname{Im} q)z} \to 0 \quad \text{as } z \to +\infty,$$

$$|e^{jq_ez}| = e^{-(\operatorname{Im} q_e)z} \to 0 \quad \text{as } z \to -\infty.$$

This surface is therefore called the *proper Riemann surface*.

In addition, however, there are three other Riemann surfaces on which the radiation condition is not satisfied, and therefore these are called the *improper Riemann surfaces*. We list these four Riemann surfaces:

   I. Proper Riemann surface: $\operatorname{Im} q < 0$, $\operatorname{Im} q_e < 0$.
  II. Improper Riemann surface: $\operatorname{Im} q > 0$, $\operatorname{Im} q_e < 0$.
 III. Improper Riemann surface: $\operatorname{Im} q < 0$, $\operatorname{Im} q_e > 0$.
 IV. Improper Riemann surface: $\operatorname{Im} q > 0$, $\operatorname{Im} q_e > 0$.

The first two belong to $\operatorname{Im} q_e < 0$ and the last two belong to $\operatorname{Im} q_e > 0$. Let us consider the first two surfaces. We first examine the complex q plane (Fig. 15.B.1). We label each quadrant as B_2, B_1, T_2, and T_1. T_1 and T_2 belong to the top surface (I) and B_1 and B_2 belong to the bottom surface (II) in the Λ plane. The corresponding α plane can be obtained by noting that

$$\lambda = k \sin \alpha \quad \text{and} \quad q = k \cos \alpha.$$

As we go around in the q plane from A to B to C to D, we can trace the corresponding locus in surfaces I and II of the Λ plane and the α plane (Fig. 15.B.1). Note that in

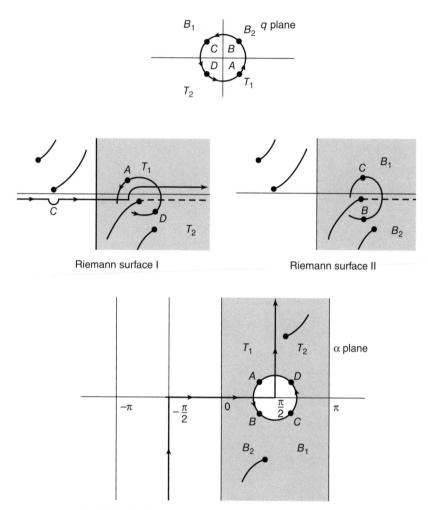

FIGURE 15.B.1 q plane, α plane, and Riemann surfaces.

the α plane, T_1, T_2, B_1, and B_2 are placed side by side, and the branch point $\lambda = +k$ in the λ plane is the regular point $\alpha = \pi/2$ in the α plane.

Now let us picture the four Riemann surfaces in the Λ plane (Fig. 15.B.2). Close examination of the Sommerfeld poles reveals that under this branch cut, the poles are on Riemann surfaces I and IV. In the α plane there are only two surfaces, one corresponding to surfaces I and II and the other corresponding to surfaces III and IV. The Sommerfeld poles are then on the top and bottom sheets in the α plane as shown in Fig. 15.B.3. We now describe one of the confusions concerning the location of the Sommerfeld pole. We write two different branch cuts below (case A and case B in Fig. 15.B.4).

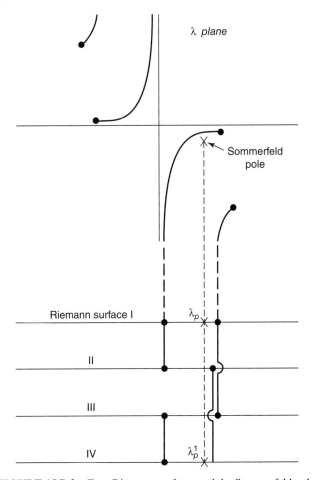

FIGURE 15.B.2   Four Riemann surfaces and the Sommerfeld pole.

For case A, the branch cuts are along Im $q = 0$, Im $q_e = 0$. For case B, the branch cuts are along Re $\lambda = $ Re k, Re $\lambda = $ Re k_e. Then for A, the pole is on I and IV, but for B, the pole is on II and III. Thus for A, the original integral along C is

$$\int_C = \int_{C_\infty} + \int_{B_{r\,A1}} + \int_{B_{r\,A2}} +2\pi j \cdot \text{residue at } P_A,$$

where $B_{r A1}$ and $B_{r A2}$ are the branch cut integrations. On the other hand, for case B,

$$\int_C = \int_{C_\infty} + \int_{B_{r\,B1}} + \int_{B_{r\,B2}},$$

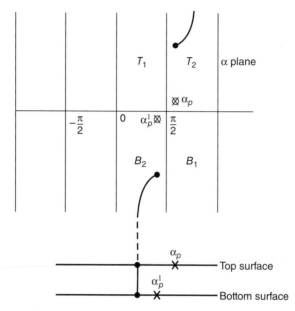

FIGURE 15.B.3 Sommerfeld poles at α_p and α_p^1.

and there is no contribution from the pole P_B, because P_B is on II. Thus it *appears* that case A gives the extra term due to the pole, and case A and case B give two different answers. Note also that the residue has all the characteristics of the Zenneck wave and thus it *appears* that case A yields the Zenneck wave, whereas case B does not.

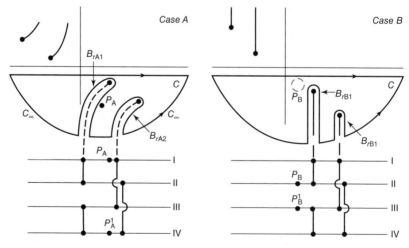

FIGURE 15.B.4 Two different branch cuts.

All these apparent contradictions are of course nonexistent. The branch cuts are drawn for convenience only, and both case A and case B should give the identical result. In fact, we can show that (the integral along C_∞ is zero) there is a difference between the branch cut integration for A and for B and this difference is exactly equal to the residue term, that is,

$$\int_{B_r A1} + \int_{B_r A2} +2\pi j \cdot \text{residue} = \int_{B_r B1} + \int_{B_r B2} .$$

In the text, the integration was performed in the α plane, where Riemann surfaces I and II are placed side by side and thus the confusion between locations P_A and P_B never arises in the α plane.

15.C MODIFIED SADDLE-POINT TECHNIQUE

Let us consider an integral

$$I_0 = \int_c F(\alpha) \exp[zf(\alpha)] \, d\alpha, \tag{15.C.1}$$

where z is a large positive number and $F(\alpha)$ has a pole at α_p, which may be located at any part of the α plane.

We first find the saddle point $\alpha = \alpha_s$ from

$$\left. \frac{\partial f(\alpha)}{\partial \alpha} \right|_{\alpha_s} = 0, \tag{15.C.2}$$

and write (15.C.1) as follows:

$$I_0 = e^{zf(\alpha_s)} \int_c F(\alpha) e^{z[f(\alpha)-f(\alpha_s)]} \, d\alpha. \tag{15.C.3}$$

Let us transform (15.C.3) into the following form using the transformation from α to s:

$$f(\alpha) - f(\alpha_s) = -\frac{s^2}{2}, \tag{15.C.4}$$

$$I_0 = e^{zf(\alpha_s)} \int_c F(s) e^{-z(s^2/2)} \left(\frac{d\alpha}{ds} \right) ds. \tag{15.C.5}$$

The transformation (15.C.4) is possible if the saddle point is an isolated one. We note that in (15.C.5) the saddle point is at $s = 0$, and the steepest descent contour (SDC) is along the real axis of s. This is pictured in Fig. 15.8 using the following example:

$$f(\alpha) = -j \cos(\alpha - \theta),$$

$$s = 2e^{-j(\pi/4)} \sin \frac{\alpha - \theta}{2},$$

$$\frac{d\alpha}{ds} = \left(1 - j\frac{s^2}{4}\right)^{-1/2} e^{j(\pi/4)}.$$

(15.C.6)

Because of $d\alpha/ds$, two branch points (B_{r1} and B_{r2}) appear at $s = \pm 2e^{-j(\pi/4)}$ in the s plane, but these branch points are far from the saddle point and have little effect on the integral.

Now we evaluate (15.C.5) for the following two cases:

1. $F(\alpha)$ *has no pole.* Let us first obtain the solution for a simpler case, where there is no pole in the integrand. In this case we can deform the contour from C to SDC.

$$\int_C ds = \int_{-\infty}^{\infty} ds.$$

(15.C.7)

Now we expand $F(s)d\alpha/ds$ in a Taylor's power series.

$$F(\alpha)\frac{d\alpha}{ds} = \sum_{n=0}^{\infty} A_{2n} s^{2n} + \sum_{n=1}^{\infty} A_{2n-1} s^{2n-1},$$

(15.C.8)

where we separated the even and odd powers for convenience.

Substituting (15.C.8) into (15.C.5), we note that the odd terms of (15.C.8) become zero. Thus

$$I_0 = e^{zf(\alpha_s)} \int_{-\infty}^{\infty} \sum_{n=0}^{\infty} A_{2n} s^{2n} e^{-z(s^2/2)} ds.$$

(15.C.9)

The integration can be performed by using the gamma function.

$$\Gamma(x) = \int_0^{\infty} t^{x-1} e^{-t} dt.$$

(15.C.10)

We also let

$$t = z\frac{s^2}{2}, \quad x = n + \frac{1}{2},$$

and note that

$$\Gamma\left(\tfrac{1}{2}\right) = \sqrt{\pi},$$

$$\Gamma\left(n+\tfrac{1}{2}\right) = \frac{1\cdot 3\cdot 5\cdots(2n-1)}{2^n}\sqrt{\pi},$$

$$\int_{-\infty}^{\infty} s^{2n}e^{-z(s^2/2)}\,ds = \frac{\Gamma\left(n+\tfrac{1}{2}\right)2^{n+1/2}}{z^{n+1/2}} \qquad (15.C.11)$$

$$= \frac{1\cdot 3\cdot 5\cdots(2n-1)}{z^{n+1/2}}\sqrt{2\pi}.$$

We then obtain

$$I_0 = e^{zf(\alpha_s)}\sum_{n=0}^{\infty} A_{2n}\frac{1\cdot 3\cdot 5\cdots(2n-1)}{z^{n+1/2}}\sqrt{2\pi}. \qquad (15.C.12)$$

The first term,

$$e^{zf(\alpha_s)}A_0\sqrt{\frac{2\pi}{z}},$$

is the result obtained from the usual saddle-point technique shown in Appendix 11.C, and (15.C.12) is the asymptotic series in inverse power of z. The series on the right side of (15.C.12), which was obtained by term-by-term integration (15.C.10), is in general divergent. However, it is an asymptotic series having the following characteristics.

If we take the partial sum

$$f_N = \sum_{n=0}^{N} A_n,$$

then the error $(I_0 - f_N)$ is smaller in absolute value than the first term neglected (A_{N+1}):

$$|I_0 - f_N| < |A_{N+1}|.$$

Usually, $|A_n|$ becomes smaller at first and then eventually gets larger and the series diverges.

Alternatively, I_0 in (15.C.6) can be evaluated numerically along the real axis of s using the Gauss quadrature technique.

2. *F(α) has a pole.* Now we come to our main problem. When $F(\alpha)$ has a pole at α_p, or $s = s_p$ in the s plane, Taylor's expansion (15.C.9) is valid only within a

circle whose radius (called the *radius of convergence*) is the distance from the origin to $|s_p|$.

$$|s| < |s_p|.$$

If the pole is close to the saddle point, $|s_p|$ is very small. In order to extend the region of Taylor's expansion, we first write $F(s) \, d\alpha/ds$ as a sum of the term with a pole and the term that is regular at $s = s_p$.

$$F(s)\frac{d\alpha}{ds} = \frac{R_1(s_p)}{s - s_p} + G_1(s). \tag{15.C.13}$$

$R_1(s_p)$ is the residue at $s = s_p$. Note also that the residue of the integrand $F(\alpha)\exp[zf(\alpha)]$ at the pole $\alpha = \alpha_p$ is given by

$$R(A_p) = R_1(s_p) \, \exp[zf(\alpha_p)]. \tag{15.C.14}$$

To show this, let $F(\alpha) = N(\alpha)/D(\alpha)$ and note that

$$R_1(s_p) = \left[\frac{N(s)}{\partial D/\partial s}\frac{d\alpha}{ds} \right]_{s=s_p} = \frac{N(\alpha)}{\partial D/\partial \alpha}\bigg|_{\alpha=\alpha_p}.$$

The evaluation of the integral must now be done differently depending on whether the pole is below or above the original contour.

It is clear that the integral along the original contour C can now be deformed along the saddle-point contour (SDC). If the pole is located below the SDC, then

$$\int_C = \int_{SDC}. \tag{15.C.15}$$

But if the pole is located above the SDC, we have

$$\int_C = \int_{SDC} -2\pi jR(s_p), \tag{15.C.16}$$

where $R(s_p)$ is the residue of $F(\alpha)\exp[-zf(\alpha)]$ at s_p shown in (15.C.14). We combine (15.C.15) and (15.C.16) and write

$$\int_C = \int_{SDC} -2\pi jR(s_p) \, U \, (\mathrm{Im}s_p), \tag{15.C.17}$$

where $U(x)$ is a unit step function. Next we evaluate the integral along the SDC.

$$\int_{SDC} = e^{zf(\alpha_s)} \int_{-\infty}^{\infty} F(s)\frac{d\alpha}{ds}e^{-z(s^2/2)}\,ds$$

$$= I_p + I_1,$$

(15.C.18)

where

$$I_p = e^{zf(\alpha_s)} \int_{-\infty}^{\infty} \frac{R_1(s_p)}{s - s_p}e^{-z(s^2/2)}\,ds,$$

(15.C.19)

$$I_1 = e^{zf(\alpha_s)} \int_{-\infty}^{\infty} G_1(s)e^{-z(s^2/2)}\,ds.$$

(15.C.20)

The evaluation of (15.C.19) can be made as follows. Write

$$I_p = e^{zf(\alpha_s)}R_1(s_p)y(z),$$

(15.C.21)

where

$$y(z) = \int_{-\infty}^{\infty} \frac{e^{-z(s^2/2)}}{s - s_p}\,ds.$$

(15.C.22)

To evaluate $y(z)$ it is necessary to reduce the integral (15.C.21) to a known integral. In this case because of the form of (15.C.22), it is possible to convert (15.C.22) into an error integral. To do this, we first multiply $y(z)$ by $\exp[zs_p^2/2]$ and take the derivative with respect to z.

$$\frac{\partial}{\partial z}\left[y(z)e^{z(s_p^2/2)}\right] = \int_{-\infty}^{\infty} \frac{s^2 - s_p^2}{2(s - s_p)}e^{-(z/2)(s^2-s_p^2)}\,ds$$

$$= -\frac{s_p}{2}e^{z(s_p^2/2)} \int_{-\infty}^{\infty} e^{-z(s^2/2)}\,ds$$

$$= -\sqrt{\pi/2}s_p\frac{e^{z(s_p^2/2)}}{\sqrt{z}}.$$

We now integrate (15.C.22) over z from $z = 0$ to z.

$$y(z)e^{z(s_p^2/2)} = y(0) - \sqrt{\pi/2}s_p \int_0^z \frac{e^{z(s_p^2/2)}}{\sqrt{z}}\,dz.$$

(15.C.23)

Using the following definition of the error function,

$$\mathrm{erf}(z) = \frac{2}{\sqrt{\pi}} \int_0^z e^{-t^2}\, dt,$$

(15.C.23) becomes

$$y(z) = e^{-z(s_p^2/2)} y(0) + j\pi e^{-z(s_p^2/2)} \mathrm{erf}\left(j\sqrt{z/2}s_p\right).\qquad (15.\mathrm{C}.24)$$

However, $y(0)$ is given by

$$y(0) = \int_{-\infty}^{\infty} \frac{ds}{s - s_p} = \begin{cases} j\pi & \text{for Im } s_p > 0, \\ 0 & \text{for Im } s_p = 0, \\ -j\pi & \text{for Im } s_p > 0. \end{cases}\qquad (15.\mathrm{C}.25)$$

Thus we write

$$y(z) = e^{-z(s_p^2/2)}\left\{ 2\pi j U\,(\mathrm{Im}\ s_p) - j\pi\left[1 - \mathrm{erf}\left(j\sqrt{\frac{z}{2}}s_p\right)\right]\right\}.\qquad (15.\mathrm{C}.26)$$

Using the complementary error function, $\mathrm{erfc}(z)$,

$$\mathrm{erfc}(z) = 1 - \mathrm{erf}(z),$$

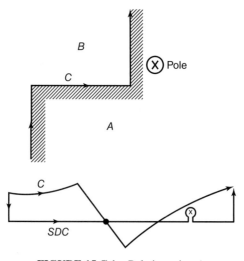

FIGURE 15.C.1   Pole in region A.

we finally obtain I_p, using $f(\alpha_s) - (s_p^2/2) = f(\alpha_p)$,

$$I_p = R(s_p) \left[2\pi j U(\text{Im } s_p) - j\pi \text{ erfc} \left(j\sqrt{\frac{z}{2}} s_p \right) \right]. \tag{15.C.27}$$

The original integral I_0 in (15.C.5) is, therefore, given by

$$I_0 = I_1 + I_p - 2\pi j R(s_p) U(\text{Im } s_p). \tag{15.C.28}$$

Note that the discontinuity in I_0 in (15.C.27) will be exactly canceled by the last term of (15.C.28), and thus we finally obtain the following.

(a) The pole is located below the original contour in region A (Fig. 15.C.1).

$$I_0 = \int_C F(\alpha) e^{zf(\alpha)} \, d\alpha$$
$$= I_1 - j\pi R(s_p) \text{ erfc} \left(j\sqrt{\frac{z}{2}} s_p \right), \tag{15.C.29}$$

where

$$I_1 = e^{zf(\alpha_s)} \int_{-\infty}^{\infty} \left[F(s)\frac{d\alpha}{ds} - \frac{R_1(s_p)}{s - s_p} \right] \exp\left(-z\frac{s^2}{2} \right) ds,$$

$$R(s_p) = R_1 (s_p) \exp[zf(\alpha_p)]$$

$$= \text{residue of } F(\alpha) \exp[zf(\alpha)] \text{ at the pole } \alpha = \alpha_p,$$

$$f(\alpha) - f(\alpha_s) = -\frac{s^2}{2} \quad \text{and } \alpha_s \text{ is the saddle point for } f(\alpha).$$

Note that the integrand for I_1 has no pole, and therefore this can be evaluated either by the expansion as in (15.C.10) or numerically. The series expansion is given by

$$I_1 = e^{zf(\alpha_s)} \sum_{n=0}^{\infty} B_{2n} \left(\frac{2}{z} \right)^{n+1/2} \Gamma(n+1/2), \tag{15.C.30}$$

and B_{2n} is the coefficient of the expansion

$$\left[F(s) \frac{d\alpha}{ds} - \frac{R_1(s_p)}{s - s_p} \right] = \sum_{n=0}^{\infty} B_{2n} s^{2n} + \sum_{n=1}^{\infty} B_{2n-1} s^{2n-1},$$

$$B_0 = A_0 + \frac{R_1(s_p)}{s_p},$$

$$B_n = A_n + \frac{R_1(s_p)}{s_p^{n+1}}, \tag{15.C.31}$$

$$A_0 = \left[F(s) \frac{d\alpha}{ds} \right]_{s=0},$$

$$A_n = \frac{1}{n!} \frac{d^n}{ds^n} \left[F(s) \frac{d\alpha}{ds} \right]_{s=0}.$$

(b) *The pole is located above the original contour (in region B)* (Fig. 15.C.1). We follow a similar procedure and obtain

$$I_0 = I_1 + j\pi R(s_p) \, \mathrm{erfc} \left(-j\sqrt{\frac{z}{2} s_p} \right). \tag{15.C.32}$$

The integral I_0 is given by (15.C.29) if the pole is located below the original contour and by (15.C.32) if the pole is located above the original contour.

APPENDIX TO CHAPTER 16

16.A   HILBERT TRANSFORM

The Hilbert transform is defined by the following (Bracewell):

$$F_h(t) = \frac{1}{\pi} \int_{-\infty}^{\infty} \frac{f(t')}{t' - t} dt', \tag{16.A.1}$$

where the integral is Cauchy's principal value. It is the convolution integral of $f(t')$ and $(-1)/(\pi t)$. The inverse Hilbert transform is given by

$$f(t) = -\frac{1}{\pi} \int \frac{F_h(t')}{t' - t} dt'. \tag{16.A.2}$$

To show this, we take the Fourier transform of (16.A.1). Noting that this is a convolution integral, we get

$$\begin{aligned} \bar{F}_h(s) &= \int F_h(t) e^{-ist} dt \\ &= K(s)F(s), \end{aligned} \tag{16.A.3}$$

Electromagnetic Wave Propagation, Radiation, and Scattering: From Fundamentals to Applications,
Second Edition. Akira Ishimaru.
© 2017 by The Institute of Electrical and Electronic Engineers, Inc. Published 2017 by John Wiley & Sons, Inc.

where

$$K(s) = -\int \frac{1}{\pi t} e^{-ist} dt,$$

$$F(s) = \int f(t) e^{-ist} dt.$$

The integral of $K(s)$ is Cauchy's principal value, and therefore

$$K(s) = \lim_{\varepsilon \to 0} \left(-\frac{1}{\pi}\right) \left(\int_{-\infty}^{-\varepsilon} + \int_{\varepsilon}^{\infty}\right) \frac{e^{-ist}}{t} dt$$

$$= \lim \frac{1}{\pi} \int_{\varepsilon}^{\infty} \frac{2i \sin st}{t} dt.$$

Noting that

$$\int_0^{\infty} \frac{\sin x}{x} dx = \frac{\pi}{2},$$

we get

$$K(s) = -\int \frac{1}{\pi t} e^{-ist} dt = i \operatorname{sgn}(s), \qquad (16.\text{A}.4)$$

where $\operatorname{sgn}(s) = 1$ if $s > 0$ and -1 if $s < 0$.

Taking the inverse transform, we get

$$-\frac{1}{\pi t} = \frac{1}{2\pi} \int i \operatorname{sgn}(s) e^{ist} ds. \qquad (16.\text{A}.5)$$

Taking the inverse Fourier transform of (16.A.3), we write

$$F_h(t) = \frac{1}{2\pi} \int i \operatorname{sgn}(s) F(s) e^{ist} ds. \qquad (16.\text{A}.6)$$

On the other hand, from (16.A.3) and (16.A.4), we get

$$f(t) = \frac{1}{2\pi} \int F(s) e^{ist} ds$$

$$= \frac{1}{2\pi} \int [-i \operatorname{sgn}(s) \bar{F}_h(s)] e^{ist} ds. \qquad (16.\text{A}.7)$$

This has a form identical to (16.A.6), and therefore

$$f(t) = -\frac{1}{\pi} \int \frac{F_h(t')}{t' - t} dt'. \qquad (16.A.8)$$

16.B RYTOV APPROXIMATION

Consider a wave field $U(\bar{r})$ satisfying the wave equation

$$(\nabla^2 + k^2 n^2)U = 0. \qquad (16.B.1)$$

We let

$$U(\bar{r}) = \exp[\Psi(\bar{r})] \qquad (16.B.2)$$

and get Riccati's equation for $\Psi(\bar{r})$

$$\nabla^2 \Psi + \nabla \Psi \cdot \nabla \Psi + k^2 n^2 = 0. \qquad (16.B.3)$$

Also we consider U_0 when $n = 1$.

$$U_0 = \exp(\Psi_0),$$
$$\nabla^2 \Psi_0 + \nabla \Psi_0 \cdot \nabla \Psi_0 + k^2 = 0. \qquad (16.B.4)$$

We now let

$$\Psi = \Psi_0 + \Psi', \qquad (16.B.5)$$

and subtracting (16.B.4) from (16.B.3), we get

$$\nabla^2 \Psi' + 2\nabla \Psi_0 \cdot \nabla \Psi' = -[\nabla \Psi' \cdot \nabla \Psi' + k^2(n^2 - 1)]. \qquad (16.B.6)$$

Now using the identity

$$\nabla^2[U_0 \Psi'] = [\nabla^2 U_0]\Psi' + 2U_0 \nabla \Psi_0 \cdot \nabla \Psi' + U_0 \nabla^2 \Psi',$$

the left side of (16.B.6) becomes

$$\frac{1}{U_0}[\nabla^2(U_0 \Psi') + k^2 U_0 \Psi'],$$

and we obtain

$$(\nabla^2 + k^2)(U_0 \Psi') = [\nabla \Psi' \cdot \nabla \Psi' + k^2(n^2 - 1)]U_0. \qquad (16.B.7)$$

This can be converted to the integral equation using Green's function G:

$$\Psi' = \frac{1}{U_0(\bar{r})} \int G(\bar{r} - \bar{r}')[k^2(n^2 - 1) + \nabla\Psi' \cdot \nabla\Psi']U_0(\bar{r}')dV'. \quad (16.B.8)$$

The first iteration for Ψ' is obtained by letting $\nabla\Psi' = 0$ in the integral. This is called the *first Rytov solution* and is given by

$$U(\bar{r}) = U_0(\bar{r})\exp[\Psi_1(\bar{r})],$$

$$\Psi_1(\bar{r}) = \frac{1}{U_0(\bar{r})} \int G(\bar{r} - \bar{r}')k^2(n^2 - 1)U_0(\bar{r}')dV'. \quad (16.B.9)$$

16.C ABEL'S INTEGRAL EQUATION

Consider the following Volterra integral equation of the first kind, known as *Abel's integral equation*

$$g(t) = \int_0^t \frac{f(\tau)}{(t - \tau)^\alpha}d\tau, \quad 0 < \alpha < 1, \quad (16.C.1)$$

where $g'(t)$ is continuous. $g(t)$ is a given function of t and $f(\tau)$ is the unknown function to be determined. We will show that the solution to (16.C.1) is given by

$$f(\tau) = \frac{\sin \pi\alpha}{\pi} \frac{g(0)}{\tau^{1-\alpha}} + \frac{\sin \pi\alpha}{\pi} \int_0^\tau \frac{g'(t)}{(\tau - t)^{1-\alpha}}dt. \quad (16.C.2)$$

Furthermore, we will show that

$$\int_0^\tau f(\tau)d\tau = \frac{\sin \pi\alpha}{\pi} \int_0^\tau \frac{g(t)}{(\tau - t)^{1-\alpha}}dt. \quad (16.C.3)$$

To prove this, we first note that $g(t)$ is a convolution integral of $K(\tau)$ and $f(\tau)$.

$$g(t) = \int_0^t K(t - \tau)f(\tau)d\tau, \quad (16.C.4)$$

where $K(\tau) = \tau^{-\alpha}$.

We can solve this by using the Laplace transform:

$$\bar{g}(s) = +g(t),$$

$$\bar{K}(s) = +K(t), \quad (16.C.5)$$

$$\bar{f}(s) = +(f(t)).$$

We then get

$$\bar{g}(s) = \bar{K}(s)\bar{f}(s). \tag{16.C.6}$$

Therefore, we get

$$f(t) = +^{-1}\left[\frac{\bar{g}(s)}{\bar{K}(s)}\right].$$

The Laplace transform $\bar{K}(s)$ is given by

$$\bar{K}(s) = +(\tau^{-\alpha}) = \frac{\Gamma(1-\alpha)}{s^{1-\alpha}} \quad \text{for } \alpha < 1. \tag{16.C.7}$$

Therefore,

$$f(t) = +^{-1}\left[\frac{s^{1-\alpha}}{\Gamma(1-\alpha)}\bar{g}(s)\right] = \frac{1}{\Gamma(1-\alpha)} +^{-1}\left[\frac{1}{s^{\alpha}}s\bar{g}(s)\right]. \tag{16.C.8}$$

We use the following to evaluate (16.C.8):

$$+^{-1}(s^{-\alpha}) = \frac{t^{\alpha-1}}{\Gamma(\alpha)}, \quad \alpha > 0,$$
$$+^{-1}(s\bar{g}(s)) = g'(t) + \delta(t)g(0), \tag{16.C.9}$$
$$\frac{1}{\Gamma(1-\alpha)\Gamma(\alpha)} = \frac{\sin \pi\alpha}{\pi}.$$

We can then write (16.C.8) in the following convolution form:

$$f(\tau) = \frac{\sin \pi\alpha}{\pi}\frac{g(0)}{\tau^{1-\alpha}} + \frac{\sin \pi\alpha}{\pi}\int_0^{\tau}\frac{g'(t)}{(\tau-t)^{1-\alpha}}dt. \tag{16.C.10}$$

To show (16.C.3), we integrate (16.C.8) and get

$$\int_0^{\tau}f(\tau)d\tau = \frac{1}{\Gamma(1-\alpha)} +^{-1}\left[\frac{1}{s^{\alpha}}\bar{g}(s)\right], \tag{16.C.11}$$

and therefore this is given by the convolution integral (16.C.3).

APPENDIX TO CHAPTER 23

23.A COMPLEX GAUSSIAN VARIABLES, CIRCULARITY, AND MOMENT THEOREM

In most problems of propagation and scattering, we deal with complex random variables. We also make the assumption that these complex variables are Gaussian. Even though real Gaussian variables are well known and defined (Davenport and Root 1958), the characteristics of complex Gaussian variables are often not carefully stated. Normally, complex Gaussian variables are expressed by the covariance matrix.

Consider a complex random function u, its average $\langle u \rangle$ and its fluctuation v.

$$u = \langle u \rangle + v, \quad \langle v \rangle = 0. \tag{23.A.1}$$

The covariance C of the fluctuation is given by

$$C = \langle v_1 v_2^* \rangle . \tag{23.A.2}$$

However, the covariance cannot completely describe the properties of $\underline{v}$. We need another quantity C_p given by

$$C_p = \langle v_1 v_2 \rangle . \tag{23.A.3}$$

Electromagnetic Wave Propagation, Radiation, and Scattering: From Fundamentals to Applications,
Second Edition. Akira Ishimaru.
© 2017 by The Institute of Electrical and Electronic Engineers, Inc. Published 2017 by John Wiley & Sons, Inc.

C_p is called the pseudo-covariance. The function C_p is rarely introduced in many problems, and is often assumed to be zero. The processes with vanishing pseudo-covariance are called "proper" or "circular" (Goodman, 1985; Neeser, 1993; Ollida, 2008; Picinbono, 1996).

If a complex Gaussian, zero-mean stationary process is assumed to be circular, we have

$$\langle v_1 v_2^* \rangle \neq 0 \quad \text{and} \quad \langle v_1 v_2 \rangle = 0. \tag{23.A.4}$$

For this process, a general moment theorem can be stated (Reed, 1962). The moment theorem for a real Gaussian zero-mean process z is well known (Middleton, 1960).

$$\langle z_1 z_2 z_3 z_4 \rangle = \langle z_1 z_2 \rangle \langle z_3 z_4 \rangle + \langle z_2 z_3 \rangle \langle z_1 z_4 \rangle + \langle z_1 z_3 \rangle \langle z_2 z_4 \rangle. \tag{23.A.5}$$

For the complex Gaussian, zero-mean process v, we have

$$\langle v_1 v_2 v_3^* v_4^* \rangle = \langle v_1 v_3^* \rangle \langle v_2 v_4^* \rangle + \langle v_1 v_4^* \rangle \langle v_2 v_3^* \rangle. \tag{23.A.6}$$

In addition, we note

$$\langle (v_1 v_2^*)^n \rangle = n! \left(\langle v_1 v_2^* \rangle \right)^n,$$
$$\langle |v|^{2n} \rangle = n! \left(\langle |v|^2 \rangle \right)^n. \tag{23.A.7}$$

Furthermore, we have

$$\langle v_1 v_2 \rangle = 0, \quad \langle v_1 v_2 v_3^* \rangle = 0.$$

Making use of (23.A.6) and (23.A.7), we have

$$\langle u_1 u_2 u_3^* u_4^* \rangle = \langle u_1 u_3^* \rangle \langle u_2 u_4^* \rangle + \langle u_1 u_4^* \rangle \langle u_2 u_3^* \rangle - \langle u_1 \rangle \langle u_2 \rangle \langle u_3^* \rangle \langle u_4^* \rangle, \tag{23.A.8}$$
$$\langle u_1^2 u_2^{*2} \rangle = 2 \langle u_1 u_2^* \rangle^2 - \langle u_1 \rangle^2 \langle u_2^* \rangle^2,$$
$$\langle I^2 \rangle = 2 \langle I \rangle^2 - |\langle u \rangle|^4,$$
$$\langle I \rangle = \langle u \rangle \langle u^* \rangle + \langle v v^* \rangle, \quad I = u u^*,$$
$$\langle u_1 u_2 \rangle = \langle u_1 \rangle \langle u_2 \rangle.$$

It should be noted that while the circular Gaussian assumption is useful, it is not exact because of the assumption $\langle v_1 v_2 \rangle = 0$.

APPENDIX TO CHAPTER 26

26.A WAVE PROPAGATION IN ELASTIC SOLID AND RAYLEIGH SURFACE WAVES

The equations governing the linearized theory of elasticity for a homogeneous, isotropic medium are given by Achenbach (1980), Redwood (1960), and Brekhovskikh (1970). Here we give a summary of the equations in the Cartesian coordinate system.

The wave propagation in elastic solids such as the earth can be expressed by the scalar potential ϕ representing the pressure wave (P-wave, longitudinal wave), and the vector potential $\hat{\psi}$ representing the shear wave (S-wave, transverse wave). These two potentials satisfy the wave equation

$$\nabla^2 \phi = \frac{1}{c_p^2} \frac{\partial^2}{\partial t^2} \phi,$$

$$\nabla^2 \hat{\psi} = \frac{1}{c_s^2} \frac{\partial^2}{\partial t^2} \hat{\psi}, \qquad (26.A.1)$$

Electromagnetic Wave Propagation, Radiation, and Scattering: From Fundamentals to Applications,
Second Edition. Akira Ishimaru.
© 2017 by The Institute of Electrical and Electronic Engineers, Inc. Published 2017 by John Wiley & Sons, Inc.

where

$$c_p = \left[\frac{\lambda + 2\mu}{\rho}\right]^{1/2} = \text{pressure wave velocity,}$$

$$c_s = \left[\frac{\mu}{\rho}\right]^{1/2} = \text{shear wave velocity,}$$

λ and μ are Lamé constants (N/m^2), and ρ is density (kg/m^3). Clearly $c_p > c_s$.

Note that in a fluid, $\mu = 0$ and only a P-wave is possible. The displacement vector $\bar{u}$ can be given in terms of ϕ and $\hat{\psi}$.

$$\begin{aligned} \bar{u} &= u\hat{x} + v\hat{y} + w\hat{z} \\ &= \nabla\phi + \nabla \times \hat{\psi}. \end{aligned} \tag{26.A.2}$$

The stress tensor τ is related to the strain tensor ε and can be given in terms of ϕ and $\hat{\psi}$. In terms of ϕ and $\hat{\psi}$, we have the displacement vector $\bar{u}$.

$$\begin{aligned} u &= \frac{\partial\phi}{\partial x} + \frac{\partial\psi_z}{\partial y} - \frac{\partial\psi_y}{\partial z}, \\ v &= \frac{\partial\phi}{\partial y} - \frac{\partial\psi_z}{\partial x} + \frac{\partial\psi_x}{\partial z}, \\ w &= \frac{\partial\phi}{\partial z} + \frac{\partial\psi_y}{\partial x} - \frac{\partial\psi_x}{\partial y}. \end{aligned} \tag{26.A.3}$$

The stress tensor τ can be expressed in terms of ϕ and $\hat{\psi}$.

$$\begin{aligned} \tau_x &= \lambda\nabla^2\phi + 2\mu\left[\frac{\partial^2\phi}{\partial x^2} + \frac{\partial}{\partial x}\left(\frac{\partial\psi_z}{\partial y} - \frac{\partial\psi_y}{\partial z}\right)\right], \\ \tau_y &= \lambda\nabla^2\phi + 2\mu\left[\frac{\partial^2\phi}{\partial y^2} - \frac{\partial}{\partial y}\left(\frac{\partial\psi_z}{\partial x} - \frac{\partial\psi_x}{\partial z}\right)\right], \\ \tau_z &= \lambda\nabla^2\phi + 2\mu\left[\frac{\partial^2\phi}{\partial z^2} + \frac{\partial}{\partial z}\left(\frac{\partial\psi_y}{\partial x} - \frac{\partial\psi_x}{\partial y}\right)\right], \\ \tau_{xy} = \tau_{yx} &= \mu\left[\frac{2\partial^2\phi}{\partial x\partial y} + \frac{\partial}{\partial y}\left(\frac{\partial\psi_z}{\partial y} - \frac{\partial\psi_y}{\partial z}\right) - \frac{\partial}{\partial x}\left(\frac{\partial\psi_z}{\partial x} - \frac{\partial\psi_x}{\partial z}\right)\right], \\ \tau_{yz} = \tau_{zy} &= \mu\left[\frac{2\partial^2\phi}{\partial y\partial z} - \frac{\partial}{\partial z}\left(\frac{\partial\psi_z}{\partial x} - \frac{\partial\psi_x}{\partial z}\right) + \frac{\partial}{\partial y}\left(\frac{\partial\psi_y}{\partial x} - \frac{\partial\psi_x}{\partial y}\right)\right], \\ \tau_{zx} = \tau_{xz} &= \mu\left[\frac{2\partial^2\phi}{\partial x\partial z} + \frac{\partial}{\partial z}\left(\frac{\partial\psi_z}{\partial y} - \frac{\partial\psi_y}{\partial z}\right) + \frac{\partial}{\partial x}\left(\frac{\partial\psi_y}{\partial x} - \frac{\partial\psi_x}{\partial y}\right)\right]. \end{aligned} \tag{26.A.4}$$

Let us now consider the boundary conditions between two media. We first start with the fluid–fluid interface. We have

  1) Fluid–fluid interface

  (a) Pressure = Continuous
  (b) Normal displacement = Continuous

  (26.A.5)

  2) Fluid–vacuum interface

  (a) Pressure = 0

  (26.A.6)

  3) Fluid–rigid wall interface

  (a) Normal velocity = 0

  (26.A.7)

These are well-known boundary conditions that have been discussed for acoustic propagation with a fluid–fluid interface (Section 2.12).

For interface with a solid, we need to include the displacement and the stress. We have

  1. Solid–solid interface

  (a) All displacement = continuous
              (normal and tangential displacement)
  (b) All stress = continuous
              (normal stress τ_{zz} and tangential stress τ_{xz} and τ_{yz})

  (26.A.8)

  2. Solid–fluid interface

  (a) Tangential stress = 0
  (b) Normal component of stress = continuuous
  (c) Normal component of stress displacement = continuous

  (26.A.9)

  3. Solid–vacuum interface

  (a) Normal stress = 0 ($\tau_{zz} = 0$)
  (b) Tangential stress = 0 ($\tau_{xz} = \tau_{yz} = 0$)

  (26.A.10)

Here, a Cartesian system shown in Fig 26.B.1 is used.

26.B TWO-DIMENSIONAL CASE, RAYLEIGH SURFACE WAVE

For the two-dimensional case ($\partial/\partial y = 0$), there is no displacement in the y direction; thus $v = 0$ and therefore $\psi_x = \psi_z = 0$.

A surface wave can exist along the solid–vacuum interface. This is called the Rayleigh surface wave. - See Fig 26.B.1. The boundary conditions are given in Eq.

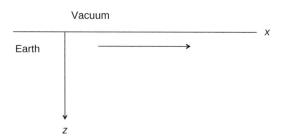

FIGURE 26.B.1 Rayleigh surface wave $(\partial/\partial y = 0)$.

(26.A.10). We then have, at $z = 0$,

$$\tau_z = \lambda \nabla^2 \phi + 2\mu \left[\frac{\partial^2 \phi}{\partial z^2} + \frac{\partial}{\partial z}\frac{\partial}{\partial x}\psi_y \right] = 0,$$

$$\tau_{zx} = \tau_{xz} = \mu \left[\frac{2\partial^2 \phi}{\partial x \partial z} - \frac{\partial^2}{\partial z^2}\psi_y + \frac{\partial^2}{\partial x^2}\psi_y \right] = 0. \tag{26.B.1}$$

We write ϕ and ψ_y in the following form:

$$\phi = A \exp[ik_l z + ik_x x],$$

$$\psi_y = B \exp[ik_t z + ik_x x], \tag{26.B.2}$$

where

$$k_t^2 = k_s^2 - k_x^2,$$
$$k_l^2 = k_p^2 - k_x^2.$$

Note that the Rayleigh surface wave exists if the velocities c_p, c_s, and c_r satisfy the following:

$$c_p > c_s > c_r, \tag{26.B.3}$$

where $k_p = \dfrac{\omega}{c_p}$, $k_s = \dfrac{\omega}{c_s}$, and $k_x = \dfrac{\omega}{c_r}$ are the wave numbers of the pressure wave (P-wave), the shear wave (S-wave), and the Rayleigh surface wave.

In view of (26.B.3), we get

$$k_p < k_s < k_r. \tag{26.B.4}$$

Therefore k_t^2 and k_l^2 in (26.B.2) are negative and k_t and k_l are imaginary.

The Rayleigh wave decays exponentially away from the surface.

$$e^{ik_l z} = e^{-\alpha_l z},$$
$$e^{ik_t z} = e^{-\alpha_t z}.$$

(26.B.5)

We now substitute (26.B.2) into (26.B.1). We get

$$\alpha A + \beta B = 0,$$
$$\delta A + \gamma B = 0,$$

(26.B.6)

where

$$\alpha = \lambda(-k_p^2) + 2\mu(-k_l^2)$$
$$= -\mu(k_s^2 - 2k_x^2),$$
$$\beta = -2\mu k_t k_x,$$
$$\delta = 2\mu[-k_x k_l],$$
$$\gamma = \mu\left[k_s^2 - 2k_x^2\right].$$

To obtain the constants α, β, δ, and γ, we make use of the following:

$$(\lambda + 2\mu) k_p^2 = \mu k_s^2.$$

(26.B.7)

A nonzero solution of (26.B.6) is obtained if the determinant of the coefficients is zero

$$\begin{vmatrix} \alpha & \beta \\ \delta & \gamma \end{vmatrix} = 0.$$

(26.B.8)

From this we can obtain the equation to determine the phase velocity of the Rayleigh wave

$$\left(2 - \frac{c_r^2}{c_s^2}\right)^2 - 4\left(1 - \frac{c_r^2}{c_s^2}\right)^{1/2}\left(1 - \frac{c_r^2}{c_p^2}\right)^{1/2} = 0.$$

(26.B.9)

Note that c_r = velocity of the Rayleigh wave, and c_s and c_p are the velocities of the shear wave and the pressure wave shown in (26.A.1). The frequency does not appear in (26.B.9) and therefore the Rayleigh surface wave is nondispersive.

REFERENCES

Abarbanel, H. D. I. (1980). Scattering from a random surface. *J. Acoust. Soc. Am.*, 68, 1459–1466.

Abramowitz, M., and I. A. Stegun, eds. (1964). *Handbook of Mathematical Functions.* Washington, D.C.: U.S. Government Printing Office.

Achenbach, J. D. (1960). *Wave Propagation in Elastic Solids.* North-Holland Publishing.

Adler, P. M. (1996). Transports in fractal porous media. *J. Hydrol.*, 187, 195–213.

Agrawal, G. P. (1989). *Nonlinear Fiber Optics.* New York: Academic Press

Andersen, J. B. (2000). Antenna arrays in mobile communications: gain, diversity, and channel capacity. *IEEE Antenna Propagat. Mag.*, 42(2), 12–16.

Andersen, J. B. (2000). Array gain and capacity for known random channels with multiple element arrays at both ends. *IEEE J. Sel. Areas Commun.*, 18(11), 2172–2178.

Anderson, P. W. (1958). Absence of diffusion in certain random lattices. *Phys. Rev.*, 109, 1492–1505.

Anderson, P. W. (1985). The questions of classical localization, a theory of white paint? *Philos Mag. B*, 52(3), 505–509.

Arii, M., J. J. van Zyl, and Y. Kim. (2010). A general characterization for polarimetric scattering from vegetation canopies. *IEEE Trans. Geosci. Remote Sens.*, 48(9), 3349–3357.

Arvas, E., and R. F. Harrington. (1983). Computation of the magnetic polarizabilityof conducting disks and the electric polarizability of apertures. *IEEE Trans. Antennas Propag.*, 31(5), 719–724.

Electromagnetic Wave Propagation, Radiation, and Scattering: From Fundamentals to Applications, Second Edition. Akira Ishimaru.
© 2017 by The Institute of Electrical and Electronic Engineers, Inc. Published 2017 by John Wiley & Sons, Inc.

Atkins, P. R., W. C. Chew, M. Li, L. E. Sun, Z.-H. Ma, and L. J. Jiang. (2014). Casimir force for complex objects using domain decomposition technique. *Prog. Electromagn. Res.*, 149, 275–280.

Ayappa, K. G., H. T. Davis, E. A. Davis, and J. Gordon. (1991). Analysis of microwave heating of materials with temperature-dependent properties. *AIChE J.*, 37(3), 313–322.

Bahl, I. J., and P. Bhartia. (1980). *Microstrip Antennas*. Dedham, MA: Artech House.

Baker, C. J., H. D. Griffiths, and I. Papoutsis. (2005). Passive coherent location radar systems. Part 2: Waveform properties. *IEE Proc. Radar Sonar Navig.*, 152(3), 160–168.

Balanis, C. A. (1982). *Antenna Theory*. New York: Harper & Row.

Balanis, C. A. (1989). *Advanced Engineering Electromagnetics*. New York: Wiley.

Banos, A., Jr. (1966). *Dipole Radiation in the Presence of a Conducting Half-Space*. Oxford: Pergamon Press.

Bass, F. G., and I. M. Fuks. (1979). *Wave Scattering from Statistically Rough Surfaces*. Oxford: Pergamon Press.

Bassiri, S., C. H. Papas, and N. Engheta. (1988). Electromagnetic wave propagation through a dielectric-chiral interface and through a chiral slab. *J. Opt. Soc. Am., A*, 5(9), 1450–1459.

Baum, C. E. (1976). Emerging technology for transient and broad-band analysis and synthesis of antennas and scatterers. *Proc. IEEE*, 64(11), 1598–1616.

Beilis, A., and F. D. Tappert. (1979). Coupled mode analysis of multiple rough surface scattering. *J. Acoust, Soc. Amer.*, 66(3), 811–826.

Blaunstein, N., and C. Christodoulou. (2007). *Radio Propagation and Adaptive Antennas for Wireless Communication Links*. New York: John Wiley.

Boas, D. A., M. A. O'Leary, B. Chance, and A. G. Yodh. (1994). Scattering of diffuse photon density waves by spherical inhomogeneities within turbid media: analytic solution and applications. *Proc. Natl. Acad. Sci. USA*, 91(11), 4887–4891.

Boerner, W.-M., ed. (1985). Inverse methods in electromagnetic imaging. In: *NATO-ASI Series C: Mathematical and Physical Sciences*. Dordrecht, The Netherlands: D. Reidel.

Boerner, W. M. (2003). Recent advances in extra-wide-band polarimetry, interferometry and polarimetric interferometry in synthetic aperture remote sensing and its applications. *IEE Proc. Radar Sonar Navig.*, 150(3), 113–124.

Bojarski, N. N. (1982a). A survey of the near-field far-field inverse scattering inverse source integral equation. *IEEE Trans. Antennas Propag.*, 30, 975–979.

Bojarski, N. N. (1982b). A survey of the physical optics inverse scattering identity. *IEEE Trans. Antennas Propag.*, 30(5), 980–989.

Booker, H. G., and W. E. Gordon. (1950). A theory of radio scattering in the troposphere. *Proc. IRE*, 38, 401–412.

Borcea, L., G. Papanicolaou, C. Tsogka, and J. Berryman. (2002). Imaging and time reversal in random media. *Inverse Problems*, 18(5), 1247–1279.

Born, M., and E. Wolf. (1964). *Principles of Optics*. Cambridge: University Press.

Bowman, J. J., T. B. A. Senior, and P. L. E. Uslenghi. (1969). *Electromagnetic and Acoustic Scattering by Simple Shapes*. Amsterdam: North-Holland.

Braunish, H., C. O. Ao, K. O'neill, and J. A. Kong. (2001). Magnetoquasistatic response of conducting and permeable spheroid under axial excitation. *IEEE Trans. Geosci. Remote Sens.*, 39, 2689–2701.

Brebbia, C. A., J. C. F. Telles, and L. C. Wrobel. (1984). *Boundary Element Techniques*. New York: Springer-Verlag.

Brekhovskikh, L. M. (1960). *Waves in Layered Media*. New York: Academic Press.

Bremmer, H. (1949). *Terrestrial Radio Waves*. Amsterdam: Elsevier.

Bringi, V. N., and V. Chandrasekar. (2001). *Polarimetric Doppler Weather Radar: Principles and Applications*. New York: Cambridge University Press.

Brookner, E., ed. (1977). *Radar Technology*. Dedham, MA: Artech House.

Brovelli, A., and G. Cassiani. (2010). A combination of the Hashin–Shtrikman bounds aimed at modeling electrical conductivity and permittivity of variably saturated porous media. *Geophysical Journal International*, 180(1), 225–237.

Butler, C. M., and D. R. Wilton. (1980). General analysis of narrow strips and slots. *IEEE Trans. Antennas Propag.*, S28(1), 42–48.

Butler, C. M., D. R. Wilton, Y. Rahamt-Samii, and R. Mittra. (1978). Electromagnetic penetration through apertures in conducting surfaces. *IEEE Trans. Antennas Propag.*, 26(1), 82–93.

Caloz, C., and T. Itoh. (2006). *Electromagnetic Metamaterials: Transmission Line Theory and Microwave Applications*. New York: Wiley Interscience.

Capon, J. (1969). High-resolution frequency-wavenumber spectrum analysis. *Proc. IEEE*, 57(8), 1408–1418.

Carcione, J. M., and G. Seriani. (2000). An electromagnetic modeling tool for the detection of hydrocarbons in the subsoil. *Geophysical Prospecting*, 48(2), 231–256.

Chan, A. K., R. A. Sigelmann, A. W. Guy, and J. F. Lehmann. (1973). Calculation by the method of finite differences of the temperature distribution in layered tissues. *IEEE Trans. Biomed. Eng.*, BME-20(2), 86–90.

Chandrasekhar, S. (1960). *Radiative Transfer*. New York: Dover. Reprint of the Oxford University Press, Oxford, 1950 edition.

Chen, P.-Y., and A. Alu. (2011). Atomically thin surface cloak using graphene monolayers. *ACS Nano*, 5(7), 5855–5863.

Chen, P.-Y., and A. Alu. (2013). Terahertz metamaterial devices based on graphene nanostructures. *IEEE Trans. Terahertz Sci. Technol.*, 3(6), 748.

Chen, Y., and D. Or. (2006). Geometrical factors and interfacial processes affecting complex dielectric permittivity of partially saturated porous media *Water Resources Research*, 42(6), WO6423. DOI: 1029 2005 WR004714.

Chen, X. D., and X. Peng. (2005). Modified Biot number in the context of air drying of small moist porous objects. *Dry. Technol.*, 23(1–2), 83–103.

Cheng, D. K. (1983). *Field and Wave Electromagnetics*. Reading, MA: Addison-Wesley.

Cheong, W., S. A. Prahl, and A. J. Welch. (1990). A review of the optical properties of biological tissues. *IEEE J. Quant. Electron.*, 26(12), 2166–2185.

Cheung, R. L.-T., and A. Ishimaru. (1982). Transmission, backscattering, and depolarization of waves in randomly distributed spherical particles. *Appl. Opt.*, 21(20), 3792–3798.

Chizhik, D., G. J. Foschini, M. J. Gans, and R. A. Valenzuela. (2002). Keyholes, correlations, and capacities of multielement transmit and receive antennas. *IEEE Trans. Wirel. Commun.*, 1, 361–368.

Cloude, S. R. (1985). Target decomposition theories in radar scattering. *Electron. Lett.*, 21, 22–24.

Cloude, S. R. (2009). *Polarization: Applications in Remote Sensing*. Oxford, UK: Oxford University Press.

Cloude, S. R., and E. Pottier. (1996). A review of target decomposition theorems in radar polarimetry. *IEEE Trans. Geosci. Remote Sens.*, 34(2), 498–518.

Cohen, J., and N. Bleistein. (1979). A velocity inversion procedure for acoustic waves. *Geophysics*, 44(6), 1077–1087.

Collin, R. E. (1966). *Field Theory of Guided Waves*. New York: McGraw-Hill.

Collin, R. E., and F. J. Zucker, eds. (1969). *Antenna Theory*. New York: McGraw-Hill.

Corbella, I., N. Duffo, M. Vall-Llossera, A. Camps, and F. Torres. (2004). The visibility function in interferometric aperture synthesis radiometry. *IEEE Trans. Geosci. Remote Sens.*, 42(8), 1677–1682.

de Hoop, A. T. (1960). A modification of Cagniard's method for solving seismic pulse problems. *Applied Scientific Research, Section B*, 8(1), 349–356.

de Hoop, A. T., and Jos H. M. T. van der Hijden. (1985). Seismic waves generated by an impulsive point source in a solid/fluid configuration with a plane boundary. *Geophysics*, 50(7), 1083–1090.

Dehong, L., J. Krolik, and L. Carin. (2007). Electromagnetic target detection in uncertain media: time-reversal and minimum-variance algorithms. *IEEE Trans. Antennas Propag.* 45(4), 934–944.

DeSanto, J. A., and G. S. Brown. (1986). Analytical techniques for multiple scattering from rough surfaces. In: *Progress in Optics XXIII*, ed. E. Wolf. Elsevier Science Publishers B.V.

Deschamps, G. A., J. Boersma, and S.-W. Lee. (1984). Three-dimensional half-plane diffraction: exact solution and testing of uniform theories. *IEEE Trans. Antennas Propag.*, 32(3), 264–271.

Devaney, A. J. (1982). A filtered back propagation algorithm for diffraction tomography. *Ultrason. Imaging*, 4(4), 336–350.

Devaney, A. J. (2005). Time reversal imaging of obscured targets from multistatic data. *IEEE Trans. Antennas Propag.*, 53(5), 1600–1610.

Devaney, A. J. (2012). *Mathematical Foundations of Imaging, Tomography and Wavefield Inversion*. Cambridge, UK: Cambridge University Press.

Devaney, A. J., E. A. Marengo, and F. K. Gruber. (2005). Time-reversal-based imaging and inverse scattering of multiply scattering point targets. *J. Acoust. Soc. Am.* 118(5), 3129–3138.

Devaney, A. J., and R. P. Porter. (1985). Holography and the inverse source problem: part II. *J. Opt. Soc. Am., A*, 2(11), 2006–2011.

Dodd, R. K., J. C. Eilbeck, J. D. Gibbon, and H. C. Morris. (1982). *Solitons and Nonlinear Wave Equations*. New York: Academic Press.

Donohue, J., and J. R. Kuttler. (2000). Propagation modeling over terrain using the parabolic wave equation. *IEEE Trans. Antennas Propag.*, 48, 260–277.

Doviak, R. J., and D. S. Zrnić. (1984). *Doppler Radar and Weather Observations*. New York: Academic press.

Dyson, F. (1993). George Green and physics. *Phys. World*, 6(8), 33.

Eleftheriades, G. V., and K. G. Balmain. (2005). *Negative Refraction Metamaterials: Fundamental Principles and Applications*. New York: John Wiley & Sons.

Elliott, R. S. (1966). *Electromagnetics*. New York: McGraw-Hill.

Elliott, R. S. (1981). *Antenna Theory and Design*. Englewood Cliffs, NJ: Prentice-Hall.

Engheta, N., and R. W. Ziolkowski, eds. (2006). *Metamaterials Physics and Engineering Explanations*. New York: John Wiley & Sons.

Falconer, K. (1990). *Fractal Geometry. Mathematical Foundations and Applications*. John Wiley & Sons.

Felsen, L. B. (1976). Transient electromagnetic fields. In: *Topics in Applied Physics*, Vol. 10. New York: Springer.

Felsen, L. B., and N. Marcuvitz. (1973). *Radiation and Scattering of Waves*. Englewood Cliffs, NJ: Prentice-Hall.

Feng, S., C. Kane, P. Lee, and A. D. Stone. (1988). Correlations and fluctuations of coherent wave transmission through disordered media. *Phys. Rev. Lett.*, 61(7), 834–837.

Feng, S. and P. N. Sen. (1985). Geometrical model of conductive and dielectric properties of partially saturated rocks. *J. Appl. Phys*, 58(8), 3236–3243.

Fink, M. (1997). Time reversed acoustics. *Phys. Today*, 50(3), 34–40.

Fink, M., D. Cassereau, A. Derode, C. Prada, P. Roux, M. Tanter, J. Thomas, and F. Wu. (2000). Time-reversed acoustics. *Rep. Prog. Phys.*, 63, 1933–1995.

Finlayson, B. A. (1972). *The Method of Weighted Residuals and Variational Principles*. New York: Academic Press.

Flinn, E. A., and C. H. Dix. (1962). *Reflection and Refraction of Progressive Seismic Waves*. McGraw-Hill.

Foschini, G. J., and M. J. Gans. (1998). On limits of wireless communications in a fading environment when using multiple antennas. *Wireless Personal Communications*, 6(3), 311–335.

Freeman, A., and S. L. Durden. (1998). A three-component scattering model for polarimetric SAR data. *IEEE Trans. Geosci. Remote Sens.*, 36(3), 963–973.

Frisch, V. (1968). Wave propagation in random medium. In: *Probabilistic Methods in Applied Mathematics*, ed. A. T. Bharucha-Reid. New York: Academic Press.

Fujimoto, J. G. (2001). Optical coherence tomography. *C. R. Acad. Sci. Paris*, t.2, Série IV, 1099–1111.

Fung, A. K. (1994). *Microwave Scattering and Emission Models and Their Applications*. Boston, MA: Artech House.

Gabriele, M. L., G. Wollstein, H. Ishikawa, L. Kagemann, J. Xu, L. S. Folio, and J. S. Schuman. (2011). Optical coherence tomography: history, current status, and laboratory work. *Invest. Ophthalmol. Vis. Sci.*, 52(5), 2425–2436.

Gao, G., A. Abubakar, and T. M. Habashy. (2012). Joint petrophysical inversion of electromagnetic and full-waveform seismic data. *Geophysics*, 77(3), WA 3–WA 18.

Ghoshal, U. S., and L. N. Smith. (1988). Skin effects in narrow copper microstrip at 77K. *IEEE Trans. Microwave Theory Tech.*, 36(12), 1788–1795.

Goldstein, H. (1981). *Classical Mechanics*. Reading, MA: Addison-Wesley Publishing Company.

Goodman, J. W. (1985). *Statistical Optics*. New York: John Wiley & Sons.

Gradshteyn, I. S., and I. M. Ryzhik. (1965). *Tables of Integrals, Series, and Products*. New York: Academic Press.

Green, G. (1828, March). *An Essay on the Application of Mathematical Analysis to the Theories of Electricity and Magnetism*. Sneinton near Nottingham.

Griffiths, H. D., and C. J. Baker. (2005). Passive coherent location radar systems. Part 1: performance prediction. *IEE Proc.-Radar Sonar Navig.*, 152(3), 153–159.

Hansen, R. C. (1966). *Microwave Scanning Antennas*, Vols. 1, 2, and 3. New York: Academic Press.

Hansen, R. C., ed. (1981). *Geometric Theory of Diffraction*. New York: IEEE Press.

Hardy, A., and W. Streifer. (1986). Coupled mode solutions of multiwaveguide systems. *IEEE J. Quantum Electron*, 22(4), 528–534.

Hardy, W. H., and L. A. Whitehead. (1981). Split-ring resonator for use in magnetic resonance from 20–2000 MHz. *Rev. Sci. Instrum.*, 52, 213–216.

Harrington, R. F. (1961). *Time-Harmonic Electromagnetic Fields*. New York: McGraw-Hill.

Harrington, R. F. (1968). *Field Computation by Moment Methods*. New York: Macmillan.

Hasegawa, A. (1989). *Optical Solitons in Fibers*. New York: Springer-Verlag.

Herman, G. T. (1979). Image reconstruction from projections. In: *Topics in Applied Physics*, Vol. 32. Berlin: Springer-Verlag.

Huang, D., E. A. Swanson, C. P. Lin, J. S. Schuman, W. G. Stinson, W. Chang, M. R. Hee, T. Flotte, K. Gregory, C. A. Puliafito, and J. G. Fujimoto. (1991). Optical coherence tomography. *Science*, 254(5035), 1178–1181.

Huynen, J. R. (1978). Phenomenological theory of radar targets. In: *Electromagnetic Scattering*, ed. P. L. E. Uslenghi. New York: Academic Press.

Ikuno, H., and K. Yasuura. (1978). Numerical calculation of the scattered field from a periodic deformed cylinder using the smoothing process on the mode-matching method. *Radio Sci.*, 13(6), 937–946.

Ishimaru, A. (1978). Diffusion of a pulse in densely distributed scatterers. *J. Opt. Soc. Am.*, 68(8), 1045–1049.

Ishimaru, A. (1991). Backscattering enhancement: from radar cross sections to electron and light localizations to rough surface scattering. *IEEE Trans. Antennas Propag.*, 33(5), 7–11.

Ishimaru, A. (1997). *Wave Propagation and Scattering in Random Media*. Piscataway, NJ: Wiley-IEEE Press.

Ishimaru, A. (2001). Acoustical and optical scattering and imaging of tissues: an overview. *Proc. SPIE 4325*, Medical Imaging. DOI: 10.1117/12.428184.

Ishimaru, A., S. Jaruwatanadilok, and Y. Kuga. (2001). Polarized pulse waves in random discrete scatterers. *Appl. Opt.*, 40(30), 5495–5502.

Ishimaru, A., S. Jaruwatanadilok, and Y. Kuga. (2004). Multiple scattering effects on the radar cross section (RCS) of objects in a random medium including backscattering enhancement and shower curtain effects. *Waves Random Media*, 14(4), 499–511.

Ishimaru, A., S. Jaruwatanadilok, and Y. Kuga. (2007). Time reversal effects in random scattering media on superresolution, shower curtain effects, and backscattering enhancement. *Radio Sci.*, 42(6), 1–9.

Ishimaru, A., S. Jaruwatanadilok, and Y. Kuga. (2007). Imaging of a target through random media using a short-pulse focused beam. *IEEE Trans. Antennas Propag.*, 55(6), 1622–1629.

Ishimaru, A., S. Jaruwatanadilok, and Y. Kuga. (2012). Imaging through random multiple scattering media using integration of propagation and array signal processing. *Waves in Random and Complex Media*, 22(2), 24–39.

Ishimaru, A., S. Jaruwatanadilok, J. A. Ritcey, and Y. Kuga. (2010). A MIMO propagation channel model in a random medium. *IEEE Trans. Antennas Propag.*, 58(1), 178–186.

Ishimaru, A., Y. Kuga, and J. Liu. (1999). Ionospheric effects on synthetic aperture radar at 100 MHz to 2 GHz. *Radio Sci.*, 34(1), 257–268.

Ishimaru, A., Y. Kuga, J. Liu, Y. Kim, and T. Freeman. (1999). Ionospheric effects on synthetic aperture radar at 100 MHz to 2 GHz. *Radio Sci.*, 34(1), 257–268.

Ishimaru, A., C. Le, Y. Kuga, L. Ailes-Sengers, and T. K. Chan. (1996). Polarimetric scattering theory for high slope rough surfaces. In: *Progress in Electromagnetic Research*. Cambridge, MA: Elsevier Science Publication.

Ishimaru, A., S.-W. Lee, Y. Kuga, and V. Jandhyala. (2003). Generalized constitutive relations for metamaterials based on the quasi static Lorentz theory. *IEEE Trans. Antennas Propag.*, 51(10), 2550–2557.

Ishimaru, A., J. D. Rockway, Y. Kuga, and S.-W. Lee. (2000). Sommerfeld and Zenneck wave propagation for a finitely conducting one-dimensional rough surfaces. *IEEE Trans. Antennas Propag.*, 48(9), 1475–1484.

Ishimaru, A., J. R. Thomas, and S. Jaruwatanadilok. (2005). Electromagnetic waves over half-space metamaterials of arbitrary permittivity and permeability. *IEEE Trans. Antennas Propag.*, 53(3), 915–921.

Ishimaru, A., and C. W. Yeh. (1984). Matrix representations of the vector radiative-transfer theory for randomly distributed nonspherical particles. *J. Opt. Soc. Am. A*, 1(4), 359–364.

Ishimaru, A., C. Zhang, M. Stoneback, and Y. Kuga. (2013). Time reversal imaging of objects near rough surfaces based on surface flattening transform. *Waves Random Complex Media*, 23, 306–317.

Itoh, T. (1980). Special domain immitance approach for dispersion characteristics of generalized printed transmission lines. *IEEE Trans. Microwave Theory Tech.*, 28(7), 733–736.

Itoh, T., ed. (1987). *Planar Transmission Line Structures*. New York: IEEE Press.

Itoh, T. (1989). *Numerical Techniques for Microwave and Millimeter Wave Passive Structures*. New York: Wiley.

Izatt, J., and M. A. Choma. (2008). Theory of optical coherence tomography. In: *Biological and Medical Physics, Biomedical Engineering*, eds. W. Drexler and J. G. Fujimoto. Berlin, Germany: Springer.

Jackson, J. D. (1975). *Classical Electrodynamics*. New York: Wiley.

Jahnke, E., F. Emde, and F. Losch. (1960). *Tables of Higher Functions*, 6th ed. New York: McGraw-Hill.

James, G. L. (1976). *Geometrical Theory of Diffraction for Electromagnetic Waves*. Stevenage, Hertfordshire, England: Peter Peregrinus.

Jaruwatanadilok, S., A. Ishimaru, and Y. Kuga. (2002). Photon density wave for imaging through random media. *Waves in Random Media*, 12(3), 351–364.

Jensen, M. A., and J. W. Wallace. (2004). A review of antennas and propagation for MIMO wireless communications. *IEEE Trans. Antennas Propag.*, 52(11), 2810–2823.

Jin, Y.-Q., ed. (2004). *Wave Propagation, Scattering and Emission in Complex Media*. Beijing: World Scientific.

John, S. (1990). The localization of waves in disordered media. In: *Scattering and Localization of Classical Waves in Random Media*, ed. P. Sheng. Singapore: World Scientific Publishing Company.

John, S. (1991). Localization of light. *Physics Today*, 44(5), 32–40.

Johnson, C. C., and A. W. Guy. (1972). Nonionizing electromagnetic wave effects in biological materials and systems. *Proc. IEEE*, 60(6), 692–718.

Jones, D. S. (1964). *The Theory of Electromagnetism*. New York: Macmillan.

Jones, D. S. (1979). *Methods in Electromagnetic Wave Propagation*. Oxford: Clarendon Press.

Jordan, E. C., and K. G. Balmain. (1968). *Electromagnetic Waves and Radiating Systems*. Englewood Cliffs, NJ: Prentice-Hall.

Jull, E. V. (1981). *Aperture Antennas and Diffraction Theory*. Stevenage, Hertfordshire, England: Peter Peregrinus.

Kak, A. C. (1979). Computerized tomography with x-ray, emission, and ultrasound sources. *Proc. IEEE*, 67(9), 1245–1272.

Kanamori, H. (1977). The energy release in great earthquakes. *Journal of Geophysical Research*, 82(20), 2981–2987.

Kantorovich, L., and V. I. Krylov. (1958). *Approximate Methods of Higher Analysis*. New York: Interscience.

Keller, J. B. (1962). Geometric theory of diffraction. *J. Opt. Soc. Am.*, 52(2), 116–130.

Keller, J. B. (1964). Stochastic equations and wave propagation in random media. *Proc. Symp. Appl. Math.*, 16, 145–170.

Kerker, M. (1969). *The Scattering of Light and Other Electromagnetic Radiation*. New York: Academic Press.

Khaled, A.-R. A., K. Vafai. (2003). The role of porous media in modeling flow and heat transfer in biological tissues. *International Journal of Heat and Mass Transfer*, 46(26), 4989–5003.

King, D. D. (1970). Passive detection. In: *Radar Handbook*, ed. M. I. Skolnik. New York: McGraw-Hill, Chapter 39.

Kleinman, R. E. (1978). Low frequency electromagnetic scattering. In: *Electromagnetic Scattering*, ed. P. L. E. Uslenghi. New York: Academic Press, Chapter 1.

Knott, E. F., and T. B. A. Senior. (1974). Comparison of three high-frequency diffraction techniques. *Proc. IEEE*, 62(11), 1468–1474.

Kong, J. A. (1972). Theorems of bianisotropic media. *Proc. IEEE*, 60(9), 1036–1046.

Kong, J. A. (1974). Optics of bianisotropic media. *J. Opt. Soc. Am.*, 64(10), 1304–1308.

Kong, J. A. (1981). *Research Topics in Electromagnetic Wave Theory*. New York: Wiley-Interscience.

Kong, J. A. (1986). *Electromagnetic Wave Theory*. New York: Wiley.

Kostinski, A. B., and W.-M. Boerner. (1986). On foundations of radar polarimetry. *IEEE Trans. Antennas Propag.*, 34(12), 1395–1404.

Kouyoumjian, R. G., and P. H. Pathak. (1974). A uniform geometric theory of diffraction for an edge in a perfectly conducting surface. *Proc. IEEE*, 62(11), 1448–1461.

Kraus, J. (1966). *Radio Astronomy*. New York: McGraw-Hill.

Kuga, Y., and A. Ishimaru. (1984). Retroreflectance from a dense distribution of spherical particles. *J. Opt. Soc. Am. A.*, 1(8), 831–835.

Kwon, D. H., and D. H. Werner. (2010). Transformation electromagnetics: an overview of the theory and applications. *IEEE Trans. Antennas Propag.*, 52(1), 24–46.

Lakhtakia, A., V. V. Varadan, and V. K. Varadan. (1988). Field equations, Huygens' principle, integral equations, and theorems for radiation and scattering of electromagnetic waves in isotropic chiral media. *J. Opt. Soc. Am.*, *A*, 5(2), 175–184.

Landau, L. M., and E. M. Lifshitz. (1960). *Electrodynamics of Continuous Media*. Reading, MA: Addison-Wesley.

Le, C., Y. Kuga, and A. Ishimaru. (1996). Angular correlation function based on the second-order Kirchhoff approximation and comparison with experiments. *J. Opt. Soc. Am. A.*, 13(5), 1057–1067.

Lee, S.-W. (1977). Comparison of uniform asymptotic theory and Ufimtsev's theory of electromagnetic edge diffraction. *IEEE Trans. Antennas Propag.*, 25(2), 162–170.

Lee, S.-W. (1978). Uniform asymptotic theory of electromagnetic edge diffraction. In: *Electromagnetic Scattering*, ed. P. L. E. Uslenghi. New York: Academic Press, pp. 67–119.

Lee, H. Y., and T. Itoh. (1989). Phenomenological loss equivalence method for planar quasi-TEM transmission lines with a thin normal conductor or superconductor. *IEEE Trans. Microwave Theory Tech.*, 37(12), 1904–1909.

Lee, S. Y., and N. Marcuvitz. (1984). Quasiparticle description of pulse propagation in a lossy dispersive medium. *IEEE Trans. Antennas Propag.*, 32(4), 395–398.

Lehman, K., and A. J. Devaney. (2003). Transmission mode time-reversal super-resolution imaging. *J. Acoust. Soc. Amer.*, 113(5), 2742–2753.

Leonhardt, U., and T. Philbin. (2010). *Geometry and Light, the Science of Invisibility*. Dover Publications.

Lewin, L., D. C. Chang, and E. F. Kuester. (1977). *Electromagnetic Waves and Curved Structures*. Stevenage, Hertfordshire, England: Peter Peregrinus.

Lewis, R. M. (1969). Physical optics inverse diffraction. *IEEE Trans. Antennas Propag.*, 17(3), 308–314.

Lighthill, J. (1978). *Waves in Fluids*. Cambridge: Cambridge University Press.

Lin, J. C. (2012). *Electromagnetic Fields in Biological Systems*. New York: CRC Press.

Livesay, D. E., and K. M. Chen. (1974). Electromagnetic fields induced inside arbitrarily shaped biological bodies. *IEEE Trans. Microwave Theory Tech.*, 22(12), 1273–1280.

Lo, Y. T., and S.-W. Lee, eds. (1988). *Antenna Handbook*. New York: Van Nostrand Reinhold.

Lo, Y. T., D. Solomon, and W. F. Richards. (1979). Theory and experiment on microstrip antennas. *IEEE Trans. Antennas Propag.*, 27(2), 137–145.

Lonngren, K., and A. Scott. (1978). *Solitons in Action*. New York: Academic Press.

López-Martínez, C., E. Pottier, and S. R. Cloude. (2005). Statistical assessment of eigenvector-based target decomposition theorems in radar polarimetry. *IEEE Trans. Geosci. Remote Sens.*, 43(9), 2058–2074.

Loyka, S. L. (2001). Channel capacity of MIMO architecture using the exponential correlation matrix. *IEEE Communications Letters*, 5(9), 369–371.

Luneburg, E. (1995). Principles of radar polarimetry. *IEICE Trans. Electron.*, E78-C(10), 1339–1345.

Ma, M. T. (1974). *Theory and Application of Antenna Arrays*. New York: Wiley.

Maanders, E. J., and R. Mittra, eds. (1977). *Modern Topics in Electromagnetics and Antennas*. Stevenage, Hertfordshire, England: Peter Peregrinus, Chapter 1.

Magnus, W., and F. Oberhettinger. (1949). *Special Functions of Mathematical Physics*. New York: Chelsea.

Mailloux, R. J. (1982). Phased array theory and technology. *Proc. IEEE*, 70(3), 246–291.

Mandelbrot, B. B. (1977). *The Fractal Geometry of Nature*. New York: W. H. Freeman and Company.

Marcuse, D. (1982). *Light Transmission Optics*. New York: Van Nostrand Reinhold.

Marcuvitz, N. (1951). Waveguide handbook. In: *MIT Radiation Laboratory Series*, Vol. 10. New York: McGraw-Hill.

Martinez, A., and A. P. Byrnes. (2001). Modeling dielectric-constant values of geologic materials: an aid to ground-penetrating radar data collection and interpretation. Current Research in Earth Sciences, Bulletin 247, Part 1.

Mendelssohn, K. (1966). *The Quest for Absolute Zero*. New York: McGraw-Hill.

Meng, Z. Q., and M. Tateiba. (1996). Radar cross sections of conducting elliptic cylinders embedded in strong continuous random media. *Waves Random Media*, 6, 335–345.

Meyer, M. G., J. D. Sahr, and A. Morabito. (2004). A statistical study of subauroral e-region coherent backscatter observed near 100 MHz with passive radar. *Journal of Geophysical Research*, 109(A7), 1–19.

Mittra, R., ed. (1973). *Computer Techniques for Electromagnetics*. Elmsford, NY: Pergamon Press.

Mittra, R. (1975). Numerical and asymptotic techniques in electromagnetics. In: *Topics in Applied Physics*, Vol. 3. New York: Springer-Verlag.

Molisch, A. F. (2005). *Wireless Communication*. Piscataway, NJ: IEEE Press.

Montgomery, C. G., R. H. Dicke, and E. M. Purcell. (1948). Principles of microwave circuits. In: *MIT Radiation Laboratory Series*, Vol. 8. New York: McGraw-Hill.

Moore, J., and R. Pizer. (1984). *Moment Methods in Electromagnetics*. Chichester, England: Research Studies Press.

Morgan, M. A., ed. (1990). *Finite Element and Finite Difference Methods in Electromagnetic Scattering*. New York: Elsevier.

Morse, P. M., and H. Feshbach. (1953). *Methods of Theoretical Physics*. New York: McGraw-Hill Book Company.

Morse, P. M., and K. U. Ingard. (1968). *Theoretical Acoustics*. New York: McGraw-Hill.

Nakayama, A., and F. Kuwahara. (2008). A general bioheat transfer model based on the theory of porous media. *International Journal of Heat and Mass Transfer*, 51(11–12), 3190–3199.

Neeser, F. D., and J. L. Massey. (1993). Proper complex random processes with applications to information theory. *IEEE Transactions on Information Theory*, 39(4), 1293–1302.

Noble, B. (1958). *Methods Based on the Wiener-Hopf Technique*. Elmsford, NY: Pergamon Press.

Novoselov, K. S., A. K. Geim, S. V. Morozov, D. Jiang, M. I. Katsnelson, I. V. Grigorieva, S. V. Dubonos, and A. A. Firsov. (2005). Two-dimensional gas of massless Dirac fermions in graphene. *Nature*, 438, 197–200.

Ogilvy, J. A. (1991). *Theory of Wave Scattering from Random Rough Surfaces*. London: IOP Publishing.

Oguchi, T. (1983). Electromagnetic wave propagation and scattering in rain and other hydrometers. *Proc IEEE*, 71, 1029–1078.

Ollila, E. (2008). On the circularity of a complex random variable. *IEEE Signal Processing Letters*, 15, 841–844.

Olsen, M. T., T. Komatsu, P. Schjonning, and D. E. Rolston. (2001). Tortuosity, diffusivity, and permeability in the soil liquid and gaseous phases. *Soil Sci. Sec. Am. J*, 65(3), 613–623.

Ozgun, O., and M. Kuzuoglu. (2010). For invariance of Maxwell's equations: the pathway to novel metamaterial specifications for electromagnetic reshaping. *IEEE Trans. Antennas Propag.*, 52(3), 51–65.

Pasyanos, M. E. (2010). A general method to estimate earthquake moment and magnitude using regional phase amplitudes. *Bulletin of the Seismological Society of America*. 100(4), 1724–1732.

Paulraj, A., R. Nabar, and D. Gore. (2003). *Introduction to Space-Time Wireless Communications*. Cambridge, UK: Cambridge University Press.

Pendry, J. B., A. J. Holden, D. J. Robbins, and W. J. Stewart. (1999). Magnetism from conductors and enhanced nonlinear phenomena. *IEEE Trans. Microw. Theory Tech.*, 47, 2075–2084.

Pendry, J. B., D. Schurig, and D. R. Smith. (2006). Controlling electromagnetic fields. *Science*, 312(5781), 1780–1782.

Pennes, H. H. (1948). Analysis of tissue and arterial blood temperatures in the resting human forearm. *Journal of Applied Physiology*, 1(2), 93–122.

Phu, P., A. Ishimaru, and Y. Kuga. (1994). Co-polarized and cross-polarized enhanced backscattering from two-dimensional very rough surfaces at millimeter wave frequencies. *Radio Sci.*, 29(5), 1275–1291.

Picinbono, B. (1996). Second-order complex random vectors and normal distributions. *IEEE Transactions on Signal Processing*, 44(10), 2637–2640.

Porter, R. P., and A. J. Devaney. (1982). Holography and the inverse source problem. *J. Opt. Soc. Am.*, 72(3), 327–330.

Prada, C., and M. Fink. (1994). Eigenmodes of the time reversal operator: a solution to selective focusing in multiple-target media. *Wave Motion*, 20(2), 151–163.

Prada, C., S. Manneville, D. Spoliansky, and M. Fink. (1996). Decomposition of the time reversal operator: detection and selective focusing on two scatterers. *J. Acoust. Soc. Amer.*, 99(4), 2067–2076.

Prada, C., and J.-L. Thomas. (2003). Experimental subwavelength localization of scatterers by decomposition of the time reversal operator interpreted as a covariance matrix. *J. Acoust. Soc. Amer.*, 114(1), 235–243.

Pruppacher, H. R., and R. L. Pitter. (1971). A semi-empirical determination of the shape of cloud and raindrops. *J. Atmos. Sci.*, 28, 86–94.

Rahmat-Samii, Y., and R. Mittra. (1977). Electromagnetic coupling through small apertures in a conducting screen. *IEEE Trans. Antennas Propag.*, 25(2), 180–187.

Ralston, T. S., D. L. Marks, P. S. Carney, and S. A. Boppart. (2006). Inverse scattering for optical coherence tomography. *J. Opt. Soc. Am. A*, 23(5), 1027–1037.

Ramo, S., J. R. Whinnery, and T. Van Duzer. (1965). *Fields and Waves in Communication Electronics*. New York: Wiley.

Ray, P. S. (1972). Broadband complex refractive indices of ice and water. *Appl. Opt.*, 11(8), 1836–1844.

Rayleigh, J. W. S. (1898). *The Theory of Sound*. Dover Publications.

Redwood, M. (1960). *Mechanical Waveguides. The Propagation of Acoustic and Ultrasonic Waves in Fluids and Solids with Boundaries*. Pergamon Press.

Reed, I. S. (1962). On a moment theorem for complex Gaussian processes. *IRE Transactions on Information Theory*, 8(3), 194–195.

Revil, A., and L. M. Cathles, III. (1999). Permeability of shaly sands. *Water Resources Research*, 35(3), 651–662.

Rice, S. O. (1951). Reflections of electromagnetic waves from slightly rough surfaces. *Commun. Pure Appl. Math.*, 4, 351–378.

Rino, C. L. (2011). *The Theory of Scintillation with Applications in Remote Sensing*. Wiley-IEEE Press.

Ruck, G. T., D. E. Barrick, W. D. Stuart, and C. K. Krichbaum. (1970). *Radar Cross Section Handbook*, Vols. 1 and 2. New York: Plenum Press.

Rumsey, V. H. (1954). The reaction concept in electromagnetic theory. *Phys. Rev., Ser. 2*, 94(6), 1483–1491.

Rumsey, V. H. (1966). *Frequency Independent Antennas*. New York: Academic Press.

Sahr, J. D., and F. D. Lind. (1997). The Manastash Ridge radar: a passive bistatic radar for upper atmospheric radio science. *Radio Sci.*, 32(6), 2345–2358.

Sailing, H., and T. Chen. (2013). Broadband THz absorbers with graphene-based anisotropic metamaterial films. *IEEE Trans. Terahertz Sci. Technol.*, 3(6), 757–763.

Sarabandi, K., and D. E. Lawrence. (2003). Acoustic and electromagnetic wave interaction: estimation of Doppler spectrum from an acoustically vibrated metallic circular cylinder. *IEEE. Trans. Antennas Propag.*, 51(7), 1499–1507.

Sato, H., and M. C. Fehler. (1988). *Seismic Wave Propagation and Scattering in the Heterogeneous Earth*. New York: Springer.

Sato, H., M. C. Fehler, and T. Maeda. (2012). *Seismic Wave Propagation and Scattering in the Heterogeneous Earth*, 2nd ed. Berlin/Heidelberg: Springer-Verlag.

Schelkunoff, S. A. (1965). *Applied Mathematics for Engineers and Scientists*. New York: Van Nostrand Reinhold.

Schmidt, R. O. (1986). Multiple emitter location and signal parameter estimation. *IEEE Trans. Antennas Propag.*, 276–280.

Schmitt, J. M. (1999). Optical coherence tomography (OCT): a review. *IEEE Journal of Selected Topics in Quantum Electronics*, 5(4), 1205–1215.

Schmitt, J. M., and A. Knüttel. (1997). Model of optical coherence tomography of heterogeneous tissue. *J. Opt. Soc. Am. A*, 14(6), 1231–1242.

Schurig, D., J. J. Mock, B. J. Justice, S. A. Cummer, J. B. Pendry, A. F. Starr, and D. R. Smith. (2006). Metamaterial electromagnetic cloak at microwave frequencies. *Science*, 314(5801), 977–980.

Schuster, A. (1905). Radiation through a foggy atmosphere. *Astrophys. J.*, 21, 1–22.

Scott, A. C. (1980). The birth of a paradigm. In: *Nonlinear Electromagnetics*. Academic Press.

Scott, C. (1989). *The Spectral Domain Method in Electromagnetics*. Norwood, MA: Artech House.

Scott, A. C., F. Y. F. Chu, and D. W. McLaughlin. (1973). The soliton: a new concept in applied science. *Proceedings of the IEEE*, 61(10), 1443–1483.

Seliga, T. A., and V. N. Bringi. (1976). Potential use of radar differential reflectivity measurements at orthogonal polarizations for measuring precipitation. *J. Appl. Meteorol.*, 15, 69–76.

Sen, P. N., C. Scala, and M. H. Cohen. (1981). A self-similar model for sedimentary rocks with application to the dielectric constant of fused glass beads. *Geophysics*, 46(5), 781–795.

Senior, T. B. A. (1979). Scattering by resistive strips. *Radio Sci.*, 14(5), 911–924.

Senior, T. B. A., and M. Noar. (1984). Low frequency scattering by a resistive plate. *IEEE Trans. Antennas Propag.*, 32(3), 272–275.

Shamonin, M., E. Shamonina, V. Kalinin, and L. Solymar. (2004). Properties of a metamaterial element: analytical solutions and numerical simulations for a singly split double ring. *J. Appl. Phys.*, 95(7), 3778–3784.

Shen, L. C., and J. A. Kong. (1987). *Applied Electromagnetism*. Boston, MA: PWS Publishers.

Sheng, P. (1990). *Scattering and Localization of Classical Waves in Random Media*. Singapore: World Scientific Publishing Company.

Shung, K. K, J. Cannata, Q. Thou, and J. Lee. (2009). High frequency ultrasound: a new frontier for ultrasound. In: IEEE EMBS Conference, Minneapolis, MN, September 2009.

Shung, K. K., and G. A. Thieme. (1993). *Ultrasonic Scattering in Biological Tissues*. Boca Raton: CRC Press.

Sihvola, A. (2007). Metamaterials in electromagnetics. *Metamaterials*, 1, 2–11.

Sinclair, G. (1950). The transmission and reception of elliptically polarized waves. *Proc. IRE*, 38, 148–151.

Skolnik, M. I., ed. (1970). *Radar Handbook*. New York: McGraw-Hill.

Skolnik, M. I. (1980). *Introduction to Radar Systems*. New York: McGraw-Hill.

Smith, D. R., P. Kolinko, and D. Schurig. (2004). Negative refraction in indefinite media. *J. Opt. Soc. Am. B*, 21, 1032–1043.

Solymar, L., and E. Shamonina. (2011). *Waves in Metamaterials*. Oxford, UK: Oxford University Press.

Sommerfeld, A. (1949). *Partial Differential Equations in Physics*. New York: Academic Press.

Sommerfeld, A. (1954). *Optics*. New York: Academic Press.

Stark, L. (1974). Microwave theory of phase-array antennas—a review. *Proc. IEEE*, 62(12), 1661–1701.

Stauffer, D., and A. Aharony. (1991). *Introduction to Percolation Theory*. CRC Press.

Stevenson, A. F. (1953). Solutions of electromagnetic scattering problems as power series in the ratio (dimensions of scatterer)/wavelength). *J. Appl. Phys.*, 24(9), 1134–1142.

Stoneback, M., A. Ishimaru, C. Reinhardt, and Y. Kuga. (2013). Temperature rise in objects due to optical focused beam through atmospheric turbulence near ground and ocean surface. *Opt. Eng.*, 52(3), 36001–36008.

Stratton, J. A. (1941). *Electromagnetic Theory*. New York: McGraw-Hill.

Stutzman, W. L., and G. A. Thiele. (1981). *Antenna Theory and Design*. New York: Wiley.

Taflove, A., and M. E. Brodwin. (1975). Computation of the electromagnetic fields and induced temperatures within a model of the microwave-irradiated human eye. *IEEE Trans. on Microwave Theory and Techniques*, MTT-23(11), 888–896.

Tai, C. T. (1971). *Dyadic Green's Functions in Electromagnetic Theory*. New York: Intext Educational Publishers.

Tamir, T., ed. (1975). Integrated optics. In: *Topics in Applied Physics*, Vol. 7. New York: Springer-Verlag.

Tamir, T., and M. Blok, eds. (1986). Propagation and scattering of beam fields. Special Issue, *J. Opt. Soc. Am., A*, 3(4), 462–588.

Tatarskii, V. I. (1961). *Wave Propagation in a Turbulent Medium*. New York: McGraw-Hill.

Tatarskii, V. I. (1971). The effects of the turbulent atmosphere on wave propagation. US. Department of Commerce, Springfield, VA, TT-68-50464.

Tatarskii, V. I., A. Ishimaru, and V. U. Zavorotny. (1993). *Wave Propagation in Random Media*. Washington: SPIE.

Tesche, F. M. (1973). On the analysis of scattering and antenna problems using the singularity expansion technique. *IEEE Trans. Antennas Propag.*, 21(1), 52–63.

Thomas, J. R., and A. Ishimaru. (2005). Wave packet incident on negative-index media. *IEEE Trans. Antennas Propag.*, 53(5), 1591–1599.

Thrane, L., H. T. Yura, and P. E. Andersen. (2000). Analysis of optical coherence tomography systems based on the extended Huygens–Fresnel principle. *J. Opt. Soc. Am. A*, 17(3), 484–490.

Tsang, L., C. H. Chan, K. Pak, and H. Sangani. (1995). Monte Carlo simulations of large-scale problems of random rough surface scattering and applications to grazing incidence with BMIA/canonical grid method. *IEEE Trans. Antennas Propag.*, 43(8), 851–859.

Tsang, L., and S. L. Chuang. (1988). Improved coupled-mode theory for reciprocal anisotropic waveguides. *J. Lightwave Technol.*, 6(2), 304–311.

Tsang, L., S. L. Chuang, A. Ishimaru, R. P. Porter, and D. Rouseff. (1987). Holography and the inverse source problem: part III. *J. Opt. Soc. Am., A*, 4(9), 1783–1787.

Tsang, L., S. L. Chuang, A. Ishimaru, R. P. Porter, D. Rouseff, J. A. Kong, and R. T. Shin. (1985). *Theory of Microwave Remote Sensing*. New York: Wiley.

Tsang, L., K.-H. Ding, X. Li, P. N. Duvelle, J. H. Vella, J. Goldsmith, C. L. H. Devlin, and N. I. Limberopoulos. (2015). Studies of the influence of deep subwavelength surface roughness on fields of plasmonic thin film based on Lippmann–Schwinger equation in the spectral domain. *J. Opt. Soc. Am. B*, 32(5), 878–891.

Tsang, L., and A. Ishimaru. (1984). Backscattering enhancement of random discrete scatterers. *J. Opt. Am. A.*, 1(8), 836–839.

Tsang, L., and J. A. Kong. (2001). *Scattering of Electromagnetic Waves*. New York: John Wiley & Sons.

Tsang, L., J. A. Kong, and R. T. Shin. (1985). *Theory of Microwave Remote Sensing*. New York: Wiley-Interscience.

Tuchin, V. V., and B. J. Thompson. (1994). *Selected Papers on Tissue Optics Applications in Medical Diagnostics and Therapy*, Vol. MS 102. Bellingham, WA: SPIE Optical Engineering Press.

Turcotte, D. L., M. M. Eldridge, and J. B. Rundle. (2014). Super fracking. *Physics Today*, 67(8), 34–39.

Twersky, V. (1964). On propagation in random media of discrete scatterers. *Proc. Symp. Appl. Math.*, 16, 84–116.

Ufimtsev, P. Y. (1975). Comparison of three high frequency diffraction techniques. *Proc. IEEE*, 63(12), 1734–1737.

Ulaby, F. T., R. K. Moore, and A. K. Fung. (1981). *Microwave Remote Sensing*, Vols. 1, 2, and 3. London: Addison-Wesley.

Unger, H. G. (1977). *Planar Optical Waveguides and Fibers*. Oxford: Clarendon Press.

Uslenghi, P. L. E., ed. (1978). *Electromagnetic Scattering*. New York: Academic Press.

Uslenghi, P. L. E. (1980). *Nonlinear Electromagnetics*. New York: Academic Press.

Van Bladel, J. (2007). *Electromagnetic Fields*. New York: IEEE Press-Wiley.

Van Bladel, J. (1968). Low-frequency scattering by hard and soft bodies. *J. Acoust. Soc. Am.*, 44(4), 1069–1073.

van De Hulst. (1957). *Light Scattering by Small Particles*. New York: Wiley.

van den Berg, P. M., A. T. De Hoop, A. Segal, and N. Praagman. (1983). A computational model of the electromagnetic heating of biological tissue with application to hyperthermic cancer therapy. *IEEE Trans. on Biomedical Engineering*, BME-30(12), 797–805.

Van Duzer, T., and C. W. Turner. (1981). *Principles of Superconductive Devices and Circuits*. New York: Elsevier.

van Zyl, J. J., H. A. Zebker, and C. Elachi. (1987). Imaging radar polarization signatures: theory and observation. *Radio Sci.*, 22(4), 529–543.

Veselago, V. G. (1968). The electrodynamics of substances with simultaneously negative values of ε and μ. *Sov. Phys. Usp.*, 10, 509–514.

Voronovich, A. G. (1994). *Wave Scattering from Rough Surfaces*. New York: Springer-Verlag.

Wait, J. R. (1959). *Electromagnetic Radiation from Cylindrical Structures*. New York: Pergamon Press.

Wait, J. R. (1962). *Electromagnetic Waves in Stratified Media*. Oxford: Pergamon Press.

Wait, J. R. (1981). *Wave Propagation Theory*. Elmsford, NY: Pergamon Press.

Wait, J. R. (1982). *Geo-Electromagnetism*. New York: Academic Press.

Wait, J. R. (1986). *Introduction to Antennas and Propagation*. Stevenage, Hertfordshire, England: Peter Peregrinus.

Wait, J. R. (1989). Complex resistivity of the earth. In: *Progress in Electromagnetics Research*, ed. J. A. Kong. New York: Elsevier, pp. 1–174.

Waterman, P. C. (1969). New formulations of acoustic scattering. *J. Acoust. Soc. Am.*, 45(6), 1417.

Watson, J., and J. Keller. (1984) Rough surface scattering via the smoothing method. *J. Acoust. Soc. Am.*, 75, 1705.

Wedge, S. W., and D. B. Rutledge. (1991). Noise waves and passive linear multiports. *IEEE Microw. Guided Wave Lett.*, 1(5), 117–119.

Weiglhofer, W. S. (2002). Constitutive relations. *Proc. SPIE*, 4806, 67–80.

Weiglhofer, W. S., and A. Lakhtakia. (1995). A brief review of a new development for constitutive relations of linear bi-anisotropic media. *IEEE Trans. Antennas Propag.*, 37(3), 32–35.

Wu, K. (2005). Two-frequency mutual coherence function for electromagnetic pulse propagation over rough surfaces. *Waves in Random and Complex Media*, 15(2), 127–143.

Xiong, J. L., M. S. Tong, P. Atkins, and W. C. Chew. (2010). Efficient evaluation of Casimir force in arbitrary three-dimensional geometries by integral equation methods. *Phys. Lett. A*, 374, 2517–2520.

Yaghjian, A. D. (1980). Electric dyadic Green's functions in the source region. *Proc. IEEE*, 68(2), 248–263.

Yamaguchi, Y., T. Moriyama, M. Ishido, and H. Yamada. (2005). Four-component scattering model for polarimetric SAR image decomposition. *IEEE Trans. Geosci. Remote Sens.*, 43(8), 1699–1706.

Yamashita, E., and R. Mittra. (1968). Variational method for the analysis of microstrip line. *IEEE Trans. Microwave Theory Tech.*, 16(4), 251–256.

Yao, Y., M. A. Kats, P. Genevet, N. Yu, Y. Song, J. Kong, and F. Capasso. (2013). Broad electrical turning of graphene-loaded plasmonic antennas. *Nano Lett.*, 1257–1264.

Yasuura, K., and Y. Okuno. (1982). Numerical analysis of diffraction from grating by the mode matching method with a smoothing procedure. *J. Opt. Soc. Am.*, 72(7), 847–852.

Yavuz, M. E., and F. L. Teixeira. (2006). Full time-domain DORT for ultrawideband electromagnetic fields in dispersive, random inhomogeneous media. *IEEE Trans. Antennas Propag.*, 54(8), 2305–2315.

Yavuz, M. E., and F. L. Teixeira. (2008). Space-frequency ultrawideband time-reversal imaging. *IEEE Trans. Geosci. Remote Sens.*, 46(4), 1115–1124.

Yavuz, M. E., and F. L. Teixeira. (2008). On the sensitivity of time-reversal imaging techniques to model perturbations. *IEEE Trans. Antennas Propag.*, 56(3), 834–843.

Yee, K. S. (1966). Numerical solution of initial boundary value problems involving Maxwell's equations in isotropic media. *IEEE Trans. Antennas Propag.*, 14(3), 302–307.

Yeh, K. C., K. H. Lin, and Y. Wang. (2001). Effects of irregular terrain on waves: a stochastic approach. *IEEE Trans. Antennas Propag.*, 49, 250–259.

Yeh, K. C., and C. H. Liu. (1972). Propagation and application of waves in the ionosphere. *Rev. Geophys.*, 10(2), 631–709.

Yeh, K. C., and C. H. Liu. (1972). *Theory of Ionospheric Waves*. New York: Academic Press.

Yeh, C., and F. I. Shimabukuro. (2008). *The Essence of Dielectric Waveguides*. New York: Springer.

Yueh, S. H., W. J. Wilson, S. J. Dinardo, and F. K. Li. (1999). Polarimetric microwave brightness signatures of ocean wind directions. *IEEE Trans. Geosci. Remote Sens.*, 37(2), 949–956.

Yura, H. T. (1972). Mutual coherence function of a finite cross section optical beam propagating in a turbulent medium. *Appl. Opt.*, 11(6), 1399–1406.

Yura, H. T., C. C. Sung, S. F. Clifford, and R. J. Hill. (1983). Second-order Rytov approximation. *J. Opt. Soc. Am.*, 73(4), 500–502.

Zebker, H. A., and J. J. van Zyl. (1991). Imaging radar polarimetry: a review. *Proc. IEEE*, 79(11), 1583–1606.

Zhang, G. (2016). *Weather Radar Polarimetry*. Taylor & Francis.

Zhang, G., R. J. Doviak, D. S. Zrnić, R. Palmer, L. Lei, and Y. Al-Rashid. (2011). Polarimetric phased-array radar for weather measurement: a planar or cylindrical configuration? *J. Atmos. Oceanic Technol.*, 28(1), 63–73.

Zienkiewicz, O. C. (1977). *The Finite Element Method*. New York: McGraw-Hill.

INDEX

ABCD matrix, 211
Abel's integral equation, 572, 574, 902
absorption coefficient, 667
absorption cross section, 320
absorption efficiency, 320
absorptivity, 594, 644
acoustic Brewster's angle, 715
acoustic reflection, 50
acoustic scattering, 344
acoustic wave, 262
active radar, 625
adiabatic process, 260
Airy disk, 180
Airy integral, 826
Airy pattern, 180
ambiguity function, 644
Ampère's law, 7
analytic continuity, 408
analytic function, 873
analytic signal, 106
Anderson localization, 789
angle of incidence, 39
angular momentum, 263
angular spread, 760

anisotropic impedance matched media, 716
anisotropic medium, 12
anisotropy factor, 667, 751
antenna pattern synthesis, 299
antenna temperature, 591, 599
aperture distribution, 602–603
aperture integral equations, 427
aperture synthesis, 607
apertures, 473
Appleton–Hartree formula, 238
Archie's law, 807
array factor, 297
atomic polarization, 234
attenuation constant, 37
axial dipole near a conducting cylinder, 364
axial resolution, 676
azimuthal waveguides, 119

Babinet's principle, 433, 435
back projection of the filtered projection, 559
backscattering cross section, 319, 324, 389
backscattering enhancement, 612
backward lateral waves, 698, 704
backward leaky wave, 53

Electromagnetic Wave Propagation, Radiation, and Scattering: From Fundamentals to Applications,
Second Edition. Akira Ishimaru.
© 2017 by The Institute of Electrical and Electronic Engineers, Inc. Published 2017 by John Wiley & Sons, Inc.

IEEE PRESS SERIES ON
ELECTROMAGNETIC WAVE THEORY

Field Theory of Guided Waves, Second Edition
Robert E. Collin

Field Computation by Moment Methods
Roger F. Harrington

Radiation and Scattering of Waves
Leopold B. Felsen, Nathan Marcuvitz

Methods in Electromagnetic Wave Propagation, Second Edition
D. S. J. Jones

Mathematical Foundations for Electromagnetic Theory
Donald G. Dudley

The Transmission-Line Modeling Method: TLM
Christos Christopoulos

The Plane Wave Spectrum Representation of Electromagnetic Fields,
Revised Edition
P. C. Clemmow

General Vector and Dyadic Analysis: Applied Mathematics in Field Theory
Chen-To Tai

Computational Methods for Electromagnetics
Andrew F. Peterson

Plane-Wave Theory of Time-Domain Fields: Near-Field Scanning Applications
Thorkild B. Hansen

Foundations for Microwave Engineering, Second Edition
Robert Collin

Time-Harmonic Electromagnetic Fields
Robert F. Harrington

Antenna Theory & Design, Revised Edition
Robert S. Elliott

Differential Forms in Electromagnetics
Ismo V. Lindell

Conformal Array Antenna Theory and Design
Lars Josefsson, Patrik Persson